Shinsaku Fujita

Organic Chemistry of Photography

Springer-Verlag Berlin Heidelberg GmbH

Shinsaku Fujita

Organic Chemistry of Photography

With 480 Figures and 32 Tables

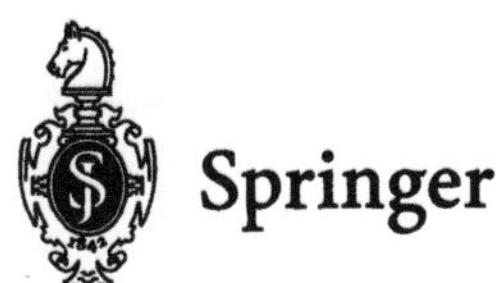

Springer

Professor Dr. Shinsaku Fujita
Kyoto Institute of Technology
Department of Chemistry
and Materials Technology
Matsugasaki, 606-8585
Kyoto, Japan
e-mail: fujitas@chem.kit.ac.jp

DOI 10.1007/978-3-662-09130-2

springeronline.com

Originally published by Springer-Verlag Berlin Heidelberg New York in 2004
MyCopy version of the original edition 2004

Typesetting: Camera ready by author
Cover-Design: Künkel & Lopka, Heidelberg / Design & Production GmbH, Heidelberg
Printed on acid free paper 02/3020/kk - 5 4 3 2 1 0
www.springer.com/mycopy

About the author:

Shinsaku Fujita was born in Kita-Kyushu City, Japan in 1944. He received his undergraduate training at Kyoto University. After earning a Master's degree in 1968, he started as a research instructor and received a Dr. Eng. degree at Kyoto University under the guidance of Prof. Hitosi Nozaki. In 1972, he joined Ashigara Research Laboratories, Fuji Photo Film Co., Ltd., where he was engaged in the R&D of organic compounds for instant color photography and in the R&D of the organic reaction database until 1997. From 1997, he has been Professor of Information Chemistry and Materials Technology at the Kyoto Institute of Technology. He was awarded the Synthetic Organic Chemistry Award in 1982 and the Society of Computer Chemistry Japan Award in 2002. His research interests have included reactive intermediates (nitrenes), synthetic organic chemistry (cylophanes, strained heterocycles, and organic compounds for photography), organic photochemistry, organic stereochemistry (theoretical approach), mathematical organic chemistry (combinatorial enumeration), and the organic reaction database (imaginary transition structures). He is the author of *Symmetry and Combinatorial Enumeration in Chemistry* (Springer-Verlag, 1991), *XΥMTEX—Typesetting Chemical Structural Formulas* (Addison-Wesley Japan, 1997), *Computer-Oriented Representation of Organic Reactions* (Yoshioka Shoten, 2001), and several books on TEX/LATEX. His homepage on World Wide Web is located at http://imt.chem.kit.ac.jp/fujita/fujitas/fujita.html

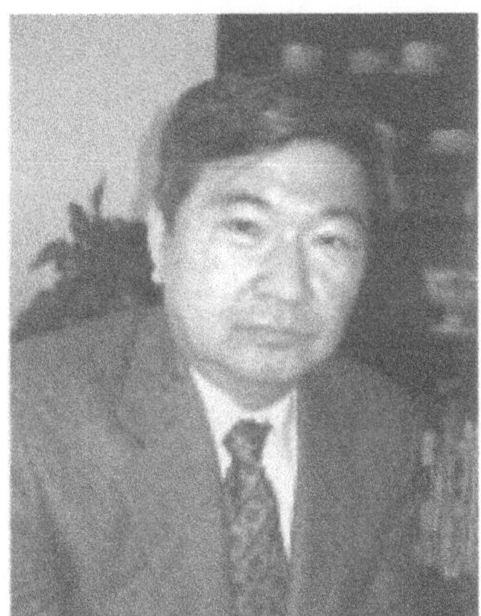

Preface

Scope and Aims of This Book

Color photography is a masterpiece of chemistry and materials technology, where organic, inorganic, and polymeric compounds dynamically participate in chemical and physical processes for producing, exposing, and processing color photographic materials. Among such compounds, organic chemicals play particularly important roles in the formation of color images of high quality. In fact, sensitizing dyes accept lights exposed, various organic additives control color reproduction, image-forming compounds (such as couplers, dye releasers, and bleachable dyes) produce image dyes, light and dark stabilizers preserve dye images, and so on, before and after we finally observe the image dyes as a completed picture.

In spite of the importance of organic chemicals, they have been described rather subsidiarily in most previous books on photography, even though some chapters have been concerned with them. Moreover, only the functions of organic chemicals have been described on the basis of general formulas (e.g., Markush's formulas), so that there have been few discussions on how the organic chemicals are designed and developed, how they are synthesized, and how things omitted from the general formulas support such functions. In other words, a viewpoint based on organic chemistry has been missing in the conventional descriptions on photography.

This book, as the title "Organic Chemistry of Photography" indicates, is intended to make up for such a viewpoint from organic chemistry. More concretely speaking, this book is devoted to comprehensive discussions on organic compounds used in color photography, especially, on their structures, functions, dynamic processes, R&D, syntheses, and related items.

The book grew directly out of lectures given at the Kyoto Institute of Technology for advanced undergraduates and graduate students (from 1997 to now) and indirectly out of my experiences at Ashigara Research Laboratories of Fuji Photo

Film Co. Ltd. (1972–1997). The words "directly" and "indirectly" are used here according to the following situations.

Although I privately tried to summarize an overview on organic chemistry of photography during the period at Fuji, it was difficult for me to complete a balanced overview, because my knowledge was restricted to organic compounds for instant color photography. In 1997, I changed my position to the Kyoto Institute of Technology (Professor of Information Chemistry and Materials Technology, Department of Chemistry and Materials Technology), where I have given lectures on "photographic chemistry" (now renamed to "photo-sensitive materials science and technology") and "information materials technology" in the undergraduate course as well as a lecture on "image-forming materials" in the graduate course. In preparing the lecture notes, I have had to begun to enlarge and reconstruct my knowledge on the whole fields of photography from a viewpoint of organic chemistry. Although this task has required troublesome efforts, it has been very challenging to me, because I have continuously kept in mind an expectation that the task would put the knowledge on photography into a universal and systematic format on the basis of organic chemistry.

This book comprises five parts, among which Part III to Part V are devoted to the main theme, i.e., "Organic Chemistry of Photography". Hence, it is possible to read only these three parts for busy or impatient readers to gain an appreciation of organochemical aspects of photography. However, because organic compounds interact and/or react with silver halide and because they are used to reproduce colors in photographic materials, well-selected pieces of information on the silver halide (Part I) and on the color reproduction (Part II) would be desirable to understand Part III to Part V more intimately.

Part I is an introduction to photography, which deals with fundamental knowledge on black-and-white and color photography (Chapter 1), photographic emulsions (Chapter 2), silver halide crystals (Chapters 3 and 4), and developers (Chapter 5). If the reader is not interested in the details of silver halide crystals, he or she can skip Part I except Chapter 1, because Chapter 1 also involves an overview of multilayer structure of photographic films and papers as well as a brief classification of chemicals for photography. Part II describes color reproduction (Chapter 6) and spectral sensitization (Chapter 7), which are common to all of the color photographic processes discussed in Part III to Part V. Part III is devoted to chromogenic photography, which is usually referred to as conventional color photography. Since this process is most popular, seven chapters (Chapter 8 to 14) are concerned with various topics of this process. Part IV deals with diffusion transfer photography, which gives a basis to so-called instant color photography. Six chapters (Chapter 15 to 20) are devoted to various functionalized compounds which are used in commercialized systems. Part V consists of one chapter (Chapter 21) concerning silver dye bleach photography.

Throughout this book, I have intended to show concrete structural formulas and to exclude Markush's formulas as far as possible, because even terminal substituents contained in the formulas have important roles in color photography. According to this policy, I have tried to depict a most plausible compound selected from many examples disclosed in a patent, although this selection has not always been an easy task.

For the convenience of the reader, the corresponding United States patents have been cited in place of Japanese patents if the original applications were done in Japan. This causes a disadvantage that the order of original Japanese application dates is missed, since the order may irregularly be changed in the publication dates of the corresponding US patents.

Additional Aims

As found in the brief summary of the five parts, this book aims at covering the fields of color photography as comprehensively as possible. At the same time, it has two additional aims with respect to my own accomplishments.

First, this book is an account of my work in collaboration with my colleagues at Ashigara Research Laboratories of Fuji Photo Film Co. Ltd., where I have engaged in the R&D of organic compounds for instant color photography and in other related tasks from 1972 to 1997. Thus, Chapter 19 involves a revised collection of our previous accounts [Fujita S (1992) R&D of Dye Releasers for Instant Color Photography: Design, Evaluation, and Synthesis. In: Noyori R (ed) *Organic Synthesis in Japan. Past, Present, and Future.* Tokyo Kagaku Dozin, Tokyo. pp 89–96; Fujita S, Koyama K, Ono S (1982) *Nikkakyo Geppo.* 35(11):29; Fujita S, Koyama K, Ono S (1991) *Nippon Kagaku Kai Shi.* 1; and Fujita S, Koyama K, Ono S (1992) *Rev Heteroatom Chem.* 7:229]. Since these accounts have been published in Japanese or they have, even in English, appeared in a rather specified book or in a domestic journal issued by a Japanese publisher, it would be worthwhile to revise and reissue them in English as one chapter of this book.

Second, this book serves as a field test of XΥMTEX, which has been developed by myself as a software for drawing structural formulas [Fujita S (1997) *XΥMTEX—Typesetting Chemical Structural Formulas.* Addison-Wesley Publishers Japan, Tokyo; Fujita S (1994) Typesetting Chemical Structural Formulas with TEX/LATEX. *Comput Chem.* 18:109; Fujita S (1995) XΥMTEX for Drawing Chemical Structures. *TUGboat.* 16:80; Fujita S, Tanaka N (2000) XΥMTEX (Version 2.00) as Implementation of the XΥM Notation and the XΥM Markup Language. *TUGBoat.* 21:1; and Fujita S, Tanaka N (2001) Size Reduction of Chemical Structural Formulas in XΥMTEX (Version 3.00). *TUGBoat.* 22:285].[1] The XΥMTEX is an implement based on XΥM Notation and XΥMML (XΥM Markup Language), which we have designed as a linear notation and as a markup language for representing chemical structural formulas [Fujita S, Tanaka N (1999) XΥM Notation for Electronic Communication of Organic Chemical Structures. *J Chem Inf Comput Sci.* 39:903; and Fujita S (1999) XΥM Markup Language (XΥMML) for Electronic Communication of Chemical Documents Containing Structural Formulas and Re-

[1] XΥMTEX is available from my homepage:
`http://imt.chem.kit.ac.jp/fujita/fujitas/fujita.html`.

action Schemes. *J Chem Inf Comput Sci.* 39:915]. All of the structural formulas and the reaction schemes contained in this book have been typeset by means of XΥMTEX. Thereby, the usefulness of the XΥMTEX system has been demonstrated to the utmost extent, even though there still remain several points to be improved.

Acknowledgments

I would like to thank the colleagues who have collaborated with me at the Ashigara Research Laboratories from 1972 to 1997. I would also like to thank Mr. Nobuya Tanaka for reading the entire script and passing on his valuable comments. I gratefully acknowledge the financial support given to our recent project by the Japan Society for the Promotion of Science: Grant-in-Aid for Scientific Research B(2) (No. 14380178, 2002–2003), since XΥMTEX has been grown into a practical tool for publishing this book.

I wish to dedicate this book to Professor Hitosi Nozaki, Professor Emeritus, Kyoto University, Japan, thanking him for guiding me to this area of organic chemistry. I am also grateful to Professor Hisashi Yamamoto of the University of Chicago for suggesting the publication of this book and to the staff of Springer-Verlag for their able cooperation.

Shinsaku Fujita

Kyoto, Japan
December 2003

Table of Contents

III Chromogenic Photography 135

[illegible]

Part I

Fundamentals of Photography

Chapter 1

Photography Based on Silver Halides. An Overview

1.1 Photographic Processes

Color photography uses silver halides (AgX) just as black-and-white (B&W) photography does. However, the silver halides in color photography work only as mediators for transforming light into organic image dyes. In fact, final pictures obtained by color photography are composed of dye images but not of silver images; no traces of silver derivatives remain in the final pictures. To begin with, a color photographic process is briefly compared with a B&W photographic process.

1.1.1 Black-and-White Photographic Process

Although black-and-white (B&W) photography for amateur use has currently been replaced by color photography, it is worth taking a brief view on B&W photographic films in order to comprehend color photographic processes based on silver halides.

Layers in a B&W photographic film B&W photographic films have a multilayer structure in which a silver halide emulsion layer[1] and a protective layer are at least coated on a polymer support. For the sake of simplic-

[1] According to the convention in photography, we use the term "emulsion" to designate a photographic dispersion system. However, such a photographic system contains silver halide grains as a suspension as well as organic compounds as an emulsion.

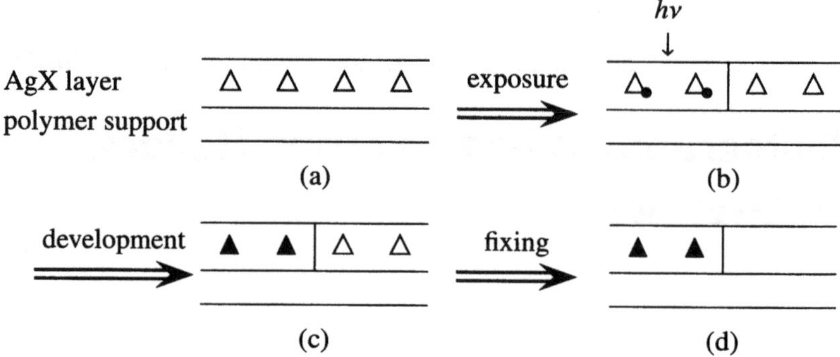

Figure 1.1. Schematic cross-sections of a B&W film during the B&W photographic process. (a) B&W film set for exposure, (b) exposed film, (c) developed film, and (d) fixed film; △: unexposed AgX, △•: exposed AgX with latent image, and ▲: developed silver.

ity, Fig. 1.1(a) schematically illustrates an AgX emulsion layer only on a polymer support.[2]

Silver halide grains are prepared from silver nitrate ($AgNO_3$) and adequate alkali halides (MeX, where Me = Na and K; X = Cl, Br, and I) in the presence of gelatin as a protective colloid. The resulting AgX emulsion displays fluidity at about 40℃, where the colloid protection of gelatin maintains suspension of AgX grains. The emulsion is then coated on a polymer support and cooled below room temperature so as to force sol-to-gel transition, which stabilizes the AgX layer generated.[3] As polymer supports, transparent PET[4] and TAC[5] are used for manufacturing photographic films, while polymer-laminated paper is used for preparing photographic papers.

Exposure and latent image formation A B&W film placed in a camera is exposed to light for taking a picture. Although intensities of light depend upon a picture taken, Fig. 1.1(b) shows an extreme case of exposure in which the left part is fully exposed ($h\nu$)[6] and the right part accepts no light. Such a light exposure forms a latent image (•), which is an invisible precursor that is generated on a surface of an AgX grain exposed to light (△•).

Development Such a latent image is invisible but is capable of giving a visible image upon photographic development. A developing agent such as hydroquinone is used to reduce silver halides into silver metal images (Fig. 1.1(c)). Since a silver halide grain with a latent image is reduced faster

[2]The cross-section of each layer is appropriately drawn, disregarding its real dimension.

[3]This process is conventionally called "setting".

[4]Polyethylene terephthalate.

[5]Cellulose triacetate or triacetylcellulose.

[6]In this book, an exposure to light is expressed by the symbol $h\nu$, where the symbol h represents Planck's constant and ν represents frequency of light.

than a grain without a latent image, an exposed area is developed to give a silver metal image (▲) in accord with eq. 1.1.

$$AgX \xrightarrow[\text{(reduction)}]{\text{developer}} Ag \tag{1.1}$$

Fixing and washing After the developing process, silver halide grains remain unchanged in an unexposed area. They should be removed to assure preservation for a long time (Fig. 1.1(d)). This process is called *fixing* (or fixation), in which silver halide is converted into a water-soluble complex by using sodium thiosulfate[7] according to eq. 1.2. The resulting complex is washed out with water.

$$AgX + 2Na_2S_2O_3 \rightarrow Na_3Ag(S_2O_3)_2(\text{soluble}) + NaX \tag{1.2}$$

1.1.2 Color Photographic Process

Color photography is also based upon the photo-sensitivity of silver halides, but produces organic dyes as color images in place of silver images. Various processes for converting silver images into dye images have been reported and commercialized: i.e., the chromogenic process (so-called conventional color photography), the dye-bleaching process, the dye transfer process (so-called instant color photography), and so on.[8] Among such processes, the chromogenic process, in which a color developer and couplers are used to form imaging dyes, is most popular and thus placed on the market in the forms of color films, color papers, and color reversal films.[9]

Since color photography is based on subtractive color mixing, at least three photo-sensitive layers (blue-, green-, and red-sentitive emulsion layers) are necessary according to three subtractive primaries (yellow, magenta, and cyan dyes). Hence, a color photographic material has a multilayer structure, where such three sets of photo-sensitive units are separated by virtue of interlayers and totally covered by a protective layer.[10]

Layers in a color photographic film For simplicity's sake, Fig. 1.2(a) schematically illustrates an AgX emulsion layer of a single photo-sensitivity on a

[7]The pentahydrate of this compound ($Na_2S_2O_3 \cdot 5H_2O$) is called "hypo", which has been used as a fixing agent.

[8]The term *photographic material* is used to designate various manufactured goods of conventional color photography (e.g., color photographic films, color papers, and color reversal films) as well as those of unconventional photography (e.g., instant color films and dye-bleaching papers).

[9]The total process using color films and color papers is frequently called a *negative-positive process* for the purpose of differentiating it from the color reversal process.

[10]In this book, the term *unit* is used to designate a set of layers that correspond to each photo-sensitivity (blue, green, or red). A photo-sensitive unit may consist of one layer as found in most color papers, or may comprise two or more layers as found in most color negative films of high sensitivity. However, the differentiation between the terms "unit" and "layer" is not so rigorous.

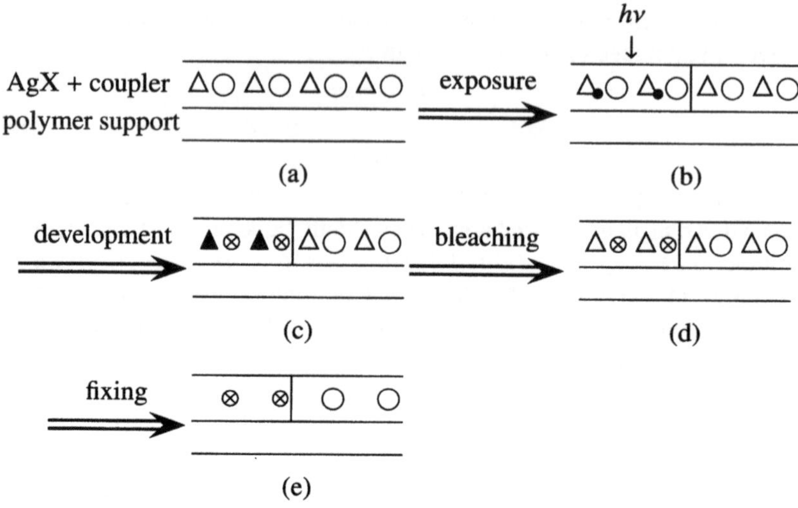

Figure 1.2. Schematic cross-sections of a color film during the color photographic process. One unit among a multilayer structure is illustrated. (a) Color film set for exposure, (b) exposed film, (c) developed film, (d) bleached film, and (e) fixed film; △: unexposed AgX, △•: exposed AgX with a latent image, ▲: developed silver, ○: coupler, and ⊗: dye.

polymer support.[11] The layer contains a dye-forming material called a *coupler* (○), which is colorless but forms a dye image (⊗) during the color development process.

Exposure and latent image formation Figure 1.2(b) shows an extreme case which contains exposed and unexposed areas. AgX grains (△•) in the exposed area produce a latent image (•) on their surface as a result of light exposure.

Color development In color development shown in Fig. 1.2(c), a *p*-phenylene diamine derivative (Dv) is used as a developing agent, which reduces silver halide grains in the exposed area, giving silver metal images (▲) and the corresponding oxidized developer (Dv_{ox}) according to eq. 1.3.

$$\mathrm{AgX} + \mathrm{Dv} \rightarrow \mathrm{Ag} + \mathrm{Dv}_{ox} \tag{1.3}$$

$$\mathrm{Cp}(\bigcirc) + \mathrm{Dv}_{ox} \rightarrow \mathrm{Dye}(\otimes) \tag{1.4}$$

The resulting oxidized developer (Dv_{ox}) couples with a coupler (Cp ○, e.g., a yellow coupler) to give an imaging dye (Dye ⊗, e.g., a yellow dye), as shown in eq. 1.4.

[11]The AgX emulsion is sensitized to capture blue, green, or red light. It is combined with a coupler for generating the corresponding complementary color image, i.e., yellow, magenta, or cyan.

Bleaching Once image dyes are formed, silver metal images along with unchanged silver halides are no longer necessary. Hence, they should be removed to observe a picture (Fig. 1.2(d)). For this purpose, silver metal is returned into a silver cation (Ag^+) by oxidation. This process is called *bleaching*, where potassium hexacyanoferrate(III) ($K_3[Fe(CN)_6]$)[12] or recently an EDTA complex of iron(III)[13] is used as an oxidizing agent.

Fixing and washing The fixing process applied to Ag^+ species in color photography is essentially the same as the one adopted for B&W photography (eq. 1.2). The resulting complex is washed out with water. As a result, dye images (⊗) in an exposed area are ready for observation, as shown in Fig. 1.2(e). Note that unchanged couplers (○) in an unexposed area have no harmful effects on color reproduction because they are colorless.

1.2 Multilayer Structure of Photographic Materials

Although Fig. 1.2 has illustrated only one photo-sensitive unit for the sake of simplicity, any color photographic materials necessitate at least three units in order to produce color images on the basis of subtractive color reproduction. Hence, they have an appropriate multilayer structure according to their uses: color negative films, color papers, and color reversal films.[14] Each of the layers contains various functionalized organic compounds in addition to the fundamental materials described above (i.e., silver halide grains, couplers, and gelatin).

This section has two purposes. One is to show the multilayer structures of photographic materials (a color paper and two color negative films). The other is to show a variety of organic compounds involved in photographic materials. The resulting catalog with respect to their functions (Table 1.1) will turn out future references for the detailed discussions of this book.

1.2.1 Multilayer Structure of Color Papers

As a typical example of multilayer structures, we choose a seven-layered color paper disclosed in a patent [1], as illustrated in Fig. 1.3.[15] This patent contains a minimum set of units, where each unit is composed of one layer and additional layers such as interlayers are also coated.

[12]This is also called "potassium ferricyanide" or "red prussiate".

[13]The term "EDTA complex" is the abbreviation of "ethylenediaminetetraacetato complex".

[14]The manufacture of photographic materials is based on the technology of multilayer coating, where 5 or more layers are coated once by using various surfactants and other additives so as not to disturb laminar flows of respective layers [2].

[15]This patent was assigned to Fuji Photo Film. Example 1 cited here is a test sample to show the superiority of the patent.

Each layer contains organic compounds having various functions. Before we proceed to discussing these functions, their structures are listed here for a brief overview.

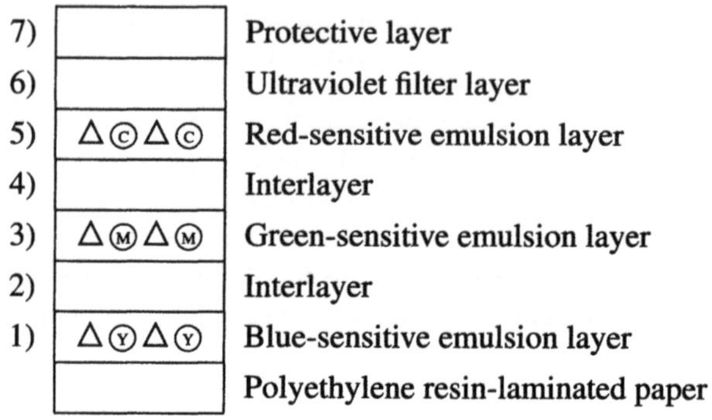

7)		Protective layer
6)		Ultraviolet filter layer
5)	△Ⓒ△Ⓒ	Red-sensitive emulsion layer
4)		Interlayer
3)	△Ⓜ△Ⓜ	Green-sensitive emulsion layer
2)		Interlayer
1)	△Ⓨ△Ⓨ	Blue-sensitive emulsion layer
		Polyethylene resin-laminated paper

Figure 1.3. Schematic cross-section of a color paper having multilayer structure. The support is paper laminated with polyethylene resin, which contains a white pigment (TiO_2/ZnO) and a fluorescent whitening agent (4,4′-bis(5-methylbenoxazolyl)stilbene). This diagram illustrates silver halide grains and couplers only. △: Silver halide grain; Ⓨ: Yellow coupler; Ⓜ: Magenta coupler; Ⓒ: Cyan coupler.

It should be noted that the multilayer structure illustrated in Fig. 1.3 has an arrangement of layers: (support) B–G–R (top), whereas that of a color negative film shown later (e.g., Fig. 1.19) has a reverse arrangement of layers: (support) R–G–B (top). The former arrangement is possible because the color photographic paper uses a silver chloride or silver chlorobromide emulsion with a rich content of chloride ions, which has less absorption within the blue region of light than bromide-based emulsions.

Layer 1 [Blue-sensitive emulsion layer] This layer mainly comprises a silver chloride emulsion, a mixture of yellow couplers, and gelatin.
The blue-sensitive emulsion cited here is spectrally sensitized by using sensitizing dyes shown in Fig. 1.4 (**1**, **2**, and **3**). Each sensitizing dye molecule is adsorbed on the surface of a silver halide grain and absorbs blue light (about 400–500nm). Note that blue is the complementary color to yellow and that the sensitizing dye that absorbs blue light is yellow in color.
The structures of the yellow couplers incorporated in Layer 1 are shown in Fig. 1.5 (**4** and **5**).[16] The yellow couplers are dispersed in a gelatin colloidal solution by the help of high-boiling solvents (oils) and surfactants.
In general, such a silver halide emulsion (dispersion) and an oil-dispersion of such couplers and other organic chemicals are prepared separately as

[16]The term "yellow coupler" is used to designate a yellow-dye forming coupler. It should be noted that a yellow coupler is colorless in itself but forms a yellow dye on the action of an oxidized color developer. The same situation is true for the terms "magenta coupler" and "cyan coupler".

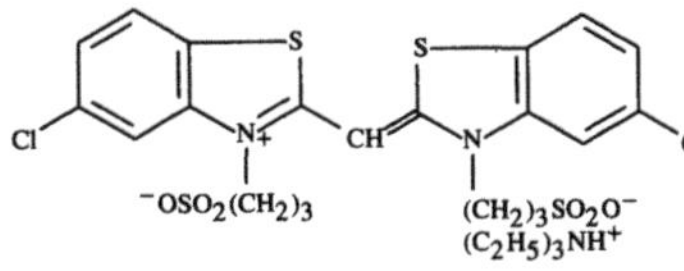

Figure 1.4. Sensitizing dyes for a blue-sensitive emulsion. Each sensitizing dye is adsorbed on the surface of a silver halide grain. The dye absorbs blue light and transfers the resulting excited electron into the silver halide grain so as to produce a latent image on the surface of the grain.

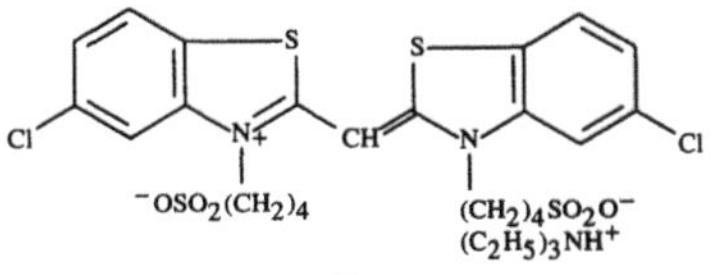

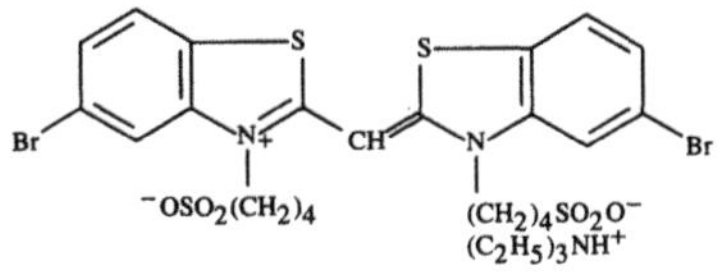

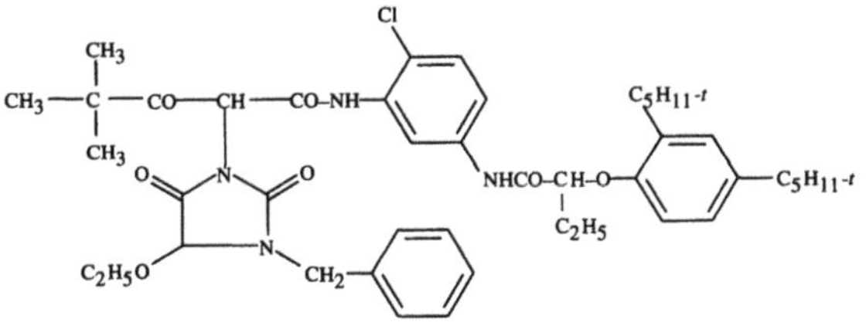

Figure 1.5. Yellow couplers for color papers. They are two-equivalent couplers of pivaloylacetanilide type, where a five-membered moiety is removed as a heterocyclic anion during color development.

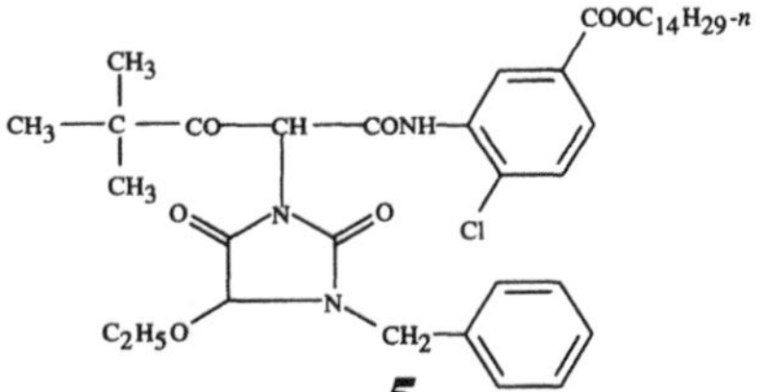

aqueous gelatin dispersions. They are then combined and coated on a support (a polymer-laminated paper or a polymer film).

Layer 1 also contains stabilizers (**6–9** listed in Fig. 1.6), which have been added to the oil-dispersion of the yellow couplers (**4** and **5**) for ensuring the light or heat stability of dyes derived from the couplers. One or more high-boiling solvents (oils) are used to disperse these organic compounds

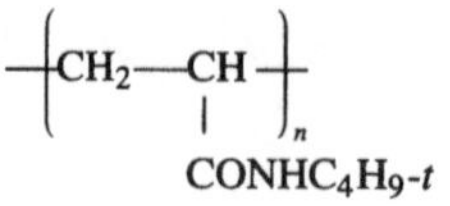

Figure 1.6. Dye stabilizers. Dye stabilizers are used to enhance the light or heat stability of dyes formed from couplers. Since there are several mechanisms for dye fading, various types of compounds are disclosed as dye stabilizers in patents. The compounds **10** and **11** are used as anti-staining agents for magenta azomethine dyes.

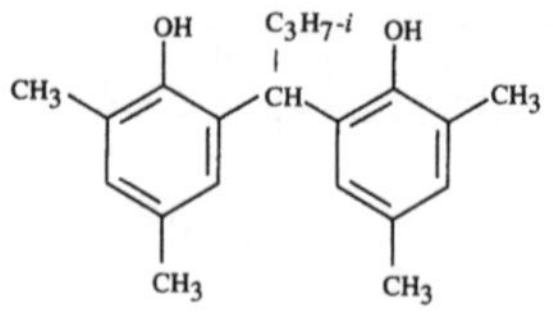

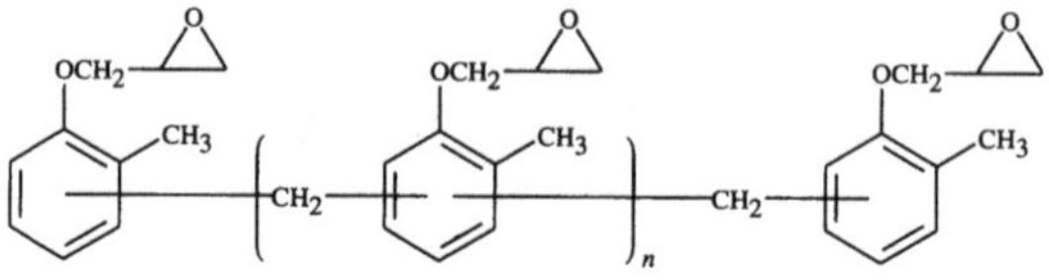

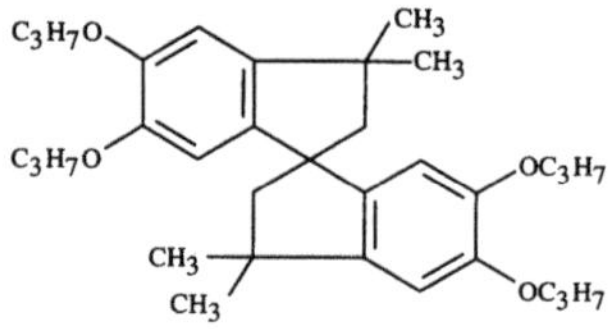

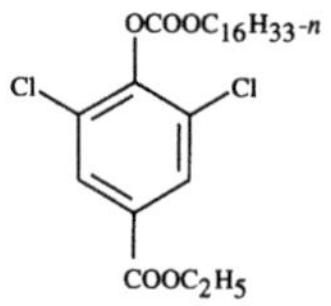

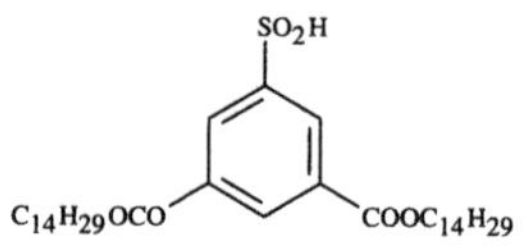

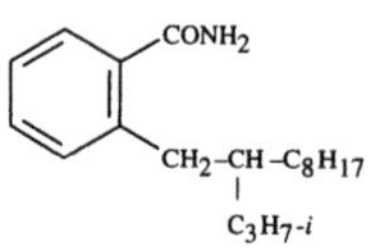

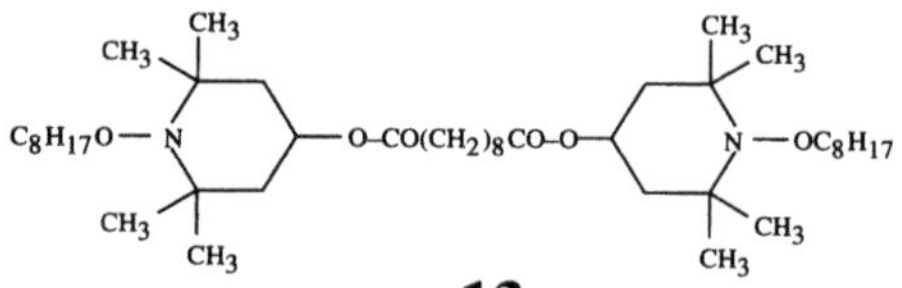

in the gelatin solution for Layer 1. As a result, Layer 1 involves **14** as such a solvent (Fig. 1.7).

Layer 2 [Interlayer] Layer 2 is an interlayer for inhibiting color mixing between the blue-sensitive layer and the green-sensitive one. It contains scavengers of an oxidized developer (**19** and **20**) as well as hexadecyl 4-hydroxybenzoate as a color-mixing inhibitor auxiliary.

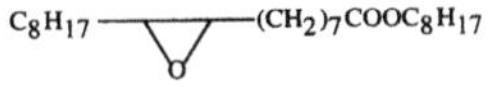

C_8H_{17}—(epoxide)—$(CH_2)_7COOC_8H_{17}$

14

i-C_3H_7—C_6H_4—O—P(=O)(—O—C_6H_4—C_3H_7-i)—O—C_6H_4—C_3H_7-i

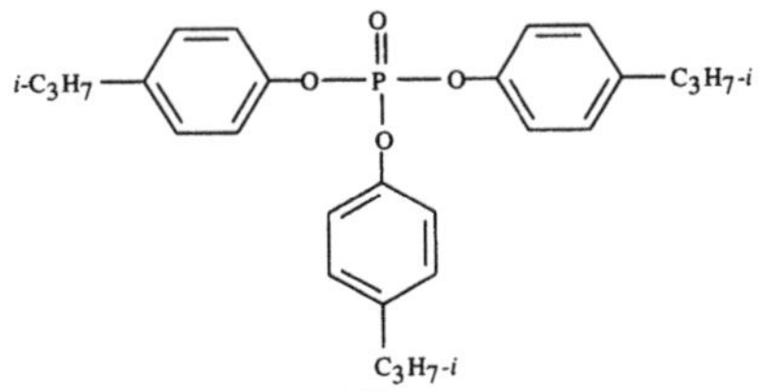

15

$C_4H_9OCO–(CH_2)_8–COOC_4H_9$

16

$O{=}P(OC_6H_{13}\text{-}n)_3$

17

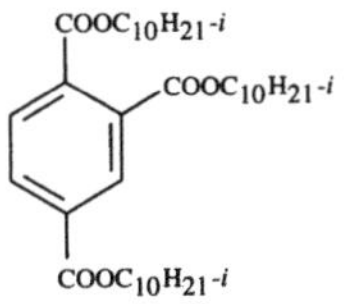

$COOC_{10}H_{21}$-i, $COOC_{10}H_{21}$-i, $COOC_{10}H_{21}$-i

18

Figure 1.7. Solvents for color photographic papers. The substitution position of each isopropyl group in **15** is tentatively presumed to be the *p*-position, though it has not been specified in the original patent. Organic compounds having a ballast group (e.g., couplers) are dispersed as oil droplets by using various high-boiling solvents (oils) and appropriate surfactants. This technique is called *oil-protection.*

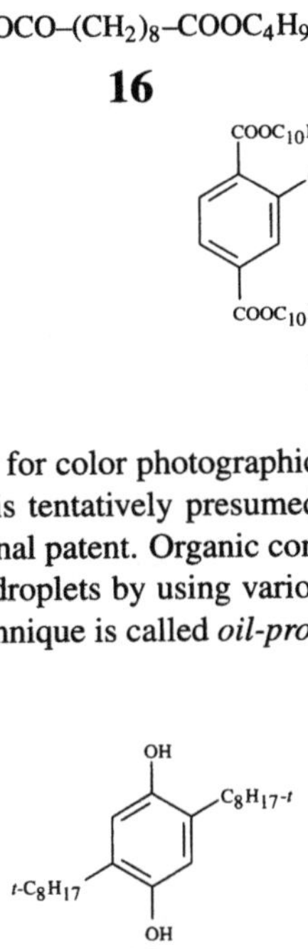

OH, C_8H_{17}-t, t-C_8H_{17}, OH

19

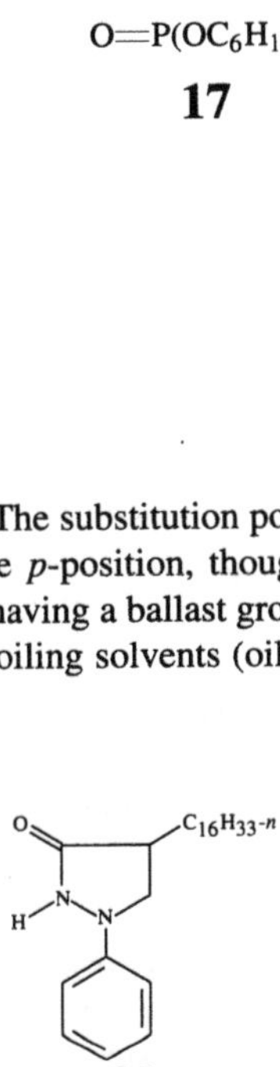

O, $C_{16}H_{33}$-n, H, N, N

20

OH, $C(CH_3)_2$-CH_2CH_2-$COOC_6H_{13}$, $C_6H_{13}OCO$-CH_2CH_2-$C(CH_3)_2$, OH

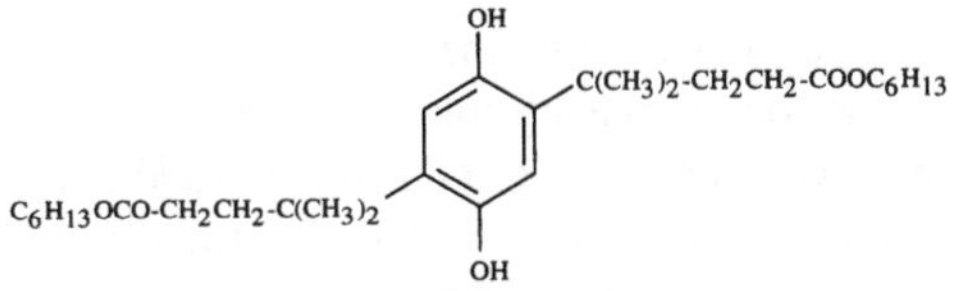

21

Figure 1.8. Scavengers for color photographic papers. Most of them are reducing agents with ballast groups to scavenge an excess of oxidized color developers (quinone diimides) or an additional portion generated from color developers remaining after processing.

Layer 2 contains a mixture of UV absorbents (**22**, **23**, **26**, and **27** in Fig. 1.9), a high-boiling solvent (**15**), and gelatin.

Layer 3 [Green-sensitive emulsion layer] Layer 3 contains a green-sensitive emulsion and gelatin. The emulsion is a mixture of a large-size emulsion

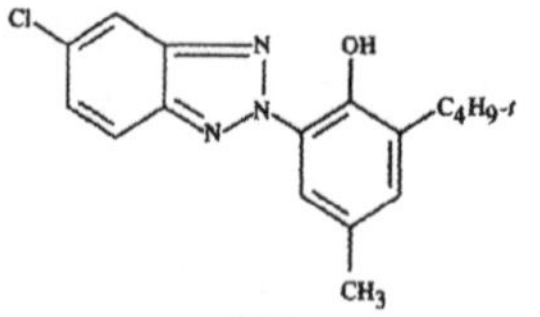

Figure 1.9. UV absorbents for color photographic papers. UV absorbents used in Layers 2, 4, and 6 in Fig. 1.3 absorb ultraviolet light to prevent the light fading of dye images or to prevent undesired UV exposure of each emulsion layer.

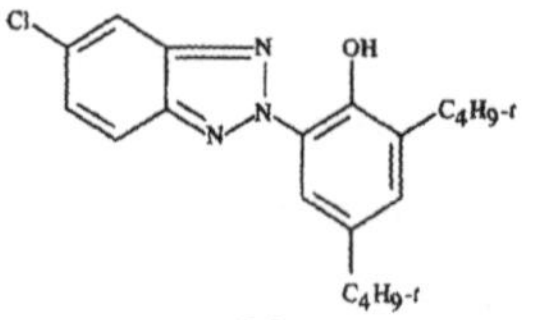

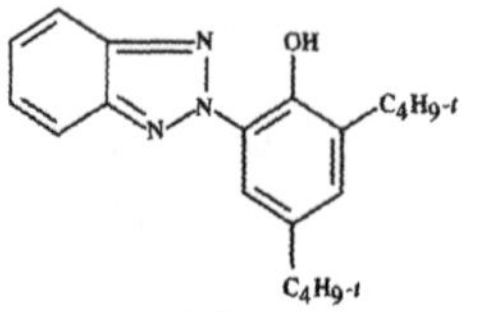

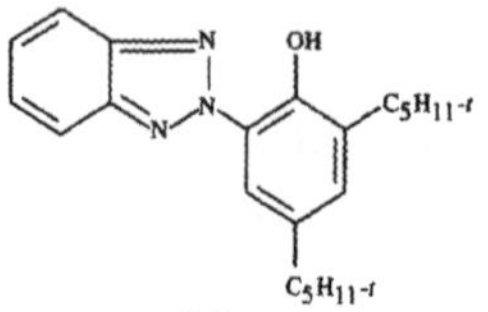

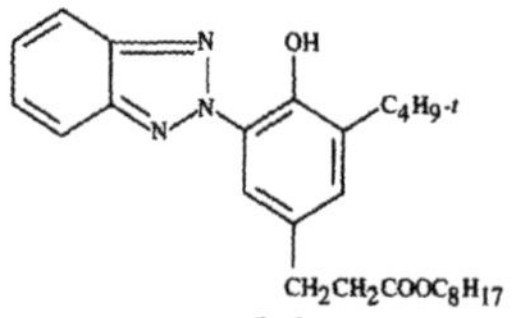

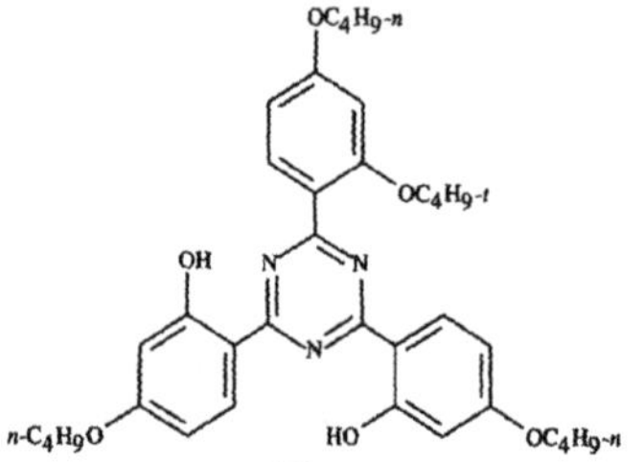

and a small-size emulsion, both of which are sensitized by sensitizing dyes (**28** and **29** in Fig. 1.10). These sensitizing dyes absorb green light (about 500–600 nm) and exhibit magenta color.

Layer 3 contains a mixture of magenta couplers (**30**–**32**) shown in Fig. 1.11. Magenta dyes derived from the couplers absorb green light. Note that magenta is the complementary color to green.

Layer 3 contains dye stabilizers (**10**, **11**, and **33**) in addition to **7**, **9**, and **20**. Layer 3 also contains high-boiling solvents (**15**, **16**, and **17**) for dispersing the magenta couplers. UV absorbents used in Layer 3 are **24** and **25** in addition to **22** and **23**.

Layer 4 [Interlayer] Layer 4 is an interlayer for inhibiting color mixing between the green-sensitive layer and the red-sensitive one. It contains oxidized developer scavengers (**20** and **21**) as well as hexadecyl 4-hydroxybenzoate as a color-mixing inhibitor auxiliary.

Layer 4 further contains the same mixture of UV absorbents as Layer 2 (**22**, **23**, **26**, and **27** in Fig. 1.9), a high-boiling solvent (**15**), and gelatin.

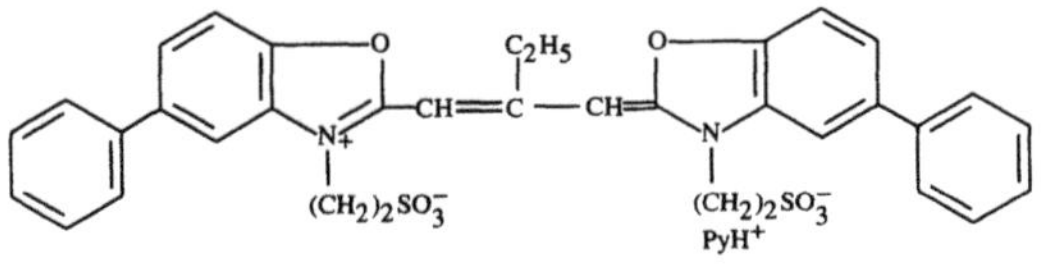

28

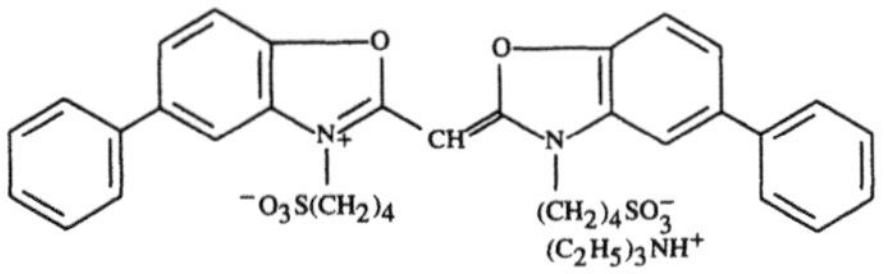

29

Figure 1.10. Sensitizing dyes for a green-sensitive emulsion. They are used in Layer 3 shown in Fig. 1.3.

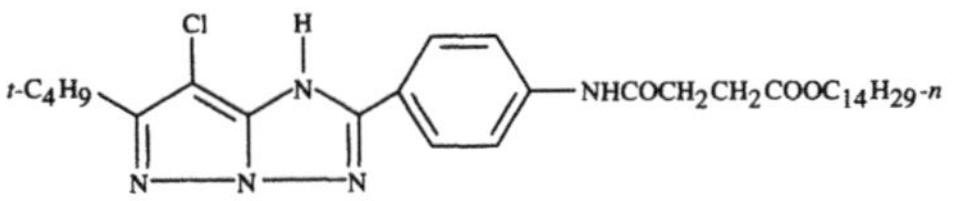

30

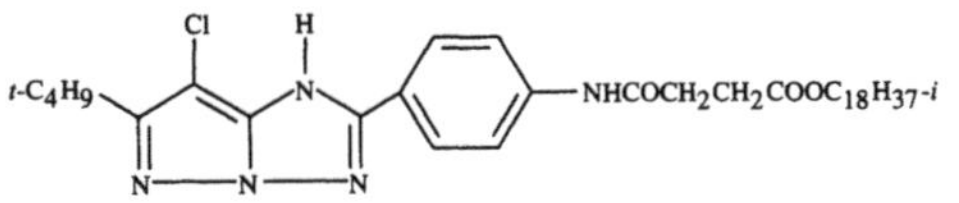

31

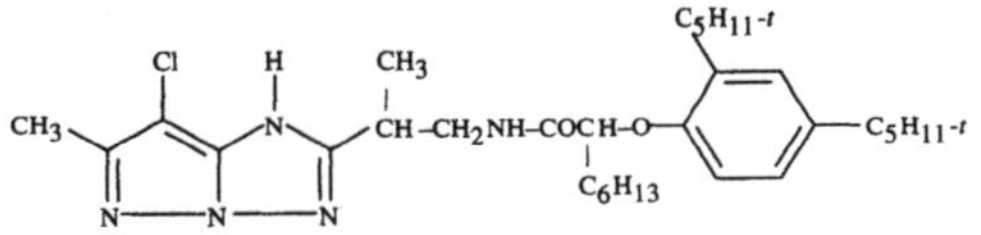

32

Figure 1.11. Pyrazolotriazole magenta couplers for color photographic papers. These couplers are two-equivalent couplers, from which a chloride anion is released during coupling with an oxidized color developer. They are incorporated in Layer 3 of Fig. 1.3.

Layer 5 [Red-sensitive emulsion layer] Layer 5 contains a red-sensitive emulsion and gelatin. The emulsion is a mixture of a large-size emulsion and a small-size emulsion, both of which are sensitized by sensitizing dyes (**34** and **35** in Fig. 1.13).

Layer 5 contains a mixture of cyan couplers (**36**–**39**) shown in Fig. 1.14. Cyan dyes derived from the couplers absorb red light. Note that cyan is the complementary color to red.

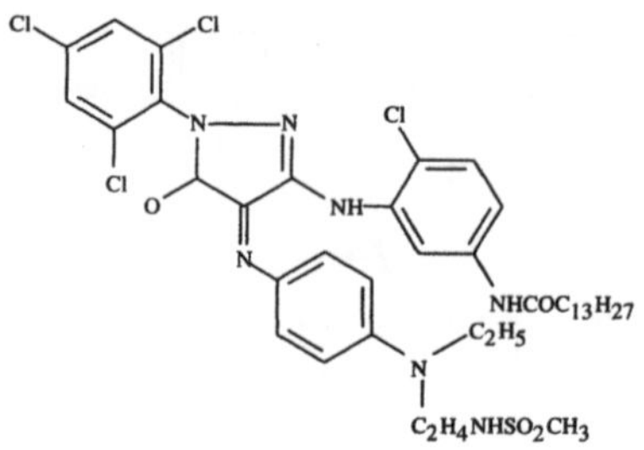

33

Figure 1.12. Dyes for color photographic papers. The dye (**33**) is used as a dye stabilizer in Layer 3.

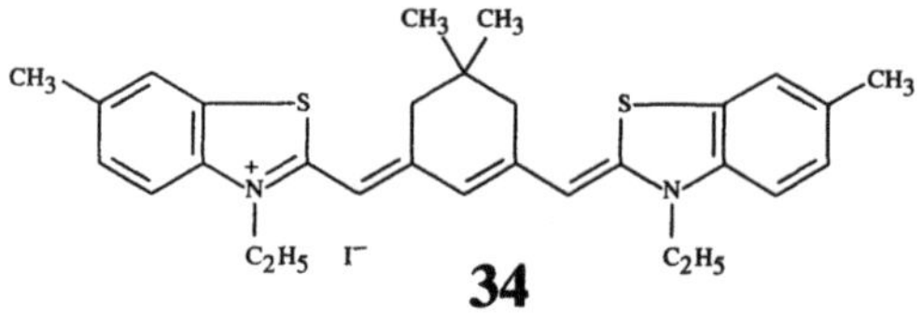

34

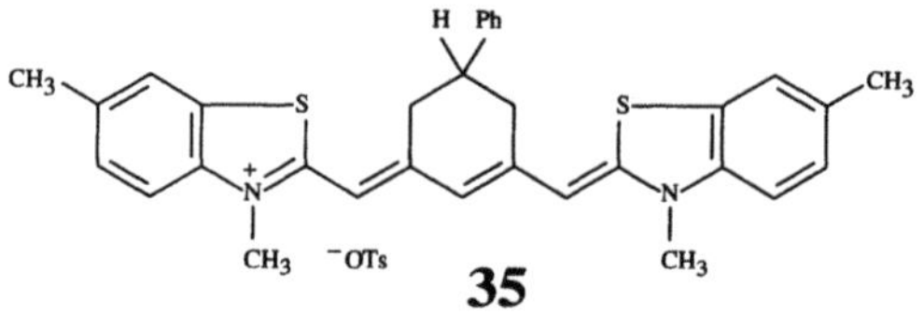

35

Figure 1.13. Sensitizing dyes for a red-sensitive emulsion. They are used in Layer 5. The symbol TsO^- represents a *p*-toluenesulfonate anion.

It contains the oxidized developer scavenger (**20**), the dye stabilizer (**10**), and the high-boiling solvent (**15**). Additional dye stabilizers (**12** and **13** in Fig. 1.6) are involved in this layer.

Layer 6 [UV filter layer] Layer 6 contains the same mixture of UV absorbents as Layer 2 does (**22**, **23**, **26**, and **27** in Fig. 1.9), as well as a high-boiling solvent (**18**) and gelatin.

Layer 7 [Protective layer] This layer contains gelatin, acrylic modified copolymer of polyvinyl alcohol (modification degree 17%), liquid paraffin, surfactants (**40** and **41** in Fig. 1.15), polydimethylsiloxane, and silicon dioxide.

Light scattering on the surface of a silver halide grain reduces the sharpness of dye images. This effect is called *irradiation*. One or more dyes such as **42–45** (Fig. 1.16) are added to inhibit the irradiation by absorbing scattered light. These

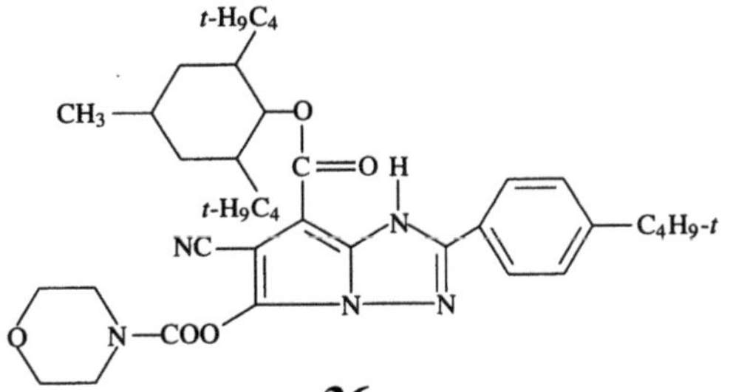

36

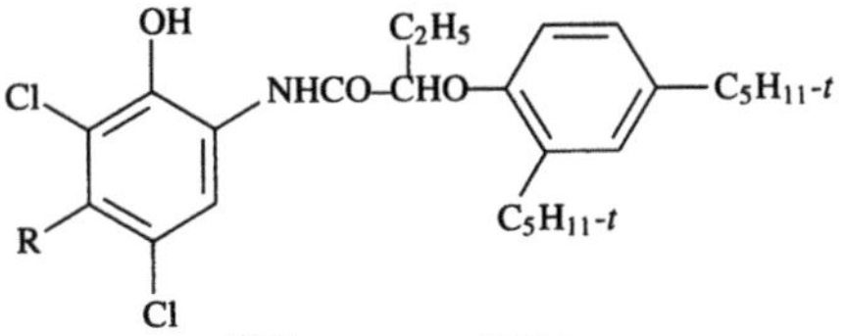

37 (R = CH_3)
38 (R = C_2H_5)

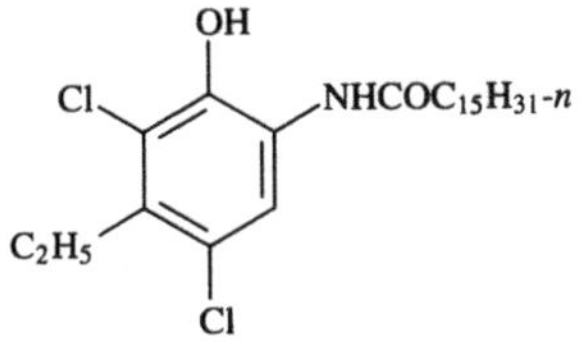

39

Figure 1.14. Cyan couplers for color papers. The coupler (**36**) at the top of this figure is a 1H-pyrrolo[1,2-*b*][1,2,4]triazole coupler. The others are cyan couplers of 2-amidophenol type. They are incorporated in Layer 5 of Fig. 1.3.

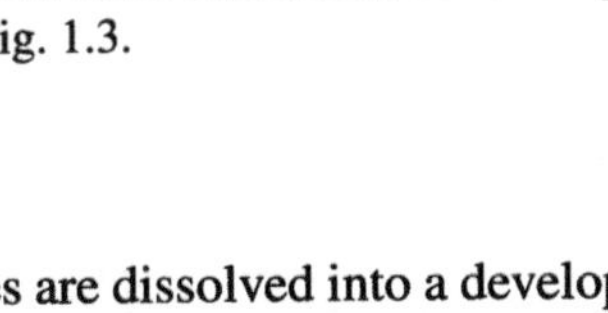

40

41

Figure 1.15. Surfactants for color photographic papers. They are used in Layer 7 shown in Fig. 1.3.

dyes are dissolved into a developing solution or one of several other photographic processing solutions.

In general, a color photographic material must maintain its multilayer structure during development processes. But it must be capable of swelling to some extent so as to permit the penetration of a developer and other reagents. The term "to some extent" is important, since over-swelling to destruct the gelatin network is harmful to yield a completed picture of high quality. To prevent such over-swelling, the photographic material is hardened by appropriate cross-linking agents (hardeners). The color paper cited here [1] has used gelatin hardeners listed in Fig. 1.17. Obviously, they are bifunctional organic compounds which link gelatin chains together.

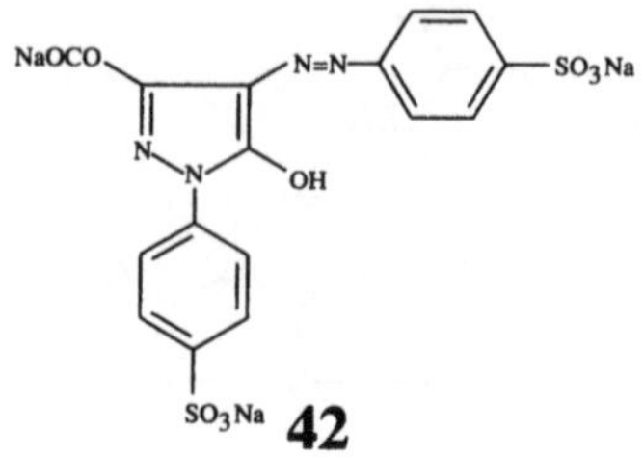

Figure 1.16. Anti-irradiation dyes for inhibiting irradiation. These dyes are added to each of the emulsion layers in order to prevent light scattering due to silver halide grains.

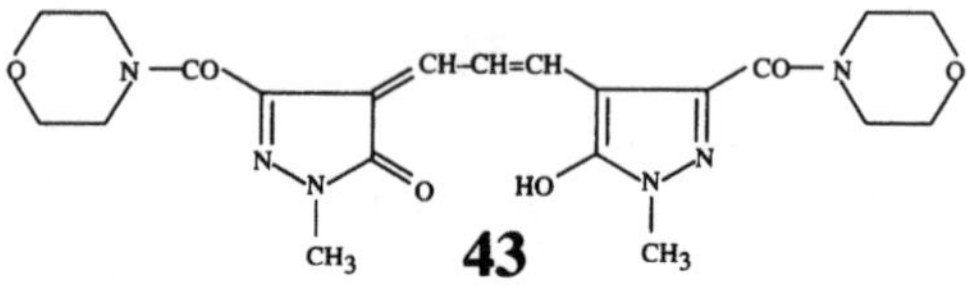

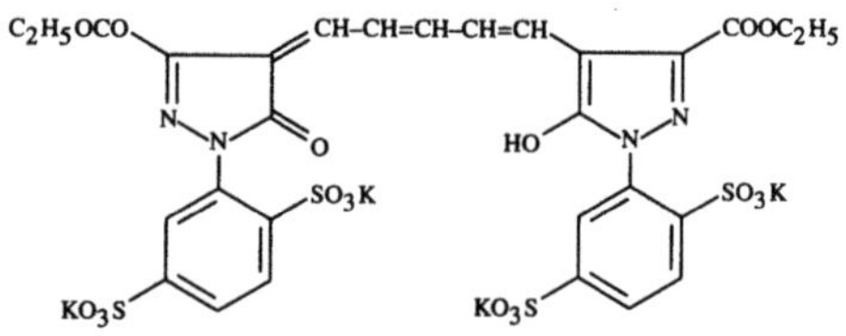

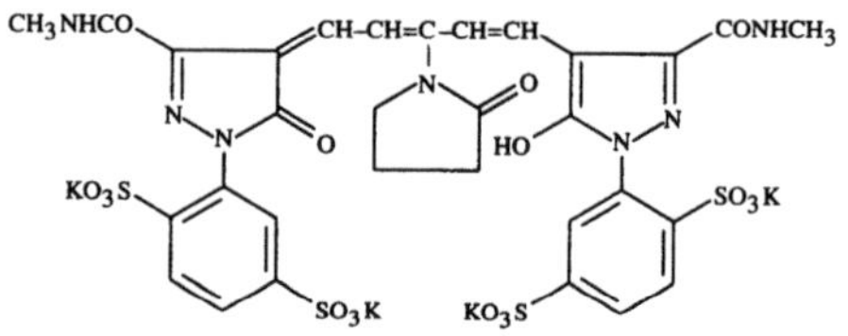

$CH_2{=}CHSO_2{-}CH_2CONH{-}CH_2CH_2{-}NHCOCH_2{-}SO_2CH{=}CH_2$

46

$CH_2{=}CHSO_2{-}CH_2CONH{-}CH_2CH_2CH_2{-}NHCOCH_2{-}SO_2CH{=}CH_2$

47

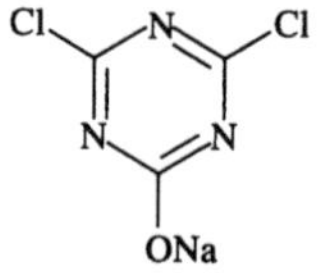

48

Figure 1.17. Gelatin hardeners for color photographic papers. These compounds are used to prevent the over-swelling of gelatin coating.

Each layer of a photographic material uses gelatin as a protective colloid. Since gelatin is a natural product, it suffers from the attack of microorganisms for long-time storage. Hence, antiseptics shown in Fig. 1.18 are added to each layer.

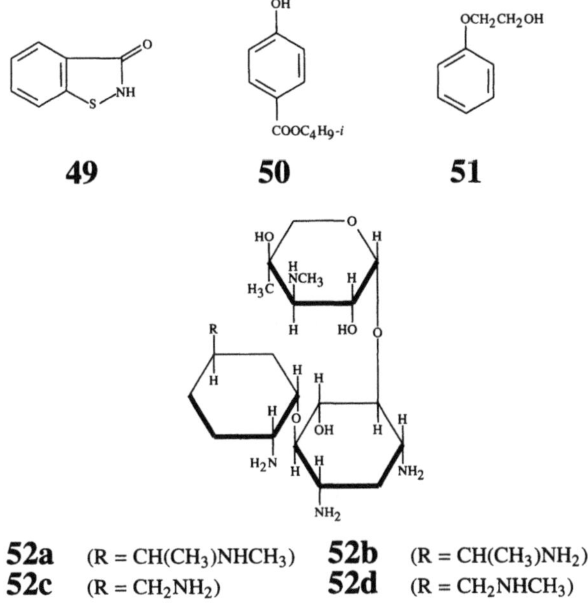

Figure 1.18. Antiseptics for gelatin.

1.2.2 Multilayer Structure of Color Films

In modern color photography, a more complicated multilayer structure is adopted as exemplified by the color negative film shown in Fig. 1.19. Thus, each photosensitive unit (blue-, green-, or red-sensitive unit) is divided into two or three subunits (layers) in order to bring about high sensitivity with maintaining other image qualities such as granularity.

Various organic compounds have been used in a 13-layered negative film disclosed in a patent [3], which is regarded as a typical example to our knowledge.[17] The ID number of each compound used in the patent is also cited for the sake of convenience.

Layer 1 [Antihalation layer][18] This layer contains UV absorbents (UV-1 and UV-2 in Fig. 1.20), an oxidized developer scavenger (DOXS-2 in Fig. 1.21), dyes for compensatory printing density (CD-2, MD-1, MM-1, and YD-1

[17]This patent was assigned to Eastman Kodak. Example 3 cited here is a comparative control at that time to show the superiority of the patent.

[18]Some of the light through a lens is not perpendicular to the interface between the surface of a film base and the lowest silver halide emulsion layer. Such slanted light is reflected at the back of the film base and irradiates the coated emulsion again. This effect that causes a blur of the image is called *halation*. An antihalation layer prevents the halation by means of antihalation dyes, black colloidal silver sol, or other light absorbents. On the other hand, the blurring effect due to the surface reflections of silver halide grains within an emulsion layer is called *irradiation*.

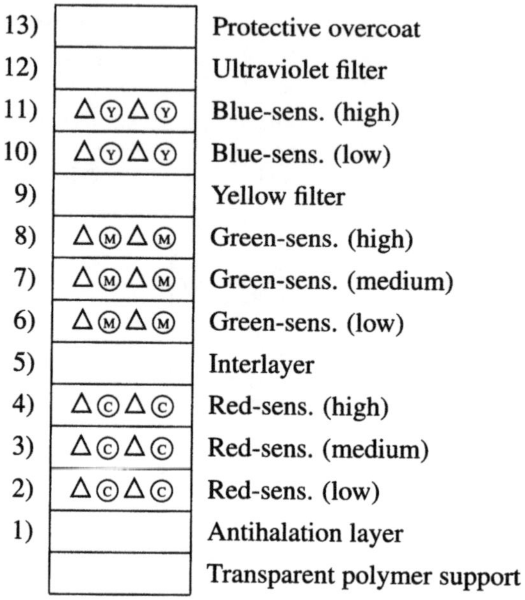

13)		Protective overcoat
12)		Ultraviolet filter
11)	△Ⓨ△Ⓨ	Blue-sens. (high)
10)	△Ⓨ△Ⓨ	Blue-sens. (low)
9)		Yellow filter
8)	△Ⓜ△Ⓜ	Green-sens. (high)
7)	△Ⓜ△Ⓜ	Green-sens. (medium)
6)	△Ⓜ△Ⓜ	Green-sens. (low)
5)		Interlayer
4)	△Ⓒ△Ⓒ	Red-sens. (high)
3)	△Ⓒ△Ⓒ	Red-sens. (medium)
2)	△Ⓒ△Ⓒ	Red-sens. (low)
1)		Antihalation layer
		Transparent polymer support

Figure 1.19. Schematic cross-section of a multilayer structure of a color negative film. The support is a polyester (or triacetylcellulose film) for color films. This scheme illustrates silver halide grains and couplers only. △: Silver halide grain; Ⓨ: Yellow coupler; Ⓜ: Magenta coupler; Ⓒ: Cyan coupler.

$C(CN)_2=CH-CH=CH-N(C_6H_{13})_2$

53 (UV-1)

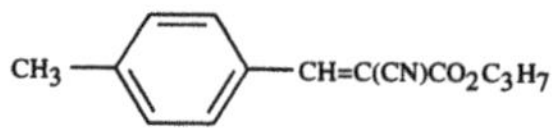

$CH_3-C_6H_4-CH=C(CN)CO_2C_3H_7$

54 (UV-2)

Figure 1.20. UV absorbents for color photography. UV absorbents used in Layers 1 and 12 (Fig. 1.19) absorb ultraviolet light to prevent the light fading of dye images.

in Fig. 1.22), solvents (S-4 and S-9 in Fig. 1.23), disodium salt of 3,5-disulfocatechol, gelatin, and black colloidal silver sol.

Layer 2 [Low sensitivity red-sensitive layer] This layer comprises a blend of a lower sensitivity emulsion (a tabular silver iodobromide emulsion: 1.3 mol% iodide; average grain diameter, 0.53 μm; and thickness, 0.09 μm thick) and a higher sensitivity emulsion (a tabular silver iodobromide emulsion: 4.1 mol% iodide; average grain diameter, 1.04 μm; and thickness, 0.09 μm thick). They are spectrally sensitized to absorb red light but sensitizing dyes have not been disclosed. As organic functionalized compounds, it contains a bleach accelerator (BAR-1 in Fig. 1.24), a cyan coupler (C-1 in Fig. 1.25), and a magenta-colored cyan coupler (CM-1 in Fig. 1.25). The oxidized developer scavenger (DOXS-3 in Fig. 1.21) and the solvents (S-1

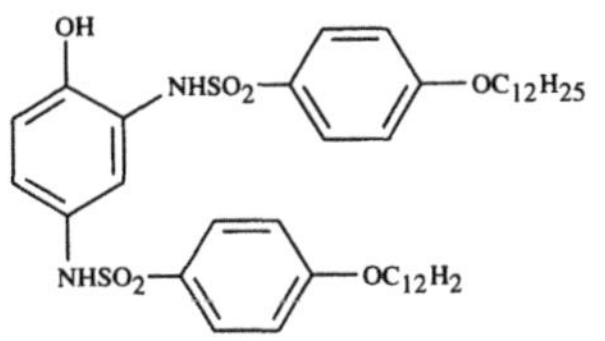

55 (DOXS-2)

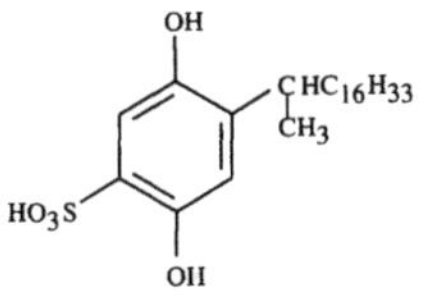

56 (DOXS-3)

Figure 1.21. Scavengers for color photography. They scavenge an excess of oxidized color developers (quinone diimides) or an additional portion generated from color developers remaining after processing.

and S-2 in Fig. 1.23) are also used here. 4-Hydroxy-6-methyl-1,3,3a,7-tetrazaindene (TAI) is a universal stabilizer for silver halide emulsions based on gelatin (Fig. 1.26).

Layer 3 [Medium sensitivity red-sensitive layer] This layer contains a red-sensitive emulsion (a tabular silver iodobromide emulsion: 4.1 mol% iodide; average grain diameter, 1.39 μm; and thickness, 0.12 μm thick). This layer contains a cyan DIR coupler (D-1 in Fig. 1.27).[19] The magenta-colored cyan coupler (CM-1 in Fig. 1.25), the cyan coupler (C-1 in Fig. 1.25), the solvents (S-1 and S-3 in Fig. 1.23), TAI (Fig. 1.26) and gelatin are also used in this layer.

Layer 4 [High sensitivity red-sensitive layer] This layer contains a red-sensitive emulsion (a tabular silver iodobromide emulsion: 4.1 mol% iodide; average grain diameter, 2.93 μm; and thickness, 0.13 μm thick).
In addition to the DIR coupler (D-1), this layer contains another cyan DIR coupler (D-2 in Fig. 1.27). The magenta-colored cyan coupler (CM-1 in Fig. 1.25), the cyan coupler (C-1 in Fig. 1.25), the solvents (S-1, S-3, and S-4 in Fig. 1.23), TAI (Fig. 1.26), and gelatin are also involved.

Layer 5 [Interlayer] This layer contains gelatin to prevent undesired interaction between the red-sensitive unit and the green-sensitive unit.

Layer 6 [Low sensitivity green-sensitive layer] This layer comprises a blend of a lower sensitivity green-sensitized emulsion (a tabular silver iodobromide emulsion: 1.3 mol% iodide; average grain diameter, 0.53 μm; and thickness, 0.09 μm thick) and a higher sensitivity green-sensitized emulsion (a tabular silver iodobromide emulsion: 4.1 mol% iodide; average grain diameter, 1.04 μm; and thickness, 0.09 μm thick).
As organic compounds, the layer involves a magenta coupler (M-1 in Fig. 1.28) and a yellow-colored magenta coupler (MM-2 in Fig. 1.28). An oxidized developer scavenger (DOXS-3 in Fig. 1.21), a solvent (S-4 in Fig. 1.23), TAI (Fig. 1.26), and gelatin are also contained.

[19] DIR is the acronym of "development inhibitor releasing".

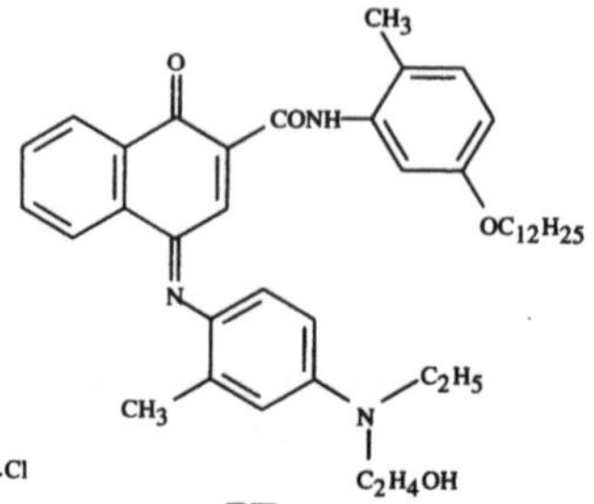

57 (CD-2)

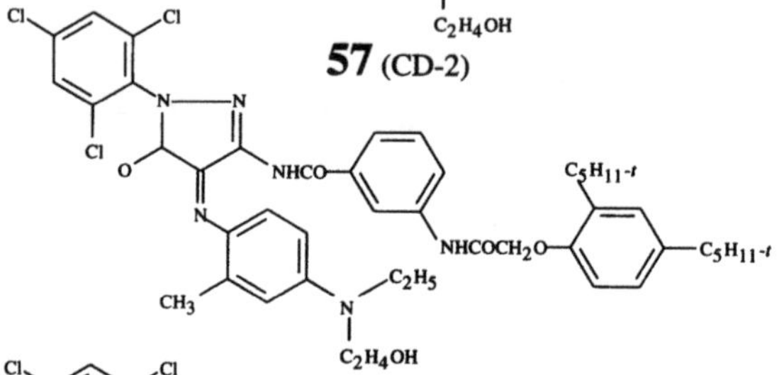

58 (MD-1)

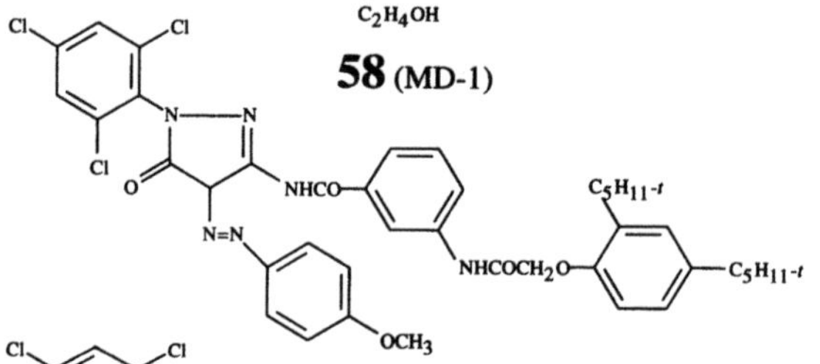

59 (MM-1)

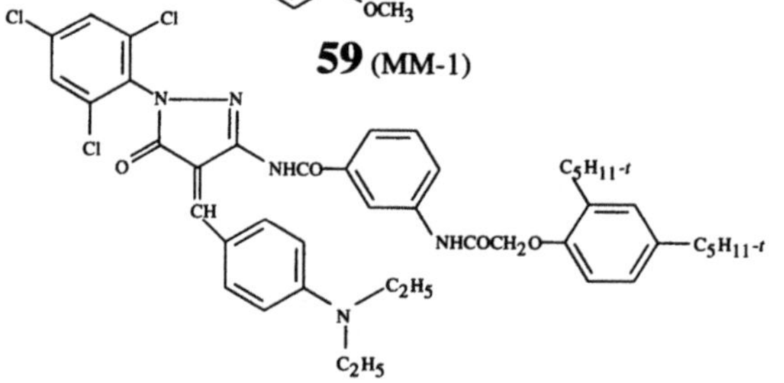

60 (YD-1)

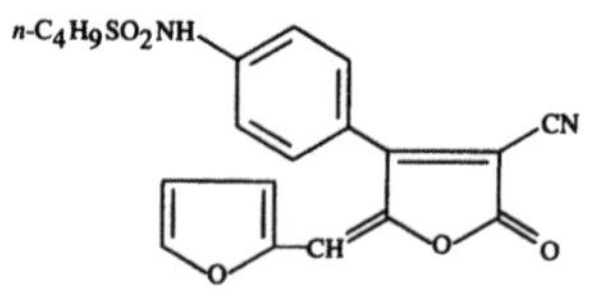

61 (YD-2)

Figure 1.22. Dyes for color photography. The dyes (**57–60**) are used for compensatory printing density in Layer 1 (Fig. 1.19). The dye (**61**) is used as a yellow filter in Layer 9.

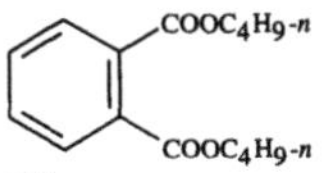

Figure 1.23. Solvents for color photography. Organic compounds having a ballast group are dispersed as oil droplets by using various high-boiling solvents (oils) and appropriate surfactants. This technique is called *oil-protection.*

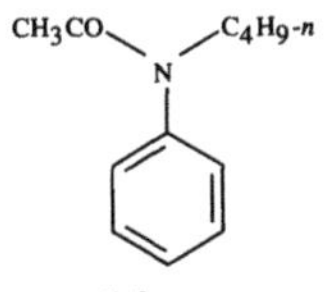

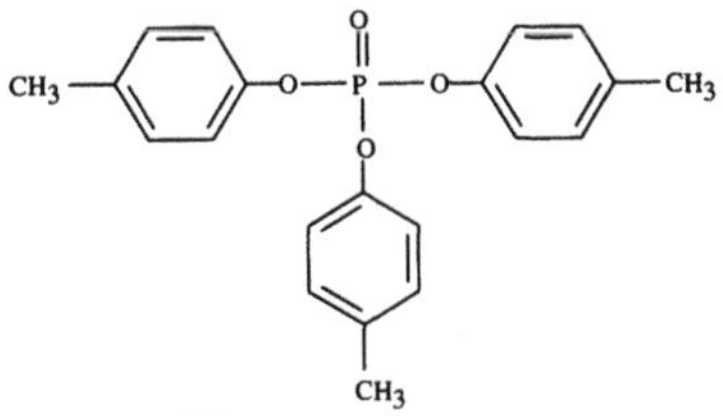

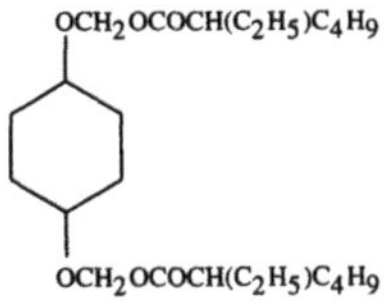

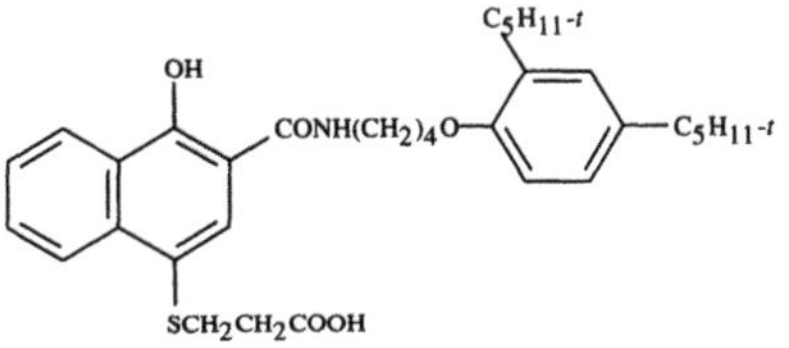

Figure 1.24. Bleach accelerator releasing (BAR) coupler for color photography. Layers 2, 7, 8, 10, and 11 (Fig. 1.19) contain this compound, which releases a sulfur-containing bleach accelerator ($HSCH_2CH_2COOH$).

Layer 7 [Medium sensitivity green-sensitive layer] This layer contains a green-sensitized emulsion (a tabular silver iodobromide emulsion: 4.1 mol% iodide; average grain diameter, 1.23 μm; and thickness, 0.12 μm thick). The layer involves the DIR coupler (D-1), the yellow-colored magenta coupler (MM-2), the magenta coupler (M-1), the oxidized developer scavenger (DOXS-3 in Fig. 1.21), a solvent (S-4 in Fig. 1.23), TAI (Fig. 1.26), and gelatin, which have been described above for the other layers.

Layer 8 [High sensitivity green-sensitive layer] This layer contains a green-sensitized emulsion (a tabular silver iodobromide emulsion: 4.1 mol%

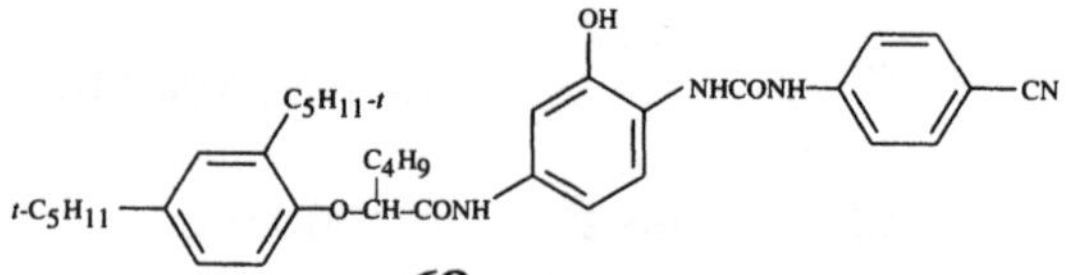

68 (C-1)

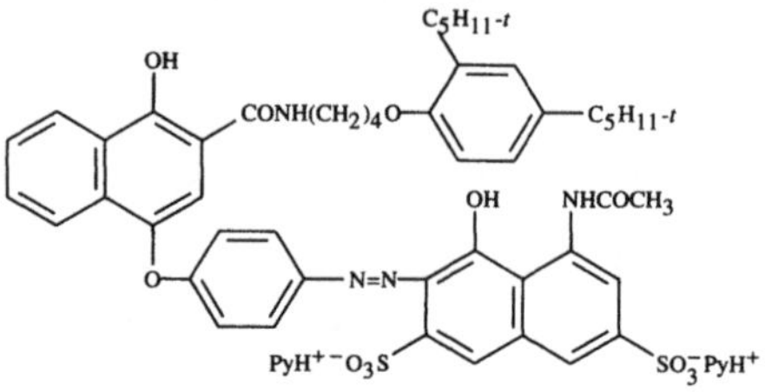

69 (CM-1)

Figure 1.25. Cyan couplers for color photography. The top compound (**68**) is a cyan coupler of 2-ureidophenol type. The bottom compound (**69**) is a magenta-colored cyan coupler.

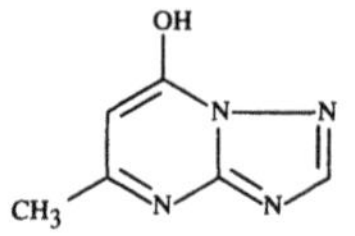

70 (TAI)

Figure 1.26. Heterocyclic stabilizer for a silver halide emulsion. A representative is 4-hydroxy-6-methyl-1,3,3a,7-tetrazainden (TAI), which is used as a stabilizer in all of the emulsion layers.

iodide; average grain diameter, 2.93 μm; and thickness, 0.13 μm thick). Two DIR couplers (D-3 and D-4 in Fig. 1.27) are involved in addition to the yellow-colored magenta coupler (MM-2), the magenta coupler (M-1), the oxidized developer scavenger (DOXS-3 in Fig. 1.21), solvents (S-1 and S-4 in Fig. 1.23), TAI (Fig. 1.26), and gelatin, which have been described above for the other layers.

Layer 9 [Yellow filter layer] This layer contains a yellow filter dye (YD-2 in Fig. 1.22) and gelatin.

Layer 10 [Low sensitivity blue-sensitive layer] This layer comprises a blend of a lower sensitivity blue-sensitized emulsion (a tabular silver iodobromide emulsion: 1.3 mol% iodide; average grain diameter, 0.53 μm; and thickness, 0.09 μm thick), a medium sensitivity blue-sensitized emulsion (a tabular silver iodobromide emulsion: 4.1 mol% iodide; average grain diameter, 0.80 μm; and thickness, 0.09 μm thick), and a higher sensitiv-

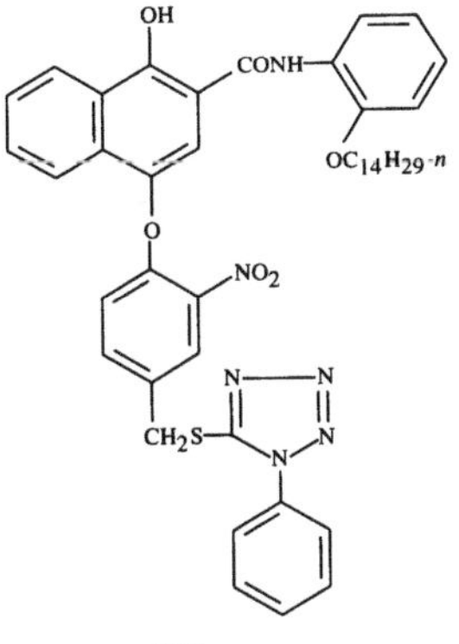

71 (D-1)

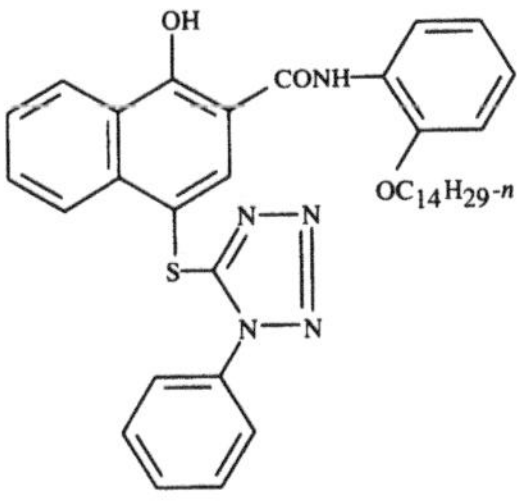

72 (D-2)

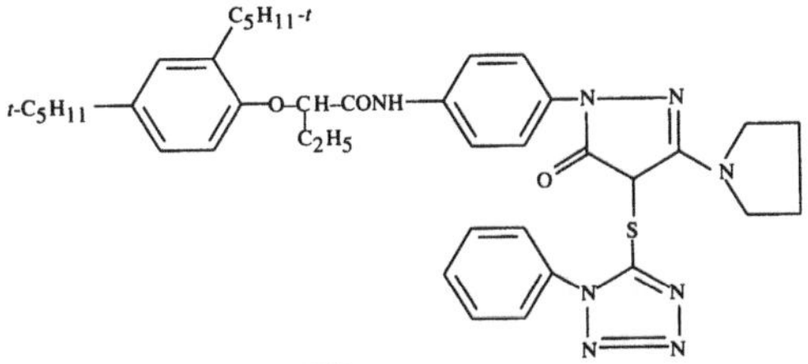

73 (D-3)

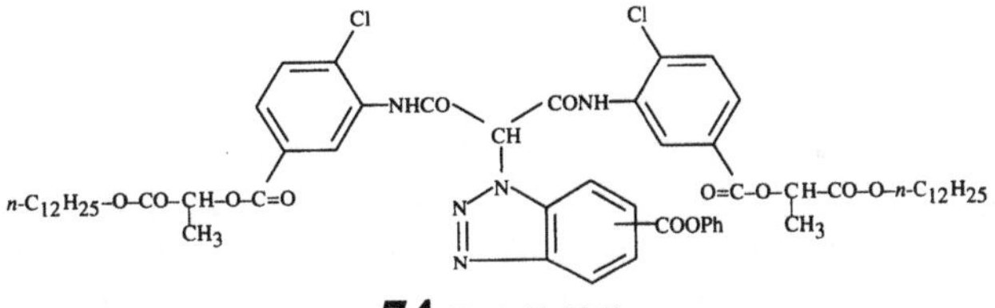

74 (D-4, ExY-1)

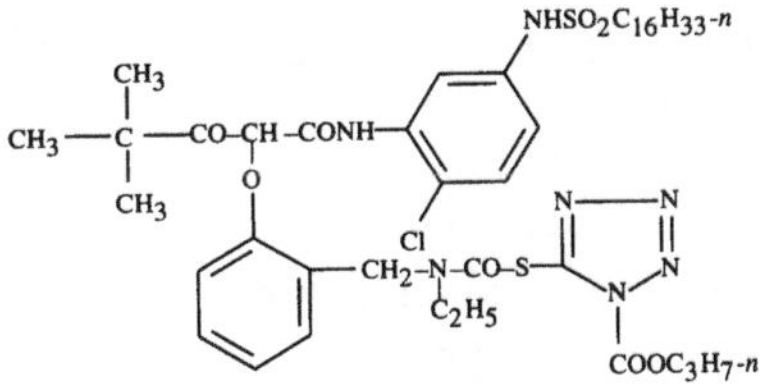

75 (D-5)

Figure 1.27. DIR couplers for color photography. The couplers **71** and **75** are DIR (development inhibitor releasing) couplers of timing-controlling type, which respectively form a cyan dye and a yellow dye. The coupler **72** is a DIR coupler of direct-releasing type, which forms a cyan dye. The couplers **73** and **74** are DIR couplers of direct-releasing type, which respectively form magenta and yellow dye images.

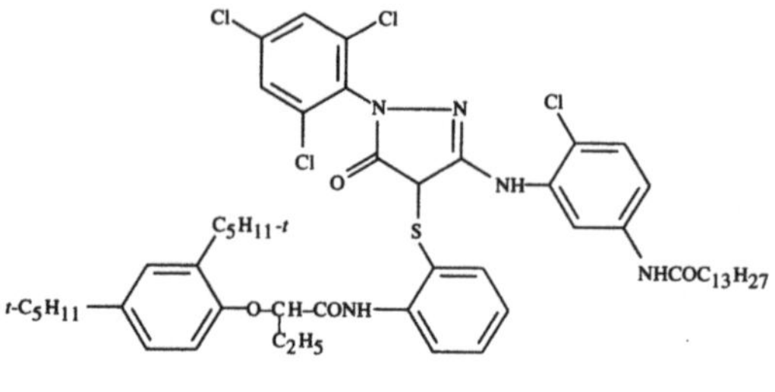

76 (M-1)

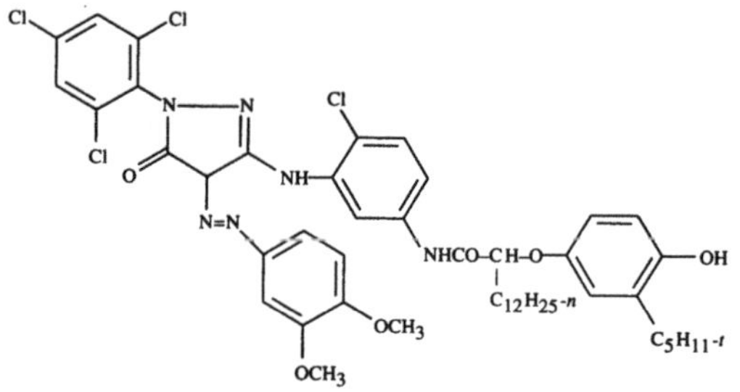

77 (MM-2)

Figure 1.28. Magenta couplers for color photography. The top compound (**77**) is a two-equivalent magenta coupler of 3-anilino-5-pyrazolone type. The bottom compound (**76**) is a yellow-colored magenta coupler.

ity green-sensitized emulsion (a tabular silver iodobromide emulsion: 6.0 mol% iodide; average grain diameter, 0.96 μm; and thickness, 0.26 μm thick). The layer contains a DIR coupler (D-5 in Fig. 1.27) and two yellow couplers (Y-1 and Y-2 in Fig. 1.29). It also involves the bleach-accelerator releasing coupler (BAR-1 in Fig. 1.24), the cyan coupler (C-1 in Fig. 1.25), the oxidized developer scavenger (DOXS-3 in Fig. 1.21), solvents (S-1 and S-2 in Fig. 1.23), TAI (Fig. 1.26), and gelatin, which have been described above for the other layers.

Layer 11 [High sensitivity blue-sensitive layer] This layer comprises a blend of a lower sensitivity blue-sensitized emulsion (a silver iodobromide emulsion: 9.0 mol% iodide; and average grain diameter, 1.06 μm) and a higher sensitivity blue-sensitized emulsion (a tabular silver iodobromide emulsion: 4.1 mol% iodide; average grain diameter, 3.37 μm; and thickness, 0.14 μm thick). It also involves the two yellow couplers (Y-1 and Y-2 in Fig. 1.29), the DIR coupler (D-5 in Fig. 1.27), the bleach-accelerator releasing coupler (BAR-1 in Fig. 1.24), the cyan coupler (C-1 in Fig. 1.25), the oxidized developer scavenger (DOXS-3 in Fig. 1.21), solvents (S-1 and S-2 in Fig. 1.23), TAI (Fig. 1.26), and gelatin, which have been described above for the other layers.

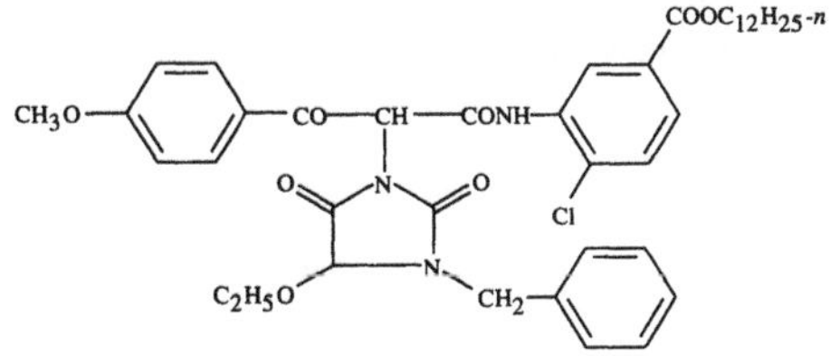

78 (Y-1, ExY-2)

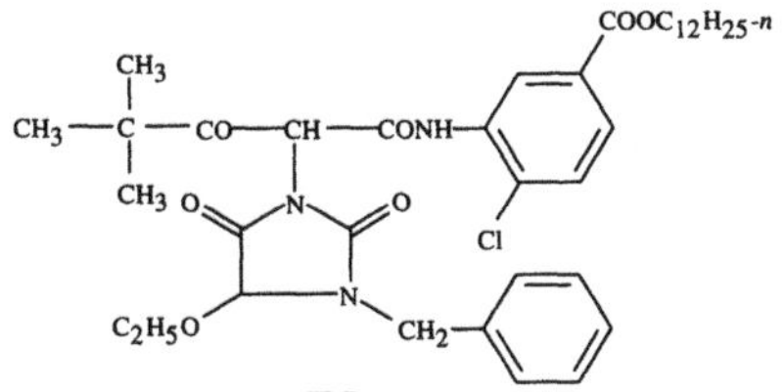

79 (Y-2)

Figure 1.29. Yellow couplers for color photography. They are two-equivalent couplers of benzoylacetanilide and pivaloylacetanilide type.

$CH_2{=}CHSO_2{-}CH_2{-}SO_2CH{=}CH_2$

80 (H-1)

Figure 1.30. Gelatin hardener for color photography. For the action of gelatin hardeners, see the caption of Fig. 1.17.

Layer 12 [Ultraviolet filter layer] This layer involves an unsensitized silver bromide Lippman emulsion as well as organic components such as the UV absorbents (UV-1 and V-2 in Fig. 1.20), solvent (S-9 in Fig. 1.23), and gelatin.

Layer 13 [Protective overcoat layer] This layer contains polymethylmethacrylate matte beads, silica gel particles, a silicon lubricant, and gelatin.
This film has been hardened by coating with 1.75% by weight of total gelatin of hardener (H-1 in Fig. 1.30).

1.2.3 Other Multilayer Structure of Color Films

Another embodiment of a negative film has used another set of organic compounds, where the structures of sensitizing dyes have been disclosed in addition to other functionalized compounds [4].[20] These compounds are used in the 15

[20]This patent was assigned to Fuji Photo Film. Example 3 cited here is a comparative control at that time to show the superiority of the patent.

layers shown in Fig. 1.31. The ID number of each compound used in the patent is also cited for the sake of convenience.

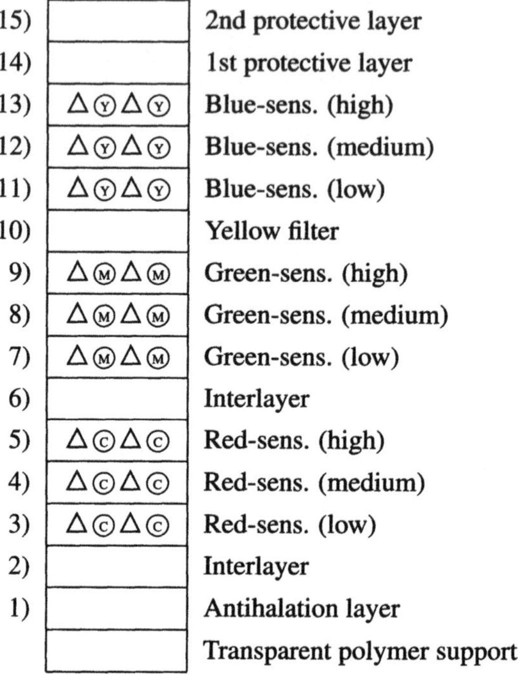

Figure 1.31. Schematic cross-section of the multilayer structure of another color negative film. This diagram illustrates silver halide grains and couplers only. The support is a polyester (or tetraacetylcellulose film) for color films. △: Silver halide grain; Ⓨ: Yellow coupler; Ⓜ: Magenta coupler; Ⓒ: Cyan coupler.

Layer 1 [Antihalation layer] This layer contanis a yellow-colored magenta coupler (ExM-1 in Fig. 1.39) and a sensitizing dye (ExF-1 in Fig. 1.33), which are used as antihalation dyes apart from their original functions. Their structures are shown later. The layer also contains tricresyl phosphate (HBS-1) as a solvent, which has been shown as **65** in Fig. 1.23. Other components such as black colloidal silver and gelatin are also involved.

Layer 2 [Interlayer] This layer contains a magenta-colored cyan coupler (ExC-2) as an antihalation dye. The structure is shown later (Fig. 1.34). UV absorbents (F-UV-1, F-UV-2, and F-UV-3 in Fig. 1.32) are involved to cut ultraviolet light.
The layer also contains di-*n*-butyl phthalate (HBS-2) as a solvent, which has been shown as **62** in Fig. 1.23. Other components such as 2,5-di-*t*-pentadecylhydroquinone, a silver halide emulsion, and gelatin are also involved.

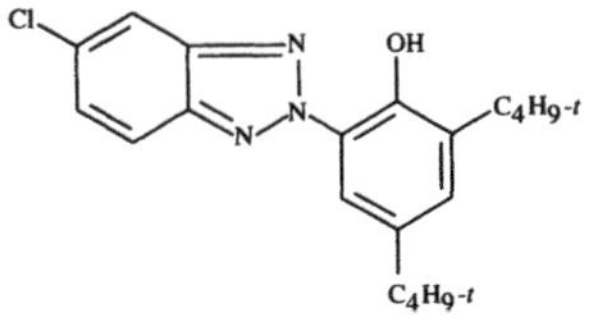

81 (F-UV-1) 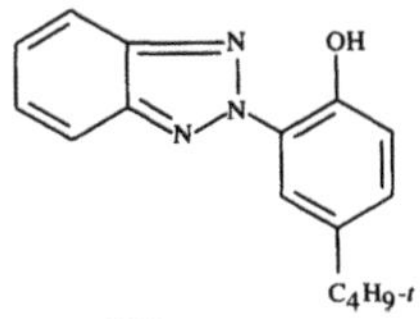 **82** (F-UV-2) 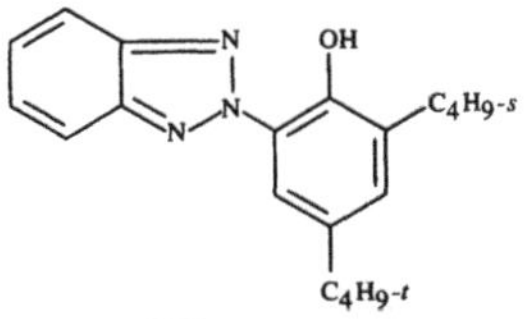 **83** (F-UV-3)

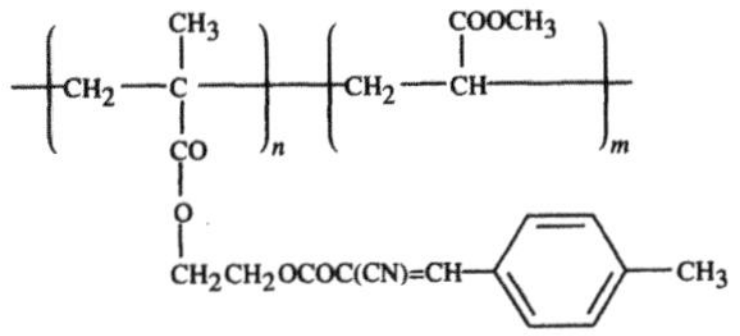

84 (F-UV-4) 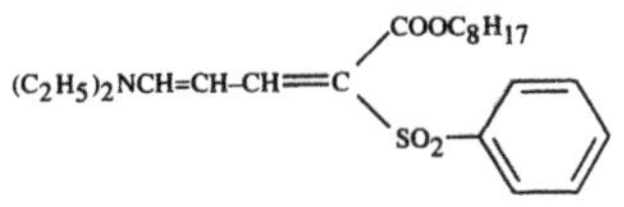 **85** (F-UV-5)

Figure 1.32. UV absorbents for color photography. UV absorbents used in Layers 2 and 14 absorb ultraviolet light to prevent the light fading of dye images. Compare these UV absorbents with those listed in Fig. 1.20.

Layer 3 [Low sensitivity red-sensitive emulsion layer] This layer comprises a blend of Emulsion A (double octahedral grains: 4.0 mol% iodide; average grain diameter, 0.45 μm; diameter/thickness ratio, 1) and Emulsion B (double octahedral grains, 8.9 mol% iodide; average grain diameter, 0.70 μm; and diameter/thickness ratio, 1). These emulsions have been subject to reduction sensitization by using thiourea dioxide and thiosulfonic acid during the preparation of the grains. They have further been subject to gold sensitization and then to sulfur and spectral sensitization in the presence of sodium thiocyanate and red-sensitizing dyes (ExS-1, ExS-2, and ExS-3 in Fig. 1.33).

As couplers for dye formation, the layer contains cyan couplers (ExC-1 and ExC-4 in Fig. 1.34), a two-equivalent cyan coupler (ExC-3 in Fig. 1.35), and a yellow-colored cyan coupler (ExC-5 in Fig. 1.34).

The layer also involves cyan DIR couplers (ExC-7 and ExC-8 in Fig. 1.36) in addition to a dye stabilizer (Cpd-2 in Fig. 1.37), tricresyl phosphate (HBS-1), and gelatin.

Layer 4 [Medium sensitivity red-sensitive emulsion layer] This layer contains Emulsion D (triple tabular grains: 9.0 mol% iodide; average grain diameter, 0.65 μm; and diameter/thickness ratio, 6), which has been subject to reduction sensitization, gold sensitization, sulfur sensitization, and spectral sensitization due to the red sensitizing dyes (ExS-1, ExS-2, and ExS-3 in Fig. 1.33).

As couplers for dye formation, the layer contains the cyan couplers (ExC-1, ExC-2, ExC-3, and ExC-4 in Fig. 1.34) and the yellow-colored cyan coupler (ExC-5 in Fig. 1.34). The layer also involves cyan DIR couplers

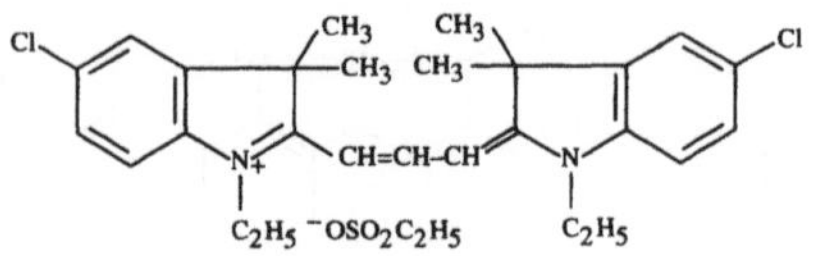

86 (ExF-1)

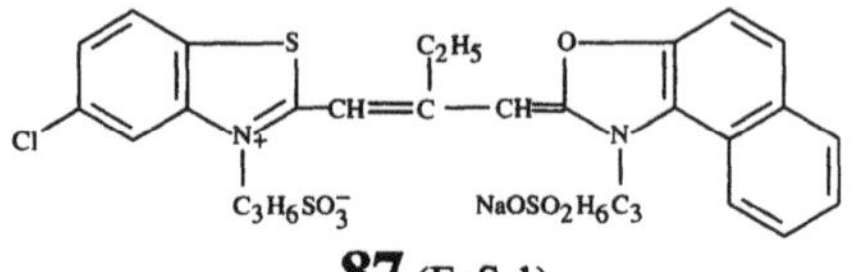

87 (ExS-1)

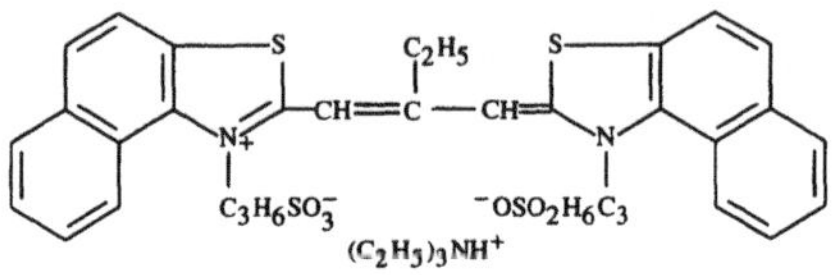

88 (ExS-2)

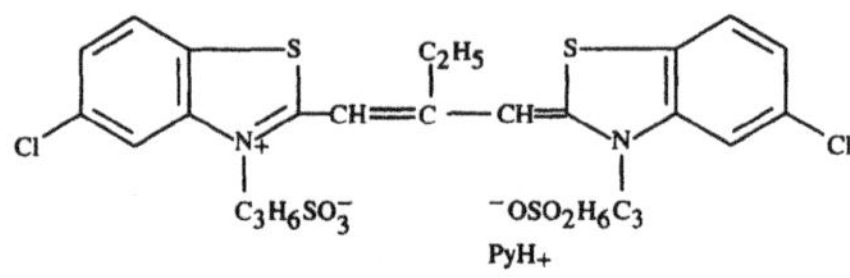

89 (ExS-3)

Figure 1.33. Sensitizing dyes. The dye (**86**) is used in Layer 1 as a dye for antihalation. The others are used for either of the red-sensitive emulsions of Layers 3–5 of Fig. 1.31.

(ExC-7 and ExC-8 in Fig. 1.36) in addition to dye stabilizer (Cpd-2 in Fig. 1.37), tricresyl phosphate (HBS-1) and gelatin.

Layer 5 [High sensitivity red-sensitive emulsion layer] This layer contains Emulsion E (triple tabular grains: 9.0 mol% iodide; average grain diameter, 0.85 μm; and diameter/thickness ratio, 5), which has been subject to reduction sensitization, gold sensitization, sulfur sensitization, and spectral sensitization in the presence of the red-sensitizing dyes (ExS-1, ExS-2, and ExS-3 in Fig. 1.33). The layer also contains a two-equivalent cyan coupler releasing a sulfur-containing compound (ExC-6 in Fig. 1.35). The other components are the same as described for Layer 4 or other layers (ExC-1, ExC-3, ExC-8, Cpd-2, HBS-1, HBS-2, and gelatin).

Layer 6 [Interlayer] This layer contains gelatin, solvent (HBS-1), and an oxidized developer scavenger (Cpd-1 in Fig. 1.37) to prevent undesired interaction between the red-sensitive unit and the green-sensitive unit.

Layer 7 [Low sensitivity green-sensitive emulsion layer] This layer contains Emulsion C (homogeneous tabular grains: 2.0 mol% iodide; average grain diameter, 0.55 μm; diameter/thickness ratio, 7), which has been subject to

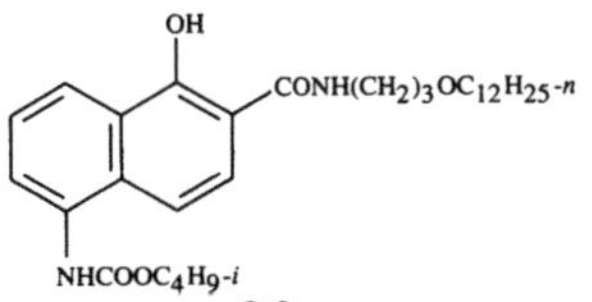

90 (ExC-1) 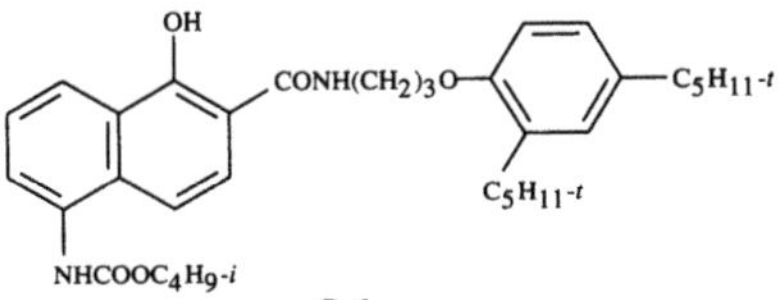 **91** (ExC-4)

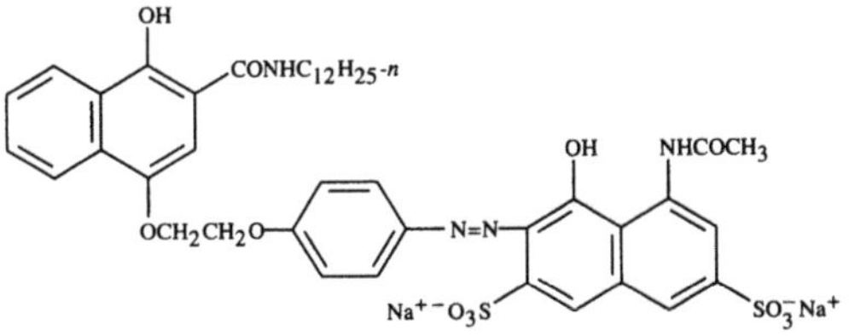

92 (ExC-2)

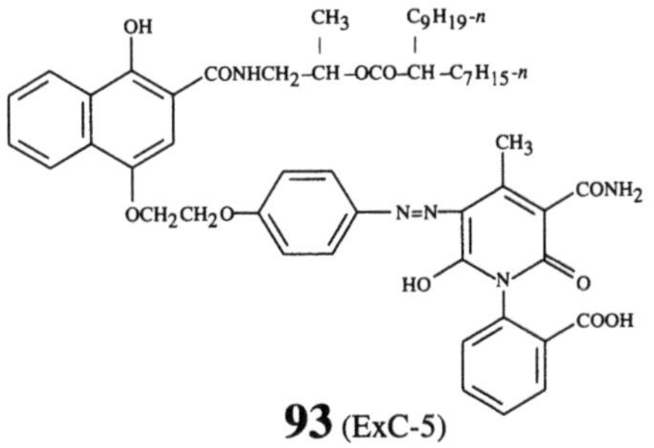

93 (ExC-5)

Figure 1.34. Cyan couplers. The top two (**90** and **91**) are four-equivalent cyan couplers. The coupler **92** is a magenta-colored cyan coupler and the coupler **93** is a yellow-colored cyan coupler.

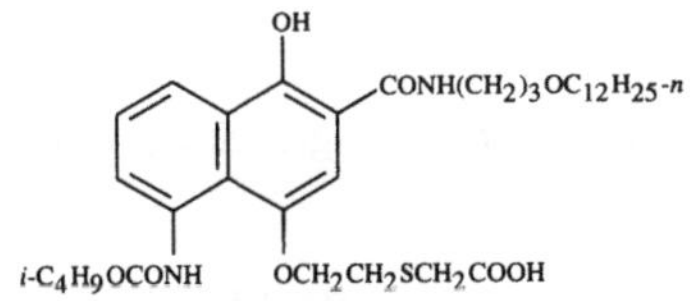

94 (ExC-3) 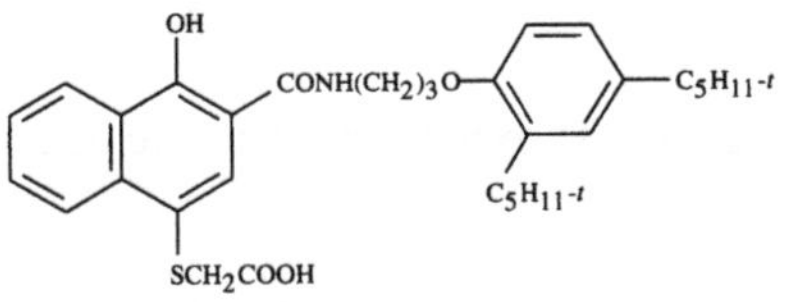 **95** (ExC-6)

Figure 1.35. Cyan couplers. These couplers are two-equivalent coupler, from which a sulfur-containing compound (maybe, a bleach accelerator) is released.

reduction sensitization by using thiourea dioxide and thiosulfonic acid during the preparation of grains. The emulsion has further been subject to gold sensitization and then to sulfur and spectral sensitization in the presence of sodium thiocyanate and green-sensitizing dyes (ExS-4, ExS-5, and ExS-6 in Fig. 1.38).

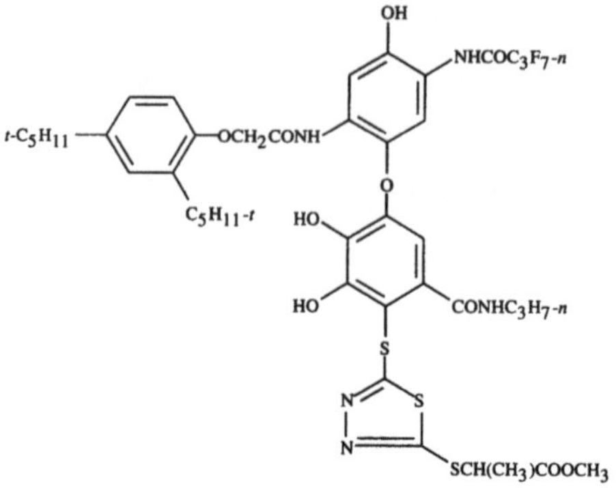

96 (ExC-7)

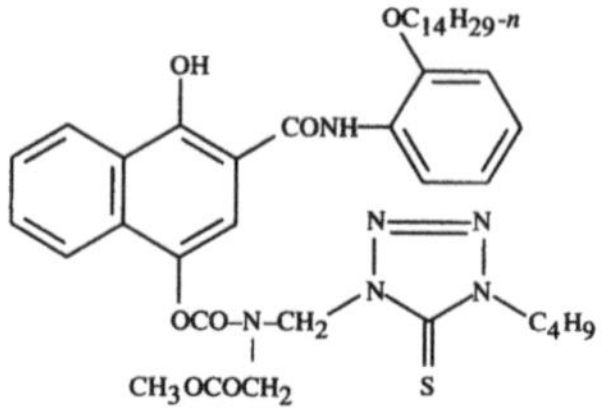

97 (ExC-8)

Figure 1.36. Cyan DIR couplers. Both are DIR couplers of timing controlling type.

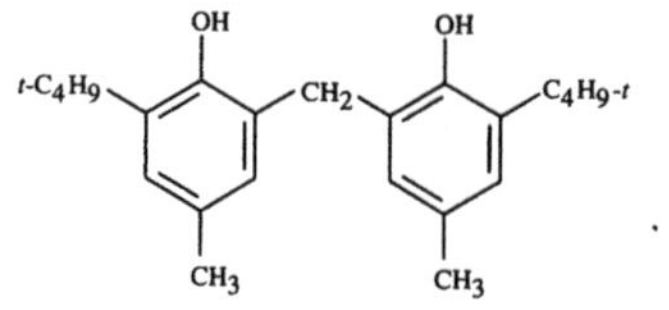

98 (Cpd-2)

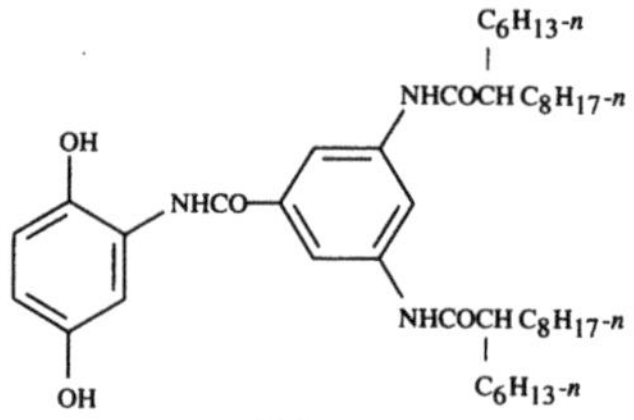

99 (Cpd-1)

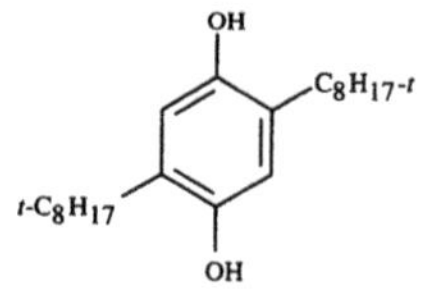

100 (Cpd-3)

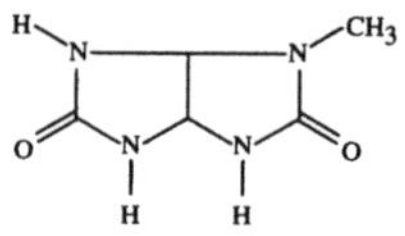

101 (S-1)

Figure 1.37. Additives. These are oxidized developer scavengers and stabilizers.

The layer contains a polymeric magenta coupler (ExM-2 in Fig. 1.39), yellow-colored magenta couplers (ExM-1 and ExM-3 in Fig. 1.39), and a yellow DIR coupler (ExY-1), the last of which has the same structure as **74** shown in Fig. 1.27.

The layer has been coated by means of a high-boiling-point solvent (HBS-3 in Fig. 1.40) as well as the solvent (HBS-1) and gelatin.

Layer 8 [Medium sensitivity green-sensitive emulsion layer] This layer contains Emulsion D described above, which is spectrally sensitized by the green-sensitizing dyes (ExS-4, ExS-5, and ExS-6) shown in Fig. 1.38. The layer involves the polymeric coupler (ExM-2 in Fig. 1.39), the yellow-colored

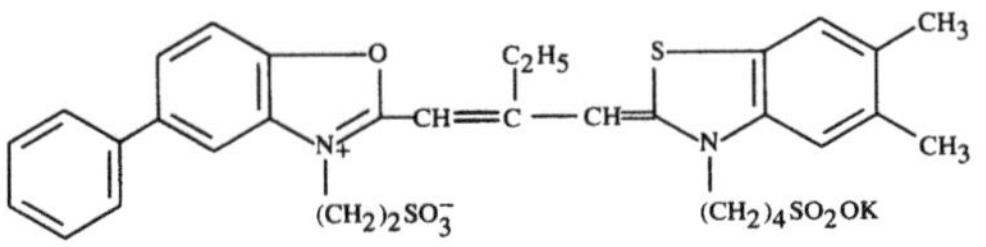

102 (ExS-4)

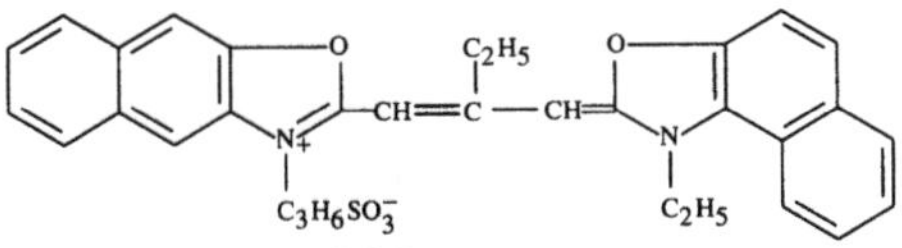

103 (ExS-5)

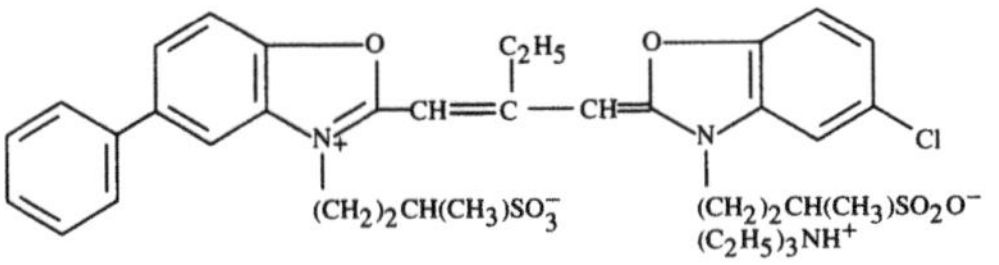

104 (ExS-6)

Figure 1.38. Sensitizing dyes for a green-sensitive emulsion. They are used in Layers 7–9 of Fig. 1.31.

magenta coupler (ExM-3 in Fig. 1.39), the yellow DIR coupler (ExY-1, the same as **74**), solvents (HBS-1 and HBS-3), and gelatin.

Layer 9 [High sensitivity green-sensitive emulsion layer] This layer contains Emulsion E described above, which is spectrally sensitized by the green-sensitizing dyes (ExS-4, ExS-5, and ExS-6) shown in Fig. 1.38.
In addition to the cyan coupler (ExC-1 in Fig. 1.34) and the yellow-colored magenta coupler (ExM-1 in Fig. 1.39), the layer involves pyrazolotriazole magenta couplers (ExM-4 and ExM-5 in Fig. 1.41). They have been coated in the presence of a scavenger (Cpd-3 in Fig. 1.37) as a dispersion in the solvents (HBS-1 and HBS-2) and gelatin.

Layer 10 [Yellow filter layer] This layer contains gelatin, solvent (HBS-1), and an oxidized developer scavenger (Cpd-1 in Fig. 1.37) to prevent undesired interaction between the green-sensitive unit and the blue-sensitive unit. The layer also involves yellow colloidal silver as a yellow filter.

Layer 11 [Low sensitivity blue-sensitive emulsion layer] This layer contains Emulsion C described above, which is spectrally sensitized by the blue-sensitizing dye (ExS-7) shown in Fig. 1.42.
The layer contains a two-equivalent yellow coupler (ExY-3 in Fig. 1.43) in addition to the yellow coulper (ExY-2), the structure of which has been described as **78** in Fig. 1.29.

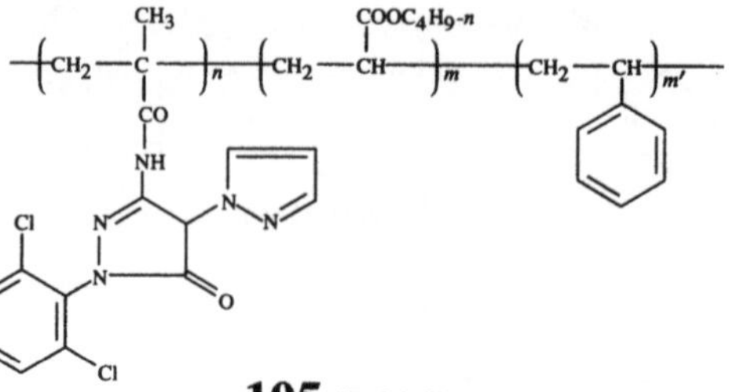

105 (ExM-2)

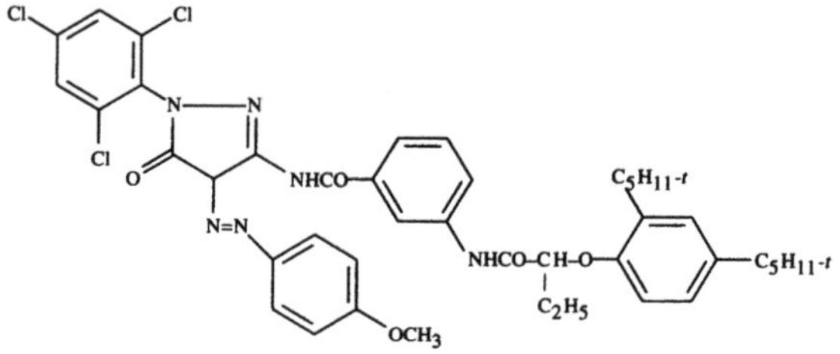

106 (ExM-1)

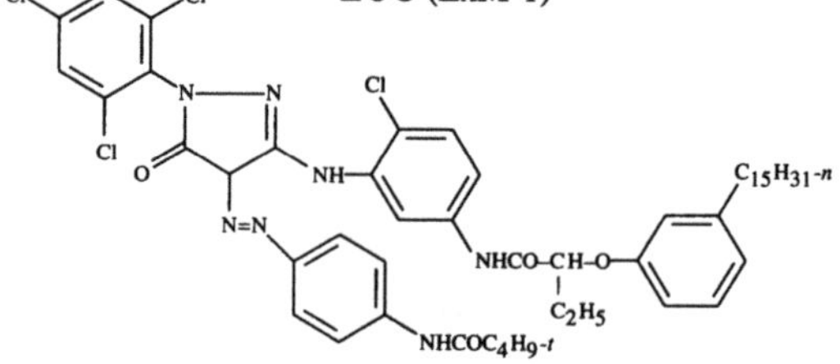

107 (ExM-3)

Figure 1.39. 5-Pyrazolone magenta couplers for color photography. The polymer magenta coupler (**105**) has $n = 50$, $m = 25$, $m' = 25$, and a molecular weight of about 20,000. The couplers **106** and **107** are yellow-colored magenta couplers.

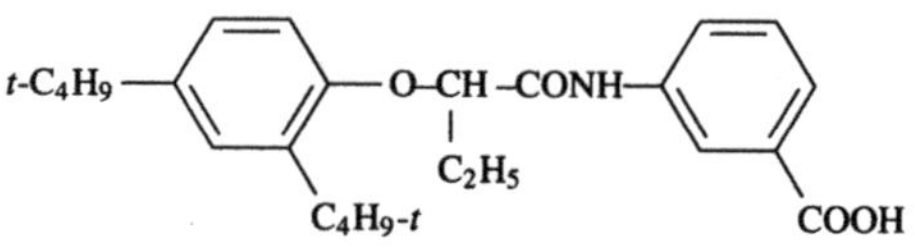

108 (HBS-3)

Figure 1.40. High-boiling solvent. For tricresyl phosphate (HBS-1 or S-4) and di-*n*-butyl phthalate (HBS-2 or S-1), see Fig. 1.23.

In addition to the DIR coupler (ExY-1) whose structure has been described as **74** in Fig. 1.27, another yellow DIR coupler (ExY-4 in Fig. 1.44) is involved in Layer 11. Solvent (HBS-1) and gelatin are also contained.

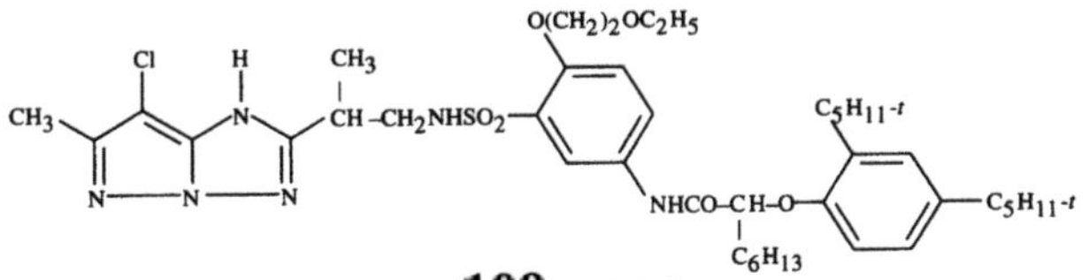

109 (ExM-4)

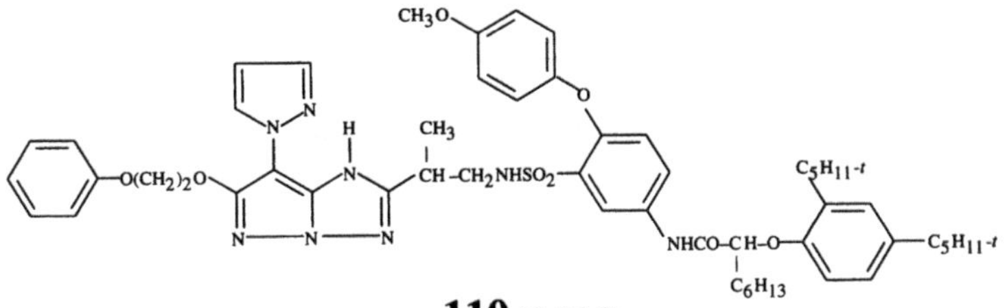

110 (ExM-5)

Figure 1.41. Pyrazolotriazole magenta couplers for color photography. Both of these couplers are two-equivalent couplers, from which a chloride anion or an imidazolyl anion is released during coupling with an oxidized color developer. Compare these couplers with those listed in Fig. 1.28 and in Fig. 1.39.

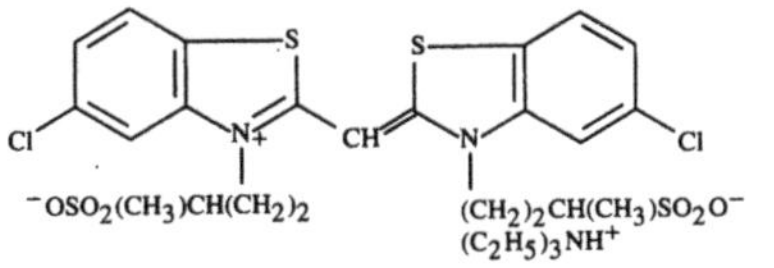

111 (ExS-7)

Figure 1.42. Sensitizing dye for a blue-sensitive emulsion.

Layer 12 [Medium sensitivity blue-sensitive emulsion layer] This layer contains Emulsion D described above, which is spectrally sensitized by the blue-sensitizing dye (ExS-7) shown in Fig. 1.42. The layer involves the yellow couplers (ExY-2 and ExY-3), cyan DIR coupler (ExC-7, the same as **96**), solvent (HBS-1), and gelatin.

Layer 13 [High sensitivity blue-sensitive emulsion layer] This layer contains Emulsion F described above, which is spectrally sensitized by the blue-sensitizing dye (ExS-7) shown in Fig. 1.42. The layer involves the yellow couplers (ExY-2 and ExY-3), cyan DIR coupler (ExC-7, the same as **96**), solvent (HBS-1), and gelatin.

Layer 14 [1st protective layer] This layer contains UV absorbents (F-UV-4 and F-UV-5 in Fig. 1.32), solvent (HBS-1), and gelatin.

Layer 15 [2nd protective layer] This layer contains polymers for matte finish (B-1 and B-2 in Fig. 1.45) and a silicon lubricant (B-3 in Fig. 1.45).

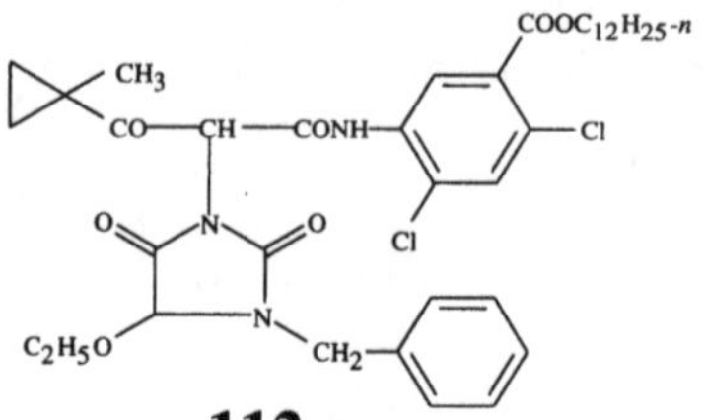

112 (ExY-3)

Figure 1.43. Yellow coupler for color photography. Note that a *t*-butyl group (see **79** in Fig. 1.29) is replaced by a 1-methyl-1-cyclopropyl group.

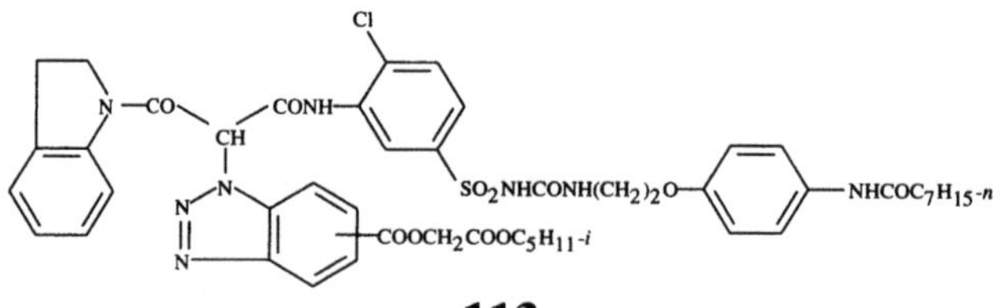

113 (ExY-4)

Figure 1.44. DIR coupler for color photography. This coupler is contained in Layer 11. See also **74** in Fig. 1.27.

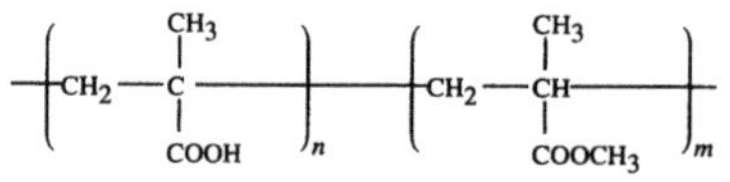

114 (B-1) ($n/m = 10/90$)
115 (B-2) ($n/m = 40/60$)

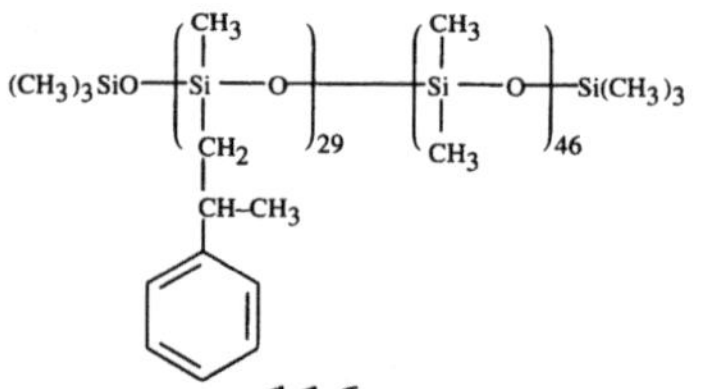

116 (B-3)

Figure 1.45. Polymers for color photography. Polymers **114** (B-1) and **115** (B-2) are polymethylmethacrylate matte beads (diameter 1.7 μm), where they are different in solubilities according to the n/m ratio. Polymer **116** (B-3) is a silicon lubricant.

> Layer 15 further contains a stabilizer (S-1 in Fig. 1.37) and gelatin. Finally, this film has been hardened by coating with a hardener (**46** in Fig. 1.17).

The couplers and other ballasted components[21] in each layer are dissolved in high-boiling solvents and dispersed in a gelatin solution in the presence of appro-

[21]In color photography, ballast groups are particularly important to control the diffusibilities of molecules.

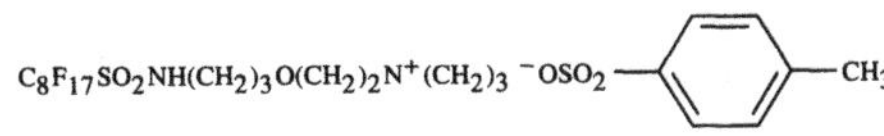

$C_8F_{17}SO_2NH(CH_2)_3O(CH_2)_2N^+(CH_2)_3{}^-OSO_2$ — CH_3

117 (W-1)

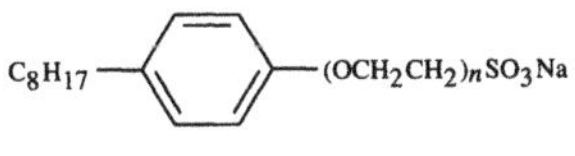

C_8H_{17} — $(OCH_2CH_2)_nSO_3Na$

118 (W-2) ($n = 2-4$)

$NaOSO_2$ — C_4H_9-*t*, C_4H_9-*t*

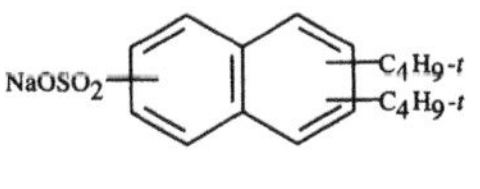

119 (W-3)

Figure 1.46. Surfactants for color photography. For the action of surfactants, see the caption of Fig. 1.23.

priate surfactants. For this purpose, cationic surfactants (e.g., W-1) and anionic ones (e.g., W-2 and W-3) are disclosed in this patent [4], as shown in Fig. 1.46.

Before closing this section, it is worthy to compare the two embodiments described above (Subsections 1.2.2 and 1.2.3). Although the filed dates of the patents are different by about 3 years, the comparison between them gives us a hint to understand the R&D policies and accomplishments of the respective companies. In particular, we should point out differences in the structures of cyan couplers (Figs. 1.24 and 1.25 vs. Figs. 1.34 and 1.35) and in those of magenta couplers (Fig. 1.28 vs. Figs. 1.39 and 1.41).[22]

1.3 Chemicals for Color Photography

As described in the preceding section, various chemicals are used in photographic materials. They are classified into several categories shown in Table 1.1, where each category of chemicals is classified according to a process to which it is closely related: the production of silver halides, the production of oil dispersions, multilayer coating, photographing process, dye-image forming processes (dye forming, dye transfer, dye bleaching), and fastness of final images. The "Examples" column of Table 1.1 contains the figure numbers shown in the preceding section.

It should be noted that one compound may be used for various purposes and that one name may be used in various meanings. For example, a thiol compound can be used as an emulsion stabilizer, an antifoggant, a development inhibitor, etc. according to the processes in which it participates. Although the word "stabilizer" is used in such terms as "emulsion stabilizer", "dye image stabilizer", etc., it denotes different phenomena which have no chemical and physical connections with each other.

[22]Respective multilayer embodiments have been adopted by other companies, although we have restricted our discussion only on a negative film of Eastman Kodak and that of Fuji Photo Film in this section.

Table 1.1. Chemicals for Silver Halide Color Photography

Process	Organic compounds used	Examples
Fundamentals	• Silver halides • Gelatin	
Producing AgX	• Silver halide solvents* • Chemical sensitizers* • Spectral sensitizers (sensitizing dyes)* • Stabilizers (antifoggants, emulsion stabilizers, etc.)*	 Figs. 1.4, 1.10, 1.13, 1.33, 1.38, 1.42 Figs. 1.26, 1.37,
Producing oil dispersions	• High-boiling solvents • Surfactants	Fig. 1.7, 1.23, 1.40 Figs. 1.15, 1.46
Film coating	• Gelatin hardeners • Binders • Plasticizers	Figs. 1.17, 1.30
Photographing	• Light absorbents (antihalation dyes, anti-irradiation dyes, filter dyes) • Lubricants • Antistatic agents • Matting agent • Brightening agents	Figs. 1.16, 1.22 Fig. 1.45 Fig. 1.45
Dye forming	• Couplers† • Color developers† • Color-mixing inhibitor	Figs. 1.5, 1.11, 1.14, 1.24, 1.25, 1.27, 1.28, 1.29, 1.34, 1.35, 1.36, 1.39, 1.41, 1.43, 1.44 Figs. 1.8, 1.21, 1.37
Dye transfer	• Dye developers† • Dye releasers†	
Dye bleaching	• Azo dyes†	
Image fastness	• Ultraviolet light absorbents • Stain inhibitors • Dye image stabilizers • Antiseptics	Figs. 1.9, 1.20, 1.32 Figs. 1.6, 1.12, 1.37 Fig. 1.18

Since the most fundamental chemicals are silver halides and gelatin, the next chapter is devoted to a brief introduction of their chemical properties. The remaining chapters of Part I deal with crystalline properties and light-sensitive properties of silver halides as well as organic compounds that interact with silver halides during the production of silver halide emulsions and during the developing process of B&W photography, as marked with an asterisk in Table 1.1.

Part III to Part V are devoted to organic compounds for generating dye images, as marked with a dagger in Table 1.1. They are couplers and color developers for chromogenic process (conventional color photography), dye developers and dye releasers for dye transfer process (instant color photography), and azo dyes for silver dye bleach process.

The other compounds without an asterisk or a dagger in Table 1.1 have been omitted in order of keeping the compactness of this book. The reader should refer to reviews [5,6] and books [7,8,9] for more detailed information, although several examples of these compounds have been given in the preceding section, as referred to in the "Examples" column of Table 1.1.

References

[1] Yoneyama H, Ohzeki K, Deguchi Y, Takada K (2001) US Patent 6 291 151 B1

[2] Miyamoto K (1990) Fuji Film Res Dev. 35:40

[3] Zengerle PL, Sowinski AF (1998) US Patent 5 726 003

[4] Sato K, Ishii Y, Yamakawa K (1995) US Patent 5 401 624

[5] Arai A (1979) Nippon Settyaku Kyokai Shi. 15:64

[6] Fujita S (1984) Sci Pub Fuji Photo Film. 29:18

[7] Arai A, Tanaka M, Kogure M, Futagi K, Oohashi M, Furutachi N, Itoh I, Adachi K, Sato R, Koboshi S, Fujita S, Nakamura T, Masuda T, Tani T, Hayami M, Ogo K (1986, 2002) Functionalized Organic Chemicals for Silver Halide Color Photographic Materials, CMC, Tokyo

[8] James TJ (ed) (1977) The Theory of the Photographic Process, 4th edn. Macmillan, New York London

[9] Keller K (ed) (1993) Science and Technology of Photography. VCH, Weinheim

Chapter 2

Photographic Emulsions

2.1 Silver Halides—Fundamental Chemical Properties

Since photo-sensitivity is mainly based on physical and chemical properties of silver halides (AgX), it is prerequisite to have a minimum set of knowledge on these properties, even though this book places emphasis on organic compounds used in photography. This section deals with essentials of such chemical properties relevant to silver halides.

2.1.1 Precipitation of Silver Halides

Suppose that an AgX bulk is placed in water. Although AgX is very slightly soluble in water, this process generates a saturated aqueous solution of AgX belonging to the equilibrium:

$$AgX(s) \rightleftharpoons AgX_{soln} \rightleftharpoons Ag^{+} + X^{-} \tag{2.1}$$

where the symbol AgX(s) represents AgX in bulk and AgX_{soln} designates a dissolved species of AgX. When we focus our attention on the latter part of eq. 2.1, the equilibrium constant is represented by

$$\frac{[Ag^{+}][X^{-}]}{[AgX_{soln}]} = K \tag{2.2}$$

Because the concentration of dissolved AgX (i.e., $[AgX_{soln}]$) is essentially constant, the solubility product (K_{sp}) of silver halide can be represented as

$$[Ag^+][X^-] = K_{sp} \quad mol^2L^{-2} \tag{2.3}$$

The solubility product can also be expressed by the logarithmic relation,

$$pK_{sp} = -\log K_{sp} \tag{2.4}$$

Hence, for silver chloride, we obtain

$$K_{sp} = [Ag^+][Cl^-] = 10^{-9.75} \quad mol^2L^{-2} \tag{2.5}$$

$$pK_{sp} = 9.75 \tag{2.6}$$

The solubility products of silver halides at 25℃ are listed as follows:

	AgCl	AgBr	AgI
pK_{sp}	9.75	12.31	16.09

The concentration of halide ions in an aqueous solution of silver halide (strictly speaking, the free halide ion activity, $[X^-]$) is determined by using silver/silver halide electrode. The potential of the electrode is given as follows:

$$E = E^0_{AgX} - \frac{RT}{F}\ln[X^-] = E^0_{AgX} - 2.30259 \times \frac{RT}{F}\log[X^-] \tag{2.7}$$

where E^0_{AgX} represents the standard potential of the silver salt electrode, R the gas constant ($8.314510\ J \cdot K^{-1} \cdot mol^{-1}$), T the absolute temperature, and F the Faraday constant ($9.6485309 \times 10^4\ C \cdot mol^{-1}$).

On the other hand, the value of pAg is frequently used to control processes for preparing silver halide emulsions. The pAg is defined by analogy with the pH as follows:

$$pAg = -\log[Ag^+] \tag{2.8}$$

The value of pAg is determined by using an indicating electrode of pure silver.

$$E = E^0_{Ag} + \frac{RT}{F}\ln[Ag^+] = E^0_{Ag} + 2.30259 \times \frac{RT}{F}\log[Ag^+] \tag{2.9}$$

where E^0_{Ag} represents the standard potential of the pure silver electrode.

Silver and Ag/AgX electrodes give the same potential for an emulsion containing a single silver halide, since a solution is in equilibrium with the bulk of silver halide crystals. In this case, we can obtain the pAg and the pX of the emulsion as follows:

$$pAg + pX = -\log K_{sp} = pK_{sp} \tag{2.10}$$

which is relevant to the solubility constant (eq. 2.3).

2.1.2 Silver Complexes

Ammine Complexes of Silver

An ammine complex of silver, $Ag(NH_3)_2^+$, which is soluble in water, forms a silver halide by the addition of a halide ion. Reversely, an aqueous suspension of a silver halide (AgX) forms the ammine complex of silver by adding ammonia as a complexing agent, where the consumption of Ag^+ is compensated by the dissociation of the AgX.

Dissociation Constants of Silver Complexes

The dissociation of the ammine complex ion $Ag(NH_3)_2^+$ is represented by the equilibrium:

$$Ag(NH_3)_2^+ \rightleftharpoons Ag^+ + 2NH_3 \tag{2.11}$$

Then, the dissociation constant (K_c) is expressed as follows:[1]

$$\frac{[Ag^+][NH_3]^2}{[Ag(NH_3)_2^+]} = K_c \tag{2.12}$$

From removing the intermediate term (AgX_{soln}) in eq. 2.1, we can derive the following equation:

$$Ag^+ + X^- \rightleftharpoons AgX(s) \tag{2.13}$$

By adding eq. 2.11 and eq. 2.13, we obtain

$$Ag(NH_3)_2^+ + X^- \rightleftharpoons AgX(s) + 2NH_3 \tag{2.14}$$

Since the amount of AgX(s) is large and can be considered to be constant, the term [AgX(s)] is disregarded in considering the dissociation constant of the equilibrium (eq. 2.14). By applying eqs. 2.12 and 2.3 to the resulting dissociation constant, we obtain

$$\frac{[NH_3]^2}{[Ag(NH_3)_2^+][X^-]} = \frac{[Ag^+][NH_3]^2}{[Ag(NH_3)_2^+]} \times \frac{1}{[Ag^+][X^-]} = \frac{K_c}{K_{sp}} \tag{2.15}$$

Precipitation of Silver Halides from Silver Complexes

According to eq. 2.14, the addition of X^- shifts the equilibrium from the left to the right so that AgX crystallizes. The value $[X^-]$ for initiating the crystallization depends upon the concentration of NH_3.

This dependence is called the *masking effect* of ammonia. The following treatment of the masking effect is based on the one described in Ref. [1].

No masking effect When ammonia is not present in excess, such masking effect does not work. For example, since eq. 2.11 indicates that 2 moles of NH_3

[1]The reciprocal of K_c is called *stability constant.*

correspond to 1 mole of Ag^+, we obtain the following relationship of their concentrations:

$$2[Ag^+] = [NH_3] \tag{2.16}$$

This relationship, which represents the condition of no excess of ammonia, is introduced into eq. 2.12. Then, we use the relationships, $K_c = 10^{-7.2}$ and $4 = 10^{0.60}$, so as to obtain

$$[Ag^+] = \sqrt[3]{\frac{K_c[Ag(NH_3)_2^+]}{4}} = 10^{-2.6} \times \sqrt[3]{[Ag(NH_3)_2^+]} \tag{2.17}$$

Suppose that the concentration of $Ag(NH_3)_2^+$ is equal to $10^{-2.1}$mol · L^{-1}. Since the dissociation of this ammine complex is considered to be negligible, we can place $[Ag(NH_3)_2^+] = 10^{-2.1}$. This is introduced into eq. 2.17, giving the following value:

$$[Ag^+] = 10^{-2.6} \times 10^{-2.1/3} = 10^{-3.3} \quad \text{mol} \cdot \text{L}^{-1} \tag{2.18}$$

Since the solubility products are determined to be $K_{sp} = 10^{-9.75}$ for silver chloride and $K_{sp} = 10^{-12.31}$ for silver bromide and $[Ag^+]$ is given in eq. 2.18, these values are introduced into eq. 2.3, giving the following results:

$$[Cl^-] = \frac{K_{sp}}{[Ag^+]} = \frac{10^{-9.75}}{10^{-3.3}} = 10^{-6.5} \quad \text{mol} \cdot \text{L}^{-1} \tag{2.19}$$

$$[Br^-] = \frac{K_{sp}}{[Ag^+]} = \frac{10^{-12.31}}{10^{-3.3}} = 10^{-9.0} \quad \text{mol} \cdot \text{L}^{-1} \tag{2.20}$$

Each of these equations indicates the concentration at which the crystallization of AgCl or AgBr begins. Since the value given by eq. 2.19 or eq. 2.20 is very small, the crystallization begins simultaneously after the addition of chloride or bromide ions.

Masking effect On the other hand, an excess of NH_3 brings about the masking effect. For example, suppose that $[NH_3] = 1$ mol · L^{-1} and that $[Ag(NH_3)_2^+] = 10^{-2.1}$ mol · L^{-1} in this case. Then, we obtain the following result by virtue of eq. 2.12:

$$[Ag^+] = \frac{K_c[Ag(NH_3)_2^+]}{[NH_3]^2} = 10^{-7.2} \times 10^{-2.1} = 10^{-9.3} \quad \text{mol} \cdot \text{L}^{-1} \tag{2.21}$$

Thereby, the concentration of Cl^- is calculated to be:

$$[Cl^-] = \frac{K_{sp}}{[Ag^+]} = \frac{10^{-9.75}}{10^{-9.3}} = 10^{-0.5} \approx 0.3 \text{ mol} \cdot \text{L}^{-1} \tag{2.22}$$

It follows that the addition of Cl^- below the concentration given by eq. 2.22 causes no crystallization of AgCl (masking effect!). Since the corresponding value for AgBr can be calculated to be smaller than the value of eq. 2.22, a higher excess of ammonia is necessary to make the masking effect work in the case of AgBr.

Dissolution of Silver Halides by Complexation

When ammonia is in turn added to silver halide, the silver halide is dissolved to form an ammine complex, $[Ag(NH_3)_2^+]$. Suppose that silver halide is fully dissolved according to eq. 2.14. Then, we obtain

$$[X^-] = [Ag(NH_3)_2^+] \tag{2.23}$$

The term $[Ag^+]$ is cancelled from eqs. 2.3 and 2.12, giving the following equation:

$$[NH_3]^2 = \frac{K_c[Ag(NH_3)_2^+]}{[Ag^+]} = \frac{K_c[Ag(NH_3)_2^+][X^-]}{K_{sp}} \tag{2.24}$$

By introducing eq. 2.23 into eq. 2.24, we obtain

$$[NH_3] = \sqrt{\frac{K_c[Ag(NH_3)_2^+][X^-]}{K_{sp}}} = \sqrt{\frac{K_c}{K_{sp}}} \times [X^-] \tag{2.25}$$

To evaluate this equation, suppose that silver halide is fully dissolved to give a concentration of 10^{-2} mol · L^{-1}. Since the solubility product of AgCl is $K_{sp} = 10^{-9.75}$ and the dissociation constant of the ammine complex is $K_c = 10^{-7.2}$, we obtain

$$[NH_3] = \sqrt{\frac{10^{-7.2}}{10^{-9.75}}} \times 10^{-2} = 10^{-0.73} \approx 0.2 \text{ mol} \cdot L^{-1} \tag{2.26}$$

This indicates that the presence of 0.2 mol · L^{-1} of ammonia is sufficient to dissolve silver chloride. Similarly, the concentration of ammonia to dissolve silver bromide is calculated to be about 4 mol · L^{-1}. In general, eq. 2.25 indicates that, if K_{sp} is smaller (i.e., the solubility of silver halide is smaller), $[NH_3]$ is larger.

Fixation of Silver Halides

The fixation represented by eq. 1.2 of Chapter 1 is essentially a complexation process, which is similar to the formation of the ammine complex described in the preceding paragraphs. Hence, eq. 1.2 of Chapter 1 is rewritten by analogy of eq. 2.11 as follows:

$$Ag(S_2O_3)_2^{3-} \rightleftharpoons Ag^+ + 2S_2O_3^{2-} \tag{2.27}$$

The dissociation constant (K_c) for this case (cf. eq. 2.12) is expressed by

$$\frac{[Ag^+][S_2O_3^{2-}]^2}{[Ag(S_2O_3)_2^{3-}]} = K_c \tag{2.28}$$

By analogy of eq. 2.23, we can obtain the following equation for the complete dissolution of AgX:

$$[X^-] = [Ag(S_2O_3)_2^{3-}] \tag{2.29}$$

The term $[Ag^+]$ is cancelled from eqs. 2.3 and 2.28 to give:

$$[S_2O_3^{2-}]^2 = \frac{K_c[Ag(S_2O_3)_2^{3-}]}{[Ag^+]} = \frac{K_c[Ag(S_2O_3)_2^{3-}][X^-]}{K_{sp}} \quad (2.30)$$

Equation 2.30 is combined with eq. 2.29 to give

$$[S_2O_3^{2-}] = \sqrt{\frac{K_c[Ag(S_2O_3)_2^{3-}][X^-]}{K_{sp}}} = \sqrt{\frac{K_c}{K_{sp}}} \times [Ag(S_2O_3)_2^{3-}] \quad (2.31)$$

This equation essentially has the same meaning as eq. 2.25 because of eq. 2.29. According to the present purpose, we convert eq. 2.31 into the following one:

$$[Ag(S_2O_3)_2^{3-}] = \sqrt{\frac{K_{sp}}{K_c}} \times [S_2O_3^{2-}] \quad (2.32)$$

Since K_c and $[S_2O_3^{2-}]$ are common to any silver halides, eq. 2.32 indicates that if K_{sp} is larger (i.e., if the solubility of a silver halide is larger), the concentration $[Ag(S_2O_3)_2^{3-}]$ is concluded to be larger.

2.2 Gelatin—Fundamental Chemical Properties

2.2.1 Merits of Gelatin as Binder

A suspension of silver halides is unstable without a binder and incapable of being coated on a polymer support. Gelatin has long been used as a binder for stabilizing the suspension of silver halides, since no alternatives have been found to substitute for gelatin [2,4].

Protective colloid Gelatin is superior as a protective colloid for silver halides. The presence of gelatin enables to prepare a stable suspension of silver halide grains of various shapes and sizes.

Sol-gel transition Gelatin is soluble in warm water (about 40℃) so that it serves as a colloidal support for suspending silver halide grains during the process of preparing photographic emulsions. On cooling after coated on a polymer support, sol-to-gel transition occurs to give stable coated layers even after drying.

Reactions with hardening agents Since gelatin is prepared by partial denaturation of collagen, it has residual functional groups that are capable of reacting with hardening agents (hardeners). Such hardening agents or hardeners are bifunctional organic compounds to cause cross-linkage between adjacent gelatin molecules. The resulting cross-links inhibit the sol-to-gel transition of gelatin. Instead, it turns out to have the property of swelling for easy processing during photographic processes.

Ion permeability Hardened gelatin is permeable to ions when swelling. The easy diffusion of such ionic species is important for photographic development, diffusion transfer process, and so on.

Influence on photographic properties Gelatin usually contains residues of nucleic acids, amino acids, and so on, which are effective as sensitizing agents, physical retarders, etc. Gelatin has been reported to reduce silver ions into silver clusters on silver halide grains, where alkali-pretreated gelatin, acid-pretreated gelatin, a thioether-containing copolymer, and an imidazole-containing copolymer have been examined as protective colloids [5]. The effects of pAg and pH on the reduction of silver ions by gelatin have been studied with respect to reduction sensitization centers [6]. The effects of the addition of sulfur-containing amino acids have been studied in the reduction of silver ions by gelatin [7]. It has been well-known that the imidazole group of gelatin (especially, histidine) interacts with silver halide grains to cause physical retardance (i.e., retardance during Ostwald's ripening). Such interactions have been studied by titration methods, where the imidazole group is blocked or not by an *N*-ethoxyformyl group [8,9]. Since the effects of gelatin on photographic properties vary to a great extent in accord with gelatin sources (see below), they should be replaced by the effects of additives having specified activities. Thus, an appropriate combination of inactive gelatin and such additives is preferable in order to control the total photographic properties of silver halide emulsions.

Stable supply Since gelatin comes from collagen that is one of the major components of animals (bones, skins, etc.), its supply is very stable as a natural material.

2.2.2 Manufacture of Gelatin

Gelatin is manufactured from collagen, which is the most abundant protein of higher animals and appears mainly in bones, skin, cartilage, tendons, and ligaments. In industrial manufacture of gelatin, cattle bones, hides, and pigskins are used as main sources [2,3]. Since hydroxyapatite makes up about 75% of cattle bones, it is removed by acid leaching to produce ossein (demineralized bone). The resulting ossein as well as cattle hides and pigskins are collagen sources, which are subjected to alkali (for about two or three months) or acid pretreatment (for several days), as shown in Fig. 2.1.

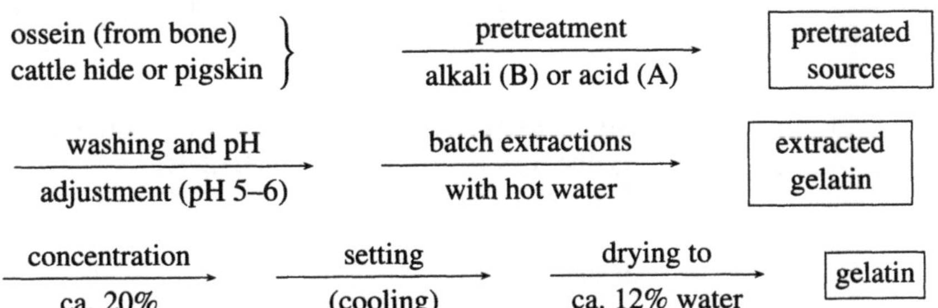

Figure 2.1. Manufacturing alkali- and acid-pretreated gelatins.

Collagen sources (ossein, etc.) are swelled and softened under prolonged alkali pretreatment (B), where peptide bonds in collagen are partially cleaved and the side-chain amide groups of asparagine and glutamine are completely hydrolyzed to form carboxylic acid groups. Thereby, the isoelectric points of alkali-pretreated gelatins range from 4.8 to 5.1.

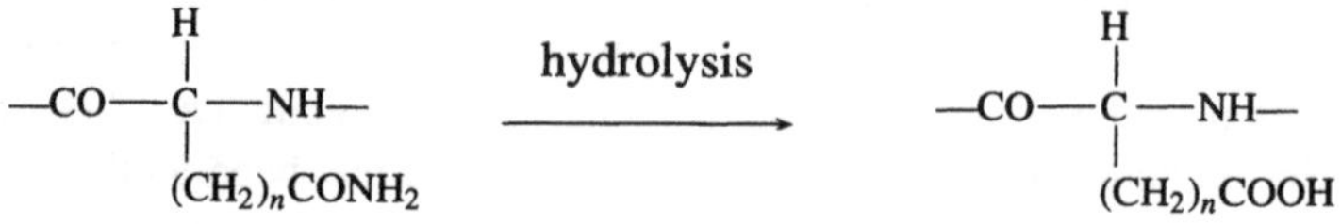

$$-CO-\underset{(CH_2)_nCONH_2}{\overset{H}{C}}-NH- \xrightarrow{\text{hydrolysis}} -CO-\underset{(CH_2)_nCOOH}{\overset{H}{C}}-NH-$$

$n = 1$ asparagine residue
$n = 2$ glutamine residue

$n = 1$ aspartic acid residue
$n = 2$ glutamic acid residue

On the other hand, the acid pretreatment (A) causes no hydrolysis of side chains so that the isoelectric points of the resulting gelatins range about 9.

2.2.3 Side-Chain Groups of Gelatin

According to the amino acid analysis reported in Ref. [2], an alkali-pretreated bone gelatin contains the following residues in decreasing order of residue number per 1000 residues: glycine (Gly 347), proline (Pro 120), alanine (Ala 113), 4-hydroxyproline (Hyp 96), glutamic acid (Glu 74), arginine (Arg 48), aspartic acid (Asp 44), serine (Ser 31)!$ lysine (Lys 27), leucine (Leu 24), valine (Val 22), threonine (Thr 17), isoleucine (Ile 12), phenylalanine (Phe 12), hydroxylysine (Hyl 5), methionine (Met 4), histidine (His 4), and tyrosine (Tyr 1). Among these residues, glutamic acid (Glu 74), arginine (Arg 48), aspartic acid (Asp 44)!$lysine (Lys 27), hydroxylysine (Hyl 5)!$and histidine (His 4) are important since they have carboxylic acid and amide groups in the side chains that undergo the attack of a hardening agent (hardener). Glutamic acid (Glu) and aspartic acid (Asp) are respectively generated from glutamine and asparagine residues as shown in the preceding subsection. The remaining residues contain the following side chains having possible reactivity.

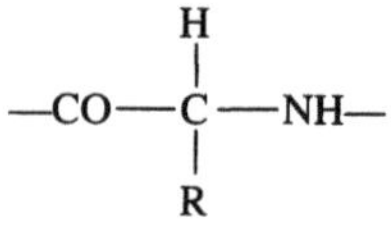

$$-CO-\underset{R}{\overset{H}{C}}-NH-$$

arginine (Arg): $R = -(CH_2)_2NH-C(NH_2)=NH$
lysine (Lys): $R = -(CH_2)_4NH_2$
hydroxylysine (Hyl): $R = -(CH_2)_2CH(OH)CH_2NH_2$

histidine (His): R = $-CH_2$–(imidazole ring: N, N–H)

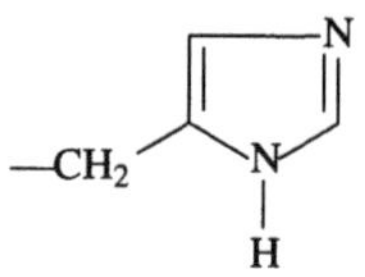

2.3 Preparation of Photographic Emulsions

Silver halide is prepared by adding an aqueous solution of silver nitrate (with or without ammonia) to an aqueous solution of alkali halide in the presence of gelatin as a protective colloid [10]. The preparation process consists of the following steps:

1. **Precipitation**: The mixing of the two solutions forms a suspention of silver halide grains in a gelatin solution, where formation of crystal nuclei, physical ripening (Ostwald's ripening) and related processes occur. The properties of the resulting silver halide grains (size, shape, light sensitivity, etc.) are influenced by various factors such as the concentrations of a $AgNO_3$ solution and of alkali halides, the rate of addition, the temperature (40–70°C), and the concentration of gelatin.

2. **Washing**: Soluble salts are washed out from the suspension. The resulting suspension (so-called AgX emulsion) is chilled and stored as a gel until later processes.

3. **After-ripening**: The AgX emulsion is dissolved on warming to give a sol. After chemical sensitizers are added to the resulting sol, this undergoes after-ripening in order to enhance light sensitivity.

4. **Addition of additives**: Sensitizing dyes, antifoggants, stabilizers, hardening agents are added to the AgX emulsion.

Then, the resulting photographic emulsion is coated on a polymer support (PET, TAC, or laminated paper), chilled to set, and dried.

2.3.1 Preparation With and Without Ammonia

The presence or absence of ammonia in preparing silver halide grains influences the properties of the resulting photographic emulsions. Hence, manufacturing procedures are usually categorized into an acid procedure (without ammonia) and an ammoniacal procedure (with ammonia).

1. **Acid procedures**: In one procedure (a single-jet method), an aqueous solution of silver nitrate is added dropwise to an aqueous solution of alkali halide (plus gelatin). In the other procedure (a double-jet method), a solution of $AgNO_3$ and a solution of alkali halide are simultaneously added to a gelatin solution. To enhance Ostwald's ripening, an excess of halide ions may be preferred.

2. **Ammoniacal procedures**: An ammoniacal aqueous solution of $AgNO_3$ is added to an aqueous solution of alkali halide in the presence of gelatin. Ammonia may be added after the addition of the silver and the halide solutions is completed. Because of the formation of an ammine silver complex, Ostwald's ripening is enhanced.

2.3.2 Examples of Procedures

As a representative of ammoniacal procedures, we should cite the preparation of the photographic emulsion named "Brovira Soft", which was reported in Fiat Final Report [11]. This emulsion was once used in all of the red-, green-, and blue-sensitive layers in Agfacolor Paper.[2]

Solution I	water	16 L
	KBr	3.06 kg
	KI (10%)	1.25 L
	ammonia (specific gravity 0.910)	260 mL
	gelatin	4 kg
	NH_4Br (25%)	2 L

Solution II	water	50 L
	$AgNO_3$	5 kg
	NH_4NO_3 (10%)	1.125 L

Solution III	gelatin	3.75 kg
	water	12 L

These three solutions are prepared in advance and treated in a darkroom as follows:

1. **Precipitation**: Solution II is added dropwise into solution I at 42°C. And solution III is added to the mixture. The mixture is allowed to stand at 48°C for 20 min and digested at 48°C for 20 min to undergo Ostwald's ripening.

2. **Washing**: The mixture is transferred to a dish and chilled to set. The set emulsion (gel) is forced through a perforated screen to form "noodles", which are washed in running water for about 8 hr. The endpoint of washing is controlled by the electric conductivity ($< 1800\ \mu S \cdot cm^{-1}$).

3. **After-ripening**: The noodles are added to the following solution prepared in advance.

gelatin	9 kg
water	60 L
acetic acid-water (1:2)	275 mL

The resulting emulsion is stirred at 75°C for 50–60 min. Then, it is chilled to set and kept in cold storage (The addition of sensitizing dyes etc. is carried out just before the successive coating process).

[2]The exact methods for preparing commercial photographic emulsions are strictly guarded as trade secrets. The example recipe cited in this text is one of the exceptional cases in which such detailed procedures were disclosed as a result of World War II.

2.4 Precipitation

2.4.1 Crystal Growth

Processes involved in the crystallization or precipitation of silver halides have been investigated in detail, as summarized in Ref. [12]. Basic processes are nucleation, growth, Ostwald's ripening, coalescence ripening, and recrystallization.

1. **Nucleation**: New crystals (nuclei) are created in this process, where the number of crystals increases explosively.
2. **Growth**: New layers are added to crystalline nuclei.
3. **Ostwald's ripening**: In this process, which occurs at higher temperatures, smaller crystals are dissolved by virtue of silver-complexing agents (halide ions, ammonia, and other additives) and consumed to make larger crystals grow even larger.
4. **Coalescence ripening**: Two or more crystals form an aggregate by direct contact, which welds together to form a larger crystal.
5. **Recrystallization**: When two or more halide species are present, repeated dissolution and crystallization occurs to form grains of different halide composition.

Several additional comments should be mentioned to characterize the features of these processes.

Nucleation

Under the assumption that a very slow addition of a silver nitrate solution suppresses the subsequent processes, a stable nucleus of AgCl has been determined to contain five silver species [13]. However, many investigations have indicated that the differentiation between the nucleation process and the subsequent growth process is difficult.

Growth

By sampling silver bromide crystals at successive times during the precipitation and by measuring the shapes and sizes, the rates of growth of the crystals have been determined precisely. They have been analyzed and explained by four different equations for grain growth, i.e., $\mathrm{d}x/\mathrm{d}t = Kx^n$ ($n = -1, 0, +1$, or $+2$), where x is a sphere diameter or a cube edge length of a grain and t represents the time of growth [14].

Ostwald's Ripening and Coalescence Ripening

In the acid procedure of a single-jet method, a solution during the precipitation of AgBr contains bromide ions in excess. Hence, the precipitated AgBr is dissolved to form a further silver complex such as $AgBr_2^-$. The complex diffuses to another AgBr crystal and deposits again on its surface. This process (dissolution→diffusion→deposition) is called *Ostwald's Ripening*.

Since small crystals are more soluble than larger crystals, such larger crystals grow larger and larger at the expense of the smaller ones. This phenomenon is explained by Ostwald's solubility formula [14]:

$$\ln \frac{S_r}{S_\infty} = \frac{2M\gamma}{RT\rho r}, \tag{2.33}$$

where the symbol S_r represents the solubility of spherical particles of radius r, S_∞ the solubility of crystals of infinite dimension, M the molecular weight of the crystal, ρ the density of the crystal, and γ the surface energy of the solid in the liquid.[3] Although the small crystals approximately can be regarded as spherical particles, their larger solubility can be explained, since eq. 2.33 indicates that S_r is larger than S_∞.

Ostwald's ripening can be enhanced in the presence of agents for complexing silver halides (so-called silver halide solvents) at higher temperature. Ammonia in the ammoniacal procedure works as such a complexing agent (ripener).

Various complexing agents other than ammonia, e.g., amines, thioethers, and thiones, are also used as ripeners for enhancing Ostwald's ripening. The following list shows representative complexing agents [15].

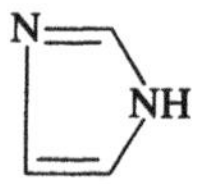

1 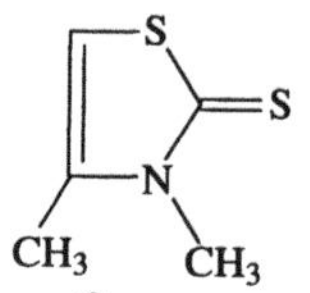 2 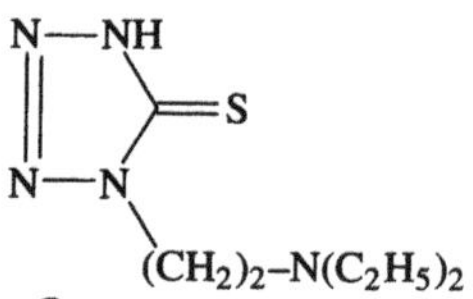 3

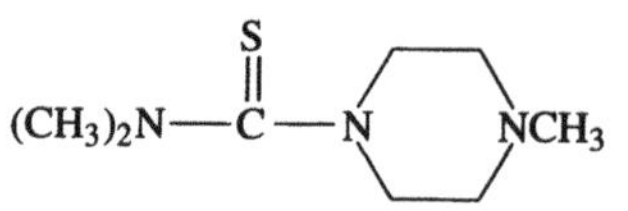

4 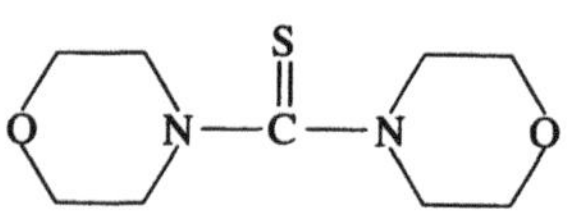 5

In contrast, the suppression of Ostwald's ripening is sometimes desirable in order to produce silver halide emulsions of high resolution and of high contrast,

[3]The Ostwald equation was originally formulated to explain the phenomenon that the solubility of small crystals in a liquid droplet was larger than that of the usual solubility of the solid, where γ represents the surface tension of such a droplet.

where it is necessary to prepare silver halide grains of small sizes that have a narrow size distribution. For this purpose, the use of gelatin containing active components (e.g., adenine (**6**) and guanine (**7**)) or the addition of strongly complexing agents is considered to suppress Ostwald's ripening.

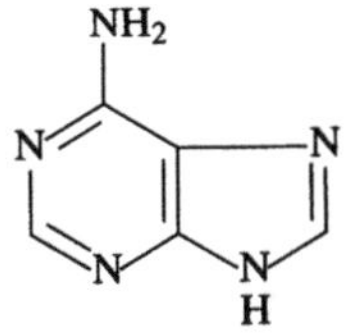

6 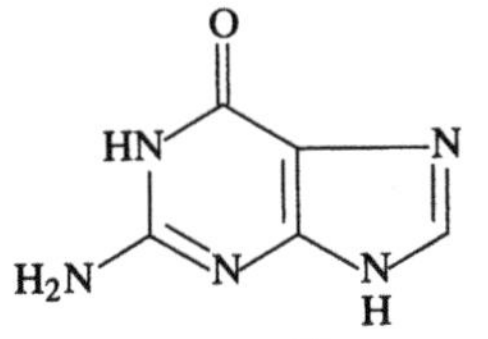 **7**

Synthesized heterocyclic compounds (**8**–**11**) are also used to suppress Ostwald's ripening, since they exhibit strong complexation (chelation) [16]. Among them, 5-mercapto-1-phenyl-1H-tetrazole (**9**) is widely used as the inhibitors of Ostwald's ripening, development, and so on, since it has a strong interaction with silver halides; its silver salt has the solubility product of $\mathrm{p}K_{sp} = 16.2$ [17].

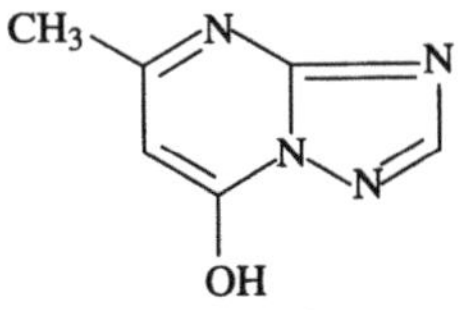

8 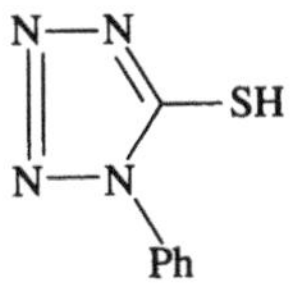 **9**

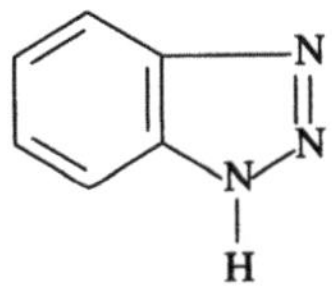

10 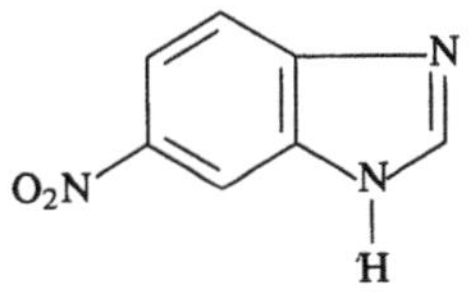 **11**

On the other hand, coalescence ripening is another type of grain growth, in which two or more crystals weld together to form a larger crystal. This process and Ostwald's ripening are alternatively called "physical ripening", whereas the after-ripening described later is called "chemical ripening".

Recrystallization

Silver chloride and silver bromide form a solid solution in all proportions, since they crystallize in the same crystalline structure (face-centered cubic or NaCl type). Such a solid solution is called "silver bromochloride". Silver iodide and other silver halides up to about 30% of iodide content can exist in solid solution [17]. When a silver nitrate solution is added dropwise into a solution of Br^- and I^-, AgI and then AgBr crystallize owing to the difference of their solubilities. The precipitated AgI is dissolved by the action of Br^- to give a silver iodobromide [12]. This process of rearrangement is called "recrystallization".

2.5 Washing

To stop the growth and ripening of silver halides, ionic species such as NO_3^-, excess halides, ammonium salts, and by-products are removed from the dispersion medium by washing. Although industrial methods other than the historical noodle washing method are trade secrets, a dialysis method and a precipitation or flocculation method have been reported in patent specifications.

1. The noodle washing method: The emulsion is chilled to a gel, which is forced through a perforated screen to form noodles. These noodles are washed in running water.
2. The dialysis method: The emulsion is dialyzed by using membrane filtration techniques.
3. The precipitation or flocculation method: A coagulant or flocculant is added to the emulsion, and the supernatant solution is repeatedly replaced with water. As such coagulants, inorganic salts, anionic surfactants, anionic polymers (such as polystyrenesulfonic acid), and gelatin derivatives (such as acylated gelatins or carbamoyl derivatives of gelatin) have been reported [18].

2.6 After-Ripening and Chemical Sensitization

Since the raw emulsion is not so highly light-sensitive, it undergoes the after-ripening process, where chemical sensitizers are added to enhance its light sensitivity. This process is also called "chemical ripening". Such chemical sensitizers are classified as follows:

1. **Sulfur sensitizers**: thiosulfates, thioureas, thiazoles, rhodanine, etc.
2. **Reduction sensitizers**: stannous salts, amines, hydrazine derivatives, formamidinesulfinic acids, silane compounds, etc.
3. **Gold sensitizers**: gold complex salts and complex salts of metals of Group VIII in the Periodic Table such as platinum, iridium and palladium.

A sensitization center generated by sulfur sensitizers or gold sensitizers work as a trigger of latent image formation, which traps a photo-induced electron. On the other hand, a center generated by reduction sensitizers traps a positive hole and inhibits the recombination of an electron and a positive hole, enhancing the efficiency of the photo-process [19].

2.6.1 Sulfur Sensitization

It had been well known that some kinds of gelatin enhanced the light-sensitivity of silver halide emulsions. However, the real source of the enhancement had not been clarified until silver sulfide generated from allylthiourea was identified by Sheppard [20].

$$S{=}C\begin{matrix} NH(CH_2CH{=}CH_2) \\ NH_2 \end{matrix} + 2Ag^+$$

$$\longrightarrow Ag_2S + NC\text{–}NH\text{–}CH_2CH{=}CH_2 + 2H^+ \tag{2.34}$$

Since allylthiourea was a non-natural sulfur compound that possessed the capability of sensitization, considerable efforts were expended in order to seek sulfur-containing components of gelatin sources as natural precursors of sulfur sensitizers. As a result, cystine ((-S-$CH_2CH(NH_2)COOH)_2$) in the gelatin sources was found to form thiosulfates, which worked as such precursors [21].

Nowadays, various organic compounds with labile sulfur atoms are known as sulfur sensitizers. They are cited in most patent specifications as obvious examples.

S, NaO—S—ONa, O

12

CH_3, S, S—ONa, O

13

S, NH—C—NH

14

CH_3, N, S, S, NH—C—NHC_2H_5

15

S, $(C_2H_5)_2N$—C—NHC_2H_5

16

S, CH_3—C—NH

17

O, NH, S, S

18

O, NH, CH, S, S

19

O, N, C_2H_5, CH, S, S

20

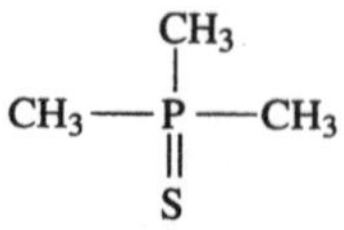

21 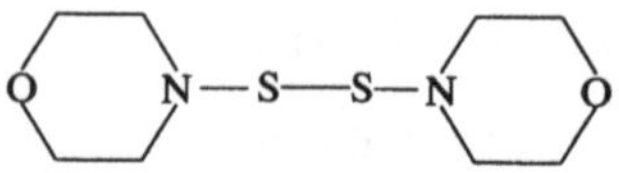 **22**

Examples of sulfur sensitizers include thiosulfates (e.g., **12** and **13**), thioureas (e.g., **14** to **16**), thioamides (e.g., **17**), rhodanines (e.g., **18**, **19**, and **20**), phosphine sulfides (e.g., **21**), disulfides (e.g., **22**), and so on. The decomposition of these sensitizers produces Ag_2S (or a cluster of Ag_2S) that is capable of trapping electrons to form a latent image. The amount of a sulfur sensitizer is generally from 1×10^{-7} to 5×10^{-3} mol per mol of silver halide.

2.6.2 Reduction Sensitization

Reduction sensitization improves the quantum yield of a latent image, where a reducing agent as a reduction sensitizer forms silver metal specks (e.g., Ag_2), which trap positive holes (or halogen atoms) and thereby prevent electron-hole recombination. Reduction sensitizers include stannous salts (e.g., tin(II) chloride: $SnCl_2$), hydrazines (e.g., *p*-tolylhydrazine: **23**), borane compounds (e.g., dimethylamine-borane: **24**), aminoiminomethanesulfinic acid (thiourea dioxide: **25**), reductones (e.g., ascorbic acid: **26**), polyamine compounds (e.g., triethylenetetramine: **27**), and so on.

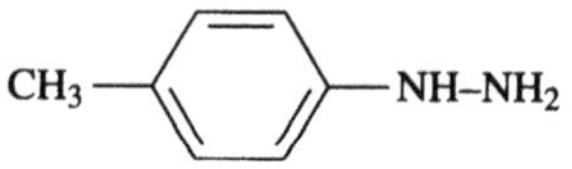

$(CH_3)_2NH{\cdot}BH_3$

23 **24**

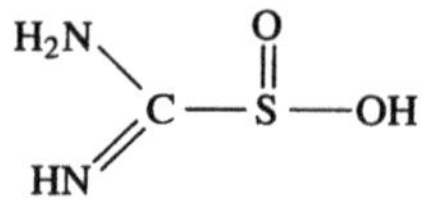

25 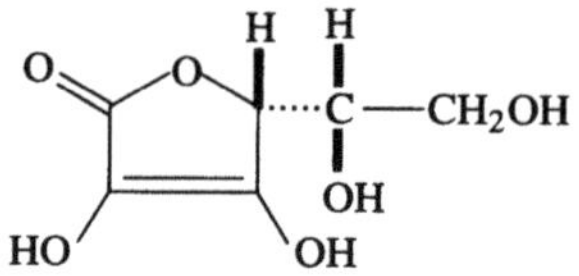 **26**

$H_2NCH_2CH_2NHCH_2CH_2NHCH_2CH_2NH_2$

27

2.6.3 Gold Sensitization

Gold sensitization was discovered by Koslowsky in 1936 [22] and was used as a fundamental technology for manufacturing photographic films of Agfa. The details of the gold sensitization was disclosed after World War II by PB reports and stimulated the advance of silver halide photography to a great extent.

Examples of gold sensitizers include chloroauric acid ($HAuCl_4$), potassium chloroaurate ($KAuCl_4$), potassium aurithiocyanate ($KAu(SCN)_4$), and gold complexes with heterocyclic ligands (e.g., **28** [23] and **29** [24]). They are usually added to silver halide emulsions during an after-ripening process. The amount of such a gold sensitizer is preferably from 1×10^{-7} to 1×10^{-2} mol per mol of silver halide.

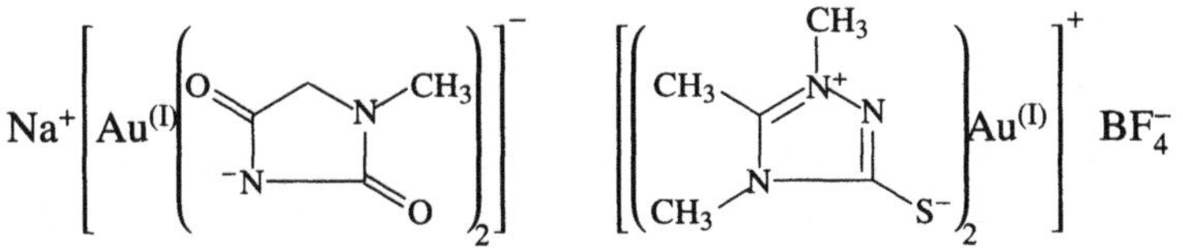

28 **29**

A combination of the sulfur sensitization with the gold sensitization is particularly preferred, because of the superadditivity of the two sensitizations. For example, a sensitizing center due to the combination is produced by the following equation [25]:

$$3AgBr + Au(S_2O_3)_2 + 2H_2O \rightarrow Ag_3AuS_2 + 3Br^- + 2H_2SO_4, \qquad (2.35)$$

where the sensitizing center is ascribed to the species Ag_3AuS_2. When gelatin is present, a further reduction can occur to form gold atom clusters that work as fog centers though light sensitivity increases [5,26].

2.6.4 Example of Chemical Sensitization

An example of an ammoniacal method using gold sensitization is cited from Fiat Final Report [11]. The resulting emulsion was called "Isopan Finegrain Color", which was used in a blue-sensitive layer of Agfacolor Negative (Type B & G) after the addition of the corresponding sensitizing dye.

Solution I	water	44 L
	KBr	9.50 kg
	KI	0.230 kg
	gelatin	2.5 kg
Solution II	water	2 × 44.0 L
	$AgNO_3$	2 × 5.25 kg
	NH_3	2 × 6.25 L
Solution III	gelatin (dry)	12.5 kg

These three solutions are prepared in advance and treated in a darkroom as follows:

1. **Precipitation**: Portions of Solution II are added four times into solution I at 25.5°C:

2 × 12.5 L added for 1 min	then allowed to stand for 4 min
2 × 12.5 L added for 1 min	then allowed to stand for 4 min
2 × 12.5 L added for 1 min	then allowed to stand for 4 min
the remaining part added	

Finally, Solution III is added to the mixture. The mixture is digested for 30 min to undergo Ostwald's ripening and then is cooled for 8 min. After the addition of acetic acid (AcOH 6.25 L and water 6.25 L) for 5 min, the mixture is stirred for 5 min.

2. **Washing** The mixture is transferred to a dish and chilled to set. The set emulsion (gel) is cut to form noodles, which are washed in running water for about 8 hr. The endpoint of washing is decided by the electric conductivity. Phenol as an antifungal agent is added to the noodles.

3. **After-Ripening** The procedure described above is repeated twice and the resulting noodles are combined and added to the following solution.

gelatin	14.9 kg
	+ MgO
KBr (10%)	1.68 L
$PhSO_2Na$ (20%)	2.4 L
urea (56 kg/6 L water)	4.8 L
V130	1.44 L

where V130 is prepared by mixing 1 ml of $AuCl_3$ (40% Au) and 1 L of a 1% aqueous solution of sodium thiocyanate; and MgO is used to adjust the pH of the gelatin solution. The sodium phenylsulfinate is added as an antifoggant or a stabilizer. The time and temperature of the after-ripening were not described in the report.

Nowadays, mixed sensitization, especially, a combination of the gold sensitization (e.g., $AuCl_3$ and KSCN) and the sulfur sensitization (e.g., *p*-toluenethiosulfonate), is used in most photographic materials of high sensitivity.

2.7 Industrial Manufacture of Silver Halide Emulsions

Industrial manufacture of silver halide emulsions has mainly been based on double-jet methods. Various apparatuses have been proposed to control grain shapes and size distributions of a silver halide emulsion. One of recent examples is a double-jet apparatus having two addition vessels with pressure-regulated feeding systems [27].

References

[1] Charlot G (1969) Les Reactions Chimiques en Lolution. L'Analyse Qualitative Minérale, 6e éd., Masson et Cie, Paris

[2] Rose PI (1977) Gelatin. In: James TH (ed) The Theory of the photographic process. Macmillan, New York London, Chapter 2

[3] http://www.nitta-gelatin.co.jp/gel-jiten/gairon/gairon4.html

[4] Kogure K (1986) In: High-Function Photochemicals. CMC, Tokyo, Chapter 2

[5] Urabe S, Sano T (1999) J Soc Photogr Sci Technol Jpn. 62:295

[6] (a) Tani T (2002) In: International Congress of Imaging Science 2002, Tokyo, pp. 41–42 (b) Tani T (1997) J Imaging Sci Technol. 41:577

[7] Tani S, Tani T (2002) In: International Congress of Imaging Science 2002, Tokyo, pp. 54–55

[8] De Clercq M, Legat JC, Rolin D (1995) J Photogr Sci. 43:154

[9] Suzuki K, Mimasu M (2002) In: International Congress of Imaging Science 2002, Tokyo, pp. 52–53

[10] Umberger JQ (1973) (ed) Photographic Chemistry. In: Thomas Jr. W (ed) SPSE Handbook of Photographic Science and Engeneering. John Wiley & Sons, New York, Section 9

[11] Gluck B (1947) Fiat Final Reoprot. No. 943

[12] Berry CR (1977) Precipitation and Growth of Silver Halide Emulsion Grains. In: James TJ (ed) The Theory of the Photographic Process, 4th edn. Macmillan, New York London, Chapter 3

[13] Klein DH, Gordon L, Walnut TH (1959) Talanta 3:177

[14] Berry CR (1974) Photogr Sci Eng. 18:4

[15] Niki K, Ohashi M (1986) In: High-Function Photochemicals. CMC, Tokyo, Chapter 2, Section 3

[16] Nakashima T (1979) Yuki Gosei Kagaku Kyokai-Shi. 37:423

[17] Pouradier J, Pailliotet, Berry CR (1977) Properties of Silver Halides. I. Structure and Thermodynamic Properties. In: James TJ (ed) The Theory of the Photographic Process, 4th edn. Macmillan, New York London, Chapter 1, Section I

[18] For example, see Shibahara Y, Ikenoue S, Watanabe T (1992) Eur Patent 0 476 327

[19] Kanzaki H (1990) J Photogr Sci Technol Jpn. 53:529

[20] Sheppard, SE (1925) Photogr J. 65:380

[21] Harbison JM, Spencer, HE (1977) Chemical Sensitization and Environmental Effects. I. Chemical Sensitization. In: James TJ (ed) The Theory of the Photographic Process, 4th edn. Macmillan, New York London, Chapter 5

[22] Koslowski R (1951) Z Wiss Phot. 26:65

[23] Deaton JC (1991) US Patent 5 049 484

[24] Deaton JC (1991) US Patent 5 049 485

[25] Faelens PA (1968) Photogr Korresp. 104:137

[26] Tani T (1995) Photographic Sensitivity: Theory and Mechanisms. Oxford Univrsity Press, New York, Chapter 6

[27] Ichikawa Y, Nagasawa H, Nomiyama T (2002) US Patent 6 406 821 B1

Chapter 3

Silver Halide Crystals. Fundamental Properties

3.1 Crystal Structures of Silver Halides

3.1.1 Dimension of Crystal Lattices

According to X-ray diffraction analysis [1], silver chloride and silver bromide form face-centered cubic structures (the rock salt type), as shown in Fig. 3.1 [2]. The lattice dimension data obtained by electron diffraction analysis [3] are also shown in Fig. 3.1.[1]

Alkali halides such as NaCl form crystals of face-centered cubic structure (the rock salt type), where the distance (a) between M^+ and X^- is found to be equal to the sum of the respective ionic radii [4]. In this book, the corresponding unit cell dimension α_{MX} is used for further discussions:

$$\alpha_{MX} = 2a = 2(\gamma_M + \gamma_X), \tag{3.1}$$

where the ionic radius of M^+ and that of X^- are respectively represented by the symbols γ_M and γ_X.

The ionic radii of ionic species are calculated on the basis of the corresponding oxides and fluorides, where the ions are considered as rigid, undeformable spheres with an electronic charge [5]. The following data are cited from the textbook of Greenwood et al. [6]:

[1] Each value a is half of the corresponding unit cell dimension. Note that 100 pm = 10^{-10} m = 1 Å.

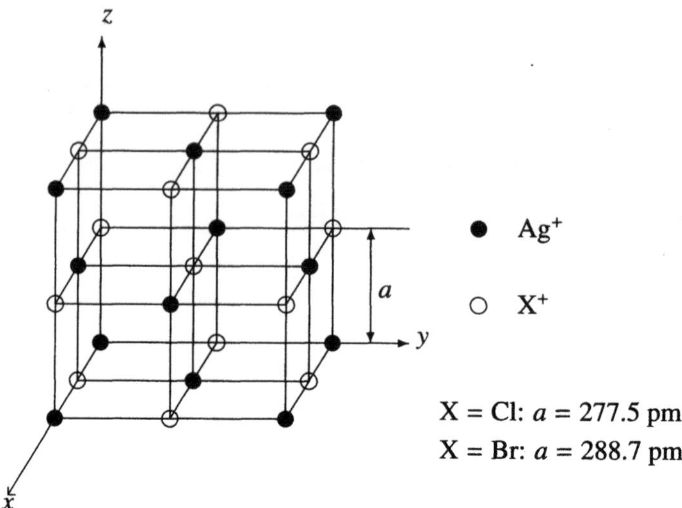

Figure 3.1. Schematic representation of silver halide crystals (face-centered cubic structure).

Ionic Radii (pm)

Li^+	76	Na^+	102	K^+	138	Ag^+	115
F^-	133	Cl^-	184	Br^-	196	I^-	220

By using these data, eq. 3.1 gives calculated values of unit cell dimensions. The corresponding observed values are cited from a data book [7].

Calculated and Observed Values of α_{MX} (pm)

NaX			AgX		
	observed	calculated		observed	calculated
NaF	462.9	470	AgF	493	496
NaCl	563.98	572	AgCl	555.8	598
NaBr	597.32	596	AgBr	577.45	622
NaI	647.28	644			

As found in this table, the calculated values for alkali halide crystals are in agreement with the observed ones, since an alkali ion and a halide ion are completely ionic and can be regarded as rigid spheres.

On the other hand, the agreement between calculated and observed values is not so good for AgCl or AgBr in contrast to AgF. Such deviation from the simple ionic model can be ascribed to the nature of the silver halide bonds that are more covalent than alkali halide bonds.

(more ionic) AgF — AgCl — AgBr — AgI (more covalent)

Silver iodide is crystallized in the hexagonal wurtzite structure (β-AgI) or the cubic zinc blende structure (γ-AgI) under the normal conditions of preparing photographic emulsions. In general, γ-AgI is preferably precipitated in the excess of silver ions while β-AgI is formed in the excess of iodide ions [8]. The selective formation of γ-AgI is accomplished when the reactant solutions are added slowly without an excess of either silver or iodide ions [9]. Two further forms have been known to be the face-centered cubic structure (δ-AgI above 300–400 MPa) and the body centered cubic structure (α-AgI above 146℃).

3.1.2 Solid Solutions

Silver halides employed mainly in photography are silver bromide and silver chloride, which are used separately or mixed (e.g., silver bromochloride or AgClBr).[2] Silver iodide is usually combined with silver bromide or silver chloride (e.g., silver iodobromide and silver iodochloride).

Silver chloride and silver bromide form a solid solution in any proportions.[3] The resulting face-centered cubic structure has the lattice parameter ($2a$ in Fig. 3.1) represented by the following experimental equation [1]:

$$\text{Silver bromochloride} \quad \alpha = 555.02 + 0.2246[\text{Br}] \text{ pm} \tag{3.2}$$

where the concentration of bromide ions is represented by [Br] mole% and the α (at 25℃) is obtained in pm units.

Although the normal crystal structures of AgI (γ-AgI and β-AgI) are not face-centered cubic, silver iodobromide and silver iodochloride form face-centered cubic crystals up to a concentration of 30% iodide ions. The lattice parameter ($2a$ in Fig. 3.1) is represented by the following experimental equations [10]:

$$\text{Silver iodobromide} \quad \alpha = 577.48 + 0.368[\text{I}] \text{ pm} \tag{3.3}$$

$$\text{Silver iodochloride} \quad \alpha = 555.02 + 0.635[\text{I}] \text{ pm} \tag{3.4}$$

where the concentration of iodide ions is represented by [I] mole% and the α (at 25℃) is obtained in pm units.

3.2 Shapes of Silver Halide Crystals

3.2.1 Representation of Shapes

The shapes of silver halide crystals play an important role to discuss the photosensitive properties of the crystals. Hence, it is convenient to make an introductory comment on the faces of such crystals. In crystallography, the faces of a crystal are represented by virtue of Millerian indices, as shown in Fig. 3.2 [11].

---

[2]The term "silver bromochloride" and the formula "AgClBr" represent a solid solution in which silver chloride as a main component contains silver bromide as a minor component.

[3]A solid solution is alternatively called *mixed crystal* for ionic crystals.

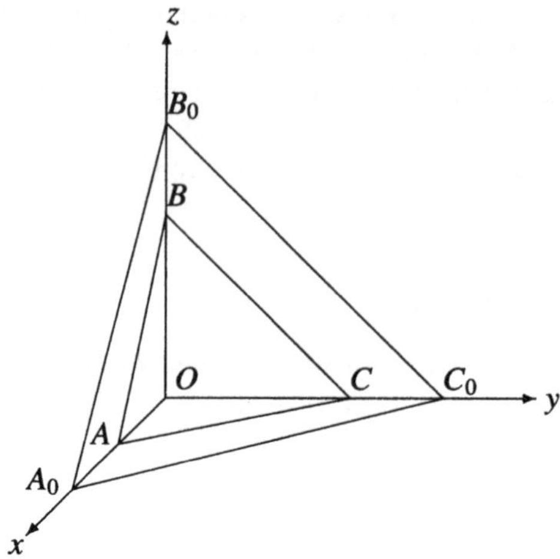

Figure 3.2. Millerian indices.

Suppose that an appropriate set of crystallographic axes (x, y, z)[4] and a parametral plane P_0 are selected. Then the plane P_0 intercepts the rectangular x, y, and z axes at A_0, B_0, and C_0 respectively. When another plane P to be described intercepts the axes at A, B, and C, there exist a set of three numbers h, k, and l satisfying the following experimental equation:

$$\frac{OA_0}{OA} : \frac{OB_0}{OB} : \frac{OC_0}{OC} = h : k : l, \tag{3.5}$$

where the three numbers are small integers that have no common divisors. The set of integers h, k, l is called the Millerian index of the plane P. The plane P is called the plane (hkl). For example, the plane that intercepts x, y, and z axes at 1/2, 2/3, and 2/3 unit lengths is described to be the (433)-plane. Obviously, the parametral plane P_0 is the plane (111). If a plane intercept an axis, the corresponding integer is defined to be zero. If an interception is negative, the index is modified with an overbar (e.g., $(11\bar{1})$).

It should be noted that even a face-centered cubic crystal does not always have a cubic shape. Cubic shaped crystals appear, when the plane (100) is predominant. When the plane (111) is predominant, octahedral or tetrahedral crystals appear. In fact, cubic crystals and octahedral crystals appear frequently in conventional procedures for preparing silver halide crystals.

On the other hand, hexagonal tabular crystals are widely used in modern photographic technology. Each of the crystals has top and bottom planes (111) and six side faces corresponding to the index (100).

---

[4] A set of rectangular axes is used in this book, since silver halide crystals usually have face-centered cubic structures.

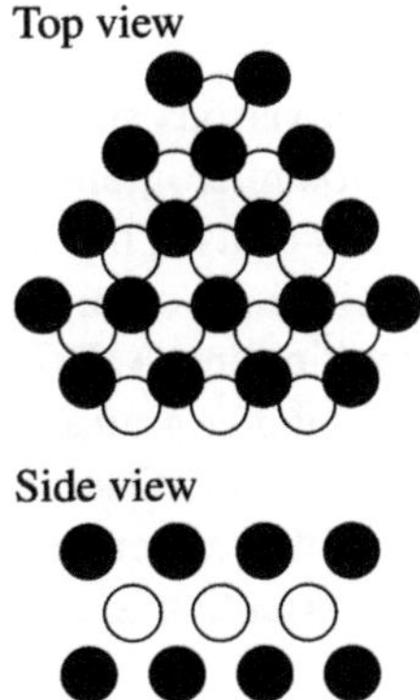

Figure 3.3. Schematic representation of a hexagonal tabular crystal.

The (111) plane of a silver halide crystal can be considered to have the structure in which planes containing silver ions and planes containing halide ions are alternately superimposed. A schematic representation[5] is shown in Fig. 3.3.

3.2.2 Controlling Crystal Shapes of Silver Halides

The double-jet method has an advantage that crystal shapes and size distribution are easily controlled by virtue of the values of pAg, etc. Thereby, crystal shapes of silver halides have been examined in detail [12].

1. In the absence of silver halide solvents, the controlled precipitation at pAg 7.9 gives small cubic grains having (100) planes. After standing for 20 hr, their vertices are truncated to give round grains.

2. In the presence of ammonia as a silver halide solvent, the controlled precipitation at pAg 7.9 gives large cubic grains having (100) planes.

3. In the presence of excess KBr,[6] tabular crystals having (111) planes form predominantly. In the condition of $[Br^-] > 10^{-3} mol \cdot L^{-1}$, the bromide ions are preferably adsorbed by the (111) planes than by the (100) planes [13]. The electronic repulsion prevents access of a complexed anion $AgBr_3^{2-}$ to the (111) planes. As a result, the $AgBr_3^{2-}$ is attached to the (100) planes, growing a tabular crystal.

[5]To show essential features, this schematic representation is simplified as far as possible. As described in the next chapter, such hexagonal tabular crystals of silver halides have been found to be twinned crystals.

[6]Although this condition is satisfied in the single-jet method, a more elaborate procedure is necessary to prepare mono-sized tabular grains that are used in modern photographic materials.

4. In the presence of ammonia and a small excess of KBr as silver halide solvents, the controlled precipitation at pAg 9.2 generates octahedral crystals having (111) planes.

Nowadays, silver halide crystals of various shapes are used according to their photographic properties. Their electron micrographs have been reported, showing the shapes classified by their purposes of use [14]: AgCl cubic grains for a color paper, AgBrCl hexagonal tabular grains for a color negative film, AgBr octahedral grains for an instant color photographic film, AgBrI cubo-octahedral grains for a color reversal film, AgBr hexagonal tabular grains for a medical X-ray film, and AgBrCl cubic grains for a graphic arts film. One of the most important features found by the electron micrographs is that silver halide grains of the same shape have been prepared in narrow size distribution according to each purpose.

3.3 Lattice Defects in Crystals

3.3.1 Kinds of Lattice Defects

Many photographic properties of silver halide (e.g., high sensitivity) stem from imperfections in the structure of silver halide crystals. Such imperfections (called *lattice defects*) are frequently incorporated in the processes of preparing photographic emulsions of respective purposes. They are categorized as follows:

- Intrinsic point defects
  - Frenkel defects
  - Schottky defects
- Extrinsic point defects
  - Defects due to an impurity with a single charge
  - Defects due to an impurity with two or more charges
- Linear defects
  - Edge dislocation
  - Screw dislocation
- Areal defects
  - Twin plane
  - Grain boundary

3.3.2 Intrinsic Point Defects

Schottky and Frenkel Defects

Even at any ordinary temperature, several ions in a silver halide crystal can be removed from their normal sites. As a result, two types of lattice defects occur, i.e., Frenkel and Schottky defects, as shown schematically in Fig. 3.4.

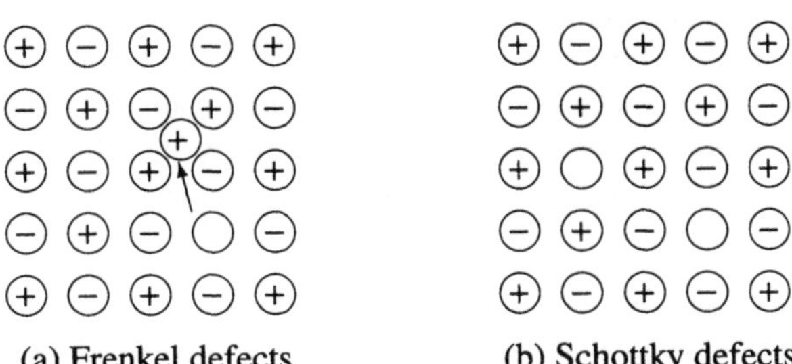

Figure 3.4. Schematic representations of intrinsic lattice defects. An open circle represents a vacancy (hole) of a cation or anion.

Frenkel defects are based on the movement of ions from their normal lattice sites into interstitial positions (Fig. 3.4a). The original lattice sites are left vacant. Since cations are usually smaller than anions in size, a pair of a shifted cation (marked with an arrow) and a negative vacancy is preferably generated.

Frenkel defects play an important role in the photographic process of silver chloride and silver bromide. Figure 3.5 illustrates a silver cation moved into an interstitial position, which is called a *tetrahedral vacancy* because it is surrounded by four adjacent atoms of the same kind (e.g., silver ions or halide ions) that construct a tetrahedron. Note that the domain of Fig. 3.5 is the eighth part of the cell shown in Fig. 3.1.

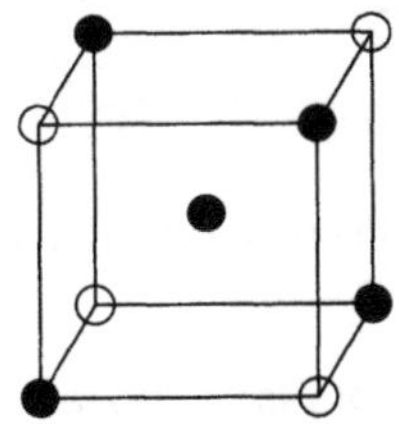

Figure 3.5. Ag^+ ion at an interstitial position (● located at the center of the cube). The vertices of the cube represent normal lattice positions, which are occupied by four silver ions (●) and four halide ions (○).

On the other hand, Schottky defects contain a pair of an anionic vacancy and a cationic one (Fig. 3.4b), where the electrical neutrality of each crystal is preserved.

Ionic Conduction

The electrical conductions of materials are classified into electronic conduction (electrons as carriers) and ionic conduction (ionic species as carriers). The conduction of a metal is electronic and gives the conductivity ranging from 10^3 to 10^7 $S \cdot m^{-1}$.[7] On the other hand, the ionic conduction is predominant in ionic crystals. Because the conductivity of typical ionic crystals is smaller than 10^{-2} $S \cdot m^{-1}$, they can be regarded as electric insulators. However, ionic crystals with point defects exhibit high electric conductivity (e.g., 131 $S \cdot m^{-1}$ for α-AgI).

Point defects are characterized by holes (for both Schottky and Frenkel defects) and interstitial ions (for Frenkel defects). The movement of these charged species generates ionic conduction, where the following mechanisms can be considered [15]:

1. Vacancy mechanism. An adjacent ion is moved to fill a vacancy (hole). This process occurs repeatedly to move the vacancy further (the hopping model).
2. Interstitial mechanism. An ion at an interstitial position moves through interstitial positions.

The electric conduction for Schottky defects is generally based on the vacancy mechanism. Since cations are usually smaller than anions in size, the former move to fill the vacancy. This process is discussed by a simple geometrical consideration (Fig. 3.6).

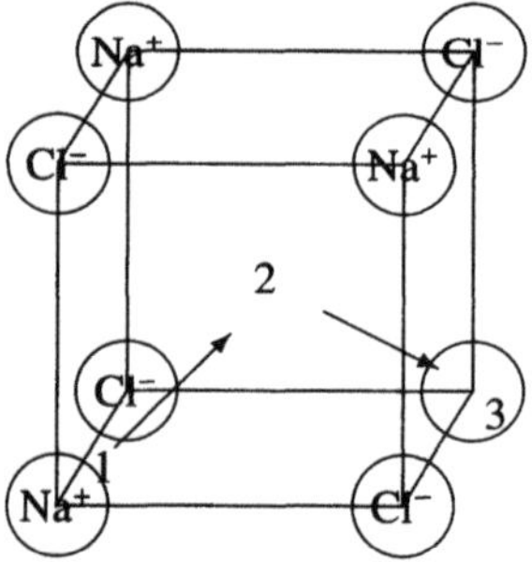

Figure 3.6. Ionic conduction of NaCl (vacancy mechanism).

Let us consider the movement of Na^+ at position 1 into the vacancy at position 3. Whereas the shortest route should cross a line between adjacent Cl^- and Cl^-, the route 1→2→3 (through a tetrahedral vacancy) is sterically preferable, since Na^+ goes through the looser interstice surrounded by three chloride ions.

In the light of a geometrical consideration similar to Fig. 3.6, the interstitial mechanism for Frenkel defects requires that an interstitial Ag^+ crosses a line

[7] $1\ S = 1\ \mho = 1\ \Omega^{-1}$.

between Cl^-–Cl^- when it moves to an adjacent tetrahedral vacancy. It is more plausible for the interstitial Ag^+ to eject the adjacent silver ion located at the normal lattice position (a pseudo-interstitial mechanism), since several experimental evidences have been reported [16].

Whereas large crystals of silver bromide gave the dark conductivity of ca. 10^{-8} S · m^{-1} at room temperature, the corresponding valuc for small crystals was estimated to be ca. 10^{-6} S · m^{-1} [17,18]. The larger conductivity of the latter was ascribed to interstitial Ag^+ ions. Since the dark conductivity was found to be proportional to the ratio of surface/volume, it was concluded that the interstitial ions were supplied from the kink positions of surfaces.

3.3.3 Extrinsic Point Defects

Extrinsic point defects are essential to provide silver halide crystals with high light-sensitivity. When a monovalent metal ion (e.g., Cu^+) replaces a silver halide in the crystal, it does not disturb the electronic balance of Frenkel defects. On the other hand, a di- or more-valent metal ion (e.g., Pb^{2+} and Cd^{2+}) can affect the balance of Frenkel defects. This means that, if the electrical neutrality of each crystal is preserved, the metal ion should accompany a vacancy of a monovalent cation, as shown in Fig. 3.7a. In other words, a Frenkel defect is generated by force. Hence, the silver halide crystal has an enhanced capability of capturing a mobile electron in the long distance.

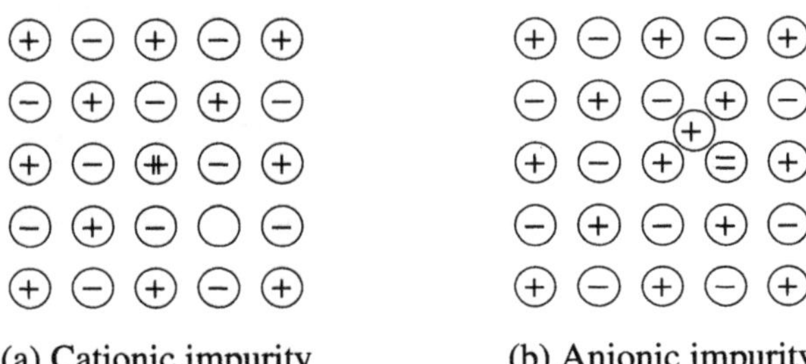

Figure 3.7. Schematic representations of extrinsic point defects by divalent ions. The symbol ⊕ represents a divalent cation, while the symbol ⊜ stands for a divalent anion.

A divalent anion S^{2-} can work in a similar way to Fig. 3.7a, where the resulting vacancy of a monovalent anion is capable of capturing a mobile hole. However, another mechanism (Fig. 3.7b) is considered for silver halide crystals, where interstitial Ag^+ ions are presumed to preserve the electrical neutrality.

3.3.4 Linear Defects—Dislocations

Another type of imperfections of crystals are dislocations, which are also recognized as active sites for photographic phenomena. For example, dislocations on silver bromide crystals have been observed as the site of silver formation by means of an optical microscope [19].

Edge dislocations and screw dislocations are known as linear defects, since the resulting imperfections give lines. Figure 3.8 schematically illustrates an edge dislocation, where a shearing stress is applied to the top half. The split plane is a horizontal one represented by a thick line. In general, such a split plane is a plane with maximal packing of atoms. For a face-centered cubic crystal (e.g., a silver halide crystal), the plane with maximal packing is a (111) plane.

As a result of the edge dislocation, an extra plane is inserted so as to give a dislocation line, which is shown as the line perpendicular to this page through the small circle in Fig. 3.8.

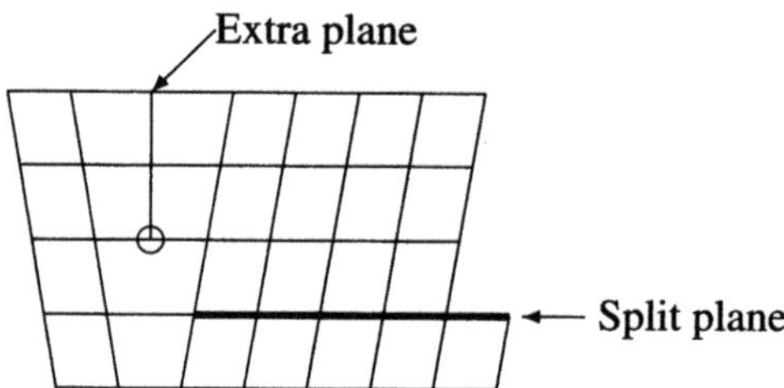

Figure 3.8. Schematic representation of edge dislocation.

On the other hand, a dislocation line for a screw dislocation is parallel to the direction of an applied stress (Fig. 3.9). The stress is applied from the right-top face of the crystal, where the resulting triangular face schematically represents a part of split plane [20].

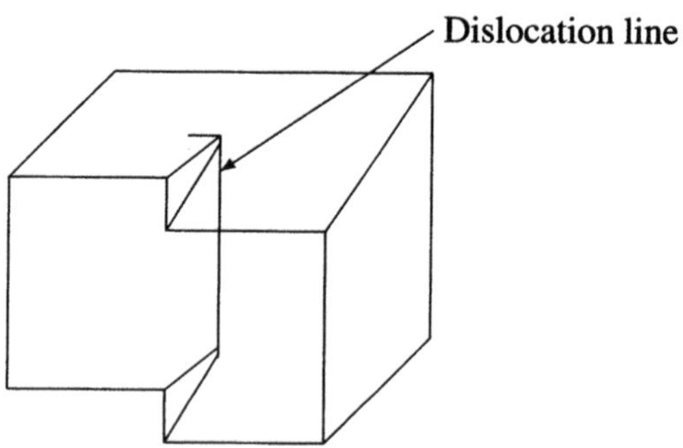

Figure 3.9. Schematic representation of screw dislocation. The vertical line is a screw dislocation line.

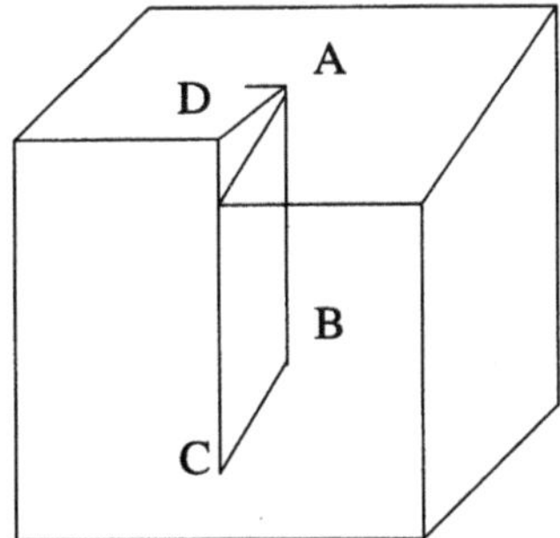

Figure 3.10. Cooperation of edge and screw dislocations. Plane ABCD: a split plane; line AB: a screw dislocation line; line BC: an edge dislocation line; and the domain around point B: a cooperative region of the two types of dislocations.

The edge dislocation (Fig. 3.8) and the screw dislocation (Fig. 3.9) can occur cooperatively, as shown schematically in Fig. 3.10. The plane denoted by ABCD is a split plane, the line AB is a screw dislocation line, and the line BC an edge dislocation line. The domain around the point B represents a cooperative region of the two types of dislocations.

The dislocations influence the modes of crystal growth. For example, the triangular split plane shown in Fig. 3.9 grows to give renewed faces. Since the screw dislocation line is fixed, the growth occurs around the line and results in the formation of a screw.

3.3.5 Kink and Jog Sites

It is important to know how line dislocations appear at the surface of a silver halide crystal, since such surface appearances influence the mechanism of the latent-image formation and the image development of photographic silver halide emulsions (dispersions).

In photographic chemistry, two kinds of surface imperfections are taken into consideration, i.e., kinks and jogs. A kink is an imperfect site (vertex or cavity) which three surfaces are incident to, while a jog is another imperfect site (convexed or concaved edge) which two surfaces are incident to.

Figure 3.11 shows schematic diagrams for three kinds of kink sites, where they are classified in terms of the modes of the incidental surfaces. Diagram A of Fig. 3.11 illustrates a neutral kink site (marked by the symbol 'a'), which is a dented site formed by three cross surfaces. The two lines represent steep surfaces between a lower layer and an upper one, where they cross with the lower layer (a terrace) at the kink site 'a'. Diagram B depicts a positively charged kink site (site 'b'), at which a silver ion is present. The silver ion loses the half number of surrounding bromide ions so that it carries a $+1/2$ net charge. Such silver ions at

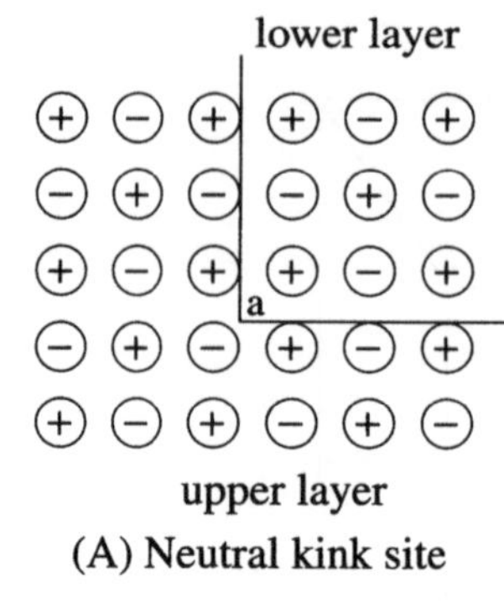

(A) Neutral kink site

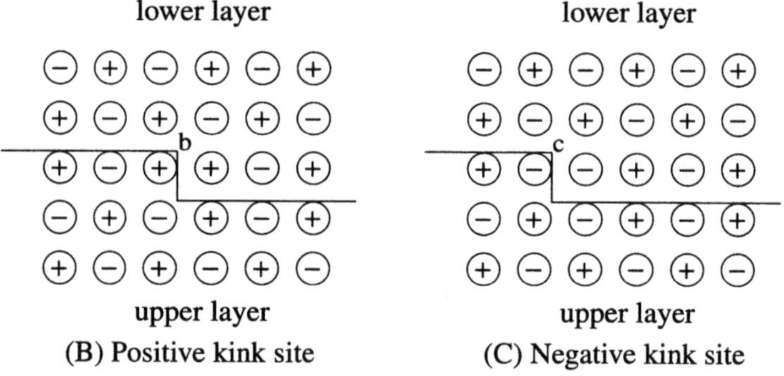

(B) Positive kink site

(C) Negative kink site

Figure 3.11. Schematic diagrams for three kinds of kink sites. A: The site designated as 'a' is a neutral kink site, which is a caved site formed by three cross surfaces. B: The sharp site marked with 'b' is a positive kink site, where the Ag^+ at the site loses the half number of surrounding bromide ions. C: The sharp site designated as 'c' is a negative kink position, where the bromide ion at the site loses the half number of surrounding silver ions. The lines represent steep surfaces between a lower layer and an upper one.

the surface kink sites can jump into interstitial positions of silver halide grains. Diagram C shows a negatively charged kink site (site 'c'), at which a halide ion (net charge $-1/2$) is located.

3.3.6 Twinned Crystals

In the formation of hexagonal tabular crystals of silver halides, twinning on a (111) plane plays an important role [21,22]. Figure 3.12 shows the schematic side view of a hexagonal tabular crystal with parallel twin planes. Each section contains alternately accumulated layers of Ag cations and of Cl anions (cf. Fig. 3.3b). Because of the reentrant angle, each position A accelerates the growth of the crystal. Since doubly twinned crystals always maintain the reentrant angles (Position A in Fig. 3.12), they show persistent growth acceleration. For more detailed discussions, see Ref. [22] and Section 3.3 of Ref. [23].

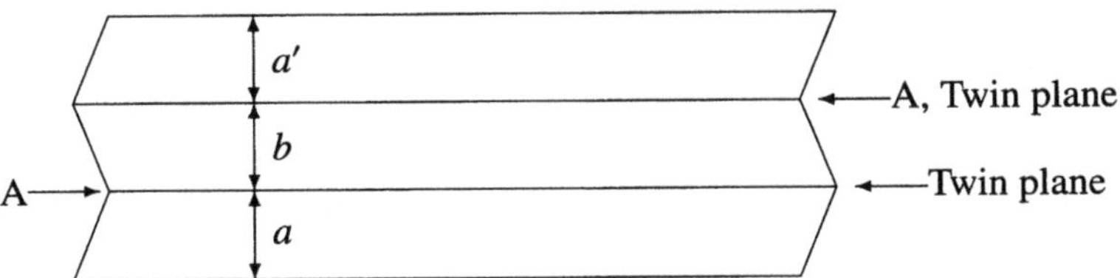

Figure 3.12. Schematic cross-section of a tabular crystal with parallel twin planes (double twinning). Positions A represent the positions of preferred growth. The ratios of the section thicknesses a, b, and a' are controlled to obtain desired photographic properties.

3.4 Spectral and Electronic Properties of Silver Halide Crystals

3.4.1 Band Theory

Spectral and electronic properties of a solid are usually explained in the light of the band theory (Fig. 3.13) [24]. Figure 3.13 shows a schematic illustration of energy bands, where the vertical axis represents the energy level of the solid in an arbitrary unit, while the horizontal axis means the spacial dimension in an abstract fashion. Each of the bands consists of many energy levels, which can be totally regarded as a quasicontinuous "band" because of close energetical values.

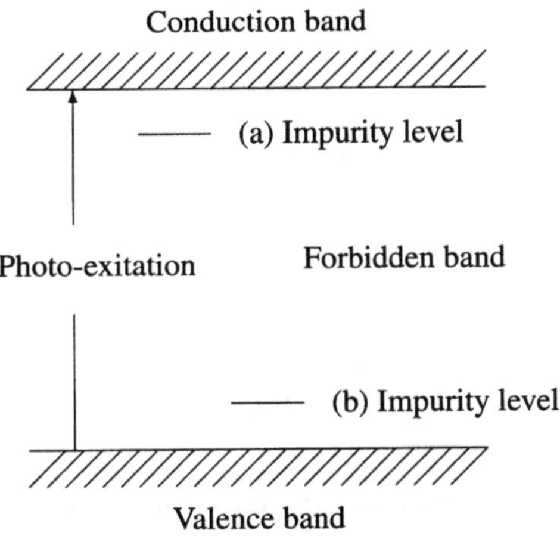

Figure 3.13. Energy level diagram of solid.

The highest occupied band is called a *valence band*, the lowest unoccupied band is called a *conduction band*, and the gap between them is designated as a *forbidden band*. Electrons in the valence band are raised to the conduction band by photo-excitation. The excited electrons contribute to the electrical conduction by moving as quasifree particles in the conduction band.

The vacancy left in the valence band during the photo-excitation is called a *hole*, which can be hypothetically considered to behave as a positively charged particle. Such photo-excited electrons and relevant holes are called electron charge carriers. Other localized energy levels can appear according to impurities and dislocations (Fig. 3.13a and b).

3.4.2 Optical Absorption

Optical absorption of silver halides has been studied extensively. Figure 3.14, which is cited from [24], shows the absorption spectra of silver halides, where the spectra of AgCl and AgBr are a composite of data on large crystals, thin sheet crystals, and evaporated layers. These spectra indicate that the optical absorption of silver halides is shorter than or at most within the blue region of visible light.

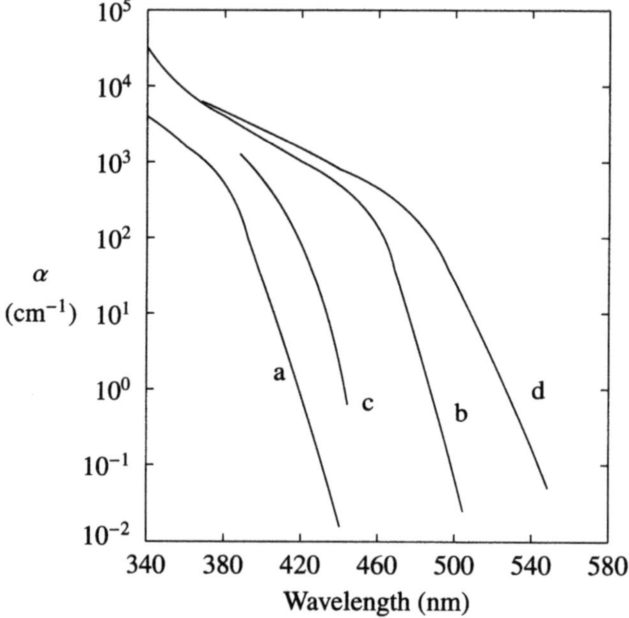

Figure 3.14. Absorption coefficients (α) of pure and mixed silver halide crystals at about 293K [24]. (a) Pure AgCl; (b) Pure AgBr; (c) AgClBr (AgBr 10 mol%); (d) AgBrI (AgI 3 mol%).

When a silver halide crystal absorbs light, an electron in the valence band of the crystal is excited to give a free electron (photo-electron) in the conduction band and to leave a positive hole in the valence band (Fig. 3.13). Since the photo-electron and the positive hole can move as electron charge carriers, their drift mobilities and Hall mobilities are determined as measures of electrical conductivity.

The drift mobilities of photo-electrons for AgBr were estimated to be 60 $cm^2 \cdot V^{-1} \cdot s^{-1}$ (large crystals) and 0.2 $cm^2 \cdot V^{-1} \cdot s^{-1}$ (small crystals) [18]! %On the other hand, the drift mobilities of holes for AgBr were estimated to be 1.1 $cm^2 \cdot V^{-1} \cdot s^{-1}$ (large crystals) and 0.001 $cm^2 \cdot V^{-1} \cdot s^{-1}$ (small crystals) [18]! %The smaller values of the small crystals indicate that both the photo-electrons and the holes are effectively trapped by shallow traps of the small crystals in pure states.[8] As a result, small crystals in pure states are considered to be incapable of generating effective latent images.

It should be noted that photographic emulsions consist of such small crystals as described above, except that they are treated to involve impurities as sensitized centers. This strongly implies that the impurities themselves play an essential role in the formation of latent images.

References

[1] (a) Wilsey RW (1921) Phil Mag. (6)42:262; (b) Wilsey RW (1923) Phil Mag. (6)46:487; (c) Wilsey RW (1925) J Franklin Inst. 200:739

[2] Pouradier J, Pailliotet, Berry CR (1977) Properties of Silver Halides. I. Structure and Thermodynamic Properties. In: James TJ (ed) The Theory of the Photographic Process, 4th edn. Macmillan, New York London, Chapter 1, Section I.

[3] Berry CR (1955) Phys Rev. 97:676

[4] Pauling L (1960) The Nature of the Chemical Bond, 3rd ed., Cornell Univ. Press, Ithaca

[5] Shannon RD (1976) Acta Cryst. A32:751

[6] Greenwood NN, Earnshaw A (1984) Chemistry of the Elements, Pergamon, Oxford, Appendix 5

[7] The Chemical Society of Japan (1966) Kagaku Binran, Maruzen, Tokyo, Table 11.29

[8] Kolkmeijer NH, van Hengel JWA (1934) Z Kristallogr. 88:317

[9] Berry CR (1967) Phys Rev. 161:848

[10] (a) Chateau H (1959) Compt Rend. 248:1950, 249:1887 (b) Chateau H, Moncet MC, Pouradier J (1958) In: Eichler W, Frieser H, Helwich O (eds) Wissenschaftliche Photographie (Ergebnisse der Intern. Konf.

[8] The drift mobility is influenced by a shallow trap that can capture an electron charge carrier. It is the accepted knowledge that silver ions at kink and interstitial positions can act as shallow traps for photo-electrons, while Frenkel defects due to such multi-valent impurities as S^{2-} can act as shallow traps for positive holes.

für Wissenshaftliche Photographie, Köln, 1956). Verlag Dr. O. Helwich, Darmstadt, p. 16.

[11] Ladd MFC (1989) Symmetry in Molecules and Crystals. Ellis Horwood, Chichester, Chapter 2

[12] Berry CR, Marino SJ, Oster Jr CF (1961) Photogr Sci Eng. 5:332

[13] Moisar E, Klein E (1963) Ber Bunsenges Phys Chem. 67:949

[14] Takada, S (1998) J Photogr Sci Technol Jpn. 61:3

[15] Smart L, Moore E (1995) Solid State Chemistry. An Introduction. 2nd ed, Chapman & Hall

[16] Friauf RJ (1957) Phys Rev. 105:843

[17] Hamilton JF, Brady LE (1964) Photogr Sci Eng. 8:189

[18] Tani T (1987) In: Technical Photograph Handbook (New Edition). Corona, Tokyo, pp. 275–277

[19] Hedges JM, Mitchell JW (1953) Phil Mag. (7) 44:223, 357

[20] Kittel G (1986) Introduction to Solid State Physics. 6th ed, John Wiley & Sons, New York

[21] (a) Urabe S (1996) J Soc Photogr Sci Technol Jpn. 59:277 (b) Urabe S (2002) In: International Congress of Imaging Science 2002, Tokyo, pp. 5–6

[22] Urabe S (2003) J Soc Photogr Sci Technol Jpn. 66:168

[23] Soc Photogr Sci Technol Jpn (ed) (1998) Principles of Photographic Science and Engineering. Corona Publishing, Tokyo, Chapter 3, Section 3.3.

[24] Moser F, Ahrenkiel RK (1977) Properties of Silver Halides. IV. Optical and Electronic Properties of Silver Halide Crystals. In: James TJ (ed) The Theory of the Photographic Process, 4th edn. Macmillan, New York London, Chapter 1, Section IV.

Chapter 4

Silver Halide Crystals. Photo-Sensitivity

4.1 Mechanism of Latent-Image Formation

When an exposed silver halide grain is developed, the development is started from several sites on such a grain to produce visible silver images. Such sites are called *latent images*, since they are invisible but have long lifetimes after exposure. What are latent images and how do they work?

When silver halide is exposed to light, the following photo-reactions occur:

$$Br^- + h\nu \rightarrow Br + e \tag{4.1}$$

$$e + Ag^+ \rightarrow Ag \tag{4.2}$$

where a bromide ion is selected as a representative halide ion. The bromine atom generated by eq. 4.1 or the silver atom generated by eq. 4.2 is converted into a latent image. Many theories for explaining the latent-image formation have been proposed [1,2], where sulfur sensitization centers were a main subject of discussion (Table 4.1).

The Gurney-Mott theory was proposed in 1938 [3] as the first well-constructed theory on the basis of modern solid physics. The following scheme is the one in which several recent results are added to the original Gurney-Mott theory.

1. Silver sulfide (▭) formed on the surface of a silver halide grain (depicted as a deformed hexagon) works as a sulfur sensitization center, which is considered as a photo-sensitive speck (a deep electron trap), as illustrated in Fig. 4.1a. This sulfur sensitization center has been proposed to be a dimeric species of silver sulfide [7,8,9].

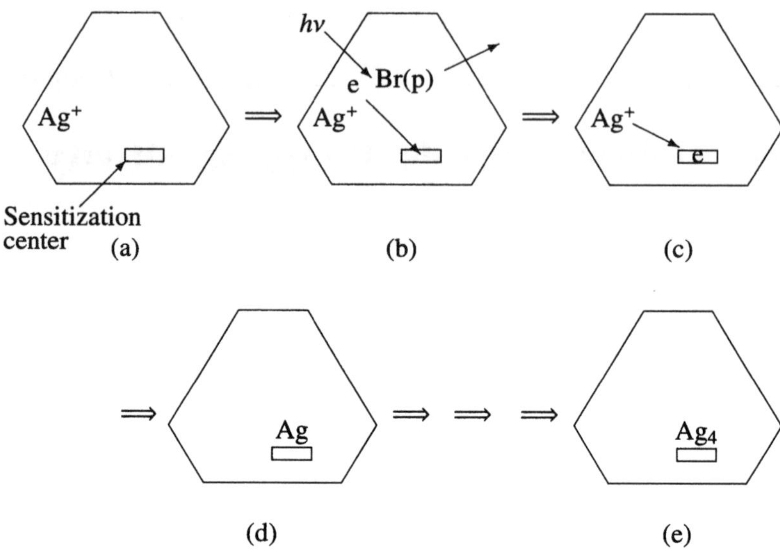

Figure 4.1. Mechanism of latent image formation. (a) Unexposed silver halide grain; (b) Electronic process: exposure of light (*hν*) and migration of resulting species (e and Br); (c) Ionic process: migration of an interstitial silver ion and combination with the trapped electron; (d) Ag atom produced by a single process; (e) Ag cluster (latent image) produced by repeated processes.

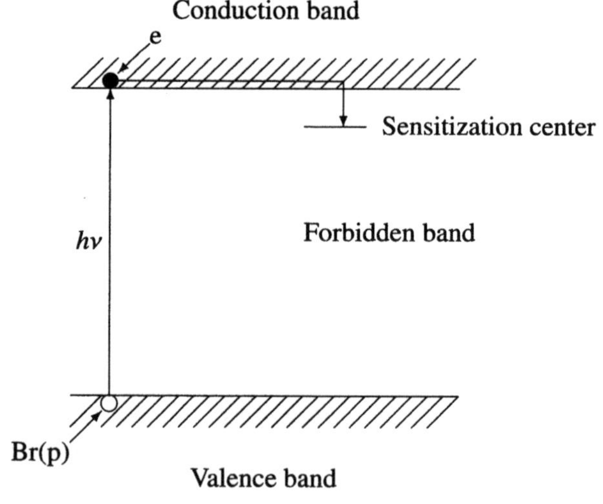

Figure 4.2. Energy level diagram of silver halide crystals for the electronic process shown in Fig. 4.1b. Photo-excitation (*hν*) promotes an electron from the valence band to the conduction band, leaving a positive hole (Br(p), ○). The excited electron (e, ●) migrates in the conduction band to be trapped by the sensitization center.

Table 4.1. Proposals for Roles of Silver-Sensitization Centers

Gurney-Mott (1938) [3]	Silver sulfide clusters (Ag_2S) on a silver halide grain work as sensitivity specks, which trap photo-electrons.
Mitchell (1957) [4]	The sensitivity speck due to silver sulfide traps a positive hole, not a photo-electron.
Hamilton (1983) [5]	The sensitivity speck due to silver sulfide does not form an electron trap, but converts shallow electron levels into deep electron traps by lattice relaxation.
Hamilton et al. (1988) [6]	The sensitivity speck due to silver sulfide forms an electron trap of 0.3 eV depth.
Kanzaki et al. (1990) [7]	A dimer of silver sulfide ($(Ag_2S)_2$) forms an electron trap of 0.5 eV depth.

2. [Electronic process] By exposure to light, a halide ion (e.g., a bromide ion) is converted into a photo-electron and a halogen atom (e.g., a bromine atom as a positive hole) in accord with eq. 4.1. Since the photo-electron has been promoted into the conduction band, it moves through the conduction band and is trapped by the photo-sensitive speck on the surface of the silver halide grain (Fig. 4.1b). The electronic process can be alternatively explained by using an energy level diagram shown in Fig. 4.2.

3. [Ionic process] An interstitial silver cation moves to the surface and combines with the trapped electron, as shown in Fig. 4.1c. As a result of eq. 4.2, a silver atom is formed, as shown in Fig. 4.1d.

4. [Clustering process] The processes described above are repeated to give a silver cluster containing about four silver atoms. The resulting cluster is regarded as a latent image (Fig. 4.1e).

5. [Photo-hole process] The bromine atom represented by the symbol Br(p) is regarded as a positive hole, which is transported to the surface of the silver halide grain and runs away as bromine (Fig. 4.1b).

For more detailed discussions, see Ref. [10].

4.1.1 Mechanism for Electronic Process

Mitchell emphasized the importance of crystalline dislocations [4]. His theory (Table 4.1) was based on knowledge of large crystals, whereas photographic emulsions are generally regarded as small crystals. Hence, to our knowledge this theory is not supported except the importance of dislocations.

The discussions by Hamilton [5,6] have reinforced the Gurney-Mott theory in considering the trapping of photo-electrons at shallow surface defects (e.g., positive kink sites or sensitizer clusters) and subsequent shallow-deep transitions (due to relaxation of surface ions).

The importance of a dimer of silver sulfide has been emphasized by several authors [7,8,9]. According to luminescence-modulation spectroscopy for different degrees of sensitization, Kanzaki et al. [7] proposed a model for silver sulfide clusters (dimers and tetramers), which played the following roles:

1. For sulfur-unsensitized AgBrI emulsions, a shallow trap (ca. 0.03 eV) due to an interstitial silver ion was observed [7]. As for monomeric Ag_2S generated by sulfur sensitization, the S^{2-} moiety required an interstitial Ag^+ in order to preserve the electrical neutrality (see Frenkel defects by divalent ions). Hence, the interstitial Ag^+ was considered to be the origin of a shallow trap in a similar way found in the unsensitized case.

2. For AgBrI emulsions of most suitable degrees of sulfur sensitization, a 0.5 eV band was observed. Note that this energy difference corresponds to the one between the conduction band and the sensitization center shown in Fig. 4.2. The sensitization center as a deep electron trap was assigned to a dimer of silver sulfide by dependence upon the amount of a sulfur sensitizer. The dimer was considered to be formed on a (111) plane, as found in Fig. 4.3 [7]. By using a technique based on a special physical developer, sulfur sensitization centers were directly observed preferably at a (111) plane rather than a (100) plane [11].

3. A further degree of sulfur sensitization caused the appearance of a 0.8 eV band, which corresponded to another deep trap [7]. The tetramer was considered to be formed on a (100) plane. Sensitometry, photo-conductivity, and oxidation-potential measurements [9] indicated that such clusters as the tetramer acted as fog centers.

Figure 4.3 shows a model structure of a silver sulfide dimer deposited on a (111) plane of silver halide, where heavy circles indicate silver ions [7]. Two of them are located in interstitial positions and the remaining two (encircled with a dashed box) are located near lattice positions. The symbol ⊜ stands for a sulfide anion (S^{2-}), which is located at an interstitial position. These atoms construct the dimer $(Ag_2S)_2$. The two Ag^+ in the dashed box are pushed out of the (111) plane to form a "bonding" orbital,[1] which is ascribed to a deep electron trap (0.5 eV).

Tani [9] has examined sulfur-sensitized tetrahedral or octahedral silver halide emulsions in detail by means of sensitometry, photo-current, and redox potential measurements. Figure 4.4 illustrates the electronic energy levels of a monomer and a dimer of Ag_2S in solid solution with silver halide [9]. An interstitial silver

[1] It should be noted that Ag^+ is considered to be a negative hole. The bonding orbital and the corresponding anti-bonding one that are formed from two Ag^+ ions are both vacant orbitals. Strictly speaking, the orbitals are non-bonding orbitals.

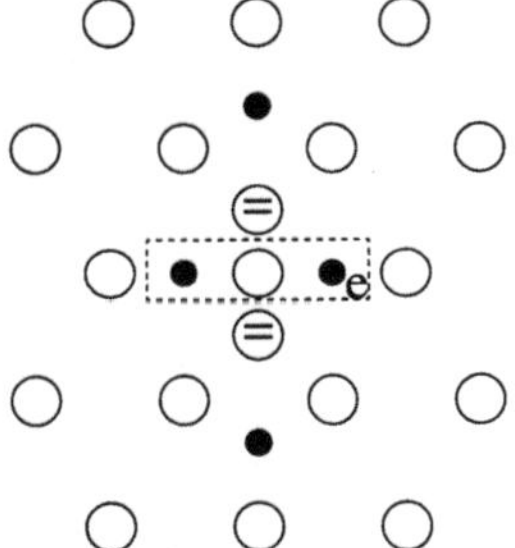

Figure 4.3. Silver sulfide dimer model for a sufur-sensitization center [7].

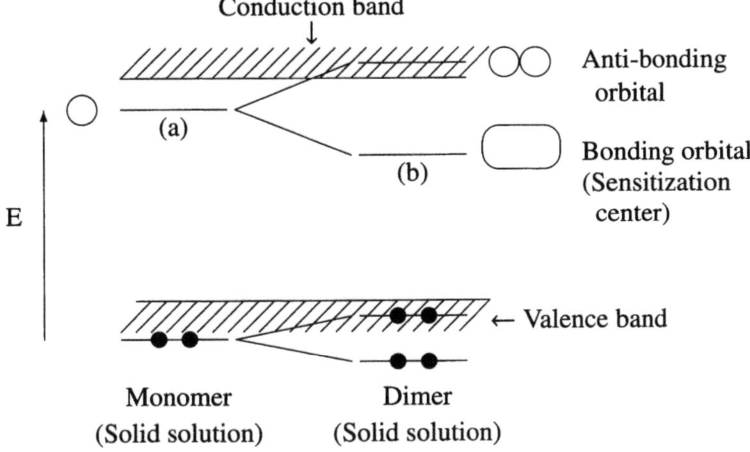

Figure 4.4. Electronic levels of silver sulfides [9]. The vertical axis (E) represents the electronic energy level, where the scale is drawn arbitrarily. The unoccupied levels represented by open circles are hydrogenic orbitals generated from interstitial silver ions, and the occupied levels represented by heavy circles are *d*-orbitals of sulfur. The level (a) indicates a shallow trap, while the level (b) indicates a deep trap. Possible energy levels for valence and conduction bands are added by hatching.

ion due to a monomeric Ag_2S provides a hydrogenic orbital (level (a) of Fig. 4.4), which is too shallow to capture a photo-electron effectively. On the other hand, a dimer $(Ag_2S)_2$ carries two interstitial silver ions, which correspond to the two electrons encircled with the dashed box in Fig. 4.3.[2] They interact with each other to form a "bonding" orbital and an "anti-bonding" orbital, the latter of which is deep enough to capture a photo-electron effectively (level (b) of Fig. 4.4).

[2]Note that the word "interstitial" was also used to designate the two Ag^+ that were pushed out of the (111) plane, as illustrated in the dashed box of Fig. 4.3.

A possible energy level for a conduction band is drawn by hatching in Fig. 4.4 according to the energy diagram of Fig. 4.2 for illustrating the electronic process. If we presume eq. 4.1 for the origin of a photo-electron, a possible energy level for a valence band should be located above the occupied levels due to sulfur atoms, as depicted in Fig. 4.2.

4.1.2 Mechanism for Ionic Process and Silver Clusters

In the ionic process (eq. 4.2), an interstitial silver ion migrates to the surface of a silver halide crystal and combines with a photo-electron trapped in the electron trap (e.g., the sulfur sensitization center), as discussed in the preceding section. This process is repeated to generate a stable silver cluster, which is composed of three or four silver atoms [12]. Such silver clusters work as latent images in the successive development process.[3]

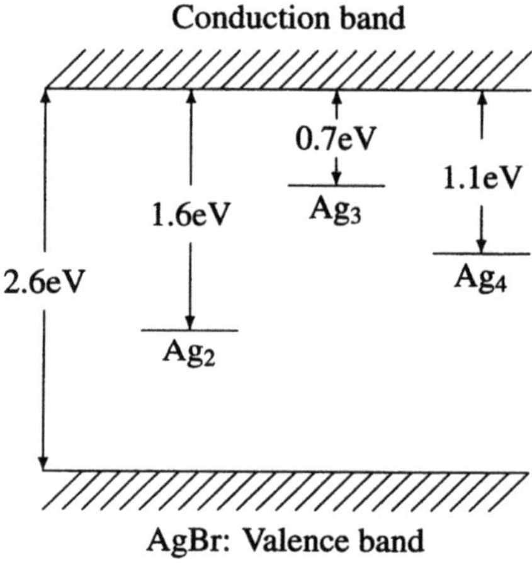

Figure 4.5. Energy level diagram of silver clusters. The highest occupied levels of Ag_2, Ag_3, and Ag_4 clusters on AgBr surface were measured by a photo-bleach technique [13].

The smallest sizes of silver clusters can be controlled by selecting appropriate conditions of developments [13]. The photo-ionization energies of such silver clusters ($Ag_n \rightarrow Ag_n^+ + e$) were determined by the photo-bleach technique, as summarized in Fig. 4.5.

[3]It should be noted that a crystal with a latent image can be reduced into a silver metal image during the development process, where Ag_4 of the latent image is magnified into Ag_n of the whole crystal. This magnification brings out the high photo-sensitivity of photographic films based on silver halide.

4.1.3 Mechanism for Photo-Hole Process

The photo-hole process is concerned with the fate of a positive hole, which is represened by h or Br· or Br(p). Although the positive hole has been tentatively considered to run away as bromine (Fig. 4.1b), the true mechanism of the photo-hole process has not been so well characterized. A hint of a possible mechanism has been obtained in investigations on reduction sensitization [14], where the role of silver clusters was emphasized.

In reduction sensitization, silver clusters on silver halide were characterized as two types of reactive centers, i.e., an image center and a reduction sensitization center [14]. The image center was recognized as a photo-electron-trapping center (P center), which was assigned to a silver cluster at a positively charged kink site [15]. On the other hand, the reduction sensitization center (R center) was assigned to a silver cluster (Ag_2) at a neutral kink site, where it works as a positive-hole-trapping center.

The dimeric silver Ag_2 reacts with a positive hole (h) to produce Ag_2^+ (eq. 4.3).

$$Ag_2 + h \rightarrow Ag_2^+ \tag{4.3}$$

$$Ag_2^+ \rightarrow Ag + Ag_{(i)}^+ \tag{4.4}$$

$$Ag \rightarrow Ag_{(i)}^+ + e \tag{4.5}$$

The resulting species Ag_2^+ is unstable and further decomposes into a single atomic silver and an interstitial silver ion $Ag_{(i)}^+$ (eq. 4.4). Thereby, the recombination of the Ag_2^+ with a photo-electron is suppressed so that the quantum efficiency of the photographic process does not decrease. The atomic silver is unstable to decompose into an interstitial silver ion and an electron (eq. 4.5). The electron according to eq. 4.5 is sometimes referred to as "Lowe's electron" [16,17]. Thus, the total process of eqs. 4.3 to 4.5 produces two interstitial silver ions and one Lowe's electron [18,19].

4.1.4 Formation of Photo-Electrons

The next problem is where a photo-electron is produced in a silver halide crystal during photo-irradiation (eq. 4.1). Note that the Gurney-Mott theory and its refined proposals have not directly aimed at the specification of a photolytic site [1]. Spectral sensitization with a sensitizing dye gave an indirect clue for specifying such a photolytic site as the site at which the sensitizing dye was adsorbed [20].

It is impartial to refer to the photo-aggregation theory proposed by Mitchell [21], where $Ag_2 ! \$Ag_2O$, Ag_2S, and (Ag, Au)S donor centers are actively involved in photochemical reactions. In other words, such donor centers are presumed to cause the photolysis, producing photo-electrons in place of the route of eq. 4.1. Mitchell's reason is that crystals of AgBr of the highest achievable purity exhibit negligible photo-conductivity and absence of photo-sensitivity. However, the photo-conductivity to be discussed is not for such AgBr crystals of high purity, but for sensitized AgBr crystals in which the photo-conductivity may be enhanced by the presence of $(Ag_2S)_2$ as an acceptor center.

4.1.5 Overall Chemical Sensitizations

Before we start our discussions on organic chemistry of photography in the next chapters, it is worthy to put the three types of chemical sensitizations (sulfur, reduction, and gold sensitizations) together into an integrated scheme. Thus the overall mechanism of latent image formation is summarized in Fig. 4.6.

1. An unexposed silver halide grain is illustrated in Fig. 4.6a. A sensitization center (S center) consists of $(Ag_2S)_2$ for sulfur sensitization or of Ag_3AuS_2 for gold sensitization. A reduction sensitization center (R center) consists of a dimeric silver species Ag_2. The species Ag^+ represents an interstitial silver ion. The thick line is a dislocation at which sensitization dyes are adsorbed.

2. Electronic process and photo-hole process (Fig. 4.6b): Exposure of light ($h\nu$) produces a photo-electron 'e' and a positive hole 'Br(p)' (bromine hole), as expressed by eq. 4.1. If a sensitization dye (Dye) is used as depicted in Fig. 4.6a, a positive hole 'Dye(p)' (dye hole) is produced in place of 'Br(p)'. The photo-electron drifts in the conduction band (Figs. 4.2 and 4.4) to arrive at the S center, where it gives a trapped electron (electronic process expressed by eq. 4.2). On the other hand, the positive hole (h) that is generated by the process, $\mathrm{Br(p)} \rightarrow \mathrm{Br^-} + \mathrm{h}$ or $\mathrm{Dye(p)} \rightarrow \mathrm{Dye} + \mathrm{h}$, reaches the R center and gives Ag_2^+ (photo-hole process expressed by eq. 4.3).

3. Ionic process (Fig. 4.6c): The interstitial silver ion moves and combines with the trapped electron to produce a latent preimage Ag at the S center. Although the timing of eq. 4.4 has not been clarified, the dimeric silver ion Ag_2^+ at the R center is decomposed to produce an unstable Ag and an interstitial Ag^+.

4. Lowe's electron-releasing process (Fig. 4.6d): The unstable Ag at the R center releases Ag^+ and a Lowe's electron.

5. Latent subimage formation (Fig. 4.6e): The processes akin to the electronic and ionic processes described above occur to form a latent subimage Ag_2 at the S center.[4]

6. Clustering process (Figs. 4.6f and 4.6g): The intermediate crystal (Fig. 4.6f) behaves in a similar way as the original crystal (Fig. 4.6a). Such repeated processes reinforce the latent subimage (Ag_2) to give a latent image (Ag_4).

The overall process requires two photons ($2h\nu$) to produce a latent image composed of Ag_4. The resulting latent image Ag_4 works as a catalyst for developing the whole part of the relevant silver halide crystal to give Ag_n (n = ca. 2×10^{10}).

---

[4]The latent subimage Ag_2 can be exchanged by Au_2 at a gold sensitization center. The resulting Au_2 can act as a latent image, not as a latent subimage. This means that the subsequent clustering process is unnecessary and that an ultimate quantum yield can be reached.

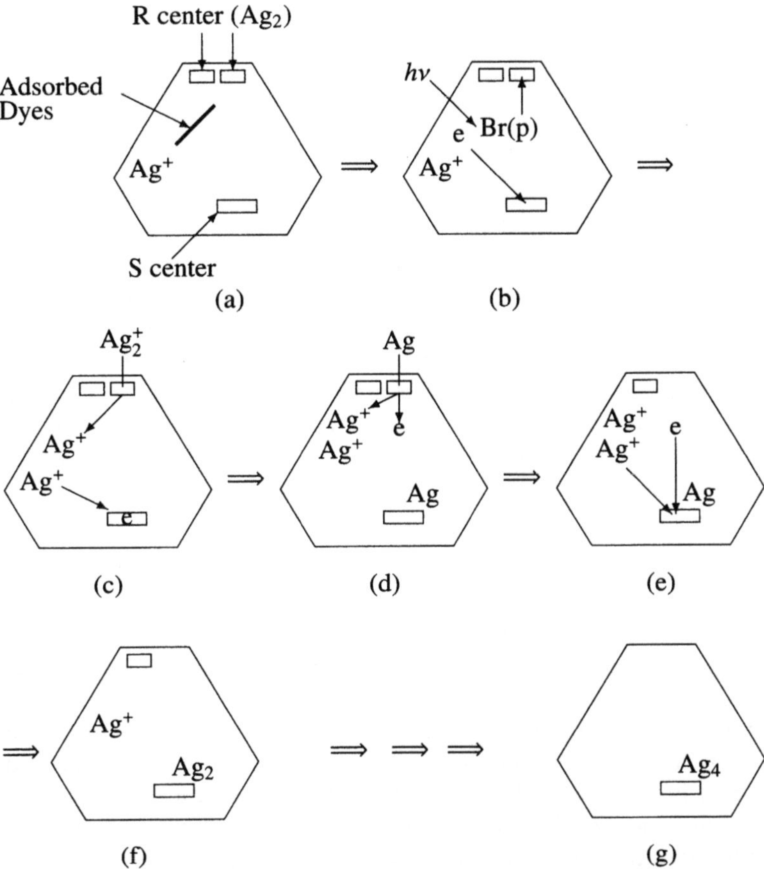

Figure 4.6. Overall mechanism of latent image formation. (a) Unexposed silver halide grain: The large frame represents a sensitization center (S center) that consists of $(Ag_2S)_2$ (sulfur sensitization center) or Ag_3AuS_2 (gold sensitization center). Each of the small frames stands for a reduction sensitization center (R center) that consists of Ag_2. The species Ag^+ represents an interstitial silver ion. The thick line is a dislocation at which sensitizing dyes are adsorbed. (b) Electronic process and photo-hole process: Exposure of light ($h\nu$) gives a photo-electron 'e' and a positive hole 'Br(p)' or 'Dye$^+$(p)'. The migration of the former produces a trapped electron at the S center and that of the latter generates Ag_2^+ at the R center. (c) Ionic process: Migration of an interstitial silver ion and combination with the trapped electron produce Ag at the S center. The dimeric silver ion Ag_2^+ produces Ag and Ag^+ at the R center. (d) Lowe's electron release: Ag produced at the R center releases Ag^+ and an electron. (e) and (f) Latent subimage formation: Ag_2 is formed at the S center. (g) Clustering process: An Ag cluster (latent image) is produced by repeated processes.

This magnification (one photon to ca. 10^{10} silver atoms) is why photographic films based on silver halide have high photo-sensitivity.

4.2 Tabular Grains

4.2.1 Tabular Grains for Color Negative Films

High photo-sensitivity is one of the most important items to be considered especially for color negative films. As found easily, tabular grains of silver halide should exhibit higher photo-sensitivity than cubic grains etc., since the former have higher covering power per photon than the latter. However, it was not until 1983 that such tabular grains were practically used in a color negative film (Kodacolor VR-1000). Although it had been well-known that tabular grains were formed during Ostwald's ripening, technical difficulties to assure stable production prevented the practical use of tabular grains. New technologies for yielding narrow distribution of grain sizes etc. made the practical usage of tabular grains (T grains) possible [22]. According to Kofron and Booms [23], T grains have the following advantages:

1. Improved minus-blue to blue sensitivity ratio: Silver halide grains have the intrinsic sensitivity at the blue region of light, which is unnecessary for emulsions in red-sensitive and green-sensitive layers. To reduce color degradation due to the intrinsic sensitivity, a blue-absorbing filter layer is usually inserted.
   Such blue sensitivity (speed) depends on crystal volume while minus-blue sensitivity depends on crystal surface. The crystal surface area of a tabular grain is larger than that of a spherical grain (e.g., a cubic grain) if their volumes are equal. Thereby, tabular grains have improved minus-blue to blue sensitivity ratio in comparison with spherical grains. It follows that the blue-absorbing filter layer can be removed so that the film structure is simplified and the coating thickness is reduced.

2. Improved sensitivity/granularity ratio: For the sake of simplicity, we select a cubic grain with 1 μm edge length (the surface area: 6 μm^2 and the volume: 1 μm^3) as a hypothetical reference (Fig. 4.7a). Let N be the number of such grains in a unit area of a coated film. Then, the granularity of the film is expressed by $\sigma = aD/\sqrt{N}$, where the root mean square (RMS) deviation for a measure of uniformity (or reversely granulality) is simplified by placing D for the term for silver coverage (density) and a for a constant. The speed/granularity ratio is expressed by S/σ, where S stands for the sensitivity (speed).
   To double the sensitivity, the surface area should be doubled to be equal to 12 μm^2, which indicates a cubic grain with 1.41 μm edge length and with 2.83 μm^3 volume (Fig. 4.7b). The number of such grains coated in a unit area is calculated to be $N/2.83$ in order to give the same silver coverage (D)

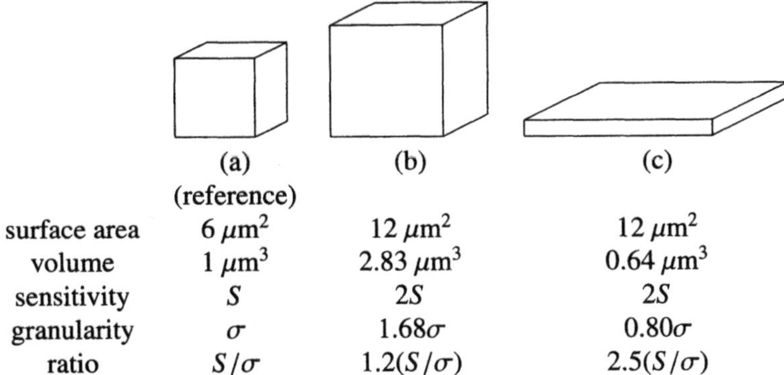

	(a) (reference)	(b)	(c)
surface area	6 μm^2	12 μm^2	12 μm^2
volume	1 μm^3	2.83 μm^3	0.64 μm^3
sensitivity	S	$2S$	$2S$
granularity	σ	1.68σ	0.80σ
ratio	S/σ	$1.2(S/\sigma)$	$2.5(S/\sigma)$

Figure 4.7. Hypothetical sensitivity/granularity characteristics: (a) for a 1 μm cubic emulsion of a 6 μm^2 surface area (reference), (b) for a 1.41 μm cubic emulsion of a 12 μm^2 surface area, and (c) for a 20:20:1 edge ratio T-Grain emulsion of a 12 μm^2 surface area (2.34 μm×2.34 μm×0.117 μm). By putting $\pi r^2 = 2.34^2$, the reduced diameter is calculated to be $2r = 2.64$ μm. Hence, the corresponding aspect ratio is calculated to be $2.64 : 0.117 = 22.6 : 1$.

after development. Thereby, the granularity of the film (σ_C) is expressed by $\sigma_C = aD/\sqrt{N/2.83} = \sqrt{2.83}\sigma = 1.68\sigma$, which indicates the worsened granularity. The hypothetical speed/granularity ratio for the cubic grain is expressed by $S_C/\sigma_C = 2S/1.68\sigma = 1.2(S/\sigma)$.

Let us now consider a square T grain with an edge ratio $20 : 20 : 1$, which corresponds to an aspect ratio $22.6 : 1$.[5] The grain has a surface area of 12 μm^2 so as to double the sensitivity (Fig. 4.7c). This T grain has the dimensions of 2.34 μm × 2.34 μm × 0.117 μm and the volume of 0.64 μm^3.[6] The number of such T grains coated in a unit area is calculated to be $N/0.64$, which gives the granularity of the film (σ_T), $\sigma_T = aD/\sqrt{N/0.64} = \sqrt{0.64}\sigma = 0.80\sigma$. The hypothetical sensitivity/granularity ratio for the T grain is expressed by $S_T/\sigma_T = 2S/0.80\sigma = 2.5(S/\sigma)$. When the sensitivity is enhanced, the granularity of T grains is not worsened so that the sensitivity/granularity ratio is improved.

[5] A tabular grain is usually characterized by an aspect ratio, which refers to the ratio of the diameter of the grain to its thickness [22]. The "diameter" of the grain is in turn defined as the diameter of a circle having an area equal to the projected area of the grain as viewed in a photo-micrograph or an electron micrograph of an emulsion sample. For the sake of simplicity, the "diameter" is here called a reduced diameter. Let $2r$ be the reduced diameter of the present grain. Then we have $\pi r^2 = 20^2$, which give $2r = 22.6$. Hence, the corresponding aspect ratio is equal to 22.6 : 1.

[6] Note that $2.34/0.117 = 20/1$. Let $1/\alpha$ be an edge ratio. Then, we can obtain x and αx as the dimensions of the edges. For the surface area, we have the expression, $2x^2 + 4\alpha x^2 = 12$. It follows that $x = \sqrt{6/(1+2\alpha)}$. When we place an edge ratio $20 : 20 : 1$ ($\alpha = 0.05$), we arrive at $x = \sqrt{6/(1+2\times 0.05)} = \sqrt{6/1.1} = 2.34$.

3. Decreased sensitivity to high-energy radiation: High-energy radiation (such as X-rays and cosmic rays) is involved in natural background radiation that has cosmic and terrestrial origin. Such high-energy radiation is one of the factors that affect change in color films after manufacture. The sensitivity for such high-energy radiation is affected by the volume of a grain, while that for visible light is related to the surface area. In parallel to the discussion described for sensitivity/granularity ratio, a T grain has a lower sensitivity to the former while a higher sensitivity to the latter because of the shape of a T grain.

4. Enhanced sharpness: Coated T grains lie with their long axis parallel to the support (polymer base) because the hydrostatic forces are involved in coating and drying processes. This arrangement decreases light scattering so that a loss in sharpness can be prevented.

5. Enhanced color sensitivity: Large surface areas are capable of adsorbing a sensitizing dye in large quantities. Thereby, higher color sensitivity can be obtained.

If the distribution of grain sizes is broad, the following disadvantages would be considered:

1. The contrast of the characteristic curve cannot be increased.

2. Since the effect of chemical sensitization depends upon grain sizes, it is difficult to realize the optimum chemical sensitization independent to such grain sizes.

Highly uniform silver bromoiodide tabular grain emulsions having an aspect ratio of 58 have been disclosed, where they have been prepared by a procedure composed of a nucleation step, a twinning step, a transition step, and a growth step [24]. It has been pointed out that the absence of the twinning step has drastically reduced the population of the tabular grains.

A further technique for yielding narrow distribution of grain sizes has been disclosed [25], where a block copolymer of ethylene oxide and propylene oxide expressed by the formula,

$$HO\text{–}(\underset{\substack{|\\CH_3}}{CH}CH_2O)_x\text{–}(CH_2CH_2O)_y\text{–}(CH_2\underset{\substack{|\\CH_3}}{CH}O)_{x'}\text{–}H$$

in which $x = 25$, $x' = 25$, and $y = 5$, has been used for preparing a monodisperse emulsion, which has been characterized by mono-shaped grains (reduced diameter: 2.20 μm; thickness: 0.113 μm; aspect ratio: 19.5; and coefficient of variation of total grains: 4.7%).

4.2.2 Σ-Grains

Silver halide grains having a core-shell structure (Σ-Grains [26]) have been used in the HR series (1983) of Fuji Photo Film Co. Ltd., where the iodine content of the silver iodobromide core is higher than that of the shell. The higher iodine content of the core has resulted in high sensitivity because of an increased amount of light absorption as well as because of high quantum sensitivity, where the recombination of a photo-electron and a positive hole is prevented [17,27]. On the other hand, the lower iodine content of the shell has assured a high efficiency of development and has brought about smaller sizes of developed silver images, where the latter effect stems from the inhibiting effect of the high iodine content of the core [27].[7]

Saito *et al.* [28] have disclosed a technique for producing mono-shaped (hexagonal shape and uniform thickness) and mono-sized (mono-dispersed) tabular grains. In the first addition step of manufacturing a silver halide emulsion, generally, there emerge nuclei having two or more parallel twin planes and other undesired nuclei of various types. The latter nuclei are dissolved in the successive step and the remaining former nuclei are grown by physical ripening to provide an emulsion of tabular grains (proportion of hexagonal tabular grains: 87%; reduced diameter: 1.4 μm; and thickness: 0.14 μm).[8] In another embodiment, such a resulting emulsion of tabular grains has been further used as a core emulsion to produce a core-shell emulsion, which has been applied to the manufacture of films for instant color photography (dye diffusion process).

Urabe [29,30] has reported transmission electromicroscopic (TEM) examination on tabular grains of core-shell AgBr-AgBrI type, which have been prepared by a double-jet method to form a silver iodobromide shell (silver iodide 10 mol %) around a tabular silver bromide core. The TEM photographs have shown that the tabular grains have striation-like imperfections parallel to the edge of the tabular grain as well as straight line imperfections perpendicular to the edge of the tabular grain (Fig. 4.8). The latter imperfections stem from edge dislocations [31].[9]

The patent itself [29] has claimed that core-shell grains without such striation-like imperfections, which are selectively prepared as mono-shaped grains (proportion of hexagonal tabular grains: 82%; reduced diameter: 2.1 μm; and thickness: 0.26 μm.), have exhibited higher sensitivity than the above-mentioned core-shell grains with striation-like imperfections. This fact shows that the striation-like imperfections are not so important and implies that imperfections due to edge dislocations may exhibit a dominant effect on high sensitivity, if they remain as fringe dislocations [32,33].

---

[7]For the techniques adopted by Konica and Agfa, see Ref. [27]. This review contains a concise overview of techniques adopted (up to 1991) by Eastman Kodak, Fuji, Konica, and Agfa for photographic negative films of high sensitivity.

[8]For a reduced diameter, see Fig. 4.7.

[9]This review [31] contains an overview of studies (up to 1996) on the growth of silver halide crystals.

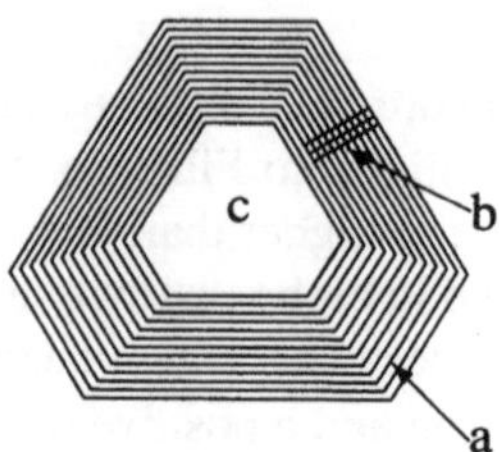

Figure 4.8. Schematic diagram of a tabular grain of core-shell AgBr-AgBrI type: (a) striation-like imperfections, (b) straight line imperfections in the shell, and (c) the uniform core. This figure has been drawn according to the TEM photograph shown in FIG. 3 of Ref. [29].

4.2.3 Example for Preparing Tabular Grains

A procedure for "extra thin" tabular silver iodobromide grain emulsion has been disclosed as an example in a patent of new sensitizing dyes [34]. The procedure consists of a nucleation step, a ripening step, and a growth step, where the specification of each step has been disclosed in detail (EXAMPLE 1 of [34]).

1. **Nucleation**: To the reaction vessel (illustrated in the original patent), water (1.0 L) and ossein gelatin of low molecular weight (2 g, average molecular weight: 10,000) were added, and the resulting solution was kept at 35℃. To the mixing vessel (having an internal volume of 2 L, as illustrated in the original patent), a 0.6 M aq. solution (50 mL) of silver nitrate and a 0.16 M aq. solution (200 mL) of potassium bromide containing 0.8% (by weight) gelatin of low molecular weight were added for 2 min, and the resulting emulsion was continuously added to the reaction vessel for 2 min. At that time, the number of stirring revolutions of the mixing vessel was 2000 rpm.

2. **Ripening**: Then, a 10% solution (300 mL) of ossein gelatin subjected to oxidation treatment (methionine content: 5 μmol/g) and KBr were added to adjust the pBr of the emulsion in the reaction vessel to 2.1, followed by elevation of the temperature to 85°C.

3. **Growth**: Thereafter, a 1.0 M aq. solution (600 mL) of silver nitrate, a 0.98 M solution (600 mL) of KBr containing 3 mol% of KI and a 5% aq. solution (800 ml) of gelatin of low molecular weight were added again to the mixing vessel at an accelerated flow rate (the flow rate at the time when addition was completed was 4 times the initial flow rate). Fine grains produced in the mixing vessel were continuously added to the reaction vessel. At that time, the number of stirring revolutions of the mixing vessel was 2000 rpm.
   During the growth of grains, at the time when 70% of silver nitrate was added, $IrCl_6$ was added in an amount of 8×10^{-8} mol/mol-Ag to dope the grains therewith. Further, before the growth of grains was completed, a solution of potassium hexacyanoferrate(II) was added to the mixing vessel.

Three percent (in terms of the amount of silver added) of shell portions of the grains were doped with potassium hexacyanoferrate(II) so as to give a local concentration of 3×10^{-4} mol/mol-Ag.

4. **Washing**: After the addition was terminated, the emulsion was cooled to 35°C, and washed with water by ordinary flocculation. Then, lime-treated ossein gelatin (70 g) was added and dissolved to adjust the pAg to 8.7 and the pH to 6.5, followed by storage thereof in a cool and dark location.
   The resulting tabular grains were extra-thin mono-dispersed tabular grains having a mean reduced diameter of 2.3 μm, a mean thickness of 0.045 μm, a mean aspect ratio of 51, and a variation coefficient in an equivalent-circle diameter of 16%.

5. **Spectral and Chemical Sensitization**: A sensitizing dye (9×10^{-4} mol/-mol-Ag) was added to the emulsion, followed by stirring at 40°C for 10 min. Then, the temperature was elevated to 60°C, and sodium thiosulfate, potassium chloroaurate, and potassium thiocyanate were added to conduct optimum chemical sensitization.

4.2.4 Crystal Habit Controlling Agents

High chloride emulsions have recently been employed because of more rapid developability and ecological advantages. In order to obtain tabular grain emulsions of high chloride contents, polymeric peptizers have been used in combination with grain growth modifiers, which are frequently called "crystal habit controlling agents".

$$-(CH_2-CH)_1-(CH_2-CH)_6- \quad \text{with side chains } COOCH_2CH_2SCH_2CH_3 \text{ and } COOCH_2CH_2CH_2SO_3Na$$

1

2, 3, 4, 5

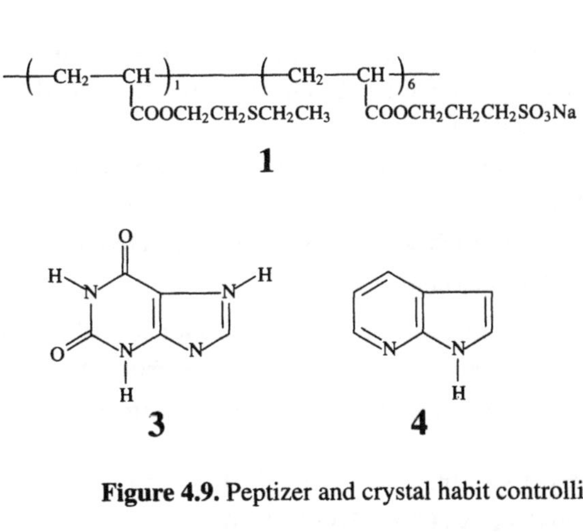

Figure 4.9. Peptizer and crystal habit controlling agents.

Thus, a copolymer having a thioether linkage (**1**) is used as a synthetic peptizer in combination with adenine (**2**) as a crystal habit controlling agent [35]. In addition, the presence of methionine generated by a partial oxidation of gelatin with hydrogen peroxide has been found to be useful to obtain a wide range of chloride ion concentration in tabular grains [36]. Xanthine (**3**) has been disclosed as another crystal habit controlling agent [37]. The combination of 7-azaindole (**4**) with 4,5,6-triaminopyrimidine (**5**) has also been proposed as a useful crystal habit controlling agent [38,39].

Specific bispyridinium salts have been disclosed as crystal habit controlling agents of another type (e.g., **6**) [40]. Monopyridinium salts (e.g., **7**) have been used as crystal habit controlling agents to produce tabular grains of a high silver chloride content for heat developable photographic materials [41,42].

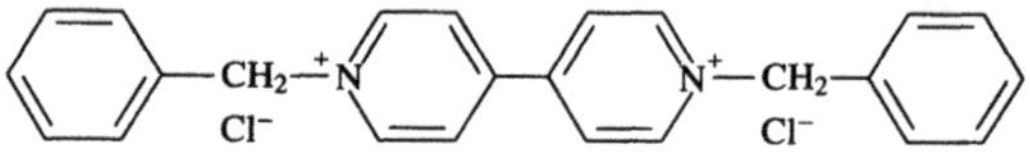

6

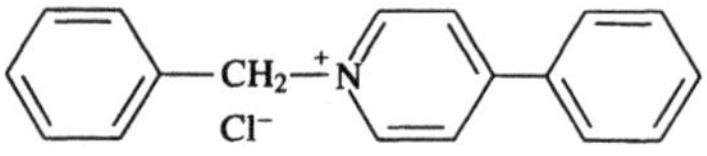

7

Figure 4.10. Another type of crystal habit controlling agents.

References

[1] Hamilton JF (1977) The Mechanism of Formation of the Latent Image. In: James TJ (ed) The Theory of the Photographic Process, 4th edn. Macmillan, New York London, Chapter 4

[2] Tani T (1995) Photographic Sensitivity: Theory and Mechanisms, Oxford Univ Press, New York

[3] Gurney RW, Mott NF (1938) Proc Roy Soc (London). A164:151

[4] Mitchell JW (1948) Sci Ind Photogr. 19:361

[5] Hamilton JF (1984) In: Balderischi A, Czaja W, Tosatti T, Tosi M (eds) The Physics of Latent Image Formation in Silver Halides, Wolrd Scientific

[6] Hamilton JF, Harbison JM, Jeanmaire DL (1988), J Imaging Sci Tech. 32:17

[7] (a) Kanzaki H (1990) J Soc Photogr Sci Technol Jpn. 53:529 (b) Kanzaki H, Tadakuma Y (1994) J Phys Chem Solids. 55:631

[8] Keevert JE, Gokhale VV (1987) J Imaging Sci Technol. 31:243

[9] Tani T (1995) J Imaging Sci Tech. 39:386

[10] Soc Photogr Sci Technol Jpn (ed) (1998) Principles of Photographic Science and Engineering. Corona Publishing, Tokyo, Chapter 2, Section 2.2

[11] Mizuno M, Mifune H, Toyama Y, Shiozawa T, Okuda J (2002) In: International Congress of Imaging Science 2002, Tokyo, pp. 74–75

[12] (a) Tani T (1990) J Soc Photogr Sci Technol Jpn. 53:501 (b) Tani T (1994) Fuji Film Res Dev 39:20 (c) Tani T (1995) J Imaging Sci Tech. 39:31

[13] Kawasaki M, Hada H (1990) J Soc Photogr Sci Technol Jpn. 53:511

[14] Spencer HE, Brady LE, Hamilton JF (1967) J Opt Soc Am. 57:1020

[15] Tani T (1997) J Imaging Sci Tech. 41:577

[16] Harbison JM, Spencer HE (1977) Chemical Sensitization and Environmental Effects. I. Chemical Sensitization. In: James TJ (ed) The Theory of the Photographic Process, 4th edn. Macmillan, New York London, Chapter 5

[17] Takada S (1998) J Soc Photogr Sci Technol Jpn. 61:3

[18] Tani T (1986) J Imaging Sci Tech. 30:41

[19] Hailstone RK, Liebert NB, Levy M, McCleary RT, Girolmo SR, Jeanmaire DL, Boda CR (1988) J Imaging Sci Tech. 32:113

[20] Tani T (1990) J Imaging Sci Tech. 34:143

[21] Mitchell JW (1995) J Imaging Sci Tech. 39:193

[22] Kofron JT, Booms RE, Jones CG, Haefner JA, Wilgus HS, Evans FJ (1984) US Patent 4 439 520

[23] Kofron JT, Booms RE (1986) J Soc Photogr Sci Technol Jpn. 49:499

[24] Antoniades MG, Daubendiek RL, Fenton DE, Hall JL, Jagannathan R (1993) US Patent 5 250 403

[25] Tsaur AK, Kam-Ng M (1993) US Patent 5 210 013

[26] Takada S, Ohshima N, Ishimaru S (1987) US Patent 4 668 614

[27] Takada S (1991) J Soc Photogr Sci Technol Jpn. 54:156

[28] Saitou S, Urabe S, Ozeki K (1989) US Patent 4 797 354

[29] Urabe S (1989) US Patent 4 879 202

[30] Urabe S (2002) In: International Congress of Imaging Science 2002, Tokyo, pp. 5–6

[31] Urabe S (1996) J Soc Photogr Sci Technol Jpn. 59:276

[32] Takada S (2001) J Soc Photogr Sci Technol Jpn. 64:9

[33] Yokota K, Tamaoki H, Suga Y, Nozawa Y, Kume Y (2001) J Soc Photogr Sci Technol Jpn. 64:298

[34] Hioki T, Morimoto K (2002) US Patent 6 365 335 B1

[35] Maskasky JE (1983) US Patent 4 400 463

[36] Maskasky JE (1983) US Patent 4 713 323

[37] Maskasky JE (1993) US Patent 5 178 998

[38] Maskasky JE (1993) US Patent 5 178 997

[39] Maskasky JE (1993) US Patent 5 185 239

[40] Ishiguro S, Morimoto K (1991) US Patent 4 983 508

[41] Ohzeki K, Asami M (2001) US Patent 6 228 565 B1

[42] Ohzeki K, Asami M, Yokokawa T, Naruse H (2001) US Patent 6 232 055 B1

Chapter 5

Photographic Development and Developers

5.1 Photographic Development

As described in Chapter 1, exposed silver halide grains in black-and-white (B&W) photography have latent images. They are reduced into silver images by means of a catalytic action of the latent images in the process of photographic development.[1]

A photographic developer is a reducing agent that reacts with silver halide:

$$DH_m + nAg^+ \rightleftharpoons nAg + D^{ox} + mH^+ \tag{5.1}$$

where the symbols DH_m and D^{ox} represent a photographic developer and its oxidized species, and where m and n are positive integers to adjust the stoichiometry of eq. 5.1. Most photographic developers take $m = n = 2$, as described in the next section.

The silver images (black) produced by eq. 5.1 appear in an exposed area, which should be white for natural image reproduction. Thus they are negative images, in which black and white are reversed. If positive images are desired, another treatment of exposure and development is necessary. In other words, "negative images of negative images" mean positive images.

[1] It should be noted that a silver halide grain with latent images is reduced as a whole under most conditions. As a result, the light-and-dark gradation is determined according to the number of developed silver grains.

5.2 Photographic Developers

5.2.1 Hydroquinones and Related Compounds

Silver halides having latent images are reduced into silver metal images by using reducing agents called *photographic developers* (simply, developers) or *developing agents*. Typical developers have structures represented by HO–(C=C)$_n$–OH, where each carbon atom is substituted by hydrogen, an alkyl group, etc. For example, hydroquinone (**1**, benzene-1,4-diol), catechol (**2**, benzene-1,2-diol), and pyrogallol (**3**, benzene-1,2,3-triol) are well known as such developers.

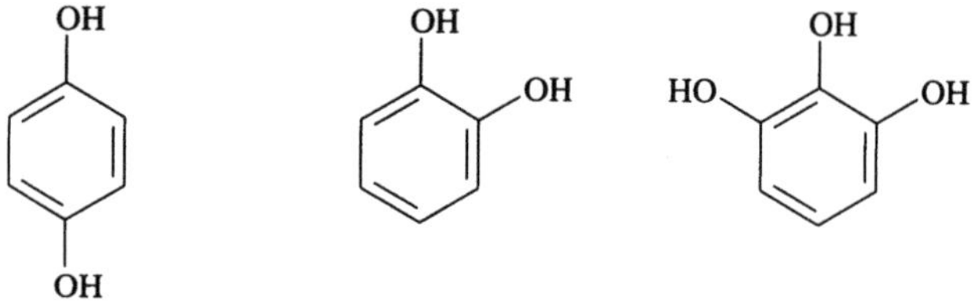

Hydroquinone (**1**) Catechol (**2**) Pyrogallol (**3**)

The reduction is represented in general as follows:

$$\begin{aligned} &\mathrm{HO{-}(C{=}C)}_n\mathrm{{-}OH} + 2\mathrm{Ag}^+ \\ &\quad \rightarrow \mathrm{O{=}C{-}(C{=}C)}_{n-1}\mathrm{{-}C{=}O} + 2\mathrm{Ag} + 2\mathrm{H}^+ \end{aligned} \qquad (5.2)$$

where 2 moles of silver ions are reduced with 1 mole of the developer. For example, hydroquinone (**1**) reacts with silver halide in agreement with the general equation shown above so as to give 1,4-benzoquinone (**4**).

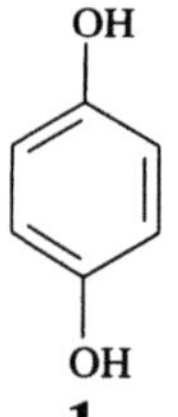

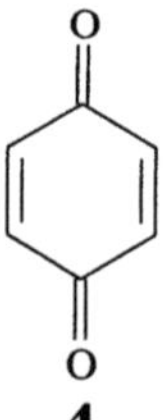

$$\mathbf{1} + 2\mathrm{Ag}^+ \rightleftharpoons \mathbf{4} + 2\mathrm{Ag} + 2\mathrm{H}^+ \qquad (5.3)$$

5.2.2 Aminophenols and Related Compounds

Developers belonging to this category have general formulas represented by HO–(C=C)$_n$–NH_2 and H_2N–(C=C)$_n$–NH_2, where each carbon atom is substituted by hydrogen, an alkyl group, etc. For example, 2- or 4-aminophenol (**5**) and 1,2- or 1,4-phenylene diamine (**6**) are representative developers.

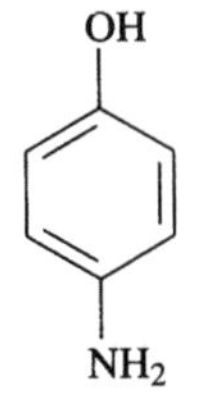

OH / NH$_2$ (para-substituted benzene ring) — 1,4-Aminophenol (**5**)

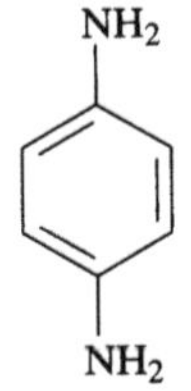

NH$_2$ / NH$_2$ (para-substituted benzene ring) — 1,4-Phenylene diamine (**6**)

5.2.3 Hydrazine Derivatives

Hydrazine is a representative inorganic developer, which reacts with silver ions according to the following equation:

$$NH_2NH_2 + 4Ag^+ \rightarrow N_2 + 4Ag + 4H^+ \quad (5.4)$$

where nitrogen gas is evolved. Mono-aryl derivatives of hydrazine are another type of photographic developers, which also evolve nitrogen gas according to the following equation:

$$Ar\text{–}NHNH_2 + 2Ag^+ \rightarrow ArH + N_2 + 2Ag + 2H^+ \quad (5.5)$$

On the other hand, di-aryl derivatives of hydrazine react with silver ions to produce azobenzene derivatives, where no nitrogen gas is evolved.

$$Ar\text{–}NHNH\text{–}Ar + 2Ag^+ \rightarrow Ar\text{–}N{=}N\text{–}Ar + 2Ag + 2H^+ \quad (5.6)$$

5.2.4 Hydroxylamine Derivatives

Hydroxylamine (NH_2OH) reacts with silver ions to evolve nitrogen gas as follows:

$$2NH_2OH + 2Ag^+ \rightarrow N_2 + Ag + 2H^+ + 2H_2O \quad (5.7)$$

N,*N*-Diethylhydroxylamine (**7**) reacts with silver ions to produce the corresponding nitrone (**8**):

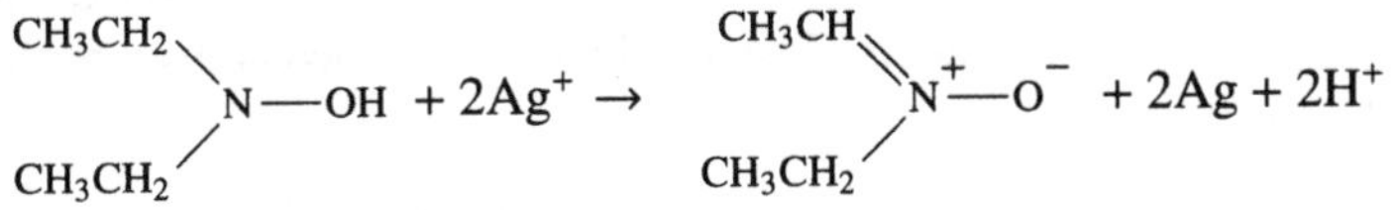

$$(CH_3CH_2)_2N\text{—}OH + 2Ag^+ \rightarrow (CH_3CH{=})(CH_3CH_2)N^+\text{—}O^- + 2Ag + 2H^+ \quad (5.8)$$

7 **8**

5.2.5 Ascorbic Acid and Related Compounds

Ascorbic acid (**9**) and hydroxytetronic acid (**10**) have a HO-C=C-OH moiety, which can be oxidized into O=C–C=O so as to exhibit reducing activity.

Ascorbic acid (**9**) Hydroxytetronic acid **10**

5.2.6 Heterocyclic Developers

1-Phenyl-3-pyrazolidone (phenidone, **11**) and its 4,4-dialkyl derivative (**12**) are another type of developers. The substitution of a hydroxymethyl group at the 4-position enhances solubility in an aqueous solution and stability to hydrolysis.

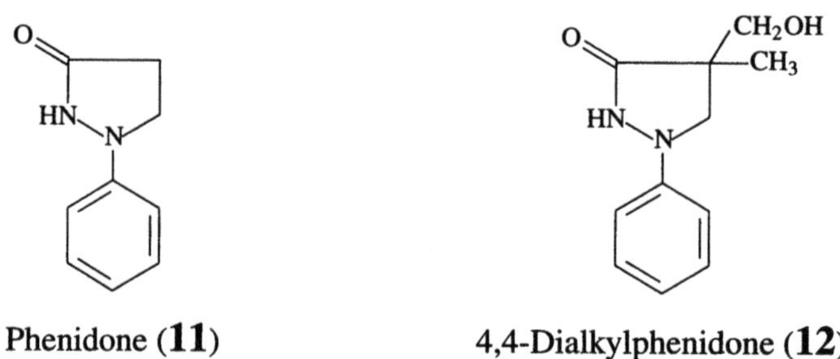

Phenidone (**11**) 4,4-Dialkylphenidone (**12**)

5.2.7 Kendall's Rule

Kendall [1] has pointed out that photographic developers are represented by a general formula:

$$A-(C=C)_n-A'$$

13

where A and A′ are selected from a set of substituents (OH, NH_2, NHR, NR_2) and n represents a non-negative integer. The developers described above are classified in terms of n as follows.

- $n = 0$: hydrogen peroxide, hydroxylamine, and hydrazine
- $n = 1$: catechols, 1,2-phenylene diamines, ascorbic acid, hydroxytetronic acid, α-ketols, α-hydroxylketones
- $n = 2$: hydroquinones, 4-aminophenols, 1,4-phenylene diamines

The heterocyclic developers **11** and **12** do not obey the general formula (**13**), but their tautomeric formulas belong to an extended formula, $A\text{-}(C{=}N)_n\text{-}A'$ [2].

5.3 Electrochemistry of Development

5.3.1 Redox Reactions

The development of silver halides (eq. 5.1) is regarded as a redox reaction in which a photographic developer (DH_m) is used as a reducing agent. It can be treated as an electrochemical process [3].

This redox reaction (eq. 5.1) can be separated into two processes (half-cells), i.e., the reduction of silver halide (eq. 5.9) and the oxidation of a developer (eq. 5.10). Note that eq. 5.10 represents the corresponding inverse reaction, i.e., the reduction of an oxidized developer.

$$E_{\mathrm{Ag}}: \qquad \mathrm{Ag^+} + \mathrm{e} \rightleftharpoons \mathrm{Ag} \tag{5.9}$$

$$E_{\mathrm{D}}: \qquad \mathrm{D^{ox}} + m\mathrm{H^+} + n\mathrm{e} \rightleftharpoons \mathrm{DH}_m \tag{5.10}$$

When the symbols E_{Ag} and E_{D} represent respective potentials, the resulting cell has the following electromotive force:

$$E_{\mathrm{cell}} = E_{\mathrm{Ag}} - E_{\mathrm{D}} \tag{5.11}$$

If $E_{\mathrm{cell}} > 0$, the reaction of eq. 5.1 proceeds rightward to promote the reduction of Ag^+. If $E_{\mathrm{cell}} < 0$, the reaction of eq. 5.1 proceeds leftward (the oxidation of Ag^+). The condition of $E_{\mathrm{cell}} = 0$ represents the equilibrium state.

If the oxidized developer (D^{ox}) is stable, the cell finally reaches an equilibrium state. If D^{ox} is unstable, eq. 5.1 proceeds rightward to promote the reduction of Ag^+.

5.3.2 The Silver Electrode Potential

From eq. 5.9, we can obtain the silver electrode potential:

$$E_{\mathrm{Ag}} = E^0_{\mathrm{Ag}} + \frac{RT}{F} \ln \frac{[\mathrm{Ag^+}]}{[\mathrm{Ag}]} \tag{5.12}$$

where R is the gas constant, T represents temperature, F is the Faraday constant, and E^0_{Ag} is the standard potential. When we place [Ag] = 1 and introduce the solubility product for AgBr, i.e., $K_{sp} = [\mathrm{Ag^+}][\mathrm{Br^-}]$, into eq. 5.12, we obtain:

$$E_{\mathrm{Ag}} = E^0_{\mathrm{Ag}} + \frac{RT}{F} \ln K_{sp} - \frac{RT}{F} \ln[\mathrm{Br^-}] \tag{5.13}$$

Equation 5.13 means that, if $[Br^-]$ is determined, we can obtain E_{Ag}. For example, when we place $E^0_{\mathrm{Ag}} = 0.808\mathrm{V}$! \$$K_{sp} = 6.3 \times 10^{-13}$ (for a bromide anion) ! \$ and $[Br^-] = 0.01$ mol/L, we obtain the following value by means of eq. 5.13:

$$\begin{aligned} E_{\mathrm{Ag}} &= 0.808 - 0.05915 \log(6.3 \times 10^{-13}) - 0.05915 \log(0.01) \\ &= 0.204\ \mathrm{V} \end{aligned} \tag{5.14}$$

5.3.3 Developer Potential

On the other hand, eq. 5.10 gives the corresponding developer potential:

$$E_{\mathrm{D}} = E_{\mathrm{D}}^{0} + \frac{RT}{nF} \ln \frac{[\mathrm{H}^{+}]^{m}[\mathrm{D}^{\mathrm{ox}}]}{[\mathrm{DH}_{m}]} \tag{5.15}$$

Let S_{ox} and S_{red} be the activity of the oxidation state and that of the reduction state. Then, we take account of the case of $m = n = 2$ in eq. 5.10 (cf. eq. 5.3)! To evaluate S_{red}, we can place

$$S_{\mathrm{red}} = [\mathrm{DH}_2] + [\mathrm{DH}^-] + [\mathrm{D}^{2-}] \tag{5.16}$$

where $[\mathrm{DH}^-]$ and $[\mathrm{D}^{2-}]$ are concerned with dissociation constants:

$$K_1 = \frac{[\mathrm{H}^+][\mathrm{DH}^-]}{[\mathrm{DH}_2]} \tag{5.17}$$

$$K_2 = \frac{[\mathrm{H}^+][\mathrm{D}^{2-}]}{[\mathrm{DH}^-]} \tag{5.18}$$

When eqs. 5.17 and 5.18 are applied to eq. 5.16, S_{red} is represented as follows:

$$S_{\mathrm{red}} = \frac{[\mathrm{DH}_2]}{[\mathrm{H}^+]^2} \times \left([\mathrm{H}^+]^2 + K_1[\mathrm{H}^+] + K_1K_2\right) \tag{5.19}$$

On the other hand, the activity of the oxidation state is evaluated to be

$$S_{\mathrm{ox}} = [\mathrm{D}^{\mathrm{ox}}] \tag{5.20}$$

The introduction of eqs. 5.19 and 5.20 into eq. 5.15 gives

$$E_{\mathrm{D}} = E_{\mathrm{D}}^{0} + \frac{RT}{2F} \ln \frac{S_{\mathrm{ox}}}{S_{\mathrm{red}}} + \frac{RT}{2F} \ln\left([\mathrm{H}^+]^2 + K_1[\mathrm{H}^+] + K_1K_2\right) \tag{5.21}$$

Let $E_{1/2}$ be the half-wave potential for polarography. Then this can be considered to be equal to the value obtained by placing $S_{\mathrm{ox}} = S_{\mathrm{red}}$ in eq. 5.21:

$$E_{1/2} = E_{\mathrm{D}}^{0} + \frac{RT}{2F} \ln\left([\mathrm{H}^+]^2 + K_1[\mathrm{H}^+] + K_1K_2\right) \tag{5.22}$$

Hence, the standard potential E_{D}^{0} can be obtained from eq. 5.22 by determining $E_{1/2}$, K_1, and K_2.

These potentials are usually represented as relative values to the standard potential of a reference electrode. A saturated calomel electrode (SCE) is used frequently because of experimental convenience. For comparison, the values due to SCE are usually converted into the values due to a normal hydrogen electrode (NHE), where we can place NHE value = SCE value + 0.246 V. As the potential value is smaller, the reducing agent has more reducing activity. The standard potential E_{Ag}^{0} is determined to be +0.7991 V (NHE), while the standard potential of hydroquinone is determined to be +0.699 V (NHE). Table 5.1 shows the potentials of representative hydroquinone derivatives [3].

Table 5.1. Potentials of Hydroquinone Derivatives

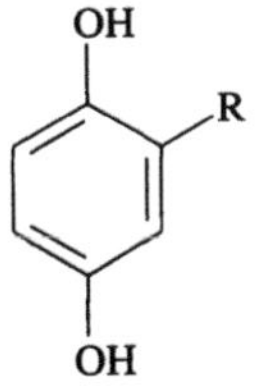

OH

R

OH

R	pK_1	pK_2	E^0_D V (NHE)
none	9.91	11.56	0.699
CH_3	10.05	11.5	0.644
OCH_3	9.91	11.9	0.602
Cl	8.90	11.0	0.712

5.3.4 Effect of Semiquinone Formation

Although eq. 5.3 represents a two-electron reaction, it consists of one-electron reactions, which should be taken into consideration in a detailed discussion [3].

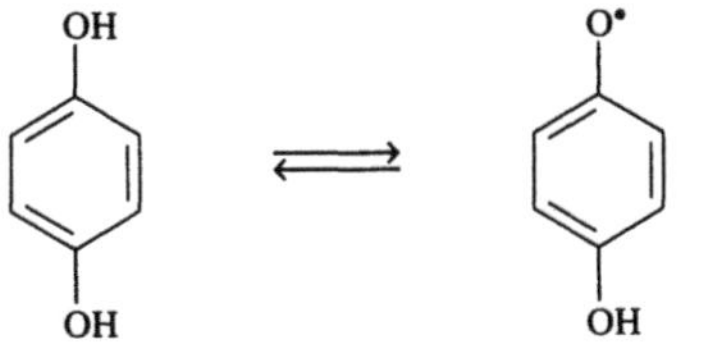

$$DH_2 \rightleftharpoons DH + e + H^+ \qquad E_1 \quad (5.23)$$

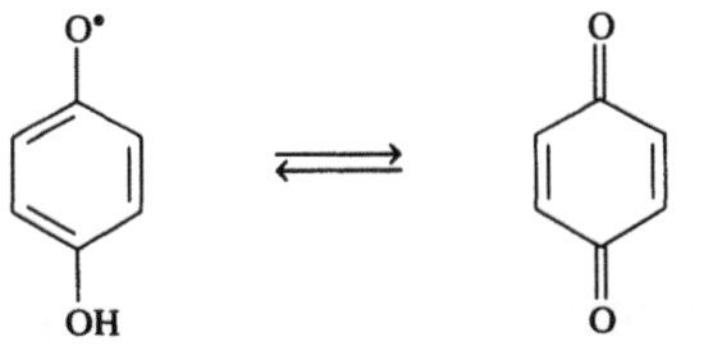

$$DH \rightleftharpoons D^{ox} + e + H^+ \qquad E_2 \quad (5.24)$$

where DH is a species derived by the protonation of a semiquinone (SQ). The intermediacy of the species in eqs. 5.23 and 5.24 results in

$$E_1 = E_1^0 + \frac{RT}{F}\ln\frac{[H^+][DH]}{DH_2} \tag{5.25}$$

$$E_2 = E_2^0 + \frac{RT}{F}\ln\frac{[H^+][D^{ox}]}{DH} \tag{5.26}$$

The equilibrium between SQ and DH is represented by

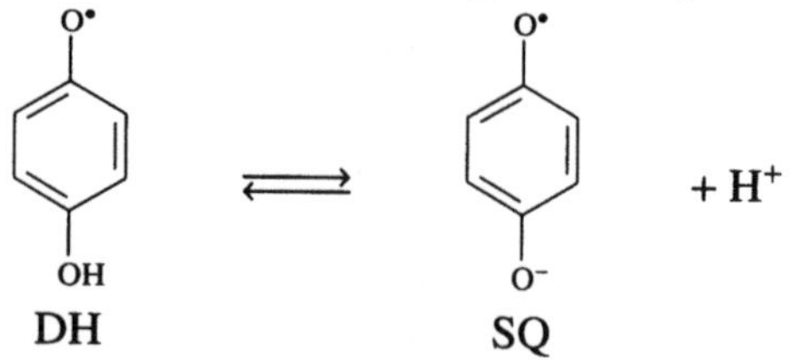

$$\text{DH (HO-C}_6\text{H}_4\text{-O}^\bullet) \rightleftharpoons \text{SQ } (^-\text{O-C}_6\text{H}_4\text{-O}^\bullet) + \text{H}^+ \qquad K_1^{SQ} \tag{5.27}$$

The equilibrium constant of eq. 5.27 is expressed by the following equation:

$$K_1^{SQ} = \frac{[\text{H}^+][\text{SQ}]}{[\text{DH}]} \tag{5.28}$$

Thereby, the activity of the oxidation state (S'_{ox}) is obtained to be

$$S'_{ox} = [\text{DH}] + [\text{SQ}] = [\text{DH}]\left(1 + \frac{K_1^{SQ}}{[\text{H}^+]}\right) \tag{5.29}$$

When eq. 5.29 and eq. 5.19 (S_{red}) are applied to eq. 5.25, we obtain

$$E_1 = E_1^0 + \frac{RT}{F}\ln\frac{S'_{ox}}{S_{red}} + \frac{RT}{F}\ln\left(\frac{[\text{H}^+]^2 + K_1[\text{H}^+] + K_1K_2}{[\text{H}^+] + K_1^{SQ}}\right) \tag{5.30}$$

The introduction of eqs. 5.20 and 5.29 into eq. 5.26 gives

$$E_2 = E_2^0 + \frac{RT}{F}\ln\frac{S_{ox}}{S'_{ox}} + \frac{RT}{F}\ln\left([\text{H}^+] + K_1^{SQ}\right) \tag{5.31}$$

When we presume $S_{red} = S'_{ox} = S_{ox}$ in eqs. 5.30 and 5.31, we obtain

$$E_1 = E_1^0 + \frac{RT}{F}\ln\left(\frac{[\text{H}^+]^2 + K_1[\text{H}^+] + K_1K_2}{[\text{H}^+] + K_1^{SQ}}\right) \tag{5.32}$$

$$E_2 = E_2^0 + \frac{RT}{F}\ln\left([\text{H}^+] + K_1^{SQ}\right) \tag{5.33}$$

By combining eq. 5.32 with eq. 5.33, we obtain

$$\frac{E_1 + E_2}{2} = \frac{E_1^0 + E_2^0}{2} + \frac{RT}{2F}\ln\left([\text{H}^+]^2 + K_1[\text{H}^+] + K_1K_2\right) \tag{5.34}$$

Compare eq. 5.34 with eq. 5.22. In the case of hydroquinone, we have E_1^0 = 1.084 V and E_2^0 = 0.324 V. Hence, we obtain $(E_1^0 + E_2^0)/2$ = 0.702 V, which is comparable with the E_D^0 value shown in Table 5.1.

References

[1] Kendall JD (1936) In: Clere LP (ed) IXe Congres Intern. de Photographie Scientifique et Appliquée, Paris, 1935, Editions Rev. D'Optique, Paris, p. 227

[2] Pelz W (1954) Angew Chem. 66:231

[3] Lee WE, Brown ER (1977) The Developing Agents and Their Reactions. In: James TJ (ed) The Theory of the Photographic Process, 4th edn. Macmillan, New York London. Chapter 11

Part II

Principles of Color Photography

Chapter 6

Color Reproduction

6.1 Color and Absorption

The wavelength of visible light ranges between 380 and 780 nm. Newton separated white light with a prism into seven rainbow colors (i.e., red, orange, yellow, green, blue, indigo, and violet),[1] which, in turn, he mixed back into white light.

On the other hand, modern color photography is based on the Helmholz theory of three primary colors, where visible light is separated into blue light (400–500 nm), green light (500–600 nm), and red light (600–700 nm), as shown in Fig. 6.1.

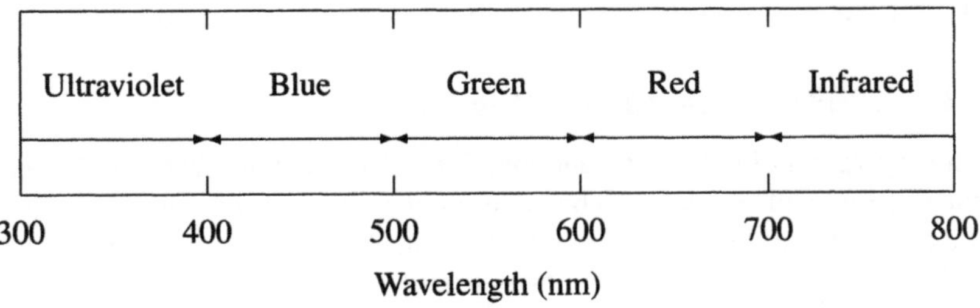

Figure 6.1. Three primary colors of visible light. The visible light is separated into blue light (400–500 nm), green light (500–600 nm), and red light (600–700 nm). This figure also contains invisible lights (ultraviolet and infrared light).

[1] To memorize the prismatic colors, say "Richard Of York Gave Battle In Vain." In Japanese, we say "seki tou ou ryoku sei ran shi."

A dye absorbs visible light of a specific region. Modern color photography uses three dyes of primary colors, i.e., yellow, magenta, and cyan dyes. A yellow (Y) dye absorbs blue light (about 400–500 nm) so that green and red light pass through transparent materials or they are reflected from a white background, as shown in Fig. 6.2a. Since a magenta (M) dye absorbs green light (about 500–600 nm), we are able see blue and red light (Fig. 6.2b). A cyan (C) dye absorbs red light (about 600–700 nm) to pass through or to reflect blue and green light, as shown in Fig. 6.2c.

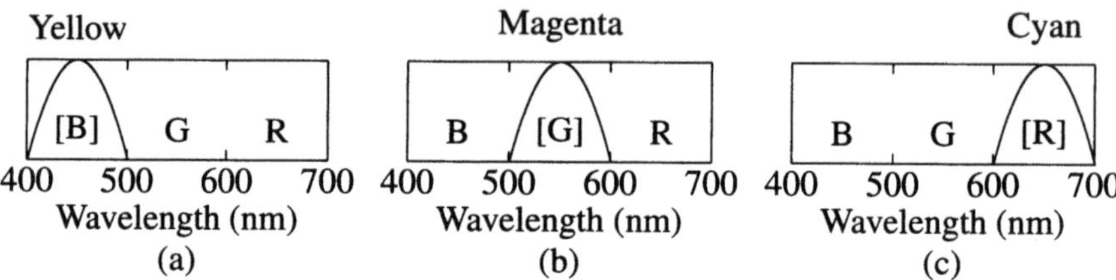

Figure 6.2. Schematic absorptions of three primary colors of dyes. (a) Yellow (Y) dye: Blue light ([B]) is absorbed. (b) Magenta (M) dye: Green light ([G]) is absorbed. (c) Cyan (C) dye: Red light ([R]) is absorbed.

By comparing Fig. 6.1 with Fig. 6.2, we can understand the relationships between lights and dye absorptions. Thus, we can say that blue is complementary to yellow (G + R), green is complementary to magenta (B + R), and red is complementary to cyan (B + G).

6.2 Additive and Subtractive Color Reproduction

6.2.1 Additive Color Reproduction

From a viewpoint of color reproduction, Fig. 6.2 shows the *additive color mixing* of blue, green, and red lights. The addition of green and red lights is equivalent to the absorption of blue light from white light. Since the latter absorption corresponds to yellow (Y), Fig. 6.2a indicates that G + R = Y. Similarly, Figs. 6.2b and 6.2c represent B + R = M and B + G = C, respectively. Note that we have B + G + R = W (white); hence, we can place Y = G + R = W − B, M = B + R = W − G, and C = B + G = W − R. These relationships are more clearly illustrated by a multilayer diagram shown in Fig. 6.3. Thereby, we are able to say yellow light to represent the light derived from G + R and so on. Then, pairs of Y and B; M and G; and C and R are called *complementary colors*, each of which gives white light in cases of the additive color mixing.

The additive color mixing is applied to the color reproduction of CRT (cathode-ray tube) or LC (liquid crystal) displays for television.

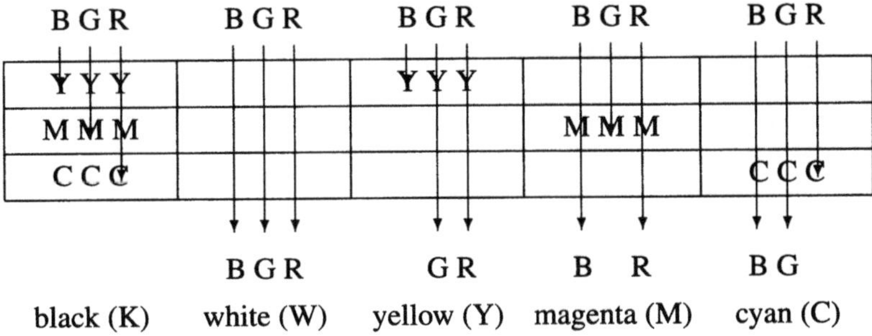

Figure 6.3. Complementary colors produced by a multilayer structure. No visible lights passing through the multilayer represent black color (K), while the arrival of all visible lights correspond to white color, i.e., B + G + R = W (white). We can place Y = G + R = W − B, M = B + R = W − G, and C = B + G = W − R.

6.2.2 Subtractive Color Reproduction

The mixing of dyes is called a *subtractive color mixing*, where yellow, magenta, and cyan dyes are used as primary colors (Fig. 6.4). The mixing of a yellow dye with a magenta dye leaves red light as an unabsorbed light, i.e., Y + M = R, as shown in Fig. 6.4a. Similarly, we have Y + C = G (Fig. 6.4b) and M + C = B (Fig. 6.4c). Note that we have Y + M + C = K (black); hence, we can place R = Y + M = K − C, G = Y + C = K − M, and B = M + C = K − Y. Thereby, we are able to say a red dye in order to represent a dye derived from Y + M and so on. Then, pairs of R and C; G and M; and B and Y are respectively called *complementary colors*, each pair of which gives black light in cases of the subtractive color mixing.

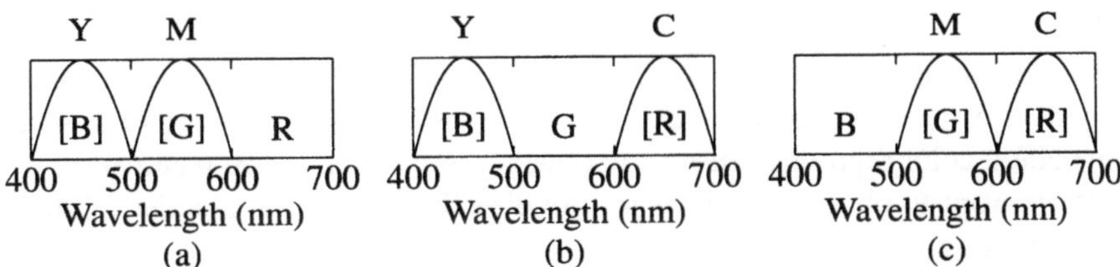

Figure 6.4. Schematic representation of subtractive color mixing. (a) Red light remains after the mixing of yellow (Y) and magenta (M) dyes. (b) Green light remains after the mixing of yellow (Y) and cyan (C) dyes. (c) Blue light remains after the mixing of magenta (M) and cyan (C) dyes. Each light in a pair of brackets is absorbed by a dye.

The subtractive color mixing is widely applied to color photography and color printing, which are the main topics of this book. Photographic films have a multilayer structure in which an appropriate color is produced in a specific photo-sensitive layer, as shown in Fig. 6.5.

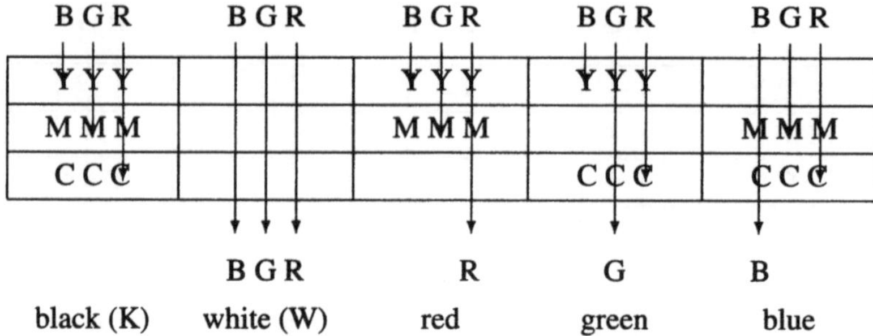

Figure 6.5. Subtractive color mixing based on a multilayer structure. No visible lights passing through the multilayer represent black color, i.e., Y + M + C = K (black). We can place R = Y + M = K − C, G = Y + C = K − M, and B = M + C = K − Y.

6.2.3 Negative and Positive Images

Most systems of color photography adopt the subtractive color mixing. In order that spectral sensitization is applied independently to three regions of visible light (B: 400–500 nm, G: 500–600 nm, R: 600–700 nm), respective sensitizing dyes should have their absorptions at these regions.[2] Hence, a sensitizing dye having its absorption at 400–500 nm is used in the blue-sensitive layer of a photographic film. The green-sensitive layer requires a sensitizing dye having its absorption at 500–600 nm and the red-sensitive layer requires a sensitizing dye having its absorption at 600–700 nm.

A usual color film is based on the following combinations:

blue-sensitive layer (B)	—	yellow dye (Y)
green-sensitive layer (G)	—	magenta dye (M)
red-sensitive layer (M)	—	cyan dye (C)

This means that the resulting picture is composed of the colors complementary to the colors of the original object. Photographically speaking, the picture is *negative*. Another treatment of exposure and development is necessary to obtain the natural color reproduction (i.e., to obtain a so-called *positive* picture).

If we want to obtain a positive picture directly, we should take the following combinations:

blue-sensitive layer (B)	—	blue dye	(or M + C)
green-sensitive layer (G)	—	green dye	(or Y + C)
red-sensitive layer (M)	—	red dye	(or Y + M)

Such blue, green, and red dyes are usually unavailable as a single compound, since the absorptions of most dyes cover about 100 nm only. If magenta and cyan dyes are used in place of a single blue dye, it is not an easy task to balance dual dye

[2]If a sensitizing dye had its absorption at two of the three regions, the resolution of spectral sensitization would be insufficient.

formations. Moreover, two dyes of any pair selected from blue, green, red dyes absorb a common region of light (e.g., red light for the pair of blue and green) so that it is impossible to construct an effective multilayer structure. Hence, several reversal processes are invented, where silver halide emulsions act reversely on the color reproduction.

6.3 Color Image Formation

6.3.1 Principles of Color Image Formation

In color photography, on the same line as black-and-white photography, photographic images are prepared by virtue of the light-sensitivity and the oxidative activity of silver halides. But the resulting final images are composed of organic dyes, but not of inorganic silver. Silver halides work only as mediators between light and dye images,[3] since unexposed silver halides and resulting silver images are all removed by successive processes. It follows that the formation of organic dyes can be discussed from an organic chemical point of view, i.e., apart from the photographic processes.

The appearance of a color image is based on the presence or absence of a dye at a site to be colored. To provide the differentiation of such presence/absence, three methodologies are conceptually possible [1]:

1. **Chromogenic (dye-forming) process:** This process is composed of the imagewise *formation of dyes* from colorless precursors.

2. **Dye-bleaching process:** This process is composed of the imagewise *decomposition of dyes* into colorless compounds.

3. **Dye transfer (diffusion transfer) process**: The color reproduction of this process is based on the imagewise *transfer (diffusion) of dyes*. Such a dye as located at one site (layer) diffuses into another site (layer) so as to give a dye image on a white background. In one embodiment, a diffusible dye is stopped imagewise; in the other embodiment, an immobile dye precursor imagewise releases a diffusible dye.

The three processes have been concisely described and compared with each other in several reviews [1,2]. All of them have been realized as commercial processes. The next subsections deal with their principal reactions of color image formation.

[3]More precisely speaking, silver halide grains mediate between photo-electrons and dye images. Note that a sensitizing dye (as an organic compound) absorbs light and the resulting photo-excited electron is injected into a silver halide grain.

6.3.2 Chromogenic (Dye-Forming) Process

Modern color photography (so-called conventional color photography) has adopted color development [3], the dye-forming reactions of which are shown in Fig. 6.6. Thus, a color developer (**1**) is imagewise oxidized by silver halide into the oxidized color developer ($\mathbf{1}_{ox}$) and reacts with respective couplers (**2**, **5**, and **8**) to form leuco intermediates (**3**, **6**, and **9**). The leuco intermediates are further oxidized into a cyan dye (**4**), a magenta dye (**7**), and a yellow dye (**10**).

Although the couplers are colorless, they are called cyan coupler (**2**), magenta coupler (**5**), and yellow coupler (**8**) in connection with the respective resulting dyes. These couplers are characterized by the presence of a ballast group (B) in their structures. Thereby, they do not migrate from one layer to another in a monopack system.[4]

Part III will be devoted to discussions on the chromogenic process.

6.3.3 Dye-Bleaching Process

Azo dyes (R–N=N–R′) are reduced by silver into colorless amines according to the following equation:

$$2Ag + R\text{–}N{=}N\text{–}R' + 4H^+ \rightarrow 4Ag^+ + RNH_2 + R'NH_2 \qquad (6.1)$$

The reactant silver on the left-hand side of eq. 6.1 is imagewise generated by the preceding development of exposed silver halide. The resulting amines on the right-hand side are removed by acidic bathing and washing with water. This reaction was applied to silver dye bleach photography [4].

Part V will deal with the dye-bleaching process.

6.3.4 Dye-Transfer Process

The first method of the dye transfer process uses dye developers (**11**), which are nondiffusible before development [5]. They are converted into the dianions (**12**), which are diffusible in an alkaline processing liquid (Fig. 6.7). If the anion encounters an exposed silver halide grain, it is imagewise oxidized into the corresponding quinone (**13**), which is changed to be nondiffusible. The unoxidized anion reaches an image-receiving layer, giving a color image. This reaction has been applied to the instant color photography of Polaroid.

The second method of the dye transfer process uses dye releasers such as *o*-sulfonamidophenols [6,7,8] and *p*-sulfonamidonaphthols [9]. Figure 6.8 shows an *o*-sulfonamidophenol dye releaser (**14**), which has been adopted in Fuji's instant color photography. The dye releaser **14** is nondiffusible because of its ballast group (B). When **14** is imagewise oxidized, the resulting compound (**15**) is hydrolyzed by an alkaline medium and releases a diffusible dye (**17**).

---

[4]The word "monopack" is rather historical, since most modern photographic films adopt the monopack system in which all of the photographic devices are coated on a film.

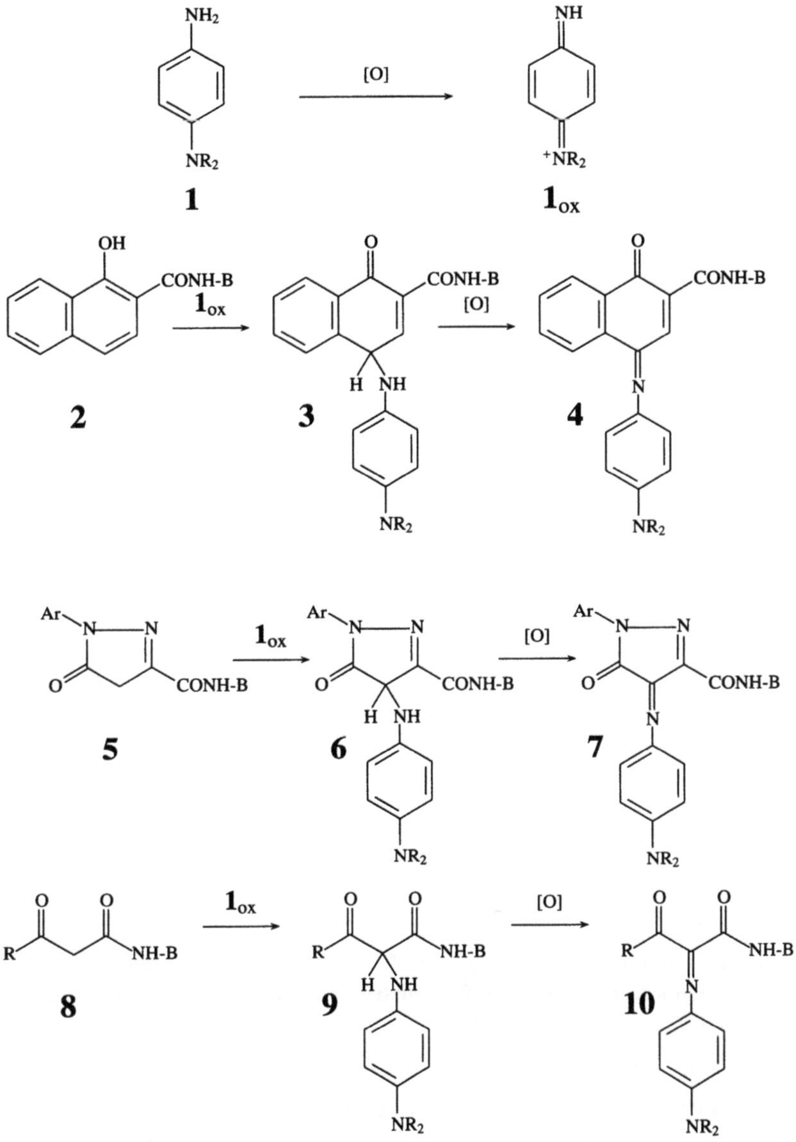

NH2 [O] NH
NR2 1 +NR2 1_{ox}
OH CONH-B 1_{ox} O CONH-B [O] O CONH-B
2 3 H NH NR2 4 N NR2
Ar N N 1_{ox} O CONH-B Ar N N [O] O CONH-B Ar N N O CONH-B
5 6 H NH NR2 7 N NR2
O O R NH-B 8 1_{ox} O O R NH-B H NH 9 NR2 [O] O O R NH-B N 10 NR2

Figure 6.6. Dye-forming processes. R: an alkyl group; Ar: an aromatic group; B: a ballast group; and [O]: an oxidizing agent such as silver halide. A color developer (**1**) is oxidized into the corresponding oxidized developer ($\mathbf{1}_{ox}$), which reacts with a cyan coupler (**2**), a magenta coupler (**5**), and a yellow coupler (**8**). Intermediate leuco dyes (**3**, **6**, and **9**) are further oxidized into a cyan dye (**4**), a magenta dye (**7**), and a yellow dye (**10**).

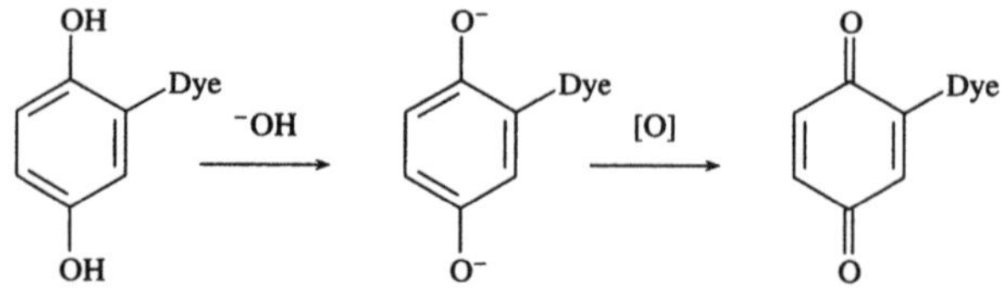

Figure 6.7. Dye transfer process (1). Dye developer process, where the symbol Dye represents a dye moiety.

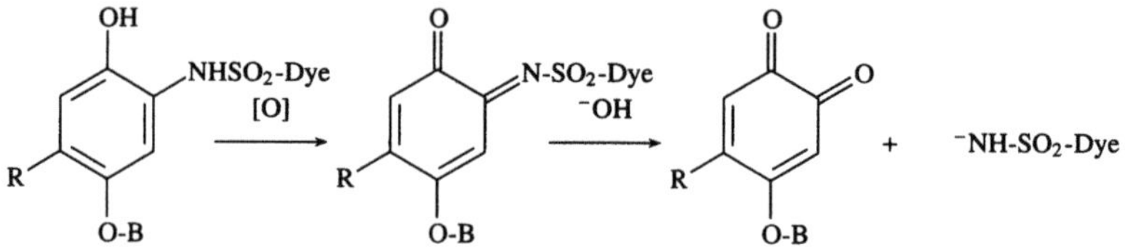

Figure 6.8. Dye transfer process (2). *o*-Sulfonamidophenol dye releaser process, where the symbols Dye and B represent a dye moiety and a ballast group.

Various systems belonging to the dye transfer process will be discussed in detail in Part IV.

References

[1] Fujita S (1984) Sensyoku Kyogyo. 32:167

[2] Thirtle JR (1979) Chemtech. 25

[3] Bailey J, Williams LA (1971) The Photographic Color Development Process. In: Venkataramn K (ed) The Chemistry of Synthetic Dyes, Vol V, Acacemic, New York London. Chapter VI

[4] Gehret EC (1964) Brit J Photogr. 111:480

[5] Rogers HG (1988) Res Technol Management. Sept-Oct, 42

[6] Fujita S, Koyama K, Ono S (1982) Nikkakyo Geppou. 35(11):29

[7] Fujita S, Koyama K, Ono S (1991) Nippon Kagaku Kaishi. 1

[8] Fujita S, Koyama K, Ono S (1991) Rev Heteroatom Chem. 7:229

[9] Hanson WT (1976) Photogr Sci Eng. 20:155

Chapter 7

Spectral Sensitization and Sensitizing Dyes

7.1 Background

Silver halides have intrinsic sensitivity only at a wavelength shorter than about 450 nm, as described in Chapters 3 and 4. This means that silver halides themselves are not sensitive to green and red lights among visible light. Hence, spectral sensitization for the green and red regions of light is indispensable to both black-and-white and color photography. Various sensitizing dyes have been used for this purpose. Comprehensive reviews on sensitizing dyes have appeared in books on photography [1–5].

Vogel [6] discovered in 1873 that a silver halide plate, when dyed with coralline (**1**, a red dye under alkaline conditions), became sensitive to light of longer wavelength. This discovery was the beginning of spectral sensitization [7].

Several sensitizing dyes of historical importance are listed in Fig. 7.1. At an earlier stage of black-and-white photography, so-called orthochromatic photographic materials were commercialized, where dyes of xanthene type (e.g., **2** and **3**) were used as spectral sensitizers. Photographic materials having light sensitivity in the entire visible range is called "panchromatic". Most dyes for panchromatic materials are cyanine dyes such as pinacyanol (**4**), which was introduced by Homolka [8]. Photographic materials having light sensitivity in the infrared region were based on sensitizing dyes such as kryptocyanine (**5**), which was discovered by Adams and Haller [9].

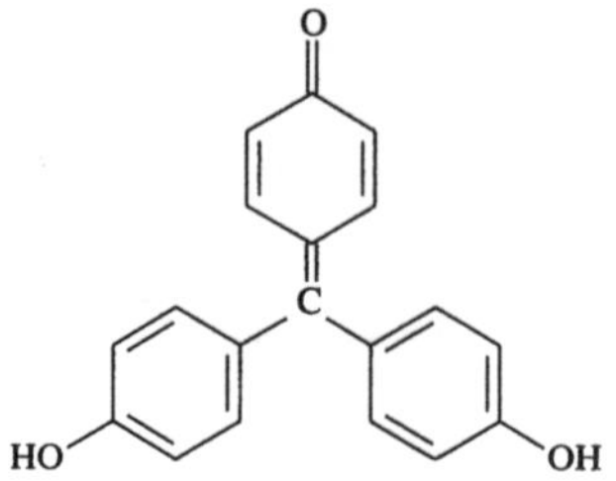

coralline (**1**)

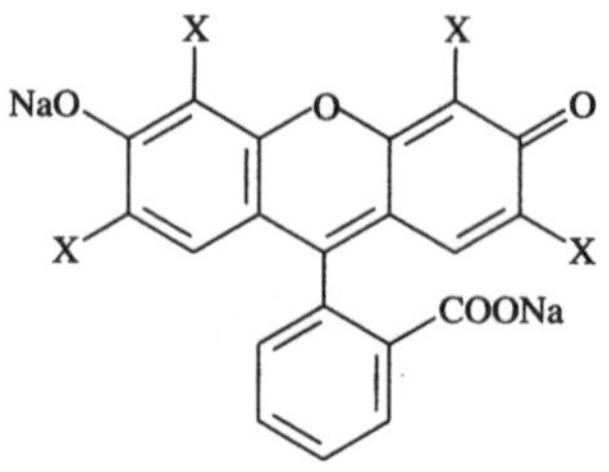

X = Br: eosine (**2**)
X = I: erythrosine (**3**)
λ_{max} 555 nm

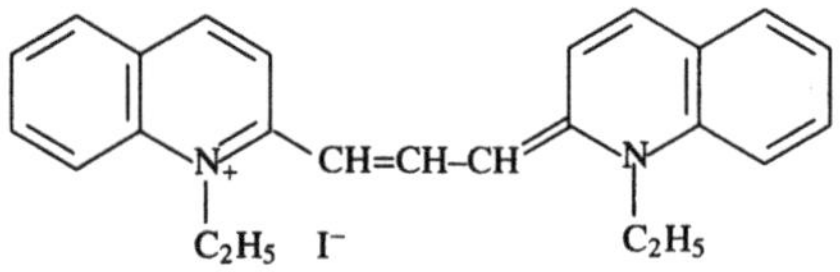

pinacyanol (**4**) λ_{max} 650 nm

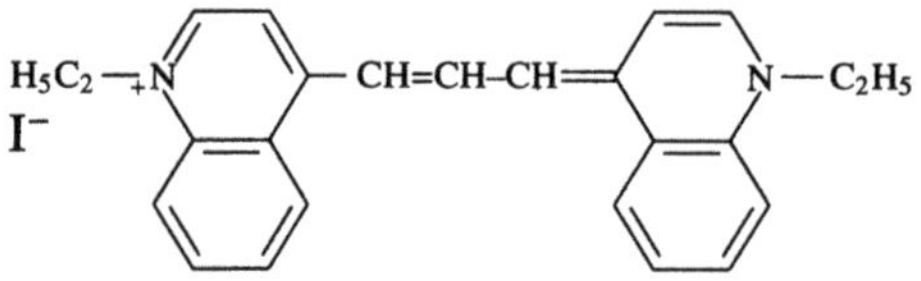

kryptocyanine (**5**) λ_{max} 740nm

Figure 7.1. Sensitizing dyes of historical importance.

7.2 Chromophoric Systems of Sensitizing Dyes

Most sensitizing dyes (spectral sensitizers) have structures in which appropriate heterocycles are placed at the two terminals of a methine chain. They are classified into the following three categories in terms of the cationic, anionic, or neutral nature of the terminal heterocycles [1–5]. The three categories are further subdivided into subclasses of more specified molecular structures.

1. [Amidinium-ion system] The amidinium-ion system contains the following structure, in which n represents the number of repeating units:

$$>\overset{+}{N}{=}CH{-}(CH{=}CH)_n{-}N< \quad \leftrightarrow \quad >N{-}(CH{=}CH)_n{-}CH{=}\overset{+}{N}< \qquad (7.1)$$

The general structure is a system of resonance, which corresponds to a cationic sensitizing dye. Note that the terminal N and N^+ can be the portions of heterocycles and that each hydrogen of the methine unit can be replaced by another substituent. Representative amidinium-ion systems are listed in Fig. 7.2.

Cyanine dyes:
(thiacarbocyanine)

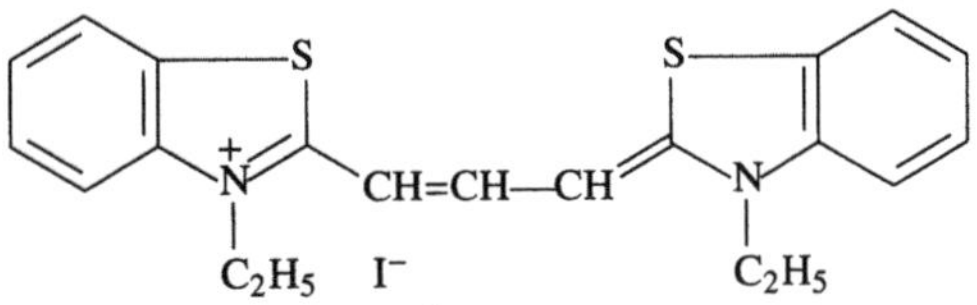

S S
$\overset{+}{N}$ CH=CH—CH N
C_2H_5 I^- C_2H_5

6

Styryl dyes:
(pinaflavol)

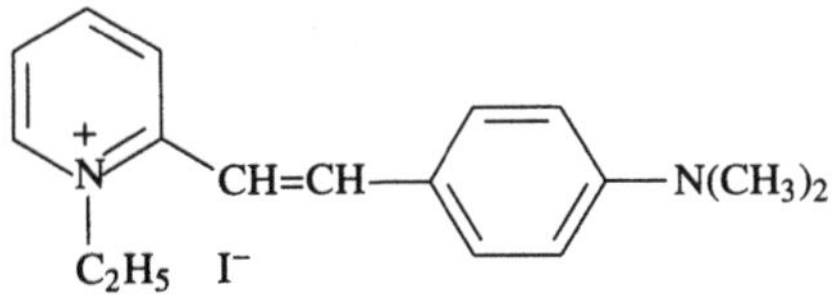

$\overset{+}{N}$ CH=CH— —$N(CH_3)_2$
C_2H_5 I^-

7

Hemicyanine dyes:

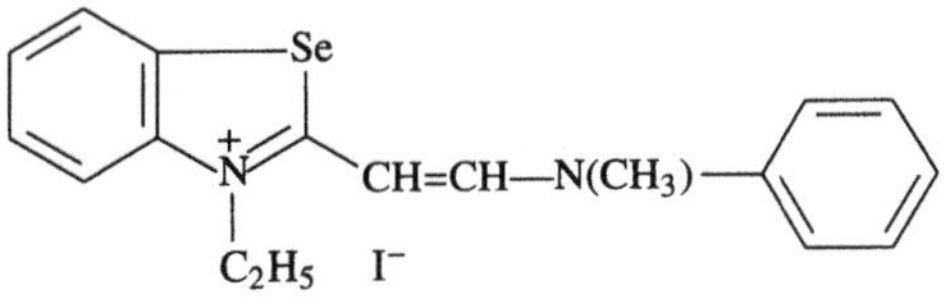

Se
$\overset{+}{N}$ CH=CH—$N(CH_3)$—
C_2H_5 I^-

8

Triphenylmethane dyes:
(malachite green)

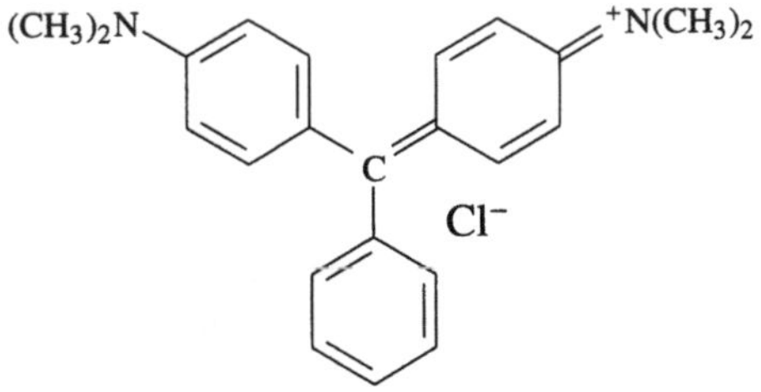

$(CH_3)_2N$ $^+N(CH_3)_2$
C
Cl^-

9

Figure 7.2. Sensitizing dyes having an amidinium system. Each of them is present in the form of a resonance hybrid shown in the general formula (eq. 7.1).

2. [Carboxyl-ion system] The carboxyl-ion system contains the following structure, in which n represents the number of repeating units:

$$\mathrm{O{=}C{-}\left(C{=}C\right)_{n}{-}O^{-}} \leftrightarrow \mathrm{O^{-}{-}\left(C{=}C\right)_{n}{-}C{=}O} \qquad (7.2)$$

This type of sensitizing dyes are anionic under dissociated conditions. Xanthene and oxonol dyes belong to this category, as shown in Fig. 7.3.

Xanthene dyes:
(eosine)

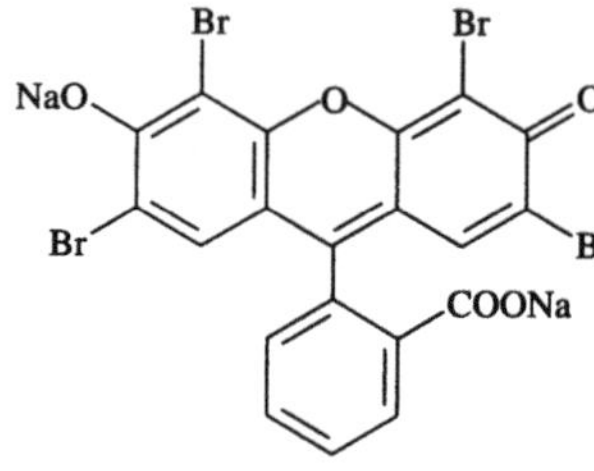

10

Oxonol dyes:

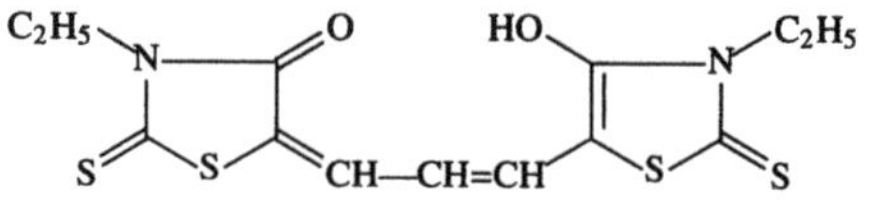

11

Figure 7.3. Sensitizing dyes of the carboxy-ion system. The NaO-group in eosine corresponds to the O^--group of the general formula (eq. 7.2). The hydroxyl group in the oxonol dye is dissociated into the O^--group.

3. [Amide system] Dyes belonging to this category are neutral, as found in the following resonance formulation so that the corresponding dipolar structure can be considered as a localized form. Merocyanine dyes are representative as listed in Fig. 7.4.

$$\mathrm{{>}N{-}\left(C{=}C\right)_{n}{-}C{=}O} \leftrightarrow \mathrm{{>}\overset{+}{N}{=}\left(C{-}C{=}\right)_{n}C{-}O^{-}} \qquad (7.3)$$

Among the sensitizing dyes mentioned above, cyanine and merocyanine dyes are mainly used in modern photography.

When sensitizing dye molecules aggregate, the maximum wavelength shifts remarkably. So-called H-aggregates that exhibit hypsochromic shifts (blue shifts) are ascribed to the formation of dimers to tetramers [2]. On the other hand, various structures (e.g., brick-stone work etc.) have been proposed to J-aggregates that exhibit bathochromic shifts (red shifts) [2].

Merocyanine dyes:

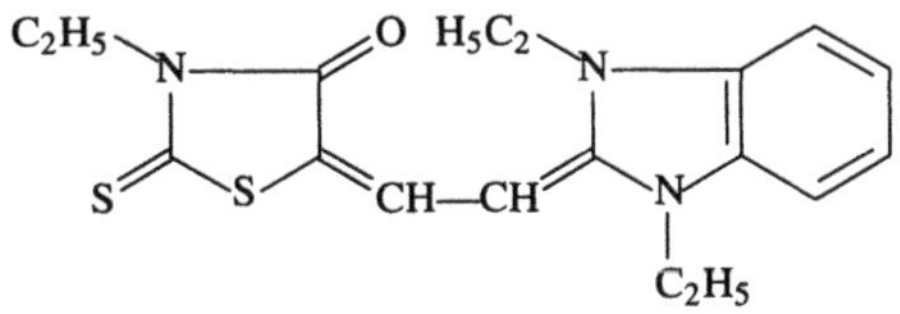

12

Figure 7.4. Sensitizing dye of the amide system. See a resonance hybrid shown in the general formula (eq. 7.3).

When a sensitizing dye is added to a silver halide emulsion, the dye molecules are adsorbed on the surface of a silver halide grain. As the concentration of the dye increases, the absorption peaks of H-aggregates or J-aggregates appear, the latter of which are important to the spectral sensitization of photographic emulsions.

7.3 Cyanine Dyes

A general formula of cyanine dyes is shown in Fig. 7.5. The nitrogen atoms of the amidinium-ion system in cyanine dyes are contained in the terminal heterocycles, which are five- or six-membered rings in most cases. The chain length can be changed by increasing the value of n to produce a series of vinylogs. They are called *simple cyanine* when $n = 0$, *carbocyanine* (or trimethine cyanine) when $n = 1$, *dicarbocyanine* (or pentamethine cyanine) when $n = 2$, and *tricarbocyanine* (or heptamethine cyanine) when $n = 3$.

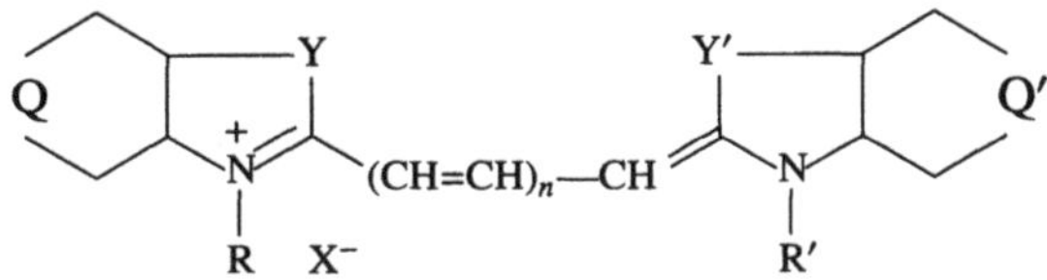

Y, Y′: O, S, Se, CH=CH, N-R etc.
R, R′: alkyl
X: Cl, Br, I, ClO_4, etc.
Q, Q′: substituents, or units for forming rings
$n = 0$ — simple cyanine
$n = 1$ — carbocyanine (or trimethine cyanine)
$n = 2$ — dicarbocyanine (or pentamethine cyanine)
$n = 3$ — tricarbocyanine (or heptamethine cyanine)

Figure 7.5. General formula for cyanine dyes.

When two terminal heterocycles are the same in a vinylogous series of sensitizing dyes of amidinium-ion or carboxyl-ion type, the values of the wavelength

S S
(CH=CH)$_n$—CH
$\overset{+}{N}$ N
C_2H_5 I^- C_2H_5

13

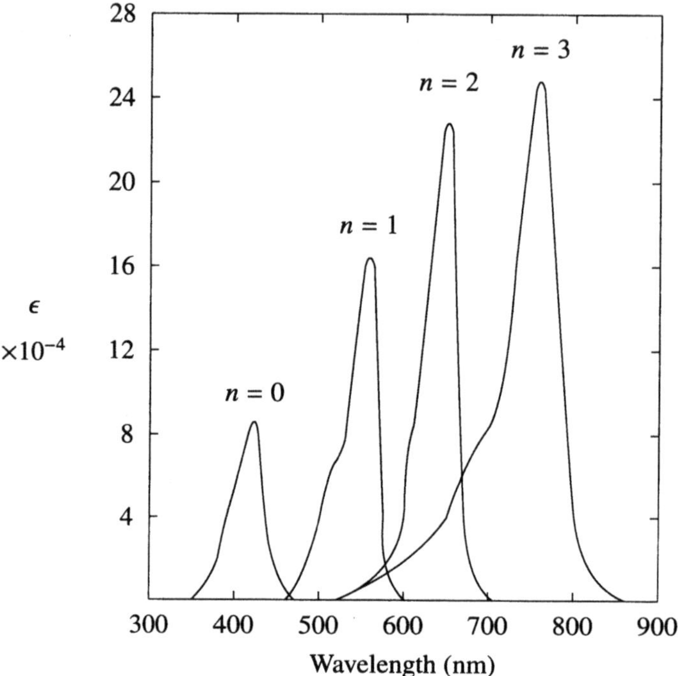

Figure 7.6. Absorption spectra of thiacarbocyanine vinylogs (**13**) in methanol. The values of the wavelength of maximum absorption (λ_{max}) were reported to be 423 nm for $n = 0$; 557.5 nm for $n = 1$; 650 nm for $n = 2$; and 758 nm for $n = 3$ [1].

of maximum absorption (λ_{max}) increase by about 100 nm as the number n of vinylene units increases by one ("vinylene shift" [10]). Such a vinylene shift of symmetrical thiacarbocyanine vinylogs are shown in Fig. 7.6 [1,11]. Color-structure relationships for cyanine dyes and for related polymethine dyes have been discussed thoroughly in a textbook [12].

Brooker [13] has shown that the light absorption of unsymmetrical cyanine dyes is blue-shifted in comparison with that of the corresponding symmetrical dyes. Such a blue shift is now well-known as "Brooker deviation", as reviewed by himself [1]. The Brooker deviation is proportional to the basicity of the heterocycles ("Brooker basicity"), which is related to the pK_a values of the sensitizing dyes [14]. Brooker [15] has examined spacial configurations of cyanine dyes, has classified them as loose dyes (which may form different conformers), compact dyes, and crowded dyes (in which steric hindrance may disturb the planar conjugate system), and has shown that only compact dyes are effective to spectral sensitization.

7.4 Mechanism of Spectral Sensitization

The mechanism of spectral sensitization has been extensively studied, as summarized in a review [16] and a book [17]. An electron transfer mechanism has become predominant over an energy transfer mechanism. A process of spectral sensitization based on the former mechanism is shown in Fig. 7.7. When a dye absorbs light, an electron on the HOMO (highest occupied molecular orbital) is excited (Fig. 7.7a) into the LUMO (lowest unoccupied molecular orbital). Note that the energy gap between the LUMO and the HOMO in the sensitizing dye is narrower than the gap between the conduction band and the valence band in the silver halide grain, so that spectral sensitization is effective. The excited electron shifts to the conduction band of a silver halide grain (Fig. 7.7b) so that it moves to an electron trap for forming a latent image (Fig. 7.7c) or is deactivated by recombination (Fig. 7.7d).

When the dye releases a photo-electron, it is converted into the corresponding radical cation Dye^+, which is usually referred to as a positive hole (eq. 7.4).[1] The fate of the photo-electron has been discussed in Chapter 4. The radical cation Dye^+ can be quenched by a bromide anion to produce a positive hole (Br), as represented by eq. 7.6.

$$\mathrm{Dye} + h\nu \rightarrow \mathrm{Dye}^+ + \mathrm{e} \quad (7.4)$$

$$\mathrm{e} + \mathrm{Ag}^+ \rightarrow \mathrm{Ag} \quad (7.5)$$

$$\mathrm{Dye}^+ + \mathrm{Br}^- \rightarrow \mathrm{Dye} + \mathrm{Br} \quad (7.6)$$

Supersensitizers are not capable of sensitizing silver halide grains but assist the spectral sensitization of sensitizing dyes, where half-wave potentials of sensitizing dyes and supersensitizers have been discussed to explain their photographic properties [19,20].

For example, a combination of a sensitizing dye (**14**) and a supersensitizer (**15**) shown in Fig. 7.8 has exhibited supersensitization. A phosphorescence band of the dye (**14**) has appeared at long wavelengths under adsorbed conditions but has been quenched on addition of the supersensitizer (**15**) [18]. As a result, the band has

[1] In the words of organic chemistry, such a radical cation is represented by the symbol $Dye\bullet^+$. The positive hole derived from Br^- is represented by Br, which corresponds to $Br\bullet$ in terms of organic chemistry.

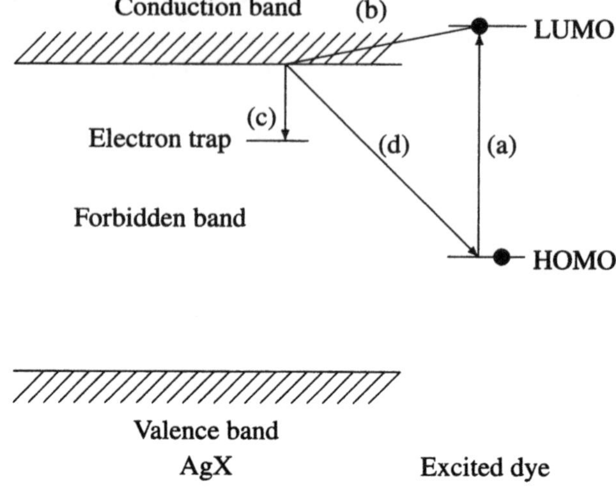

Figure 7.7. Schematic diagram of spectral sensitization. (a) Excitation of an electron by light absorption; (b) Movement to the conduction band of a silver halide grain; (c) Trapping by an electron trap to form a latent image; and (d) Deactivation by the cancellation of the positive hole.

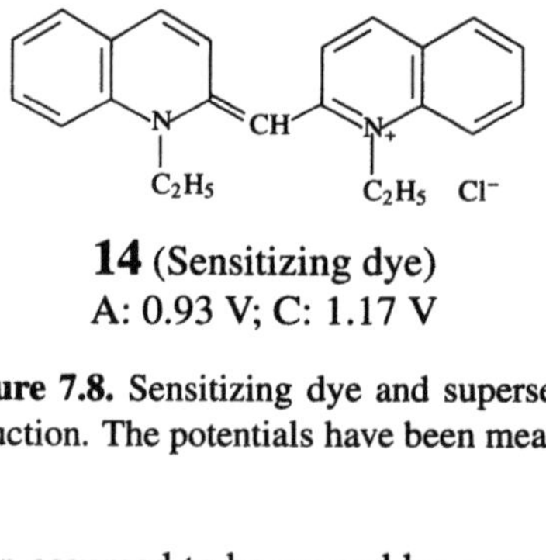

14 (Sensitizing dye)
A: 0.93 V; C: 1.17 V

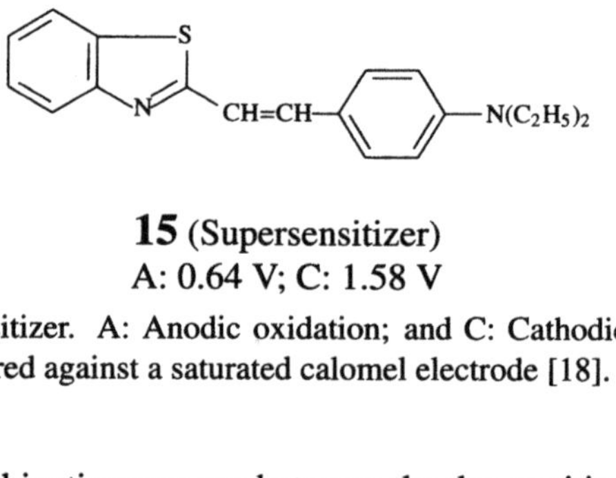

15 (Supersensitizer)
A: 0.64 V; C: 1.58 V

Figure 7.8. Sensitizing dye and supersensitizer. A: Anodic oxidation; and C: Cathodic reduction. The potentials have been measured against a saturated calomel electrode [18].

been assumed to be caused by a recombination process between the dye positive holes and the photo-electrons. This means that the supersensitizer prevents the recombination process that corresponds to the process of Fig. 7.7d. Note that the oxidation potential of the supersensitizer (**15**) is more positive than that of the sensitizer (**14**), as shown in Fig. 7.8.

To explain supersensitization by using such accumulated experimental results as described above, Gilman [21] has proposed a mechanism shown in Fig. 7.9, where an electron in the HOMO of a supersensitizer moves to the vacancy in the HOMO of a sensitizing dye (Fig. 7.9e) so that the positive hole of the sensitizing dye is cancelled. As a result, the original LUMO of the sensitizing dye is shifted up to the LUMO′ so as to accelerate the movement of a photo-electron to the

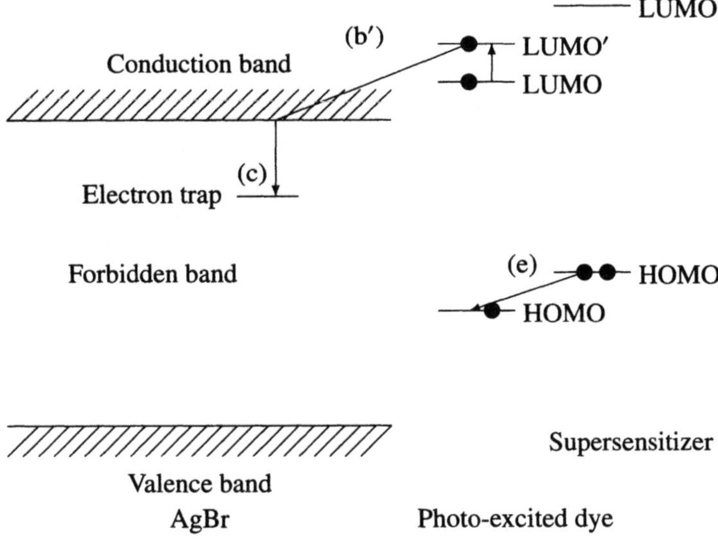

Figure 7.9. Schematic diagram of supersensitization. (b′) Movement from the shifted HOMO′ to the conduction band of a silver halide grain; (e) Cancellation of the positive hole at the expense of Fig. 7.7d. The process (a) is omitted. See the caption of Fig. 7.7.

conduction band of an silver halide grain. This means the higher efficiency of latent image formation.

7.5 Syntheses of Sensitizing Dyes

7.5.1 Symmetrically Substituted Sensitizing Dyes

Since König [10] discovered an elegant method of synthesizing carbocyanines by using triethyl orthoformate, a vast number of sensitizing dyes have been synthesized, as summarized in reviews [22,23]. The original scheme of synthesizing pinacyanol (**4**) by König is shown in eq. 7.7, where two moles of the 2-methyl heterocyclic quaternary salt react with triethyl orthoformate.

CH_3, N^+, I^-, C_2H_5 + $CH(OC_2H_5)_3$ + CH_3, N^+, I^-, C_2H_5

$\longrightarrow$ CH=CH–CH (N^+–C_2H_5 I^-; N–C_2H_5) **4** (7.7)

A sensitizing dye (**16** called Rr340; the sensitizing maximum 545–550 nm) for the green-sensitive layer of the Agfacolor Reversal film has been reported to be synthesized according to eq. 7.8, where pyridine (Py) has been used as a basic condensing agent [24]. In order to introduce an ethyl substituent on the methine chain, triethyl orthopropionate has been used as a polymethine source.

Ph, O, N^+, CH_3, C_2H_5 I^- + $CH_3C(OC_2H_5)_3$ + CH_3, O, N^+, Ph, C_2H_5 I^-

$\xrightarrow{Py}$ Ph, O, N^+, C_2H_5 I^-, CH=C(C_2H_5)–CH=, O, N, C_2H_5, Ph (7.8)

16 (Rr340)

The quaternary salt of 2-methyl-5-phenylbenzoxazole for the synthesis of Rr340 (**16**) has been synthesized according to the following scheme [24].[2]

17 (Ph, OH) $\xrightarrow[\text{2) } NaHSO_3]{\text{1) p-}Cl^-N_2^+\text{–}C_6H_4\text{–}SO_3H}$ **18** (Ph, OH, NH_2) $\xrightarrow{(CH_3CO)_2O}$

19 (Ph, O, N, CH_3) $\xrightarrow{(C_2H_5O)_2SO_2}$ **20** (Ph, O, N^+, CH_3, C_2H_5 X^-) (7.9)

As found in the preceding examples, most methods for synthesizing cyanine dyes involve the reaction of 2- (or 4-)methyl heterocyclic quaternary salt (**21**) with another reagent (a polymethylene source) in the presence of a basic con-

[2]The gegen ion (X^-) was derived from diethyl sulfate. According to the recipe [24], the quaternary salt (**20**) was used in situ in the successive reaction, although the original equation (7.8) involved an iodide ion as a gegen anion. By referring to the other recipes, the gegen anion (X^-) was presumed to be replaced by an iodide ion by adding potassium iodide during the recrystallization process of the sensitizing dye (**16**).

densing agent. The intermediacy of the corresponding methylene species (**22**) has been postulated in the dye preparations [25]. For further synthetic methods of sensitizing dyes, the reader should refer to review articles [23,26] and a book [22].

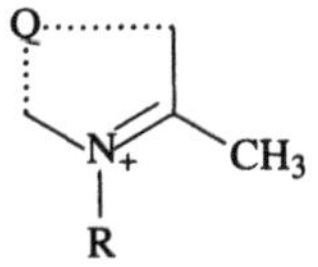

Q N+ CH_3 R **21** Q N CH_2 R **22**

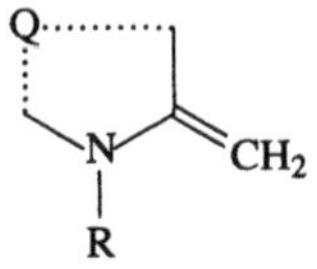

7.5.2 Unsymmetrical Sensitizing Dyes

As an example of synthesizing unsymmetrical dyes, Fig. 7.10 shows the preparation of another sensitizing dye (**36** called Kt945; the sensitizing maximum 555–560 nm), which was used in the green-sensitive layer of the Agfacolor Reversal film together with **16** [24].

The right part of **36** contains a benzoselenazole ring, which comes from an anil intermediate (**29**). To introduce a selenium atom, a diazonium chloride (**24**) is condensed with potassium isoselenocyanate with releasing nitrogen. The resulting intermediate (**25**) undergoes a reductive coupling to give a diselenide (**26**) after acetylation. The diselenide function of **26** is reduced to a selenol, which is cyclized intramolecularly to give a benzoselenazole (**27**). After quaternization, the active methyl group is functionalized to give 3-ethyl-2-methylene-benzoselenazole-anil (**29**). The polymethylene source is N, N'-diphenylformamidine used in the reaction from **27** to **28**.

The left part of **36** involves a substituted benzimidazole, which comes from the quaternary salt (**35**). The benzimidazole ring is formed by the acetylation of an *o*-diamino derivative (**32**) and the subsequent intramolecular cyclization of the acetyl derivative (**33**). The quaternary salt (**35**) is obtained by the methylation of **34** with dimethyl sulfate. Finally, the target dye (**36**) is obtained by the condensation of the anil (**29**) and the quaternary salt (**35**).

7.5.3 Chains Incorporating a Heterocyclic Ring

A cyanine dye (**46** called Rr2632; the sensitizing maximum 625–630 nm) in which a heterocyclic ring has been incorporated in the chain has been used in the red-sensitive layer of the Agfacolor Reversal film [24]. The synthetic pathway of **46** is shown in Fig. 7.11.

The left part of the sensitizing dye contains a benzothiazole ring. This ring is introduced by the oxidation of the thio-amido group of **40**, where the intramolecular cyclization takes place to give the benzoxazole (**41**). After quaternization, the active methyl group is condensed with N, N'-diphenylformamidine to give an anil intermediate (**42**).

a) "3-Ethyl-2-methylene-benzoselenazole-anil"

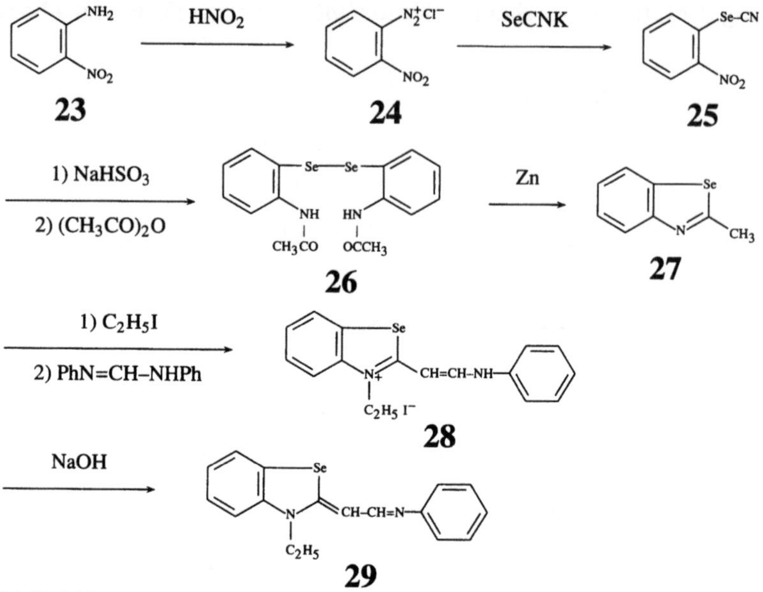

NH_2, NO_2 **23** —HNO_2→ $N_2^+Cl^-$, NO_2 **24** —SeCNK→ Se–CN, NO_2 **25**

—1) $NaHSO_3$ 2) $(CH_3CO)_2O$→ Se–Se, NH, HN, CH_3CO, $OCCH_3$ **26** —Zn→ Se, N, CH_3 **27**

—1) C_2H_5I 2) PhN=CH–NHPh→ Se, N+, CH=CH–NH–, C_2H_5 I^- **28**

—NaOH→ Se, N, CH–CH=N–, C_2H_5 **29**

b) Kt945

Cl, Cl, NO_2, Cl **30** —$NH_2(CH_2)_3OH$→ Cl, Cl, NO_2, $NH(CH_2)_3OH$ **31** —$SnCl_2$→

Cl, Cl, NH_2, $NH(CH_2)_3OH$ **32** —$(CH_3CO)_2O$→ Cl, Cl, $NHCOCH_3$, $NH(CH_2)_3OCOCH_3$ **33**

→ Cl, Cl, N, N, CH_3, $(CH_2)_3OCOCH_3$ **34** —$(CH_3O)_2SO_2$→ Cl, Cl, N+, CH_3, $^-OSO_2OCH_3$, N, CH_3, $(CH_2)_3OCOCH_3$ **35**

—**29**, NH_4CNS→ Cl, Cl, N+, CH_3, ^-CNS, N, CH=CH–CH=, Se, N, $(CH_2)_3OCOCH_3$, C_2H_5 **36** (Kt945)

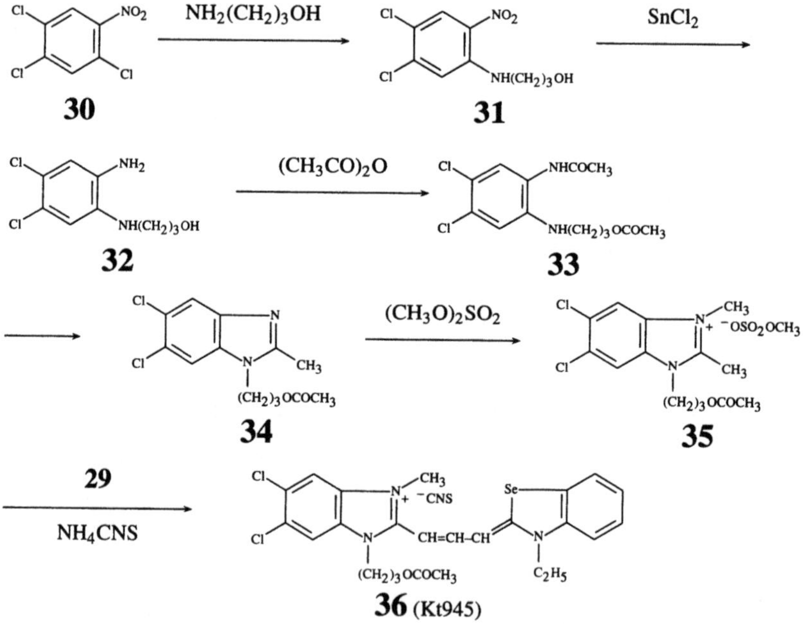

Figure 7.10. Synthesis of an unsymmetrically substituted sensitizing dye (Kt945) for the Agfacolor Reversal film [24].

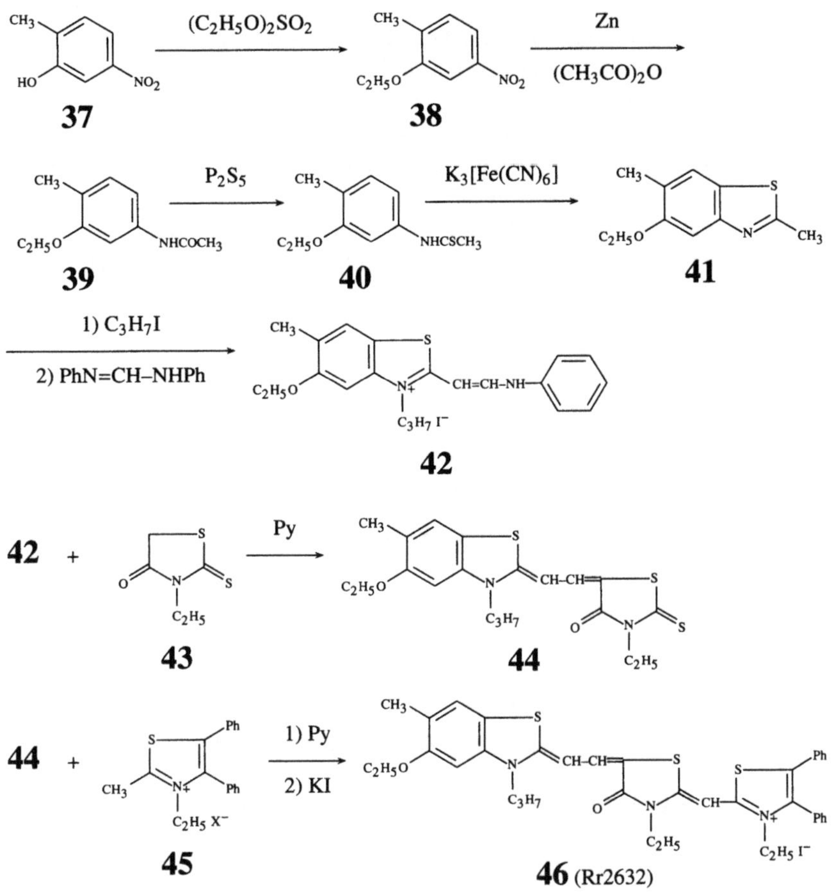

Figure 7.11. Synthesis of a sensitizing dye (Rr2632) for the Agfacolor Reversal film [24]. The gegen ion (X^-) in **45** was derived from diethyl sulfate.

Then, a merocyanine (**44**), which is a key intermediate in this preparation, is derived by the condensation between the anil (**42**) and *N*-ethylrhodanine (**43**). Thereby, the heterocyclic moiety due to the rhodanine is incorporated in the centeral chain. The final step is the condensation between the merocyanine (**44**) and a thiazole derivative (**45**) so as to produce the target molecule (**46**).

7.6 Recent Topics

7.6.1 Residual Color Due to Sensitizing Dyes

As a result of recent progress in tabular grain techniques, new sensitizing dyes having additional properties have been required to ensure high quality of color photography. Tabular grains are advantageous to realize high color sensitivity because sensitizing dyes can be adsorbed in large quantities. This advantage, in turn, provides a problem that residual color after processing increases to an undesired extent, especially when rapid procedures of processing are adopted. By elaborating the *N*-substituents of terminal heterocycles, new dyes (**47a** and **47b**; and **48a** and **48b**) that are less in residual color have been developed [27,28]. They have substituents $-CH_2CONHSO_2CH_3$ or $-CH_2SO_2NHCOCH_3$, in which the NH group can dissociate easily to give the corresponding anion. As a result, they are less in residual color than the corresponding dyes with $R = -(CH_2)_3\text{-}SO_3^-$ and a gegen ion $(C_2H_5)_3NH^+$.

47a ($R = CH_2CON^-SO_2CH_3$); **47b** ($R = CH_2SO_2N^-COCH_3$)

48a ($R = CH_2CON^-SO_2CH_3$); **48b** ($R = CH_2SO_2N^-COCH_3$)

Figure 7.12. Sensitizing dyes less in residual color.

7.6.2 Additional Sensitizing Layer

A color negative film at least consists of a blue-sensitive, a green-sensitive, and a red-sensitive emulsion layer, each of which contains one or more sensitizing dyes having an appropriate visible absorption. As shown in Fig. 7.13 (B, G, and R), the spectral sensitivities of the sensitizing dyes are designed not to overlap each other. Although this specification prevents color contamination in cases of objects having broad light absorption, it provides insufficient color reproduction on the occasion of objects having a sharp spectrum. Moreover, the addition of a DIR (development inhibitor releasing) coupler in the green-sensitive layer of a color negative film provides an interlayer effect to the red-sensitive layer so that the formation of cyan dyes is suppressed in the regions irradiated with green light. As a result, the amount of cyan dyes increases in the corresponding regions of the color print obtained from the color negative film. In particular, the region irradiated with yellow-green to yellow light (about 570 nm) suffers from the absorption of the cyan dyes so that the yellow-green to yellow color is shifted to green color. What is to be done is to make the cyan dyes increased by the interlayer effect

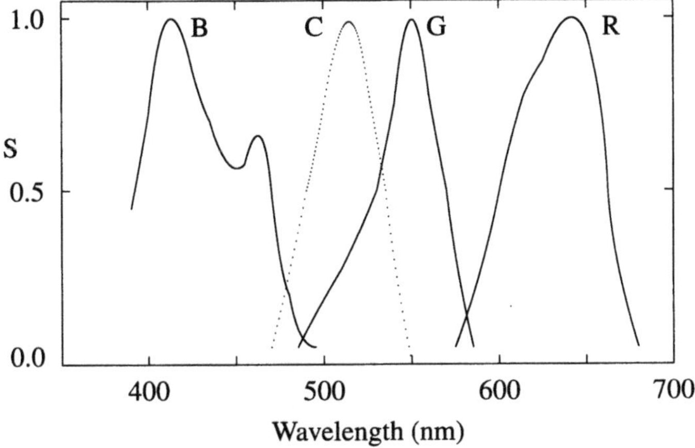

Figure 7.13. Schematic illustration of spectral sensitivities and an additional sensitivity [29]. The solid lines represent usual sensitivities: blue (B), green (G), and red (R). The dotted line with the symbol (C) represents an additional sensitivity (so-called "cyan sensitivity").

harmless in the color print. This has been realized by placing an additional sensitive emulsion layer which contains a sensitizing dye absorbing the light of ca. 520 nm (dotted line in Fig. 7.13) and a DIR coupler [29].[3]

An example of multilayer photographic films that contains such an additional sensitized layer (so-called "fourth emulsion layer" of "cyan sensitivity") has been disclosed in a patent [30]. The schematic cross-section is shown in Fig. 7.14.

The usually sensitized emulsion layers involve the sensitizing dyes that are substantially the same as those shown in Chapter 1: the sensitizing dyes of Fig. 1.33 (Chapter 1) for the red-sensitive layers (Layers 3 and 4 in Fig. 7.14), the dyes of Fig. 1.38 (Chapter 1) for the green-sensitive layers (Layers 6 and 7), and the dye of Fig. 1.42 (Chapter 1) for the blue-sensitive layers (Layers 11 and 12).

On the other hand, the additional layer of "cyan sensitivity" (Layer 9) contains a silver halide emulsion sensitized with a new sensitizing dye having an absorption at about 520 nm (**49**), where the *t*-amyl (*t*-pentyl) group has been selected by examining the STERIMOL parameters (Fig. 7.15) [30].

According to the "cyan sensitivity", Layer 9 of Fig. 7.14 contains a DIR yellow coupler (**50**),[4] a yellow-colored magenta coupler (**51**), and a polymeric magenta coupler (**52**). The DIR yellow coupler releases a development inhibitor, which diffuses into the red-sensitive layers and inhibits the development of silver halide. Note that the green-sensitive layers can be considered to be unexposed in the

[3] A color negative film employing this type of color correction has been placed on the market under the name of Fujicolor REALA (1989).

[4] This coupler is a deactivatable DIR coupler. See Chapter 14.

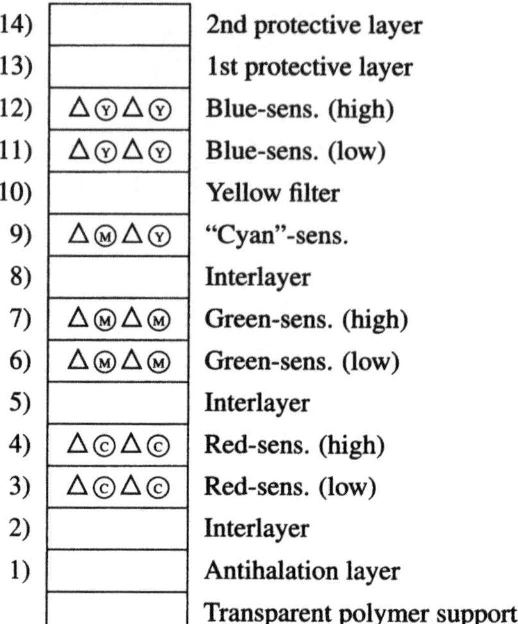

Figure 7.14. Schematic section of multilayer structure with an additional layer of "cyan sensitivity". Δ: Silver halide grain; Ⓨ: Yellow coupler; Ⓜ: Magenta coupler; ©: Cyan coupler.

region where the cyan-sensitive layer is exposed, so long as the region requires color correction by the mechanism of the cyan-sensitive layer. This means that the development inhibitor released in the cyan-sensitive layer runs through the green-sensitive layers and reaches the red-sensitive layers.[5]

7.6.3 Infrared-Sensitizing Dyes for Digital Printing

Since laser diodes exhibiting emission at the infrared region of light have been introduced as new light sources for digital imaging, renewed interests have been focused on infrared-sensitizing dyes [32]. The light sources used in a digital printer (Pictrography 3000 from Fuji Photo Film) are three laser diodes of different wavelength: 680, 750, and 810 nm, where the latter two diodes require infrared-sensitizing dyes.[6]

[5]An inverse inhibiting action from the red-sensitive layer to the blue-sensitive layer has been realized by using a DIRR (development inhibitor releaser releasing) coupler [31]. See Chapter 14.

[6]This embodiment is so-called "false color reproduction". Thus infrared-sensitizing dyes at 750 and 810 nm are incorporated in a cyan layer and a yellow layer, respectively. A sensitizing dye for 680 nm corresponds to a magenta layer. See Chapters 19 and 20.

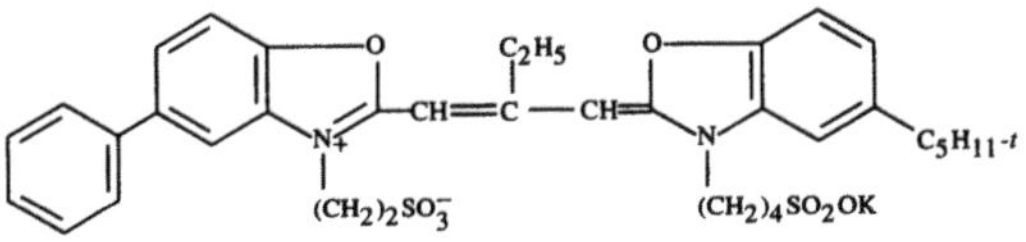

Figure 7.15. Sensitizing dye and couplers for the additional layer of "cyan sensitivity" (Layer 9 of Fig. 7.14).

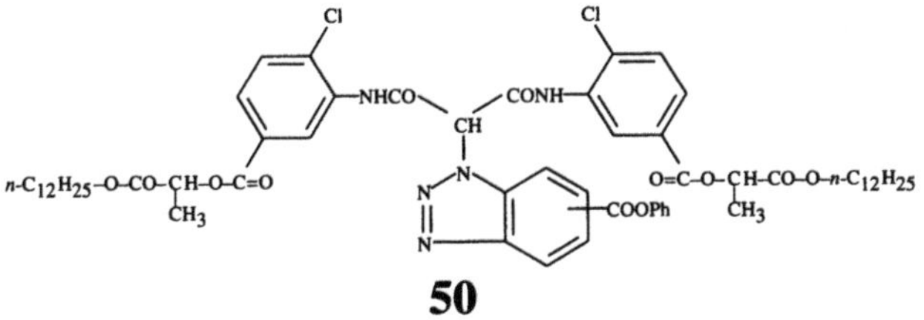

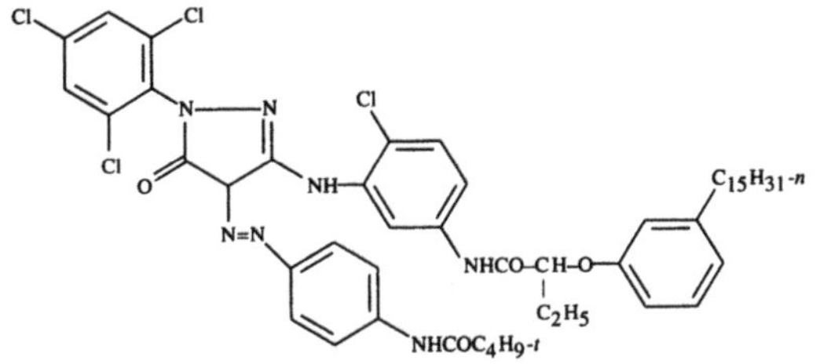

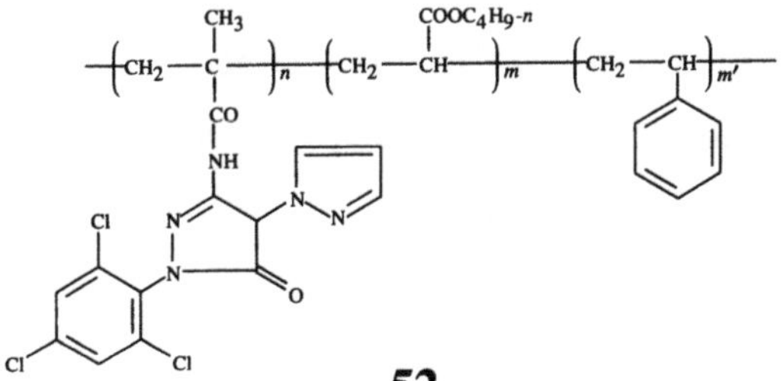

A thiatricarbocyanine dye is used as an infrared-sensitizing dye for the 810 nm light after enhancing stabilization to an oxidative degradation. A bulky *N*-substituent shown in **53b** (Fig. 7.16) has been introduced to improve the stability to oxidation.

On the other hand, a thiadicarbocyanine dye is used as an infrared-sensitizing dye for the 750 nm light after tuning J-aggregate formation. Substituents on the bridging ring of the methylene chain (e.g., a phenyl group in **54**) have been found to exhibit an essential effect on J-aggregation.

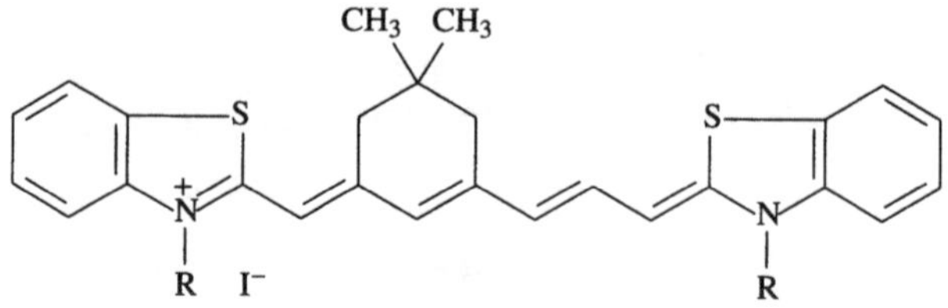

53a R = C_2H_5; **53b** R = $-CH_2CH_2O-$

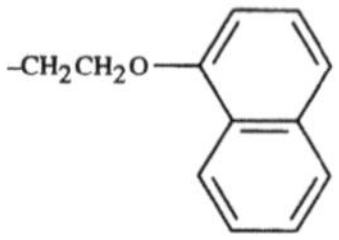

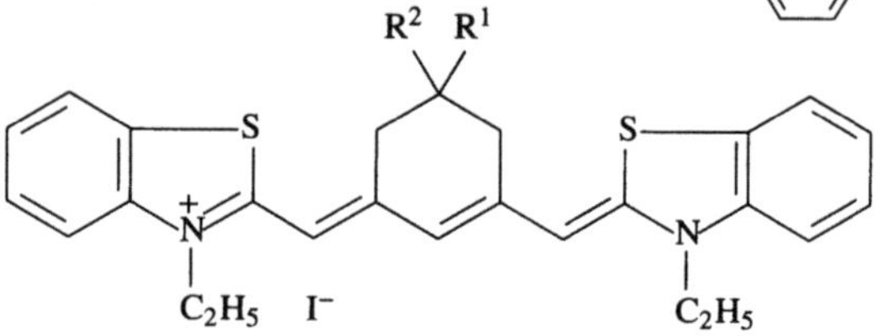

54 R^1 = Ph and R^2 = H

Figure 7.16. Infrared-sensitizing dyes for digital printing. The dye **53b** is combined with a diode emitting the 810 nm light, while the dye **54** is combined with a diode emitting the 750 nm light.

7.6.4 Sensitizing Dyes Applied to CD-R

As the quantity of information treated with computers has increased more and more, recording media having higher capacity have been developed. Among them, this subsection deals with compact disk recordable (CD-R, a write-once memory using compact disk) and digital video disks recordable (DVD-R, a write-once memory using digital video disk),[7] since one of the representative systems uses cyanines and related dyes [33,34].

Figure 7.17 shows a schematic cross-section view of a CD-R (or DVD-R) of organic ablation type, where a dye layer, a reflection layer (containing silver, gold, copper metal), and a protective layer are coated on a transparent polymer support such as polycarbonate.

A laser beam having a wavelength of about 780 nm (for CD-R) or about 635 nm (for DVD-R), which is modulated to carry information, is irradiated to the dye layer (Fig. 7.17). The resulting excited dye is decomposed exothermally or relaxed thermally so as to cause thermal transformations of the polycarbonate support and of the reflection layer. Such transformations produce a pit with which the information is stored in terms of difference in the reflection ratio.

Because cyanine dyes are unstable to environmental light, appropriate photostabilization is necessary in order to improve the lifetime of compact disks. One

[7] The term DVD is also used as the abbreviation of "digital versatile disk".

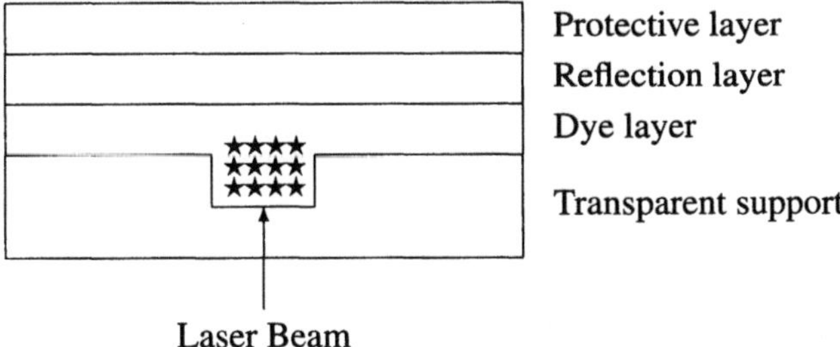

Figure 7.17. Schematic cross-section of a CD-R or DVD-R of organic ablation type. The site with the symbols ★ schematically represents a pit produced by the laser-beam ablation of organic dyes in the dye layer.

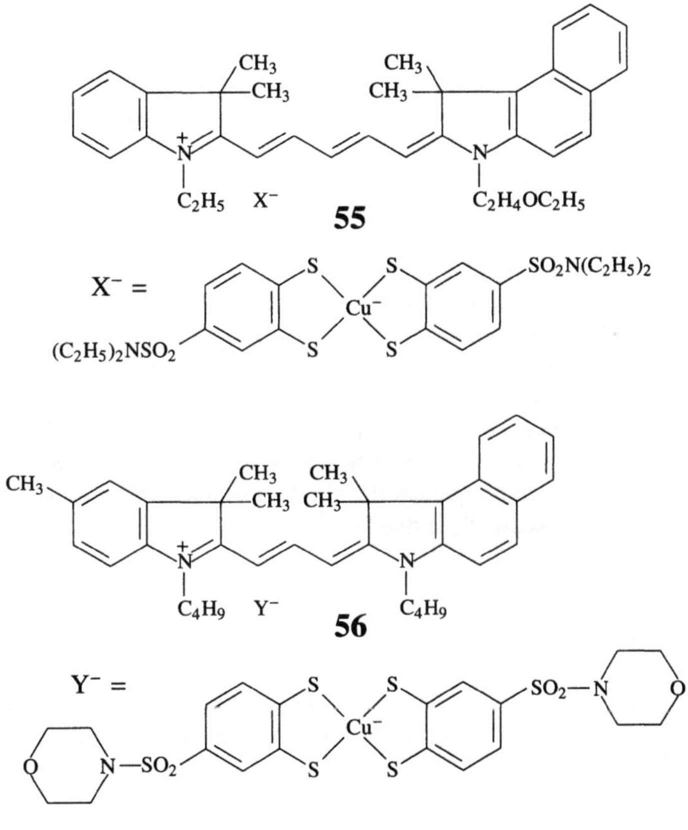

55

56

Figure 7.18. Cyanine dyes having a metal complex as a counter ion for a CD-R and a DVD-R.

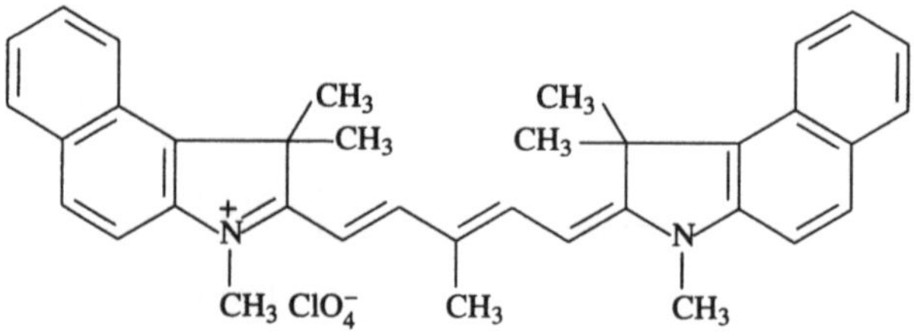

Figure 7.19. Cyanine dye (absorption maximum: 670 nm) and organic oxidizing agents for improving the life-time of a CD-R.

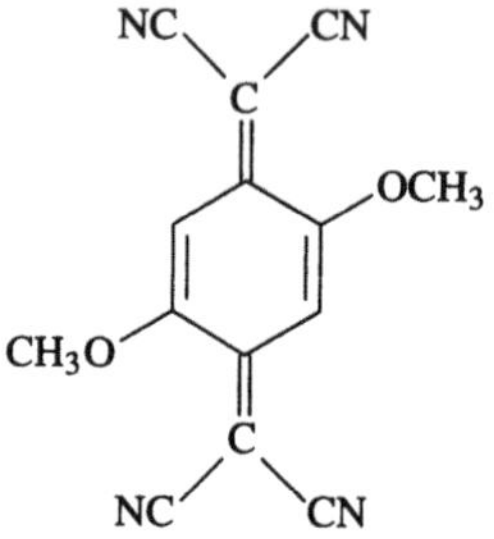

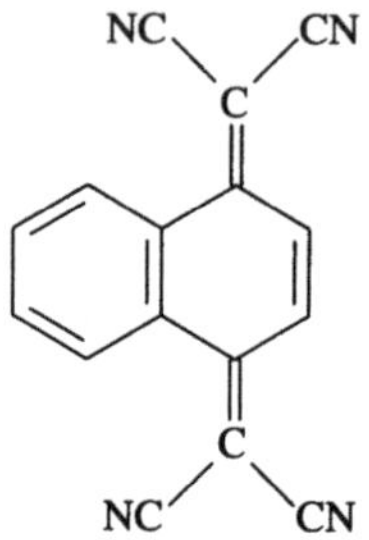

technique to enhance photo-stability is to use a cyanine dye with a specific counter ion (X^- or Y^-), as shown in Fig. 7.18, where **55** is a dye for CD-R and **56** is a dye for DVD-R [35]. Other types of counter ions have been disclosed [36].

An alternative technique for enhancing the light stability of cyanine dyes for a CD-R (e.g., **57**) is the addition of an oxidizing agent such as tetracyanoquinodimethanes (**58** or **59**), as shown in Fig. 7.19 [37,38]. Since cyanine dyes are presumed to be faded by light by virtue of auto-sensitization and oxidation due to singlet oxygen, quenchers of excited states of the dyes or those of singlet oxygen are used as light stabilizers. Tetracyanoquinodimethanes such as **58** and **59** work as electron acceptors which quench the singlet excited state of a cyanine dye to reproduce the ground state of the dye [39].[8]

References

[1] Brooker LGS (1966) Sensitizing and Desensitizing Dyes. In: James TJ (ed) The Theory of the Photographic Process, 3rd edn. Macmillan, New York London. Chapter 11

[8] Since the article [39] has described no structural formulas and no compound names of dyes and of tetracyanoquinodimethane derivatives, the data do not correspond to the structures shown in Fig. 7.19.

[2] Sturmer DM, Heseltine DW (1977) Sensitizing and Desensitizing Dyes. In: James TJ (ed) The Theory of the Photographic Process, 4th edn. Macmillan, New York London. Chapter 8

[3] Hishiki Y (1975) J Soc Photogr Sci Technol Jpn. 38:421; Hishiki Y (1981) Problems and Prospects of Sensitizing Dyes. In: Ogawara M, Kuroki N, Kitao T (ed) Chemistry of Functionalized Dyes, CMC, Tokyo. Chapter 7

[4] Kampfer H (1993) Spectral Sensitization. In: Keller K (ed) Science and Technology of Photography, VCH, Weinheim. Section 2.2

[5] Futaki K, Oohashi M (1986, 2002) In: Functionalized Organic Chemicals for Silver Halide Color Photographic Materials, CMC, Tokyo. Chapter 2, Section 3.2

[6] Vogel HW (1873) Ber. 6:1302

[7] Dähne S (1974) Photogr Sci Eng. 18:582

[8] Homolka B (1906) Ger. Patent 172 118

[9] Adams EQ, Haller HL (1920) J Am Chem Soc. 42:2661

[10] König W (1922) Ber dtsch Chem Ges. 55:3293

[11] Thirtle JR (1979) Chemtech. 25

[12] Gordon PF, Gregory P (1983) Organic Chemistry in Color, Springer-Verlag, Berlin-Heidelberg, Section 5.5

[13] Brooker LGS, Sklar AL, Cressman HWJ, Keyes GH, Smith LA, Sprague RH, VanLare E, VanZandt G, White FL, Williams WW (1945) J Am Chem Soc. 67:1875

[14] Herz H (1974) Photogr Sci Eng. 18:207

[15] Brooker LGS (1951) J Photogr Sci. 1:173

[16] Dähne S (1994) J Imaging Sci Tech. 38:101

[17] Tani T (1995) Photographic Sensitivity: Theory and Mechanisms, Oxford Univ Press, New York

[18] Gilman PB (1970) The Effects of Aggregation, Temperature, and Supersensitization on the Luminescence of 1,1'-Diethyl-2,2'-cyanine Chloride Adsorbed to Silver Chloride. In: Berg WF, Mazzucato U, Meier M, Semerano G (ed) Dye Sensitization: Symposium Bressnone 1967, Focal Press, London. pp. 218–233

[19] Tani T (1969) Nature 221:466

[20] Gilman PB (1977) Pure Appl Chem. 49:357

[21] Gilman PB (1974) Photogr Sci Eng. 18:418

[22] Hamer FM (1964) The Cyanine Dyes and Related Compounds. In: Weissberger A (ed) The Chemistry of Heterocyclic Compounds, Vol 18, Wiley Intersci, New York

[23] Ficken GE (1971) Cyanine Dyes. In: Venkataraman K (ed) The Chemistry of Synthetic Dyes, Vol IV, Academic, New York London. Chapter V

[24] Gluck B (1947) The Manufacture of Agfacolor Material. Fiat Final Report. No. 943

[25] Brooker LGS, Dent SG, Heseltine DW, VanLare E (1953) J Am Chem Soc. 75:4335

[26] Hishiki Y (1974) Yuki Gosei Kagaku Kyokai Shi. 32:971

[27] Hioki T, Morimoto K (2002) US Patent 6 365 335 B1

[28] Ikegawa A, Mihara Y (1997) US Patent 5 604 089

[29] Sasaki N, Takahasi K, Ikoma H (1989) J Soc Photogr Sci Technol Jpn. 52:41

[30] Ikegawa A, Ohashi Y, Okazaki M (1993) US Patent 5 198 332

[31] Ichijima S (1989) J Soc Photogr Sci Technol Jpn. 52:145

[32] Inagaki Y, Hioki T, Katoh T (1996) J Soc Photogr Sci Technol Jpn. 59:260

[33] Shinkai M, Namba K (1992) Senryo to Yakuhin. 37:185

[34] Inagaki Y (2002) Kagaku to Kyoiku. 50:137

[35] Namba K, Kitagawa S, Shinkai M, Suzuki M, Kimura S, Hirako K (2000) US Patent 6 071 672

[36] Kasada C, Hata Y, Kawata T, Yasui S (2002) US Patent 6 413 607 B1

[37] Morishima S, Wariishi K, Ishida T, Shibata M (1999) US Patent 5 879 772

[38] Ishida T, Shibata M, Wariishi K, Morishima S (1999) US Patent 5 998 094

[39] Morishima S, Wariishi K, Inagaki Y, Shibata M, Ishida T, Kubo H (1999) Jpn J Appl Phys. 38:1634

Part III

Chromogenic Photography

Chapter 8

History of Color Development

8.1 Fischer's Pioneering Work

The early history of color photography has been summarized in Friedman's book [1]. In the present section, we shall focus our attention on the history of couplers and color developers, since they are directly related to modern color photography.

Fischer [2,3] disclosed color photography in which coupling reactions between 1,4-phenylene diamines (as color developers) and phenols or active methylene compounds (as coupling components or couplers) occurred on the action of exposed silver halide. Representative couplers and developers that were disclosed to give cyan, magenta, and yellow dyes are shown in Fig. 8.1 [2,3]. Moreover, Fischer [4] disclosed the production of color photographs by making three selective, color-sensitive silver halide emulsions with the corresponding couplers and by superimposing the three layers one on another, where colorless intermediate layers were formed between three sensitive layers in order to prevent the diffusion of the couplers. Further, yellow filter layers were placed to reduce the blue-sensitiveness of the green- and red-sensitive layers [4].

As described in the preceding paragraph, Fischer's embodiments seemed to cover most of the essential items of modern color photography, i.e., couplers, color developers, multilayer structure, interlayers, and yellow filter layers. However, there were two major problems: (1) sensitizing dyes migrated from one layer to another and (2) proper image dyes were not formed in the proper lay-

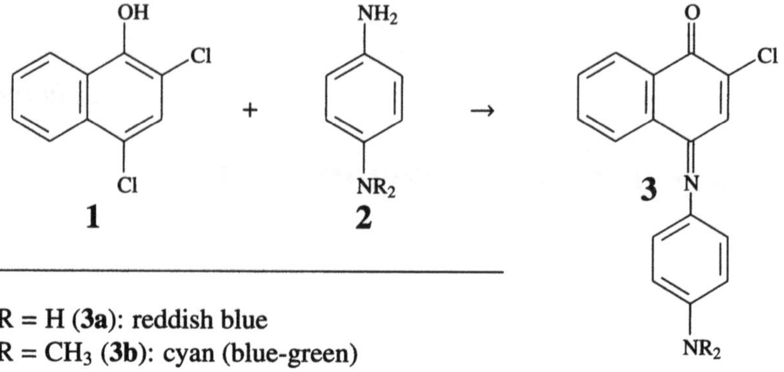

R = H (**3a**): reddish blue
R = CH_3 (**3b**): cyan (blue-green)

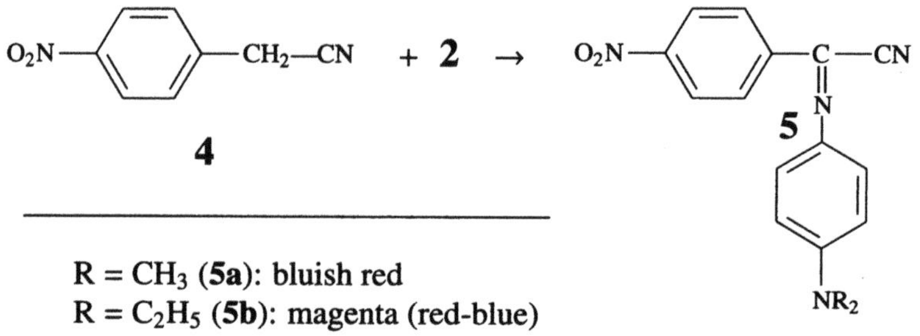

R = CH_3 (**5a**): bluish red
R = C_2H_5 (**5b**): magenta (red-blue)

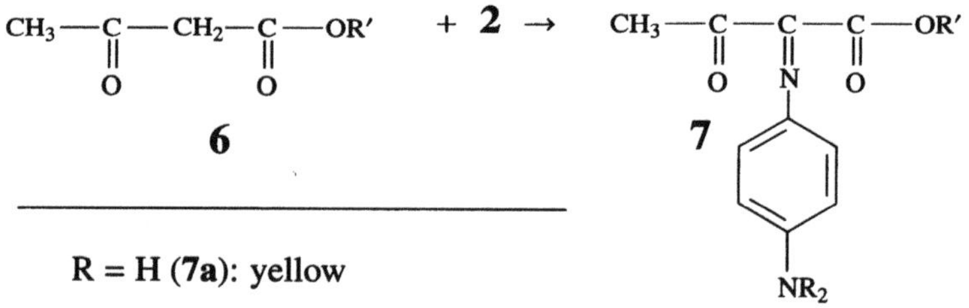

R = H (**7a**): yellow

Figure 8.1. Color development according to Fischer [2,3]. All dye formations require silver halide as an oxidizing agent. Couplers and developers selected from the patent specifications are depicted. The developer for **7a** is uncertain.

ers.[1] The first problem was solved by Brooker, who made sensitizing dyes non-diffusible by selecting appropriate substituents [5]. But the second problem was not solved until 1935 when the method of Mannes and Godowsky was applied to Kodachrome [6–10]. An alternative solution was achieved by introducing non-diffusible couplers in Agfacolor Reversal [11].

[1] The history of overcoming these problems has been described, showing a climate for innovation in Kodak [5].

8.2 Diffusible Couplers

Instead of putting couplers into emulsions (as adopted by Fischer), Mannes and Godowsky fed a coupler and a color developer from a processing liquid into an appropriate layer to form the corresponding image dye at the layer. This process was placed on the market under the name Kodachrome in 1935 [12].

Figure 8.2 shows examples of such couplers, which are here called *diffusible couplers*.[2] Diffusible couplers are soluble in aqueous alkaline media and so diffusible as to penetrate into the appropriate layers of a color reversal film. Then, they are reacted with a coupler by virtue of silver halide and deposit non-diffusible dyes in the respective layers.

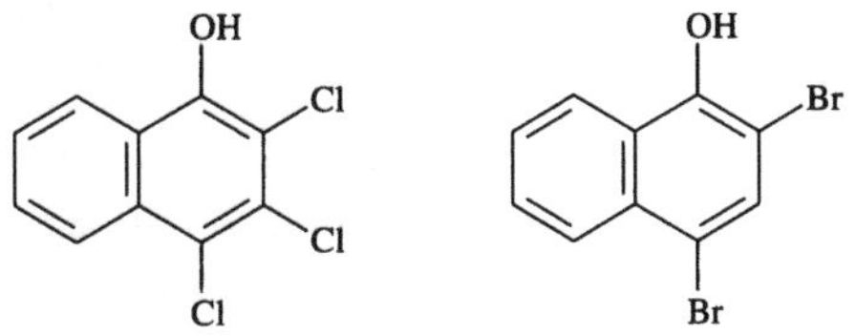

8 **9**

Diffusible cyan couplers

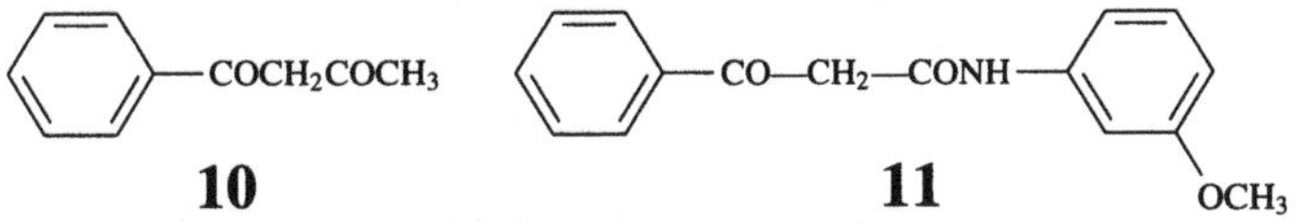

10 **11**

Diffusible yellow couplers

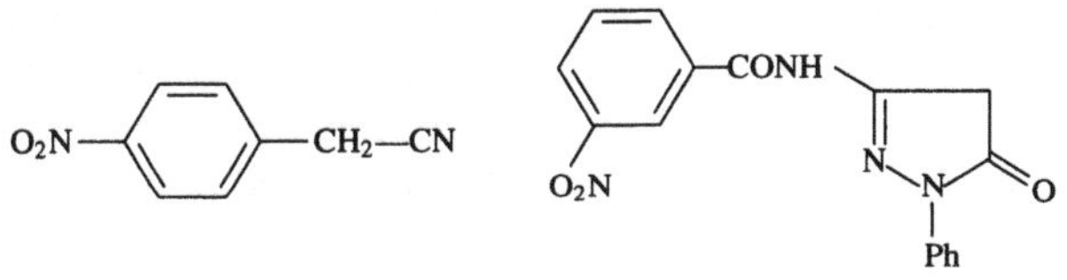

4 **12**

Diffusible magenta couplers

Figure 8.2. Diffusible couplers for color reversal photography. The formulas **8** and **10** are cited from [8]. The formula **4** is cited from [9]. The other formulas are cited from [10].

Since the original process of Kodachrome adopted the selective penetration of bleaching agents [6,7], it was slow and very clumsy. The revised process of Kodachrome adopted the selected exposure shown in Figs. 8.3 and 8.4.

1. Figure 8.3a shows a photographic film that is composed of a blue-sensitized silver halide layer, a yellow-filter layer, a green-sensitized silver halide

[2]They are also called "couplers of outer type". In contrast, non-diffusible couplers are called "couplers of inner type".

layer and a red-sensitized silver halide layer are coated on a transparent polyester support. The film is exposed by lights shown on the top of the figure. The symbol △ represents an exposed silver halide grain.

2. Figure 8.3b indicates the first development (B&W development).

3. Then the film is exposed with red light and color-developed with a cyan coupler (**9**), as shown in Fig. 8.3c. Thus the remaining silver halide grains in the red-sensitive layer activate the cyan color development, where the symbol △ represents a developed silver halide grain with cyan dyes.

4. The film is exposed with blue light and color-developed with a yellow coupler (**11**), as shown Fig. 8.3d. Thus the remaining silver halide grains in the blue-sensitive layer activate the yellow color development, where the symbol △ represents a developed silver halide grain with yellow dyes.

5. The film is then developed with a solution containing a fogging developer which makes the remaining halide developable. Thereby, a magenta coupler (**12**) is reacted with an oxidized color developer (Fig. 8.4e). The symbol △ represents a developed silver halide grain with magenta dyes.

6. Finally the developed silver is converted (bleached) to silver halide and removed (fixed) from the layers into a thiosulfate solution (Fig. 8.4f). As a result, the film reproduces blue (= M + C), green (= Y + C), red (= Y + M), white (no dyes), and black (all dyes).

As found easily, the first B&W development is essential to the process shown in Figs. 8.3 and 8.4, because an original object should be exposed only once. If the original image did not remain as the B&W image (i.e., ▲ shown in Fig. 8.3b), each of the successive color developments (Fig. 8.3c to Fig. 8.4e) would be indiscriminate in the effect on the whole of the layer at issue. This means that the process works reversal only, so that its use was limited only as slide and movie films.[3]

[3]The supply of color photographic materials based on diffusible couplers (Kodachrome or others using outer-type coupler) has been discontinued.

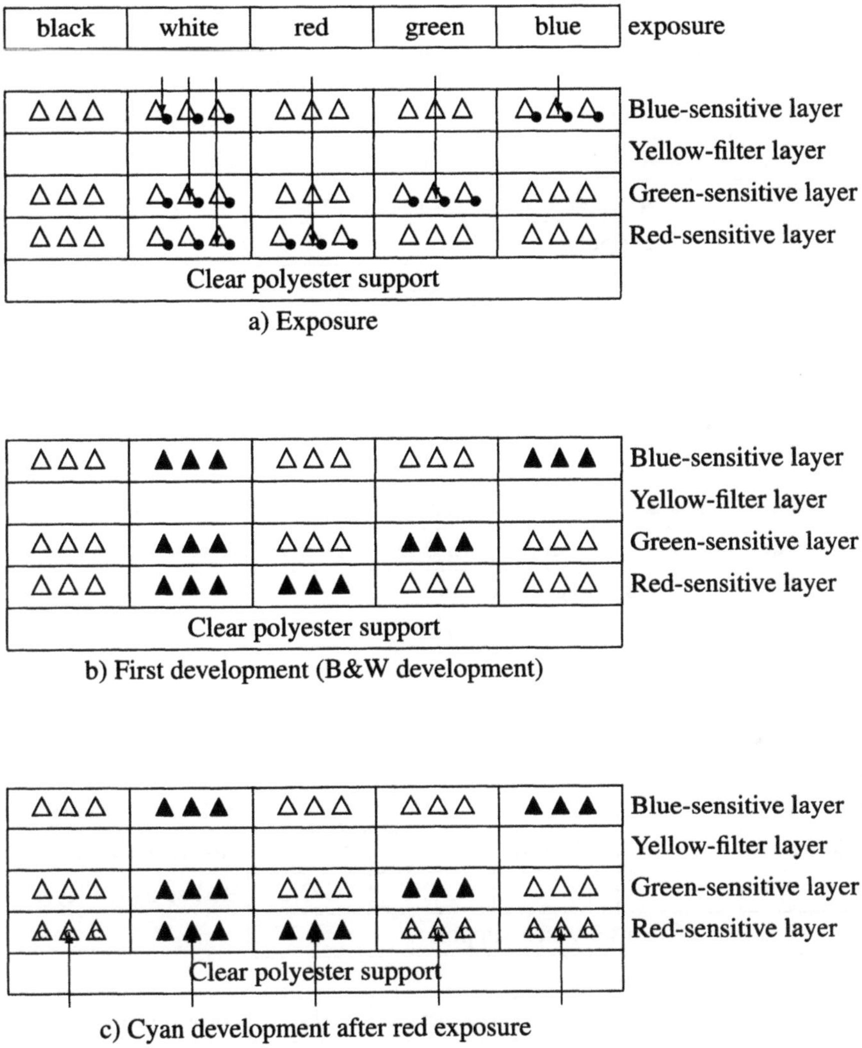

black | white | red | green | blue | exposure

Blue-sensitive layer
Yellow-filter layer
Green-sensitive layer
Red-sensitive layer
Clear polyester support

a) Exposure

Blue-sensitive layer
Yellow-filter layer
Green-sensitive layer
Red-sensitive layer
Clear polyester support

b) First development (B&W development)

Blue-sensitive layer
Yellow-filter layer
Green-sensitive layer
Red-sensitive layer
Clear polyester support

c) Cyan development after red exposure

Figure 8.3. Schematic color development of diffusible couplers for a reversal color film (to be continued to Fig. 8.4). △: Silver halide grain; △•: Exposed silver halide grain with a latent image; ▲: Developed silver halide grains without dye forming; △: Developed silver halide with cyan dyes; △: Developed silver halide with yellow dyes; △: Developed silver halide with magenta dyes.

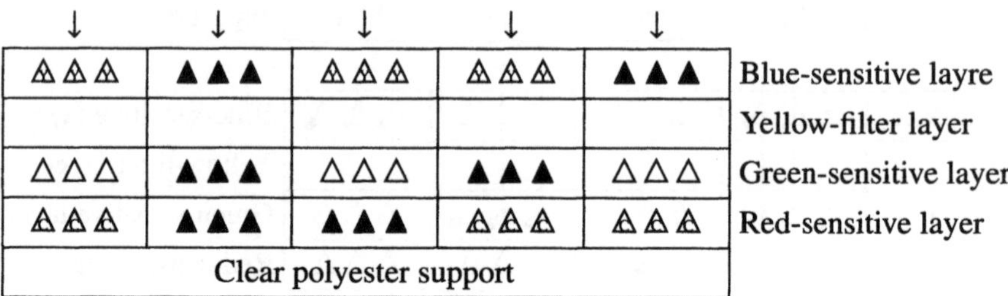

d) Yellow development after blue exposure

Blue-sensitive layer
Yellow-filter layer
Green-sensitive layer
Red-sensitive layer
Clear polyester support

e) Magenta development

Y Y Y	! !! !! !	Y Y Y	Y Y Y		Blue-sensitive layer
					Yellow-filter layer
M M M		M M M		M M M	Green-sensitive layer
C C C			C C C	C C C	Red-sensitive layer
Clear polyester support					
black	white	red	green	blue	color reproduction

f) Bleach and fixation

Figure 8.4. Schematic color development of diffusible couplers for a reversal color film (continued from Fig. 8.3). : Developed silver halide with yellow dyes; : Developed silver halide with magenta dyes; : Developed silver halide with cyan dyes; Y: Yellow dye; M: Magenta dye; C: Cyan dye.

8.3 Non-diffusible Couplers

8.3.1 Hydrophilic Couplers

In the Fischer process, each of the couplers was incorporated in an appropriate layer so that during development three different dyes were expected to be produced simultaneously in the respective layers. In other words, the couplers should not wander from one layer to another.

A natural way to make a coupler non-diffusible is to increase the molecular weight of the coupler by introducing a ballast group to its structure. However, such an organic compound as having a high molecular weight is difficult to be dispersed in aqueous media of gelatino-silver halide emulsions. To overcome this difficulty, a soap-like (or detergent-like) function was incorporated in a coupler molecule as a remarkable practical innovation. Thus, Fischer's original idea was perfected and marketed as Agfacolor in 1936, where non-diffusible hydrophilic couplers were incorporated in the three emulsion layers.

Hydrophilic couplers (**13–16**) incorporated in Agfacolor Reversal film are shown in Fig. 8.5 [11]. They are non-diffusible because of the presence of such a ballast group as having a long aliphatic chain ($C_{17}H_{35}$). Their hydrophilicity is based on the presence of an SO_3H or COOH group.

Because they are amphiphilic compounds that have both a lipophilic (hydrophobic) site and a hydrophilic site, they can be dissolved in a gelatin solution as the result of micelle formation. Thereby, they can be coated on a film so as to be incorporated in the respective photo-sensitive layers, as shown in Fig. 8.6a. After being coated, they are non-diffusible during the processes shown in Fig. 8.6.

1. Figure 8.6a shows a film coated with three photo-sensitive layers, i.e., a red-sensitive emulsion layer with a cyan coupler, a green-sensitive emulsion layer with a magenta coupler, and a blue-sensitive emulsion layer with a yellow coupler. The film is exposed by lights shown on the top of the figure. The symbols Δ and Δ• represent an unexposed silver halide grain and an exposed one with a latent image. The symbols Ⓨ, Ⓜ, and Ⓒ express a yellow, a magenta and a cyan coupler, respectively.

2. Figure 8.6b shows the first development (B&W development), in which no dyes are generated.

3. After exposure with white light (or reversal action with a fogging agent), the film is processed by color development as shown in Fig. 8.6c. Image dyes are generated at the sites where no B&W development has occurred.

4. Finally the developed silver is converted (bleached) to silver halide and removed (fixed) from the layers in a thiosulfate solution (Fig. 8.6d). As a result, the film reproduces blue (= M + C), green (= Y + C), red (= Y + M), white (no dyes), and black (all dyes). Note that each layer contains unreacted coupler (Ⓨ, Ⓜ, or Ⓒ), which is colorless and harmless to color reproduction.

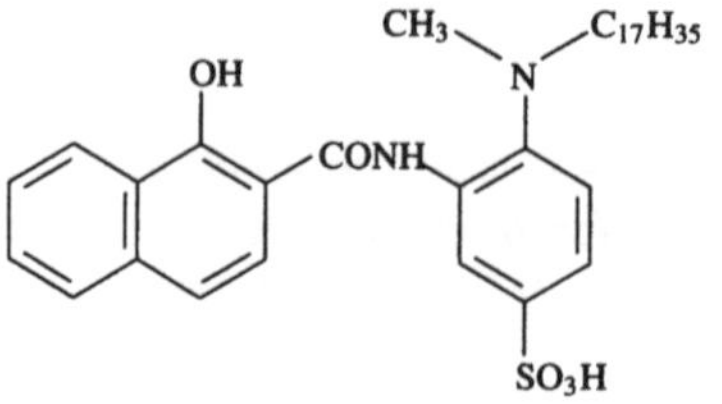

13 (F654) 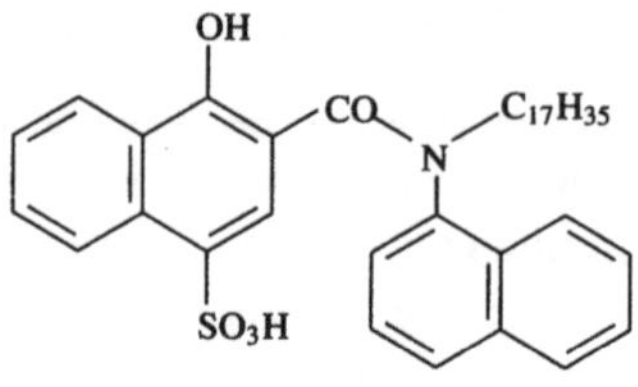 **14** (Koe308)

Non-diffusible cyan couplers

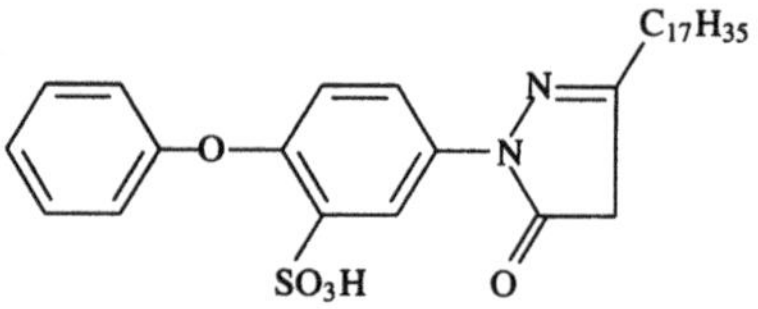

15 (Z169)

Non-diffusible magenta coupler

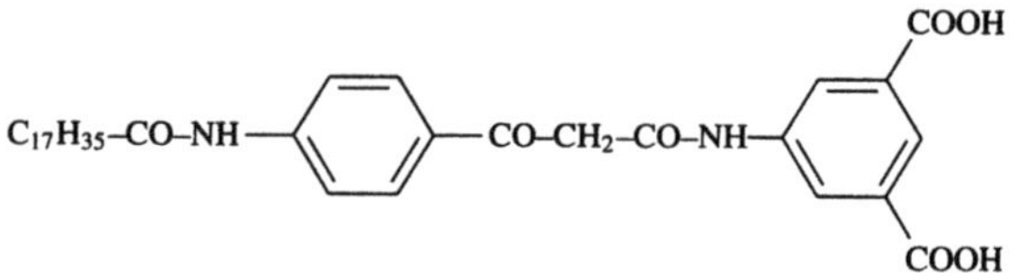

16 (F535)

Non-diffusible yellow coupler

Figure 8.5. Hydrophilic couplers incorporated in Agfacolor Reversal film. The notations, F654, Koe308, Z169, and F 535, are cited from the report [11].

8.3.2 Oil- and Polymer-Protected Couplers

Oil-Protected Dispersion

Non-diffusible couplers of another type are oil-protected couplers, which were placed on the market under the name Kodacolor in 1941 [5,12]. They had a lipophilic ballast group but no hydrophilic group so that they required suitable oils for dissolving and techniques for producing stable dispersions. However, solving these requirements was somewhat easier than overcoming the synthetic difficulties of hydrophilic couplers.[4] Hence, oil-protected couplers have been

[4]There were problems in the manufacture of hydrophilic couplers. In particular, their purification was difficult because of their detergent-like nature [5]. See Figs. 8.9 to 8.11. It should be noted that the amphiphilic function of hydrophilic couplers has been separated into a lipophilic function and a

predominantly used in the modern color photography, as will be described in the remaining parts of this book.

Figure 8.7 shows examples of oil-protected couplers disclosed later by Eastman Kodak in more elaborate embodiments: a cyan coupler **17** [13], a magenta coupler **18** [14], and a yellow coupler **19** [15].

These couplers can be dispersed as fine oil droplets by using about equal amounts of oils such as tricresyl phosphate or dibutyl phthalate, where they are kept liquid so that no crystallization occurs. For example, the magenta coupler **18** dissolved in ethyl acetate and dibutyl phthalate can be dispersed in a 10% gelatin solution by using Alkanol B (sodium triisopropylnaphthalenesulfonate) [16]. The resulting dispersions can be incorporated in a color film (e.g., a reversal color film shown in Fig. 8.6).

It should be emphasized that photographic materials based on non-diffusible couplers (whether they are hydrophilic or oil-protected ones) work well, even if the first B&W development is omitted, which is necessary to the reversal process. This has given a foundation to the negative-positive process which has been adopted widely for amateur use.

Polymer-Protected Dispersion

Hydrophobic compounds such as non-diffusible couplers of oil-protected type can be incorporated in emulsion layers by using loadable latexes in place of oils having a high boiling point. A practical method for loading (distributing or dissolving) various hydrophobic compounds (including oil-protected couplers for conventional color photography and dye-releasing compounds for instant color photography) within loadable polymer particles has been disclosed in an extensive manner [17]. For example, a solution of a yellow coupler (**20**) in acetone is blended with a loadable polymer latex (**21**) to give a stable coupler-loaded latex composition after evaporating acetone. The use of homopolymers has been reported to improve the dark stability of image dyes derived from cyan couplers of phenol type [18]. This topic will be discussed later in detail (Subsection 10.3.2 of Chapter 10).

---

hydrophilic one. The former has been succeeded by oil-protected couplers, while the latter has been shared by a surfactant used for dispersing the oil-protected couplers.

black	white	red	green	blue	Exposure
△Ⓨ△Ⓨ	△•Ⓨ△•Ⓨ	△Ⓨ△Ⓨ	△Ⓨ△Ⓨ	△•Ⓨ△•Ⓨ	Blue-sens.
					Yellow-filter
△Ⓜ△Ⓜ	△•Ⓜ△•Ⓜ	△Ⓜ△Ⓜ	△•Ⓜ△•Ⓜ	△Ⓜ△Ⓜ	Green-sens.
△Ⓒ△Ⓒ	△•Ⓒ△•Ⓒ	△•Ⓒ△•Ⓒ	△Ⓒ△Ⓒ	△Ⓒ△Ⓒ	Red-sens.
Clear polyester support					

a) Exposure

△Ⓨ△Ⓨ	▲Ⓨ▲Ⓨ	△Ⓨ△Ⓨ	△Ⓨ△Ⓨ	▲Ⓨ▲Ⓨ	Blue-sens.
					Yellow-filter
△Ⓜ△Ⓜ	▲Ⓜ▲Ⓜ	△Ⓜ△Ⓜ	▲Ⓜ▲Ⓜ	△Ⓜ△Ⓜ	Green-sens.
△Ⓒ△Ⓒ	▲Ⓒ▲Ⓒ	▲Ⓒ▲Ⓒ	△Ⓒ△Ⓒ	△Ⓒ△Ⓒ	Red-sens.
Clear polyester support					

b) First development (B&W development)

▲Y ▲Y	▲Ⓨ▲Ⓨ	▲Y ▲Y	▲Y ▲Y	▲Ⓨ▲Ⓨ	Blue-sens.
					Yellow-filter
▲M ▲M	▲Ⓜ▲Ⓜ	▲M ▲M	▲Ⓜ▲Ⓜ	▲M ▲M	Green-sens.
▲C ▲C	▲Ⓒ▲Ⓒ	▲Ⓒ▲Ⓒ	▲C ▲C	▲C ▲C	Red-sens.
Clear polyester support					

c) Color development after exposure (or reversal)

Y Y	Ⓨ Ⓨ	Y Y	Y Y	Ⓨ Ⓨ	Blue-sens.
					Yellow-filter
M M	Ⓜ Ⓜ	M M	Ⓜ Ⓜ	M M	Green-sens.
C C	Ⓒ Ⓒ	Ⓒ Ⓒ	C C	C C	Red-sens.
Clear polyester support					

d) Bleach and fixation

Figure 8.6. Schematic color development of non-diffusible couplers for reversal color film. △: Silver halide grain; △•: Exposed silver halide grain with a latent image; ▲: Developed silver halide grains without dye forming; Ⓨ: Yellow coupler; Ⓜ: Magenta coupler; Ⓒ: Cyan coupler; Y: Yellow dye; M: Magenta dye; C: Cyan dye.

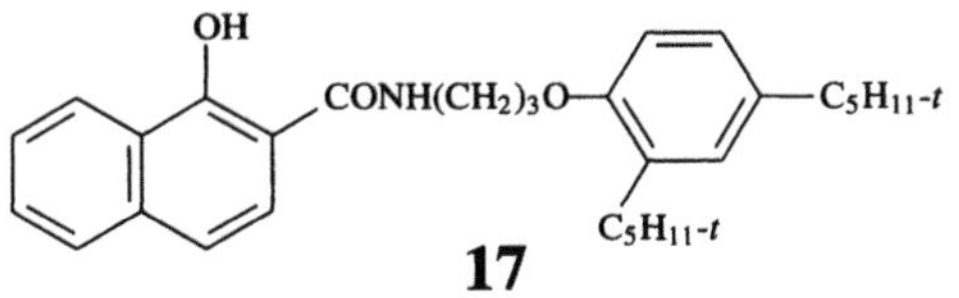

17
Oil-protected cyan coupler

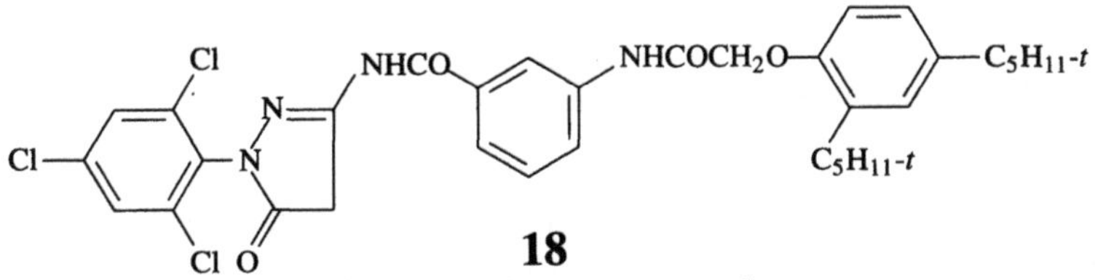

18
Oil-protected magenta coupler

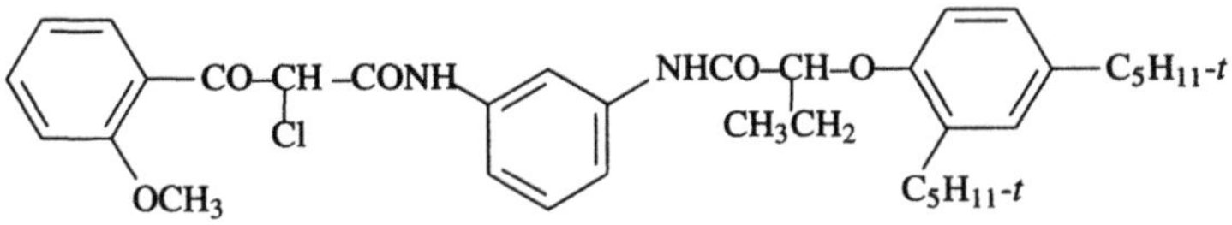

19
Oil-protected yellow coupler

Figure 8.7. Oil-protected couplers. These couplers have no hydrophilic groups. The symbol C_5H_{11}-*t* represents a 1,1-dimethyl-1-propyl group (*t*-amyl group).

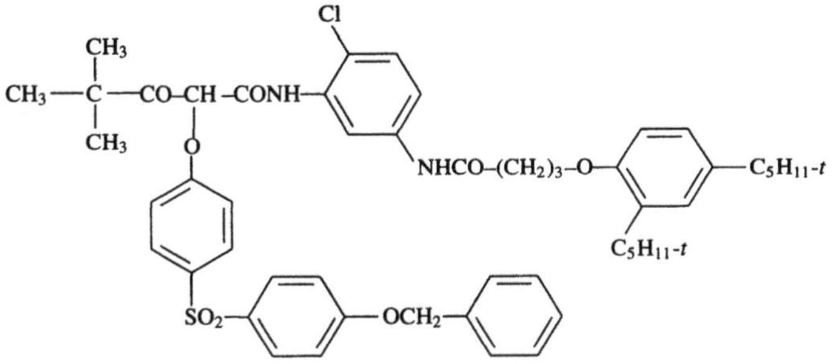

20

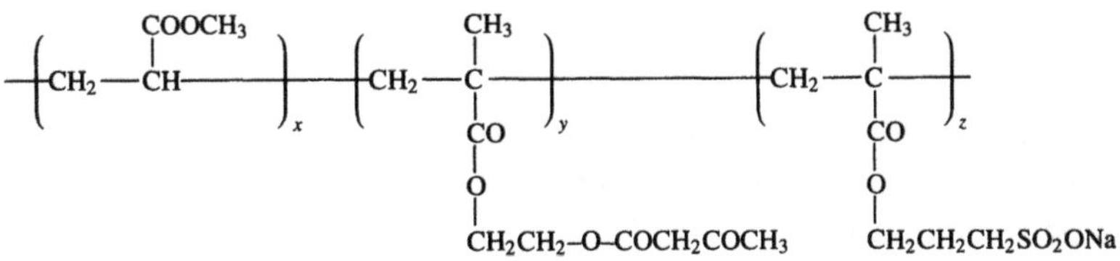

21

Figure 8.8. Example for a two-equivalent yellow coupler and a loadable polymeric latex.

8.4 Syntheses of Non-diffusible Couplers

As described in the preceding section, the relative merits of hydrophilic vs. oil-protected couplers were decided by their synthetic easiness. We here compare the syntheses of hydrophilic couplers with those of oil-protected ones in order to show an example in which the organic syntheses decided the fate of methodologies.

8.4.1 Syntheses of Hydrophilic Couplers

Figure 8.9 shows the synthetic pathway for preparing the cyan coupler **13** (F654) [11]. The ballast group ($C_{17}H_{35}$) and the hydrophilic group (SO_3Na) are introduced in the first step in which the sodium salt of 4-chloro-3-nitrobenzenesulfonic acid (**22**) and *N*-methyl-*N*-stearylamine are condensed. The presence of two electron-withdrawing groups (NO_2 and SO_3H) at the ortho- and para-positions in the substrate (**22**) enhances the activity of the 4-chloro substituent, so that the nucleophilic attack of such an amine occurs. The nitro group of the condensate (**23**) is reduced into an amino group to give **24**. The condensation of the amine (**24**) and 1-hydroxynaphthalene-2-carbonyl chloride (**25**) is conducted in pyridine. The product **13** (F654) is purified by washing the crude product with methanol, where the losses occurring in each of the purification are relatively large.

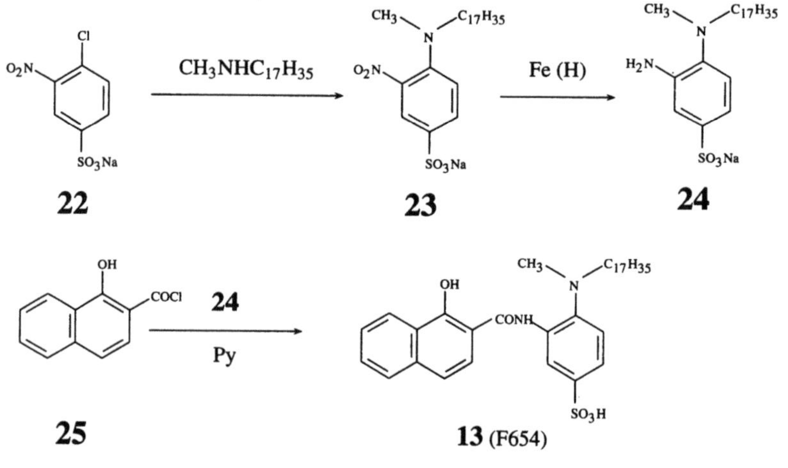

Figure 8.9. Synthetic pathway of the cyan coupler **13** (F654).

The synthetic pathway for preparing the magenta coupler **15** (Z169) is shown in Fig. 8.10 [11]. The nucleophilic attack of a phenolate anion onto the chlorine atom in the sodium salt of 2-chloro-5-nitrobenzenesulfonic acid (**27**), which is isomeric to **22**, provides a diphenyl ether intermediate (**28**). The nitro group of **28** is chemically reduced into an amino group in the presence of iron powder. The amino group is in turn converted into a hydrazine group via a diazonium salt (**28** → **29** → **30**). The key step of this synthetic pathway is the condensation of the hydrazine intermediate **30** and β-ketoester **31**, where a ring closure occurs to give the pyrazolone magenta coupler (**15**). The product is purified by salting out. The hydrophilic group (SO_3H) comes from the hydrazine intermediate (**30**), while the ballast group is introduced as a portion of the β-keto-ester (**31**).

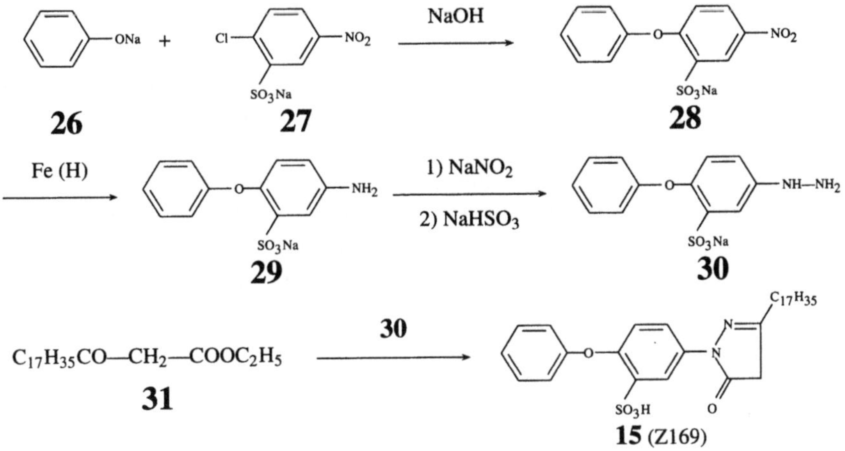

Figure 8.10. Synthetic pathway of the magenta coupler **15** (Z169).

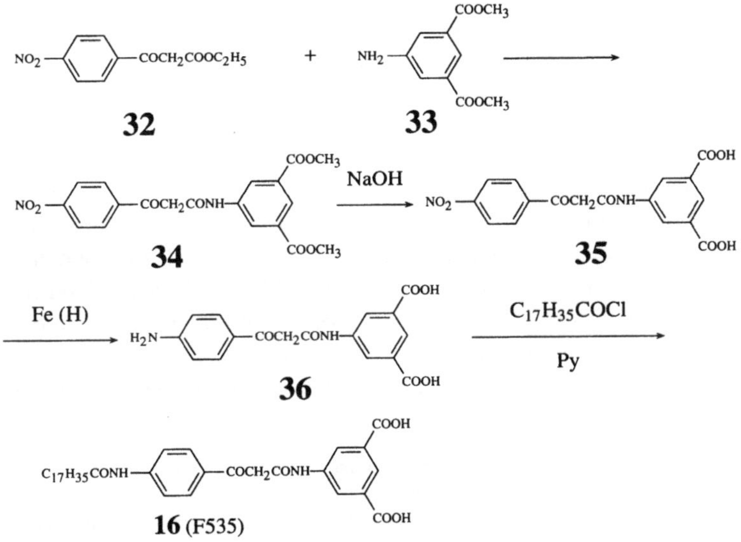

Figure 8.11. Synthetic pathway for preparing the yellow coupler **16** (F535).

The synthetic pathway for preparing the yellow coupler **16** (F535) is shown in Fig. 8.11 [11]. The ballast group coming from stearyl chloride is introduced in the last step. The crude product of **16** is purified by stirring it at room temperature with acetone, filtering it off, and washing it with acetone [11], but its melting point has not been described.

8.4.2 Syntheses of Oil-Protected Couplers

The cyan coupler **17**, which belongs to a category of oil-protected couplers, is synthesized according to the pathway shown in Fig. 8.12 [13]. The ballast group is prepared by the Michael addition of acrylonitrile, where 2,4-di-*t*-amylphenol (**37**),[5] which is commercially available, is used as a nucleophile. The cyano group of the adduct (**38**) is catalytically reduced by hydrogen in the presence of Raney's nickel to produce 3-(2,4-di-*t*-amylphenoxy)propylamine (**39**).[6] This amine intermediate[7] is condensed with phenyl 1-hydroxy-2-naphthoate (**40**),[8] giving the cyan coupler (**17**) as crystals. The coupler has a melting point at 91–93℃.

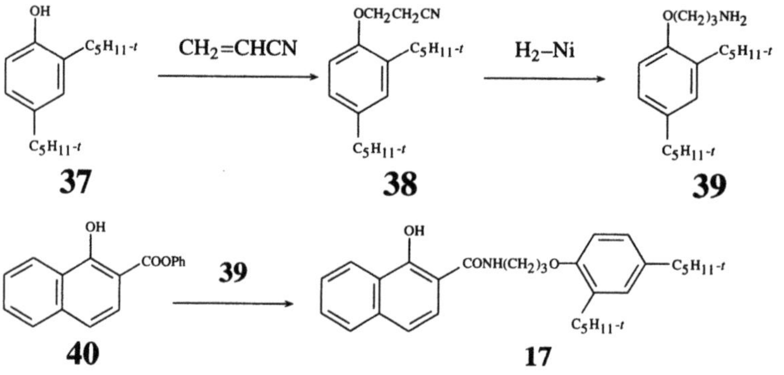

Figure 8.12. Synthetic pathway for preparing the cyan coupler **17**.

The synthetic pathway for preparing the magenta coupler **18** is shown in Fig. 8.13 [14]. The hydrazine intermediate **43** is prepared from 2,4,6-trichloroaniline (**41**) via the corresponding diazonium salt (**42**). The hydrazine (**43**) is condensed with ethyl *β*-ethoxy-*β*-iminopropionate (**44**) to give **45**, which is cyclized into a pyrazolone intermediate (**46**). The 3-amino group of **46** is converted into an amido group by the reaction with 3-nitrobenzoyl chloride. The nitro group coming from the 3-nitrobenzoyl chloride is chemically reduced by iron powder to give

[5] A 1,1-dimethyl-1-propyl group used in color photography is frequently referred to as a *t*-amyl group and is represented by the symbol C_5H_{11}-*t* in structural formulas.

[6] Compare this catalytic reduction of a nitro group (**38** → **39**) with the chemical reduction described in Fig. 8.10 (**28** → **29**).

[7] One important methodology in coupler syntheses is the introduction of ballast groups through an amido linkage. Such a ballast-involving amine as **39** is a key intermediate for preparing an amido linkage.

[8] Compare this intermediate having a phenyl ester group with the acid chloride (**25**) described in Fig. 8.9. They are used as synthetic equivalents.

an intermediate with an amino group (**47**). Then, the amino group is reacted with α-(2,4-di-*t*-amylphenoxy)acetyl chloride,[9] which has been prepared by the reaction between 2,4-di-*t*-amylphenol and chloroacetic acid and the subsequent chlorination. Thereby, the pyrazolone magenta coupler (**18**) is obtained, which has a melting point of 133–135°C.

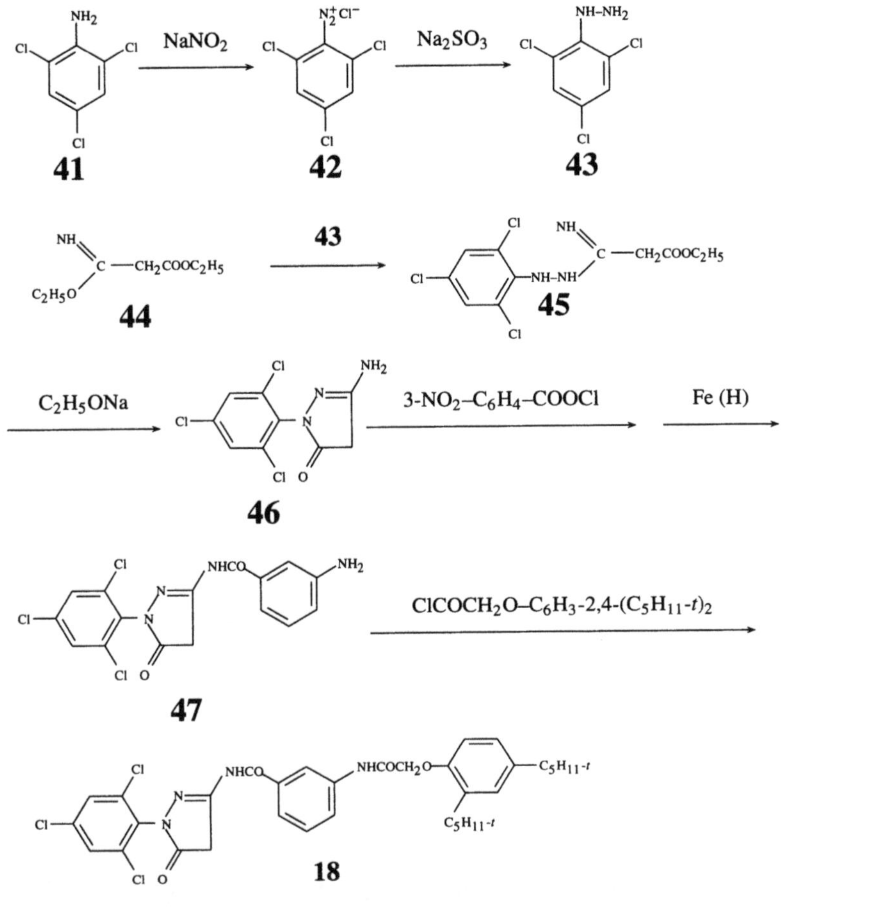

Figure 8.13. Synthetic pathway for preparing the magenta coupler **18**.

The synthetic pathway for preparing the yellow coupler **19** is shown in Fig. 8.14 [15]. The ballast group containing a 2,4-di-*t*-amylphenoxy group is introduced by virtue of the reaction of 3-nitroaniline (**48**) with a carbonyl chloride, namely, α-(2,4-di-*t*-amylphenoxy)butyryl chloride (**49**). The nitro group of the resulting intermediate (**50**) is catalytically reduced to give an amino interme-

[9] Compare this intermediate with the compound **39**. Thus, amido linkages with a ballast group can be introduced in alternative umpolung ways.

diate with a ballast group (**51**). This intermediate is condensed with ethyl (2-methoxybenzoyl)acetate so as to give the four-equivalent coupler (**53**), which is obtained as an amorphous solid melting above 60℃ but not obtained in a crystalline form. The chlorination of **53** by sulfuryl chloride produces the final product (**19**), which is also an amorphous solid. When the intervening 1,3-phenylene group of **19** was replaced by a 1,4-phenylene group, the resulting two-equivalent coupler has a melting point at 88–90℃ after recrystallization from methanol.

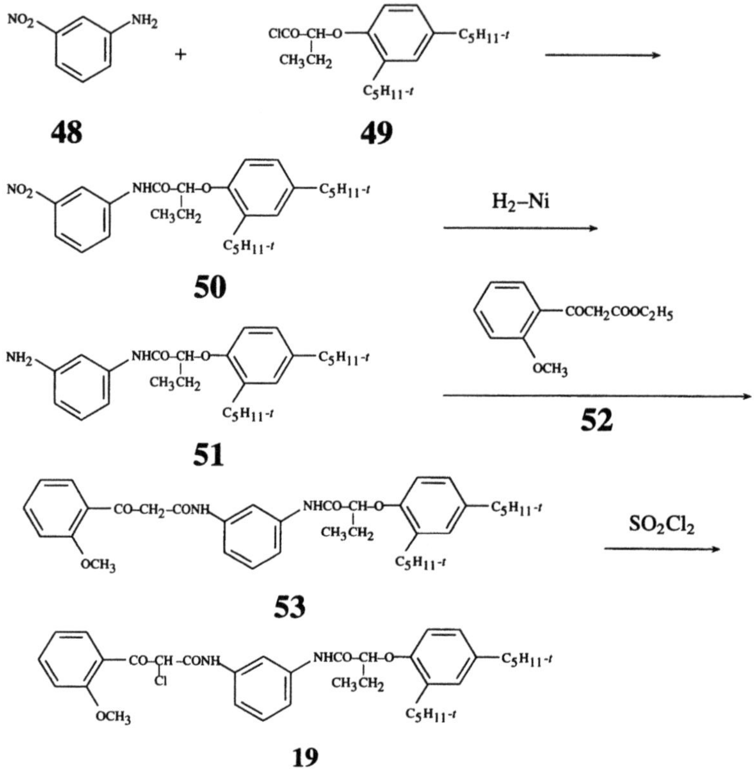

Figure 8.14. Synthetic pathway of the yellow coupler **19**.

References

[1] Friedman JS (1968) History of Color Photography. 2nd ed, Focal, London, Chapter 23

[2] Fischer R (1912) Ger Patent 253 335

[3] Fischer R, Siegrist (1914) Photogr Korres. 51:18

[4] Fischer R (1912) Eng Patent 15 055; Fischer R (1913) Brit J Photo 60:595, 712

[5] Weissberger A (1980) Chemtech. 340

[6] Mannes LD, Godowski Jr L (1924) US Patent 1 516 824

[7] Mannes LD, Godowski Jr L (1935) US Patent 2 019 718

[8] Mannes LD, Godowski Jr L (1936) US Patent 2 059 884

[9] Mannes LD, Godowski Jr L (1936) US Patent 2 059 887

[10] Kikuchi S (1968) Shashin Kagaku (Chemistry of Photography). New ed, Kyoritsu Zensho, Kyoritsu, Tokyo, Chapter 16

[11] Gluck B (1947) The Manufacture of Agfacolor Material. Fiat Final Report. No. 943

[12] Mees CEK (1942) J Franklin Inst. 233:41; Mees CEK (1942) Am Photogr. 36:8

[13] Weissberger A, Salminen IF, Vittum PW (1949) US Patent 2 474 293

[14] Loria A, Weissberger A, Wittum PW (1952) US Patent 2 600 788

[15] McCrossen FC, Wittum PW, Weissberger A (1955) US Patent 2 728 658

[16] Fierke SS, Chechak JJ (1957) US Patent 2 801 171

[17] Chen TJ (1980) US Patent 4 199 393; US Patent 4 203 716

[18] Takahashi O, Ogawa T (1991) Fuji Film Res & Dev. 36:7

Chapter 9

Color Developers

9.1 Syntheses and Basic Properties

Color developers used in modern color photography are the derivatives of 1,4-phenylene diamine (4-amino-*N*, *N*-dialkylanilines). A variety of such color developers have been synthesized by several methods, from which two representative methods are illustrated in Fig. 9.1 [1].

According to Weissberger's retrospection on the situation in the 1940s [2], there was a hitch as far as amateurs were concerned with color photography, because some individuals developed a skin sensitivity to color developers such as 4-amino-*N*, *N*-dimethylaniline (CD-1) and its 3-methyl derivative (CD-2). The acidic sulfonamido group was found not only to increase the solubility in the alkaline-processing solution but also to reduce allergenic activity. Thereby, the developer named CD-3, i.e., 4-amino-*N*-ethyl-*N*-(2-methanesulfonylethyl)-3-methylaniline, and related color developers were invented, as shown in Table 9.1 [1]. The low allergenic activity was ascribed to the reduction of the solubility in the lipids of the fat glands, the sweat glands, and the hair follicles of the skin, since these were regarded as the portals through which color developers entered into the bloodstream to cause the skin allergy [2].

Table 9.1 shows the electrochemical, photographic, and allergenic properties of several color developers that are selected from the data of about 60 compounds reported in Ref. [1].

The elementary process of photographic development is the reduction of silver ion to silver, where the electron is supplied by a color developer (**3**), as shown in Fig. 9.2. The first step is the one-electron oxidation of the color developer

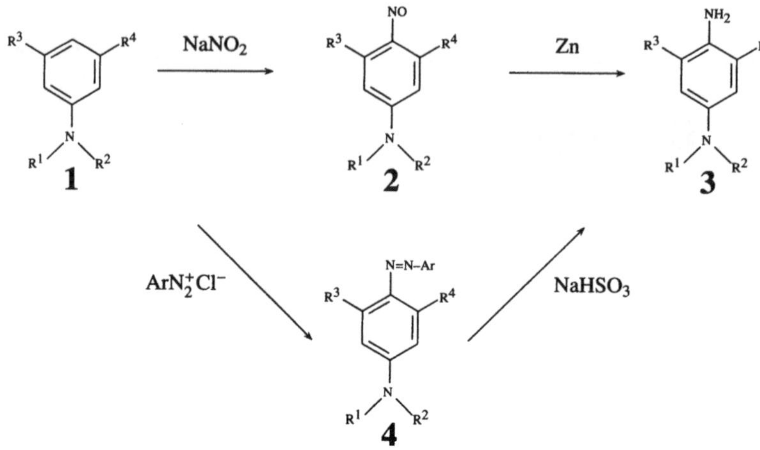

Figure 9.1. Syntheses of color developers. The symbol Ar represents a *p*-sulfophenyl group, which comes from the diazonium salt prepared by the diazotization of sulfanilic acid. The use of other diazonium salts has also been reported [1].

Table 9.1. Electrochemical, Photographic, and Allergenic Properties of Selected Color Developers [1]

Substituents				$E_{1/2}$	Develop-	Allergenic	Coupling
R^1	R^2	R^3	R^4	vs. NHE pH 11.0 (V)	ment rate (min^{-1})	activity[a]	efficiency
CH_3	CH_3	H	H	−235	0.30		++
[b]C_2H_5	C_2H_5	H	H	−222	0.44	M to H	++
n-C_3H_7	*n*-C_3H_7	H	H	−204	0.53		++
C_2H_5	C_2H_4NH-SO_2CH_3	H	H	−219	0.18	L	++
C_2H_5	C_2H_4OH	H	H	−206	0.42		++
[c]C_2H_5	C_2H_5	CH_3	H	−190	0.80	M to H	++
[d]C_2H_5	C_2H_4NH-SO_2CH_3	CH_3	H	−190	0.38	L to M	++
[e]C_2H_5	C_2H_4OH	CH_3	H	−188	0.67		++

[a]L: low; M: moderate; and H: high.
[b]Kodak code: CD-1; [c]Kodak code: CD-2; [d]Kodak code: CD-3; [e]Kodak code: CD-4.

(R: **3**) to produce the corresponding semiquinone ion (**5**). Note that the electron represented by the symbol −e is accepted by the silver ion or other oxidizing agents. The semiquinone ion (**5**) is dissociated into the semiquinone (S: **6**), which is involved in equilibria such as disproportionation (2S $\rightleftharpoons$ R + Q) and dimerization (2S $\rightleftharpoons$ Dimer). The semiquinone undergoes a further one-electron oxidation to produce a quinonediimine (Q: **7**).

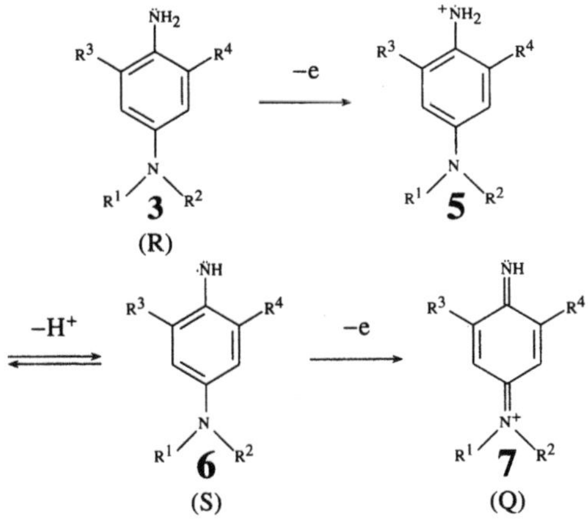

Figure 9.2. Oxidation of color developers.

The tendency of the color developer to release electrons can be estimated by its polarographic half-wave potential, $E_{1/2}$. Thus, the more positive is the oxidation half-wave potential, the greater the tendency of the color developer to release electrons. The longer aliphatic groups attached to the nitrogen atom exhibit the half-wave potentials shifted to more positive: *N,N*-dimethyl (−235 V), *N,N*-diethyl (−222 V), and *N,N*-dipropyl (−204 V), as found in Table 9.1.

The development rates ($1/t'$) of the developers, which were determined as the reciprocals of the time in minutes required to attain an optical density of 0.2 above fog ($t' = t - t_0$, where t_0 is the induction period) under coated conditions on photographic films, are collected also in Table 9.1.

The coupling efficiencies are based on the yields of the azomethine dyes formed with 2-cyanoacetylcoumarone: 0, indicating no coupling; +, slight coupling; and ++, good coupling [1], although Table 9.1 collects developers of good coupling only.

9.2 Mechanism of Reactions

9.2.1 Dye Formation

Direct analyses of photographic reactions are difficult, because they are solid-liquid reactions among three components: silver halides, color developers, and couplers. In most cases, the reaction between a quinonediimine (Q) and a coupler (C) in liquid media has been examined after the former is generated from a color developer (R) by using potassium hexacyanoferrate(III) $K_3[Fe(CN)_6]$ in place of silver halide (Fig. 9.3) [3]. Since the dye formation of the coupler shown in Fig.

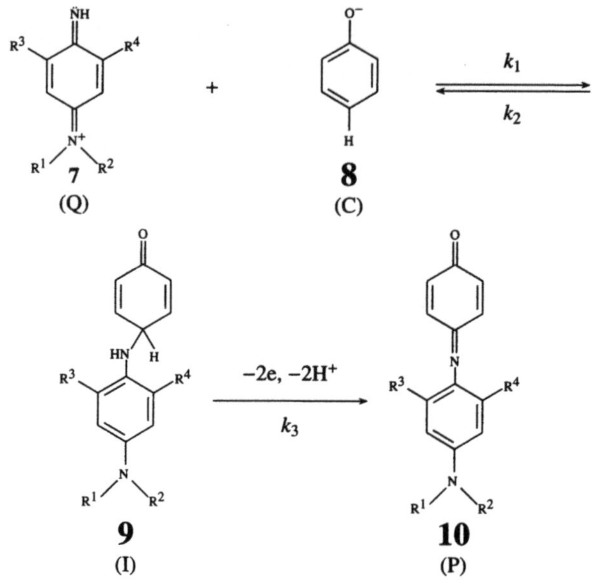

Figure 9.3. Coupling of an oxidized color developer with a four-equivalent coupler. Although phenolate ion is illustrated as a coupler anion (C), it should be replaced by appropriate couplers for discussing general cases.

9.3 requires the loss of four equivalents of electrons (two equivalents for Fig. 9.2 and two for Fig. 9.3), it is called a *four-equivalent* coupler. The intermediate (I, **9**) is usually called a *leuco* form, which is regarded as a reduced dye. Such a leuco form is oxidized into the corresponding dye in the second step by the excess of an oxidizing agent or by virtue of the disproportionation with another quinonediimine. If the second step (for k_3) is fast, the formation of the dye (P) can be regarded as the second-order reaction with a bimolecular rate constant k_C as follows:

$$\frac{d[P]}{dt} = -\frac{d[Q]}{dt} = k_C[Q][C] \tag{9.1}$$

When 6-nitro-1-naphthol is used as a coupler (C) which reacts with the quinonediimine (Q) derived from 4-amino-*N*, *N*-diethyl-3-methylaniline (CD-2), eq. 9.1 is satisfied so as to give $k_C = 8.1 \times 10^4$ L·mol^{-1}·sec^{-1} at pH 12.0 [3].

When couplers have an electro-negative releasing group at their coupling position, they are called *two-equivalent* couplers. Note that the coupling reaction for this type of dye formation requires no electrons, as shown in Fig. 9.4, while the formation of the quinonediimine (**7**) requires two equivalents of electrons (Fig. 9.2).

The mechanism of the reaction of **7** with a two-equivalent coupler (**11**) has been investigated within the framework of the scheme shown in Fig. 9.4 [4]. If the second step (for k_3) is fast, eq. 9.1 is also satisfied. When 2,6-dihydroxymethyl-4-methoxyphenol (X = OCH_3) is used as a coupler (C-X) which reacts with the

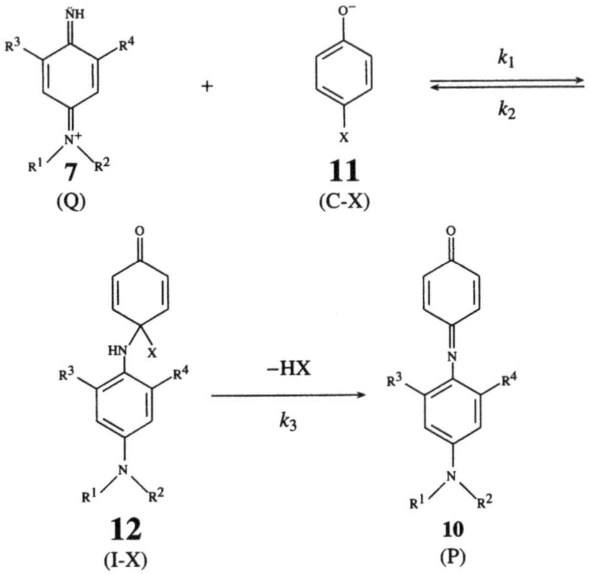

Figure 9.4. Coupling of an oxidized color developer with a two-equivalent coupler. Although a 4-X substituted phenolate ion is illustrated as a coupler anion (C), it should be replaced by appropriate couplers for discussing general cases.

quinonediimine (Q) derived from 4-amino-*N*, *N*-diethyl-3-methylaniline (CD-2), the second-order rate constant k_C is obtained to be 3.0×10^4 L·mol^{-1}·sec^{-1} at pH 10.13 [4]. For the quinonediimine derived from 4-amino-*N*, *N*-diethylaniline (CD-1), the value of k_C is determined to be 0.79×10^6 L·mol^{-1}·sec^{-1} at pH 10.13.

When the naphthol coupler (**13**, X = Cl) is used, the coupling reactions obey the second-order mechanism satisfying eq. 9.1 [4]. The quinonediimine (Q) derived from CD-2 reacts with the coupler, giving the second-order rate constant k_C = 6.3×10^3 L·mol^{-1}·sec^{-1} at pH 9.87. For the quinonediimine derived from CD-1, the value of k_C is determined to be 2.8×10^5 L·mol^{-1}·sec^{-1} at pH 8.17.

On the other hand, the naphthol coupler (**13**, X = OCH_3) obeys the first-order mechanism in which the rate-determining step is the elimination of a methoxide anion (for k_3). Thus, the k_3 values are determined to be 4.3 sec^{-1} (at pH 10.12) for the quinonediimine (Q) derived from CD-2 and 3.2 sec^{-1} (at pH 10.05) for the quinonediimine (Q) derived from CD-1.

As for an arylthio-leaving pyrazolone magenta coupler (e.g **14**), the second elimination step is reported to be very slow compared with the first coupling step [5]. The arylthio anion (ArS^-) eliminated from **14** is rapidly oxidized with a quinonediimine into the corresponding disulfide (ArS–SAr). In addition to the usual elimination mechanism of the dye formation, there is another path of the dye formation in which the leuco dye (I-X) is oxidized with a quinonediimine to produce the dye (P) and $Ar{-}S^+$, as shown in Fig. 9.6. The two-equivalent coupler (**14**) dispersed with dibutyl phthalate in a gelatin solution is reacted with

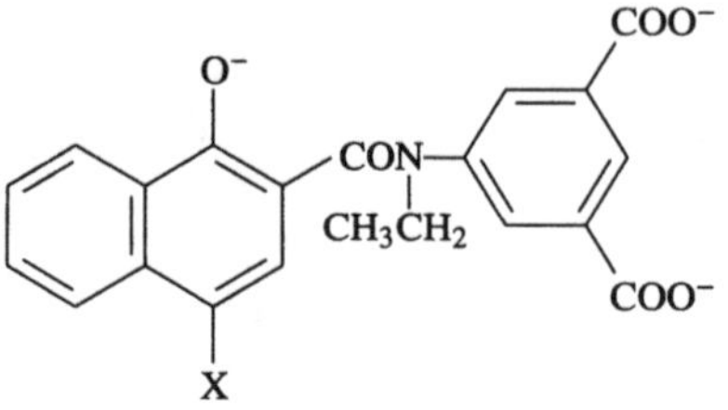

13

Figure 9.5. An alkaline-soluble cyan coupler for examining reaction mechanisms.

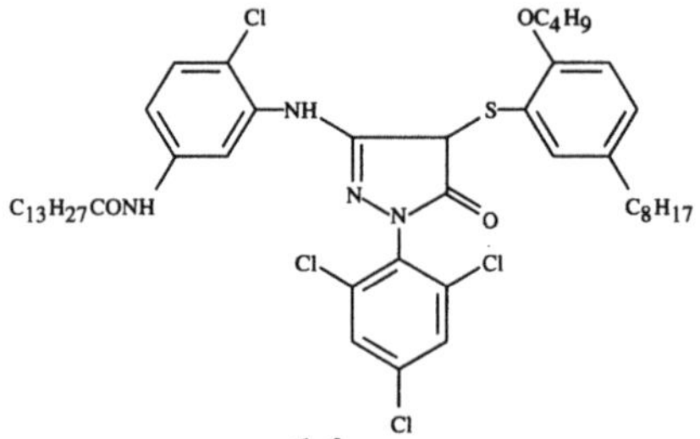

14 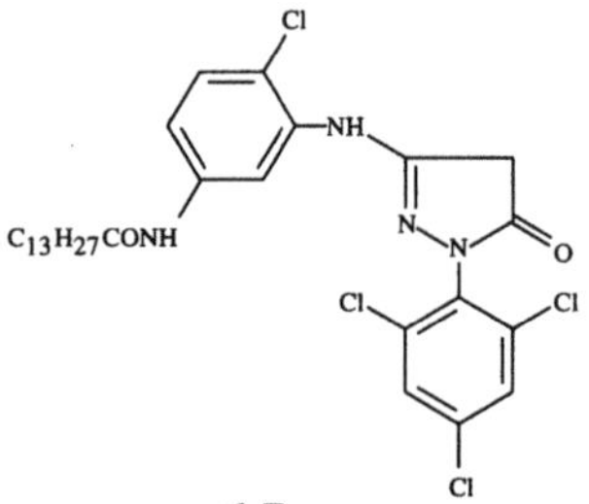 **15**

$$\mathbf{14}\ (\text{C-X}) + \mathbf{7}\ (\text{Q}) \xrightarrow{k_1} \text{Leuco dye (I-X)} \xrightarrow{k_2} \underset{\text{Elimination}}{\text{Dye (P)} + {}^{-}\text{S-Ar}}$$

$$\text{Leuco dye (I-X)} \xrightarrow[+\text{Q}]{k_2'} \underset{\text{Oxidation}}{\text{Dye (P)} + {}^{+}\text{S-Ar}}$$

$$2\,{}^{-}\text{S-Ar} + \text{Q} \xrightarrow{\text{fast}} \text{ArS–SAr}$$

Figure 9.6. Reaction scheme for the dye formation of the magenta coupler (**14**) with the quinonediimide (**7**) derived from CD-1. The symbol Ar represents the phenyl moiety released from **14**. For the symbols C-X, Q, I-X, and P, see Fig. 9.4.

the quinonediimine produced by the anodic oxidation of CD-1. The second-order rate constant for the coupling step (k_1) and the first-order rate constant of the elimination step (k_2) are determined to be $> 10^5$ L·mol^{-1}·sec^{-1} and 0.02 sec^{-1}, respectively, at pH 10.5. The k_2' value for the oxidation step is determined to be 1.2×10^3 L·mol^{-1}·sec^{-1} [6]. The four-equivalent coupler (**15**) and CD-1 gives $k_1 = 1.8 \times 10^3$ L·mol^{-1}·sec^{-1}, while the successive oxidation step is fast.

9.2.2 Side Reactions

Although the coupling reactions of color developers are sufficiently fast, there may occur several side reactions. One of such side reactions is the deamination

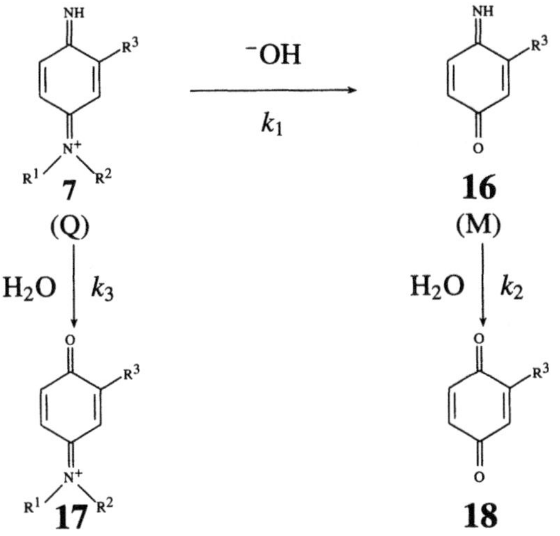

Figure 9.7. Deamination of oxidized color developers.

Table 9.2. Deamination of Color Developers [7,8]

Substituents				pH	$k\ (= k_1 + k_3)^a$	$\log\{k/[^-OH]\}$
R^1	R^2	R^3	R^4		(sec^{-1})	
CH_3	CH_3	H	H	10.02	2.53	4.40
				11.10	30.4	
				11.96	196.	
[b]C_2H_5	C_2H_5	H	H	10.10	1.21	3.72 (3.99)
				11.10	11.10	
				11.96	77.6	
C_2H_5	C_2H_4NH-SO_2CH_3	H	H	10.06	0.177	2.34
				11.10	0.421	
				11.96	2.16	
[c]C_2H_5	C_2H_5	CH_3	H	10.04	0.141	3.22
				11.08	2.00	
				11.59	5.84	
				11.93	13.2	
[d]C_2H_5	C_2H_4NH-SO_2CH_3	CH_3	H	10.00	0.166	2.48
				11.02	0.519	
				11.93	2.86	

[a]Pseudo-first order rate constants at 25°C.
[b]Kodak code: CD-1; [c]Kodak code: CD-2; [d]Kodak code: CD-3.

of oxidized color developers, the kinetics of which has been investigated in detail [3,7].

The disappearance of the quinonediimine obeys the second-order mechanism concerning [Q] and [OH^-]. Under a fixed pH (e.g., in a buffered solution), this reaction can be considered to obey the pseudo-first order mechanism, because [OH^-] can be regarded as a constant:

$$-\frac{d[Q]}{dt} = k_Q[Q][^-OH] = k[Q] \tag{9.2}$$

where $k = k_1 + k_3$ according to the notation shown in Fig. 9.7. Several data of k are cited from Ref. [7] and collected in Fig. 9.2. The second-order rate constants can be calculated by $\log k_Q = \log\{k/[^-OH]\} = \log k + 14 - \text{pH}$, which are cited from Refs. [7,8].

References

[1] Bent RL, Dessloch JC, Duennebier FC, Fassett DW, Glass DB, James TH, Julian DB, Ruby WR, Snell JM, Sterner JH, Thirtle R, Vittum PW, Weissberger A (1951) J Am Chem Soc. 73:3100

[2] Weissberger A (1980) Chemtech. 340

[3] Tong LKJ, Glesmann, MC (1957) J Am Chem Soc. 79:583

[4] Tong LKJ, Glesmann MC (1968) J Am Chem Soc. 90:5164

[5] Furutachi N (1989) Fuji Film Res & Dev. 34:1

[6] Kobayashi H, Okazawa M, Kurono H, Okawa Y, Ohno T (1994) J Imaging Sci Tech. 38:28

[7] Tong LKJ (1954) J Phys Chem. 58:1090

[8] Tong LKJ, Glesman MC, Bent RL (1960) J Am Chem Soc. 82:1988

Chapter 10

Couplers

10.1 Negative-Positive Process

Modern photographic films and papers for amateur use are based on the negative-positive process. As shown at the top of Fig. 10.1, a negative film is composed at least of three photo-sensitive layers, i.e., a red-sensitive emulsion layer with a cyan coupler, a green-sensitive emulsion layer with a magenta coupler, and a blue-sensitive emulsion layer with a yellow coupler. In contrast to the reversal process which involves black-and-white development in addition to color development, the negative-positive process undergoes color development only. Hence, the resulting picture is a negative image (i.e., a complementary color image) of the original image.

1. The film (or paper) is exposed by lights shown in Fig. 10.1a. The symbols △ and △• represent an unexposed silver halide grain and an exposed one with a latent image. The symbols Ⓨ, Ⓜ, and Ⓒ express a yellow, a magenta, and a cyan coupler, respectively.

2. The film (or paper) is processed by color development as shown in Fig. 10.1b to produce a dye image consisting of complementary color to exposed light.

3. Finally the developed silver is converted (bleached) to silver halide and removed (fixed) from the layers in a thiosulfate solution (Fig. 10.1c). As a result, the film reproduces yellow (complementary to blue), magenta (complementary to green), cyan (complementary to red), black (all dyes), and

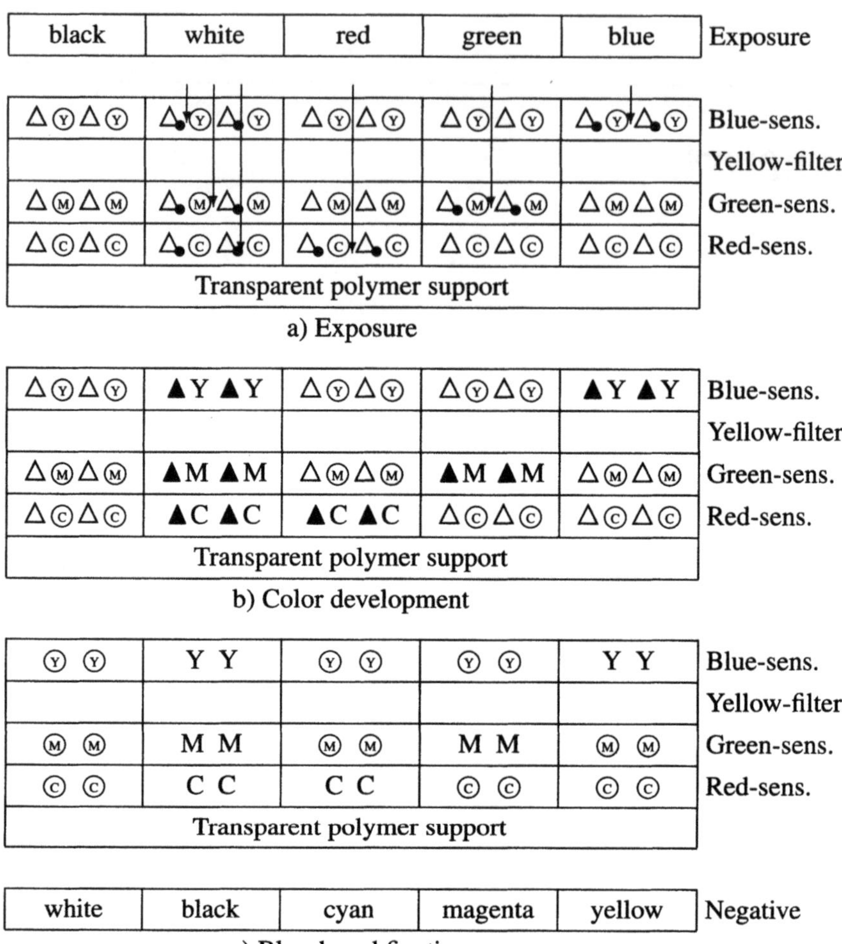

Figure 10.1. Schematic color development of non-diffusible couplers for negative photographic materials (color films). The support is a polyester (or tetraacetylcellulose film) for color films. △: Silver halide grain; △•: Exposed silver halide grain with a latent image; ▲: Developed silver halide grains with dye forming; Ⓨ: Yellow coupler; Ⓜ: Magenta coupler; Ⓒ: Cyan coupler; Y: Yellow dye; M: Magenta dye; C: Cyan dye.

white (no dyes). Note that each layer contains unreacted coupler (Ⓨ, Ⓜ, or Ⓒ), which is colorless and harmless to color reproduction.

Since the image of the developed film is negative, a further processing is necessary to obtain a positive image. Positive photographic materials (color papers) are also composed of at least three photo-sensitive layers, as shown Fig. 10.2. The negative color image shown in the bottom row of Fig. 10.1c is now the original of the successive positive process, which is shown at the top row of Fig. 10.2a. The

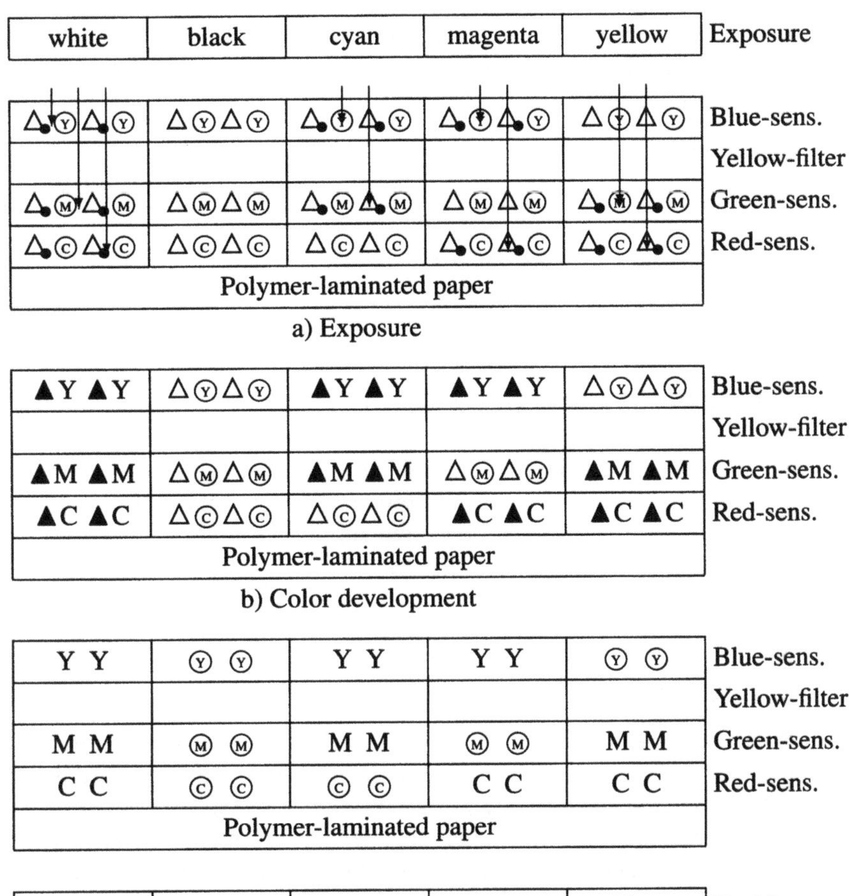

Figure 10.2. Schematic color development of non-diffusible couplers for positive photographic materials (color papers). The support is a polyethylene-laminated paper for color papers. For the symbols used in this figure, see Fig. 10.1.

process shown in Fig. 10.2 is essentially the same as that of Fig. 10.1, although the detailed procedures are somewhat different between films and papers.

The resulting positive image is shown in the bottom row of Fig. 10.2c, which is the same as the top row of Fig. 10.1a. Thus, we are able to obtain such a positive image by making a negative image from the negative image of an original object:

Original Object		Negative (Color film)		Positive (Color paper)
Blue	→	Yellow	→	Blue
Green	→	Magenta	→	Green
Red	→	Cyan	→	Red

When Fig. 10.2 is inversely considered to be the process of a negative material (color film) and Fig. 10.1 to be that of a positive material (color paper), another mode of color reproduction can be illustrated:

Original Object		Negative (Color film)		Positive (Color paper)
Cyan	→	Red (Yellow + Magenta)	→	Cyan
Magenta	→	Green (Yellow + Cyan)	→	Magenta
Yellow	→	Blue (Magenta + Cyan)	→	Yellow

10.2 Couplers and Azomethine Dyes

The R&D of couplers has been summarized in several reviews on the chemistry of couplers [2–5]. This section is devoted to brief discussions on fundamental items common to cyan, magenta, and yellow couplers. In the following sections, the properties and organic syntheses of couplers are discussed by referring to rather old but essential topics as well as by introducing recent progress of this field.

10.2.1 Dye-Forming Reactions

A coupler has a general formula (**1**) shown in Fig. 10.3, where one terminal of a double-bond conjugation takes a dissociative hetero group (HX–, frequently HO–) and the other terminal takes a hydrogen atom or a removable hetero group (Y–).[1] The coupler is dissociated in an alkaline solution of chromogenic (color) development to give a coupler anion ($^{-}$X–), which is reacted with a color developer (**2**) by the oxidative action of silver halide. The resulting intermediate called a leuco dye (**3**) is further reduced (when Y = H) or releases HY (when Y is a hetero group) to produce a dye (**4**), which is called an azomethine ($n = 0$) or indoaniline dye ($n = 1$).

Although the stoichiometry of the chromogenic reaction has already been discussed in Chapter 9, it is worthwhile to recall the concept of equivalence on the basis of Fig. 10.3. When Y is a hydrogen atom, totally four equivalents of silver halide are necessary to obtain one mole of the dye according to the scheme shown in Fig. 10.3. This type of couplers are called *four-equivalent couplers*. On the other hand, when Y represents a hetero group, the second step from **3** to **4** is an isohypsic process, which does not require the oxidative action of silver halide. Hence, the latter type of couplers are called *two-equivalent couplers*.

[1] It should be noted that the general formula can be converted into the corresponding keto form by virtue of the keto-enol tautomerism. As a matter of convenience, a coupler based on an active methylene is written as having a keto form, while a phenol or naphthol coupler is written as having an enol form. The coupler anion is regarded as a single species whether it is generated from the enol form or from the keto form.

$HX-(C=C)_n-C=C-Y$ + $NH_2-C_6H_4-NR_2$

1 2

$2Ag^+$ → 3

$-HY$ (Y = hetero atom) or $2Ag^+$ (Y = H) → 4

Figure 10.3. Dye formation by the reaction of a coupler (**1**) with a color developer (**2**) in chromogenic development. The symbol X represents a hetero atom, Y represents a hydrogen atom or a hetero atom, and n is a non-negative integer (in most cases, $n = 0$ or 1). Each vertical bond should be filled up by an appropriate substituent or a component necessary to construct a ring structure. Although the coupler (**1**) is written as an enol form, the corresponding keto form can exist in terms of a keto-enol tautomerism.

In the R&D of couplers, several items that depend mainly on a coupler nucleus should be taken into consideration:[2]

1. Activity to an oxidized color developer (or efficiency of dye forming), which is characterized by the maximum density (D_{max}). This item has become more and more important, because it allows the reduction in the amount of silver halide so that it leads to improved color image reproduction, lower cost photographic materials, as well as less potential environmental damage from the development process. The activity of a coupler may be influenced by a ballast group or a leaving group.

2. Stain-free property of unreacted couplers in the storage of developed films or papers. Since more rapid processes of development have recently become popular, compounds capable of reacting with the unreacted couplers tend to remain after the development.

[2]The term "coupler nucleus" is used to designate the coupling site and the adjacent groups having a direct influence on the reactivity of the coupler. The coupler is composed of a coupler nucleus, a ballast group, and (for a two-equivalent coupler) a leaving group.

10.2.2 Properties of Dyes

Since the final images of color photography are azomethine or indoaniline dyes derived from couplers, the properties of the dyes are essentially important in order to improve the qualities of the photographic images:

1. Excellent hue as a yellow, magenta, or cyan dye. This means that the dye should exhibit an absorption of an appropriate absorption maximum and a narrow but sufficient band shape so as to cover the blue, green, or red region of light.
2. Excellent light stability. Visible light and ultraviolet light may cause the fading of the dye. Various mechanisms (oxidative or reductive ones) work in the fading reactions according to the structures of couplers.
3. Excellent dark stability. Heat and humidity in storage may cause the fading of the dye.

The spectra, stabilities, and other properties of azomethine or indoaniline dyes have been extensively investigated in order to place photographic materials of higher quality on the market.

10.2.3 Ballast Groups for Oil-Protection

A coupler is coated on a polymer support (film or paper) as an oil-dispersion in an aqueous gelatin medium. A coupler should satisfy various requirements for manufacturing and storing photographic materials:

1. Good solubility in dispersion oils. This assures a stable oil-in-water dispersion during the manufacture.
2. No crystallization in storage. The coupler dispersion should be stable after coating so that no crystallization occurs.

These properties mainly come from a ballast group incorporated in a coupler.

Methodologies for the introduction of ballast groups have been extensively investigated concurrently with those of the other parts of a coupler, because the introduction of an appropriate ballast group into the coupler structure is important to incorporate the coupler into a photo-sensitive layer. A number of procedures for the syntheses of couplers are outlined in several reviews [6,7]

10.3 Cyan Couplers

Organic compounds for forming cyan dyes can be classified into naphthol couplers, phenol couplers, and heterocyclic couplers. The hues of the resulting indoaniline or related dyes and other properties such as light and dark stabilities have been extensively investigated by using various substituents. In this section, we do not pay explicit attention to the difference between four-equivalent couplers and two-equivalent ones. In other words, examples cited here may be two- or four-equivalent couplers, where their common features are discussed.

10.3.1 Naphthol Couplers

1-Naphthol Couplers

Dyes from 1-Naphthols

Phenol or 1-naphthol derivatives have been mainly used as cyan couplers because of easy availability. The basic data of absorption maxima of indoaniline dyes derived from 2-substituted 1-naphthols have been reported by several groups [8, 9]. The effect of anilide substituents upon absorption spectra of 1-hydroxy-2-naphthanilide has been extensively investigated, as cited in Table 10.1 [10].

(Dis)advantages of 1-Naphthol Couplers

An oil-protected 1-naphthol coupler has already been described in Chapter 8, where its synthesis is illustrated in Fig. 8.12.

1-Hydroxy-2-naphthanilide couplers exhibit various degrees of reactivities to oxidized color developers and are capable of generating cyan dyes of good hue, i.e., λ_{max} about 700 nm (Table 10.1). These merits have resulted in the long-term adoption of such 1-hydroxy-2-naphthanilide couplers in color negative films.

In spite of the above-mentioned merits of the 1-hydroxy-2-naphthanilide couplers, the dye images derived from them have some disadvantages [11]:

1. The reduction of color density takes place when development processing is carried out by using a fatigued bleaching solution. The reduction of color density is ascribed to the reduction with accumulated ferrous ions (reductive fading).[3]

2. The conversion of leuco dyes into the corresponding cyan dyes is slow when a bleaching solution with weak oxidative activity is used in development processing.

3. The cyan dyes derived from the 1-hydroxy-2-naphthanilide couplers are not good in heat stability. The heat fading is ascribed to the reduction due to several remaining compounds that have reducing activity, e.g., a ferrous ion, a thiosulfate ion, and couplers [12].

[3]Phenol couplers having a phenylureido group at the 2-position and carbonamido group at the 5-position have been disclosed to overcome this disadvantage (see below). However, they involve another type of disadvantage in that the absorption maximum of a color image varies widely, depending upon the color density.

Table 10.1. Absorption Maxima of Indoaniline Dyes (**5**) Derived by Oxidative Coupling of 4-Amino-3-methyl-*N, N*-diethylaniline (CD-2) with 1-Hydroxy-2-naphthanilides[a] [10]

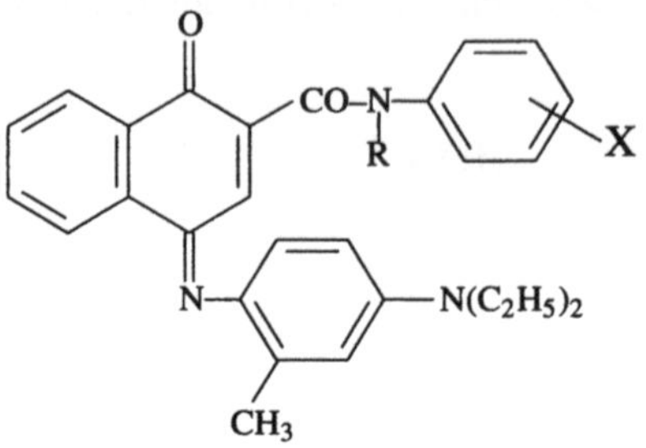

5

Substituents		λ_{max}	$\epsilon_{max} \times 10^{-4}$
R	X	(nm)	
H	H	694	3.4
CH_3	H	629	1.9
H	*o*-$N(CH_3)_2$	689	3.2
H	*m*-$N(CH_3)_2$	694	3.4
H	*p*-$N(CH_3)_2$	687	3.3
H	*o*-OCH_3	682	3.3
H	*m*-OCH_3	691	3.4
H	*p*-OCH_3	689	3.3
H	*o*-Cl	693	3.4
H	*m*-Cl	700	3.5
H	*p*-Cl	697	3.5
H	*o*-NO_2	700	3.3
H	*m*-NO_2	712	2.8
H	*p*-NO_2	713	3.9

[a] All data were determined in butyl acetate.

Various ballast groups suitable to oil-dispersion have been proposed to improve stability toward ferrous ion reduction. For example, ballast groups containing a sulfonyl group have been disclosed for this purpose [13].

5-Amido-1-naphthol Couplers

Improved Dark Stability

Since the drawbacks described above for 1-naphthol couplers are all concerned with the reduction of the corresponding indoaniline dyes, 1-naphthol couplers providing less reducible dyes would be the next promising targets. This guiding principle has been embodied by introducing a 5-amido group into the 1-naphthol nucleus (Fig. 10.4) [14].

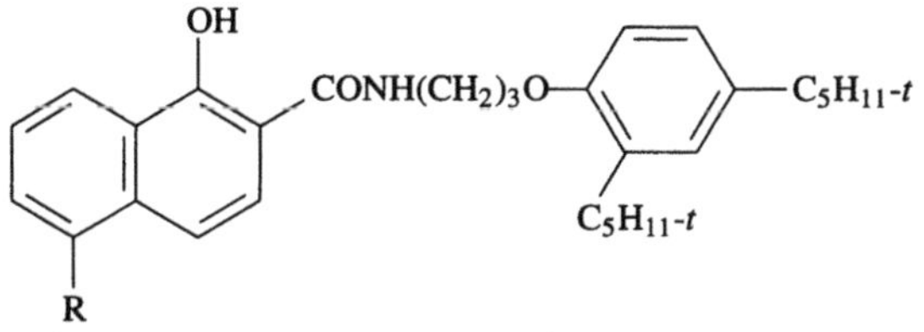

OH

$CONH(CH_2)_3O$ — C_5H_{11}-*t*

C_5H_{11}-*t*

R

6a (R = $NHSO_2CH_3$) **6b** (R = $NHCOCH_3$)
6c (R = $NHCOOC_2H_5$) **6d** (R = H)

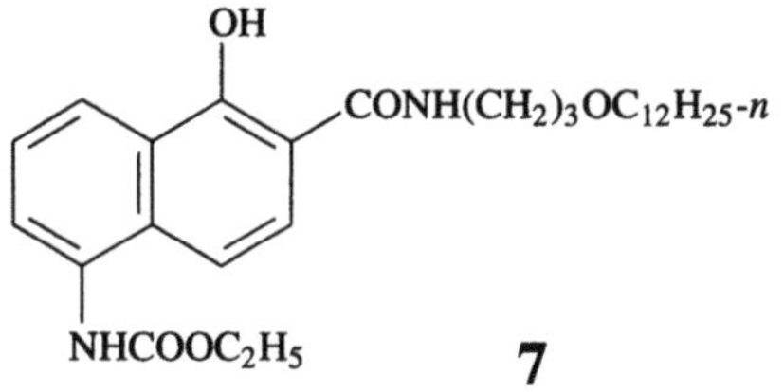

OH

$CONH(CH_2)_3OC_{12}H_{25}$-*n*

$NHCOOC_2H_5$ **7**

Figure 10.4. Non-diffusible four-equivalent couplers forming a cyan indoaniline dye (1-hydroxy-5-amido-2-naphthanilides).

Table 10.2. Fastness and Absorption of Indoaniline Dyes Derived by Oxidative Coupling (CN-16 Processing) of 1-Hydroxy-5-amido-2-naphthanilides [11]

Coupler	Remaining of dye image (%)		λ_{max} (nm)	
	Fe^{2+a}	dark[b]	in ethyl acetate	in film
6a	98	92	688	710
6b	99	74	696	708
6c	100	96	680	698
7			674	696
6d	56	50	674	698

[a]Produced by the reduction of 10% EDTA-Fe(III) in a bleach solution.
[b]Stored at 100°C for 16 days.

The dark stabilities of the indoaniline dyes derived from the couplers **6a** to **6c** have been compared with that of the dye derived from **6d** [11]. The dark stabilities have been tested by determining the remaining dye images on the reductive action of ferrous ions or on the exposure of heat in a dark place. The data cited in Table 10.2 show the superiority of the 5-amido couplers (**6a** to **6c**) over the standard coupler (**6d**) with no 5-substituent.

In general, the substitution of a 5-amido group causes the bathochromic shift of an absorption maximum. To adjust absorption maxima, 5-amino substituents and ballast groups have been varied so that a coupler **7** has been found to exhibit an equivalent absorption maximum to that of the standard cyan dye derived from **6d** (Table 10.2).

The 5-amidonaphthol couplers have a secondary absorption at the blue region of light in the same manner as other naphthol couplers. Irregular color reproduction due to the secondary absorption can be corrected by using yellow-colored cyan couplers. This topic will be later discussed in detail (Chapter 12, Subsection 12.2.2).

Syntheses

The synthetic pathway of the coupler (**6a**) is illustrated in Fig. 10.5 [14]. The key intermediate of the pathway is **12**, which has been prepared after the initial block of the amino group of **10** by a trifluoroacetyl group. The intermediate **10** has been prepared by a one-pot reaction in which the corresponding acid chloride of **9** has not been isolated. The leaving group (*p*-nitrophenoxy group) of **10** is more effective than the usual phenoxy group (see Fig. 8.12 in Chapter 8).

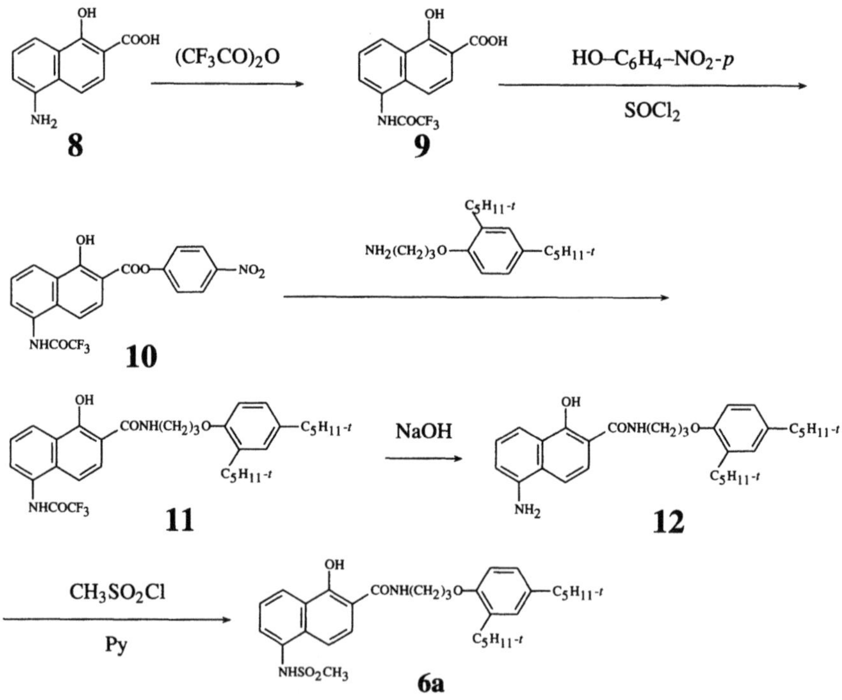

OH, COOH, NH_2 — **8**
$(CF_3CO)_2O$ →
OH, COOH, $NHCOCF_3$ — **9**
$HO–C_6H_4–NO_2$-*p*, $SOCl_2$ →

OH, COO, NO_2, $NHCOCF_3$ — **10**
$NH_2(CH_2)_3O$, C_5H_{11}-*t*, C_5H_{11}-*t* →

OH, $CONH(CH_2)_3O$, C_5H_{11}-*t*, C_5H_{11}-*t*, $NHCOCF_3$ — **11**
NaOH →
OH, $CONH(CH_2)_3O$, C_5H_{11}-*t*, C_5H_{11}-*t*, NH_2 — **12**

CH_3SO_2Cl, Py →
OH, $CONH(CH_2)_3O$, C_5H_{11}-*t*, C_5H_{11}-*t*, $NHSO_2CH_3$ — **6a**

Figure 10.5. Synthetic pathway of the cyan coupler **6a**.

10.3.2 Phenol Couplers

2-Amidophenol Couplers

Improved Dark Stability

Phenol couplers have long been used in color papers, as described in several reviews [5,15]. The coupler **13a**, which was earlier reported as an oil-protected coupler [16],[4] had been long-lived until the first half of the 1980s, because of high reactivity, the superior hue of the resulting dye, and light stability.

However, the insufficient dark stability of **13a** was pointed out in connection with the preservation of color prints. This instability was found to stem from the fact that the dye derived from **13a** was reduced back into the corresponding leuco dye [17].

Fortunately, the dark stability was dramatically improved by replacing the 5-methyl group of **13a** with an ethyl group, as exemplified by **14** and **13b** [18].[5] Since the reason of this dramatic improvement is not clear theoretically, similar experimental luckiness may be expected in other situations of R&D.

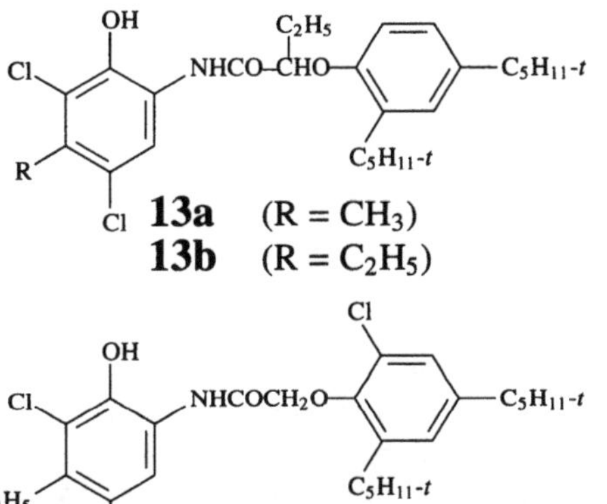

13a (R = CH_3)
13b (R = C_2H_5)

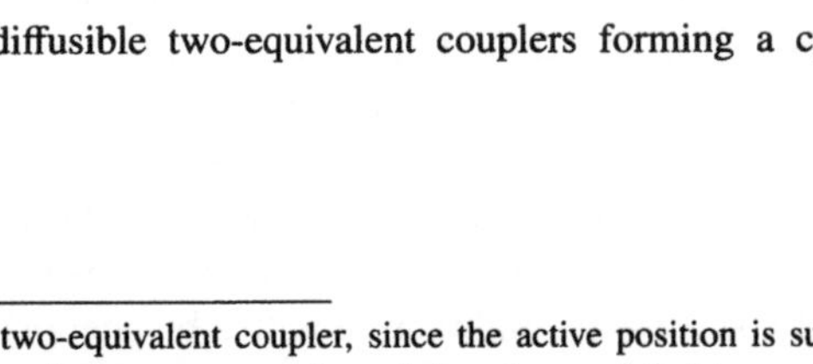

14

Figure 10.6. Non-diffusible two-equivalent couplers forming a cyan indoaniline dye (2-amidophenols).

[4] This coupler is a two-equivalent coupler, since the active position is substituted by a chlorine atom. In general, two equivalent cyan couplers of phenol or naphthol type can be easily obtained by introducing a halogen atom at the active position.

[5] This patent was assigned to Minnesota Mining and Manufacturing Co. A color paper using a coupler of 2-amido-5-ethylphenol type was first placed on the market by Eastman Kodak in the 1980s. Couplers of this type have been also used in color papers of other companies after the expiration of the original patent.

Polymer-Protected Dispersion

Couplers of 2-amido-5-alkylphenol type (e.g., **13b**), when used in usual oil-protected dispersions, have merits of giving dyes with superior hue and light stability. Although the dark stability has been dramatically improved as described above, a further improvement has been required to keep up with the dark stabilities of yellow and magenta couplers. This task has been accomplished by adopting a polymer (e.g., poly(*N*-*t*-butylacrylamide)) as a dispersion medium in place of such an oil as tricresyl phosphate [19]. According to model experiments at high temperatures, the lifetime of a polymer-protected dye has been estimated to be about 10^4 years at room temperature, whereas that of an oil-protected dye has been estimated to be about 10^2 years [20].[6] See also Section 8.3.2.

Syntheses

The synthetic pathway of **14** is illustrated in Fig. 10.7 [16]. One of the common methodologies in coupler syntheses is the introduction of ballast groups through an amido linkage. In this methodology, such an acid chloride as **17** is a key intermediate for preparing an amido linkage. The ethyl analog **14** has been synthesized by replacing 2-amino-4-chloro-5-methylphenol (**18**) with 2-amino-4-chloro-5-ethylphenol [18] except that the 6-chlorine atom is introduced in the last step.

2,5-Diamidophenol Couplers

Hues and Stabilities

2,5-Diamidophenol couplers such as **22** have been disclosed to give cyan dyes having a good hue, where the introduction of a fluoroalkyl group causes a bathochromic shift of about 20 nm as compared with couplers with a simple alkyl group [21]. Other 2,5-diamidophenol couplers have been proposed in patents, as listed in Fig. 10.8: **23** [22], **24a** [23], **25** [24], **26** [25], and **27** [26]. Although these 2,5-diamidophenol couplers are composed of a common nucleus, they contain various substituents at the 2- and 5-positions. The characteristic of **22** and **23** is a fluorine-containing sulfonamido group at the 2-position of the coupler nucleus [21,22]. The perfluorophenyl group is characteristic of the couplers **24a** and **24b** [23].[7] The characteristic of the coupler **25** is a *p*-alkylsulfonamidophenoxy terminal moiety at the 5-position [24]. The coupler **26** also has a *p*-alkylsulfonamidophenoxy moiety at the 5-position. This coupler has been disclosed to be used in combination with a bisphenol derivative as a light

[6] Color papers using this technique were placed on the market under the name Fujicolor Paper SUPER HG and SUPER FA in 1989.

[7] A color paper using **24a** was placed on the market under the name Konicacolor QA Paper Type A5 in 1991.

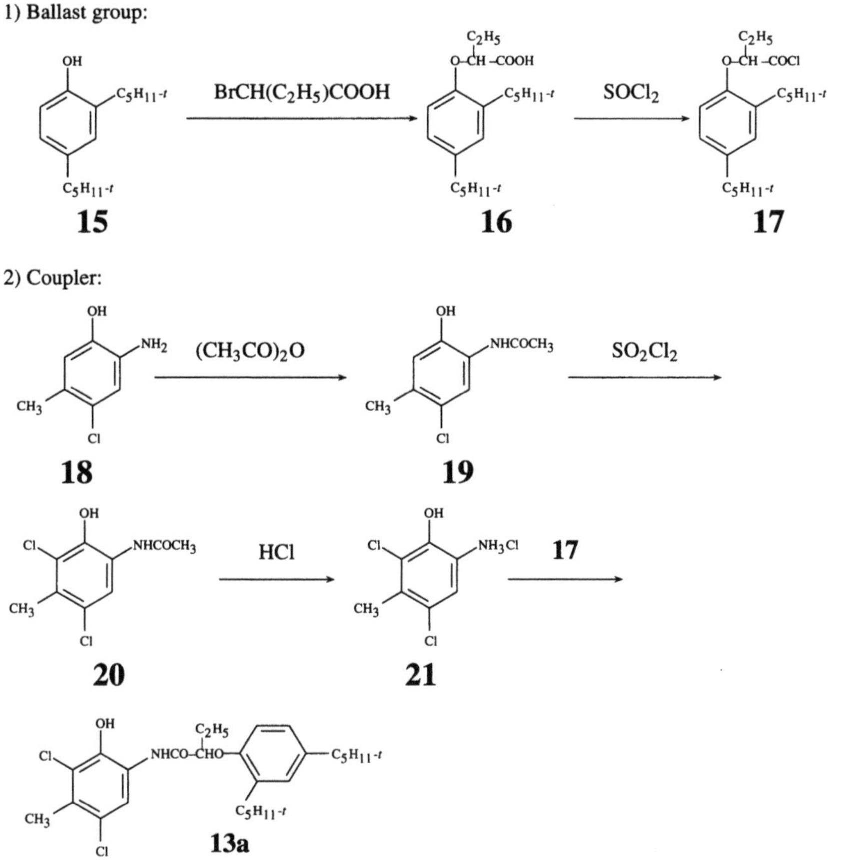

1) Ballast group:

15 — $BrCH(C_2H_5)COOH$ → 16 — $SOCl_2$ → 17

2) Coupler:

18 — $(CH_3CO)_2O$ → 19 — SO_2Cl_2 →

20 — HCl → 21 — 17 →

13a

Figure 10.7. Synthetic pathway of the cyan coupler **13a**.

and thermal stabilizer for the corresponding cyan dye [25]. The coupler **27** contains a 2-chlorophenyl moiety at the 5-position of the coupler nucleus [26]. The phenoxy group at the 5-position has been changed into a naphthoxy group, which provides high coupling activity, narrowed bandwidth, and improved image dye stability [27]

Several couplers have been compared qualitatively with respect to essential properties of the resulting dyes: (hue, light stability, and dark stability) [15]. Thus, these properties have been reported as follows: (good hue, good light stability, and poor dark stability) for **13a**;[8] (good, poor, and good) for **23**; (poor, good, and good) for **25**; and (good, good, and good) for **24a**.

Several properties of dyes derived from 2,5-diamidophenols analogous to **24b** are collected in Table 10.3 [15], where the pentafluorophenyl group of **24b** is re-

[8]The data for the ethyl analog (**13b**) have not been involved in this comparison, though **13b** has been reported to be superior to **13a** in dark stability.

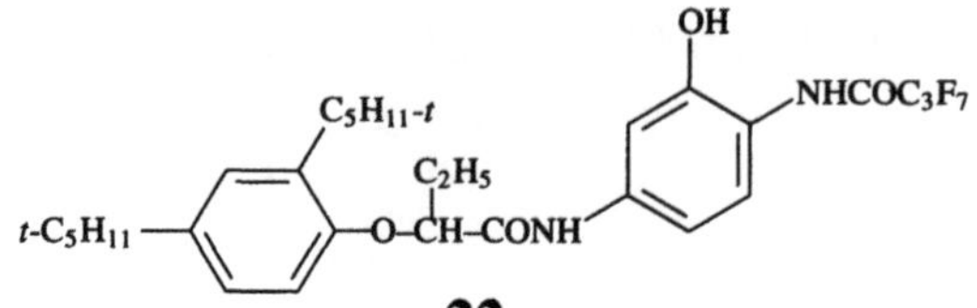

22

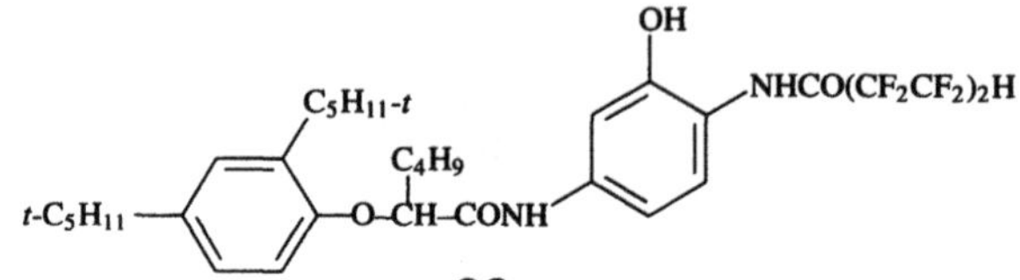

23

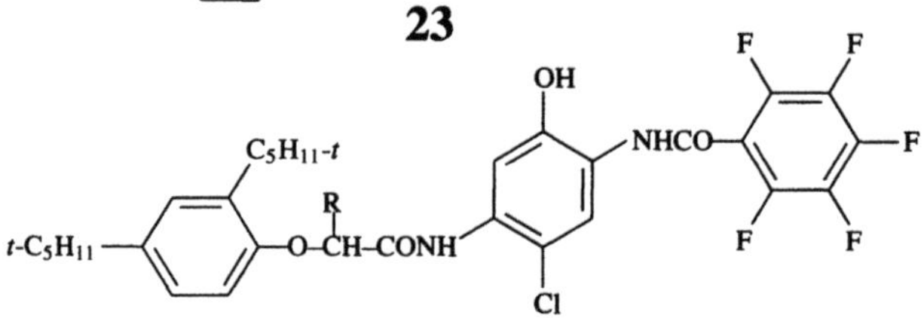

24a (R = C_3H_7-*i*)
24b (R = C_4H_9-*n*)

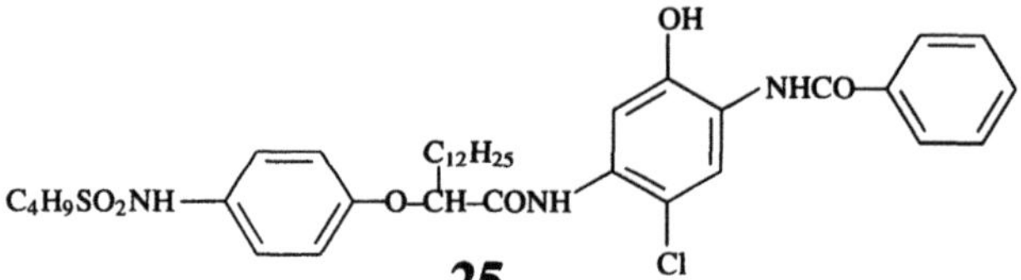

25

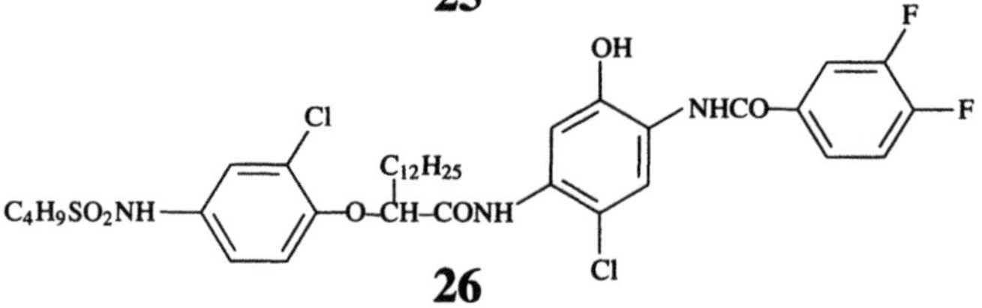

26

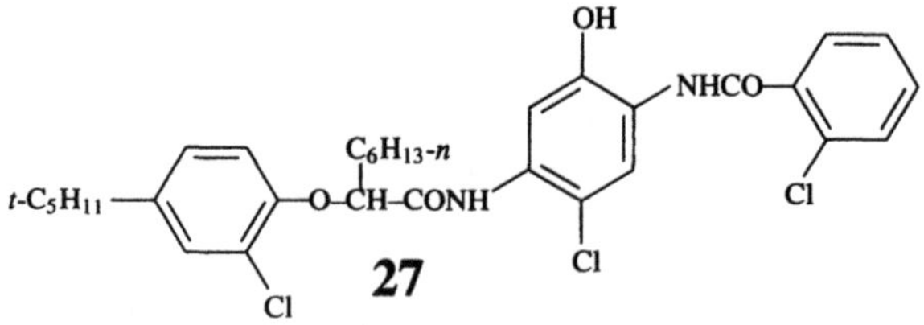

27

Figure 10.8. Non-diffusible four- and two-equivalent couplers (2,5-diamidophenols) forming a cyan indoaniline dye.

Table 10.3. Fastness and Absorption of Indoaniline Dyes Derived by Oxidative Coupling of 2,6-Diamidophenols with CD-3[a] [15]

Phenyl moiety[b]	Remaining of dye image (%)		λ_{max}
	light[c]	dark[d]	(nm)
4-F	72	100	643
2-F	55	100	639
2,6-F_2	69	100	648
2-F,6-Cl	61	96	655
4-CF_3	74	98	648
3-CF_3	75	97	641
F_5 (**24b**)	77	100	655

[a] *N*-Ethyl-*N*-(2-methanesulfonamidoethyl)-3-methyl-4-aminoaniline.

[b] A phenyl group with these substituents is used in place of the pentafluorophenyl group of **24b**.

[c] Irradiated with a xenon lamp (70 KLux) for 1 week.

[d] Stored at 100°C, 60% relative humidity for 2 weeks.

placed by a phenyl group with various substituents. The data show the superiority of **24b** over these analogs cited. To improve solubility, the butyl group in the side chain of **24b** has been changed into the *i*-propyl group, as found in **24a**.

Narrow Bandwidth Couplers

Couplers which give an indoaniline dye having an absorption with a steep slope on the short wavelength side are called "NB couplers" (narrow bandwidth couplers). In other words, the resulting cyan dyes have no unwanted absorption in the green and blue regions of the spectrum. An appropriate selection of amide substituents (e.g., *m*-alkylphenylsufonyl groups) brings about the NB effect, as found in **28** (Fig. 10.9) [28,29].

Table 10.4 lists several data of NB-couplers. The LBW (left bandwidth) has been adopted as a measure of the sharpness of the curve on the left (short wavelength) side of the absorption band. Compare the LBW of **28** (X = SO_2) and that of **29** (X = O). This effect becomes more remarkable when the 2-amido substituent is changed into a branched one, as exemplified by **30** to **32** [30]. Their data are listed also in Table 10.4, which shows remarkable NB effects, as compared with that of **13b**. A variation of 2-amido substituents has extensively been disclosed to bring about the NB effects [31,32,33].

Syntheses

The synthetic pathway for preparing **22**, which contains a perfluoroalkylsulfonamido group at the 2-position, is illustrated in Fig. 10.10 as a representative 2,5-diamidophenol coupler [21]. Note that the ballast group is introduced by using the intermediate (**17**) described in Fig. 10.7.

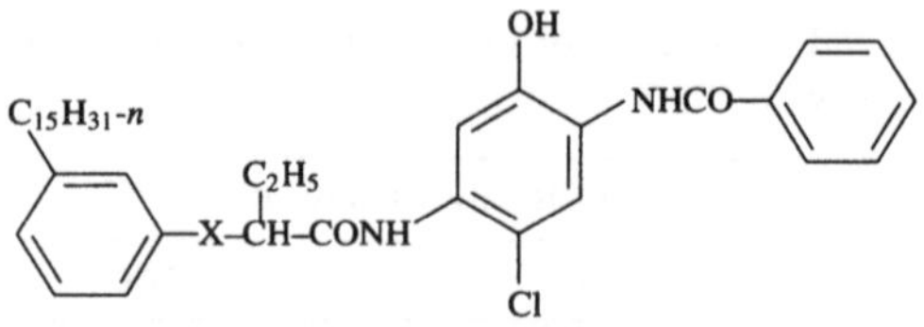

28 ($X = SO_2$)
29 ($X = O$ for comparison)

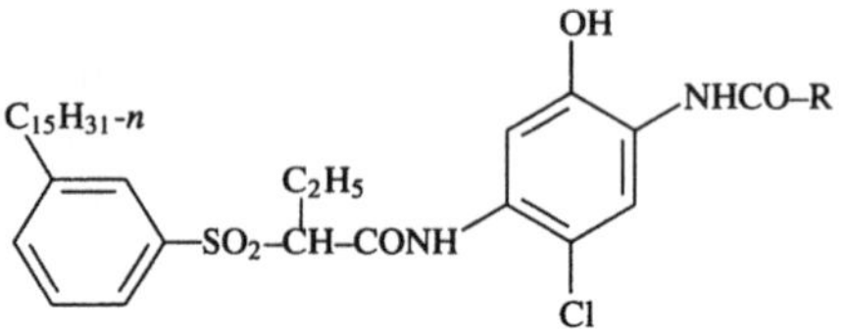

30 ($R = C_4H_9$-*t*)
31 ($R = CCl_3$)
32 ($R = CCl_2F$)

Figure 10.9. 2,5-Diamidophenol couplers forming a cyan indoaniline dye with narrow bandwidth.

Table 10.4. Absorption Properties of Dyes Derived from "NB Couplers" and CD-3 [28,30]

Coupler	λ_{max} (nm)	LBW (nm)[a]	Ref.
28	624	53	[28]
29[b]	641	81	[28]
30	559	63	[30]
31	614	40	[30]
32	630	53	[30]
13b[b]	661	80	[28]

[a] The LBW (left bandwidth) is the value obtained by subtracting the wavelength at the point on the left side of the absorption band where the normalized density is 0.5 from the λ_{max}.
[b] Couplers for comparison.

The synthetic pathway for preparing **25**, the characteristic of which is a *p*-alkylsulfonamidophenoxy terminal moiety at the 5-position, is illustrated in Fig. 10.11 [24]. The ballast group is introduced by using an acid chloride intermediate (**39**).

From a synthetic point of view, it is worthwhile to compare the scheme shown in Fig. 10.5 with the schemes shown in Figs. 10.7, 10.10, and 10.11. The scheme of Fig. 10.5 adopts a methodology of preparing an amide linkage in which Cp-COCl (or Cp-COOAr) is reacted with Ballast-NH_2. On the other hand, the schemes of Figs. 10.7, 10.10 and 10.11 prepare an amide linkage by the reaction of Cp-NH_2 with Ballast-COCl.

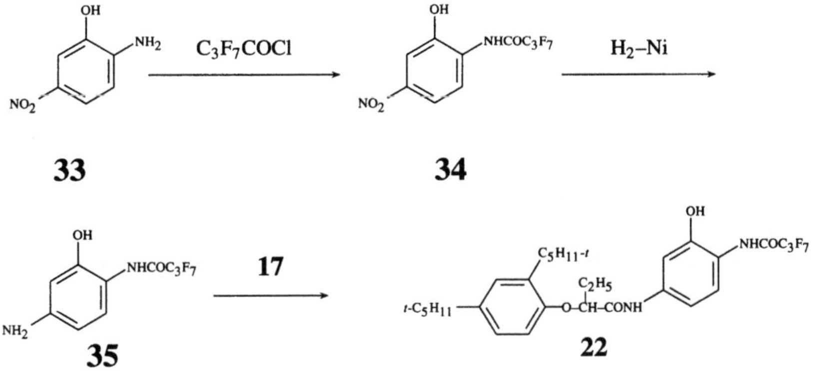

Figure 10.10. Synthetic pathway of the cyan coupler **22**. The preparation of the intermediate (**17**) has been described in Fig. 10.7.

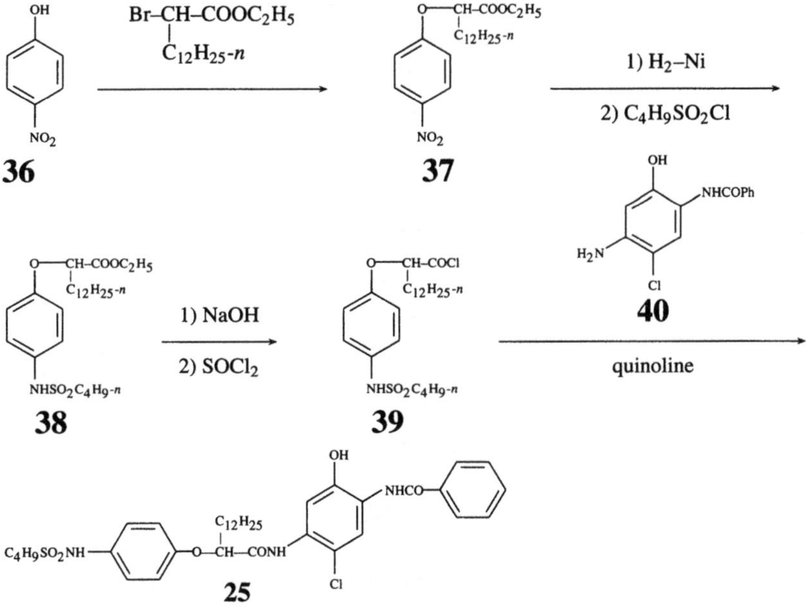

Figure 10.11. Synthetic pathway of the cyan coupler **25**.

2-Ureidophenol Couplers

Improved Hues and Activities

2-Ureidophenol cyan couplers such as **41** have been widely used in color negative photographic materials,[9] because they have good hues (absorption maxima at about 700 nm), exhibit high dark stabilities, and are less active to the reduction due to ferrous ion in bleach processing [34]. Other couplers of ureidophenol type have been disclosed, e.g., **42** [35] and **43** [36]. The modification of the 5-substituent shown in **44** [37] has been disclosed to improve the properties of the couplers described in [34] (e.g., **41**).

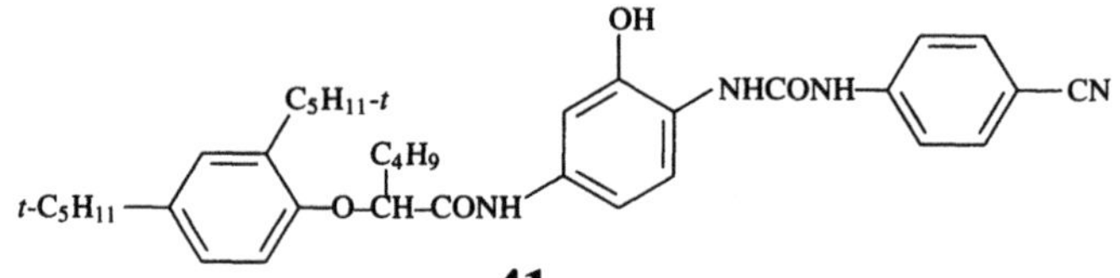

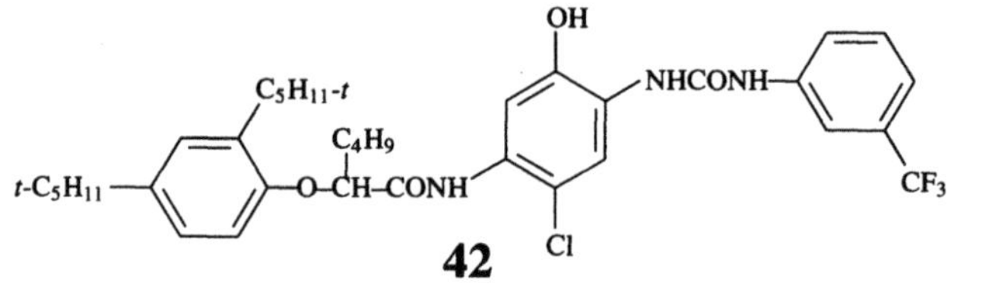

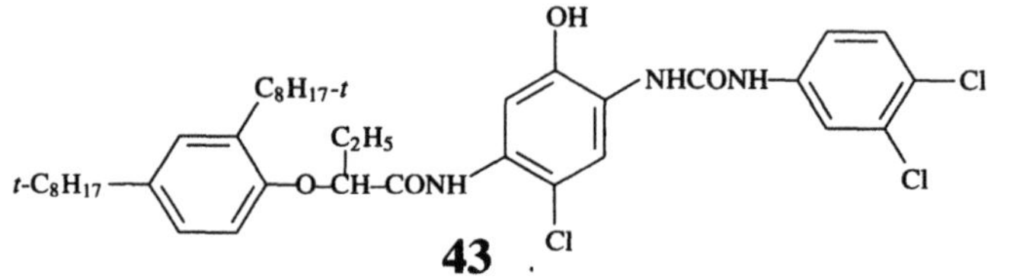

Figure 10.12. Non-diffusible couplers forming a cyan indoaniline dye (2-ureidophenols).

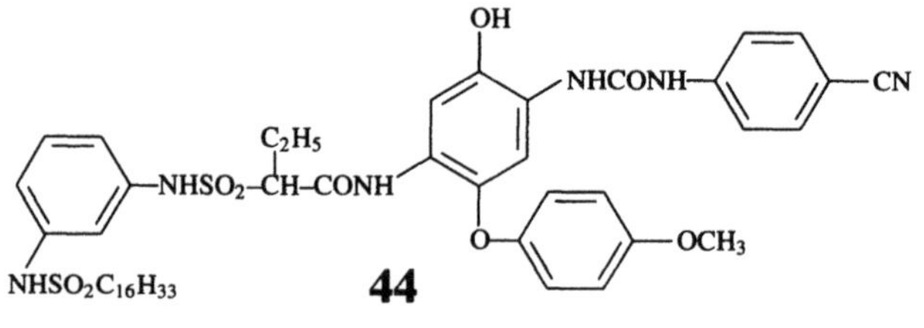

[9] 2-Amidophenol couplers and 2,5-diamidophenol couplers are used in color papers, not in color negative films. Such a 2-ureidephenol coupler as **41** is the first example of phenolic couplers used in color negative photographic materials.

However, the hue of the indoaniline dye derived from the 2-ureidophenol coupler (**41**) is sensitive to the density of the dye image as well as to environmental conditions such as dispersion oils, temperature, and other contaminants. As a result, the absorption maximum of the dye shifts in a range of 630–700 nm. The effect of dispersion oils on the hue of the dye derived from **41** and CD-4 (4-amino-3-methyl-*N*-ethyl-*N*-(2-hydroxyethyl)aniline) has been investigated by using ^{1}H-NMR spectroscopy and molecular weight measurement with a vapor pressure osmometer [38]. The dye molecules have been concluded to aggregate to form a head-to-head type dimer, in which the two hydrogens of the ureido group in one molecule form chelation with the oxygen of the indoaniline ring in the other molecule, and vice versa. The bathochromic shift due to dispersion oils has been ascribed to such dimer formation.

The activity of the 2-ureidophenol coupler **41** has been enhanced by using dibutyl sebacate, oleyl alcohol, or phenylethyl benzoate as a dispersion oil (a high boiling solvent) in place of dibutyl phthalate [39,40].

Syntheses

The synthetic pathway for preparing **41** is shown in Fig. 10.13 [34]. The formation of the ureido group comes from the isocyanate **45**. The incorporation of a ballast group is based on the amide formation with **48**.

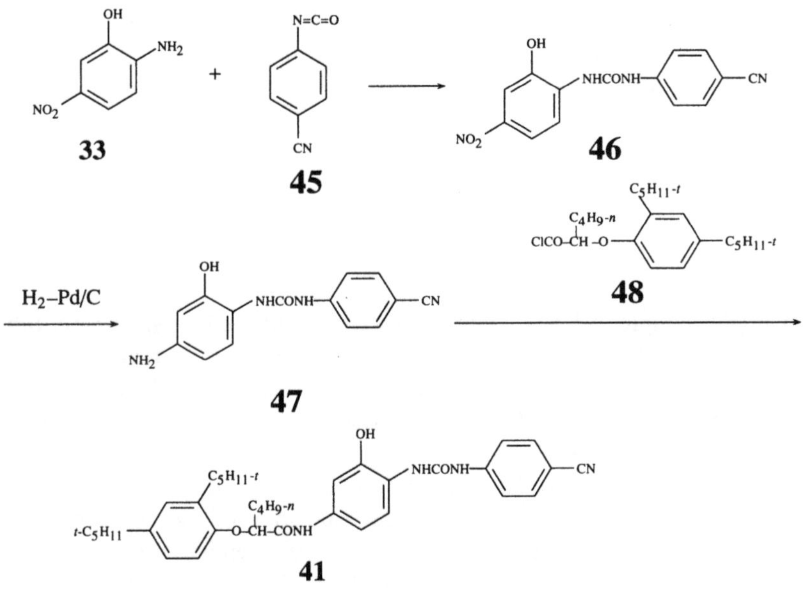

OH
NH_2
NO_2
33
+
N=C=O
CN
45
OH
NHCONH
CN
NO_2
46
C_5H_{11}-*t*
C_4H_9-*n*
ClCO–CH–O
C_5H_{11}-*t*
48
H_2–Pd/C
OH
NHCONH
CN
NH_2
47
OH
NHCONH
CN
C_5H_{11}-*t*
C_4H_9-*n*
t-C_5H_{11}
O–CH–CONH
41

Figure 10.13. Synthetic pathway of the cyan coupler **41**. The preparation of the intermediate (**48**) is analogous to that of **17** shown in Fig. 10.7.

10.3.3 Heterocyclic Cyan Couplers

Pyrazolotriazoles

In place of 5-pyrazolone magenta couplers, pyrazolotriazoles have been developed as new types of magenta couplers, as will be described in the following section.

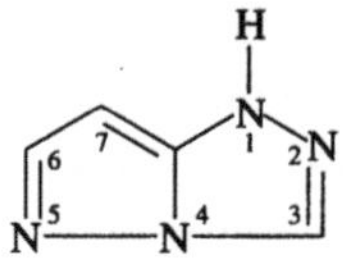

49

The success of the pyrazolotriazole magenta couplers has stimulated studies on heterocyclic cyan couplers, as summarized in a review [41]. For example, several couplers based on 1H-pyrazolo[5,1-*c*][1,2,4]triazole (**49**) have been disclosed to exhibit cyan hues when they have electron-withdrawing substituents [42,43]. However, the incorporation of such electron-withdrawing substituents inevitably causes low coupling activity with oxidized color developers. Pyrazolopyrimidines and pyrazolotriazines having a fused 5- and 6-membered ring structure have also been studied [41].

Pyrrolotriazoles

More recently, 1H-pyrrolo[1,2-*b*][1,2,4]triazole (**50**) has been disclosed as a new heterocyclic nucleus for cyan couplers which have high coupling activity with oxidized color developers [44,45,46].

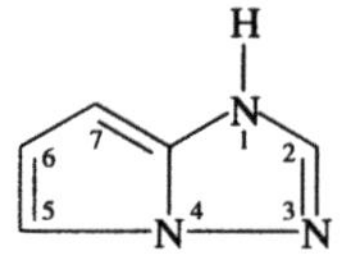

50

For example, coupler **51** has exhibited $D_{max} = 2.60$ and $D_{500} = 0.08$ (absorbance at 500 nm at the maximum absorption wavelength is normalized to 1.0) in the presence of 2-hexadacanoxybenzoic acid, whereas coupler **13b** has exhibited $D_{max} = 2.53$ and $D_{500} = 0.21$. The comparison of D_{max} indicates that coupler **51** has more improved color forming activity than **13b**. Moreover, the lower secondary absorption determined by D_{500} shows the improved color reproducibility of **51**. Note that the ballast group on the 7-position of the pyrrolotriazole ring is based on a cyclohexane ring, where the stereochemical arrangement of three substituents has not been specified. Pyrrolotriazines having a fused 5- and 6-membered ring structure have also been studied [47].

1H-Pyrrolo[1,2-*b*][1,2,4]triazoles having a ballast group on the 2-position have been disclosed, where the ballast group is an aryl group linked with an additional group through $NHSO_2$, NHCO, etc. [48]. Couplers forming cyan dyes

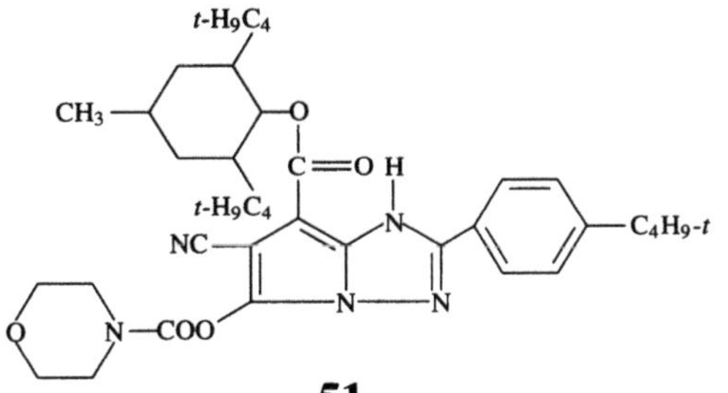

51

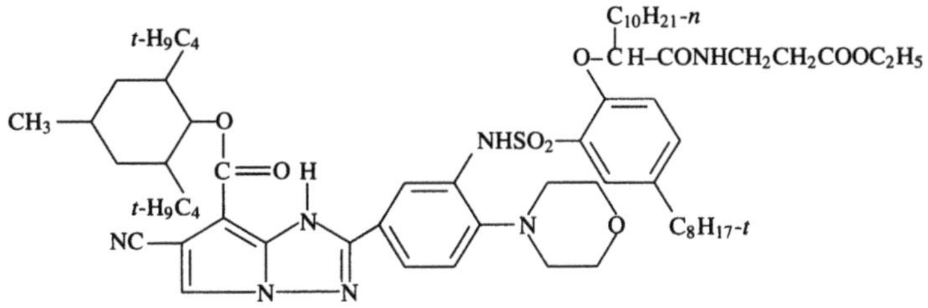

52

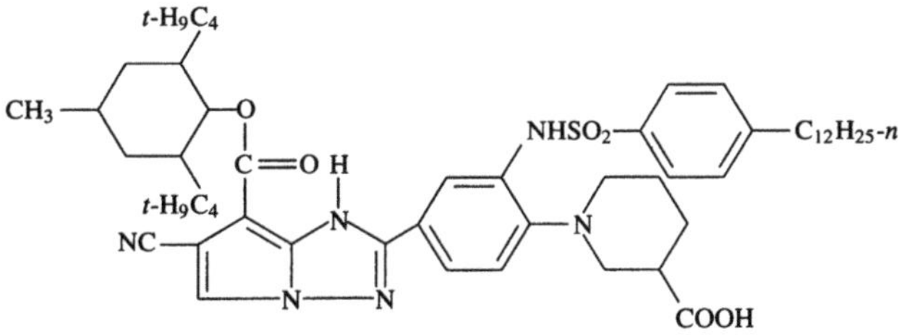

53

Figure 10.14. Non-diffusible heterocyclic couplers (1H-pyrrolo[1,2-*b*][1,2,4]triazoles) for forming a cyan dye.

of improved stability (e.g., **52**) have been disclosed, where the 1H-pyrrolo[1,2-*b*][1,2,4]triazole nucleus takes an aminophenyl group at the 2-position, where the amino group of the aminophenyl group is a morpholino moiety [49,50]. The morpholine ring has been replaced by a piperidine ring to give another type of couplers (e.g., **53**) [51]. These couplers are mainly employed in color reversal films and color papers.

Syntheses

The synthetic pathway of **52** shown in Figs. 10.15 and 10.16 contains two steps of constructing heterocycles [49]. The one is the construction of a triazole ring, where the hydrazide (**56**) is condensed with diethyl malonimidate to give the triazole derivative (**57**) after hydrolysis. The other step is the generation of a pyrrolotriazole ring, where the triazole (**59**) is reacted with acrylonitrile in the presence of DBU, giving **60**. The α-ketoester function of **59** is introduced by the oxidation of the methylene sandwiched between the triazole ring and the carbonyl group. It should be noted that the cyan coupler **52** contains a specific protective group of a saturated six-membered ring, which comes from 2,6-di-*t*-butyl-4-methylcyclohexanol.

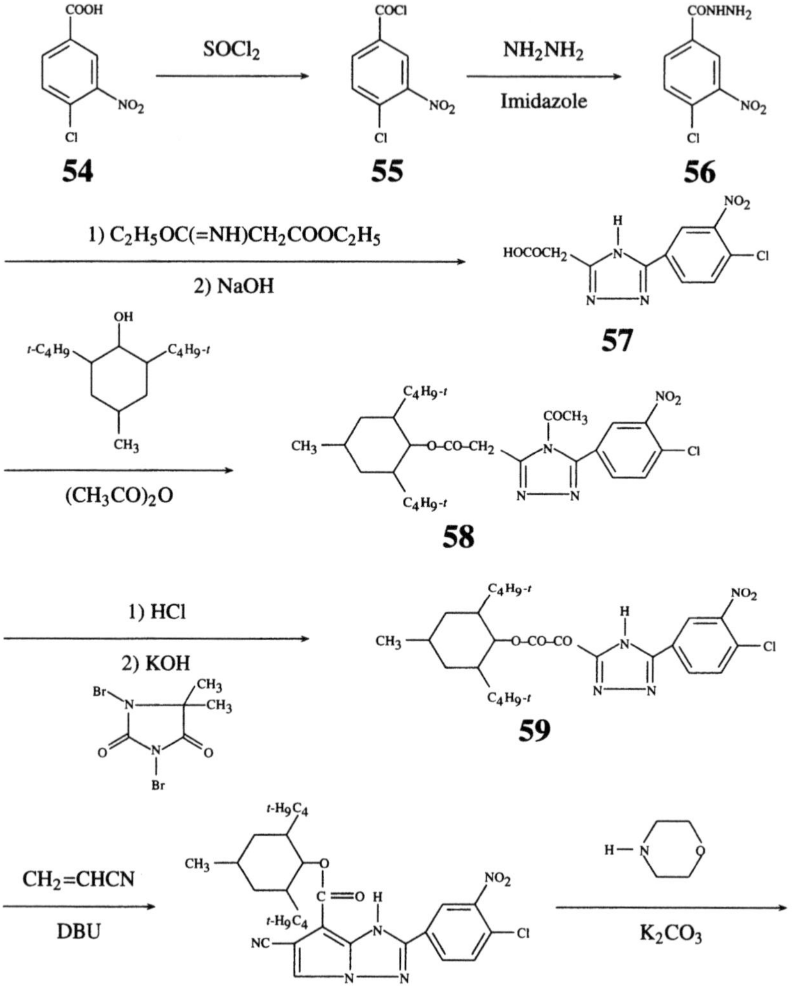

COOH
NO$_2$
Cl
54
$SOCl_2$
COCl
NO$_2$
Cl
55
NH_2NH_2
Imidazole
CONHNH$_2$
NO$_2$
Cl
56

1) $C_2H_5OC(=NH)CH_2COOC_2H_5$
2) NaOH
HOCOCH$_2$
57

(CH$_3$CO)$_2$O
58

1) HCl
2) KOH
59

CH$_2$=CHCN
DBU
60
K$_2$CO$_3$

Figure 10.15. Synthetic pathway of the cyan coupler **52** (to be continued to Fig. 10.16). DBU is the acronym of 1,8-diazabicyclo[5.4.0]undec-7-ene.

After being used as an activating group for introducing a morpholino ring (**60** → **61**), the nitro group of **61** is reduced chemically by iron powder to produce the amino intermediate. The resulting amino group is used to construct an amido linkage with a ballast group having a chlorosulfonyl function so that the final coupler **52** is obtained, as found in Fig. 10.16

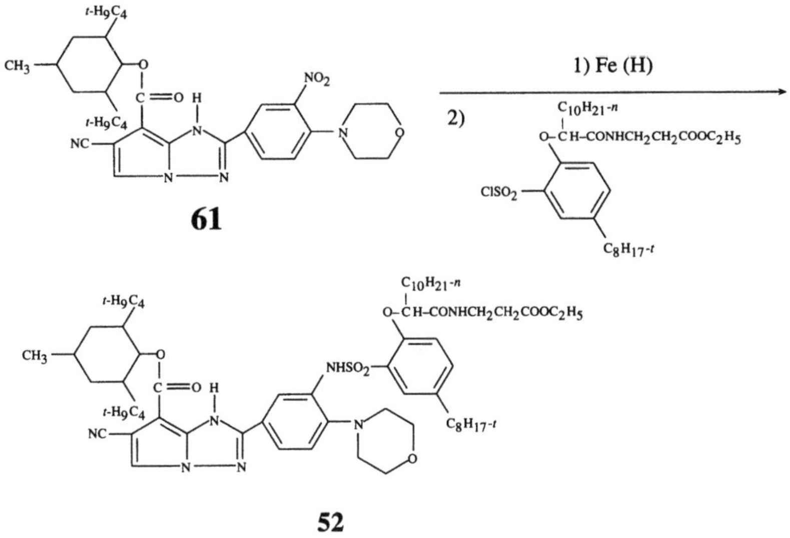

Figure 10.16. Synthetic pathway of the cyan coupler **52** (continued from Fig. 10.15).

10.4 Magenta Couplers

According to an excellent review on magenta couplers by Furutachi [54], the progress on magenta couplers consists of five generations, among which three generations are concerned with four-equivalent 5-pyrazolone couplers: 3-alkyl-5-pyrazolones (1st generation), 3-acylamido-5-pyrazolones (2nd generation), and 3-anilino-5-pyrazolones (3rd generation). The 4th generation involves two-equivalent pyrazolone couplers. Pyrazolotriazole couplers are referred to as the 5th generation. Since four-equivalent magenta couplers of oil-protected type began mainly with the 3-acylamido-5-pyrazolones of the 2nd generation, this section deals with the progress after the 2nd generation.

10.4.1 5-Pyrazolone Couplers

Dyes from 5-Pyrazolones

One of the most useful classes of magenta couplers involves 5-pyrazolone derivatives. The absorption maxima and several other properties of azomethine dyes (**62**) derived from the 5-pyrazolone derivatives have been reported by

Weissberger's group [52], as cited in Table 10.5. This list contains absorption maxima for representative dyes derived from 3-alkyl, 3-carbamoyl, and 3-anilino derivatives of 1-phenyl-5-pyrazolone.

Table 10.5. Absorption Maxima of Azomethine Dyes (**62**) Derived by Oxidative Coupling of 4-Amino-3-methyl-*N*, *N*-diethylaniline (CD-2) with 5-Pyrazolones[a] [52]

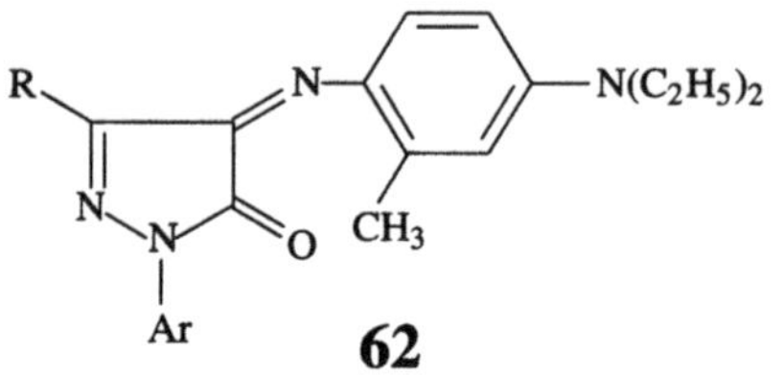

R, N, N, Ar, O, N, CH_3, $N(C_2H_5)_2$

62

Substituents		Primary band (*x*-band)		Secondary band (*y*-band)	
Ar	R	λ_{max} (nm)	ϵ_{max} $\times 10^{-4}$	λ_{max} (nm)	ϵ_{max} $\times 10^{-4}$
Ph	H	546	3.3	449	0.8
Ph	CH_3	522	3.6	448	1.6
Ph	n-$C_{15}H_{31}$	522	3.0	450	1.4
Ph	$COOC_2H_5$	557	4.0	465	1.5
Ph	$CONH_2$	565	3.8	460	1.2
Ph	$CONHCH(C_2H_5)Ph$	526	5.2	440	1.0
Ph	NH_2	506	4.1	430	1.2
Ph	NHPh	522	4.7	432	0.9

In general, azomethine dyes (**62**) derived from 5-pyrazolone derivatives have secondary absorption bands (*y*-bands) along with primary ones (*x*-bands), as collected in Table 10.5. The primary bands undergo strong bathochromic shifts of about 30 to 50 nm in more polar solvents (e.g., butyl acetate) as compared with non-polar solvents (e.g., cyclohexane) [52]. On the other hand, the secondary bands exhibit little shifts by using butyl acetate in place of cyclohexane as a solvent. Because such secondary bands influence the quality of color reproduction, various techniques (e.g., colored couplers) have been invented to overcome this drawback.

The azomethine dye **63** (R = CH_3 or NHCOPh) has been found to be rapidly faded by treating an excess of the mother pyrazolone **64** [53]. This fading has been ascribed to the mechanism shown in Fig. 10.17, where the disproportionation between the dye (**63**) and 2 moles of the couplers (**64**) produces the corresponding leuco dye (**65**) and the bis-pyrazolone (**66**). The leuco dye is further reacted with 2 moles of the couplers to cause formation of the bis-pyrazolone and regeneration of the coupler and color developer (CD-2).

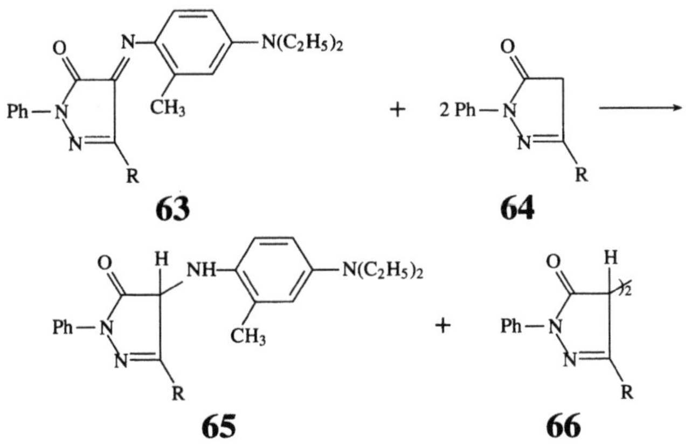

Figure 10.17. Fading of magenta azomethine dyes.

3-Acylamido-5-Pyrazolone Couplers

An oil-protected 3-acylamido-5-pyrazolone coupler (the 2nd generation) has already been described in Chapter 8, where its synthesis has been illustrated in Fig. 8.13. This coupler and related ones having at least two halogen substituents on the 1-phenyl group have been disclosed to exhibit improved stability to light and heat when incorporated in emulsion layers as well as to form dyes having absorption maxima which are shifted toward the red by about 50 to 200 nm as compared with dyes from corresponding couplers with a simple 1-phenyl or 1-chlorophenyl substituent [55]. This type of magenta couplers has now been used as two-equivalent couplers which will be described in Chapter 11.

3-Anilino-5-pyrazolone Couplers

Improved Hues

3-Anilino-5-pyrazolone couplers have been originally disclosed as colored couplers, where 3-NH–$COCH_2O$-C_6H_4-NO_2-*p* in the 5-pyrazolone ring has been reported to be converted into 3-NH–C_6H_4-NO_2-*p* on the action of sodium hydroxide [56]. This reaction has been applied to the production of chlorine-substituted 3-anilino derivatives, which provide magenta dyes exhibiting a sharp-cut profile at the longer wavelength side. Thereby, 3-anilino-5-pyrazolone couplers such as **67** (the 3rd generation) have been developed so as to improve color reproduction as well as the light stability of dye images [57]. This type of magenta couplers has been widely used after introducing various ballast groups. For example, **68** has been reported in a patent on an intermediate film producing a duplicate negative for motion picture films [58]. Such a ballast group can influence the properties of a dye generated from the coupler. For example, the ballast group in **69** has been

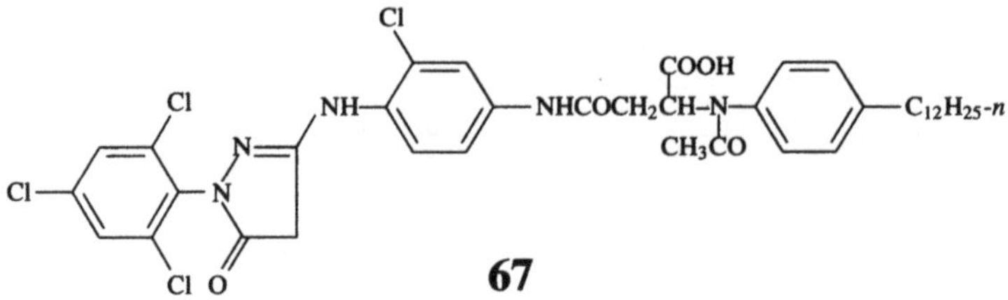

Figure 10.18. Non-diffusible four-equivalent couplers forming a magenta azomethine dye (3-anilino-5-pyrazolone couplers).

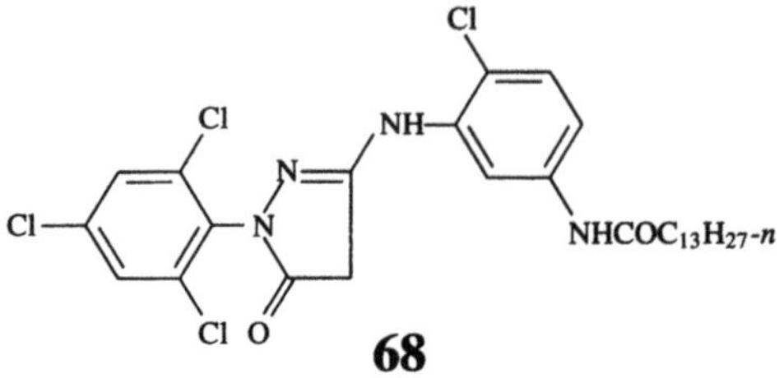

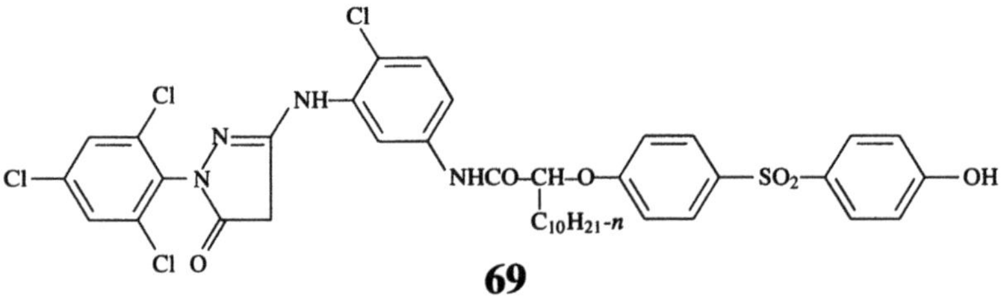

disclosed to enhance the activity of the coupler, which results in an increased maximum dye density and an improved γ-value [59].

Syntheses

Figure 10.19 shows a synthetic pathway for preparing **67** [57]. As already pointed out, the key step is the intramolecular condensation of **72** into **73** with releasing $HOCH_2COONa$, where the alkoxy group is considered to be activated by the *p*-nitro group and the nucleophilic attack of the amide anion forms a five-membered intermediate or transition state. The reduction of the nitro group of **73** is effected by hydrazine on a catalytic action of Raney's nickel.

10.4.2 Pyrazolotriazole Magenta Couplers

Among the properties of 5-pyrazolone magenta couplers described in the preceding subsections, insufficient hues of azomethine dyes derived from them have been one of the most critical drawbacks needed to be improved. As summarized in Table 10.5, there is a secondary absorption in the blue region of light along with an extra absorption in the red region. Although such unnecessary ab-

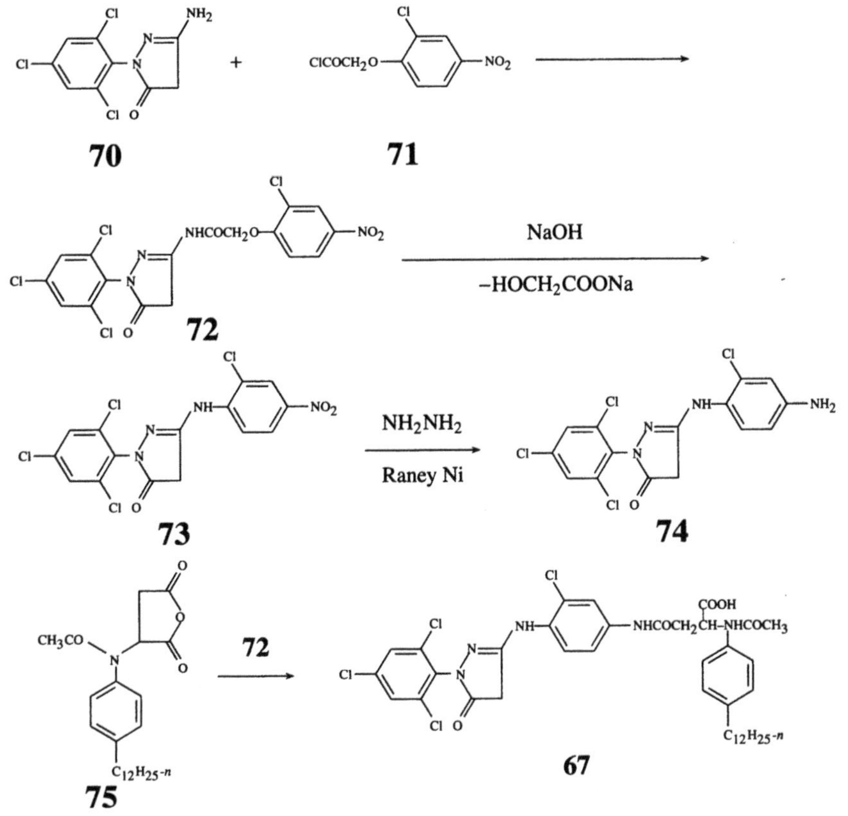

Figure 10.19. Synthetic pathway for preparing a non-diffusible coupler (**67**) of the 3rd generation (3-anilino-5-pyrazolone type).

sorptions can be compensated in negative films by appropriate techniques,[10] such techniques cannot be applied to color papers. This means that magenta couplers generating a dye of an excellent hue should be developed by starting from a new nucleus other than 5-pyrazolones so as to improve the color reproduction of color papers. Two nuclei, 1H-pyrazolo[5,1-*c*][1,2,4]triazole (**49**; EK nucleus) and 1H-pyrazolo[1,5-*b*][1,2,4]triazole (**76**; FF nucleus), have been extensively examined so that color papers using them have been placed on the market.[11] Although

[10]For example, the color-masking technique based on colored couplers has been used for color negative films, as discussed in Chapter 12. As a result of this technique, a developed negative film is orange in color.

[11]In this book, 1H-pyrazolo[5,1-*c*][1,2,4]triazole is called "EK nucleus", since Eastman Kodak has developed it. This nucleus was used in a Kodak negative film (VR 1000) in 1982. On the other hand, 1H-pyrazolo[1,5-*b*][1,2,4]triazole is called "FF nucleus", since it was developed by Fuji Photo Film. This nucleus was used in Fujichrome paper Type 34 in 1986 and in Fujicolor paper HG in 1989.

they appear similar in structure, the different modes of ring fusions have required different approaches to syntheses.

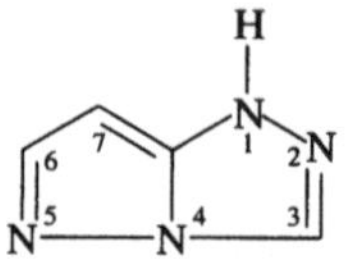

49 (EK nucleus) 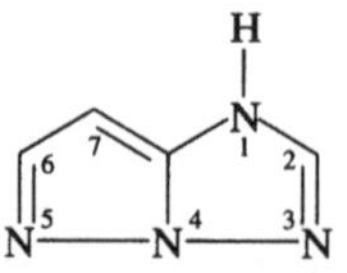 **76** (FF nucleus)

One of the common advantages is such excellent hues that azomethine dyes derived from them have no secondary absorption in the blue region of light and less extra absorption in the red region. On the other hand, the light stability of the dyes has been a main concern, since it is more critical to color papers than to color films.

1H-Pyrazolo[5,1-*c*][1,2,4]triazole Couplers

Azomethine Dyes Derived from 1H-Pyrazolo[5,1-c][1,2,4]triazoles

As one type of such new magenta couplers, 1H-pyrazolo[5,1-*c*][1,2,4]triazoles (**49**) have been reported to give a dye which exhibits a magenta hue superior to those of 5-pyrazolone azomethine dyes [60]. Selected spectral data are listed in Table 10.6. They have higher extinction coefficients than azomethine dyes derived from 5-pyrazolone couplers (see Table 10.5). Moreover, they have no secondary blue absorption and show a sharper cut-off on the long wavelength side.

Merits of 1H-Pyrazolo[5,1-c][1,2,4]triazole Couplers

The use of couplers based on the pyrazolotriazole nucleus in photographic materials has been covered by patents that have emphasized the merit of good hue [61]. The coupling activity of such pyrazolotriazole couplers as **78** also has been enhanced by using the ballast group described for **69**, where the maximum dye density and the γ-value are increased [59].

Light Stability of Dyes

Although azomethine dyes derived from 1H-pyrazolo[5,1-*c*][1,2,4]triazole couplers have good hues, their light stability is lower as compared with dyes derived from 3-anilino-5-pyrazolone couplers. For example, the light stability of the dye derived from 6-methyl-3-undecyl-1H-pyrazolo[5,1-*c*][1,2,4]triazole and CD-3 has been reported to be inferior to that of the dye derived from the 3-anilino-5-pyrazolone coupler (**68**) and CD-3 [62,63]. Thus, only about 35% of the initial density of the former dye remains after one-day irradiation with a xenon lamp (100 klx) through a UV filter, whereas about 85% remains for the latter dye derived from **68** under the same conditions.

The light stability of dyes derived from 1H-pyrazolo[5,1-*c*][1,2,4]triazole couplers has been improved when a *t*-alkyl group was incorporated at the 6-position

Table 10.6. Absorption Maxima of Azomethine Dyes (**77**) Derived by Oxidative Coupling of 4-Amino-3-methyl-*N*, *N*-diethylaniline (CD-2) with 1H-Pyrazolo[5,1-*c*][1,2,4]triazoles [60]

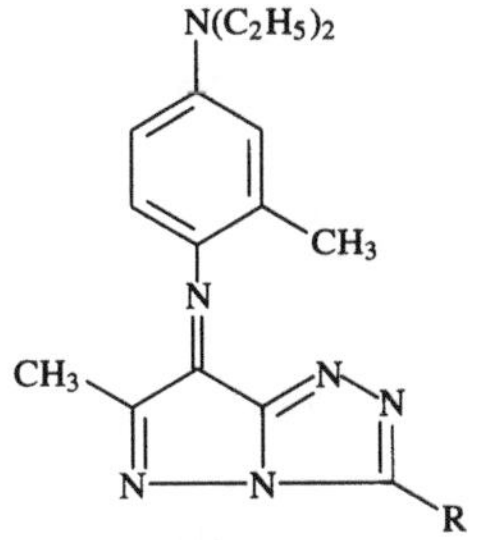

$N(C_2H_5)_2$ ... CH_3 ... N ... CH_3 ... N ... N ... N ... N ... R

77

Substituents R	Absorption[a]	
	λ_{max} (nm)	ϵ_{max} $\times 10^{-4}$
Ph	551	6.77
C_6H_4-CH_3-*o*	541	5.26
C_6H_4-CH_3-*p*	549	6.18
C_6H_4-$(CH_3)_3$-2, 4, 6	531	5.01
C_6H_4-Cl-*o*	539	5.46
C_6H_4-Cl-*p*	555	6.51
C_6H_{13}-*n*	528	4.64

[a]All data were determined in ethyl acetate.

and an aryl group placed at the 3-position, as exemplified by **79** shown in Fig. 10.20 [64]. The light stability of the dye derived from **79** and CD-3 is compared with the corresponding dye in which the 6-*t*-butyl group of **79** is replaced by a methyl group. Thus, 93% of the initial density of the former dye with the *t*-butyl group remains after 300 hr irradiation with a high-intensity fluorescent light source (13 klx), whereas only 59% remains for the latter dye with the methyl group under the same conditions. To assure light stability, a 1H-pyrazolo[5,1-*c*][1,2,4]triazole coupler (**80**) having a *t*-butyl group at the 6-position has been combined with a light stabilizer (**82**) [65,66]. 1H-Pyrazolo[5,1-*c*][1,2,4]triazole couplers having a *t*-alkyl linkage at the 3-position have been disclosed to improve dye light stability in patents [67,68]. More recently, another type of 1H-pyrazolo[5,1-*c*][1,2,4]triazole couplers (e.g., **81**) having a *t*-alkyl linkage at the 3-position as well as a *t*-butyl group at the 6-position has been disclosed to improve dye light stability [69].

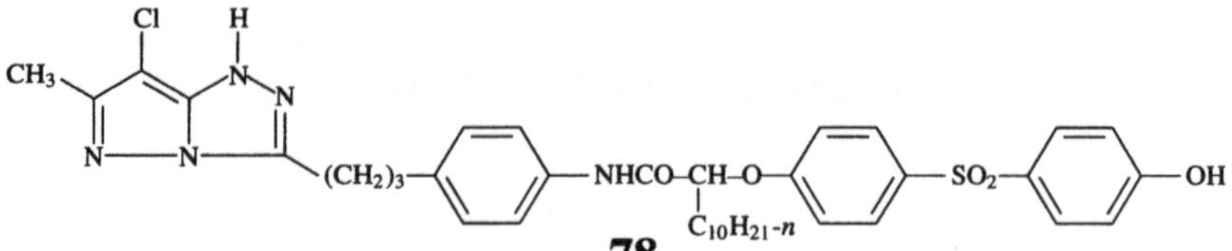

78

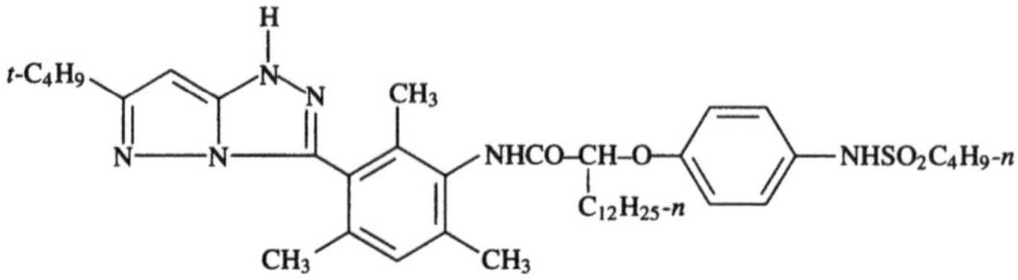

79

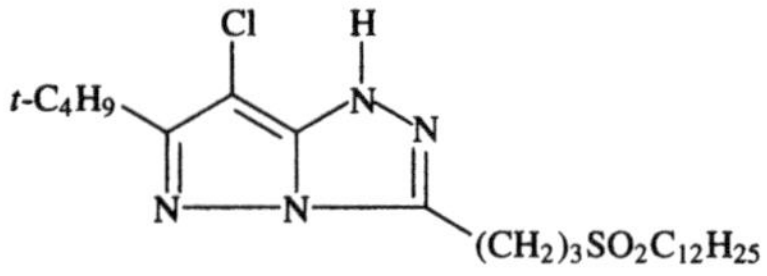

80

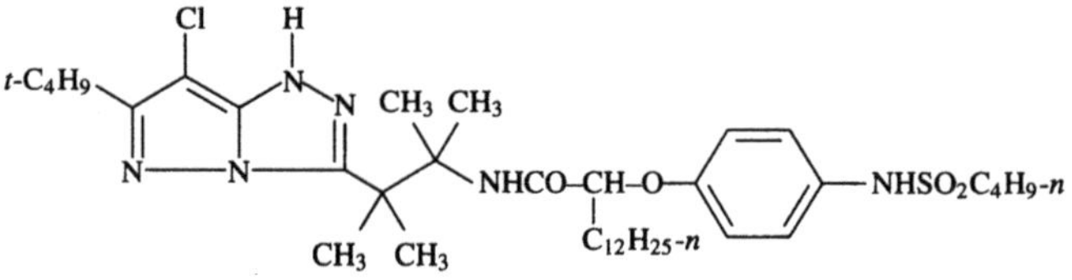

81

Figure 10.20. 1H-Pyrazolo[5,1-*c*][1,2,4]triazole couplers as non-diffusible couplers forming a magenta azomethine dye.

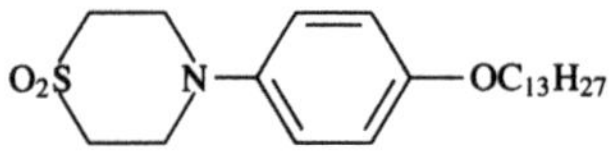

82

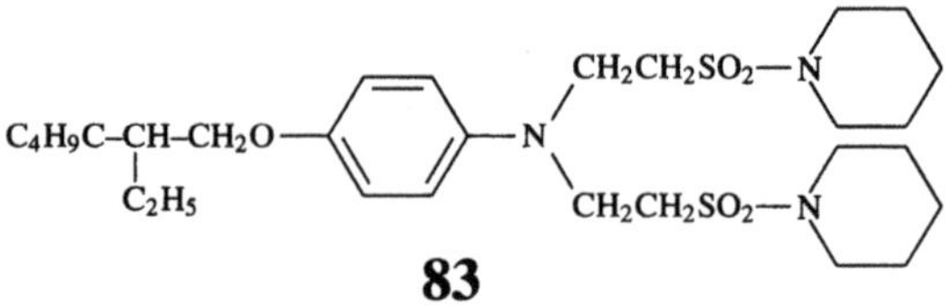

83

Figure 10.21. Light stabilizers for magenta azomethine dyes derived from 1H-Pyrazolo[5,1-*c*][1,2,4]triazole couplers. The light stability of the dye derived from **80** shown in Fig. 10.20 has been improved by adding the light stabilizer (**82**).

Activity vs. Dye Stabilities

Although the incorporation of a *t*-butyl group at the 6-position affords the merit of light stability, it sterically hinders the coupling reaction of an oxidized color developer at the adjacent 7-position of the 1H-pyrazolo[5,1-*c*][1,2,4]triazole nucleus. Even in such cases, the presence of an ionizing group such as a phenolic hydroxyl group (e.g., **78**) and a sulfonamido group (e.g., **79**) in the ballast enhances coupling activity during the development process. In several cases, ballasts containing a sulfone group (e.g., **80**) have provided useful coupling activity without the presence of an ionizing group [70,71]. However, the coupling activity should be further improved without reducing light stability in order to comply with recent advances in processing speed.

A 1H-pyrazolo[5,1-*c*][1,2,4]triazole coupler (**84**) in which the ballast group branches at the adjacent position of the triazole ring and involves a carboxylic group has been disclosed to enhance coupling activity, as shown in Fig. 10.22 [72]. However, the presence of such a water-solubilizing group as a carboxylic acid group causes undesirable color contamination as well as undesirable sensitivity to the pH of the developer solution. Thereby, the relationship between the amount of exposure and the resulting dye density varies for the coupler as a function of the pH of the processing solution. These drawbacks have been overcome by the proposal of **85** with another attaching mode of a water-solubilizing group [73]. Ballast groups having hydroxyl groups also have been disclosed to bring about high coupling activity as well as improved light stability, as exemplified by **86** shown in Fig. 10.22 [74,75]. Couplers having more complicated ballast groups (e.g., **87**) have been proposed to provide dyes with improved light stability [76].

Light Stabilizers and Mechanism of Light Fading

1H-Pyrazolo[5,1-*c*][1,2,4]triazole couplers were first used in color films, because light stability is less critical than in color papers. It has taken more time to incorporate the couplers into color papers, since the light stability is an essential requirement for color papers.

To apply the 1H-pyrazolo[5,1-*c*][1,2,4]triazole couplers to color papers, light stabilizers have been extensively studied, as described in a review [65]. Singlet oxygen quenchers such as **82** have been developed as light stabilizers [77,78], since light fading was shown to be caused by the attack of singlet oxygen to the dyes derived from these couplers [79]. To obtain information about ground-state potential energy surfaces, activation parameters for the thermal *syn-anti* isomerization of pyrazolotriazole azomethine dyes have been examined [80]. Further efforts to develop new light stabilizers have been undertaken continuously. For example, **83** shown in Fig. 10.21 has been disclosed to exhibit a good singlet oxygen quenching rate and compound stability in comparison to **82** [81].

Syntheses

Figure 10.23 shows a synthetic pathway for preparing **79** [64]. Two ring closure reactions should be mentioned as key steps. The first ring closure creates

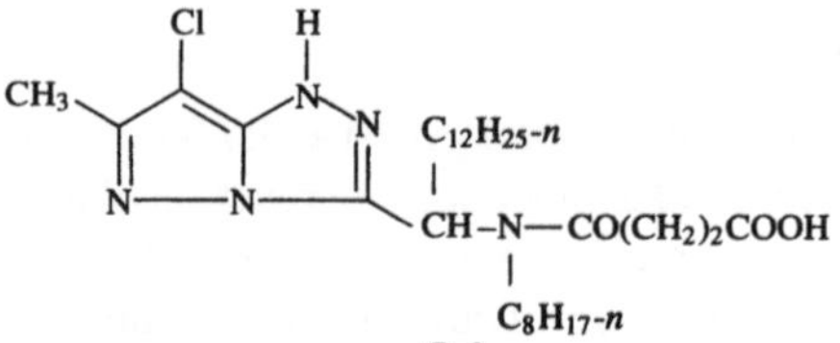

84

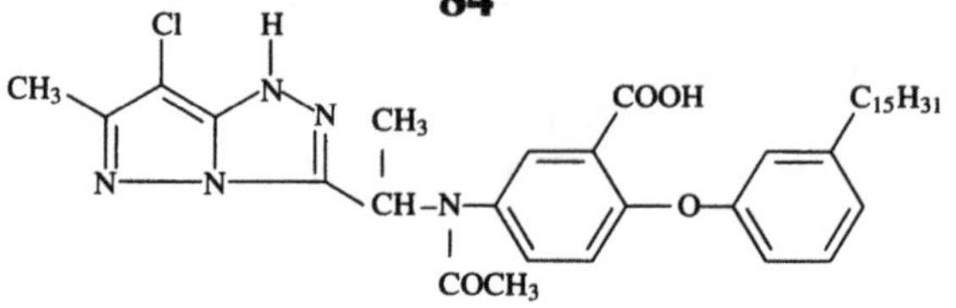

85

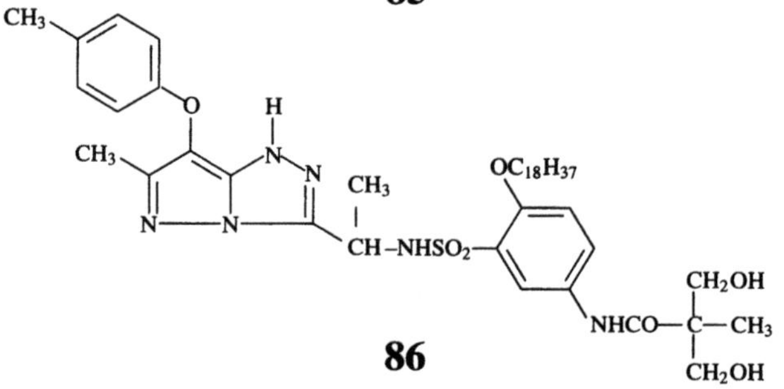

86

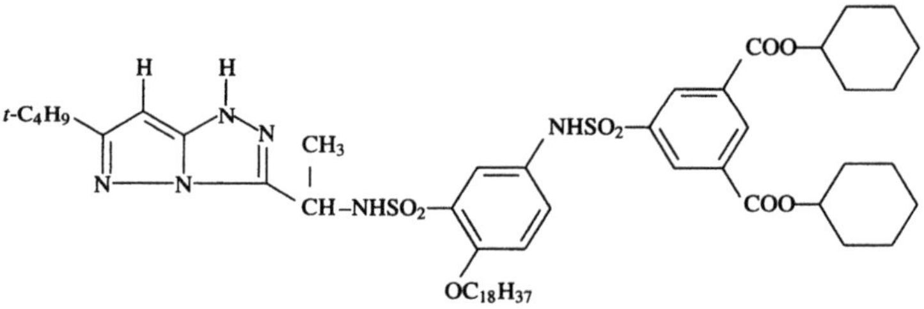

87

Figure 10.22. Various ballast groups for improving the photographic properties of 1H-pyrazolo[5,1-*c*][1,2,4]triazole couplers.

a pyrazole ring (**90** → **91**). The second ring closure is an intramolecular oxidative coupling, which generates a pyrazolotriazole ring (**93** → **94**). The cyano group of the intermediate **94** is removed by a hydrolytic decarboxylation to generate a free 7-position as a coupling site.

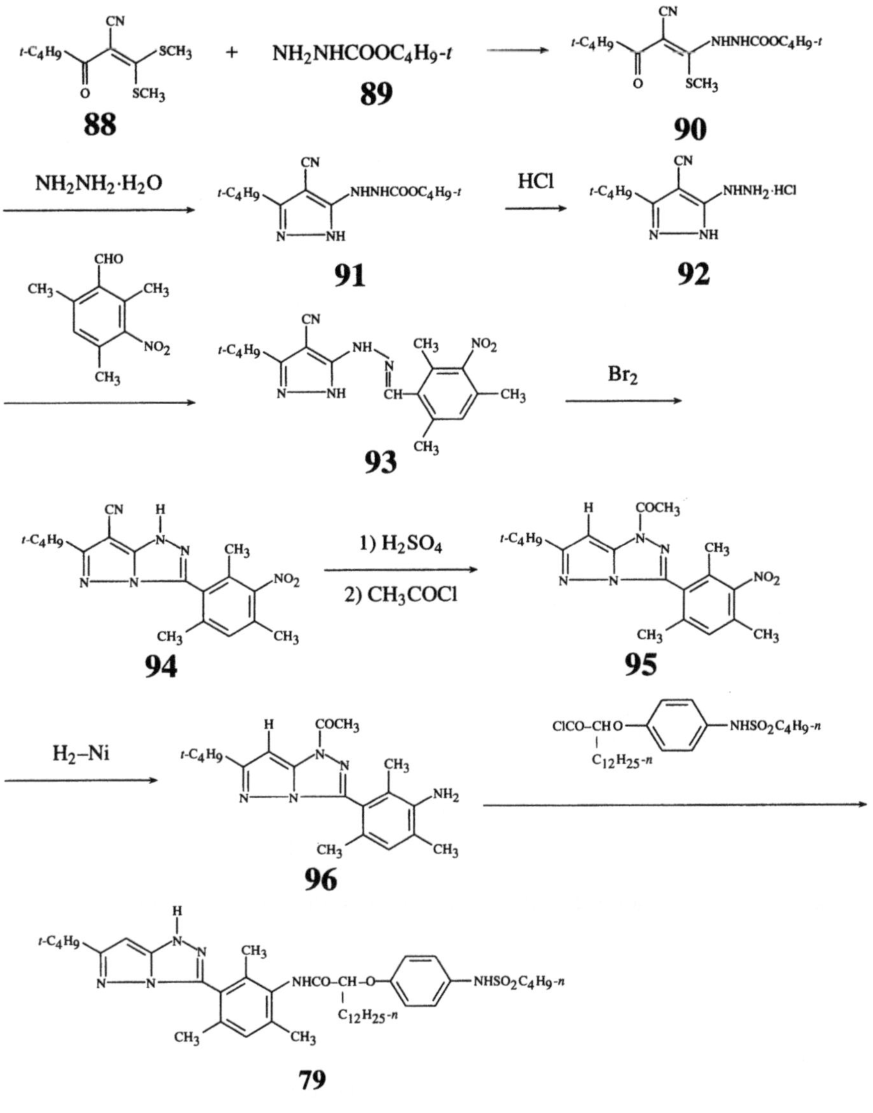

Figure 10.23. Synthetic pathway for preparing the non-diffusible magenta coupler (**79**) of 1H-pyrazolo[5,1-*c*][1,2,4]triazole type.

1H-Pyrazolo[1,5-*b*][1,2,4]triazole Couplers

Azomethine Dyes Derived from 1H-Pyrazolo[1,5-b][1,2,4]triazoles

As another type of pyrazolotriazole couplers, magenta couplers based on the 1H-pyrazolo[1,5-*b*][1,2,4]triazole nucleus (FF nucleus) have been disclosed to give a dye which exhibits a magenta hue superior to those of 5-pyrazolone azomethine dyes [62,63]. The dye (**97**) derived from 3,6-dimethyl-1H-pyrazolo[1,5-*b*][1,2,4]triazole and CD-3 exhibits no secondary absorption in the blue region of light and less absorption in the red region, as shown in Fig. 10.24 [62,82]. Figure 10.24 also shows the absorption curve of **98** derived from a 3-anilino-5-pyrazolone coupler, which exhibits a secondary absorption at about 430 nm and a rather higher absorption at about 580–620 nm as compared with that of **97**.

Merits of 1H-Pyrazolo[1,5-b][1,2,4]triazole Couplers

To develop the 1H-pyrazolo[1,5-*b*][1,2,4]triazole nucleus (FF nucleus), several pyrazoloazoles have been examined in addition to the 1H-pyrazolo[5,1-*c*][1,2,4]triazole nucleus (EK nucleus) described above. Table 10.7 collects dyes derived from them [63]. The dye **97** derived from the FF nucleus exhibits λ_{max} at a higher wavelength than the corresponding dye (**100**) derived from the EK nucleus. They have absorption curves of substantially equal profile in terms of half bandwidths and *S* values. Table 10.7 also contains data of **99** (R = $C_{13}H_{27}$), in which the *S* value is larger than those of the other data listed in the table.

Substituents at the 6-position of the FF nucleus influence the absorption maxima of azomethine dyes. The substituent effect on the absorption maxima has been investigated to show that the wavenumbers ν_{max} (= $1/\lambda_{max}$) are linearly related to Hammet's σ_p-values [63].

The use of couplers based on the FF nucleus in photographic materials has been covered by patents that have emphasized the merit of good hue, as summarized in Table 10.7 [82,83,84]. For example, a four-equivalent coupler (**103**) and the corresponding two-equivalent coupler (**104**) shown in Fig. 10.25 have been disclosed [82].

Light Stability vs. Activity

With respect to the light stability of the dye derived from 6-methyl-3-undecyl-1H-pyrazolo[1,5-*b*][1,2,4]triazole and CD-3, it has been reported that about 80% of the original density of the dye remains after one-day irradiation with a xenon lamp (100 klx) through a UV filter [63]. The data show that the light stability of the dye is substantially equal to that of the dye derived from the 3-anilino-pyrazolone coupler (**68**), since about 85% remains under the same conditions. A similar experiment has shown that only about 35% of the initial density remains for the dye from 6-methyl-3-undecyl-1H-pyrazolo[5,1-*c*][1,2,4]triazole (see page 190). Hence, we can safely say that the light stability of dyes derived from 1H-

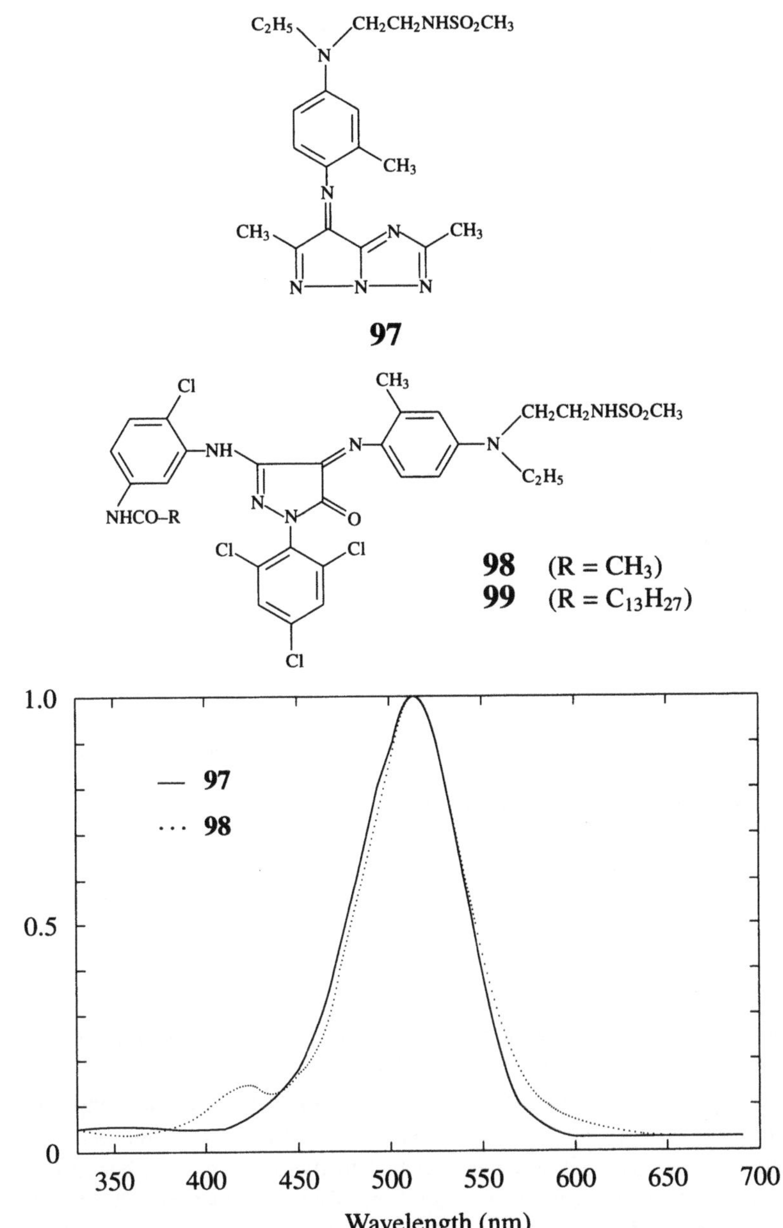

Figure 10.24. Normalized absorption spectrum of the dye (**97**) derived from 3,6-dimethyl-1H-pyrazolo[1,5-*b*][1,2,4]triazole in comparison with that of the dye (**98**) derived from a 3-anilino-5-pyrazolone coupler [82].

Table 10.7. Absorption Properties of Dyes Derived from Pyrazoloazoles [62,63]

Structure[a]	Absorption in Ethyl Acetate		
	λ_{max} (nm)	$W_{1/2}$[b] (nm)	S[c]
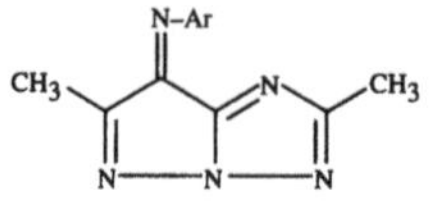 97	527	66	0.053
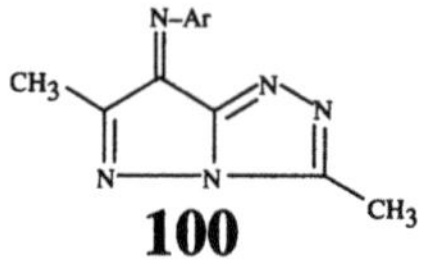 100	521	66	0.049
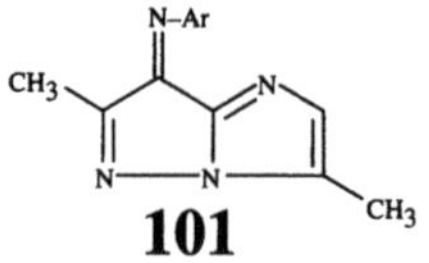 101	520	78	0.050
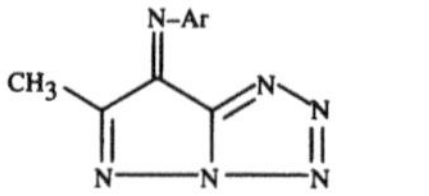 102	550	66	0.044
99 (Comparison)	527	67	0.134

[a]The substituent Ar comes from CD-3. For the full structure of **97**, see Fig. 10.24.

[b]The symbol $W_{1/2}$ represents a half bandwidth.

[c]The S value is defined by $S = D_{max+60}/D_{max}$ where D_{max} represents the optical density at λ_{max} and D_{max+60} is the optical density at the wavelength $\lambda_{max}+60$ nm. The S value was originally introduced for characterizing yellow couplers. See eq. 10.1.

pyrazolo[1,5-*b*][1,2,4]triazole couplers is superior to that of dyes derived from 1H-pyrazolo[5,1-*c*][1,2,4]triazole couplers.

The dye derived from **105** and CD-3 changes its spectral shape at higher concentrations, where a shoulder appears at the short-wavelength side of the absorption curve. This shoulder is ascribed to dye aggregation, because a higher branched alkyl group (ethyl, isopropyl, or *t*-butyl), which is substituted for the methyl group at the 6-position of **105**, exhibits a less pronounced extent of such shoulder appearance even in high concentrations [63]. In separate experiments, it has been found that dyes derived from **105** become more stable to light, as the substituent at the 6-position becomes bulkier (methyl, ethyl, isopropyl, and *t*-butyl). These facts have strongly suggested that the dye aggregation causes the dye fading. According to further experiments, the fading of a dye from a 1H-

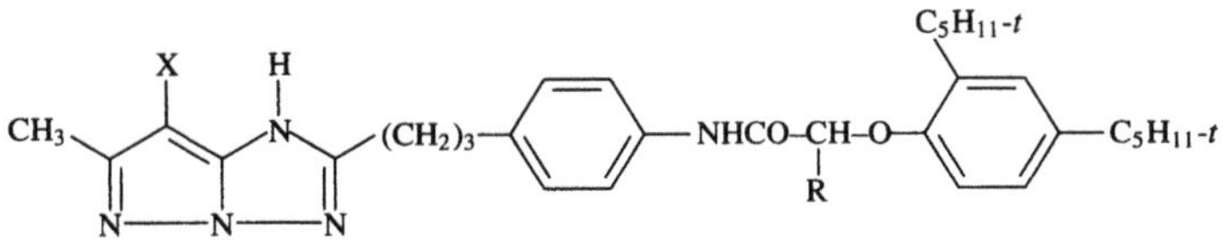

103 (X = H and R = C_6H_{13}-*n*)

104 (X = 3,5-dimethyl-1,2,4-triazol-4-yl and R = C_6H_{13}-*n*)

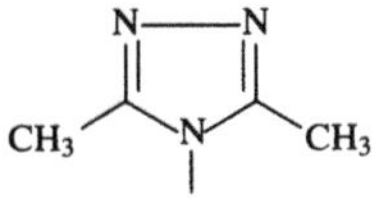

105 (X = H and R = C_4H_9-*n*)

Figure 10.25. 1H-Pyrazolo[1,5-*b*][1,2,4]triazole couplers as non-diffusible couplers forming a magenta azomethine dye.

pyrazolo[1,5-*b*][1,2,4]triazole coupler occurs in the presence of oxygen, where two molecules of the dye participate in the fading reaction.

The light stability of dyes derived from 1H-pyrazolo[1,5-*b*][1,2,4]triazole couplers has been improved when a *t*-alkyl group is incorporated at the 6-position and an aryl group is placed at the 2-position, as exemplified by **106** shown in Fig. 10.26 [85,86,87]. This coupler is capable of rapid processing and produces an image dye with excellent stability. To improve the coupling efficiency of this coupler, a water-solubilizing group such as a sulfonamido group has been incorporated in the ballast group, as exemplified by **107** [88].

The dye aggregation can be alternatively inhibited by the introduction of branching at the adjacent position of the 2-substituent (a ballast group), as embodied in the coupler (**108**) [63]. Moreover, the coupling activity can be enhanced by using high-boiling solvents having a dielectric constant not less than 4.00 (at 25℃ and 10 kHz) [89]. It follows that the light stability of couplers based on the FF nucleus can be assured without reducing the coupling activity. This methodology is parallel to that of the couplers based on the EK nucleus described in Fig. 10.22.

The coupling rates have been compared, as shown in Table 10.8 [54].[12] The pK_a value of **110** (EK nucleus) is higher than that of **109** (FF nucleus). The dye-forming rate of the fully dissociated coupler anion of **110** (EK nucleus) is larger than that of **109** (FF nucleus). However, the couplers in a coated layer are not fully dissociated under usual processing conditions of color films and color papers (at about pH 10). The dye-forming rates for such actual conditions can be estimated by determining coupling rates at pH 7.8 in aqueous solutions. Under

[12] Although the structure of the ballast group in **109** and **110** has not been disclosed in this review, see **108** as a typical example of ballast groups.

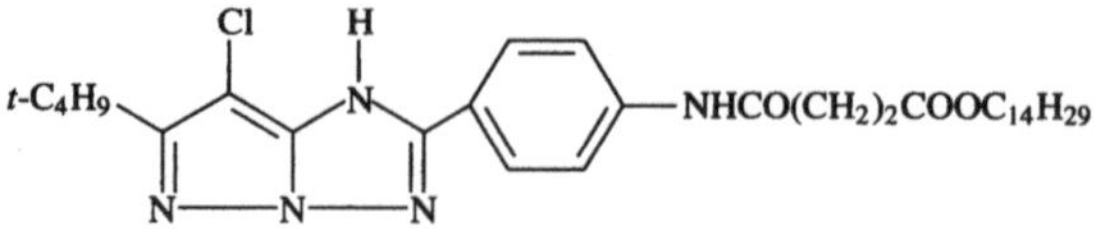

106

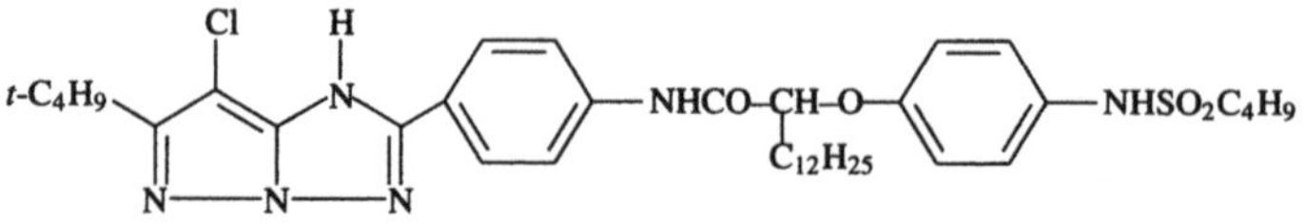

107

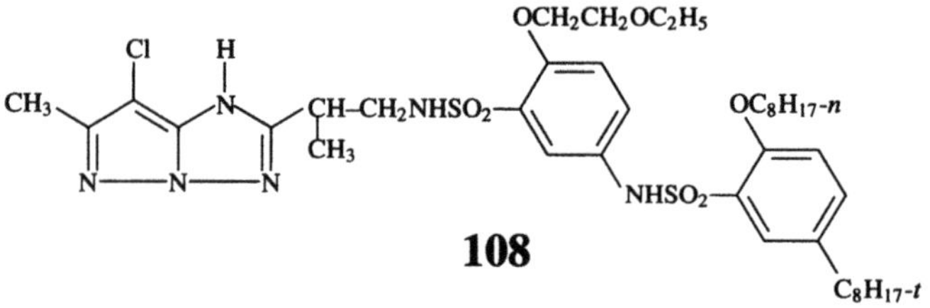

108

Figure 10.26. 1H-Pyrazolo[1,5-*b*][1,2,4]triazole couplers with various substituents.

Table 10.8. Comparison Between Magenta Couplers Based on FF and EK Nuclei [54].

Structure[a]	pK_a[b]	Dye-forming rate[c]		Absorption[d]	
		as Cp- anion	at pH 7.8	λ_{max} (nm)	ϵ $\times 10^{-4}$
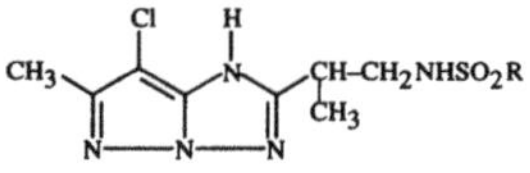 **109** (FF)	8.11	1	4.4	539.6	5.48
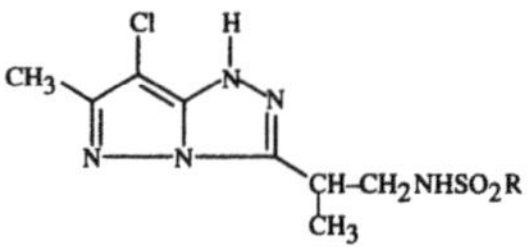 **110** (EK)	10.73	64	1	527.9	5.58

[a] The ballast group R is common to both couplers of FF and EK types.

[b] Measured in tetrahydrofuran/water (6/4).

[c] Measured in aqueous solutions by the stopped flow method by using CD-3 as a color developer and potassium hexacyanoferrate(III) as an oxidizing agent.

[d] Measured in ethyl acetate for the corresponding dyes with the moiety derived from CD-3.

this condition (at pH 7.8), the dye-forming rate of **109** (FF nucleus) is larger than that of **110** (EK nucleus).

Light Stabilizers and Mechanism of Light Fading

Since the light fading of dyes derived from 1H-pyrazolo[1,5-*b*][1,2,4]triazole couplers has been found to be caused by the attack of oxygen on (excited) dye aggregates, various antioxidative agents and related compounds have been examined in order to find effective light stabilizers. Among them, a spiroindane derivative (**111**) [90] and a hydroquinone diether (**112**) [91] are remarkably effective to inhibit the light fading of dyes derived from such couplers as **108** [63].

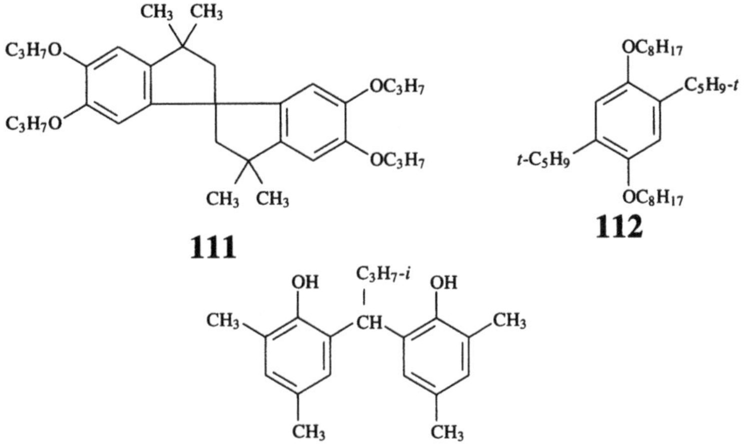

CH_3 CH_3 C_3H_7O C_3H_7O OC_3H_7 OC_3H_7 CH_3 CH_3

111

OC_8H_{17} C_5H_9-*t* *t*-C_5H_9 OC_8H_{17}

112

OH C_3H_7-*i* OH CH_3 CH CH_3 CH_3 CH_3

113

Figure 10.27. Light stabilizers for magenta azomethine dyes derived from 1H-pyrazolo[1,5-*b*][1,2,4]triazole couplers.

The spiroindane derivative (**111**) has been found to work as a quencher of an excited state of a dimeric dye or as a quencher of an excimer [63,92]. Thus the dye derived from 5-methyl-3-undecyl-1H-pyrazolo[1,5-*b*][1,2,4]triazole and CD-3 exhibits a fluorescence peak at about 746 nm at 173 K, which is ascribed to the dimeric excited state (or the excimer).[13] The dimer peak has been found to be quenched efficiently by the spiroindane derivative (**111**).

Anti-Staining Agents

Rapid processing machines (known popularly as "mini-labos") are used today for processing color negative films and color papers. They sometimes cause troubles of new type, e.g., stains due to color developers remaining in a developed film

[13] At higher temperature (213 K), the dye exhibits a fluorescence peak due to a monomer at about 636 nm.

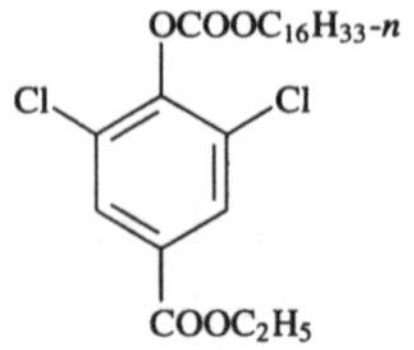

$OCOOC_{16}H_{33}$-*n*
Cl Cl
$COOC_2H_5$

114

SO_2H
$C_{14}H_{29}OCO$ $COOC_{14}H_{29}$

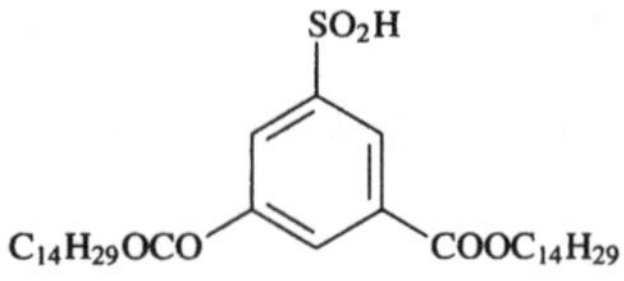

115

Figure 10.28. Anti-staining agents for magenta azomethine dyes derived from 1H-pyrazolo[1,5-*b*][1,2,4]triazole couplers.

or paper, because complete finishing is not always assured. Such stains become conspicuous during storage particularly in low magenta density regions, since more magenta couplers remain unreacted in these regions than in high magenta density regions. To prevent this type of stains, anti-staining agents such as **114** and **115** are used together with light stabilizers such as **111** and **113** [93]. The former anti-staining agent (**114**) is a scavenger of a color developers, which is converted into the corresponding carbamate. On the other hand, the latter compound (**115**) scavenges an oxidized color developer (a quinonediimide), which is converted into the corresponding sulfonamide even under neutral conditions [63].

Syntheses

The 1H-pyrazolo[1,5-*b*][1,2,4]triazole nucleus (FF nucleus) had not been synthesized before its application to photographic materials described above. The R&D for developing new synthetic routes of the FF nucleus has been reviewed in detail [94]. In order to include the FF nucleus into a coupler molecule, an intermediate which has an appropriate group (e.g., an amino group) for linking a ballast group is necessary to be synthesized. When an amino group is selected as a linking group, a ballast group is linked through a carbamido or sulfonamido group in most cases.

Figure 10.29 shows a synthetic pathway for preparing the four-equivalent coupler (**103**) and the corresponding two-equivalent coupler (**104**) [84]. One of the key steps of this pathway is the ring closure of **120** to give a 1H-pyrazolo[1,5-*b*][1,2,4]triazole (**121**), where the ring-closure reaction contains an N–N bond formation with releasing a tosyloxy (*p*-toluenesulfonyloxy) group. The nitro group on the benzene ring of the intermediate **121** is reduced into an amino group, which is the site at which a ballast group is attached, as found in the four-equivalent coupler **103**. To convert this coupler into the corresponding two-equivalent coupler, a nitroso group is placed at the 7-position of **103** and is reduced into an amino group to give **123**. Then, a triazole ring is formed by virtue of the amino group so as to yield the two-equivalent coupler (**104**).

Figure 10.30 illustrates another example of the preparation of a two-equivalent coupler that includes a chlorine atom as a leaving group [94]. The amino group

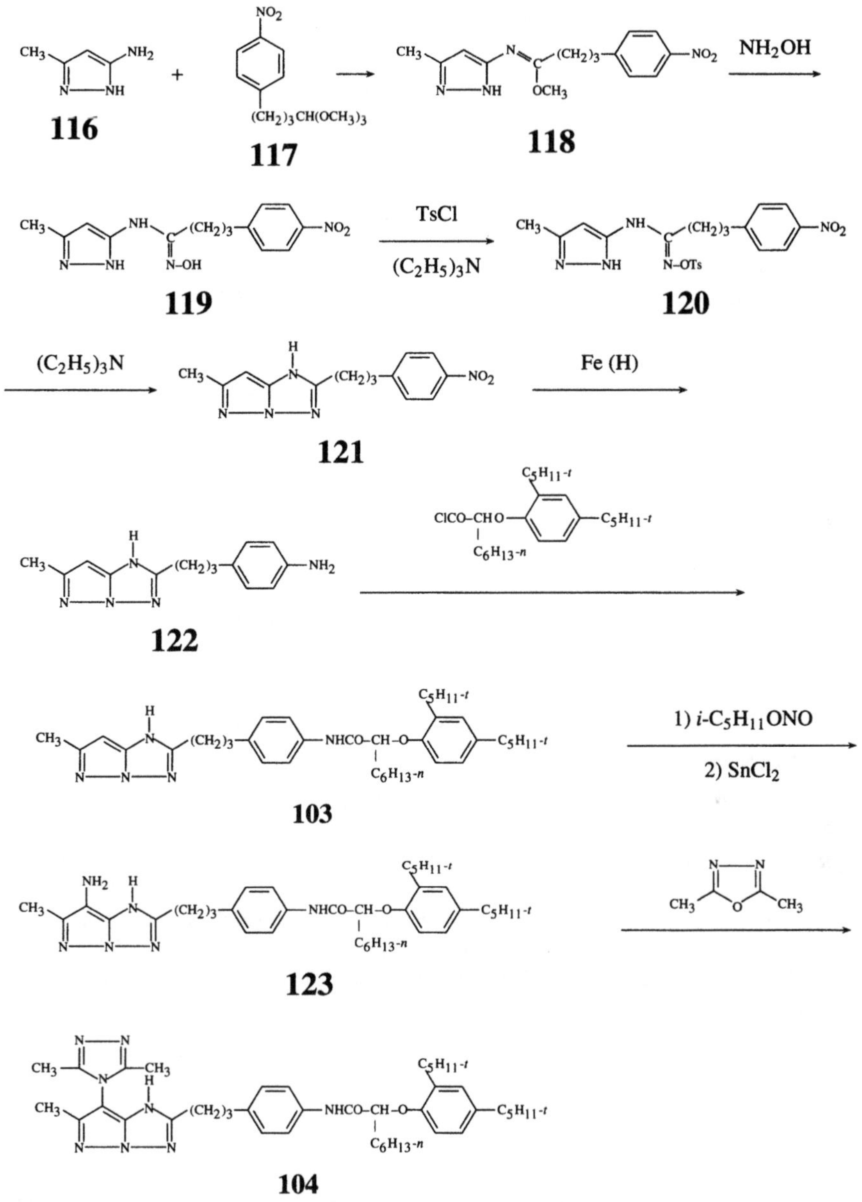

116 + 117 → 118 $\xrightarrow{NH_2OH}$ 119 $\xrightarrow[(C_2H_5)_3N]{TsCl}$ 120 $\xrightarrow{(C_2H_5)_3N}$ 121 $\xrightarrow{Fe\ (H)}$ 122 $\xrightarrow{ClCO-CH(C_6H_{13}\text{-}n)O-C_6H_3(C_5H_{11}\text{-}t)_2}$ 103 $\xrightarrow[2)\ SnCl_2]{1)\ i\text{-}C_5H_{11}ONO}$ 123 → 104

Figure 10.29. Synthetic pathway for preparing the non-diffusible magenta couplers (**103** and **104**) of 1H-pyrazolo[1,5-*b*][1,2,4]triazole type [84]. The symbol Ts represents a *p*-toluenesulfonyl group.

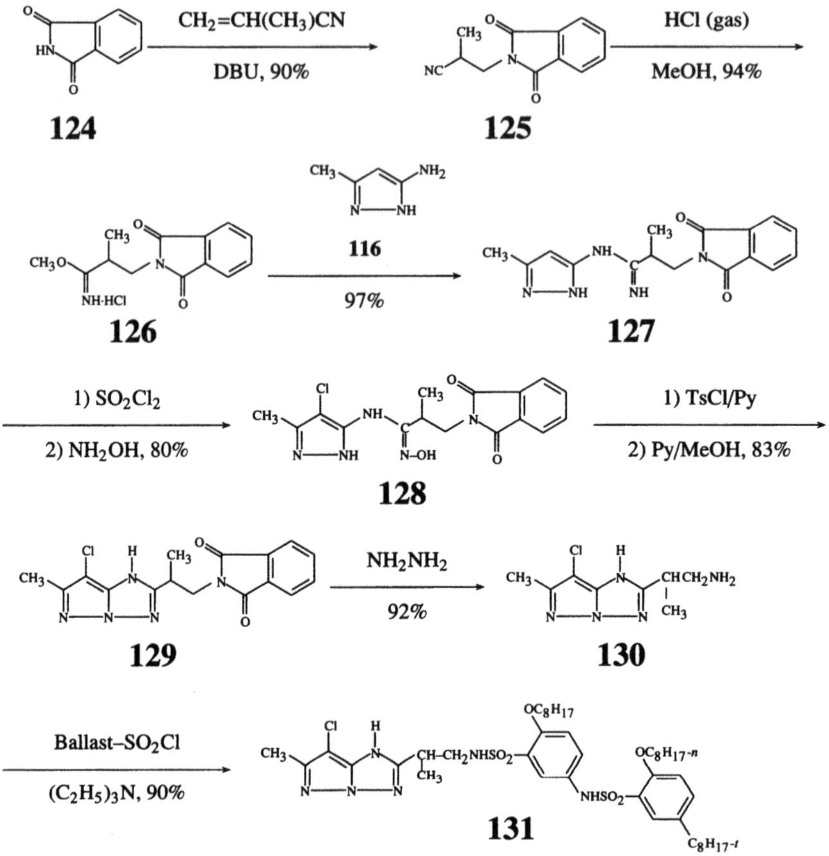

Figure 10.30. Synthetic pathway for preparing the non-diffusible magenta coupler (**131**) of 1H-pyrazolo[1,5-*b*][1,2,4]triazole type [94]. DBU is the acronym of 1,8-diaza-bicyclo[5.4.0]undec-7-ene. MeOH is the abbreviation of methanol. The symbols Ts and Py represent a *p*-toluenesulfonyl group and pyridine, respectively. The symbol Ballast represents the ballast group shown in **131**. The value shown below each arrow is the yield of the step at issue.

is protected as a phthalimide group before reaching the key intermediate **129** and is regenerated by removing the protective group. Note that 5-amino-3-methylpyrazole (**116**) contains an amino group, which should be differentiated from the protected one in the intermediate (**126**). The regenerated amino group in **130** is the site at which a ballast group is attached, as found in the two-equivalent coupler **131**.

Application to Digital Printing

Since azomethine dyes derived from 1H-pyrazolo[1,5-*b*][1,2,4]triazoles have good hue, they have been applied to digital printing, especially, as magenta dyes for a thermal transfer image recording method [95].

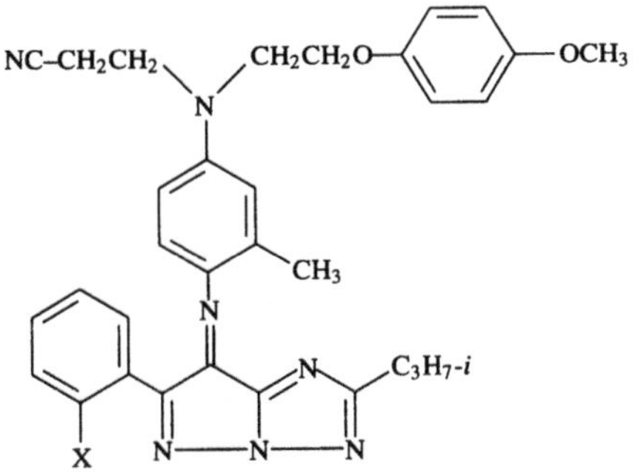

132 X = Cl
133 X = H

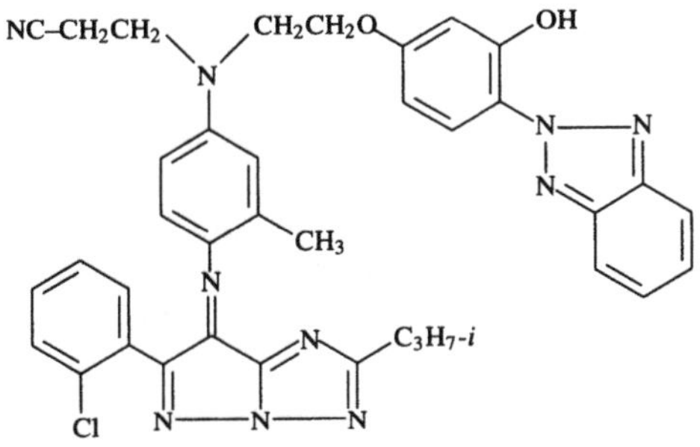

134

Figure 10.31. Azomethine dyes derived from 1H-pyrazolo[1,5-*b*][1,2,4]triazoles for digital printing.

When a 2-chlorophenyl group is placed at the 6-position of 1H-pyrazolo[1,5-*b*][1,2,4]triazole, the resulting azomethine dye (**132**) exhibits an absorption with a steep profile both at the long and the short wavelength side as compared with the azomethine dye (**133**) having an unsubstituted phenyl group [96].

In general, the addition of light stabilizers in a thermal transfer material is less effective than in the photographic material described above. Note that a light stabilizer molecule and an azomethine dye molecule are present in an oil droplet in case of the photographic material. This means that their intimate interaction may cause the quenching of fluorescence, as described above. On the other hand, the dye molecule is deposited rather separately from the light stabilizer molecule in the thermal transfer material so that the light stabilizing effect is difficult to work. To make the light stabilization more effective, a light stabilizer moiety has been incorporated into an azomethine dye molecule [96]. For example, **132** and **133** involve a *p*-methoxyphenoxy group as a light stabilizing moiety [97]. The dye **134** involves a UV-absorbing moiety according to a similar guiding principle.[14] 1H-Pyrazolo[1,5-*b*][1,2,4]triazoles generate azo dyes by coupling with appropriate diazo compounds. The resulting azo dyes have been applied to a thermal transfer material as yellow dyes [98].

10.5 Yellow Couplers

From the beginning of the chromogenic process, active methylene compounds such as acylacetamides have been used as yellow couplers. Among them, benzoylacetanilides and pivaloylacetanilides[15] are representative. This section is mainly devoted to the introduction of couplers of these types.

10.5.1 Benzoylacetanilide Couplers

Dyes from Benzoylacetanilides

A representative class of yellow couplers involves benzoylacetanilide derivatives. The basic data of absorption maxima of azomethine dyes (**135**) derived from benzoylacetanilide derivatives have been reported by Weissberger's group [1], where 4-amino-3-methyl-*N*, *N*-diethylaniline (CD-2) is used as a color developer. Several data are cited from this article and collected in Table 10.9.

In addition to the absorption maximum (λ_{max}) of each dye, the shape of the absorption curve is important to improve the quality of color reproduction. In particular, the sharp-cut slope is desirable at the longer side of the absorption curve. To estimate whether the absorption is sharp-cut or not, the S values defined by the following equation have been determined:

$$S = \frac{D_{max+60}}{D_{max}} \tag{10.1}$$

where D_{max} represents the optical density at λ_{max} and D_{max+60} is the optical density at the wavelength λ_{max} + 60 nm. By the inspection of the S values collected in Table 10.9, the substitution of a methoxy group or a chlorine atom at the *o*-position of anilide is found to make an absorption curve sharp-cut [1]. This

[14] Remember that a photographic material (e.g., a color paper) contains an ultraviolet filter layer at the top of its multilayer structure. The thermal transfer material is incapable of assuming such multilayer structure in principle.

[15] The term "pivaloyl" is commonly used to refer to a trimethylacetyl group in photographic science and technology.

Table 10.9. Absorption Maxima of Azomethine Dyes (**135**) Derived by Oxidative Coupling of 4-Amino-3-methyl-*N*, *N*-diethylaniline (CD-2) with Benzoylacetanilides[a] [1]

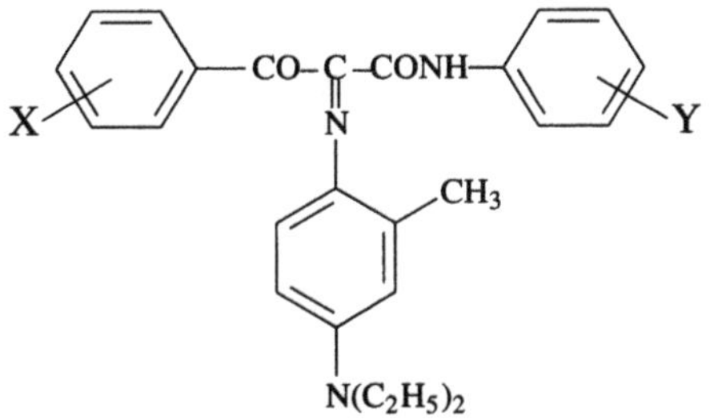

135

Substituents		λ_{max} (nm)	$\epsilon_{max} \times 10^{-4}$	*S*
X	Y			
H	H	433	1.6	0.20
H	*o*-OCH_3	432	1.9	0.17
H	*m*-OCH_3	436	1.7	0.20
H	*p*-OCH_3	430	1.5	0.23
H	*o*-Cl	442	2.2	0.18
H	*m*-Cl	438	1.6	0.22
H	*p*-Cl	438	1.7	0.22
H	*o*-NO_2	460	1.9	0.23
H	*m*-NO_2	441	1.7	0.24
H	*p*-NO_2	451	2.1	0.21
o-OCH_3	H	420	1.0	0.31
m-OCH_3	H	434	1.6	0.20
p-OCH_3	H	430	1.6	0.20
p-Cl	H	436	1.6	0.25
m-NO_2	H	443	1.4	0.38
p-NO_2	H	430	1.5	0.42

[a] All data were determined in butyl acetate.

effect is ascribed to the chelation of the lone pair of the methoxy oxygen or the chlorine atom to the *N*-hydrogen of the anilide.

Ballast Groups of Benzoylacetanilide Couplers

To incorporate benzoylacetanilides into photo-sensitive layers, a ballast group should be introduced into the structure represented by **135**. Either of the substituents X and Y can be selected as a ballast group, as shown in Fig. 10.32. Thus X is used as a ballast group for the non-diffusible coupler **136**, while Y is used as

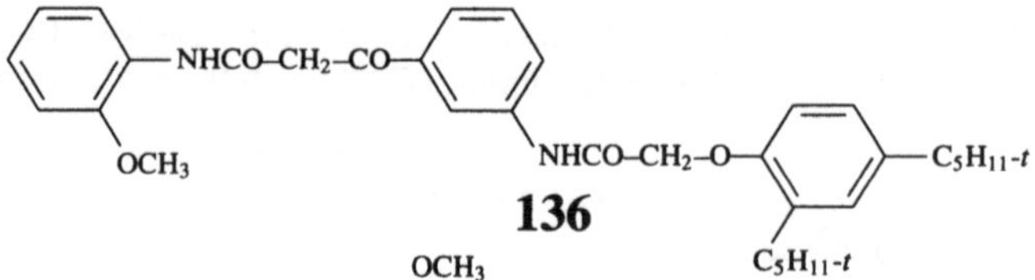

Figure 10.32. Non-diffusible four-equivalent couplers forming a yellow azomethine dye (benzoylacetanilides).

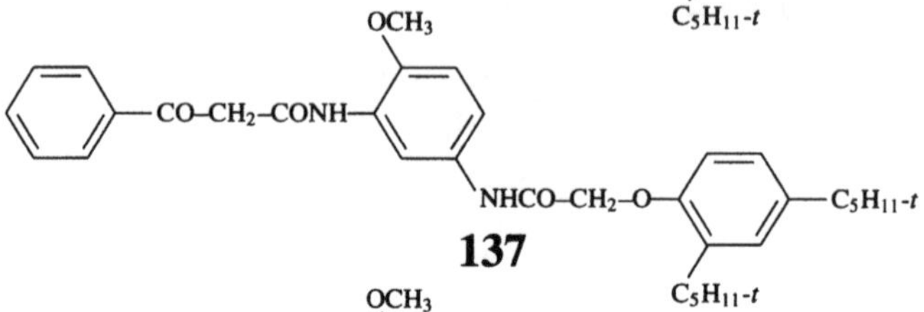

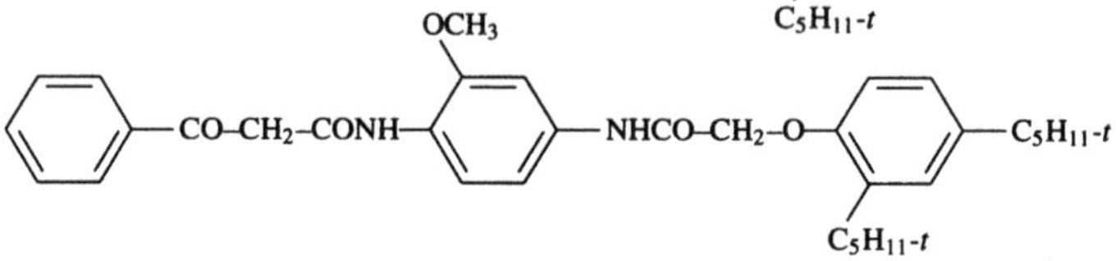

a ballast group for the non-diffusible couplers **137** and **138** [16]. Note that these couplers have a methoxy group at the *o*-position of the anilide moiety.

Syntheses

The synthetic pathways for preparing the yellow couplers (**136** and **137**) are summarized in Fig. 10.33. Key intermediates are coupler nuclei having an amino group such as **144** and **148**. The amino group affords a hold to the linkage of a ballast group, which is introduced by using α-(2,4-di-*t*-amylphenoxy)acetyl chloride (**140**).

The intermediate (**140**), which is derived from 2,4-di-*t*-amylphenol (**139**), has already appeared in several synthetic pathways as a key for introducing a ballast group. As found easily, it is a common intermediate for preparing the couplers (**136** to **138**) shown in Fig. 10.32. To construct a benzoylacetanilide chromophore, an appropriate benzoylacetate (**142** or **145**) is condensed with an aromatic amine (**141** or **146**). The nitro group of the resulting benzoylacetanilide (**143** or **147**) is reduced chemically (Fe powder) or catalytically (Raney's nickel and hydrogen) to give an amino group (**144** or **148**), which is then reacted with the ballast intermediate (**140**) to produce the final coupler (**136** or **137**).

The coupler **138** is synthesized according to the essentially same pathway as that of the coupler **137** (Fig.10.33c), except that 2-methoxy-5-nitroaniline (**146**) is replaced by 2-methoxy-4-nitroaniline as an intermediate and except that the Raney-nickel-catalyzed hydrogenation (**147** $\rightarrow$ **148**) is replaced by the reduction with iron powder (similar to the step **143** $\rightarrow$ **144**).

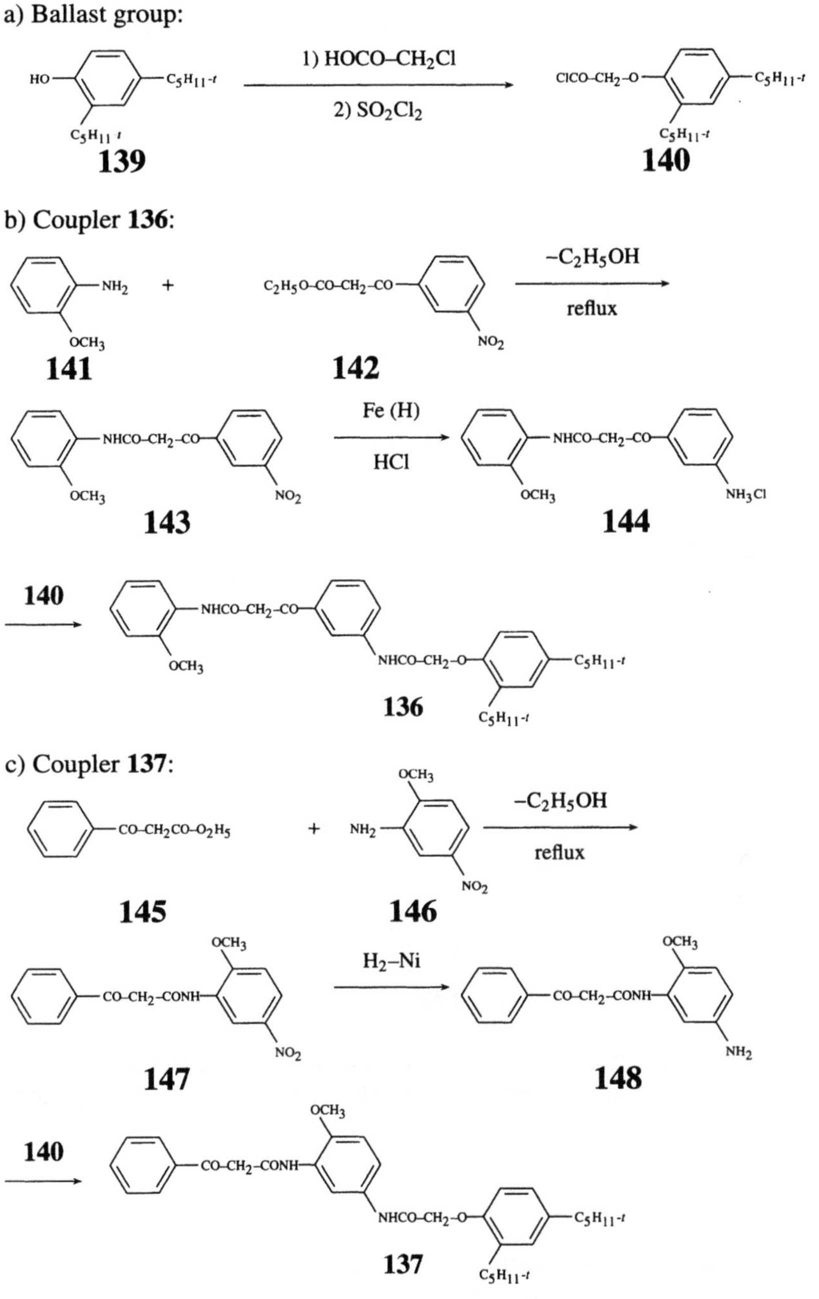

a) Ballast group:

HO– –C_5H_{11}-*t*, C_5H_{11}-*t*

139

1) HOCO–CH_2Cl

2) SO_2Cl_2

ClCO–CH_2–O– –C_5H_{11}-*t*, C_5H_{11}-*t*

140

b) Coupler **136**:

NH_2, OCH_3

141

+

C_2H_5O–CO–CH_2–CO–, NO_2

142

–C_2H_5OH

reflux

NHCO–CH_2–CO–, OCH_3, NO_2

143

Fe (H)

HCl

NHCO–CH_2–CO–, OCH_3, NH_3Cl

144

140

NHCO–CH_2–CO–, OCH_3, NHCO–CH_2–O– –C_5H_{11}-*t*, C_5H_{11}-*t*

136

c) Coupler **137**:

CO–CH_2CO–O_2H_5

145

+

OCH_3, NH_2, NO_2

146

–C_2H_5OH

reflux

CO–CH_2–CONH–, OCH_3, NO_2

147

H_2–Ni

CO–CH_2–CONH–, OCH_3, NH_2

148

140

CO–CH_2–CONH–, OCH_3, NHCO–CH_2–O– –C_5H_{11}-*t*, C_5H_{11}-*t*

137

Figure 10.33. Synthetic pathways for preparing the non-diffusible benzoylacetanilide couplers (**136** and **137**).

10.5.2 Pivaloylacetanilide Couplers

Another representative class of yellow couplers involves pivaloylacetanilide derivatives, which give yellow dyes of high stability of light and heat [99]. Figure 10.34 shows representative couplers of this class (**149** to **151**).

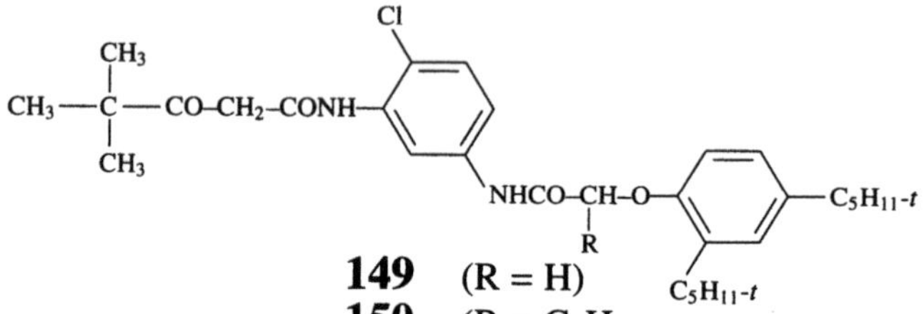

Figure 10.34. Non-diffusible pivaloylacetanilide couplers forming a yellow azomethine dye. The top two are four-equivalent couplers, while the last one is a two-equivalent coupler.

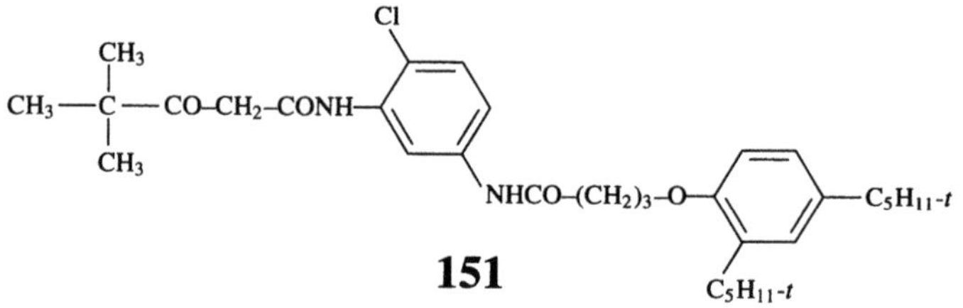

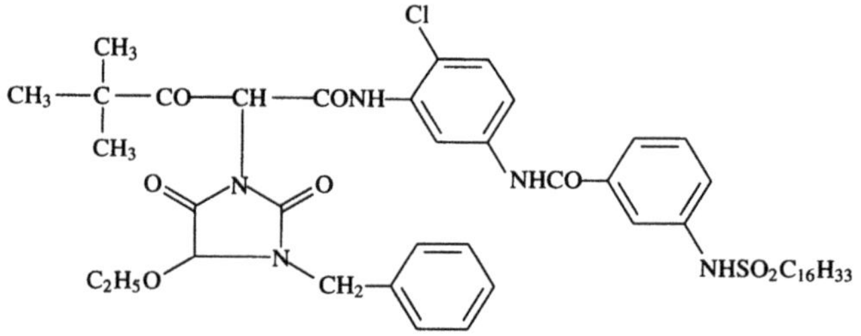

The absorptions and light stabilities of azomethine dyes derived from **149** to **151** are shown in Table 10.10 [99]. The good hues of the dyes have been rather qualitatively shown by the words "exhibiting particularly good absorption characteristics". Later, the absorption spectral diagram of a dye generated from a pivaloylacetanilide coupler with CD-3 was reported to exhibit a sharp-cut profile

Table 10.10. Absorptions and Light Stabilities of Azomethine Dyes Derived from Pivaloylacetamide Couplers (**149** to **151**) [99]

Coupler	λ_{max} of dye with		Decrease[a] in density of dye with	
	CD-2[b] (nm)	CD-3[c] (nm)	CD-2	CD-3
149	448	444	0.12	
150	446	444		0.10
151	448	444		0.12

[a]Density loss at λ_{max} after 30 hr exposure in a xeon arc fadeometer.
[b]4-Amino-3-methyl-*N*, *N*-dimethylaniline.
[c]*N*-Ethyl-*N*-(2-methanesulfonamidoethyl)-3-methyl-4-aminoaniline.

at the longer wavelength side, as compared with a dye from a benzoylacetanilide coupler [5].[16]

The improved light stabilities of the dyes (Table 10.10) have been demonstrated by comparing them with dyes derived from couplers having primary and secondary alkyls in place of the *t*-butyl group [99]. For example, when the *t*-butyl group in the pivaloyl group of **149** is replaced by a 2,2-dimethyl-1-propyl group (*t*-Bu-CH_2-, a primary alkyl group), the resulting dye with CD-3 exhibits a density loss of 0.4. In addition, the replacement of the *t*-butyl group of **151** by a 1-methyl-1-propyl group ($CH_3CH_2CH(CH_3)$-, a secondary alkyl group) causes a density loss of 0.70 in the resulting dye with CD-3.

The light and heat stabilities of pivaloylacetanilide couplers themselves[17] have been reported to be higher than those of benzoylacetanilide couplers [99]. When the pivaloyl group of **149** is replaced by a benzoyl group, the film containing the benzoylacetanilide coupler shows a 4.5% reduction in transmission, after exposure for 30 hr in a xenon arc fadeometer, as compared to the 0.5% reduction shown by the film containing **149**. The reduction in transmission due to the yellowing by storing for one week at 60°C and 75% relative humidity is improved to be 0% for **149** as compared to a 1% reduction for the benzoylacetanilide coupler.

The reactivities of pivaloylacetanilide couplers with oxidized color developers are generally lower than those of benzoylacetanilide couplers, because the pK_a's of the former are higher than those of the latter by about 1.5. This shortcoming has been overcome by developing two-equivalent pivaloylacetanilide couplers, as

[16]The sharp-cut profile may come from the steric hindrance due to the *t*-butyl group of a pivaloylacetanilide coupler, where the aggregation of azomethine dyes therefrom may be inhibited by the steric hindrance.

[17]The properties of an azomethine dye (e.g., the data in Fig. 10.10) are theoretically the same whether the dye is produced from a four-equivalent coupler or from the corresponding two-equivalent one. On the other hand, the properties of the four-equivalent coupler, such as the yellowing, is different from those of the two-equivalent coupler.

discussed later (Chapter 11). Moreover, the incorporation of a sulfonamido or sulfamoyl group into the ballast group of such a two-equivalent coupler has been disclosed to enhance the reactivity, as exemplified by the coupler **152** [100].

Syntheses

The synthetic pathway for preparing the pivaloyl-type yellow coupler (**149**) is shown in Fig. 10.35 [99], which contains essentially the same unit procedures as shown in Fig. 10.33. The key of this pathway is also a coupler nucleus having an amino group (**156**), where the amino group reacts with the same intermediate, i.e., α-(2,4-di-t-amylphenoxy)acetyl chloride (**140**).

The first step is the condensation of ethyl pivaloylacetate (**153**) with 2-chloro-5-nitroaniline (**154**) to produce an pivaloylacetanilide intermediate (**155**), the nitro group of which is then catalytically reduced into an amino group (**156**). The amino group is linked with the ballast group by using **140**.

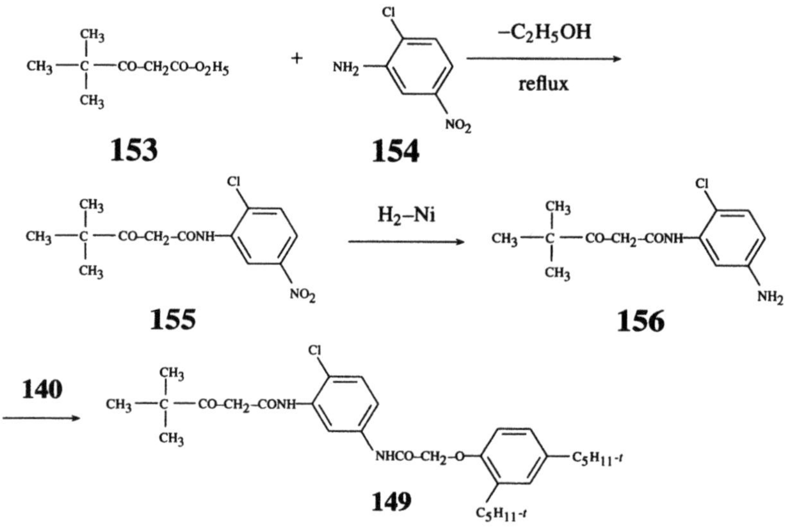

CH_3 CH_3—C—CO-CH_2CO-O_2H_5 CH_3 + NH_2 Cl NO_2 —$-C_2H_5OH$, reflux→

153 **154**

CH_3 CH_3—C—CO-CH_2-CONH Cl NO_2 —H_2–Ni→ CH_3 CH_3—C—CO-CH_2-CONH Cl NH_2

155 **156**

—**140**→ CH_3 CH_3—C—CO-CH_2-CONH Cl NHCO-CH_2-O C_5H_{11}-t C_5H_{11}-t

149

Figure 10.35. Synthetic pathways for preparing the non-diffusible pivaloylacetanilide coupler (**149**). The synthesis of the intermediate **140** has been shown in Fig. 10.33.

10.5.3 Cyclopropanecarbonylacetanilide Couplers

As shown in the preceding subsections, benzoylacetanilide couplers have a drawback of having a broad absorption peak, although they have merits in that they have high coupling reactivity with oxidized color developers and the corresponding azomethine dyes have a sufficient molar extinction coefficient. On the other hand, pivaloylacetanilide couplers are excellent in the hue of the resulting yellow azomethine dyes but they have disadvantages that they have lower coupling reactivity with oxidized color developers and the corresponding azomethine dyes have a lower molar extinction coefficient.

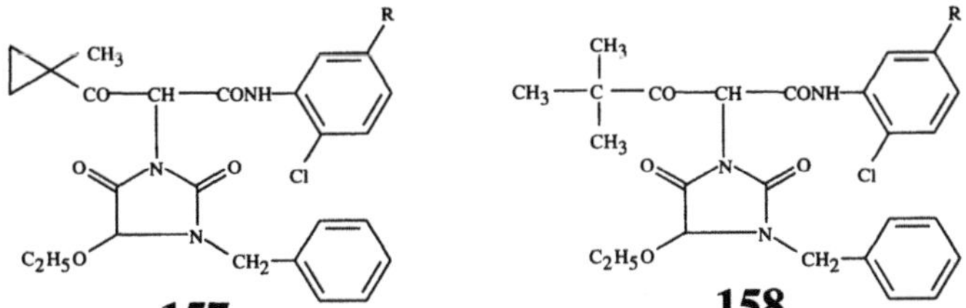

157

158

a: R = $COOC_{12}H_{25}$-*n*

b: R = NHCO–CH(C_2H_5)–O–C_6H_3(C_5H_{11}-*t*)$_2$

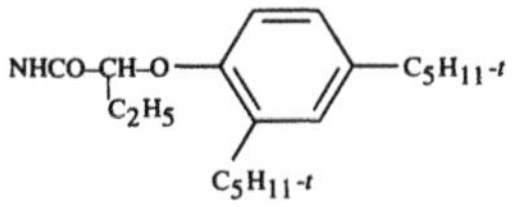

Figure 10.36. Yellow couplers. Comparison between a 1-methyl-1-cyclopropyl group (**157a** and **157b**) and a *t*-butyl group (**158a** and **158b**).

Table 10.11. Properties of 1-Methylcyclopropane-1-carbonylacetanilide Couplers [101, 102]

	Absorption of dye image[a]			Density[b]	
Coupler	λ_{max} (nm)	$\Delta\lambda_{0.5}$[c] (nm)	$\Delta\lambda_{0.1}$[d] (nm)	D_{max}	γ
157a	440	36.7	71.0	2.10	2.10
157b	440	37.7	73.2	2.04	2.01
158a	444	39.3	72.5	1.63	1.41
158b	444	40.3	74.3	1.58	1.04

[a] Developed by using *N*-ethyl-*N*-(2-methanesulfonamidoethyl)-3-methyl-4-aminoaniline sulfate (CD-3). The transmission absorption spectrum at the maximum density of each film was determined.

[b] A maximum color density D_{max} and the gradient γ of the tangent line at the concentration of the half of D_{max} were determined from the sensitometry curve of each processed sample.

[c] The value represents the difference between λ_{max} and the long-side wavelength at a half of D_{max}.

[d] The value represents the difference between λ_{max} and the long-side wavelength at 1/10 of D_{max}.

To maintain or enhance the advantages of pivaloylacetanilide couplers as well as to cover their disadvantages, the 1-methylcyclopropane-1-carbonyl group has been proposed in place of the pivaloyl group, as exemplified by **157a** and **157b** [101,102]. The properties of the couplers of 1-methylcyclopropane-1-carbonyl type have been compared with those of **158a** and **158b** of pivaloyl type (Table 10.11).

As found by inspection of the $\Delta\lambda_{0.5}$ and $\Delta\lambda_{0.1}$ values, the absorption peak of each dye of 1-methylcyclopropane-1-carbonyl type exhibits a steeper profile at the long-wavelength side as compared with that of the corresponding dye of pivaloyl type. Moreover, the comparison of the D_{max} values indicates that the couplers of 1-methylcyclopropane-1-carbonyl type react more efficiently with oxidized color developers than those of pivaloyl type.

10.5.4 Malondiamide Couplers

Although malondianilide couplers have long been known [103], the corresponding azomethine dyes have low dark stability and insufficient hue. On the other hand, malondiamide couplers in which one amido group comes from a secondary cyclic amine (e.g., **159**) have been disclosed to exhibit excellent hue as well as high dark stability [104].

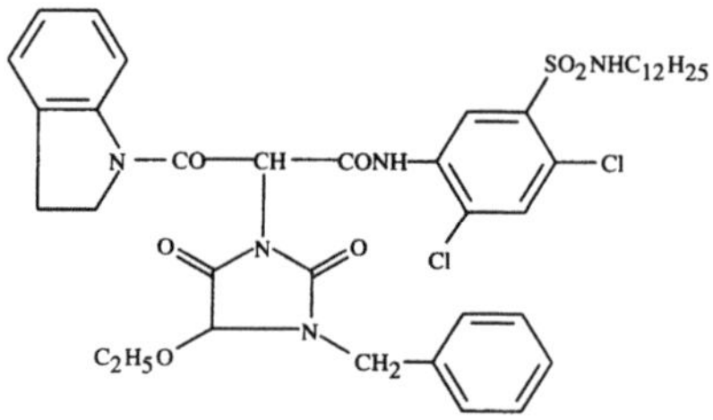

159

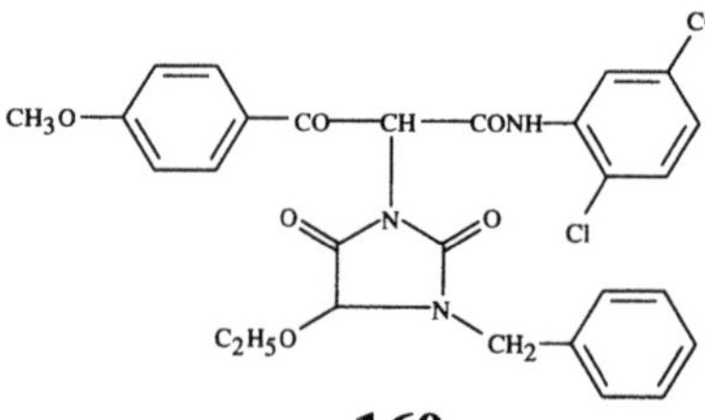

160

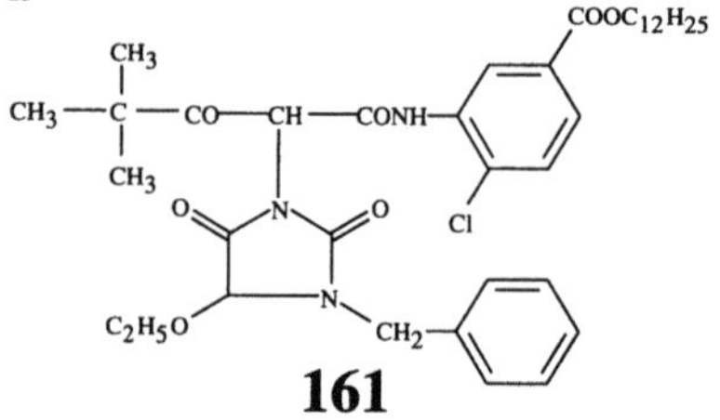

161

Figure 10.37. Yellow coupler of malondiamide type (**159**) and comparative couplers of other types (**160** and **161**).

The comparison of D_{max} values indicates that the coupler of malondiamide type (**159**) reacts more efficiently with oxidized color developers than the couplers of other types (**160** and **161**), as shown in Table 10.12. Moreover, the D_{520} value of the dye derived from **159** (malondiamide type) is smaller than the corresponding

Table 10.12. Properties of Yellow Couplers of Malondiamide Type and Comparative Data [104]

Coupler	Absorption of dye image[a]		D_{max}	Remaining dye image[c] (%)
	λ_{max} (nm)	D_{520}[b] (%)		
159	452	10.8	2.72	98
160	452	15.2	2.52	86
161	449	12.5	2.04	97

[a]Developed by using *N*-ethyl-*N*-(2-hydroxyethyl)-3-methyl-4-aminoaniline sulfate (CD-4).
[b]The ratio of the density at 520 nm to D_{max}.
[c]The dark stability (the remaining dye density) was measured after being stored for 7 days under conditions of 60°C and 70% relative humidity.

values for **160** (benzoylacetanilide type) and **161** (pivaloylacetanilide type). This means that the absorption curve of **159** has a steep profile at the long-wavelength side. The dark stability of **159** is also superior to the couplers of the other types. A malondiamide coupler that is similar to **159** but has a different ballast group has been employed in a color reversal film [105].[18]

10.5.5 Pyrroloylacetanilides and Related Couplers

Another approach to improve yellow couplers in hue and other properties has been focused on the incorporation of aromatic heterocycles. For example, pyrroloylacetanilide couplers have been disclosed in patents [106,107,108]. Yellow couplers of indolecarbonylacetanilide type were later disclosed, as exemplified in Fig. 10.38 (**162a**, **162b**, and **162c**) [109].

Comparison of the D_{max} values collected in Table 10.13 (Run 1) indicates that the couplers of 1-indolecarbonylacetanilide type react more efficiently with oxidized color developers than those of pivaloylacetanilide type (**162c** vs. **164**). This fact can be seen more clearly by the data of the "activity ratio" column (Run 1), which have been calculated by selecting the value of **161** as the standard.

The activities of the 1-indolecarbonylacetanilide couplers (**162a**, **162b**, and **162c**) have been indirectly compared with those of the coupler (**163**) of malondiamide type, where the data of Run 2 have been used together with the data of Run 1 by selecting **161** as a common standard. It is to be noted, however, that careful consideration should be given to such comparison between Run 1 and Run 2 because the original D_{max} values are extremely different.

The comparison among **162a**, **162b**, and **162c** indicates the effect of leaving groups at the coupling position of the two-equivalent couplers. The presence of a

[18]This reversal film has been placed on the market under the name of Fujichrom Provia 100 (1994).

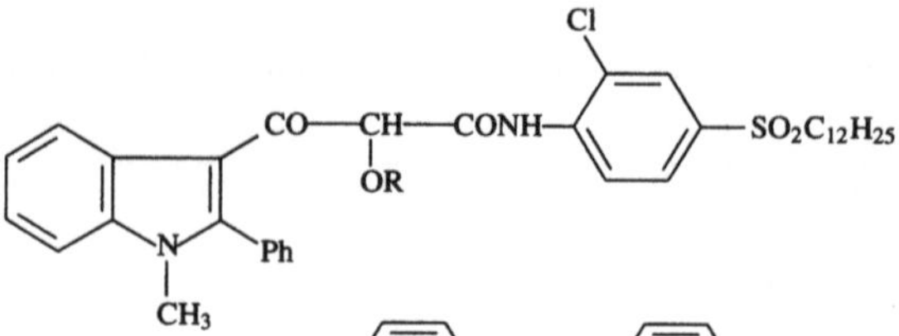

162a R =

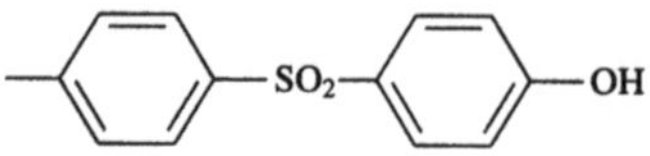

162b R =

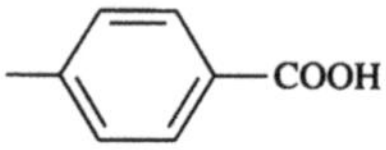

162c R =

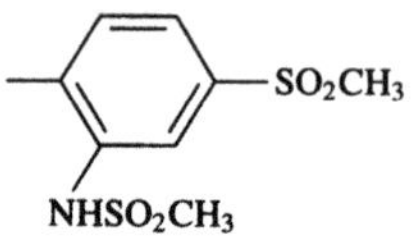

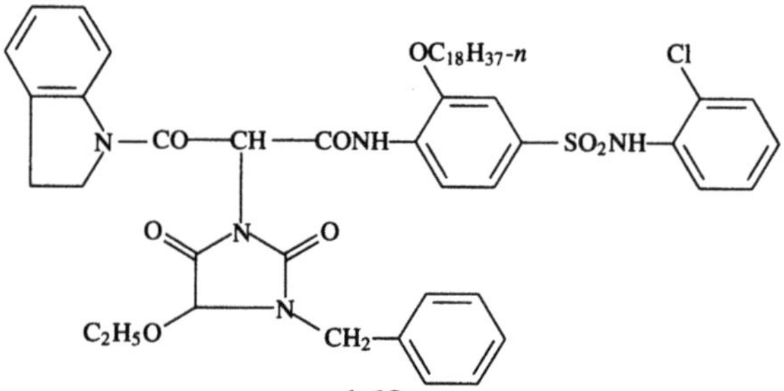

163

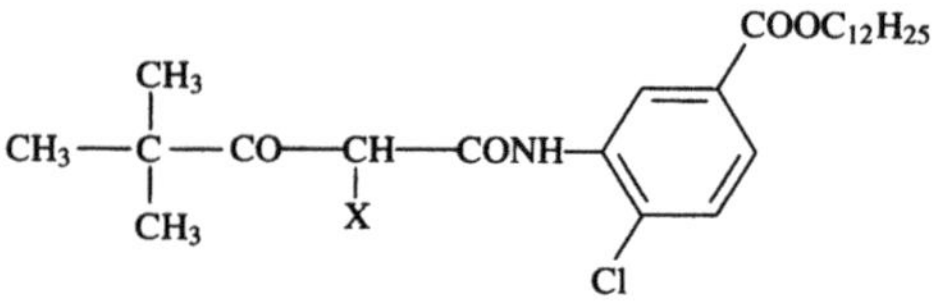

161 For X, see Fig. 10.37

164 X =

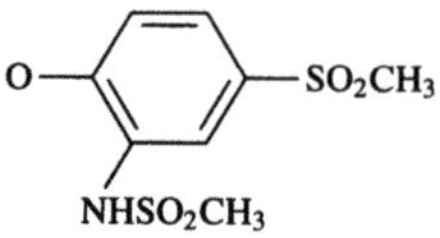

Figure 10.38. Yellow couplers of indolecarbonylacetanilide type (**162a**, **162b**, and **162c**) and comparative couplers of other types.

water-solubilizing group (e.g., a carboxylic acid group or a methanesulfonamido group) in each leaving group improves the activity of the coupler.

A further approach using another heterocyclic moiety (imidazolonoylacetanilide) has appeared, as collected in Table 10.14 [110]. The absorption spectra of the

Table 10.13. Reactivity of Indolecarbonylacetanilides and Related Couplers [109].

Coupler	D_{max} [a]		Activity ratio [b]	
	Run 1	Run 2	Run 1	Run 2
162a	2.44		1.67	
162b	2.56		1.75	
162c	2.67		1.83	
163		(4.29)		(1.53)
161	1.46	(2.80)	1.00	(1.00)
164	2.07	(3.83)	1.40	(1.37)

[a] Developed by using process E-6 solutions which probably contain *N*-ethyl-*N*-(2-methanesulfonamidoethyl)-3-methyl-4-aminoaniline sulfate (CD-3).
[b] The ratio was calculated by selecting the value of **161** as the standard.

dyes **165e** and **165f** derived from the couplers of this type have been compared with those of the comparative dyes (**165a** to **165d**) derived from couplers of other types. The sharp band shapes of the former dyes have been characterized by the $\Delta\lambda_{0.5}$ values that are as small as those of the latter comparative dyes.

Then-2-oyl- or then-3-oylacetanilide couplers have been disclosed as alternative candidates for couplers exhibiting good dye stability [111].

Table 10.14. Absorption Maxima of Azomethine Dyes (**165**) Derived by Oxidative Coupling of *N*-Ethyl-*N*-(2-hydroxyethyl)-3-methyl-4-aminoaniline sulfate (CD-3) with Pyrroloylacetamides[a] [109,110]

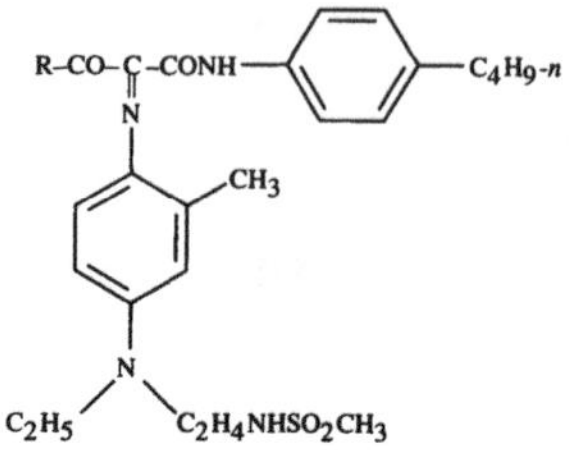

R–CO–C(=N)–CONH–C_6H_4–C_4H_9-*n*; N bonded to a 2-CH_3-4-[N(C_2H_5)($C_2H_4NHSO_2CH_3$)]phenyl ring

165

No.	Substituent R	λ_{max} (nm)	$\epsilon_{max} \times 10^{-4}$	$\Delta\lambda_{0.5}$[b] (nm)
165a	*t*-C_4H_9	428	1.18	42
165b	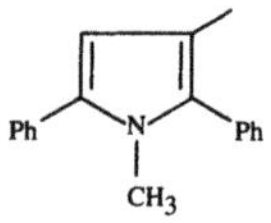 Ph, N, Ph, CH_3	426	1.64	36
165c	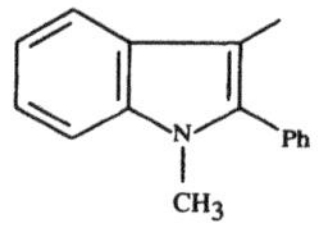 N, Ph, CH_3	426	1.97	32
165d	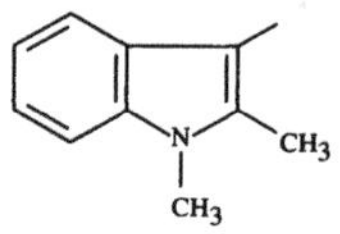 N, CH_3, CH_3	428	1.62	37
165e	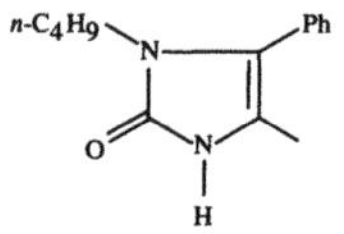 *n*-C_4H_9, N, Ph, O, N, H	435	2.09	35
165f	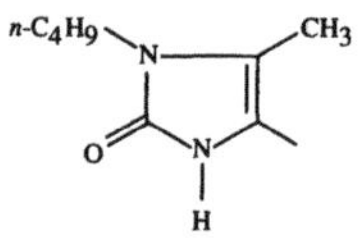 *n*-C_4H_9, N, CH_3, O, N, H	436	1.43	42

[a]All data were determined in acetonitrile.

[b]The value represents the difference between λ_{max} and the long-side wavelength at a half D_{max}.

10.6 Polymer Couplers

One more methodology for giving non-diffusibility to couplers is the incorporation of a coupling nucleus into polymeric structure. Such couplers are called *polymer couplers*, while the corresponding coupler nuclei are referred to as *monomer couplers*. For example, acrylic acid amides of 3-aminopyrazolones as monomer couplers (e.g., 1-(*p*-bromophenyl)-3-methacrylamido-5-pyrazolone) have been incorporated into polymer chains, where their copolymers with acrylamide and acrylic acid have been isolated and dissolved into solution before coating [112]. Later, emulsion polymerization has been applied to the preparation of polymer couplers. For example, emulsion polymerization methods in an aqueous gelatin phase [113] and in water [114] have been utilized to prepare polymer couplers in a latex form.

Monomer pyrazolone couplers of a specific substitution on the 1-phenyl group (e.g., **166**) have been disclosed to be used in a latex form [115]. Advantages of using polymer couplers in a latex form are as follows:

1. Polymer coupler latexes require no dispersion oils, which are in turn necessary to disperse oil-protected couplers. It follows that coated layers become thinner than those resulting from the oil dispersion.
2. The strength of the film can be maintained upon addition of latexes rather than the oil dispersion.
3. A polymer coupler contains coupler nuclei in higher concentration than an oil-protected coupler.

The vinyl function and the pyrazolone nucleus in a monomer coupler can be linked by a spacer as found in (e.g., **167**) [116].

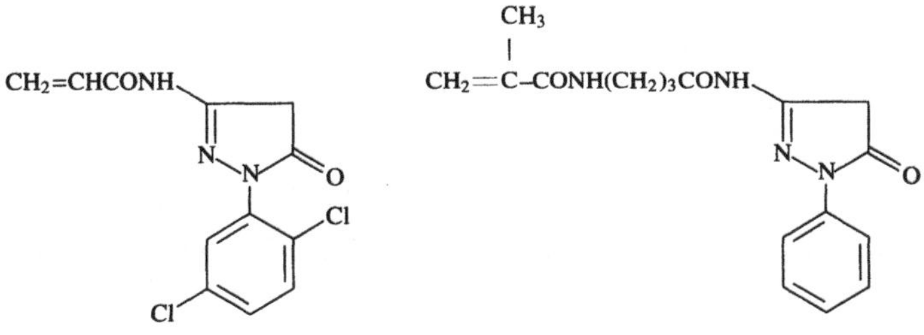

Figure 10.39. Four-equivalent magenta monomer couplers for latex formation. They serve as monomers for emulsion copolymerization with ethyl acrylate, *n*-butyl acrylate, and/or other vinyl monomers.

Four-equivalent pyrazolone couplers have several disadvantages, which have been overcome by converting them into two-equivalent ones, as will be discussed in the next chapter. Such two-equivalent pyrazolone couplers have been incorporated into polymer chains, where seed polymerization has been applied to the

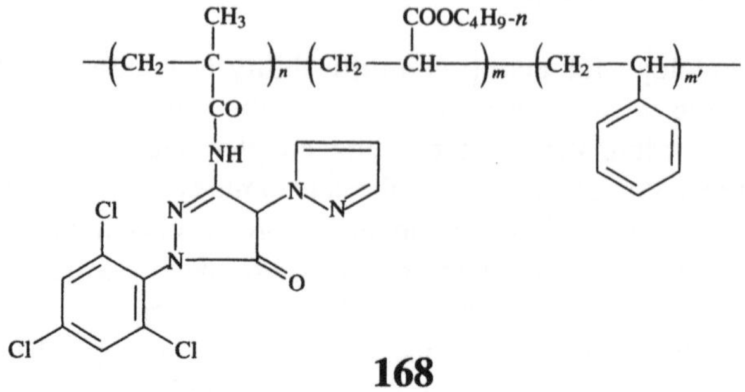

168

Figure 10.40. 5-Pyrazolone polymer magenta coupler with $n = 50$, $m = 25$, and $m' = 25$ by weight and a molecular weight of about 20,000.

preparation of polymer coupler latexes having a layered structure [117]. Thus, a monomer coupler having a 1-imidazolyl leaving group at the 4-position of a 5-pyrazolone nuclei has been copolymerized into a latex form. A copolymer latex represented by the formula **168** has been compared with the corresponding four-equivalent polymer coupler, where the presence of a phenolic cyan coupler [118] or of a DIR (developer inhibitor releasing) coupler [119] has been claimed.[19]

References

[1] Brown GH, Figueras J, Gledhill RJ, Kibler CJ, McCrossen FC, Pafmerter SM, Vittum PW, Weissberger A (1957) J Am Chem Soc 79:2919

[2] Bailey J, Williams LA (1971) The Photographic Color Development Process. In: Venkataramn K (ed) The Chemistry of Synthetic Dyes, Vol V, Acacemic, New York London. Chapter VI

[3] Furutachi N (1983) Yuki Gosei Kagaku Kyokai Shi. 41:439

[4] Kaneko U (1992) J Soc Photogr Sci Technol Jpn. 55:353

[5] Furutachi N (1986, 2002) In: Functionalized Organic Chemicals for Silver Halide Color Photographic Materials, CMC, Tokyo. Chapter 2, Section 4

[6] Kimura S, Yoshida M (1965) Yuki Gosei Kagaku Kyokai Shi. 23:989

[7] Kimura S, Yoshida M (1967) Yuki Gosei Kagaku Kyokai Shi. 25:922

[8] Portuaya BS, Spasokukotsky NS, Twitsyna NF, Bobkova TP, Arbuzov GL, Leukoev II (1956) J Gen Chem USSR. 26:2381

[19]Color negative films employing a polymer coupler of this type have been placed on the market under the name Fujicolor HR series (1984), where the polymer coupler has been called "L-coupler" after the word "latex" [120].

[9] Lurie AP, Brown GH, Thirtle JR, Weissberger A (1961) J Am Chem Soc. 83:5015

[10] Barr CR, Brown GH, Thirtle FR, Weissberger A (1961) Photogr Sci & Eng. 5:195

[11] Kobayashi H, Mihayashi K (1994) J Soc Photogr Sci Technol Jpn. 57:316

[12] Bard CC, Larson GW, Hammnd H, Packard C (1980) J Appl Photogr Eng. 6:42

[13] Lau PTS, Jozefiak TH, Welter TR (1995) US Patent 5 457 008

[14] Saito N, Aoki K, Yokota Y (1987) US Patent 4 690 889

[15] Ishii F (1989) J Soc Photogr Sci Technol Jpn. 52:156

[16] Fierke SS, Chechak JJ (1957) US Patent 2 801 171

[17] Tuite RJ (1979) J Appl Photogr Eng. 5:200

[18] Ramello P (1973) US Patent 3 772 002

[19] Ogawa T, Takahashi O (1989) US Patent 4 857 449

[20] Ogawa T, Takahashi O (1991) Fujifilm Res Dev. 36:7

[21] Salminen IF, Barr CR, Loria A (1959) US Patent 2 895 826

[22] Kojima T, Sato O, Endo T, Usui T, Horiuchi T (1978) Jpn Patent S53-2728

[23] Sasaki T, Kaneko U, Ishii F, Tsuda Y, Kimura K, Kato K (1988) Jpn Patent S63-30619

[24] Osborn HJ (1978) US Patent 4 124 396

[25] Clarke D, Leyshon LJ, Smith KE (2002) US Patent 6 444 417 B1

[26] Ogawa A, Aoki K, Tanabe O, Umemoto M (1985) Jpn Kokai S60-24547

[27] Tang P, Steele MD (2000) US Patent 6 096 494

[28] Lau PTS, Cowan SW, Rossi LJ (1997) US Patent 5 686 235

[29] Lau PT, Cowan SW, Rossi LJ (1999) US Patent 5 962 198

[30] Begley WJ, Russo GM, Curt DT (2001) US Patent 6 251 575

[31] Begley WJ, Coms FD, Russo GM (2001) US Patent 6 194 132

[32] Begley WJ, Russo GM, Curt DT (2001) US Patent 6 197 491

[33] Begley WJ, Coms FD, Russo GM (2001) US Patent 6 207 363

[34] Lau PTS (1982) US Patent 4 333 999

[35] Sato R, Katoh K, Sasaki T, Sugita H (1983) Jpn Kokai S58-213748; Sugita H, Tsuda Y, Ito K, Shimba S (1984) US Patent 4 434 225; Katoh K, Nakagawa S (1986) US Patent 4 609 619

[36] Kamio T, Ono M, Aoki K, Watanabe T, Arakawa J (1985) Jpn Kokai S60-37557

[37] Hoke D, Kilminster KN (1989) US Patent 4 849 329

[38] Kobayashi H, Furuya K, Tsukahara Z (1994) J Soc Photogr Sci Technol Jpn. 57:333

[39] Zengerle PL, Sowinski AF (1998) US Patent 5 726 003

[40] Zengerle PL, Sowinski AF (1998) US Patent 5 834 175

[41] Kaneko U (1992) J Soc Photogr Sci Technol Jpn. 55:199

[42] Bailey J, Harrow DN (1988) US 4 728 598

[43] Tachibana K, Kaneko Y, Ishii F (1989) US 4 873 183

[44] Suzuki M, Sato T, Sato K, Ishii Y, Naruse H, Shimada Y (1993) US Patent 5 256 526

[45] Takahashi O, Yoshioka Y, Soejima S, Shimada Y, Morigaki M (2000) US Patent 6 103 460

[46] Seto N, Morigaki M, Yoshioka Y, Soejima S, Takahashi O, Mikoshiba H (2000) US Patent 6 132 945

[47] Sato K, Ishii Y, Yamakawa K (1995) US Patent 5 401 624

[48] Matsuda N (2000) US Patent 6 159 671

[49] Tateishi K, Mikoshiba H, Matsuda N (2002) US Patent 6 399 291 B1

[50] Tateishi K, Mikoshiba H, Matsuda N (2002) US Patent 6 495 697 B1

[51] Kato Y, Mikoshiba H, Matsuda N, Kojima T (2003) US Patent 6 541 192 B2

[52] Brown GH, Graham B, Vittum PW, Weissberger A (1951) J Am Chem Soc. 73:919

[53] Vittum PW, Duennebier FC (1950) J Am Chem Soc. 72:1536

[54] Furutachi N (1992) J Soc Photogr Sci Technol Jpn. 55:192

[55] Loria A, Weissberger A, Wittum PW (1952) US Patent 2 600 788

[56] Beavers LE (1961) US Patent 2 983 608

[57] Greenhalgh CW, Stone KC (1964) US Patent 3 127 269

[58] Sawyer JF, Fenton DE (1995) US Patent 5 399 468

[59] Lestina GJ (1984) US Patent 4 443 536

[60] Bailey J (1977) J Chem Soc Perken 1. 2047

[61] Bailey J, Middlesex H, Knott EB, Marr PA (1973) US Patent 3 725 067

[62] Sato T (1989) J Soc Photogr Sci Technol Jpn. 52:162

[63] Ogawa T, Sato T, Takahashi O, Hasebe K, Furutachi N (1991) Nippon Kagakukai Shi. 719

[64] Buckland PR, Leyshon LJ (1988) US Patent 4 777 121

[65] Kaneko U (1993) J Soc Photogr Sci Technol Jpn. 56:301

[66] Kita H, Iizuka H, Daifuku K (1999) J Soc Photogr Sci Technol Jpn. 62:31

[67] Nakayama N, Kawakatsu S, Katoh K, Shinozaki K (1991) US Patent 5 032 497

[68] Kita H, Ishida H, Kaneko Y (1995) US Patent 5 470 697

[69] Romanet RF, Burns PA, Fischer SM, Spara PP, Balasubramanian RP, Lincoln DG (1999) US Patent 5 972 587

[70] Iijima T, Kumashiro K, Kashiwagi H, Hatta K, Hotta Y, Ohya H, Nakayama N, Kawakatsu S, Katoh K, Shinozaki K (1989) US Patent 4 840 886

[71] Burns PA, Harder JW (1991) US Patent 5 021 325

[72] Romanet RF, Chen T-H (1993) US Patent 5 183 728

[73] Bose JA, Friedrich LE, Hoke D, Potenza JC, Romanet RF, Singer SP, Valente RR (1999) US Patent 5 985 532

[74] Nakamine T, Motoki M, Kawagishi T, Matsuda N (1997) US Patent 5 656 418

[75] Nakamine T, Motoki M, Kawagishi T, Matsuda N (1999) US Patent 5 858 635

[76] Mikoshiba H, Fukuzawa H, Matsuda N (2003) US Patent 6 548 237 B2

[77] Nishijima T (1991) US Patent 5 017 465

[78] Kadokura K, Yamazaki K (1993) US Patent 5 236 819

[79] Douglas P, Townsend SM, Ratcliffe R (1991) J Imaging Sci. 35:211

[80] Douglas P, Clarke D (1991) J Chem Soc Perkin 2. 1363

[81] Franke CA, Mura Jr AJ, Eiff SL (2000) US Patent 6 013 429

[82] Sato T, Kawagishi T, Furutachi N (1985) US Patent 4 540 654

[83] Sato T, Kawagishi T, Furutachi N (1986) US Patent 4 621 046

[84] Sato T, Kawagishi T, Furutachi N (1987) US Patent 4 705 863

[85] Asami M, Yoneyama H (1996) US Patent 5 578 437

[86] Mizukawa Y, Motoki M, Sato T, Takahashi O (1995) US Patent 5 451 501

[87] Mizukawa Y, Motoki M, Sato T, Takahashi O (1996) US Patent 5 532 377

[88] Tang P-W, Cowan SW (2000) US Patent 6 143 485

[89] Nakazyo K, Hirose T, Hurusawa G, Takahashi O, Furutachi N, Kobayashi H (1990) US Patent 4 900 655

[90] Furutachi N (1986) US Patent 4 588 679

[91] Morigaki M, Kawagishi T, Nakazyo K, Seto N, Kamei S (1988) US Patent 4 735 893

[92] Furuya K, Furutachi N, Oda S, Maruyama K (1994) J Chem Soc Perkin 2. 531

[93] Morigaki M, Seto N, Aoki K (1993) US Patent 5 212 055

[94] Sato T (1991) Yuki Gosei Kagaku Kyokai Shi. 49:541

[95] Tanaka M, Kubodera S (1991) US Patent 5 034 371

[96] Mikoshiba H, Tanaka M, Kubodera S (1993) US Patent 5 227 379
[97] Mikoshiba H, Tanaka M, Kubodera S (2000) J Soc Photogr Sci Technol Jpn. 63:322
[98] Mikoshiba H, Kamio T (2002) US Patent 6 458 194 B1
[99] Weissberger A, Kibler CJ (1966) US Patent 3 265 506
[100] Usui T, Kato K, Kojima T (1982) US Patent 4 356 258
[101] Shimura Y, Kobayashi H, Yoshioka Y (1994) US Patent 5 359 080
[102] Shimura Y, Kobayashi H, Yoshioka Y (1994) US Patent 5 427 902
[103] Tanaka M, Yagihara M, Aono T, Hirose T (1979) US Patent 4 149 886
[104] Motoki M, Ichijima S, Saito N, Kamio T, Mihayashi K (1993) US Patent 5 213 958
[105] Shibahara Y, Yamada K, Ishimaru S (1995) Fuji Film Res & Dev. 40:1
[106] Clark BA, McNab H, Sommerville CC (1997) US Patent 5 674 667
[107] Welter TR, Reynolds JH (2000) US Patent 6 015 658
[108] Welter TR, Reynolds JH (2000) US Patent 6 057 087
[109] Welter TR, Reynolds JH (2000) US Patent 6 083 677
[110] Welter TR, Coms FD (2002) US Patent 6 455 241 B1
[111] Clark BA, Gourley RN, NcNab H, Sommerville CC (1997) US Patent 5 693 458
[112] Firestine JC, Umberger JQ (1964) US Patent 3 163 625
[113] Dawson GA (1968) US Patent 3 370 952
[114] Van Paesschen AJ, Priem JJ (1978) US Patent 4 080 211
[115] Yagihara M, Hirano T, Mihayashi K, Ozawa T (1983) US Patent 4 409 320
[116] Yagihara M, Mihayashi K, Ozawa T (1984) US Patent 4 474 870
[117] Hirano T, Mihayashi K, Yagihara M (1984) US Patent 4 444 870
[118] Sakanoue K, Hirano T (1988) US Patent 4 791 051
[119] Hirano T, Sakanoue K (1990) US Patent 4 937 179
[120] Itoh I, Takada S, Ikenoue S (1984) Nikkakyo Geppo. 1984(12):18

Chapter 11

Two-Equivalent Couplers

11.1 Merits of Two-Equivalent Couplers

A coupler which has no substituent at the active position is usually called a four-equivalent coupler, because totally four moles (equivalents) of silver halide are necessary to generate one mole of an azomethine dye. On the other hand, a coupler whose active position is substituted with an electro-negative atom such as chlorine, oxygen, or sulfur is called a *two-equivalent coupler*, because it theoretically requires two moles of silver halide to produce one mole of the corresponding dye. The dye-forming reaction involving a two-equivalent coupler is referred to as *elimination coupling*.

1. The reduction of the amount of silver halide provides an economic merit because silver is a noble metal, which is expensive and is frequently regarded as an object of speculation.[1]
2. The improvement of quality (e.g., sharpness) is another merit of two-equivalent couplers, because the thickness of each emulsion layer can be dramatically reduced.
3. The oil-solubility of a two-equivalent coupler can be controlled by adjusting a leaving (coupling-off) group as well as a ballast group. This property is very important, since the coupler is dispersed as oil droplets in a gelatin colloidal dispersion.

---

[1] In fact, the price of silver has once risen enormously by speculation in 1979–1980 (40,000 yen/kg → max 300,000 yen/kg).

Table 11.1. Leaving Groups for Two-Equivalent Couplers[a] [1]

Leaving group	Cyan coupler	5-Pyrazolone magenta coupler		Yellow coupler
		3-Acylamino	3-Anilino	
–Halogen	○	– (Fog)	– (Fog)	– (Fog)
–O-Alkyl	○	– (NP)	– (NP)	○
–O-Aryl	○	– (NP)	– (NP)	○
$-NHSO_2R$	○	– (US)	– (US)	○
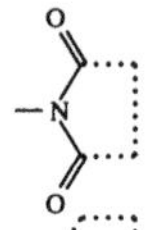 O –N O	– (LR)	– (LR)	– (LR)	○
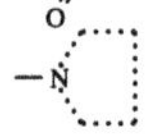 O –N	– (LR)	○	– (US)	○
–SR	(○)	(○)	○	(○)

[a] The symbol ○ indicates that a coupler having a leaving group at issue is practically usable, while – represents no practical usefulness because of Fog: fog formation; NP: no practical synthetic route; US: unstable; and LR: low coupling rate. The symbol (○) designates the usefulness as DIR couplers.

4. The coupling reactivity can be optimized by selecting an appropriate leaving group. This effect is associated with the adjustment of the pK_a value of the resulting coupler.

5. The released compound due to the leaving group is used to control qualities of photographic images. For example, colored couplers and DIR couplers are used to improve color reproduction. They will be discussed in other chapters (Chapters 12 and 13).

The selection of a leaving group (coupling-off group) is particularly important, since it greatly influences the activity of the coupling site. Table 11.1 lists various leaving groups [1]. A wide variety of leaving groups work well for cyan and yellow couplers. On the other hand, leaving groups for 5-pyrazolone magenta couplers have been selected from narrower possibilities.

11.2 Two-Equivalent Cyan Couplers

11.2.1 Two-Equivalent Cyan Couplers of Naphthol Type

Two-equivalent cyan couplers have not been so difficult to access. For example, 2,4-dichloronaphthol used as a cyan coupler in the original proposal by Fischer (see Chapter 8, Section 8.1) is a two-equivalent coupler [2]. The hydrophilic coupler Koe308 described in Chapter 8 (Subsection 8.3.1) is a kind of two-equivalent coupler [3]. Moreover, a patent disclosing oil-protected cyan couplers has already involved several two-equivalent naphthol couplers that have a chlorine atom at the coupling position (e.g., **1**) [4].

However, the above-mentioned merits of two-equivalent couplers had not been taken into explicit consideration until leaving groups with an oxygen linkage were incorporated into cyan couplers (in the 1960s). Thus, in order to control the activity of a cyan coupler to an oxidized color developer, aryloxy and alkoxy groups were incorporated as leaving groups at the reaction site of the coupler. For example, aryloxy groups such as **2** [5], alkoxy- or hydroxy-substituted alkoxy groups such as **3** [6], sulfonamido-substituted alkoxy groups such as **4** [7], and alkoxy groups of another type such as **5** [8] were disclosed in patents.

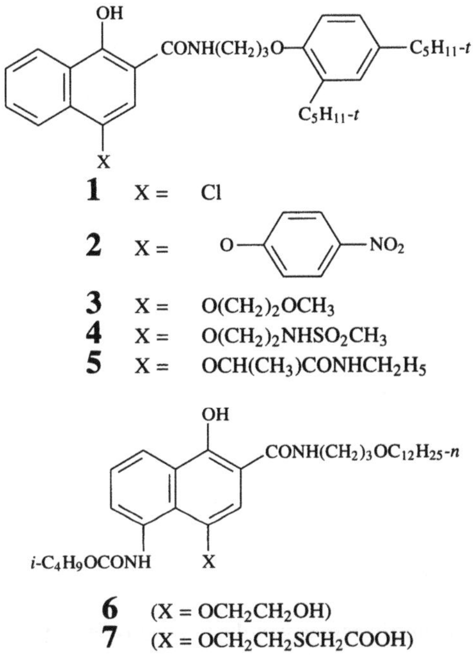

OH
$CONH(CH_2)_3O$
C_5H_{11}-*t*
C_5H_{11}-*t*
X

1 X = Cl

2 X = O–C_6H_4–NO_2

3 X = $O(CH_2)_2OCH_3$

4 X = $O(CH_2)_2NHSO_2CH_3$

5 X = $OCH(CH_3)CONHCH_2H_5$

OH
$CONH(CH_2)_3OC_{12}H_{25}$-*n*
i-C_4H_9OCONH
X

6 (X = OCH_2CH_2OH)

7 (X = $OCH_2CH_2SCH_2COOH$)

Figure 11.1. Two-equivalent cyan couplers based on 1-naphthol.

The resulting dyes derived from the 1-naphthol couplers have several disadvantages, as described in Chapter 10 (Subsection 10.3.1). To overcome the disadvantages of the 1-naphthol couplers, four-equivalent 5-amido-1-naphthol couplers have been disclosed, as also described in Chapter 10 (Subsection 10.3.1). Since the merits of two-equivalent couplers had been well known at the time of the disclosure, the patents claiming the 5-amido-1-naphthol couplers already contained the corresponding two-equivalent couplers such as **6** [9,10]. Two-equivalent couplers of the same type (e.g., **7**) have been used together with yellow-colored cyan couplers [11].

Syntheses

A synthetic pathway for preparing **2** is illustrated in Fig. 11.2 [5]. The key step to introduce an aryloxyl group (**8** → **9**) is a nucleophilic substitution of 1-fluoro-4-nitrobenzene, where the anion of 1,4-dihydroxy-2-naphthoic acid (**8**) works as a nucleophile and the fluorine atom is capable of leaving as an anion under the influence of a *p*-nitro group. Note that a fluorine atom can leave as an anion more easily than other halogen atoms. The carboxyl group of the intermediate (**9**) is converted into a chlorocarbonyl group (**10**), which is reacted with a ballast amine to give the target molecule (**2**).

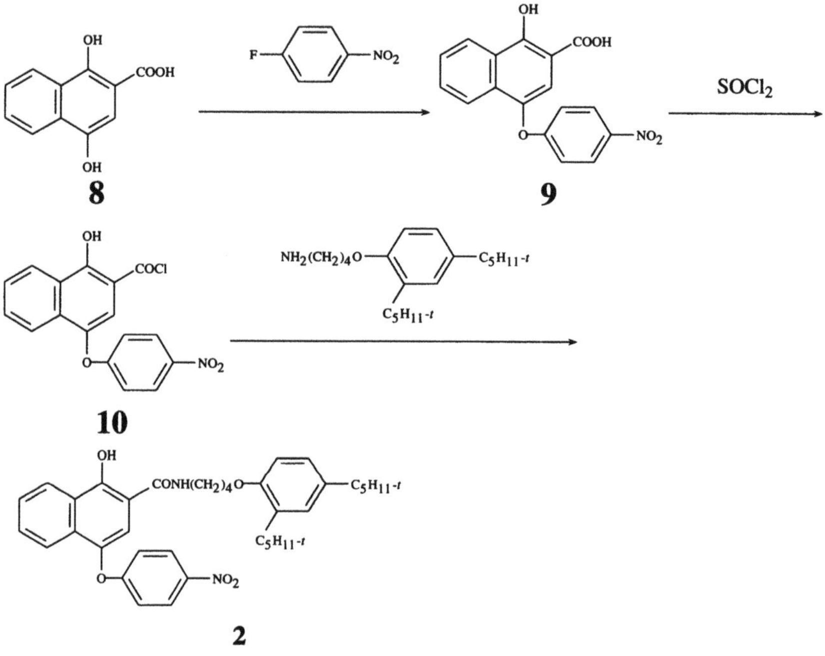

Figure 11.2. Synthetic pathway of the two-equivalent cyan coupler **2**.

The coupler **2** is important as a further intermediate for preparing colored couplers, after the nitro group is reduced into an amino group, which can be a clue for forming azo dyes. This topic will be discussed later in the chapter on colored couplers (Chapter 12).

A synthetic pathway for preparing **3** is illustrated in Fig. 11.3 [6]. The key step to introduce an alkoxyl group (**8** → **11**) is based on an acid-catalyzed ether formation between 1,4-dihydroxy-2-naphthoic acid (**8**) and 2-ethoxyethanol. Compare this step with the transformation from **8** to **9** in Fig. 11.2, where a basic condition has been employed to promote the nucleophilic displacement. The step of linking a ballast group uses a *p*-nitrophenyl ester (**12**) in place of an acid chloride (cf. **10** shown in Fig. 11.2).

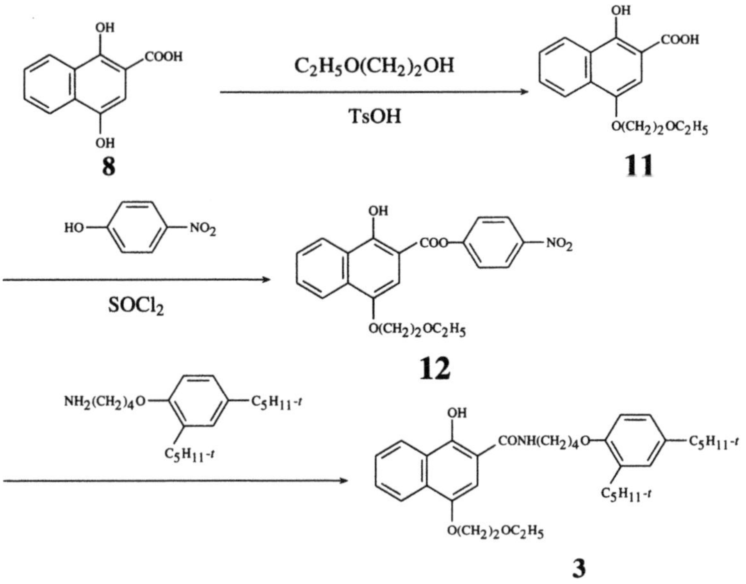

Figure 11.3. Synthetic pathway of the two-equivalent cyan coupler **3**.

A synthetic pathway for preparing **6** is shown in Fig. 11.4 [9,10]. Because of the presence of an acylamido group in the 5-position of **6**, the corresponding 5-acylamido-1,4-hydroxy-2-naphthoic acid cannot be used as a starting material.[2] The key step (**14** → **6**) is a kind of Ullman ether synthesis using a copper catalyst. Although Ullman synthesis usually requires high temperature (about 200°C), the intermediate **14** is exceptionally reactive because of the neighboring effect of the 5-amido group.[3] Thus, the reaction from **14** to **6** proceeds at room temperature in a high yield.

11.2.2 Two-Equivalent Cyan Couplers of Other Types

As described in Chapter 10 (Subsection 10.3.2), most couplers of phenol type have already been disclosed as two-equivalent couplers, where a chlorine atom is substituted at the coupling position (4-position). Several two-equivalent phenol couplers having an aryloxy leaving group will be discussed in the chapter on DIR couplers (Chapter 13).

[2]Since the hydroxyl group at the 1-position of **8** is protected by the adjacent carboxyl group, only the 4-hydroxyl group participates in the ether formations shown in Figs. 11.2 and 11.3. On the other hand, the two hydroxyl groups of a 5-substituted 1,4-hydroxy-2-naphthoic acid as a possible intermediate is presumed not to be differentiated from each other because of intramolecular chelations.

[3]The neighboring effect has been confirmed by the fact that 1-acetamido-8-bromonaphthalene reacts with an alcohol at room temperature but 1-bromonaphthalene does not react [9].

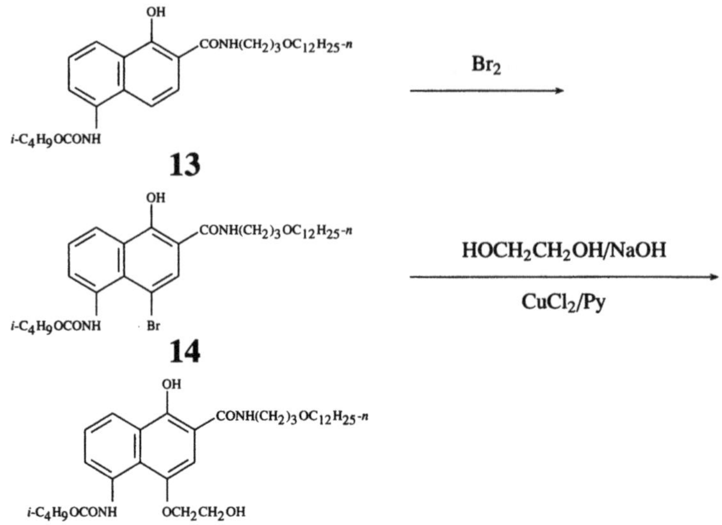

OH

$CONH(CH_2)_3OC_{12}H_{25}$-*n*

i-C_4H_9OCONH

13

Br_2

OH

$CONH(CH_2)_3OC_{12}H_{25}$-*n*

i-C_4H_9OCONH Br

14

$HOCH_2CH_2OH/NaOH$

$CuCl_2/Py$

OH

$CONH(CH_2)_3OC_{12}H_{25}$-*n*

i-C_4H_9OCONH OCH_2CH_2OH

6

Figure 11.4. Synthetic pathway of the cyan coupler **6**. The symbol Py represents pyridine.

Heterocyclic cyan couplers (Chapter 10, Subsection 10.3.3) have also been disclosed as two-equivalent couplers, where various groups such as chloro, imidazolyl, arylthio, and aryloxy are used as leaving groups.

11.3 Two-Equivalent Magenta Couplers

11.3.1 Two-Equivalent 5-Pyrazolone Couplers

Disadvantages of Four-Equivalent 5-Pyrazolone Couplers

5-Pyrazolone couplers as four-equivalent couplers have been widely used in color photography (color negative films and color papers). However, they have several disadvantages as follows:

1. The efficiency of dye-forming is so low that only about 50% of the intermediate leuco dye is used to form an azomethine dye and the remaining portion is consumed to form by-products [12]. This means that 2 moles of the coupler and 8 equivalents of silver halide are necessary for the formation of 1 mole of the azomethine dye, although the stoichiometry theoretically requires 1 mole of the four-equivalent coupler and 4-equivalents of silver halide.

2. An unreacted coupler (**15**) after development processing is easily reacted with an oxygen so as to give yellow stains in storage. The yellow stains

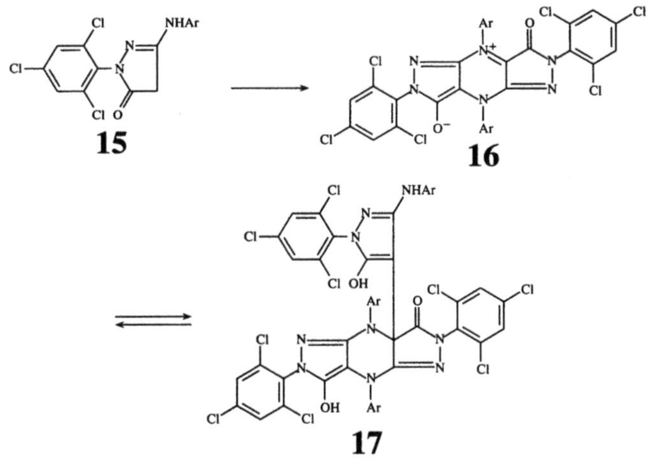

Figure 11.5. Yellow stain formation due to the oxidative self-coupling of a four-equivalent 3-anilino-5-pyrazolone coupler (**15**). The dimer (**16**) and the trimer (**17**) cause yellow stains after further decomposition. Since 3-anilino-5-pyrazolone couplers are mainly used in color papers, the improved stabilities of the resulting images (more dye fastness and less yellow stains) are primary requisites to be accomplished.

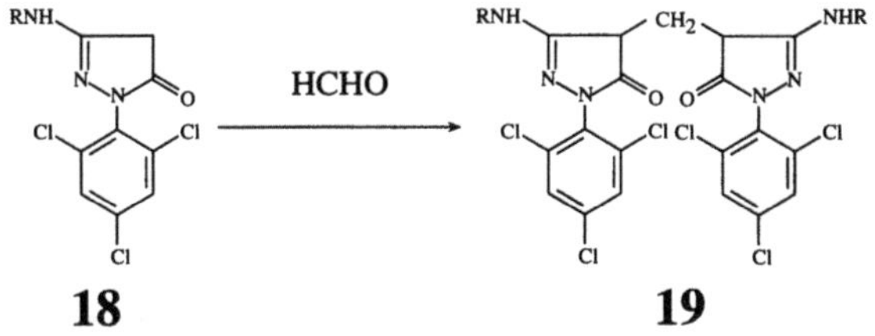

Figure 11.6. Reaction of a four-equivalent 3-acylamido-5-pyrazolone coupler (**18**) with formaldehyde. Since this type of couplers are mainly used in color negative films, an unexposed film in a camera may be exposed to formaldehyde gas if it is left in a new piece of furniture.

have been investigated to come from the dimer (**16**) and the trimer (**17**) shown in Fig. 11.5 [12].

3. A coupler (**18**) is reacted with formaldehyde (due to furniture) to form a coupling product (**19**). It follows that a portion of coated coupler molecules are consumed in vain, as shown in Fig. 11.6 [12].

As clarified by the undesired reactions shown in Figs. 11.5 and 11.6, these disadvantages come from the absence of a substituent at the coupling site (4-position) of the 5-pyrazolone nucleus. It follows that two-equivalent couplers which have an appropriate substituent at the 4-position are desirable to overcome the disadvantages.

Various groups linking through a hetero atom with the active position have been extensively investigated to develop two-equivalent couplers having improved properties. For example, a chlorine atom [13] and a thiocyano group (-SCN) [14,15] have been disclosed as leaving groups for two-equivalent magenta couplers. Two-equivalent magenta couplers of 5-pyrazolone type disclosed in a patent [16] have had aryloxy leaving groups such as O–C_6H_4-SO_2CH_3-*p* and O–C_6H_4-SO_2NR_2-*p*, while those disclosed in another patent [17] have been characterized by acyloxy leaving groups such as O–CO–alkyl and O–CO–aryl. A further type of acyloxy groups has been proposed in yet another patent [18]. Heterocyclic leaving groups such as a triazolyl group have been proposed under the title of "two-equivalent couplers for photography" [19], which will be, however, discussed as DIR couplers later (Chapter 13). Magenta couplers having various leaving groups (e.g., chloro, thiocyano, sulfo, 3-octadecylcarbomylphenylthio, acetoxy, and stearoyloxy groups) have been disclosed in a patent [20] aiming at a shortened processing (blix).[4] 5-Pyrazolone magenta couplers having $N(R)SO_2R'$ at the coupling position have also been proposed as two-equivalent couplers having improved properties [21].

The two-equivalent magenta couplers described in the preceding paragraph have not been practically used, because of one or more disadvantages in that a remarkable color fog is produced, in that the coupling reactivity is insufficient, in that the couplers per se are chemically unstable and change into compounds which cannot form dyes, or in that many difficulties are encountered in the preparation thereof (Table 11.1).

Arylthio Groups as Leaving Groups

Arylthio groups have been disclosed as leaving groups of two-equivalent magenta couplers [22]. The thiol compounds which are formed upon the coupling reaction have a severe photographic effect (i.e., DI or development inhibition) so that the couplers described in this patent are now well-known under the name of DIR couplers (development inhibitor releasing couplers), as will be discussed in Chapter 13. Hence, they have been used as additional couplers together with other main couplers.[5] To use such thiol-releasing couplers as main couplers, the DI effect, which is undesired for this purpose, should be suppressed in the light of appropriate methodology [1,12]. This task has been accomplished by means of couplers such as **20** shown in Fig. 11.7 [23,24].

The coupler **20** has been reported to give the corresponding azomethine dye in 83% yield, which is far higher than the value (about 50%) for the corresponding four-equivalent coupler [12]. Since the released group ArS^- is oxidized into the corresponding disulfide ArS–SAr in about 20% yield, an additional amount of

[4]The term "blix" is the abbreviation of "bleach-fixing".

[5]Main couplers are used to produce most of image dyes, while additional couplers are subsidiarily used to control photographic properties (e.g., color reproduction).

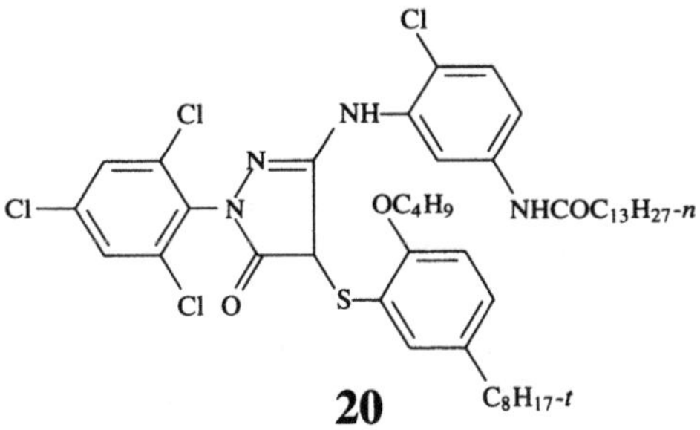

20

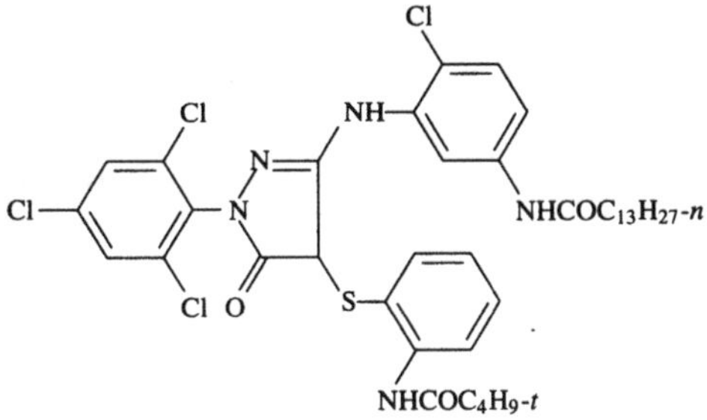

21

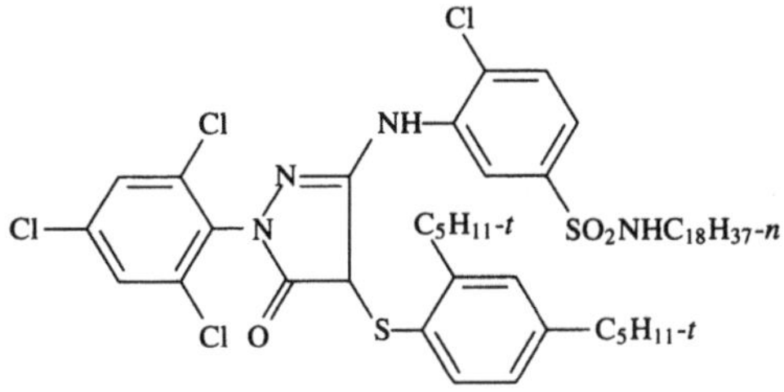

22

Figure 11.7. Non-diffusible two-equivalent couplers forming a magenta azomethine dye (3-anilino-5-pyrazolone coupler). The first structure (**20**), which is covered by the patent [23,24] described in the text, is cited from a more recent article [25]. The second structure (**21**) and the third structure (**22**) are covered by the patents [26] and [27], respectively.

the oxidized color developer (or silver halide) is consumed in the development process. Hence, precisely speaking, the coupler is an about 3.3-equivalent coupler rather than a two-equivalent coupler in its literal meaning. However, the two-equivalent coupler still exhibits twice or more efficiency than the four-equivalent coupler.

A two-equivalent coupler having a 2-alkoxy-5-alkylphenylthio group (e.g., **20**) has been reported to exhibit less yellow stain than the corresponding four-equivalent coupler [1]. Thus, about 90% of the former coupler remains at 40℃

and 70% relative moisture, while about 45% of the latter remains under the same conditions. See Fig. 11.5 for the reactions forming yellow stains.

Thiol derivatives released from magenta couplers might influence the light fastness of the resulting magenta dyes, since they are non-diffusible and remain in the neighborhood of the dyes. Such *o*-alkoxyarylthio groups as found in the coupler **20** have been reported to be exceptionally useful, since they have no harmful influence on the light fastness [12].

The presence of an acylamido group in the ortho position on the phenylthio leaving group of the coupler **21** enhances the acceptability to rapid machine processing [26]. The coupler **22** has been used in the presence of organic disulfides as light stabilizers of image dyes [27]. This type of couplers has been described in Chapter 1 (Subsection 1.2.3, Fig. 1.28) as a recent embodiment for a color negative film [28]. An analogous coupler has also been disclosed in a patent by emphasizing the solubility and dispersibility, where the substituent at the 3-position of **22** is replaced by a 2-chloro-4-dodecylsulfonylanilino group [29].

Syntheses

The synthetic pathway of a two-equivalent magenta coupler **28** is illustrated in Fig. 11.8 [23]. 2-Dodecyloxy-5-chlorobenzenesulfenyl chloride (**27**), which is derived from the thiol (**26**) by the action of sulfuryl chloride, is used to introduce a thioether function to the active position of a pyrazolone nucleus.

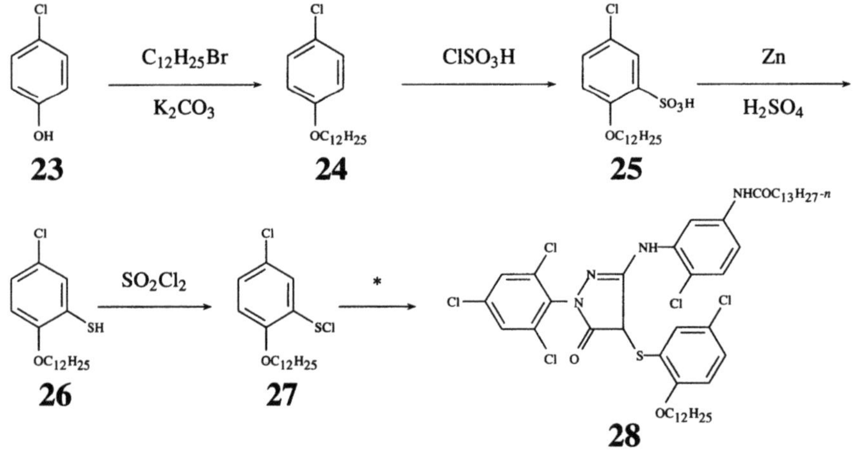

Figure 11.8. Synthetic pathway of the magenta coupler **28**. This scheme shows the synthesis of a sulfur-containing leaving group (**27**), which is reacted with the corresponding four-equivalent coupler represented by an asterisk.

The synthesis of a two-equivalent coupler having a 2-alkoxy-5-alkylphenylthio group (e.g., **20**), which has been reported to be practically used in color papers, has been described in detail [12]. Along with a desired monoarylthio product (e.g., **20**), the corresponding diarylthio derivative has been generated as a by-product, although the 2-alkoxy and the 5-alkyl groups in the arylthio group have not been specified in the reference [12]. The desired product is obtained in a high yield by using dimethylformamide as a solvent or triethylamine as a base.

Pyrazolyl Groups as Leaving Groups

3-Acetamido-5-pyrazolone couplers, which have long been used in color negative films, have required another type of leaving groups to be changed into the corresponding two-equivalent couplers. Various heterocyclic leaving groups (e.g., 1-pyrazolyl, 1-imidazolyl, phthalimido, and 2-methyl-1-imidazolyl) have been tested whether they are suitable to this purpose or not [12]. Among them, the 1-pyrazolyl group has been selected because a pyrazolyl-substituted coupler such as **29** is stable in a color film and exhibits high reactivity with an oxidized color developer [30]. Dark fading of the magenta dyes derived from two-equivalent 3-acylamido-5-pyrazolone couplers has been ascribed to the reaction of the magenta dye with the unreacted coupler, where the pK_a value due to the leaving group (including pyrazolyl groups) at the coupling position influences the fading rate of the dye [31].

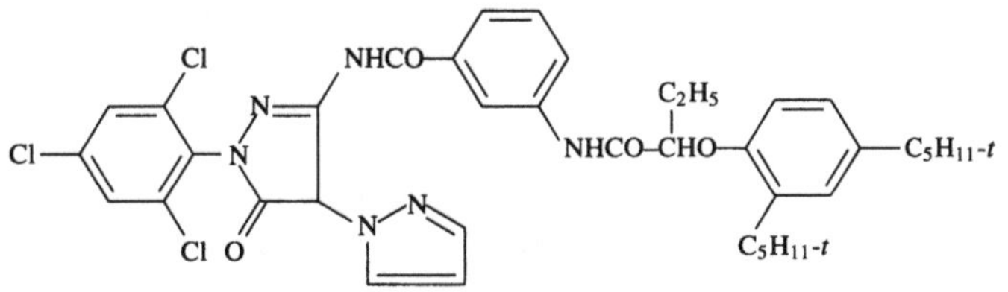

29

Figure 11.9. Non-diffusible two-equivalent coupler forming a magenta azomethine dye (3-acylamido-5-pyrazolone coupler). This structure, which is covered by the patent [30] described in the text, is cited from a recent patent [32].

Syntheses

The synthetic pathway of a two-equivalent magenta coupler **31** is illustrated in Fig. 11.10 [30]. The monobromo derivative (**30**), which is derived by the bromination of the corresponding four-equivalent magenta coupler, is condensed with 3,5-dimethylimidazole to produce the magenta coupler (**31**). The monobromo derivative (**30**) undergoes disproportionation to give the corresponding four-equivalent coupler and the dibromo derivative, the latter of which causes side-reactions to produce dimeric materials as by-products [12]. The coupler (**29**) is obtained in a similar way except that imidazole is substituted for 3,5-dimethylimidazole.

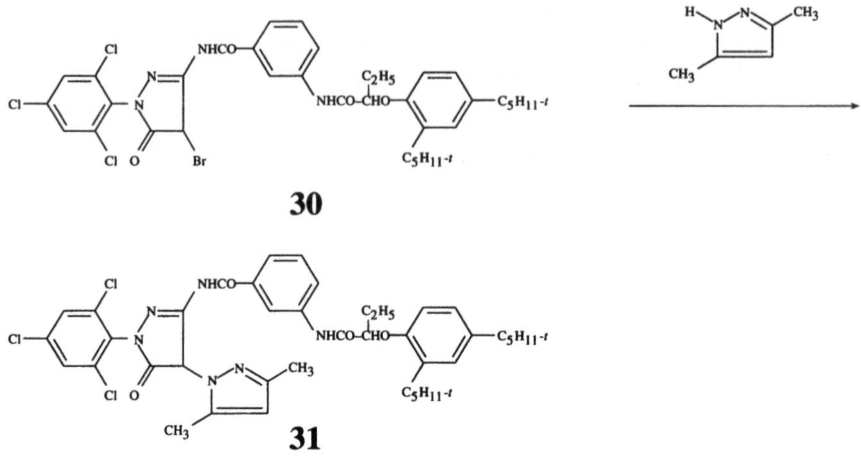

30

31

Figure 11.10. Synthetic pathway of a two-equivalent magenta coupler **31**.

Aryloxy Groups as Leaving Groups

In spite of several patents (e.g., [16]), aryloxy groups have not been practically used as leaving groups for 3-anilino- and 3-acylamido-5-pyrazolone couplers. This is mainly because the pyrazolone couplers having aryloxy leaving groups are instable during film storage and because there have been no practical synthetic routes as summarized in Table 11.1.

As a result, two-equivalent couplers which contain a leaving group having a sulfur or nitrogen atom linked to the pyrazolone nucleus have first been placed on the market, as discussed previously. However, such pyrazolone-based couplers as having an aryloxy leaving group have long been considered to be potentially attractive two-equivalent magenta couplers, because the released aryloxy compound is inactive to silver halide in comparison with the thiolic or heterocyclic compounds released.

Interest in 3-alkyl-5-pyrazolones (e.g., **32**) has been revived leading to the development of stable couplers having an aryloxy leaving group at the 4-position of the 5-pyrazolone nucleus, as exemplified in Fig. 11.11 [33]. The coupler **32** forms a magenta dye having λ_{max} at 554 nm after the C-41 processing using CD-4 (4-amino-*N*-ethyl-*N*-(2-hydroxyethyl)-3-methylaniline).

3-Aryl-5-pyrazolones (e.g., **33** in Fig. 11.11) have been disclosed to be easy to prepare and to be stable during synthesis, film manufacturing, and during film storage [34].

11.3.2 Two-Equivalent Magenta Couplers of Other Types

In Chapter 10, we have already described 1H-pyrazolo[5,1-*c*][1,2,4]triazole and 1H-pyrazolo[1,5-*b*][1,2,4]triazole couplers, most of which are two-equivalent

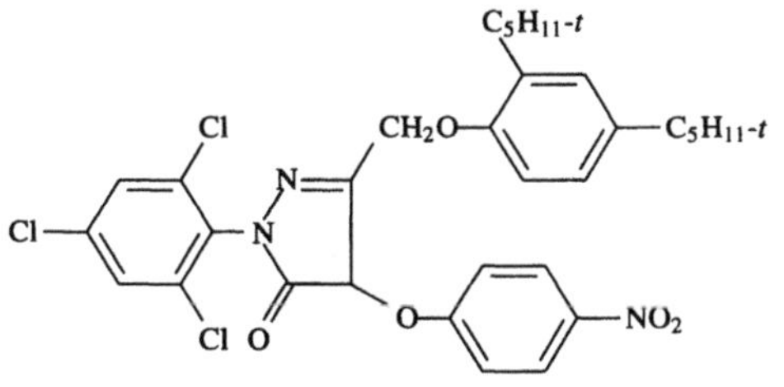

32

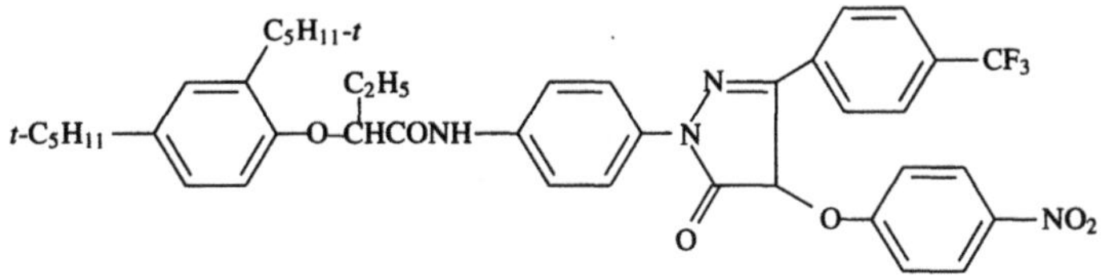

33

Figure 11.11. Non-diffusible two-equivalent couplers forming a magenta azomethine dye (3-alkyl- and 3-aryl-5-pyrazolone couplers).

couplers. For example, a chlorine atom (e.g., **34a**), a 1-pyrazolyl group (e.g., **34b**), and a 4-methylphenoxy group (e.g., **34c**) have been disclosed as leaving (coupling-off) groups for 1H-pyrazolo[1,5-*b*][1,2,4]triazole couplers (Fig.11.12) [35]. Couplers of this type have been described in Chapter 1 (Section 1.2.3, Fig. 1.41) as a recent embodiment for a color negative film [36].

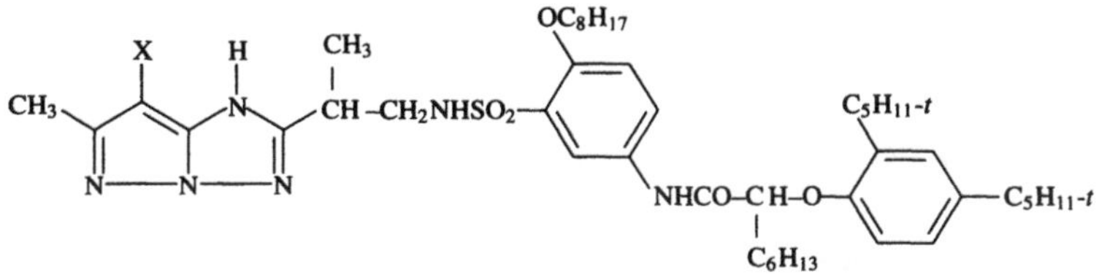

	34a	34b	34c
X =	Cl	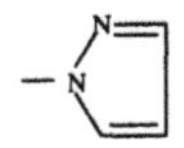 1-pyrazolyl	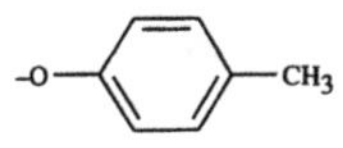 $-O-C_6H_4-CH_3$

Figure 11.12. Two-equivalent pyrazolotriazole magenta couplers.

11.4 Two-Equivalent Yellow Couplers

11.4.1 Earlier Two-Equivalent Yellow Couplers

A chlorine atom at an active methylene position of benzoylacetanilide couplers is too active to be applied to photographic materials [37]. It follows that these chlorinated couplers give rise to higher stain in the unexposed area. The formation of such stains is ascribed to one of two causes: (a) the coupler reacts directly with the unoxidized color developer to form leuco dye which is later oxidized to the dye, or (b) the chlorine substituent is so active as to oxidize a color developer, which then reacts with another mole of coupler. The effect of an *o*-methoxy group in two-equivalent couplers **35a** and **35b** was disclosed to reduce the activity of the chlorine substituent and, as a result, to prevent the formation of excessive dye stains [38]. In other words, the corresponding coupler with no chlorine atom is not as active in color development. The coupling activity is increased by the presence of the chlorine substituent on the active methylene group, where, at the same time, stain is reduced by virtue of the presence of the *o*-methoxy group. Although two-equivalent couplers having acyloxy groups [39] and sulfonyloxy groups [40] as leaving groups also have been disclosed, they are expected to be susceptible to hydrolysis.

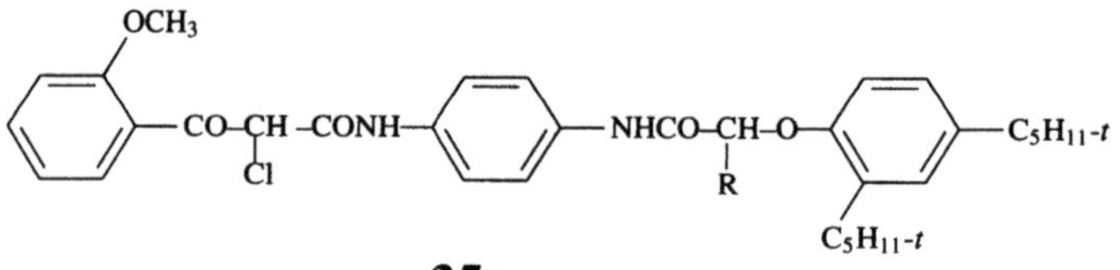

35a (X=H)
35b (X=C_2H_5)

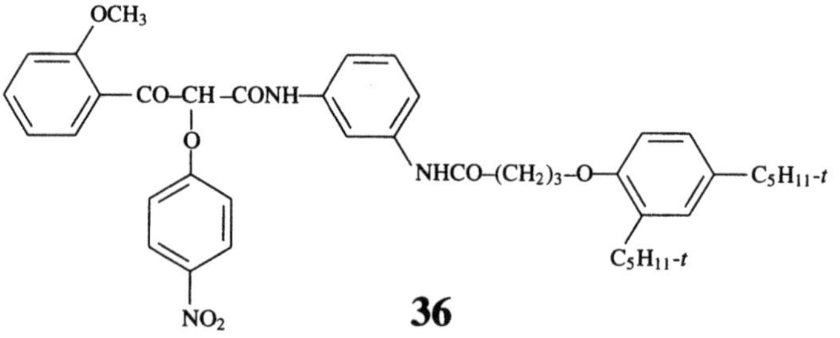

36

Figure 11.13. Two-equivalent yellow couplers (benzoylacetanilide couplers). The chloro atom and the *p*-phenoxy group are leaving groups, which are inert to silver halide grains.

11.4.2 Acylacetanilide Couplers

Aryloxy Groups as Leaving Groups

Two-equivalent yellow couplers of benzoylacetanilide type that have an aryloxy leaving group at the coupling position have been disclosed [41], as exemplified by the formula **36** (Fig. 11.13).

Various aryloxy groups have been examined as leaving groups for pivaloylacetanilide couplers. For example, 4-sulfonyl-, 4-sulfamoyl-, or 4-sulfonamidophenoxy groups (e.g., **37a**) have exhibited improved coupling activities [42]. The data of the competitive coupling test using citrazinic acid (1,2-dihydroxy-4-pyridinecarboxylic acid) can be applied to the evaluation of their coupling activities, though the original data aimed at another technical purpose.

Thus, **37a**, which has a phenoxy leaving group substituted by 4-benzyloxybenzenesulfonyl group, exhibits the coupling activity comparative to citrazinic acid, whereas such a coupler as **36** having a *p*-nitrophenoxy group shows lower coupling activity.

The coupling activity can be enhanced by incorporating a dissociative group such as sufonamido groups at the ortho position (e.g., **37b**) [43]. This coupler exhibits D_{max} 3.16 by color development with 4-amino-*N*-ethyl-*N*-(2-hydroxyethyl)-3-methylaniline sulfate (CD-3). On the other hand, the coupler having a 4-methoxycarbonylphenoxy group and the one having a 4-acetamidophenoxy group give lower D_{max} values (2.64 and 2.08 respectively).

The active position in **37c** is substituted with a phenoxy group having a branched chain alkoxycarbonyl group at the 4-position of the phenoxy group [44]. This coupler is not susceptible to the variation of pH of a color developing solution so that it forms a dye image of reduced fluctuations in its image density.

Figure 11.14 contains a two-equivalent yellow coupler (**37d**) having a sulfonamido group as a leaving (coupling-off) group, which is characterized by the presence of a *N*-pyridyl group [45]. This type of couplers gives less stain after storage in the heating cupboard. Another type of leaving groups has been disclosed to ensure sufficient color densities even with short processing times [46].

The dark stability can be improved by incorporating a bulky substituent (e.g., a pivaloylamido group) at the ortho position, as found in **38** [47]. Thus, the loss of the original density (at $D = 1.0$) for the dye derived from the coupler **38** has been determined to be −0.15 after 6 weeks at 60℃ and 70% relative humidity. When the pivaloylamido group of the leaving group in **38** is replaced by an acetamido group, the corresponding value for the resulting dye was determined to be −0.21. When the leaving group of **38** is replaced by that of **37a**, the resulting dye showed the corresponding value of −0.20. Since these couplers produce the same dye, the above-mentioned difference among the dark stabilities suggests some interaction between the dye and the released phenol derivatives.

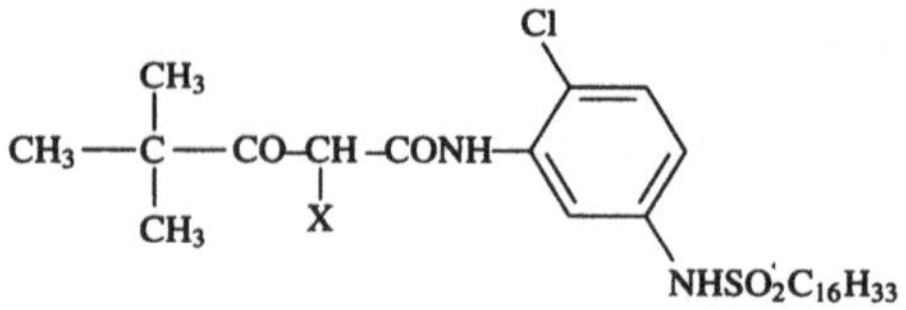

Figure 11.14. Two-equivalent couplers forming a yellow azomethine dye (pivaloylacetanilides).

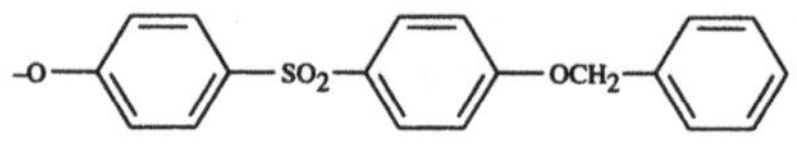

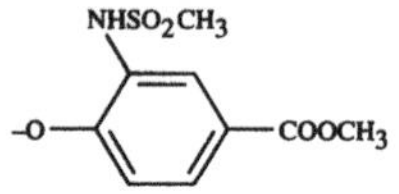

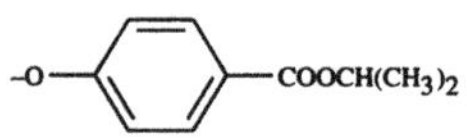

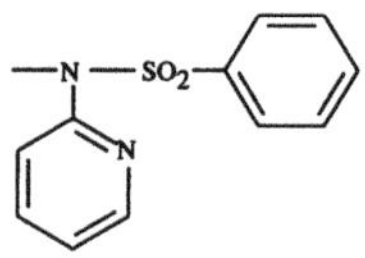

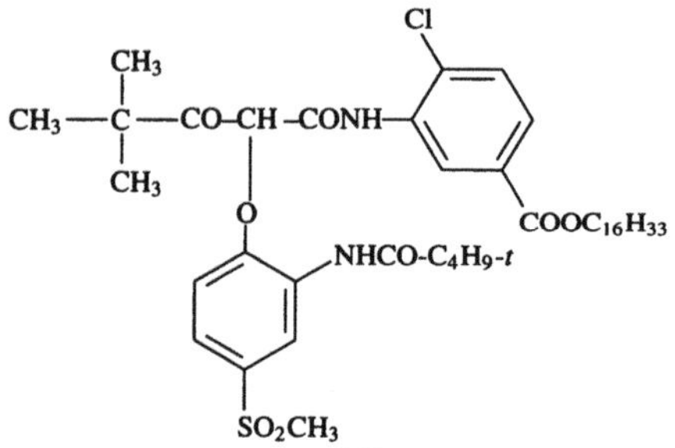

Heterocyclic Leaving Groups

Various 5-membered heterocyclic groups have been disclosed as leaving groups for two-equivalent pivaloylacetanilide couplers. For example, a hydantoin (e.g., **39a**) [48] and dimethadione (e.g., **39b**) [49] have been selected as such leaving heterocycles. One of the most convenient leaving groups is the 4-ethoxy-3-benzyl-1-hydantoinyl group, as exemplifies by the formula **39c** [50]. This group has been widely used as a leaving group for two-equivalent yellow couplers of various types. For further examples, see Figs. 1.5, 1.29, and 1.43 (Chapter 1) as well as Figs. 10.34, 10.36, 10.37, and 10.38 (Chapter 10). It has been also incorporated into benzoylacetamide couplers [51].

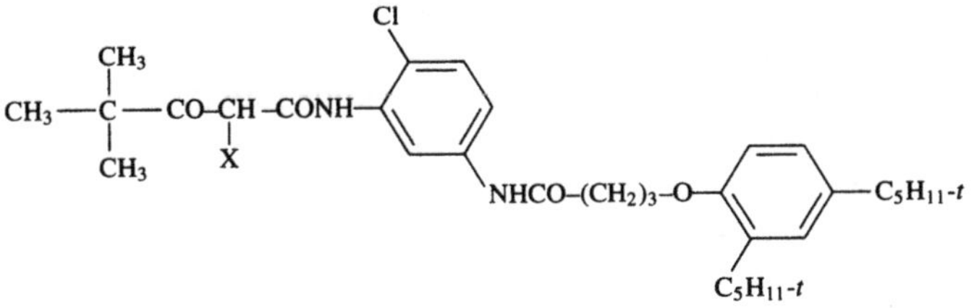

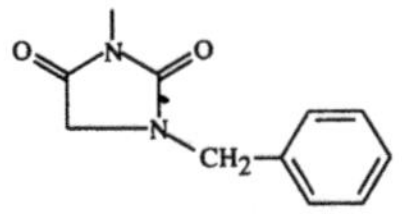

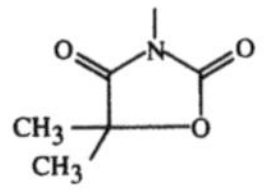

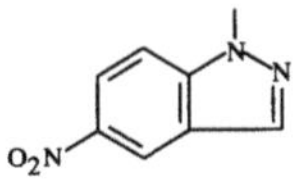

Figure 11.15. Various heterocyclic leaving groups for two-equivalent couplers forming a yellow azomethine dye (pivaloylacetanilides).

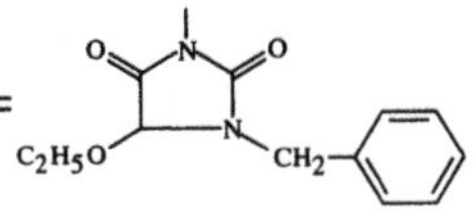

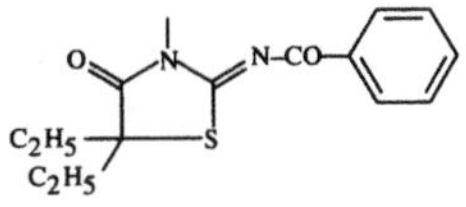

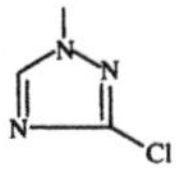

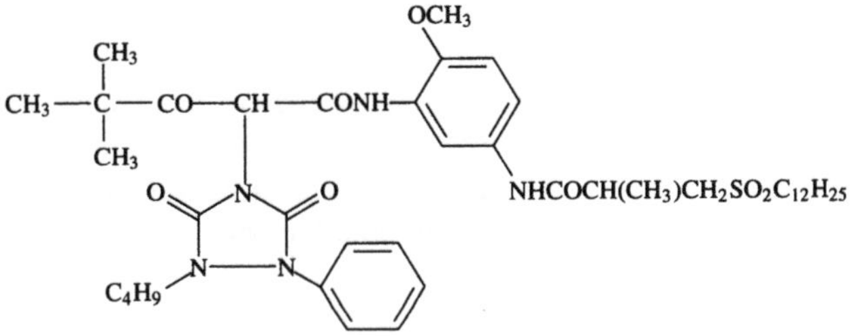

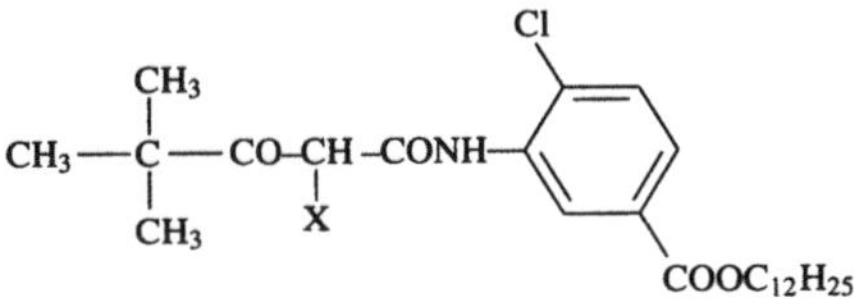

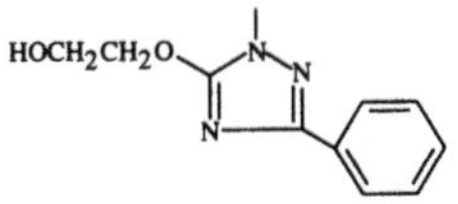

5,5-Diethyl-2-benzoylimino-1,3-thiazolidine-4-one has been used to produce a two-equivalent coupler represented by the formula **39d** [52]. 3-Chloro-1,2,4-

triazole has been used as a leaving group of **39e** [53]. These couplers have been found to exhibit a very good reactivity.

The dye derived from a two-equivalent pivaloylacetanilide coupler (e.g., **40**) having a methoxy group on the anilide ring has a lemon-yellow hue, which is more desirable than an orange-yellow hue that is characteristic of dyes derived from couplers having a chlorine atom on the anilide ring [54].

The dye-forming efficiency of **41a** has been evaluated by D_{max} 3.03 [55], while the coupler having the 4-ethoxy-3-benzyl-1-hydantoinyl leaving group exhibits D_{max} 2.20. The coupler **41b** has been disclosed to give a noticeable improvement of sharpness [56]

Syntheses

In general, a two-equivalent coupler is synthesized by the introduction of a halogen atom at the active position of the corresponding four-equivalent coupler and by the following substitution of an appropriate leaving group for the halogen atom. Figure 11.16 illustrates the preparation of a yellow coupler of the cyclopropyl type (**49**), the properties of which have been discussed in the preceding chapter [57].

1-Methylcyclopropane-1-carboxylic acid (**42**) is chlorinated by oxalyl chloride to give 1-methylcyclopropane-1-carboxyl chloride (**43**). This is condensed with the magnesium salt of ethyl acetylacetate to give ethyl α-(1-methylcyclopropane-1-carboxyl)acetylacetate (**44**), which is deacetylated under basic conditions to produce ethyl (1-methylcyclopropane-1-carboxyl)acetate (**45**). The corresponding four-equivalent coupler (**47**), which is prepared by the condensation of a β-ketoester (**45**) and a ballast amine (**46**), is chlorinated by sulfuryl chloride to give the chloro intermediate (**48**). Then, the chlorine atom is replaced by a heterocyclic anion to give the two-equivalent coupler (**49**).

1) β-Ketoester:

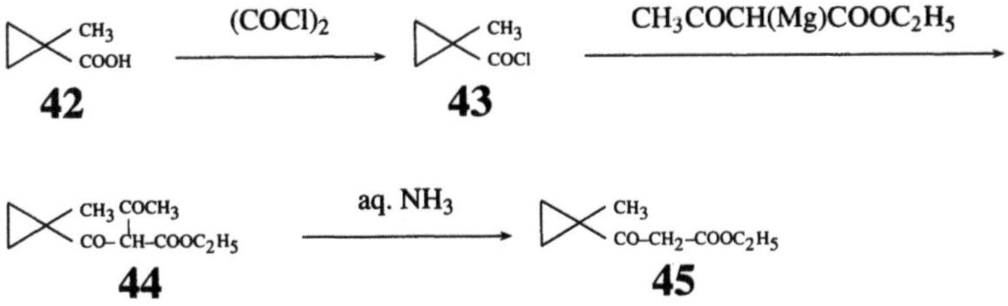

CH_3 COOH **42** —$(COCl)_2$→ CH_3 COCl **43** —$CH_3COCH(Mg)COOC_2H_5$→

CH_3 $COCH_3$ CO–CH–$COOC_2H_5$ **44** —aq. NH_3→ CH_3 CO–CH_2–$COOC_2H_5$ **45**

2) Two-equivalent yellow coupler:

Cl NH_2 C_2H_5 NHCO–CHO– –C_5H_{11}-t C_5H_{11}-t **46** —45→ **47** —SO_2Cl_2→

CH_3 CO–CH–CONH– Cl Cl C_2H_5 NHCO–CHO– –C_5H_{11}-t C_5H_{11}-t **48** —H N O O C_2H_5O N CH_2Ph, $(C_2H_5)_3N$→

CH_3 CO–CH–CONH– Cl O N O C_2H_5O N CH_2– C_2H_5 NHCO–CHO– –C_5H_{11}-t C_5H_{11}-t **49**

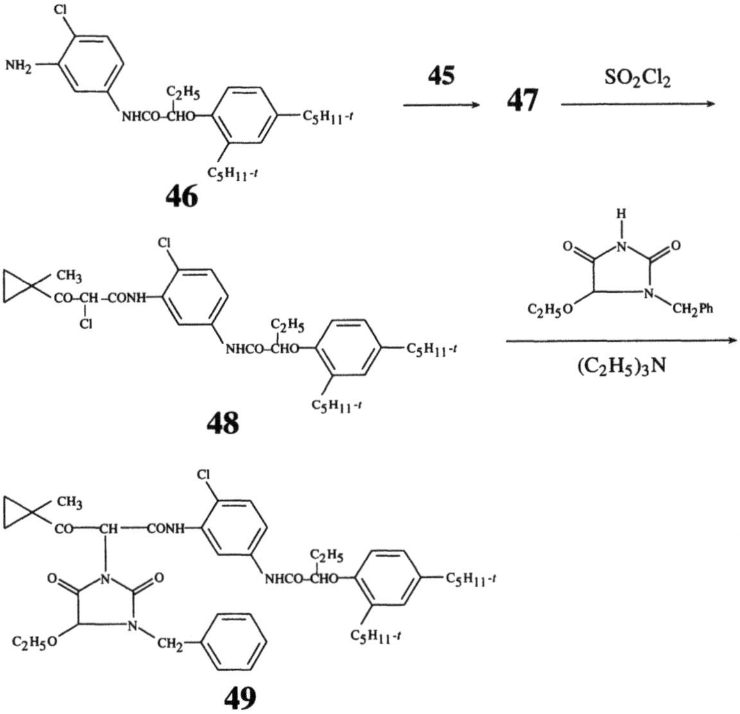

Figure 11.16. Synthetic pathway of a two-equivalent yellow coupler **49**. The intermediate **47** is the corresponding four-equivalent coupler.

References

[1] Furutachi N, Nakamura K, Ichijima S (1986) Nikkakyo Geppo. (3):10
[2] Fischer R, Siegrist (1914) Photogr Korres. 51:18
[3] Gluck B (1947) The Manufacture of Agfacolor Material. Fiat Final Report. No. 943
[4] Weissberger A, Salminen IF, Vittum PW (1949) US Patent 2 474 293
[5] Loria A (1969) US Patent 3 476 563
[6] Yagihara M, Yokota Y (1981) US Patent 4 264 722
[7] Yagihara M, Yokota Y (1980) US Patent 4 228 233
[8] Deguchi H, Endo T, Kikuchi S, Komaita T (1980) US Patent 4 205 990
[9] Kobayashi H, Mihayashi K (1994) J Soc Photogr Sci Technol Jpn. 57:316
[10] Yamakawa K, Kobayashi H, Itoh I (1988) Jpn Kokai S63-258446
[11] Ohkawa A, Kamio T, Motoki M, Mihayashi K (1992) US Patent 5 112 730
[12] Furutachi N, Nakamura K, Ichijima S (1987) Yuki Gosei Kagaku Kyokai Shi. 45:151
[13] Loria A, Reckhow WA, Salminen IF (1961) US Patent 3 006 759
[14] Loria A, Salminen IF (1965) US Patnt 3 214 437
[15] Loria A, Salminen IF (1966) US Patent 3 253 924
[16] Young DV (1968) US Patent 3 419 391
[17] Loria A (1967) US Patent 3 311 476
[18] Arai A, Oishi Y, Yamada M, Furutachi N, Nakamura K (1975) US Patent 3 926 631
[19] Sawdey GW (1971) US Patent 3 617 291
[20] Edens CO, Van Campen, JH (1971) US Patent 3 582 322
[21] Arai A, Shiba K, Yamada M, Furutachi N, Nakamura K (1980) US Patent 4 237 217
[22] Barr CR, Williams J, Whitmore KE (1966) US Patent 3 227 554
[23] Aoki K, Seto N, Yabuki Y, Morigaki M, Furutachi N, Nakamura K (1982) US Patent 4 351 897
[24] Mitsui A, Nakamura K (1983) US Patent 4 413 054
[25] Kobayashi H, Okuzawa M, Kurono H, Okawa Y, Ohno T (1994) J Imaging Sci Technol. 38:28
[26] Krishnamurthy S, Johnston BH, Kilminster KN, Vogel DC, Buckland PR (1989) US Patent 4 853 319
[27] Krishnmurthy S (1988) US Patent 4 740 438
[28] Zengerle PL, Sowinski AF (1998) US Patent 5 726 003

[29] Krishnamurthy S, Crawley MW, Bailey DS, Pawlak JL (2000) US Patent 6 015 657

[30] Ichijima S, Yabuki Y, Watanabe T, Furutachi N (1982) US Patent 4 310 619

[31] Sakanoue K, Furutachi N (1988) J Photogr Sci. 36:64

[32] Matsumoto K, Ito T (2002) US Patent 6 337 176 B1

[33] Poslusny JN, Anderson LG, Mooberry JB, Wu ZZ (1996) US Patent 5 576 167

[34] Slusarek WK, Poslusny JN, Wu ZZ (2001) US Patent 6 280 919 B1

[35] Mizukawa Y, Naruse H, Watanabe T, Sato T (1995) US Patent 5 409 808

[36] Sato K, Ishii Y, Yamakawa K (1995) US Patent 5 401 624

[37] Weissberger A, Kibler CJ (1966) US Patent 3 265 506

[38] McCrossen FC, Vittum PW, Weissberger A (1955) US Patent 2 728 658

[39] Loria A (1969) US Patent 3 447 928

[40] Porter RF (1970) US Patent 3 542 840

[41] Loria A (1968) US Patent 3 408 194

[42] Cameron RG, Neuberger D (1976) US Patent 3 933 501

[43] Lau PTS (1983) US Patent 4 401 752

[44] Ogawa A, Furuzawa H (1988) US Patent 4 770 983

[45] Kunitz F-W, Salzmann H (1980) US Patent 4 186 019

[46] Lowski D, Schranz K-W (1979) US Patent 4 146 400

[47] Tang PW, Reynolds JH, Corcoran DE (2000) US Patent 6 130 032

[48] Fujiwhara M, Matsuo S, Kojima T (1976) US Patent 3 973 968

[49] Arai A, Oishi Y, Okumura A, Nakazyo K (1983) US Patent 4 404 724

[50] Okumura A, Sugizaki A, Ichijima S, Shiba K, Nakazyo K (1976) Jpn Kokai S51-102636

[51] Arai A, Oishi Y, Nakazyo K, Sugizaki A, Okumura A (1981) US Patent 4 269 936

[52] Tchopp P (1980) US Patent 4 206 278

[53] Quaglia A (1980) US Patent 4 182 630

[54] Tsuruta M, Mizukura N, Nakagawa S (1991) US Patent 4 992 360. See also Kaneko Y (1993) J Soc Photogr Sci Technol Jpn. 56:301

[55] Dahlhause U, Langen H, Wiesen H (2001) US Patent 6 297 385 B1

[56] Crawley MW (2001) US Patent 6 190 853 B1

[57] Shimura Y, Kobayashi H, Yoshioka Y (1995) US Patent 5 427 902

Chapter 12

Colored Couplers

12.1 Color Masking

Although color photography based on the negative-positive process is popular, its mechanism is not so well-known to users. For example, it is seen that a negative film after development (in the transparent region) is orange in color. Why is this? This chapter is devoted to giving an answer to this question.

12.1.1 Azomethine Dyes From Azo Dyes

As mentioned in Chapter 10, azomethine dyes derived from 5-pyrazolone couplers as well as indoaniline dyes derived from 1-naphthol couplers have secondary absorption bands at the shorter-wavelength regions of light. These secondary absorptions cause irregular color reproduction. One way to ensure correct or faithful color reproduction is to develop couplers which give dyes without such secondary absorptions, as described in Chapter 10. Another way to compensate the secondary absorptions is color correction using colored couplers, which is nowadays called *color masking*.[1]

[1] A further way for color correction is the usage of DIR (development inhibitor releasing) couplers. This will be later discussed in detail (Chapter 13).

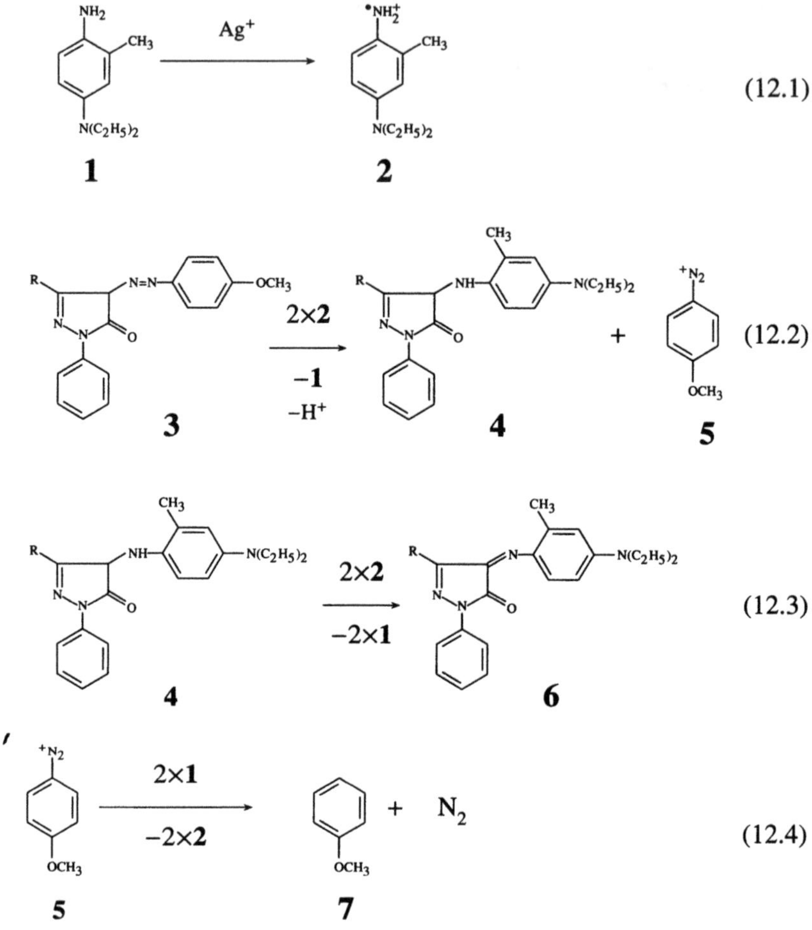

(12.1)

(12.2)

(12.3)

(12.4)

Figure 12.1. Proposed mechanism for the coupling reaction of an azo dye (**3**, R = $NHCOCH(C_2H_5)C_6H_5$) with an oxidized color developer (**2**) to produce the corresponding azomethine dye (**6**). This reaction is basic to color masking due to colored couplers [3].

The beginning of color masking was the discovery of the fact that a pyrazolone azo dye (e.g., **3**) can be reacted with an oxidized color developer (e.g., **2**) to produce a magenta azomethine dye (e.g., **6**) [1,2,3].[2]

[2] An episode of the invention of colored couplers for masking has been vividly described in Weissberger's retrospection [4] on the situation in the 1940s when Vittum, Hanson, and other scientists at Kodak played their own roles in this invention. A color negative film using colored couplers was placed on the market by Kodak in 1948.

The proposed mechanism of the reaction is shown in Fig. 12.1 [3]. According to the original paper [3], the oxidized form of a color developer (CD-2) is formulated as a semiquinone ion (eq. 12.1) rather than the quinonediimide species described in the previous chapters. The arylazopyrazolone anion generated from **3** is coupled with the semiquinone ion (**2**) to give the leuco azomethine (**4**), the diazonium ion (**5**), and a regenerated molecule of **1** (eq. 12.2). The oxidation of the leuco dye (**4**) into the azomethine dye (**6**) is formulated as an oxidation of the leuco dye by the semiquinone (**2**), as shown in eq. 12.3. The resulting diazonium ion (**5**) is reduced by two moles of the color developer into anisole and nitrogen (eq. 12.4). As an alternative mechanism, the oxidation of the leuco dye may take place by a direct interaction between the leuco dye and the diazonium ion. Another article has presumed the participation of a quinonediimide, where the corresponding adduct is considered to be further oxidized, giving the azomethine dye (**6**) and the diazonium ion (**5**) [5].

Although the mechanism shown in Fig. 12.1 is rather complex, the pyrazolone coupler is concluded to be a two-equivalent coupler as follows. The second equation (eq. 12.2) requires two moles of the semiquinone ion (**2**). The third equation (eq. 12.3) also requires two moles of **2**. Since one mole of the semiquinone ion (**2**) corresponds to one equivalent of silver halide as found in eq. 12.1, this would seem to be a four-equivalent reaction overall. However, the positively charged diazonium ion (**5**) acts as an oxidizing agent to generate two moles of the semiquinone ion (**2**) in eq. 12.4. This means the overall balance of a two-equivalent (two-electron) process.

Both the mechanisms reported in references [3,5] have presumed the intermediacy of the diazonium ion (**5**) and its reduction (eq. 12.4). However, an alternative mechanism can be considered without the intermediacy of the diazonium ion (**5**), as shown in Fig. 12.2. Thus, the quinonediimide (**8**) reacts with the anion of the colored coupler (**3**) to produce an adduct represented by the formula (**9**), which is capable of generating the azomethine dye (**6**), anisole (**7**), and nitrogen in terms of concerted electron shifts designated by bent arrows.[3] This step (eq. 12.7) is considered to be an intramolecular redox reaction in such an intermediate (**9**). The mechanism shown in Fig. 12.2 is simpler than that of Fig. 12.1, although it is still a speculation of the author. Since two equivalents of silver halide participate in the first step (eq. 12.5) and the other steps require no silver halide (no electron removal), the colored coupler (**3**) is concluded to be two-equivalent.

12.1.2 Azo Dyes as Colored Couplers

As found in the preceding subsection, a pyrazolone azo dye (yellow) is converted into an azomethine dye (magenta) by color development. The azo dye has an absorption in the range of 400–500 nm, as shown in Fig. 12.3 (Curve (a)). On the other hand, the resulting azomethine dye absorbs light of wavelength 500–

[3] Strictly speaking, the directions of these arrows may be reversed or non-determinable.

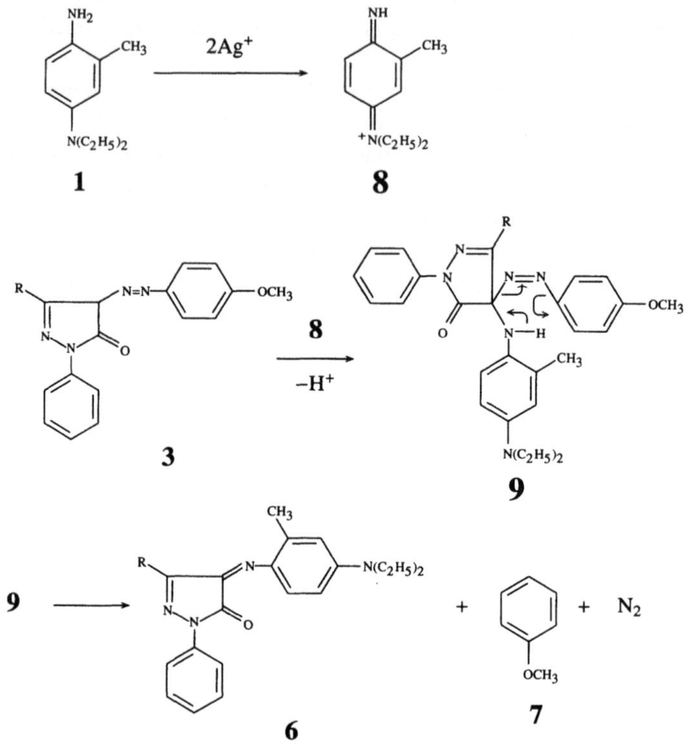

Figure 12.2. Alternative mechanism of the reaction of an azo dye (**3**, where R = $NHCOCH(C_2H_5)C_6H_5$) with an oxidized color developer (**8**) to produce the corresponding azomethine dye (**6**). This reaction is basic to color masking due to colored couplers.

600 nm (Curve (b)), where the secondary absorption band appears in the blue region (400–500 nm) of light. This conversion has been applied to color correction for compensating the undesired secondary absorption [1,6]. This type of color correction is now called *color masking* and the coupler used for the color masking is called a *colored coupler* [7,8].

When such a pyrazolone azo dye is used alone as a coupler, its absorption in the blue region is greater than undesired absorption of the magenta dye (Fig. 12.3). This results in so-called over-correction. Hence, a mixture of a colored coupler (e.g., **10**) and an uncolored coupler (main coupler, e.g., **11**) has been used to prevent such over-correction (Fig. 12.4) [1].

Figure 12.5 shows absorption spectra for such a mixture of a colored coupler (e.g., **10**) and an uncolored main coupler (e.g., **11**) upon exposure under a step tablet and development by a color developer [1]. Curve (a) of Fig. 12.5 represents the absorption spectrum of the colored coupler itself, that is, in the region where there is no exposure. The absorption in the range 400–500 nm of Curve (a) of Fig.

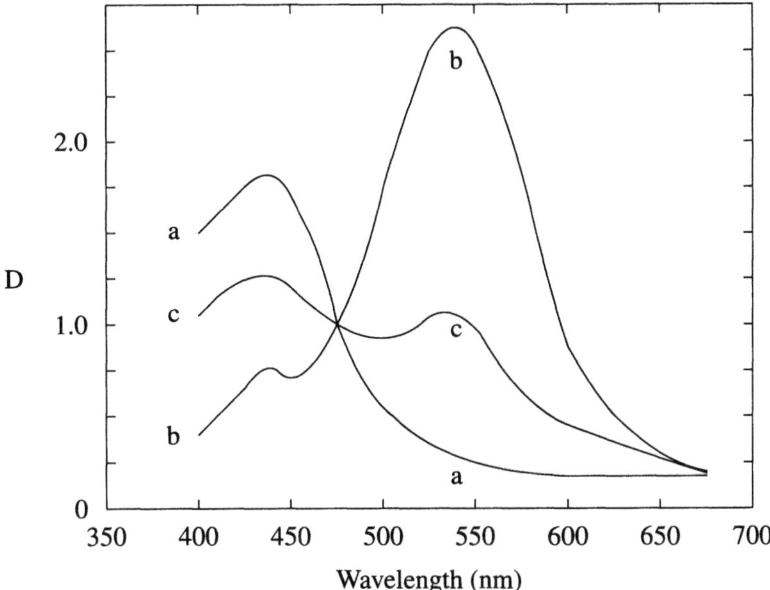

Figure 12.3. Absorption spectra of a colored coupler and the resulting azomethine dye. Curve (a) represents the absorption spectrum of the unreacted colored coupler. Curve (b) represents that of the resulting magenta azomethine dye. Curve (c) shows a mixture of the coupler and the dye [1].

12.5 is approximately equal to the secondary absorption of the resulting magenta dye, which is represented by Curve (e). The latter curve represents maximum dye density where the couplers are entirely converted into the corresponding magenta dyes (from the colored coupler and the uncolored coupler). Curves (b), (c), and (d) shown in Fig. 12.5 represent absorptions for transient regions with increasing amounts of exposure, where the absorption in the range 400–500 nm is substantially constant for all of these regions of exposure. It follows that the undesired side absorption has been cancelled out by printing a negative film based on the color masking onto a color paper.

Because Curves (a) to (e) shown in Fig. 12.5 correspond to the increasing amounts of exposure (H), the sensitometric curve ($\log H$ vs. D) can be drawn, as shown schematically in Fig. 12.6. The curve (G) obtained through a green filter shows that the magenta dye image exhibits the density increasing with increase in exposure. On the other hand, the density through the blue (B) and the red (G) filters is substantially constant regardless of exposure.

In order to improve color reproduction in modern color photography, a *yellow*-colored magenta coupler as well as a *magenta*-colored cyan coupler are incorporated in color negative films. Thereby, background absorptions in the blue and green regions of light are maintained constant by the absorptions due to the colored coupler (yellow and magenta) as well as by the secondary absorption of

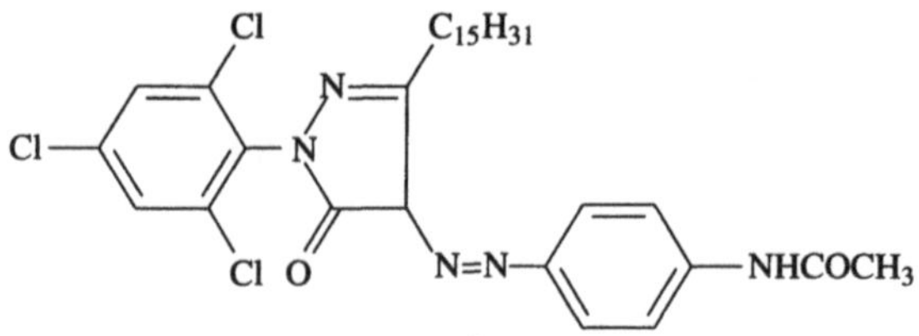

Figure 12.4. Mixture of a colored coupler (**10**) and an uncolored coupler (**11**) for forming magenta dyes [1].

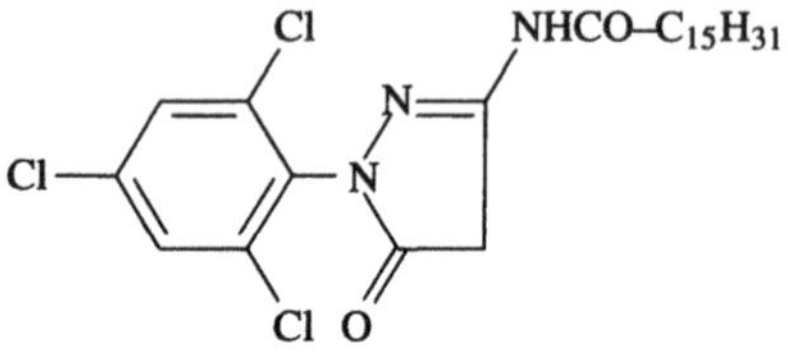

the azomethine dye derived from the main coupler. It follows that a negative film after development is orange in color, where an appropriate mixture of the yellow and the magenta dyes (the colored coupler and the generated dyes) exhibits constant orange color.

12.2 Colored Cyan Couplers

The original idea of colored couplers has been based on the conversion of an azo dye into an azomethine dye, as described in the preceding section. This methodology can be called “decomposition masking” in terms of “masking due to the decomposition of an azo dye”. Another masking method is based on the removal of an azo dye, as will be discussed in this section. This methodology can be called “removal masking”.

12.2.1 Magenta-Colored Cyan Couplers

“Decomposition Masking” by Azo Dyes

Indoaniline dyes derived from 1-naphthol cyan couplers (e.g., **13**) have undesired side absorptions in the range of 500–600 nm (and sometimes 400–500 nm), which are compensated by using colored couplers (e.g., **12**) [1]. Magenta-colored cyan couplers of this type cause the decomposition of an azo dye moiety, exhibiting masking effects in a similar way to yellow-colored magenta couplers illustrated in Fig. 12.1. Thus, such a coupler as **12**, which has an azo group at the

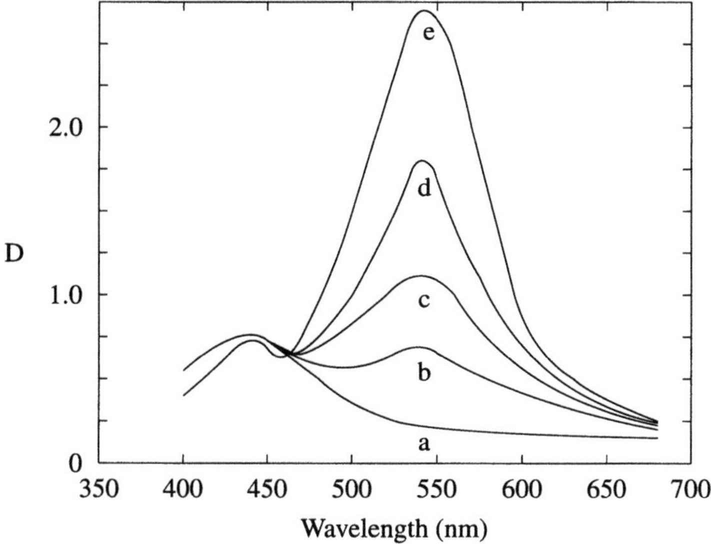

Figure 12.5. Absorption spectra for a mixture of a colored coupler and an uncolored coupler upon exposure under a step tablet and development by a color developer. Curve (a) represents the absorption spectrum of the colored coupler itself. Curve (e) represents maximum dye density where the couplers are entirely reacted to give the corresponding magenta dyes (from the colored coupler and the uncolored coupler). Curves (b), (c), and (d) represent transient regions with increasing amounts of exposure [1].

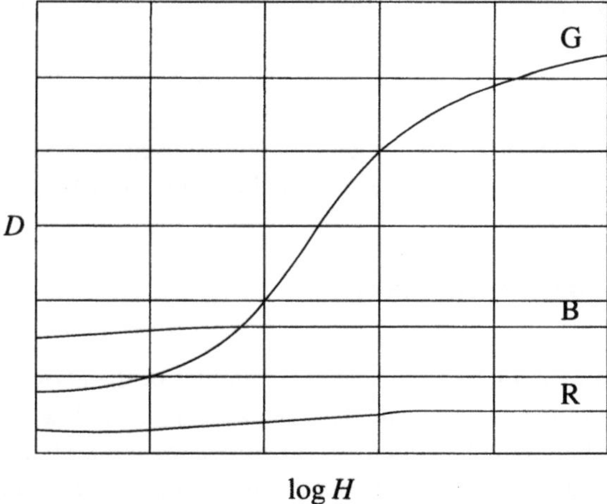

Figure 12.6. Schematic sensitometric curves (log H vs. Density) for a series of dye images such as those represented by Fig. 12.5, where the dye images were read through green (G), blue (B), and red (R) filters [1].

coupling position for an oxidized color developer, works as a magenta-colored coupler according to the methodology of "decomposition masking". This coupler is used as a mixture with a main cyan coupler (Fig. 12.7) in a similar way to a mixture of a magenta coupler/yellow-colored magenta coupler.

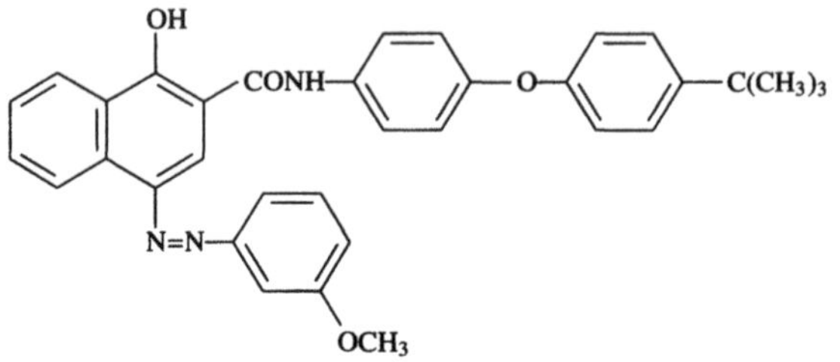

12

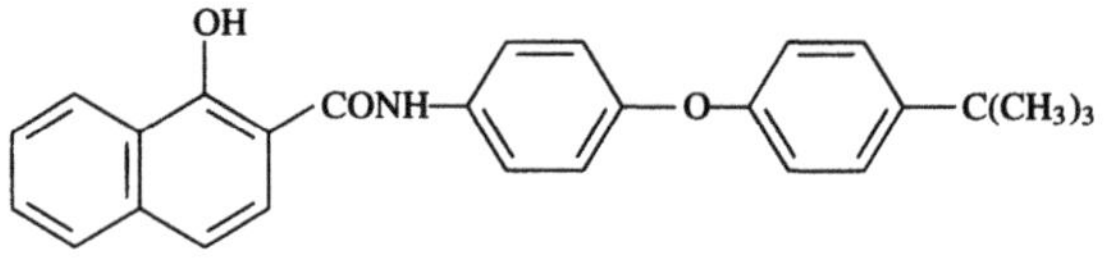

13

Figure 12.7. Mixture of a colored coupler (**12**) and an uncolored coupler (**13**) for forming cyan dyes [1].

An azo dye (i.e., a colored coupler) which is derived from a cyan coupler is usually magenta in color. As a result, the long-wavelength side of the absorption (in the red region of light) lowers the red-sensitivity corresponding to cyan color images. To prevent such reduction of sensitivity, the original color of the colored coupler can be temporarily shifted to yellow by acylating the azo moiety [9]. Thereby, the azo moiety is fixed to the azo form (e.g., **14**) rather than the corresponding hydrazone form. The coupler **14** can be used even in the blue sensitive emulsion layer without interfering with the light sensitivities of other layers.[4]

"Removal Masking" by Azo Dyes

The coupling reaction of a colored cyan coupler of the above-mentioned type, where the azo group is released and decomposed, is rather slow so as to provide an insufficient masking effect. Moreover, the unreacted colored coupler tends to be changed in color as the pH of a processing solution is varied. Colored couplers of

[4]Temporarily shifted colored couplers of the type shown in Fig. 12.8 have not been used in practical photographic materials, because the acyloxy group is so unstable as to be hydrolyzed during storage.

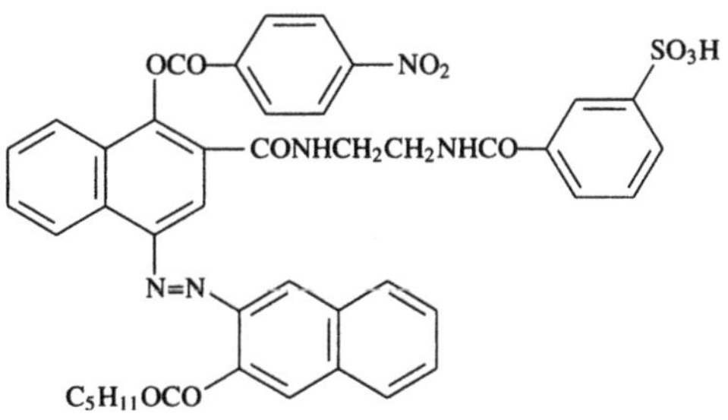

14

Figure 12.8. Colored cyan coupler of temporarily shifted type. This coupler itself is yellow in color, whereas the corresponding coupler having a free 2-hydroxy group is magenta in color [9].

another type (e.g., **15**), in which an azo dye is released without being decomposed ("removal masking"),[5] have been disclosed to overcome such disadvantages [10].

The reaction of such a colored coupler (**15**) with an oxidized color developer (**16**) is illustrated in Fig. 12.9. The unreacted coupler (**15**) and the resulting indoaniline dye (**17**) are non-diffusible and remain in a negative film because of the presence of a ballast group (Ballast). The released azo dye (**18**), on the other hand, diffuses out of a negative film and dissolves into a processing solution. An early example (**19**) is shown in Fig. 12.10 [10].

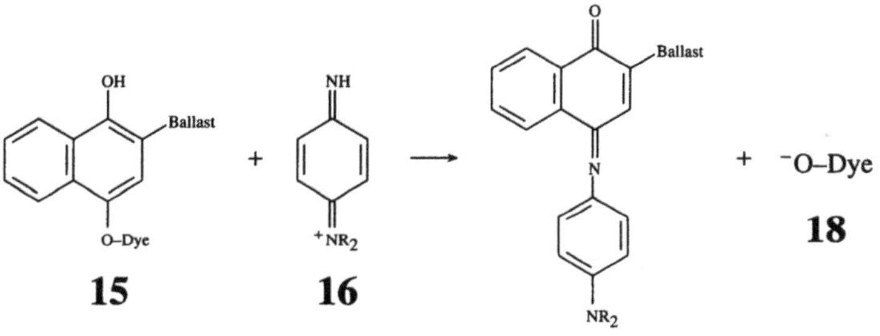

15 **16** **17** **18**

Figure 12.9. Reaction of a colored coupler to form a cyan indoaniline dye and a diffusible dye (Dye–O$^-$).

The magenta-colored cyan coupler (**20**)[6] has once been disclosed as a dye-releasing coupler for diffusion transfer photography [11,12,13]. The azo dye is attached to the 4-position through an aryloxy group and is released as an aryloxide

[5] It should be noted that the dye moiety (Dye) is not always an azo dye according to the removal masking, although an azo dye is usually selected.

[6] This formula contains pyridinium ions as gegen cations. Other cations (e.g., sodium cation) have also been reported as gegen cations in several patents.

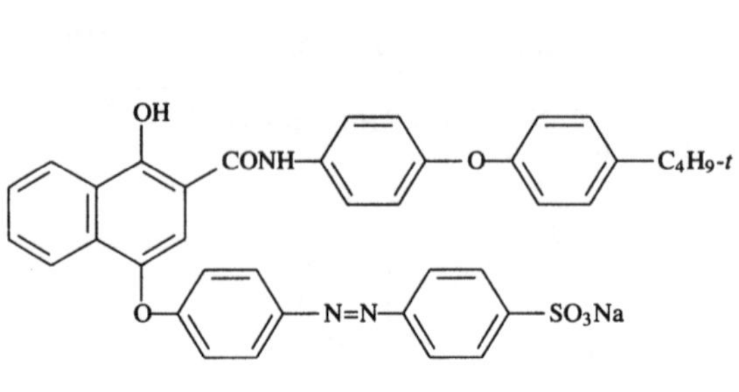

Figure 12.10. Colored couplers for forming a cyan indoaniline dye. The top is a yellow-colored cyan coupler. The other couplers are magenta-colored cyan couplers.

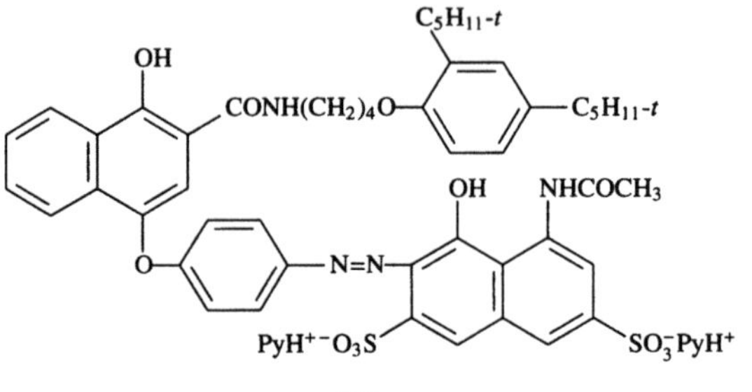

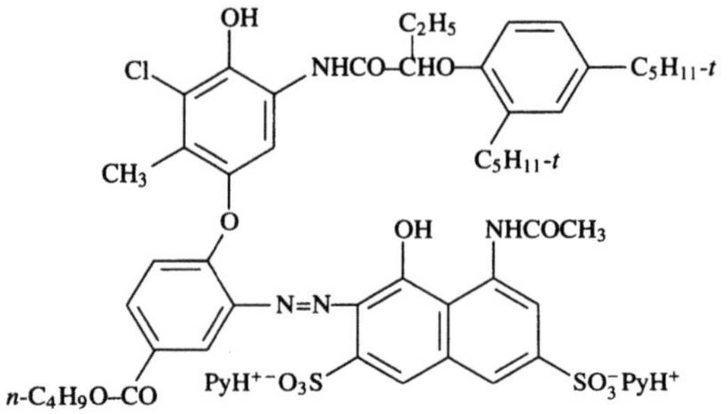

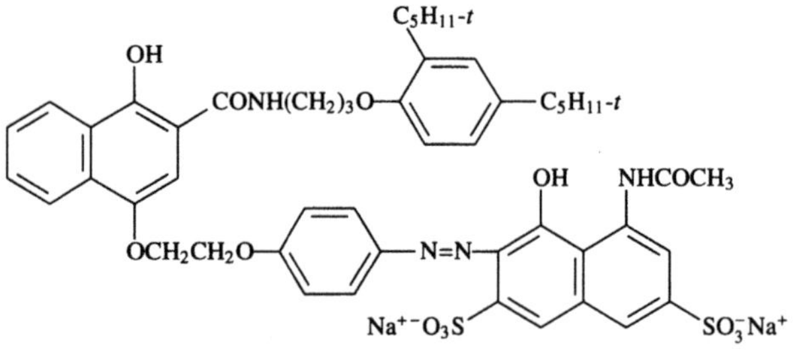

anion. This coupler has been widely used as a representative colored cyan coupler in recent patents (e.g., [14] and [15]). One of such embodiments has been already described in Chapter 1 (Fig. 1.25).

An ideal masking should give uniform absorption in the green and blue regions (particularly in the green region). This depends upon which magenta-colored cyan coupler is selected to be used together with an uncolored cyan coupler. A combination of a phenolic colored coupler such as **21** with a colorless naphtholic coupler has been disclosed to be capable of color-correcting almost perfectly [16].

As for another useful magenta-colored cyan coupler (**22**), the azo-dye moiety is attached to the 4-position through an alkoxy group and is released as an alkoxide anion [17]. Colored cyan couplers based on the same mode of dye releasing have been widely used as representative colored cyan couplers in recent patents (e.g., [18] and [19]), where the ballast group at the 2-position is replaced by a more simple one (e.g., $CONH(CH_2)_3OC_{12}H_{25}$-*n*). One of such embodiments has been already described in Chapter 1 (Fig. 1.34). Colored cyan couplers having a more complicated ballast group (e.g., $CONH$-C_6H_4-2-$CH_2CH(C_6H_{13})C_8H_{17}$) have been disclosed to improve dispersibility [20].

Syntheses

The synthetic pathway of the magenta-colored cyan coupler **21** is illustrated in Fig. 12.11 [16]. The key of this scheme is the nucleophilic attack of an aryloxy nucleophile (from **29**) to butyl 4-fluoro-3-nitrobenzoate (**25**).

One of the key intermediates (**25**), which becomes an diazo component and incorporated as a linking unit in the target molecule (**21**), is prepared by a halide-exchange reaction of butyl 4-chloro-3-nitrobenzoate (**24**).

The other key intermediate (**29**) is prepared by using a two-equivalent cyan coupler (**26**) as a starting material. Thus, the coupler (**26**) is converted into a benzoquinone derivative (**27**) by using sodium nitrite as an oxidative agent, where the *p*-chlorophenol moiety of **26** is oxidized into the benzoquinone moiety of **27**. The quinone is catalytically reduced into the corresponding hydroquinone (**28**), which is in turn mono-benzylated to give a protected intermediate (**29**).

The two key intermediates (**25** and **29**) are condensed under alkaline conditions to give a condensate (**30**). The nitro group of **30** is reduced into an amino group under catalytic hydrogenation, which, at the same time, removes the protective benzyl group catalytically.

The resulting amino intermediate (**31**) is diazotized into the corresponding diazonium salt (**32**), which is coupled with 8-acetamido-3,6-disulfonaphthol (disodium salt) in the presence of pyridine (Py), giving the target molecule (**21**).

The synthesis of a 1-naphthol coupler having a *p*-nitrophenoxy group at the 4-position [10] has been described in Chapter 11 (Fig. 11.2) to illustrate a synthesis of two-equivalent couplers. This compound, i.e., *N*-[4-(2,4-di-*t*-amylphenoxy)butyl]-1-hydroxy-4-(4-nitrophenoxy)-2-naphthamide, can be used as an intermediate for preparing the colored coupler (**20**). After the nitro group is reduced into an amino group, the resulting intermediate is diazotized and coupled

1) Linking unit:

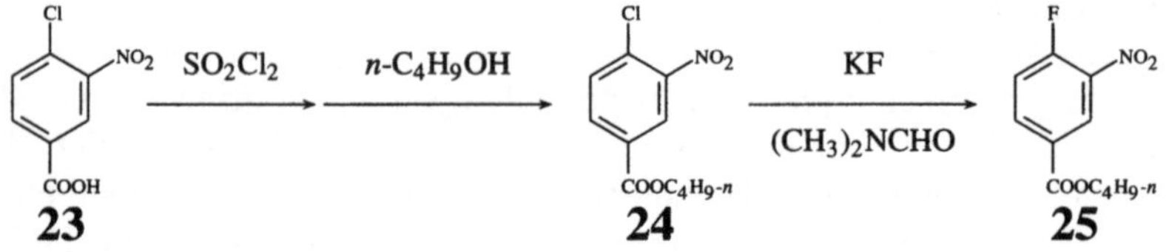

2) Magenta-colored cyan coupler:

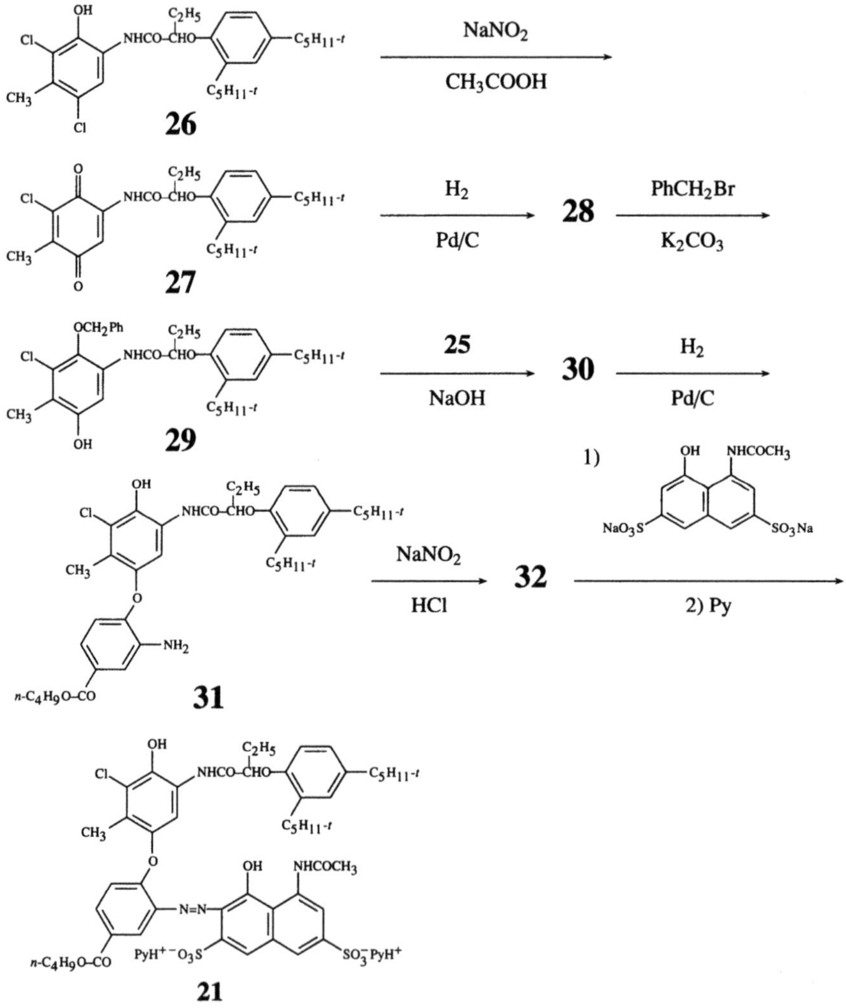

Figure 12.11. Synthetic pathway of the magenta-colored cyan coupler **21** [16].

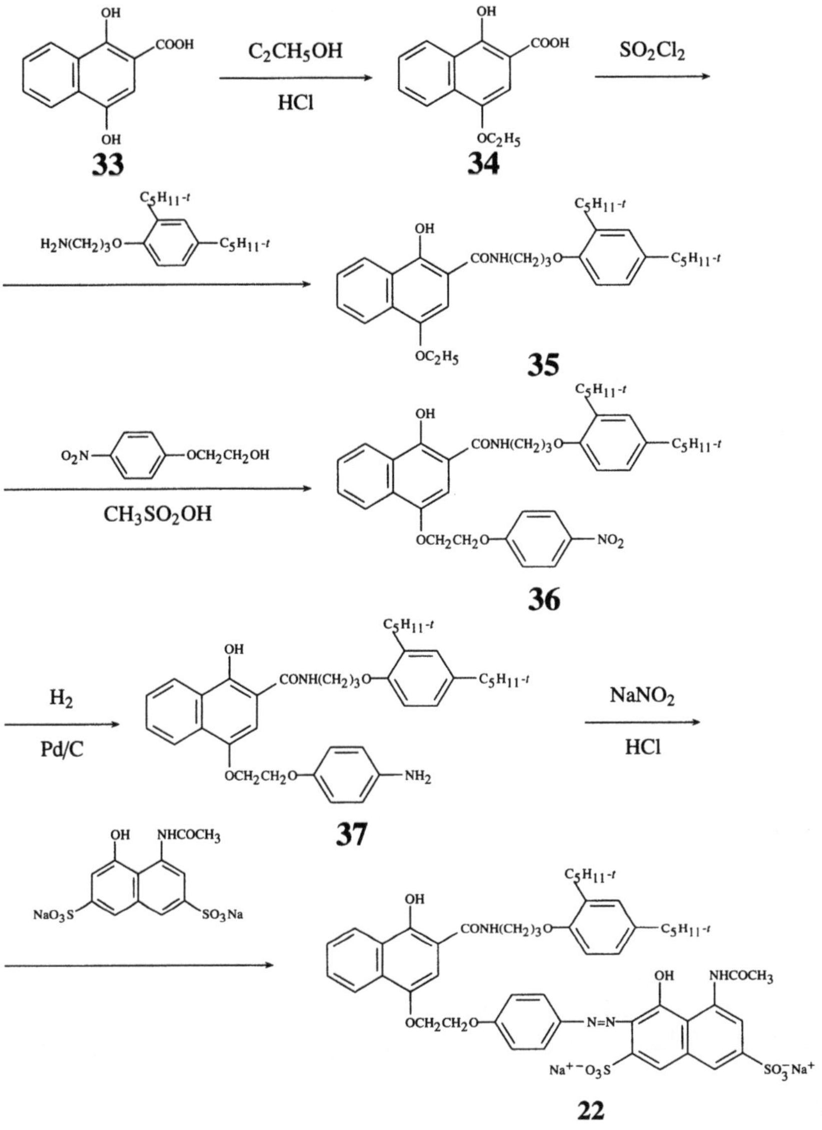

Figure 12.12. Synthetic pathway of the magenta-colored cyan coupler **22** [17].

with 8-acetamido-3,6-disulfonaphthol to yield **20** in a similar way to the processes converting **31** into **21** (Fig. 12.11).

Another synthetic pathway shown in Fig. 12.12 illustrates the preparation of the magenta-colored cyan coupler (**22**) [17]. The key step is the ether exchange of **35** under acidic conditions to yield the intermediate (**36**). Compare this intermediate with the two-equivalent coupler shown in Fig. 11.3 (Chapter 11), i.e., *N*-[4-(2,4-di-*t*-amylphenoxy)butyl]-4-(2-ethoxyethoxy)-1-hydroxy-2-naphthamide.

The intermediate (**36**, i.e., *N*-[3-(2,4-di-*t*-amylphenoxy)propyl]-4-ethoxy-1-hydroxy-2-naphthamide) is prepared by starting from 1,4-dihydoxy-2-naphthoic acid. After an acidic ether formation, the resulting intermediate (**34**) is chlorinated to produce the corresponding acid chloride, which is condensed with 3-(2,4-di-*t*-amylphenoxy)propylamine to give an intermediate **35**. The acidic ether exchange from the ethoxy group of **35** to a 2-(4-nitrophenoxy)ethoxy group produces the key intermediate (**36**). Then the nitro group of **36** is further reduced into an amino group. The resulting intermediate **37** is diazotized and coupled with 8-acetamido-3,6-disulfonaphthol to yield the colored coupler **22**.

12.2.2 Yellow-Colored Cyan Couplers

The merit of 5-amido-1-naphthol couplers such as **38** and **39** has been described in Chapter 10. This type of couplers, however, generates cyan dyes having an undesired secondary absorption band in the blue region of light [21].

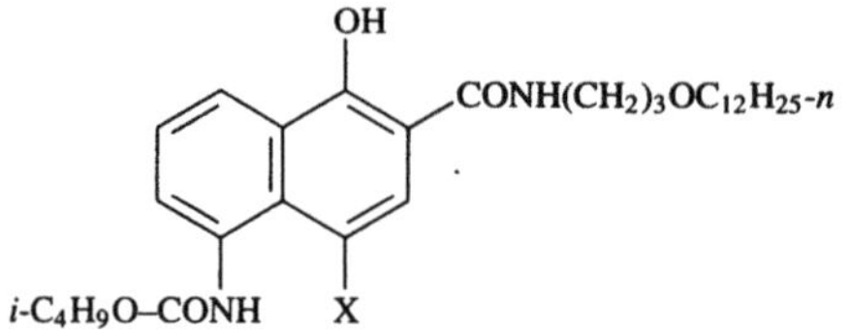

OH

$CONH(CH_2)_3OC_{12}H_{25}$-*n*

i-C_4H_9O–CONH X

38 X = H

39 $X = OCH_2CH_2OH$

Figure 12.13. 5-Amido-1-naphthol couplers for improving cyan images. These couplers are used together with yellow-colored cyan couplers listed in Fig. 12.14.

The undesired band can be compensated by adding a yellow-colored cyan coupler such as **40** as a masking coupler [22]. Another type of yellow-colored cyan couplers (e.g., **41**) has been reported to be used together with a cyan coupler such as **39** [21].

The cyan indoaniline dye derived from a coupler analogous to **39** exhibits an absorption spectrum shown in Fig. 12.15d, although the ballast group is not specified. Note that the undesired secondary absorption appears in the range of 400–500 nm of Curve (d) of Fig. 12.15. A mixture of such a coupler as **39** and such a yellow-colored cyan coupler as **41** is exposed under a step tablet and developed with a color developer. The absorptions with the increasing amounts of exposure are shown in Fig. 12.15a–c, where the absorptions in the blue region of light are maintained to be substantially constant, compensating the undesired absorption of the image coupler (Fig. 12.15d) [21]. The processing of color negative films employing a 5-amido-1-naphthol coupler (e.g., **39**) together with a yellow-colored cyan coupler (e.g., **41**) has been disclosed to be accelerated by a further

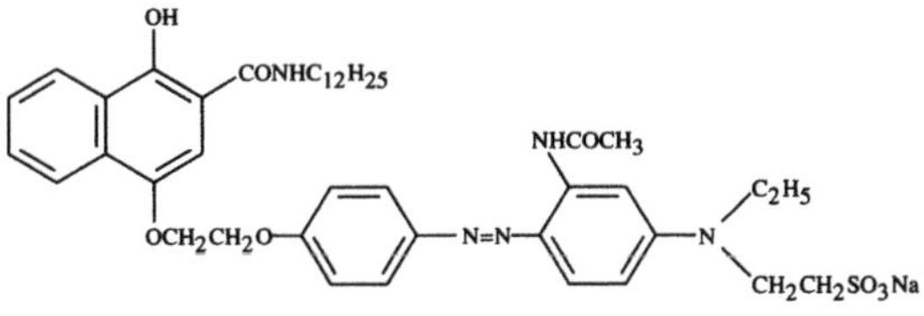

40

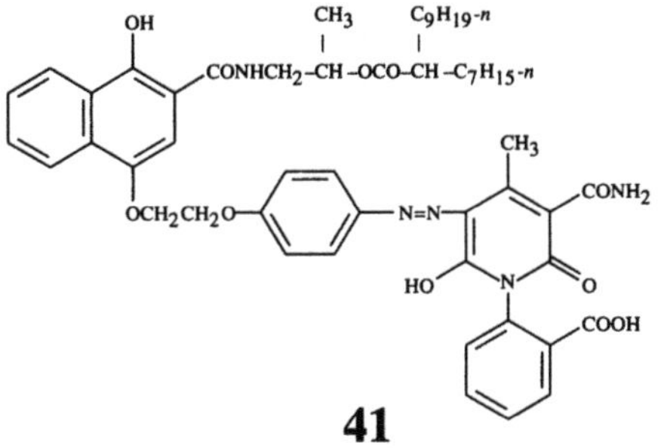

41

Figure 12.14. Yellow-colored cyan couplers. These couplers are used together with the cyan couplers listed in Fig. 12.13.

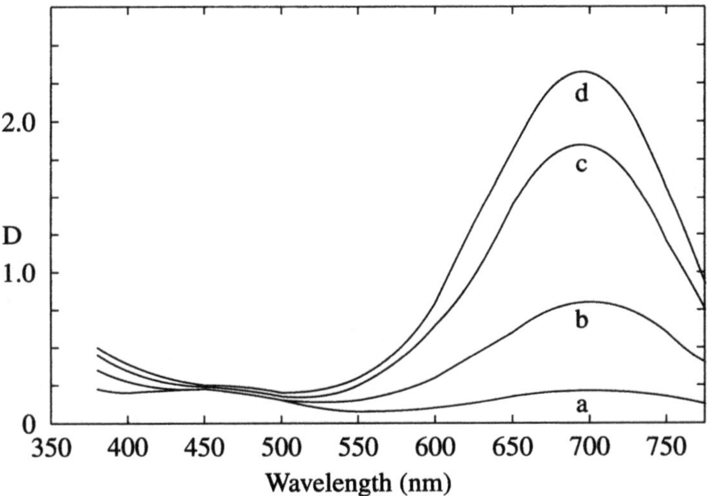

Figure 12.15. Absorption spectra for a mixture of a yellow-colored cyan coupler and an uncolored cyan coupler. Curve (d) represents a maximum dye density where the colored coupler entirely reacted to give the corresponding magenta dyes (from the colored coupler and the uncolored coupler). Curves (a), (b), and (c) represent transient regions with increasing amounts of exposure [21].

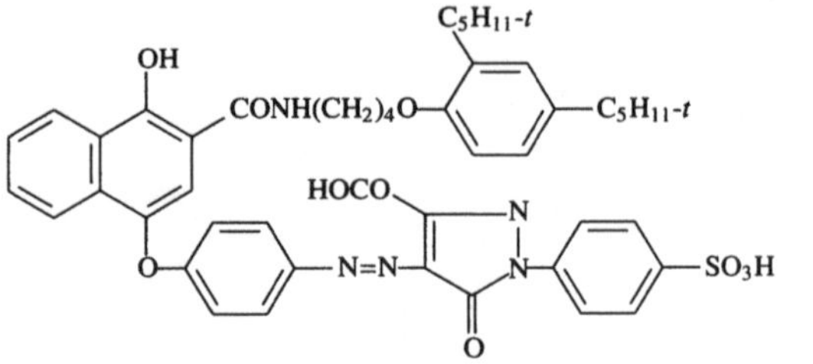

42

Figure 12.16. Yellow-colored cyan couplers.

addition of a BAR (bleach accelerator releasing) coupler to a emulsion layer [23] or by addition of an oxidizing agent to the processing solution [24].

Pyrazolone azo dyes can be used as released moieties, as exemplified by a yellow-colored magenta coupler (**42**) [25]. The coupler **42** acts as a colored cyan coupler for releasing a pyrazolone azo dye, where the pyrazolone azo dye moiety does not react with an oxidized color developer. A combination of the colored coupler (**42**) and an uncolored coupler **38** has been disclosed in a patent [26].

12.3 Colored Magenta Couplers

12.3.1 Yellow-Colored Magenta Couplers

As described in Section 12.1, the technique of color masking was first established by means of yellow-colored magenta couplers (see Figs. 12.1 and 12.4).[7] As magenta coupler nuclei have been developed in the order of 3-alkyl (the first generation), 3-amido (the second generation), and 3-anilino-5-pyrazolone (the third generation), the corresponding colored couplers have appeared for the respective generations.[8] For example, 3-amido-type colored couplers such as **43** have been derived by coupling the corresponding four-equivalent coupler with a diazotized aminoisophthalic ester or diazotized aminoisophthalic acid [27]. These couplers can be incorporated directly into usual photographic emulsions. Yellow-colored magenta couplers based on the 3-anilino-5-pyrazolone nucleus have later been

[7] As already discussed in Chapter 11, it is difficult to introduce an aryloxy or alkoxy group (which is capable of releasing an inactive molecule) at the active position of a 5-pyrazolone magenta coupler. This means that colored couplers according to the "removal masking" methodology has been difficult to be realized. Hence, colored couplers based on the "decomposition masking" remain predominant, as long as the 5-pyrazolone nuclei are used as magenta couplers.

[8] For the different generations of magenta couplers, see Chapter 10 (Section 10.4). The fourth generation is concerned with two-equivalent magenta couplers, where leaving groups attach to the active position of the 5-pyrazolone nuclei via a sulfur or nitrogen atom.

disclosed as exemplified by **44** [28]. These colored couplers have been disclosed to be more reactive than those based on the 3-amido-5-pyrazolone nucleus.

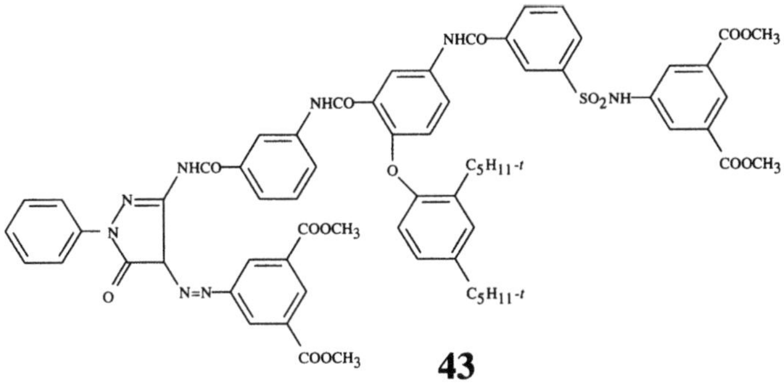

43

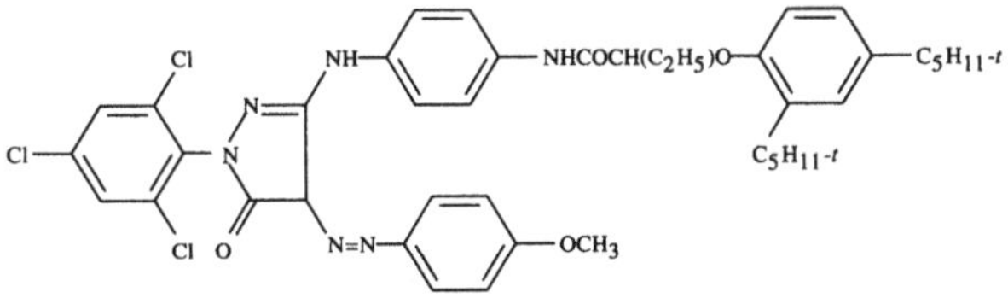

44

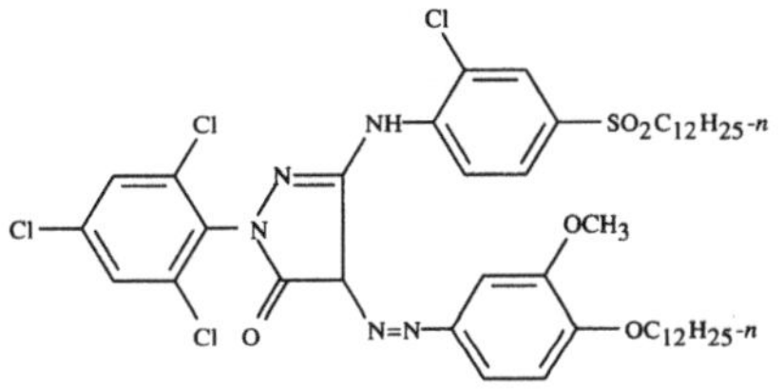

45

Figure 12.17. Yellow-colored coupler for forming magenta dyes.

As found in Fig. 12.1, the released azo moiety is decomposed into a colorless compound with evolving nitrogen gas. The colorless compound is usually washed out of the film during processing. However, the colorless compound may be non-diffusible, as exemplified by the coupler **45** [29].

The coupling activity of this class of couplers can be increased if the 4-phenylazo group is substituted by a hydroxyl group at the 2- or 4-position. For example, **46** [30], **47** [31], and **48** [32] have been disclosed in patents. An analogous colored coupler (**49**) has been used in a yellow filter layer to improve sensitivity without increase of granularity [26].

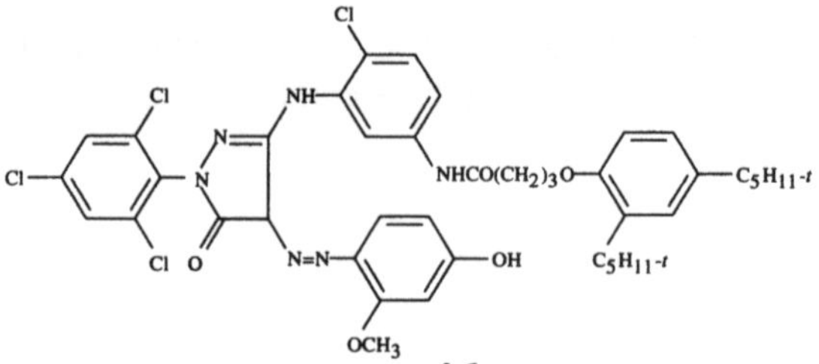

46

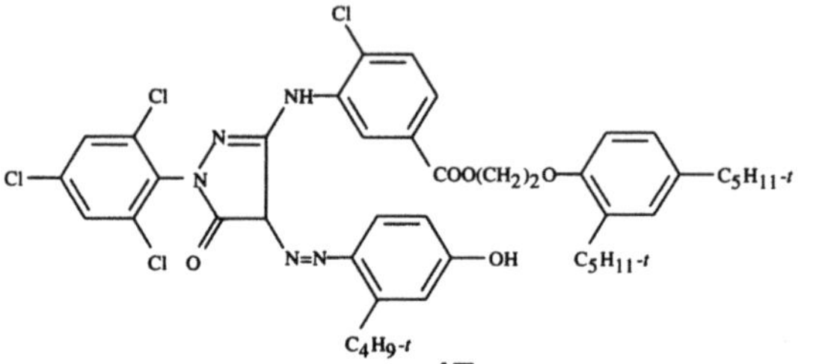

47

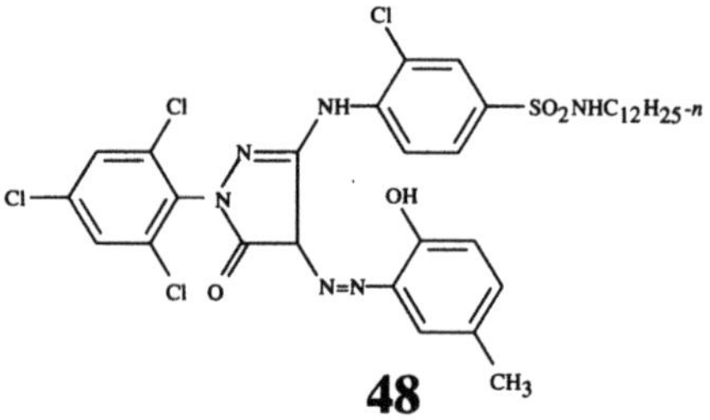

48

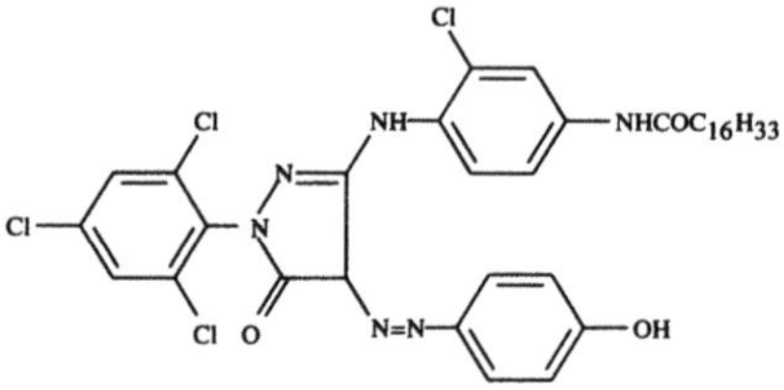

49

Figure 12.18. Yellow-colored magenta couplers having a hydroxyphenylazo moiety.

12.3.2 Hydrophilic Yellow-Colored Magenta Couplers

Oil-protected couplers have been mainly used in recent decades because of their merits (enhanced coupling reactivity, stable dispersion, synthetic easiness, etc.). However, the use of oils (high-boiling solvents) causes an increase of the thickness of photographic films and may also raise ecological concerns if high solvent levels are required. As one approach for omitting such oils, the revival of hydrophilic couplers such as **50** [33] and **51** [34] has been disclosed, because these couplers form self-assembled micellar aggregates in water and can be directly incorporated into an aqueous coating solution without the need for high-boiling solvents (Fig. 12.19). For example, the couplers **50–52** can be dissolved in water containing 2% (w/w) 2-phenoxyethanol. On the other hand, the couplers **53–56** do not possess adequate solubility under this condition. Dispersion of the latter couplers requires *n*-propanol or ethyl acetate as a solvent.

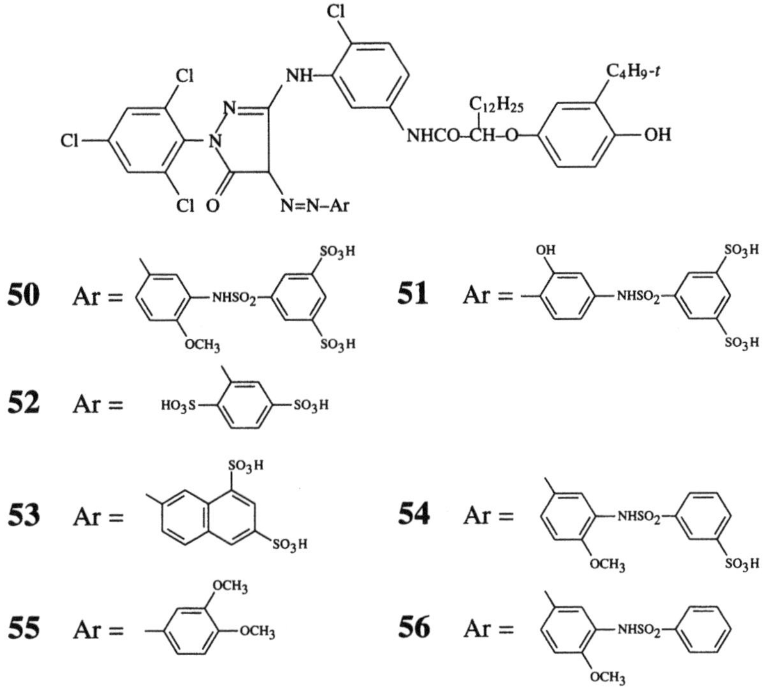

Figure 12.19. Hydrophilic yellow-colored magenta couplers (**50** and **51**) and comparative couplers.

Syntheses

A synthetic pathway for preparing the yellow-colored magenta coupler (**50**) is shown in Fig. 12.20 [33]. The strategy of this synthesis is to avoid difficulties due to sulfo groups by using fluorosulfonyl groups as precursors.

Thus, 3,5-bis(fluorosulfonyl)aniline (**57**) is diazotized to the diazonium salt (**58**), which is converted into 3,5-bis(fluorosulfonyl)benzenesulfonyl chloride (**59**) by the reaction with sulfur dioxide in the presence of cuprous chloride (the Sandmeyer reaction). This intermediate is reacted with 2-methoxy-5-nitroaniline and the resulting sulfonamide (**60**) is catalytically hydrogenated into the amino coupling component (**61**). Then, the amino group of **61** is diazotized and the resulting diazonium salt (**62**) is allowed to couple with the corresponding four-equivalent coupler. The product has two terminal fluorosulfonyl groups, which are finally hydrolyzed into sulfonic acid groups, giving the target coupler (**50**).

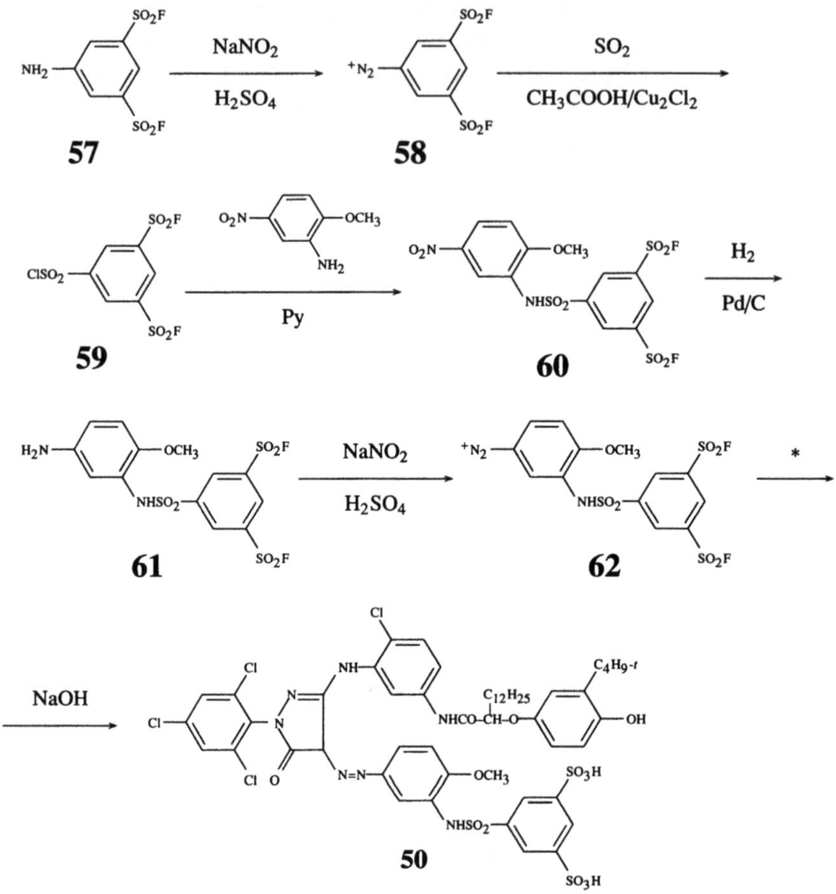

Figure 12.20. Synthetic pathway of the yellow-colored magenta coupler 50. The symbol * represents the corresponding four-equivalent magenta coupler [33].

12.4 Colored Yellow Couplers

12.4.1 Magenta-Colored Yellow Couplers for Movie Films

Colored cyan couplers and colored magenta couplers are used for compensating undesired secondary absorptions that appear in the shorter wavelength ranges than desired primary absorption bands. Thus, magenta-colored cyan couplers, yellow-colored cyan couplers, and yellow-colored magenta couplers are applied to the improvement of color reproduction for usual color negative films. In contrast, magenta-colored yellow couplers are used for duplicating color motion picture

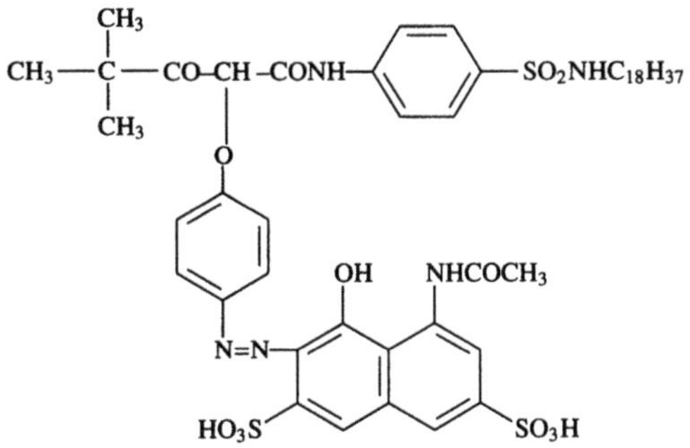

63

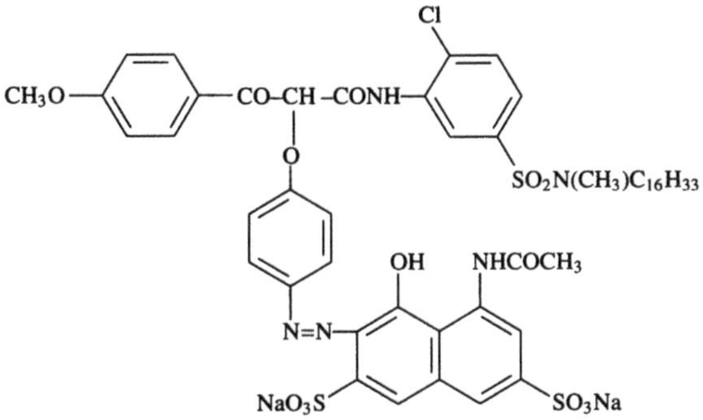

64

Figure 12.21. Magenta-colored couplers forming a yellow azomethine dye.

films, where the longer wavelength ranges are objects to be compensated.[9] For example, the magenta-colored yellow coupler (**63**) has been disclosed to be used in a blue-sensitive emulsion layer of a duplicate film for producing motion pictures [35].

12.4.2 Magenta-Colored Yellow Couplers for Color Negative Films

Magenta-colored yellow couplers such as **64** have been disclosed to improve sharpness and color reproduction for color negative films based on high-silver

[9]Most production practices of color motion pictures consist of two or four steps. The two-step procedure contains (1) the recording of the scene onto a camera negative film and (2) the production of a direct print in which this original negative is printed onto a negative-working print film. On the other hand, the four-step procedure contains (1) the recording of the scene onto a camera negative film, (2) the printing onto a negative-working intermediate film, yielding a master positive, (3) the printing onto a negative-working intermediate film, providing a duplicate negative, and finally (4) the production of a release print where the duplicate negative is printed onto a print film. The use of magenta-colored yellow couplers is concerned with such intermediate films.

chloride emulsions [36].[10] Such high-silver chloride films have become widely used for rapid processing in so-called "mini-laboratories" or "mini-labos", since silver chloride emulsions can be generally developed more rapidly than other silver halide emulsions. The incorporation of the magenta-colored yellow couplers keeps the films free from the deterioration of the image qualities even under the conditions of shortened processing time.

[10]Example 1 of this patent involves an embodiment of a color negative film which gives a full repertoire of funtionalized organic compounds in terms of specified structural formulas. The comparison between this example (2002) and the example (1995) described in Chapter 1 would provide useful pieces of information on the R&D of photography during this period.

References

[1] Vittum PW, Arnold RJ (1947) US Patent 2 428 054
[2] Glass DB, Vittum PW, Weissberger A (1948) US Patent 2 455 170
[3] Vittum PW, Sawdey GW, Herdle RA, Scholl MK (1950) J Am Chem Soc. 72:1533
[4] Weissberger A (1980) Chemtech. 340
[5] Barr CR, Thirtle JR, Vittum PW (1969) Photogr Sci Eng. 13:74
[6] Hanson Jr WT (1948) US Patent 2 449 966
[7] Hanson Jr WT, Vittum PW (1947) J Photo Soc Am. 13:94
[8] Hanson Jr WT (1950) J Opt Soc Am. 40:166
[9] Williams J (1958) US Patent 2 860 974
[10] Loria A (1969) US Patent 3 476 563
[11] Barr CR, Bush WM, Thomas LJ, Cole HE (1973) UK Patent GB 1 330 524
[12] Barr CR (1972) US Patent T 900029
[13] Cole HE (1974) Jpn Patent S49-21660
[14] Chari K, Keech JT, Sawyer JF, Schmoeger JW, Zengerle (1993) US Patent 5 190 851
[15] Visconte GW, Fant AB, Wang Y, Olsen RG (1999) US Patent 5 928 847
[16] Orvis RL (1977) US Patent 4 004 929
[17] Hirose T, Shiba K, Yokota Y, Inouye K, Okumura A (1979) US Patent 4 138 258
[18] Yoneyama H (1998) US Patent 5 744 292
[19] Hioki T, Ihama M (1998) US Patent 5 763 152
[20] Ichijima S, Shimada Y, Arakawa J (1989) US Patent 4 837 136
[21] Kobayashi H, Mihayashi K (1994) J Soc Photogr Sci Technol Jpn. 57:316
[22] Ohkawa A, Kamio T, Motoki M, Mihayashi K (1992) US Patent 5 112 730
[23] Ishii Y, Fujita Y, Mihayashi K (1995) US Patent 5 459 022
[24] Fujita Y, Mihayashi K (1993) US Patent 5 178 993
[25] Renner G, Langen H, Klein U (1993) US Patent 5 241 058
[26] Bell P, Borst H-U, Büscher R, Willsau J, Siegel J (1998) US Patent 5 853 971
[27] Graham B, Vittum PW, Weissberger A (1953) US Patent 2 657 134
[28] Beavers LE (1961) US Patent 2 983 608
[29] Kapp DL, Ross RJ, Singer SP, Clark BA (1996) US Patent 5 492 799
[30] Imamura H, Sato S, Kojima T, Endo T (1978) US Patent 4 070 191
[31] Shiba K, Hirose T, Arai A, Okumura A, Yokota Y (1979) US Patent 4 163 670

[32] Kapp DL, Younathan JN, Ross RJ, Merrill JP (1997) US Patent 5 622 818
[33] Crawley MW, Chari K, Sowinski AF (2000) US Patent 6 010 839
[34] Younathan JN, Crawley MW, Chari K (2000) US Patent 6 132 943
[35] Sawyer JF, Fenton DE (1995) US Patent 5 399 468
[36] Matsumoto K, Ito T (2002) US Patent 6 337 176 B1

Chapter 13

DIR Couplers and Related Compounds

13.1 Development-Inhibitor Releasing

The term "DIR" is the acronym of "development-inhibitor releasing". To discuss the functions of the DIR couplers, the present section is devoted to explaining the terms "development inhibitor" and "releasing".

13.1.1 Development Inhibitors

Solubilities

Photographic materials tend to suffer from fogs during manufacture, storage, and processing. Various stabilizers are used to prevent such fogs. They are called antifoggants, emulsion stabilizers, inhibitors of Ostwald's (physical) ripening, development inhibitors, etc. according to the conditions or processes in which they are used [1].[1] Since they more or less interact with silver halide crystals, the same compounds are frequently used as development inhibitors if their activities are

[1] Stabilizers used in the developer solution are called "development restrainers" or "antifogging agents (antifoggants)". If they are used to delay the onset of fog during the after-ripening of the emulsion making process, or if they are added to emulsions prior to the coating, they are called "emulsion stabilizers". If they are added during the process of Ostwald's ripening, they are referred to as "inhibitors of physical ripening" or "physical ripening control agents". If they are added to film layers to inhibit the development of emulsion, they are frequently called "development inhibitor". However, these terms are not absolutely distinct.

sufficient to suppress development. This subsection deals with antifoggants or emulsion stabilizers that can be regarded as candidates of development inhibitors.

Table 13.1. Solubility Products of Thiolic Antifoggants [2]

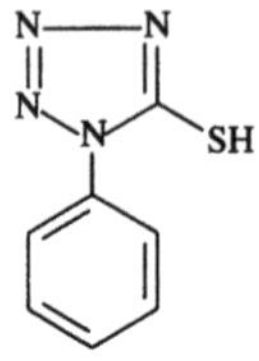

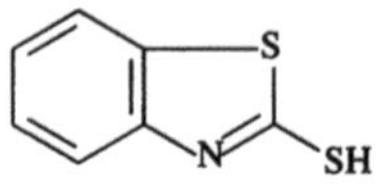

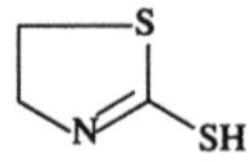

Compound	pK_{sp} (pH 5.6)[a]	pK_{sp} (pH 9.6)[a]
1	14.5–14.3	15.0–15.3
2	14.5–16.2	16.7–15.3
3	12.05	15.7
4	10.95	13.8
5		15.1–16.3
6	12.15	14.6
7	14.6	14.8
8	16.35	16.95
9	14.6	15.0

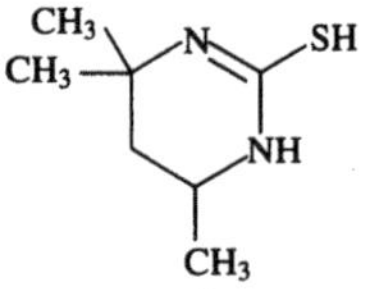

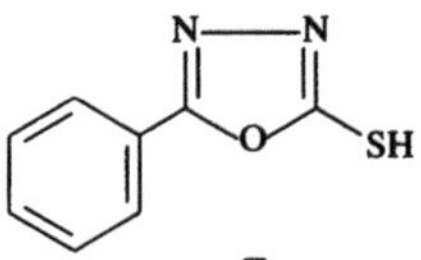

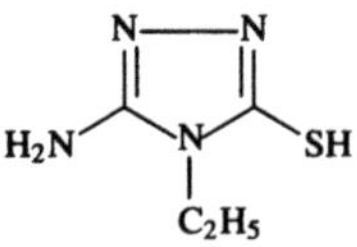

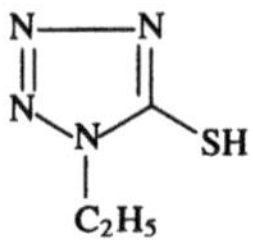

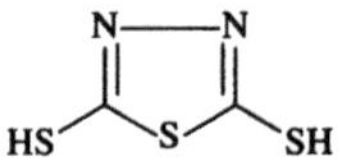

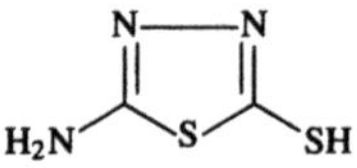

[a]Measured under conditions of excess of mercaptide anions at 30°C, p[AgSR] = 4.00.

A representative class of antifoggants is composed of heterocyclic thiol derivatives, as listed in Table 13.1. To evaluate their interactions with silver halides, Table 13.1 collects the solubility constants of the corresponding silver salts in the form of logarithms (pK_{sp}) [2]. These values should be compared with those of silver halides (pK_{sp}: 9.75 for AgCl, 12.31 for AgBr, and 16.09 for AgI). Thus, the silver salts are substantially insoluble in water.

Examples of heterocyclic antifoggants other than those having thiol groups are shown in Fig. 13.1: 4-hydroxy-6-methyl-1,3,3a,7-tetrazaindene (**10**) and 1,2,3-benzotriazole (**11**). Further examples have been listed in a reference [3].

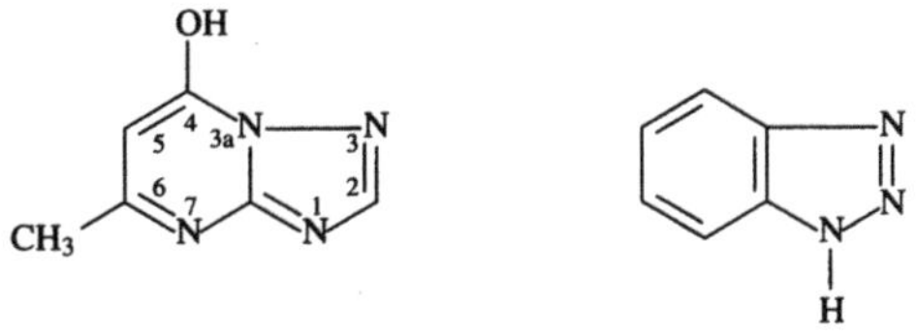

Figure 13.1. Heterocyclic antifoggants.

The adsorption of development inhibitors (e.g., 1-phenyl-5-mercapto-1H-tetrazole, 1,2,3-benzotriazole, and potassium iodide) to metallic silver surface has been measured to evaluate their effects on physical development [4]. The phenylmercaptotetrazole has been found to cover the silver surface completely with an apparent monolayer, while the maximum coverage of the benzotriazole is only 10 to 15% of the surface. These data have been related to their inhibiting effects on physical development.

Diffusibilities

The interlayer and intralayer effects of development inhibitors depend upon their diffusibilities. A practical test has been reported to evaluated diffusibilities, as shown in Fig. 13.2 [13]. The amount of silver bromide in Layer 2 of the test material shown in Fig. 13.2 varies from 0 to 6 g/m^2. After exposure, each test material is processed by a chromogenic development in the presence of a developer inhibitor (**12**) to be tested. The density of the resulting dye image derived from the couplers contained in Layer 1. The densities for each development inhibitor are determined by varying the amount of AgBr in Layer 2, where D_0 denotes the density in the absence of the development inhibitor, $D_{0.7}$ denotes the value for 0.7 g/cm^2 AgBr, and D denotes the value for 6.0 g/cm^2 AgBr. Thereby, the diffusibility D_f and the inhibiting efficiency I_n are defined by the following equations:

$$D_f = \frac{10(D_0 - D_{0.7})}{D_0} \tag{13.1}$$

$$I_n = \frac{100(D_0 - D)}{D_0} \tag{13.2}$$

For 5-substituted 1,2,3-benzotriazoles as development inhibitors, the diffusibility D_f can be estimated by the following equation:

$$D_f = -0.0099(\pm 0.0036)M - 0.0474(\pm 0.0103)I_n + 9.33 \tag{13.3}$$

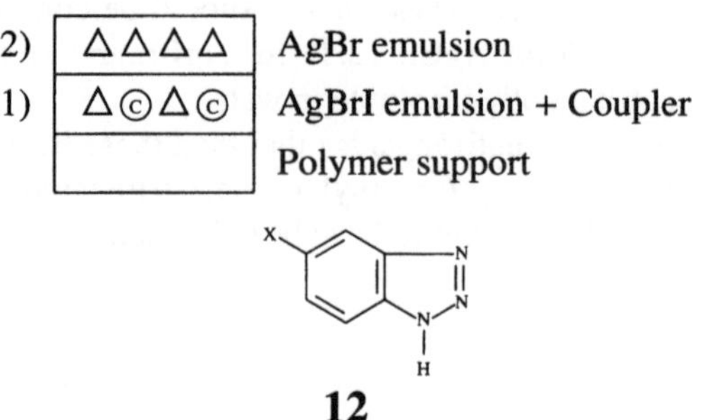

Figure 13.2. Test material for evaluating the diffusibilities of development inhibitors [13]. The amount of silver bromide in Layer 1 varies from 0 to 6 g/m^2. This material is developed in the absence or presence of each development inhibitor (**12**).

where M denotes the molecular weight of a benzotriazole derivative ($n = 28$, $r = 0.941$, $s = 0.41$, and each value in parentheses represents 95% confidence interval). In terms of eq. 13.3, the molecular weights (M) of the benzotriazoles influence the D_f values in the same order as I_n.

To evaluate diffusibilities, another practical test has been reported in a patent, as shown in Fig. 13.3 [5]. Test materials A and B shown in Fig. 13.3 have been developed in the absence or presence of development inhibitors. The diffusibility D_f is determined by the following equation:

$$D_f = \frac{(D_{A_0} - D_A)/D_{A_0}}{(D_{B_0} - D_B)/D_{B_0}} \tag{13.4}$$

The color densities of test materials A and B after development without addition of an inhibitor are respectively denoted by the symbols D_{A_0} and D_{B_0}. On the other hand, D_A and D_B represent the color densities of test materials A and B after development with the addition of an inhibitor, where the level of the concentration is conformed to $(D_{B_0} - D_B)/D_{B_0} = 0.5$.

Heterocyclic development inhibitors and their diffusibilities (D_f) are listed in Fig. 13.4 [5]. Preferred inhibitors have been disclosed to exhibit D_f of larger than or equal to 0.4 so that this patent has claimed DIR couplers releasing a heterocyclic inhibitor such as **23**. The inhibitors **25** and **26** are a kind of timing releasers which release a development inhibitor.

13.1.2 Timing and Imagewise Releasing

The action of development inhibitors described in the preceding subsection is direct or immediate after adding to processed solutions. Such direct action is undesirable for the purpose of incorporating the inhibitors into a photographic film. Hence, some structural modification is required to retard the inhibiting action.

In general, there are two modes of modification, i.e., timing releasers and imagewise releasers [6]. They are differentiated in terms of the ways how the con-

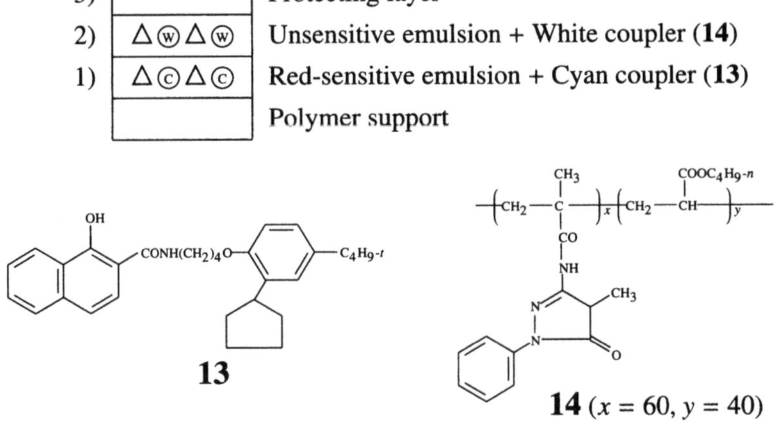

Figure 13.3. Test materials for evaluating the diffusibilities of development inhibitors [5]. A test material A contains 0.754 g of the cyan coupler **13** and 0.38 g of the white coupler **14**, while another test material B contains 0.754 g of the cyan coupler **13** and 0.346 g of the white coupler **14**. Tested development inhibitors are listed in Fig. 13.4.

cealed inhibiting function is regenerated, or in other words, how a development inhibitor is released.

Timing Releasing

The timing releasing is not related to an image to be pictured and occurs in all regions of the film. For example, let us consider a precursor **27** [7], which releases a developer inhibitor anion (**28**) as the result of a retro-Michael reaction (Fig. 13.5). The timing of releasing depends upon the rate of the retro-Michael reaction. Various timing releasers have been used in instant color photography, since it is a crucial technique to stop the development of silver halide at the time when sufficient images appear to be observed [8]. This point will be discussed later in Chapters 16–20 for instant color photography.

Timing releasers such as **29** and **30** have recently been disclosed to be useful in photographic reversal films that are push-processed because they tend to release their development inhibitors after extended development times. The precursor **29** forms a quinone-methide-like intermediate to release a thiolic inhibitor [9]. On the other hand, the precursor **30** suffers from the attack of a nucleophile at the position to which the sulfur atom attached [10]. The two nitro groups and the carbamoyl group are all electron-withdrawing so that such a nucleophilic attack occurs easily.

Imagewise Releasing

The imagewise releasing corresponds to an image to be pictured so that the concentration of a released compound is proportional (or inverse proportional) to the

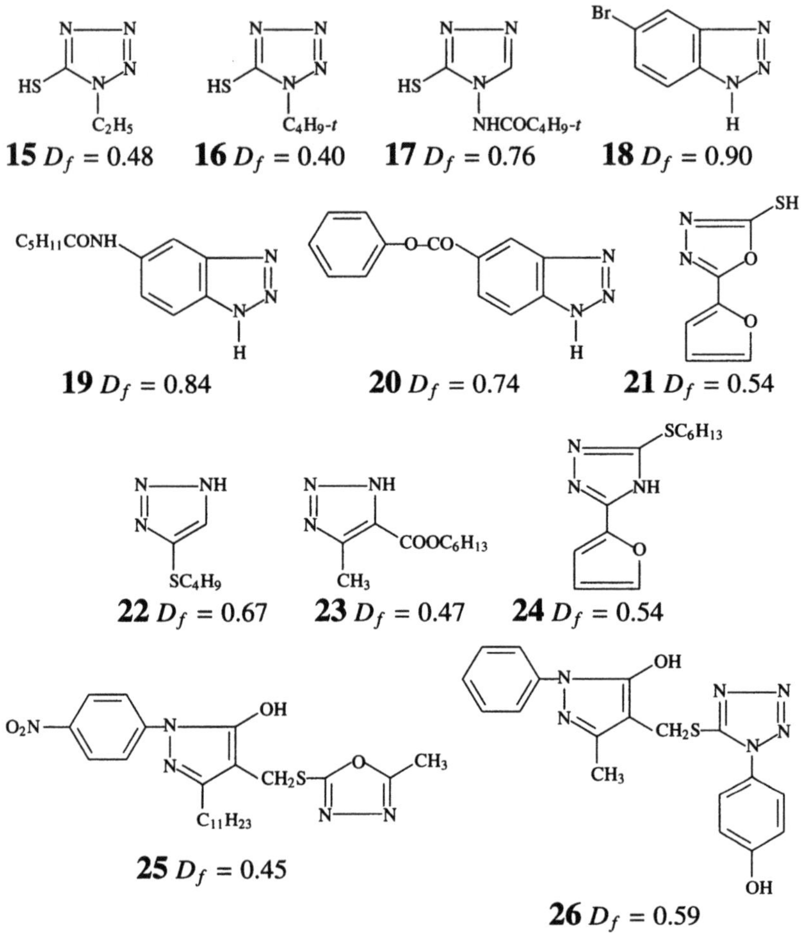

15 $D_f = 0.48$ **16** $D_f = 0.40$ **17** $D_f = 0.76$ **18** $D_f = 0.90$

19 $D_f = 0.84$ **20** $D_f = 0.74$ **21** $D_f = 0.54$

22 $D_f = 0.67$ **23** $D_f = 0.47$ **24** $D_f = 0.54$

25 $D_f = 0.45$ **26** $D_f = 0.59$

Figure 13.4. Heterocyclic development inhibitors. Each D_f value has been obtained for the test materials shown in Fig. 13.3 by using eq. 13.4 [5].

image gradient. Simply speaking, two-equivalent couplers discussed in Chapter 12 can be regarded as imagewise releasers if the compounds released from them are photographically active.

For example, the coupler **33** reacts with an oxidized color developer (**32**) to give an azomethine dye (**34**) and a development inhibitor (**35**) as shown in Fig. 13.6. Because the oxidized developer (**32**) is generated by the color development according to the image gradient, the concentration of the inhibitor (**35**) is proportional to the image gradient through the amount of the oxidized developer.

There are several mechanisms of imagewise releasing. Although the present chapter is devoted to the discussion on DIR couplers used in conventional color photography, instant color photography uses a variety of imagewise releasers

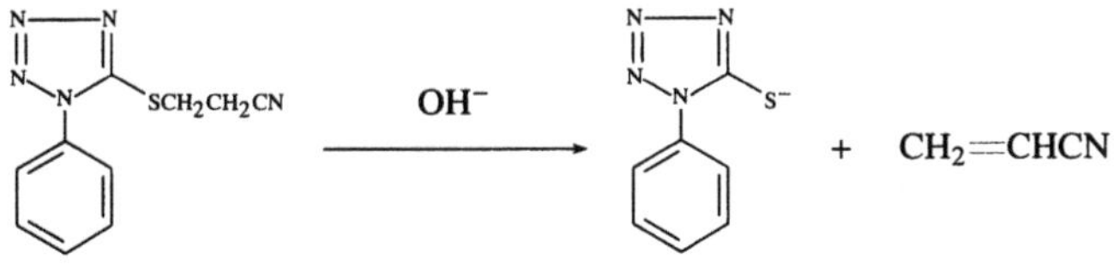

Figure 13.5. Timing releasing of development inhibitors. The precursor **27** is based on a retro-Michael reaction, as shown in the scheme of the first row. On the other hand, the precursors **29** and **30** use a quinone-methide-like formation and a nucleophilic displacement, respectively, to release thiolic inhibitors.

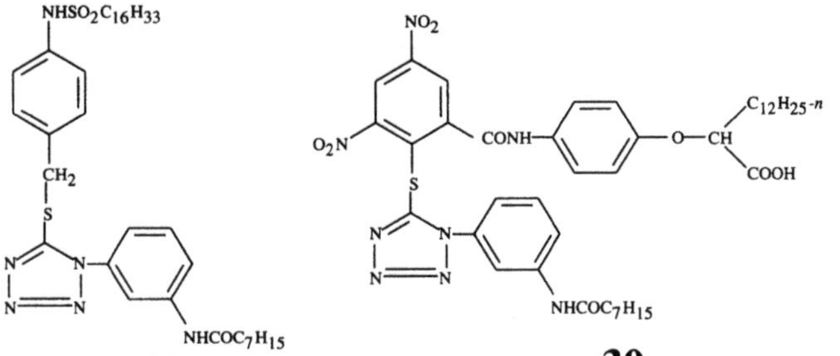

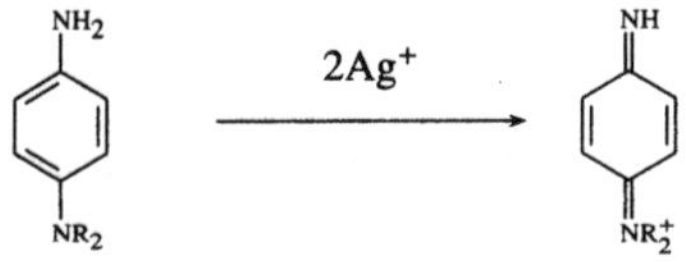

Figure 13.6. Imagewise releasing of a development inhibitor. This scheme is the same as that of a two-equivalent coupler but characterized by the fact that the released thiol (**35**) is so photographically active as to work as a development inhibitor.

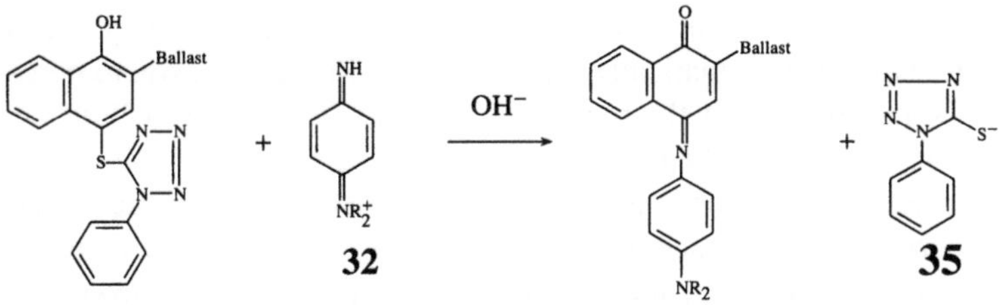

such as hydroquinone releasers, sulfonamide releasers, etc., which will be later discussed in detail (Chapters 16–20 of Part IV).

13.2 Effects of DIR Couplers

DIR couplers are imagewise-releasing compounds, which release a developer inhibitor (DI) as a function of developed silver halides (Fig. 13.6). The merits of DIR couplers have been discussed in an earlier review by Barr [11] and more recent reviews by others [12,13,14], where the effects of the DIR couplers are classified into intralayer effects and interlayer ones.

1. intralayer effects
    (a) improved sharpness
    (b) reduced granularity
2. interlayer effects

As will be found in the following discussions, these effects depend largely upon the diffusibilities of released inhibitors (or their precursors) as well as upon their powers of inhibition.

13.2.1 Intralayer Effects

Improved Sharpness

One of the most important effects is the improvement of sharpness, which is based on the intralayer diffusion of a development inhibitor released from a DIR coupler. Consider the boundary between a low exposure area and a high exposure area in a photographic material (Fig. 13.7).

Without DIR couplers or any other development inhibitors (e.g., iodide ion), the density curve is schematically represented by Fig. 13.7A, where point (a) and point (b) show the blur of the boundary. Thus the scattering of light lowers the sharpness remarkably at the boundary.

In the presence of a DIR coupler, the development inhibitors are released from the couplers in higher concentration in the high exposure area than in the low exposed area. As a result, the development inhibitors diffuse from the former area to the latter one across the boundary. As for the high exposure area, the concentration of development inhibitors becomes lower near the boundary than in the main part of the high exposure area, as found at point (a) in Fig. 13.7B. As for the low exposure area, the diffused development inhibitors suppress the development near the boundary as found at point (b) in Fig. 13.7B. The total feature of the curve of Fig. 13.7B shows the improvement of sharpness in terms of the edge effect.

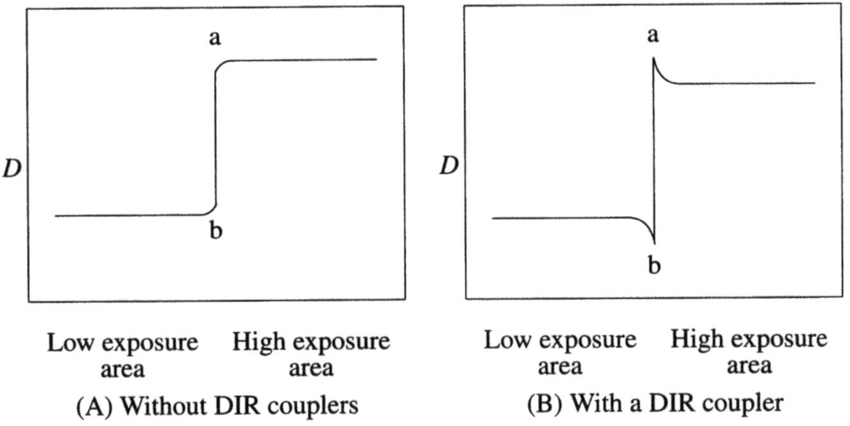

Figure 13.7. Edge effect by means of a DIR coupler. (A) No edge effect without DIR couplers. (B) Edge effect with a DIR coupler.

Reduced Granularity

A development inhibitor released from a DIR coupler reduces the granularity[2] of the developed image within a layer [11]. A dye cloud is larger than the corresponding silver image in the absence of DIR couplers. On the other hand, the presence of a DIR coupler causes the generation of dye clouds of smaller size. Thus the development has been considered to take place at a number of development centers in each silver halide grain and causes the generation of two or more silver grains (or filamentary silver appendages), around which the dye image clouds are formed during development [15,16].

13.2.2 Interlayer Effects

Development inhibitor molecules released from a DIR coupler also diffuse into the adjacent layers, where they suppress the development of the other-light sensitive emulsions [11]. The top of Fig. 13.8 shows a coating for demonstrating interlayer inhibition by DIR couplers, where Layer 1 contains a magenta coupler, Layer 2 is a gelatin interlayer, and Layer 3 contains a usual cyan coupler for comparison or a DIR coupler for a test. The coatings with the usual cyan coupler with the cyan DIR coupler are flashed with a uniform green-light exposure and a gradient red-light exposure. They are processed by color development, fixed, bleached, fixed, washed, and dried. The sensitometric curve obtained from the comparison coating is shown in Fig. 13.8A, which shows that the density due to the magenta dye is constant in agreement with the uniform green-light exposure. On the other hand, the sensitometric curve obtained from the test coating (Fig. 13.8B) shows

[2]The term "graininess" is also used in place of the term "granularity".

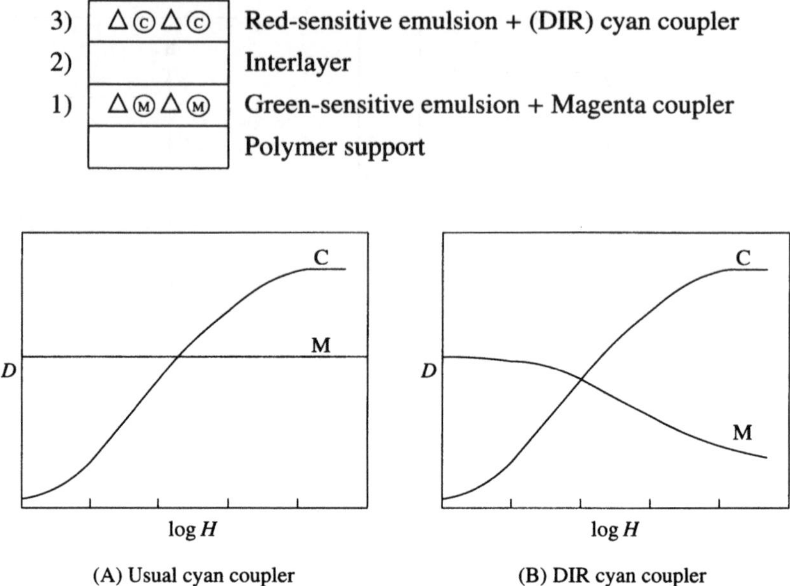

Figure 13.8. Schematic sensitometric curves (log H vs. Density) for a usual cyan coupler (A) and for a DIR cyan coupler (B). Each of the couplers is incorporated in Layer 3 of the test coating illustrated at the top of this figure. Curve C represents the density of the cyan dye vs. the log exposure (log H). Curve M represents the density of the magenta dye vs. the log exposure (log H).

that the magenta dye is formed in inverse proportion to the amount of the cyan dye formed in the red-sensitive layer, although uniform green-light exposure is applied. This is because the development inhibitor released from the DIR coupler in the red-sensitive emulsion layer (Layer 3) diffuses into the green-sensitive layer (Layer 1) so as to cause the imagewise suppression of the development in Layer 1.

Although the adjacent effect occurs together with the development suppression of the layer in which the DIR coupler is incorporated, the DIR coupler can be used to correct color reproduction in place of colored couplers.

13.3 DIR Couplers of Direct-Releasing Type

There are several ways of imagewise releasing development inhibitors. They are categorized into direct releasing and indirect releasing. The term "direct releasing" is used to refer to cases in which a released molecule is a development inhibitor itself. On the other hand, the term "indirect releasing" denotes the releasing of a precursor, which afterwards releases a development inhibitor. This section deals with the direct releasing of development inhibitors.[3]

13.3.1 Releasing Thiolic Development Inhibitors

Whitmore et al. have disclosed the imagewise releasing of development inhibitors (DIs) in chromogenic development, where the DIR couplers are classified into several classes [15]. As one of the predominant classes, they have disclosed DIR couplers that release thiolic development inhibitors, e.g., **36** (cyan) and **37** (magenta) [15,16,17]. These couplers have been and are widely used as DIR couplers. For example, the DIR coupler in which the ballast group at the 2-position of **36** is replaced by 2-tetradecyloxyphenylcarbamoyl group has been used in recent color negative films, as described in Chapter 1 (Fig. 1.27).

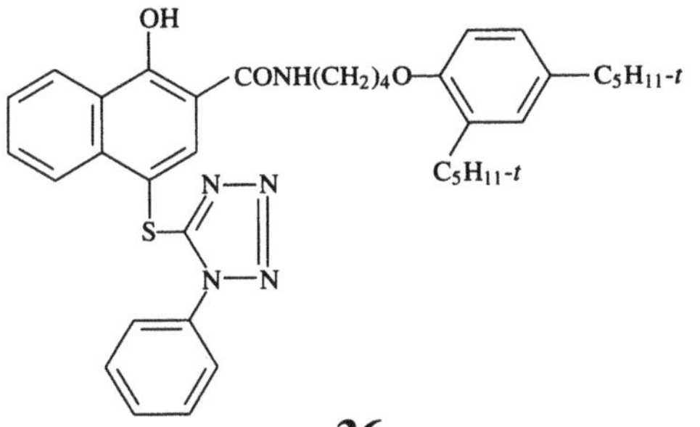

36

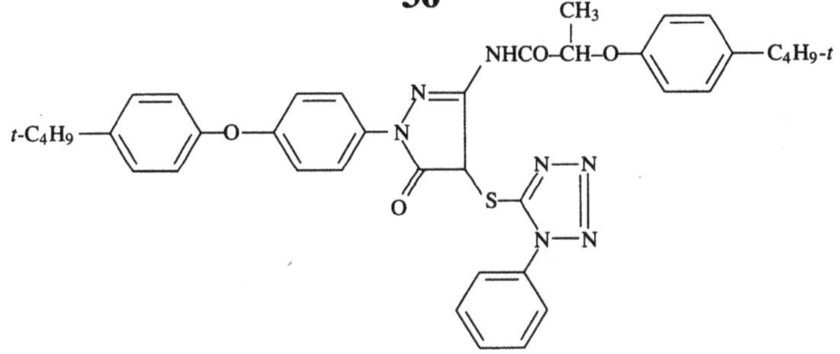

37

Figure 13.9. DIR couplers. The coupler (**36**) is a DIR cyan coupler. The coupler (**37**) is a DIR magenta coupler.

The use of a 3-primary amido-substituted 5-pyrazolone DIR coupler such as **37** in an admixture with a usual (main) coupler results in decreased contrast, some loss in speed of the silver halide emulsion, and image-dye formation which is not

[3] A color negative film using DIR couplers has been placed on the market for the first time under the name of "Kodacolor II" (1972).

linear to the exposure received. These disadvantages can be overcome by substituting a secondary amino group for the 3-primary amido group, as exemplified by **38** [18]. The coupler **38** has been still used in a recent embodiment described in Chapter 1 (Fig. 1.27).

DIR couplers such as **39** are sometimes called "DIR white couplers", because they react with an oxidized color developer to release a development inhibitor but does not generate an image dye [19]. This type of couplers can be incorporated in any light-sensitive emulsion layers.

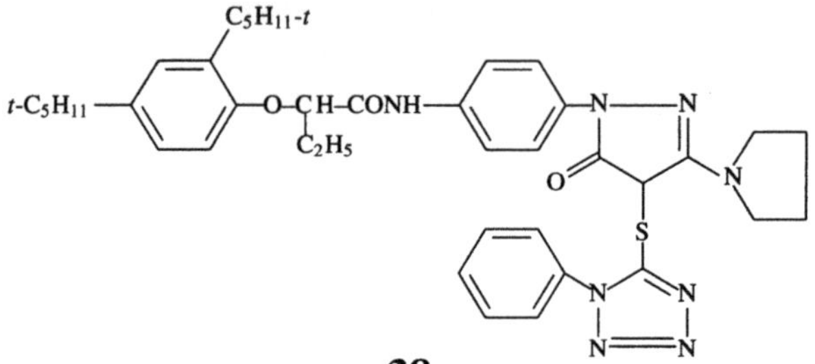

Figure 13.10. DIR couplers. The coupler (**38**) is a DIR magenta coupler. The coupler (**39**) is a DIR white coupler, which produces a colorless adduct after a chromogenic development.

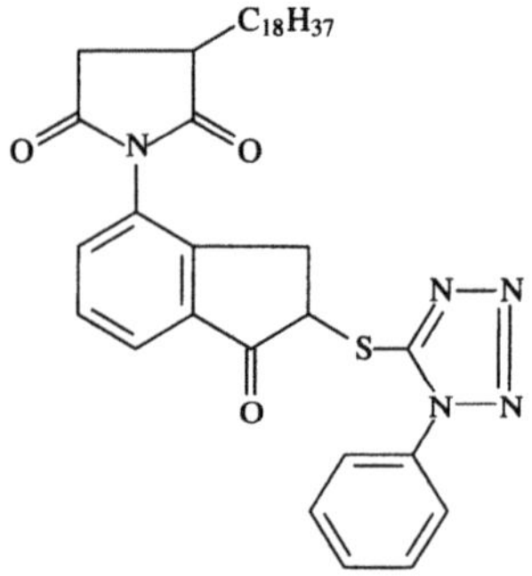

Syntheses

A synthetic pathway for preparing **38** is illustrated in Fig. 13.11 [18]. A pyrrolidino group as a secondary amino group is introduced into the 3-position of the pyrazolone nucleus, where the ethoxy group of **40** is replaced by a pyrrolidino group to produce **41**. After the hydrogenation of the nitro group of **41** into an amino group, the acid chloride of a ballast group, i.e., α-(2,4-di-*t*-amylphenoxy)butyryl chloride, is condensed with the amino group to give the corresponding four-equivalent coupler (**43**). A developer inhibitor moiety is incorporated in the last step (**43** $\rightarrow$ **38**) by using a sulfenyl chloride, which is prepared by 1-phenyl-2-tetrazoline-5-thione (or 1-phenyl-5-mercapto-1H-tetrazole) by chlorine gas.

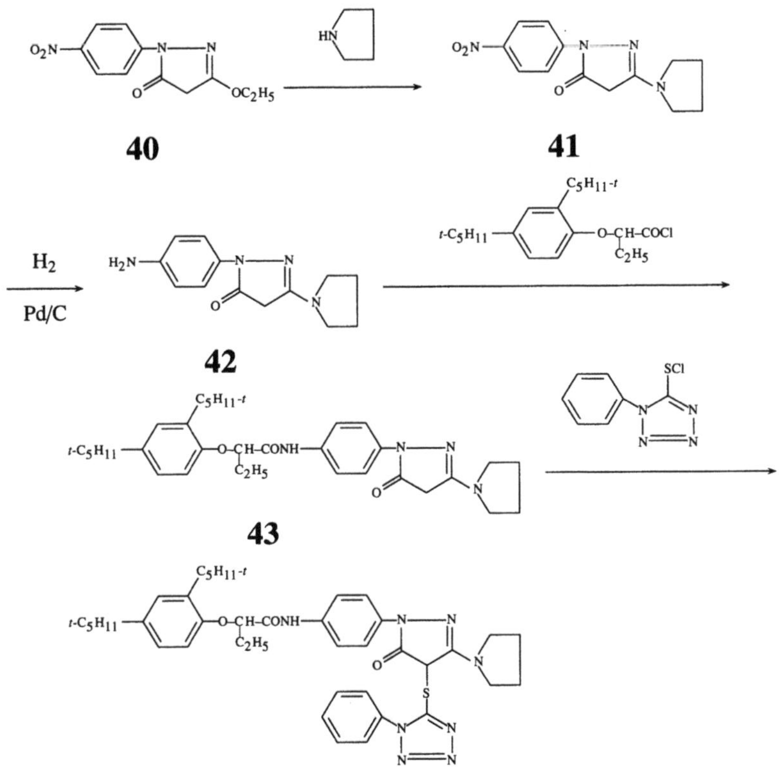

Figure 13.11. Synthetic pathway for preparing the DIR magenta coupler (**38**).

13.3.2 Releasing Heterocyclic Development Inhibitors

Whitmore et al. have disclosed couplers that release an intermediate *o*-aminobenzenediazonium ion [15]. The intermediate has been reported to cyclize subsequently into a benzotriazole,[4] which works as a development inhibitor.

---

[4]The intermediacy of the *o*-aminobenzenediazonium ion has been adopted by the review by Barr et al. [11]. However, the diazonium ion has not been isolated and, instead, only a 1,2,3-benzotriazole has been isolated in a solution reaction, so long as we inspect the original patent by Whitmore et al. [15]. The intermediacy of the *o*-aminobenzenediazonium ion is not always necessary to be postulated as a discrete compound, if we consider that the participation of the *o*-amino group takes place in the step in which the coupling of an oxidized developer at the active position gives the corresponding

Benzotriazoles have later been incorporated as a substituent at the active position of cyan (e.g., **44**), magenta (e.g., **45**), and yellow couplers (e.g., **46**), as illustrated in Fig. 13.12 [20]. These couplers absorb ultraviolet radiation and fluorescence in the blue region of the visible spectrum by virtue of the presence of the benzotriazole ring.

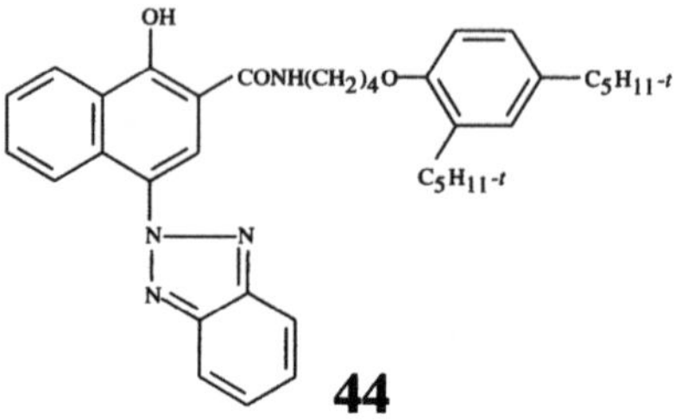

Figure 13.12. DIR couplers releasing a developer inhibitor of benzotriazole type.

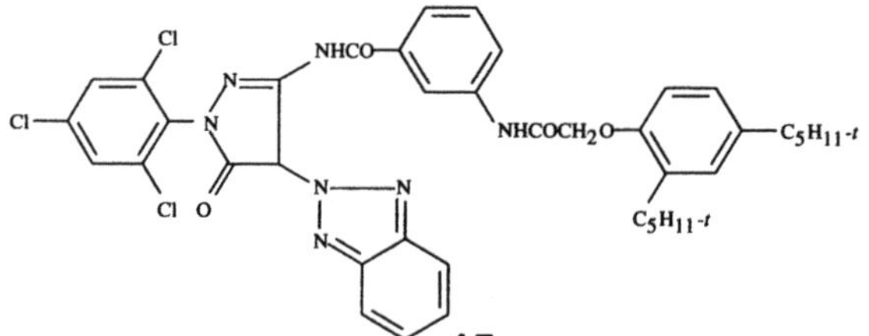

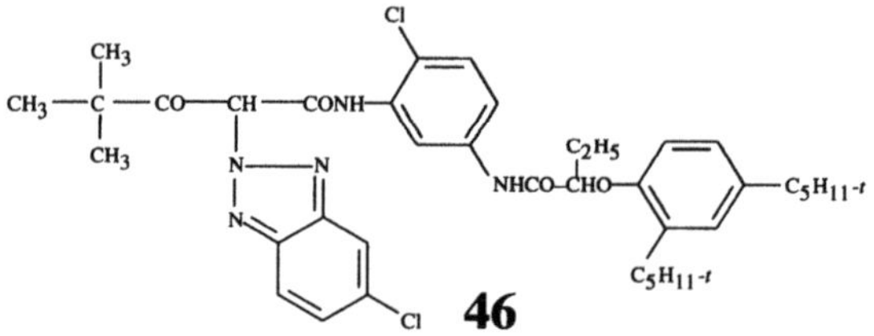

5-Bromo-1,2,3-benzotriazole has been reported to be more diffusible than 1-phenyl-5-mercapto-1H-tetrazole [13]. Since the improved sharpness depends upon the interlayer diffusion of a released development inhibitor, various substituents have been examined to control the diffusibility (Fig. 13.13) [21]. The specific substituent represented by **47a** has been selected to provide more improved image sharpness and more reduced granularity than the representative

leuco dye. This participation may cause a cyclization to give a 1,2,3-benzotriazoline attached at the coupling position, at which a developer moiety is also attached. Then, a further oxidation may cause the release of 1,2,3-benzotriazole from the leuco intermediate, giving an indoaniline cyan dye. In this connection, we have discussed in Chapter 12 the validity of the intermediacy of a diazonium ion in the reaction of a colored coupler (Fig. 12.1 vs. Fig. 12.2). These problems on the intermediacy of diazonium ions are still open to further experimental examinations.

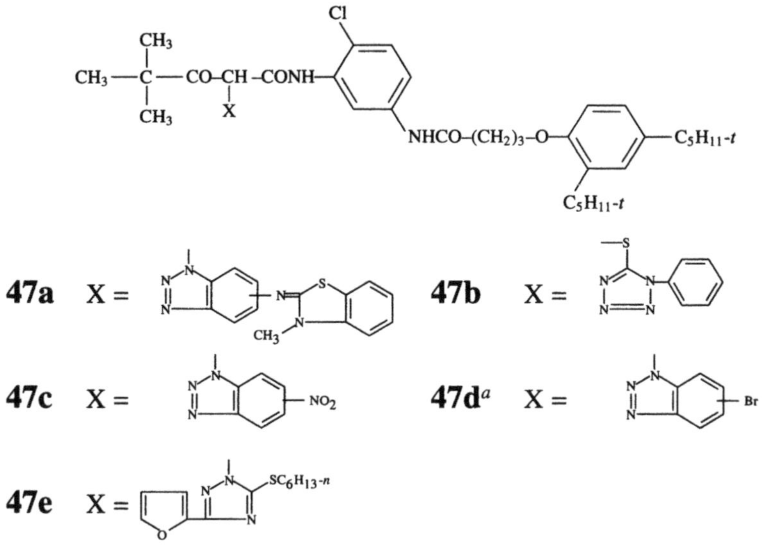

Figure 13.13. Pivaloylacetanilide couplers leaving inhibitors of benzotriazole type (**47a**, **47c**, and **47d**) and a comparison coupler leaving a thiolic inhibitor (**47b**) [21]. [a]The corresponding benzoylacetamide coupler has been used for comparison.

releasing group (**47b**) and the other substituents (**47c** and **47d**). The same leaving group as found in **47a** has been also incorporated in malondiamide couplers [22].

Yellow DIR couplers with a 5-furyl-1,2,4-triazole coupling-off group (e.g., **47e**) have been disclosed to produce comparatively powerful interlayer effects [23], where the furyl substituent is compared with an amino group, a phenyl group, etc.

Monocyclic 1,2,3-triazoles have been used as leaving groups (Fig. 13.14). 1,2,3-Triazoles having an alkoxycarbonyl group have been incorporated in DIR yellow couplers, as exemplified by **48** [24]. The substitution positions of the methyl and butoxycarboxyl groups may be altered, because this DIR has been synthesized by the condensation between the corresponding chloro intermediate and butyl 4-methyl-1H-1,2,3-triazole-5-caboxylate. The DIR coupler (**48**) has been compared with another DIR coupler leaving 5-hexylthio-1,2,4-triazole.

The same development inhibitor as found in **48** has been incorporated in a DIR yellow coupler (**49**) of another type, although the other mode of substitution of the leaving group has been adopted [5]. The merit of an acyloxymethyl group on the 1,2,3-triazole nucleus has been disclosed, as found in **51** [25], which has been compared with **50** having the same leaving group as **49**. The coupler **52** is a so-called white coupler, which gives a colorless coupling product and a 1,2,4-triazole development inhibitor [26].

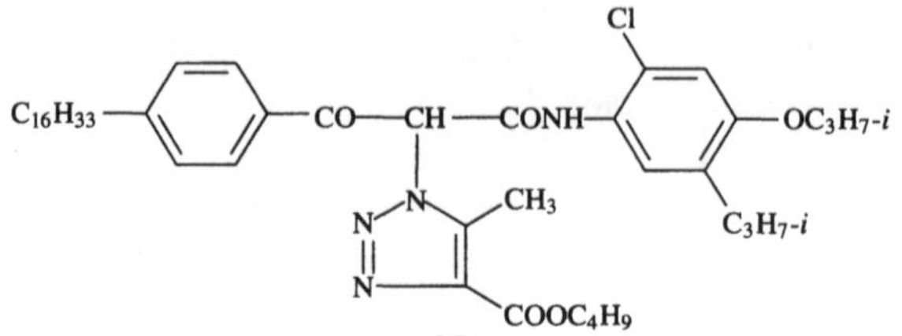

Figure 13.14. DIR yellow couplers (**48–51**) leaving a 1,2,3-triazole development inhibitor and a white coupler (**52**) leaving a 1,2,4-triazole development inhibitor.

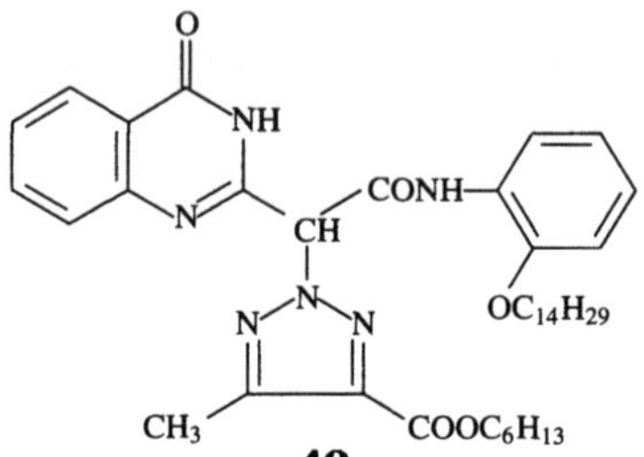

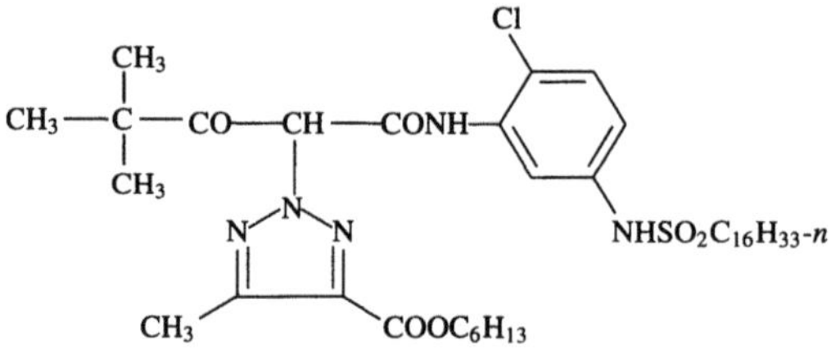

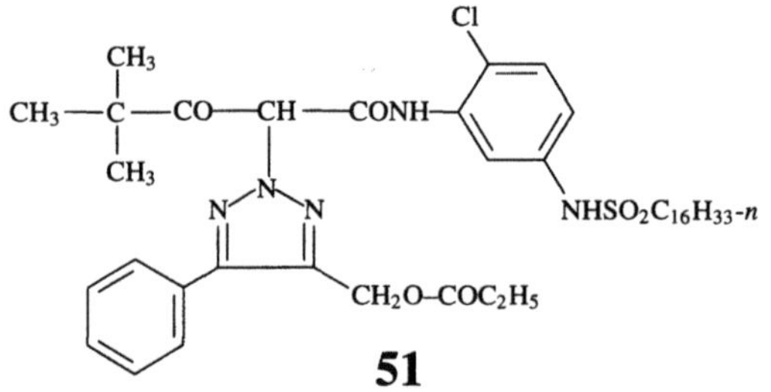

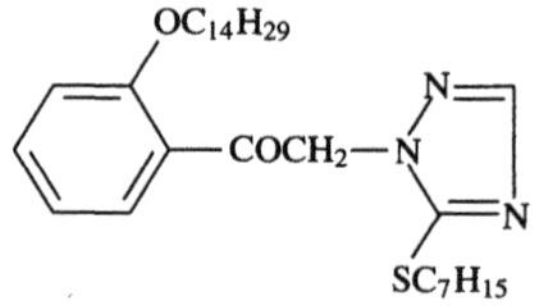

Syntheses

The coupler **45** (Fig. 13.12) can be prepared by the intramolecular cyclization of a nitro-azo starting material (**47a**), as shown in Fig. 13.15, where the reduction of the nitro group with zinc dust is employed [20].

The synthetic pathway of **56** (Fig. 13.16) has been disclosed as an analogous but simpler DIR coupler than **47a** [21]. Thus, the corresponding four-equivalent

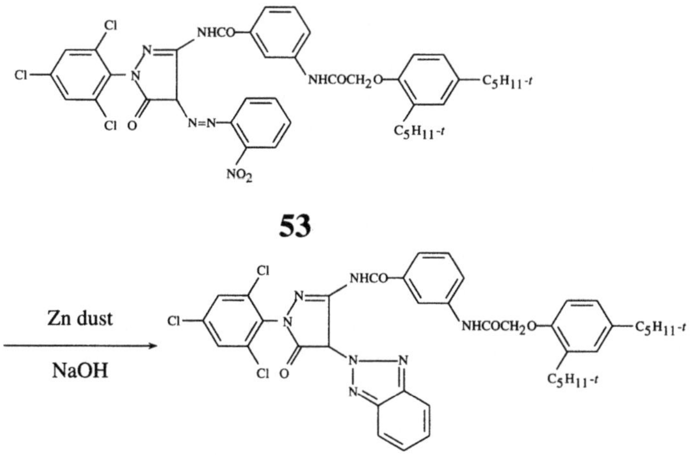

53

Zn dust / NaOH

45

Figure 13.15. Synthetic pathway for preparing the DIR magenta coupler (**45**).

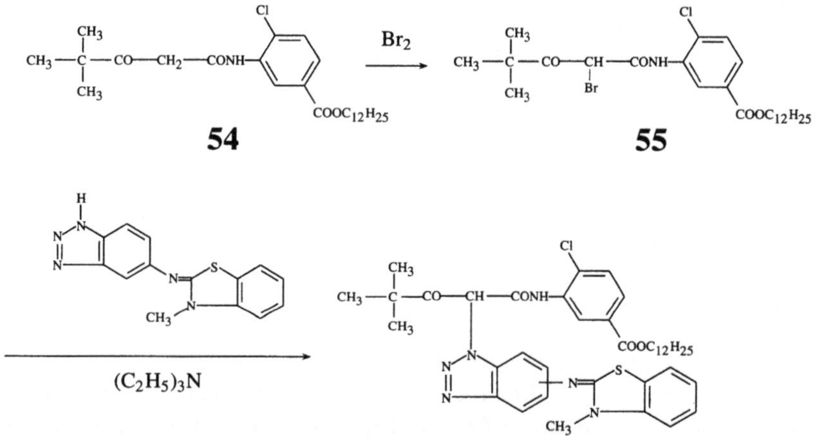

Br_2

54 → 55

$(C_2H_5)_3N$

56

Figure 13.16. Synthetic pathway for preparing the DIR yellow coupler (**56**).

coupler (**54**) is brominated at the active methylene position to give a bromo derivative **55**. Since the 1,2,3-benzotriazole intermediate used in the step from **55** to **56** may be either 5-(3-methylbenzothiazolinylidene)amino-1,2,3-benzotriazole or the 6-substituted counterpart in terms of tautomerism, the final product (**56**) is not specified with respect to the substitution position.

13.3.3 Deactivatable DIR Couplers

Development inhibitor molecules released from DIR couplers during film processing would diffuse from a photographic film into the processing solution. The inhibitor molecules accumulated in the solution cause the inhibition of subsequent development, since commercial processes of film processing repeatedly use such a processing solution as contaminated with the inhibitor molecules, as discussed in a review by Ichijima [13]. This problem, commonly referred to as "seasoning", can be overcome, if a development inhibitor molecule is converted into a species which is inactive as a development inhibitor. This type of inhibitors is here called "deactivatable development inhibitor" or "deactivatable DI". Deactivatable DIR couplers incorporating a deactivatable DI have been disclosed, where the deactivation process is the hydrolysis of an ester function, as exemplified by **57** [27]. The 1,2,3-benzotriazole-5-carboxylic acid ester released from the coupler (**57**) works as a development inhibitor in a photographic film. On the other hand, 1,2,3-benzotriazole-5-carboxylic acid, which is produced soon after contact with the processing solution, is inactive as a development inhibitor. A similar coupler has been used in a more recent embodiment, as discussed in Chapter 1 (Fig. 1.27). Magenta and cyan couplers that release deactivatable development inhibitors have also been disclosed in the patent [27].[5] Deactivatable DIR couplers of another malondiamide type (e.g., **58**) have been disclosed more recently in a patent [28] and used in recent color negative films, as discussed already in Chapter 1 (Fig. 1.44)

The same guideline has been applied to the incorporation of thiolic development inhibitors [29]. The phenyl group of 1-phenyl-2-tetrazoline-5-thione (or 1-phenyl-5-mercapto-1H-tetrazole), which has been discussed in Chapter 1 (Fig. 1.27) and in this chapter (**36** and **38**) is replaced by an alkoxycarbonylalkyl group, as exemplified by **59** and **60**.

Syntheses

A synthetic pathway for preparing the DIR magenta coupler (**57**) is illustrated in Fig. 13.19 [27]. The corresponding four-equivalent coupler **62** has been prepared by the condensation of the amine **61** and diethyl malonate. Then, bromination and substitution at the active methylene position of the malondimide (**62**) yield the target molecule **57**.

α-Halo intermediates such as **63** easily replace their halogen atom by an *O*- or *N*-nucleophile under suitable reaction conditions so as to produce DIR yellow couplers (e.g., **57**), as shown in Fig. 13.19. On the other hand, 4-monohalo intermediates derived from magenta couplers of 5-pyrazolone type (e.g., **64**) frequently suffer from further bromination, producing 4,4-dibromo intermediates. Monohalo intermediates, even if they can be isolated, disproportionate easily un-

[5] A set of such deactivatable DIR couplers as **57** has been placed on the market from Fuji Photo Film (Super HR etc., 1983) [13].

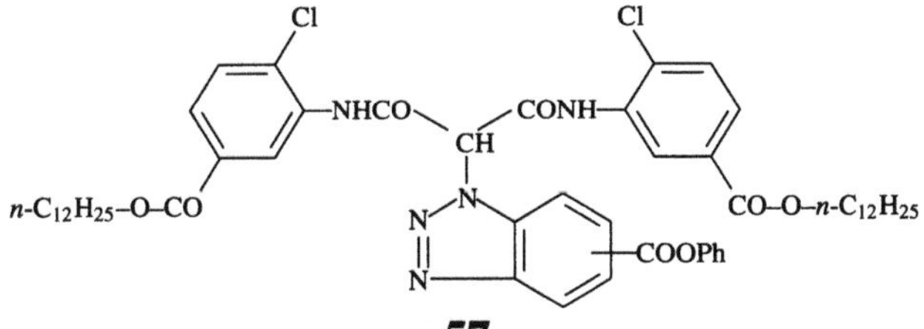

Figure 13.17. Deactivatable DIR couplers of symmetrical and unsymmetrical malondiamide types, which release a 1,2,3-benzotriazole inhibitor.

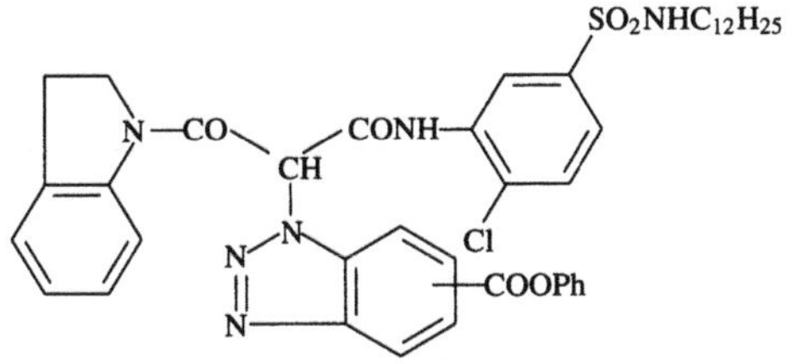

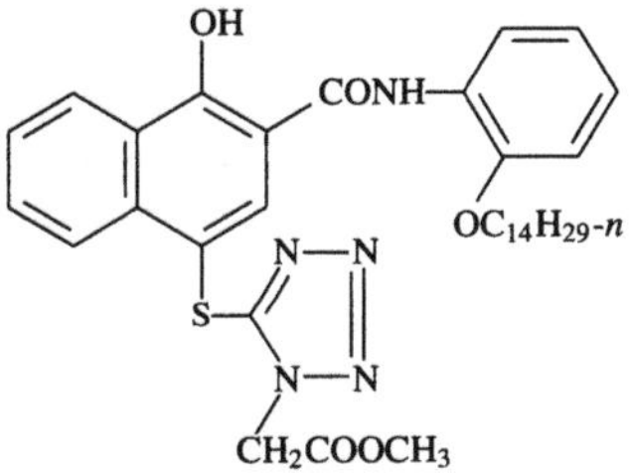

Figure 13.18. Deactivatable DIR couplers releasing a thiolic development inhibitor.

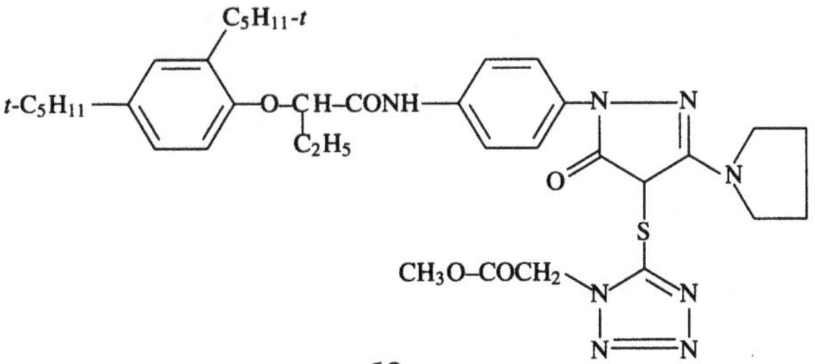

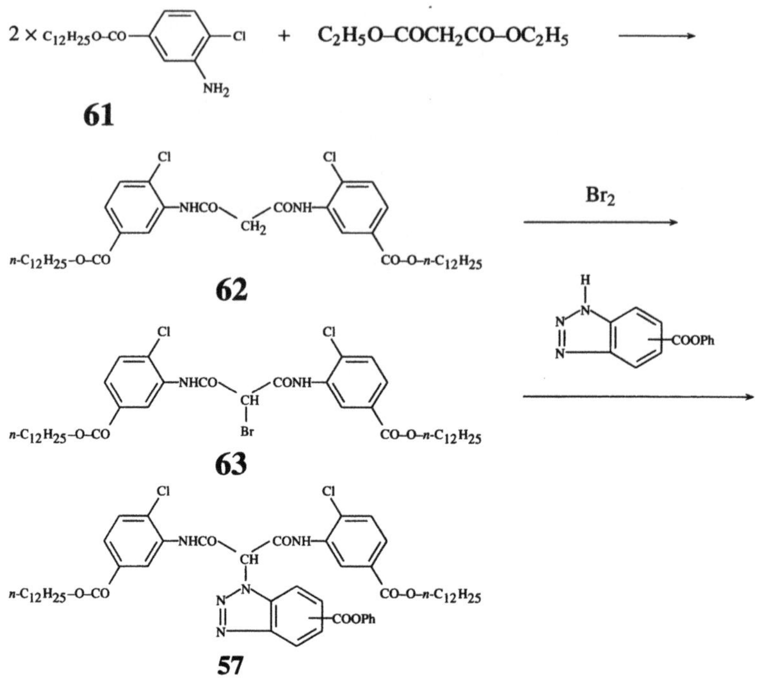

Figure 13.19. Synthetic pathway for preparing the DIR yellow coupler (57).

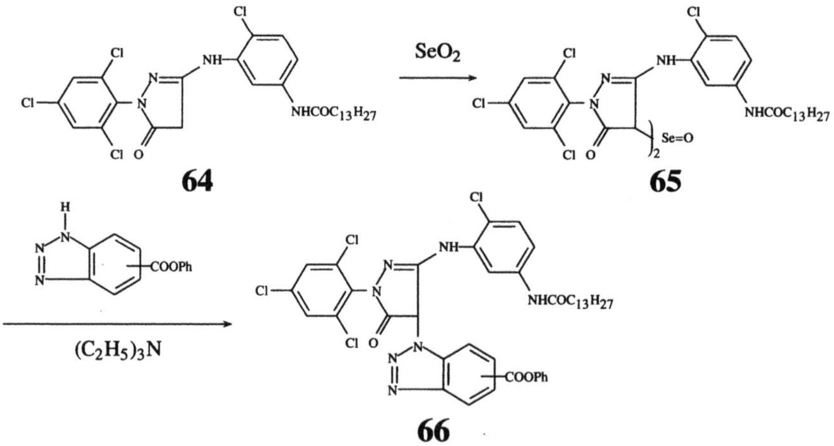

Figure 13.20. Synthetic pathway for preparing the DIR magenta coupler (66).

der subsequent attacks by basic nucleophiles. See Chapter 11 for a discussion of two-equivalent couplers.

A bispyrazolylselenium oxide (e.g., 65) has been reported to react with a 1,2,3-benzotriazole, producing a DIR magenta coupler (66) (Fig. 13.20) [30]. The bispyrazolyselenium oxide (**65**) can be prepared as a stable intermediate by the reaction of selenium dioxide with the four-equivalent coupler (**64**).

13.4 Timing DIR Couplers

As has been discussed in Subsection 13.1.2, there are two primary ways of releasing development inhibitors, i.e., timing releasing (cf. Fig. 13.5) and imagewise releasing (cf. Fig. 13.6). These two approaches can be combined to create additional ways of releasing development inhibitors. Thereby, timing DIR couplers have been invented as another important category of DIR couplers.[6] In contrast to the DIR couplers of direct-releasing type described in the preceding section, a timing DIR coupler releases a precursor of a development inhibitor, which permits a more elaborate control over imagewise releasing rate of the precursor, diffusion rate of the precursor, and DI-releasing timing from the precursor.

13.4.1 Timing Due to Hydrolysis

The first type of timing DIR couplers contain a timing group which, after releasing, is hydrolyzed to leave the active form of a development inhibitor, as shown in Fig. 13.21. This mechanism of DI releasing is here called "hydrolysis mechanism". The developer inhibitor moiety (DI) is linked to the active position of the naphthalene nucleus through the timing group O–CH_2 (**67**) [31]. This timing DIR cyan coupler can release a precursor HO–CH_2–DI or the corresponding anion (**69**), which is hydrolyzed into DI (**70**) and formaldehyde.[7] The DI (**70**) can be deactivated by hydrolysis, giving a non-effective free acid (**71**).

Such a deactivation function as shown in the bottom scheme of Fig. 13.21 is not always incorporated in a timing DIR coupler. For example, the coupler **72** (Fig. 13.22) is a timing DIR cyan coupler of this type, where the developer inhibitor moiety is a non-deactivatable one [31].

The merit of deactivated development inhibitors has been already discussed in Subsection 13.3.3. They can be incorporated into timing DIR couplers of this type, as shown in Fig. 13.21. As an additional example, the timing DIR coupler **73** releases a deactivatable development inhibitor, which is hydrolyzed after a coupling-off reaction [32]. An analogous coupler has been used in color films placed on the market, as referred to in a review [13].

The methylene group of the linkage can be substituted by an alkoxycarbonyl group, as found in another timing DIR coupler **74** [33]. The alkoxycarbonyl group in the timing group of **74** is so electron-withdrawing as to control the rate of the

[6]Fortunately, various mechanisms of timing releasing have been invented for instant color photography during the 1970s and disclosed in patents, as summarized in a review [6].

[7]The discrete release of the precursor (or its anion) is not always necessary. The intermediate leuco dye can form the DI molecule and formaldehyde spontaneously at the time of the coupling-off reaction. This mechanism should require another explanation to rationalize improved sharpness etc.

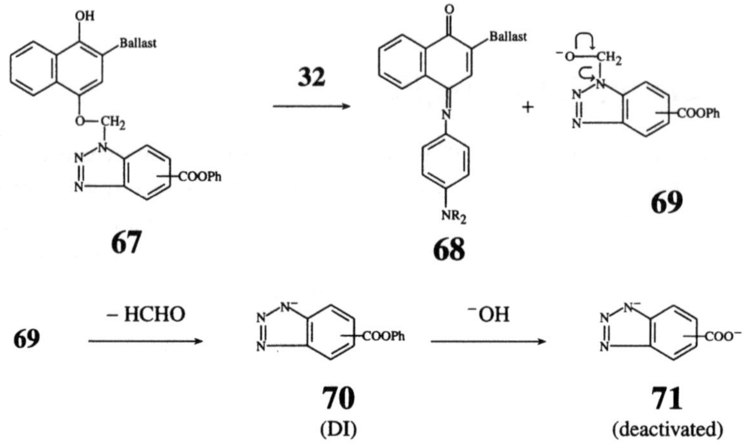

OH Ballast O—CH2 N N N COOPh **67** —32→ O Ballast N NR2 **68** + ⁻O CH2 N N N COOPh **69**

69 —(– HCHO)→ N N⁻ N COOPh **70** (DI) —(⁻OH)→ N N⁻ N COO⁻ **71** (deactivated)

Figure 13.21. Releasing reactions (hydrolysis mechanism) of a timing DIR coupler, where the intermediate **69** undergoes electron shifts to release a development inhibitor (**70**) and formaldehyde. The DI (**70**) is deactivated by hydrolysis.

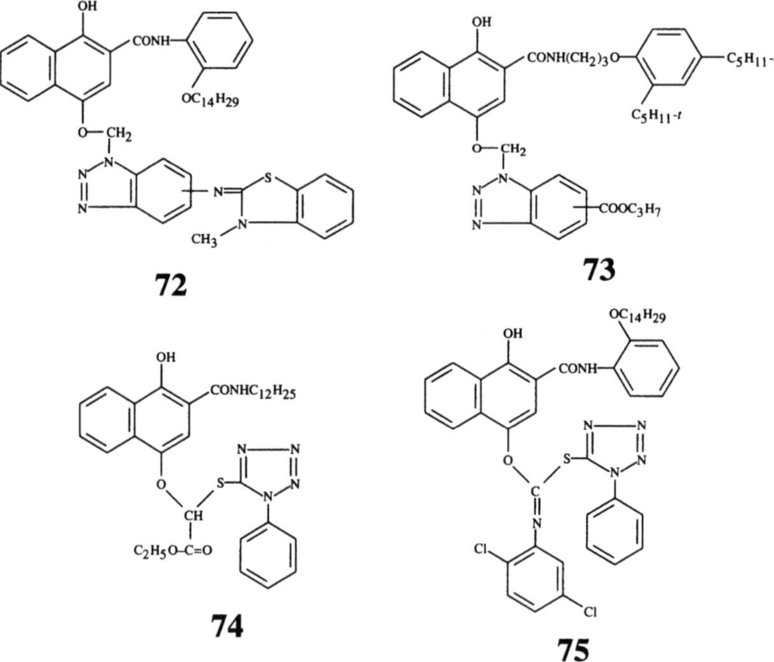

Figure 13.22. Timing DIR couplers with a hydrolyzable timing group.

coupling-off reaction and the stability and diffusibility of the resulting precursor anion ($^{-}$O–CH(COOR)–DI).

A further linking group for timing releasing has been disclosed, as exemplified by the coupler **75** [34]. The linking group has an arylimido group (=NAr), which influences the speed of diffusion of the primary product ($^{-}$O–C(=NAr)–DI) released from the coupler as well as the speed of the subsequent cleavage to release DI. A merit of linking groups of this type is that these speeds can be controlled by varying the substitution on the arylimide group.

Syntheses

To illustrate synthetic routes for preparing timing DIR couplers of this type, Fig. 13.23 shows the synthesis of **72** [31]. The developer inhibitor **76** is hydroxymethylated with formaldehyde and then converted into the chloromethyl intermediate (**77**). This is reacted with 1,4-dihydroxy-*N*-(2-tetradecyloxyphenyl)-2-naphthamide (**78**) in the presence of sodium ethoxide.

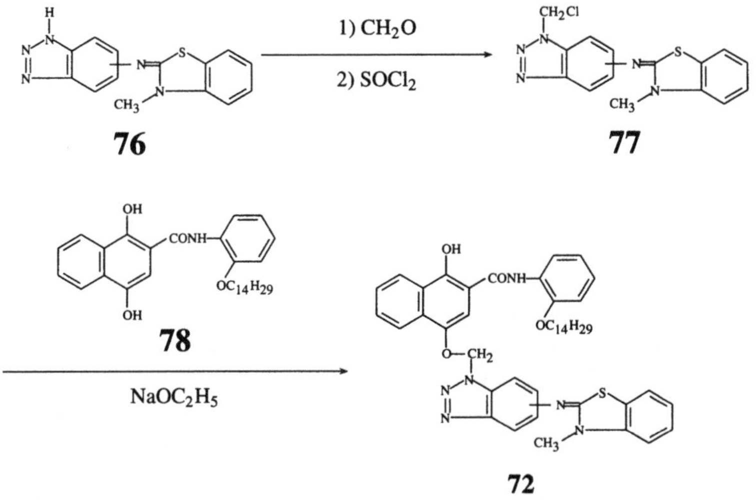

1) CH_2O
2) $SOCl_2$
76 **77**
78
$NaOC_2H_5$
72

Figure 13.23. Synthetic pathway for preparing **72**. The developer inhibitor **76** is represented by a Markush-type structural formula, because it is unspecified in terms of tautomerism.

13.4.2 Timing Due to Intramolecular Attacks

The second type of timing DIR couplers contains a timing group which is displaced from the coupler as a result of reaction of the coupler with an oxidized color developer and subsequently undergoes an intramolecular nucleophilic substitution to release a development inhibitor, as illustrated in Fig. 13.24 [35]. Thus, the timing coupler (**79**) reacts with the oxidized color developer (**32**) to give a cyan indoaniline dye **68** as well as to release a precursor (**80**). The phenoxide anion of **80** attacks the intramolecular thiocarbamate group as shown by curved arrows.[8] Thereby, a cyclization occurs to leave a six-membered molecule **81** and to release a developer inhibitor molecule **35**.[9] This mechanism of DI releasing is here called an "intramolecular cyclization mechanism" or briefly a "cyclization mechanism".

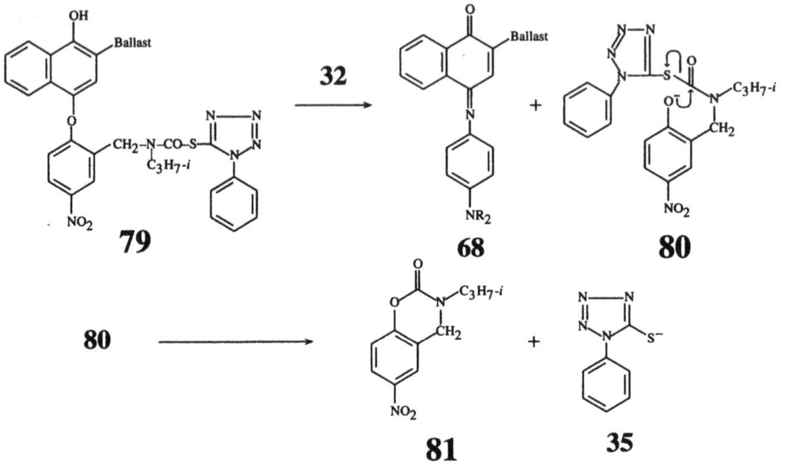

Figure 13.24. Releasing reaction (cyclization mechanism) of a timing DIR coupler, where the intermediate **80** undergoes intramolecular nucleophilic cyclization to release a development inhibitor.

As found easily, the *N-i*-propyl group on the thiocarbamate group can be replaced by another *N*-alkyl group in order to control the cyclization rate of such an intermediate as **80**. This type of timing DIR couplers with various *N*-alkyl groups and various ballast groups have been disclosed in patents, as summarized in reviews [12,14]. For example, a timing DIR cyan coupler (**82**) and a yellow one (**83**) are shown in Fig. 13.25 [35].

To control transportability, the simple phenyl group of the tetrazole inhibitor (e.g., **83** in Fig. 13.25) has been replaced by a *p*-methoxylbenzyl moiety, which is exemplified by 1-(4-methoxybenzyl)-2-tetrazoline-5-thione (or 1-(4-methoxybenzyl)-5-mercapto-1H-tetrazole), involved in **84** (Fig. 13.26). The released precursor yields a lower degree of interlayer effect but causes a high degree of "image acutance" (sharpness) due to an intralayer effect [36].

The development inhibitor released from a yellow coupler **85** in the mechanism similar to that of Fig. 13.24 is 1-propyloxycarbonylmethyl-2-tetrazoline-5-thione

[8]The attack of the phenoxide ion may first generate an intermediate having a tetrahedral carbon. This plausible intermediate is omitted for the sake of simplicity.

[9]It should be noted that the discrete release of the phenoxide anion is not always necessary. The intermediate leuco dye can undergo the coupling-off reaction and the cyclization reaction spontaneously. This mechanism should require another explanation to rationalize improved sharpness etc.

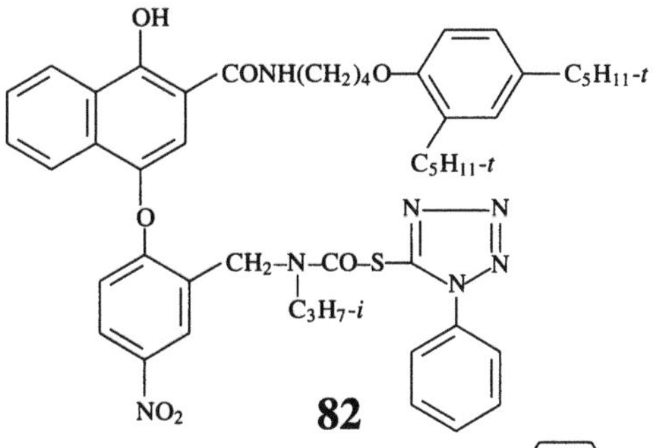

Figure 13.25. Timing DIR couplers releasing a thiolic development inhibitor on the basis of the intramolecular-cyclization mechanism. The released development inhibitor is 1-phenyl-5-mercapto-1H-tetrazole, which is a typical development inhibitor.

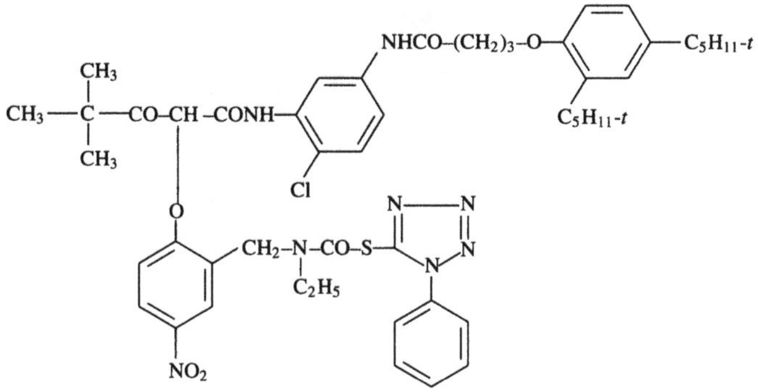

(or 1-propoxycarbonylmethyl-5-mercapto-1H-tetrazole), which can be deactivated by the hydrolysis of the ester group [29].[10] Compare the phenyl group of the development inhibitor from **83** with the propoxycarbonylmethyl group of the development inhibitor from **85**, where the latter is hydrolyzed (deactivated) after diffusing into a processing solution.

Tabular silver halide emulsions that have high chloride contents have been used in order to realize faster and easier processability (i.e., faster and easier development, beaching, and fixing), where various timing DIR couplers are incorporated to provide high sensitivity and high dye density formation [37].

Syntheses

As an example of syntheses of timing DIR couplers, Fig. 13.27 illustrates the synthesis of **83** [35]. The chloro intermediate **86** is condensed with the phenolic timing unit, i.e., 4-nitro-2-(*N*-ethyl-*N*-trifluoroacetylaminomethyl)phenol (**87**), in

[10]The yellow timing DIR coupler shown at the bottom of Fig. 1.27 (Chapter 1) may be the same compound as **85**.

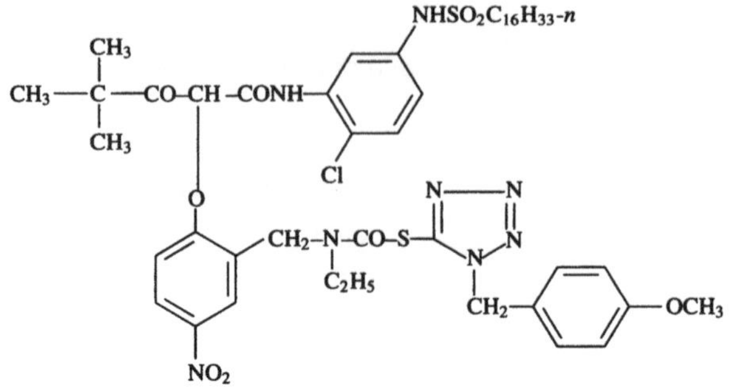

84

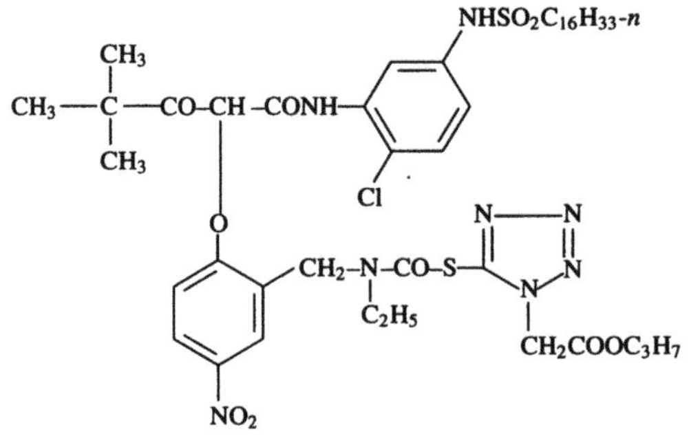

85

Figure 13.26. Additional examples of timing DIR couplers on the basis of the intramolecular-cyclization mechanism. The thiolic development inhibitors released therefrom have a substituent other than a phenyl group.

the presence of diisopropylethylamine to yield a condensed product. Then the protective group (CF_3CO–) is removed by hydrolysis to give the intermediate **88**. This intermediate is reacted with S,S'-carbonyldi-(1-phenyl-5-mercapto-1H-tetrazole) in the presence of sodium bicarbonate to give the final timing DIR coupler (**83**). The S,S'-carbonyldi-(1-phenyl-5-mercapto-1H-tetrazole) is prepared by the reaction of 1-phenyl-5-mercapto-1H-tetrazole with phosgene.

13.4.3 Timing Due to Quinone-Methide Formation

A timing releasing due to a quinone-methide mechanism has been well-known in instant color photography, as summarized in a review [6]. This mechanism has been incorporated into timing DIR couplers of the third type [38]. For example, the timing coupler (**89**) reacts with the oxidized color developer (**32**) to give a leuco-dye intermediate, which in turn gives a cyan indoaniline dye **68** and a phenoxide species (**90**). The phenoxide anion (**90**) undergoes electron shifts shown by curved arrows so as to release a developer inhibitor molecule **35** as

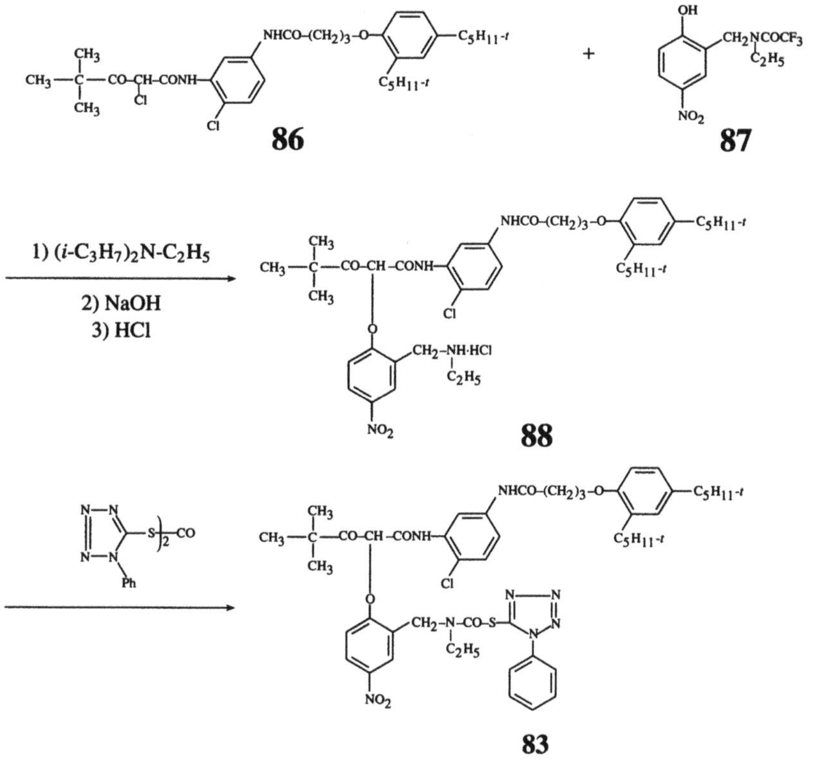

Figure 13.27. Synthetic pathway for preparing **83**.

well as to leave an *o*-quinone methide **91**.[11] This mechanism of DI releasing is here called "quinone-methide mechanism" or more precisely "*o*-quinone-methide mechanism".

As an example of timing DIR couplers based on the quinone-methide mechanism, the coupler **92**, which is covered by the patent [38], has been reported to have improved properties, which stem from the presence of a phenyl group on the methide moiety [14].

Pyrazolone magenta couplers (especially, 3-aninilo- or 3-acylamido-5-pyrazolones) having an aryloxy leaving group are unstable during film storage. It follows that pyrazolone-based timing DIR couplers have been difficult to be developed. As discussed in Chapter 11 (Section 11.3), 3-alkyl-5-pyrazolones can be used to overcome the instability due to the substitution of an aryloxy group at the active 4-position of a pyrazolone nucleus [39]. This methodology has been applied to develop timing DIR magenta couplers (e.g., **93**), which are characterized by a

[11] The intermediate leuco dye can undergo the coupling-off reaction and the DI-releasing reaction spontaneously, where the discrete release of the intermediate phenoxide species is not postulated. This mechanism should require another explanation to rationalize improved sharpness etc.

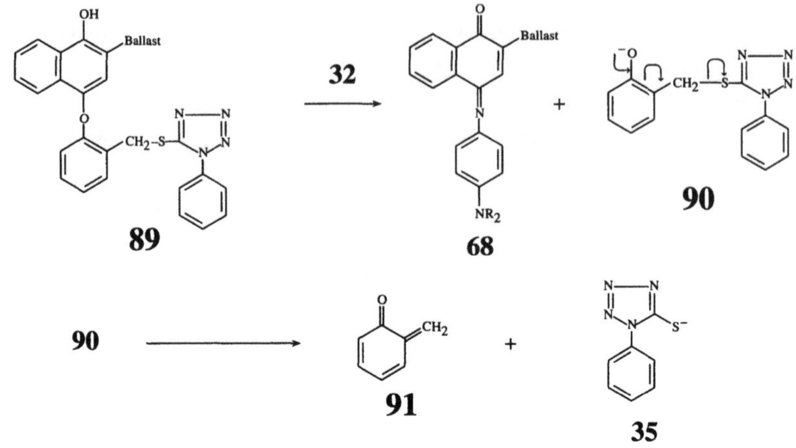

Figure 13.28. Releasing reaction (quinone-methide mechanism) of a timing DIR coupler, where the intermediate **90** undergoes electron shifts to release a development inhibitor and an *o*-quinone methide.

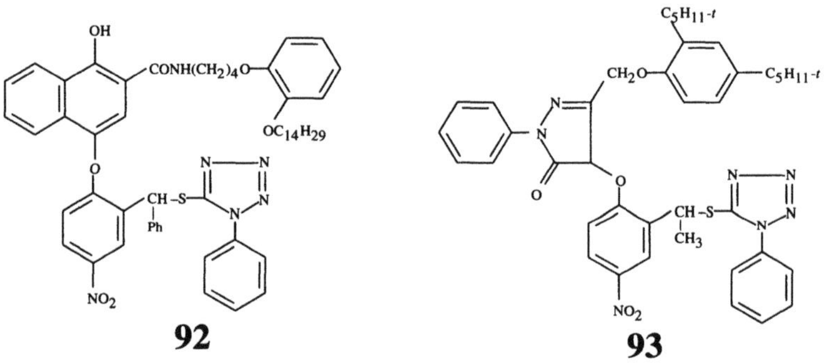

Figure 13.29. Timing DIR couplers releasing a thiolic development inhibitor on the basis of the *o*-quinone-methide mechanism.

substituted alkyl group such as a 2,4-di-*t*-amylphenoxymethyl group at the 3-position [40].

The quinone-methide mechanism is based on a timing moiety represented by the formula O^-—(C═C)$_n$—CR_2—DI, which is incorporated in a carbocylic aromatic ring (*n*: non-negative integer).[12] An analogous mechanism is also effective when the moiety is incorporated in a heterocyclic aromatic ring, as discussed in a review [14]. For example, a pyrazolone ring has been selected as one of such timing moieties, as exemplified by **94** [41].

[12] The present cases correspond to $n = 1$ and 2. For the cases of $n = 0$, see Subsection 13.4.1.

The timing DIR coupler (**95**) of this type gives a cyan indoaniline dye, which diffuses from a photographic film into the processing solution [42].[13] This means that the coupler can be used not only in a red-sensitive emulsion layer but also in any other layers in order to make the released developer inhibitor effective. To make the coupler non-diffusible, a ballast group ($C_{11}H_{23}$) is attached at the linking unit in place of the naphthol nucleus. Compare between **94** and **95** with respect to the ballast groups.

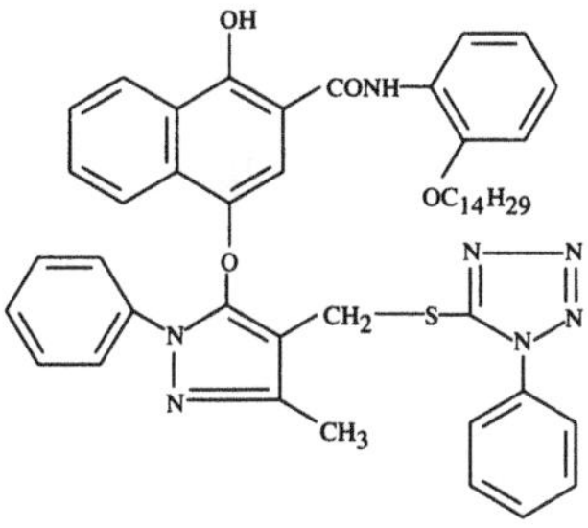

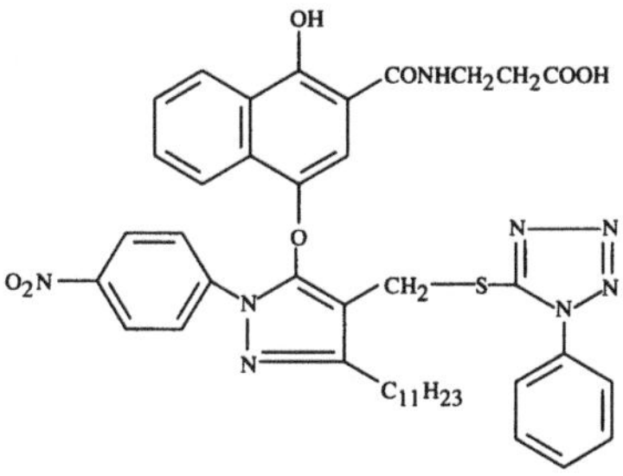

Figure 13.30. Timing DIR couplers releasing a thiolic development inhibitor on the basis of heterocyclic analogs of the quinone-methide mechanism.

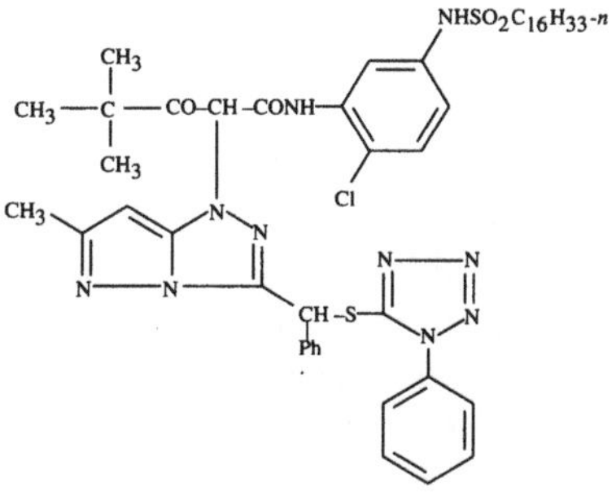

Another mode of timing releasing is embodied in **96**, where a precursor containing a moiety N^{-}—$(N{=}C)_n$—CR_2—DI ($n = 1$) is released by the coupling-off reaction with an oxidized color developer [43]. Note that the moiety is fully embedded in a heterocyclic ring.

Timing DIR couplers giving a cyan indoaniline dye (e.g., **97** and **98**) have been used together with 2-ureidophenol couplers [44,45]. The coupler **97** has been used in a more recent embodiment of a color film, as described in Chapter 1 (Fig. 1.27). The DI releasing of these couplers is based on the *p*-quinone-methide mechanism, which is a vinylog to the mechanism shown in Fig. 13.28.

[13] Since the cyan dye derived from the DIR coupler is removed into the processing solution, the DIR coupler works as a kind of white coupler.

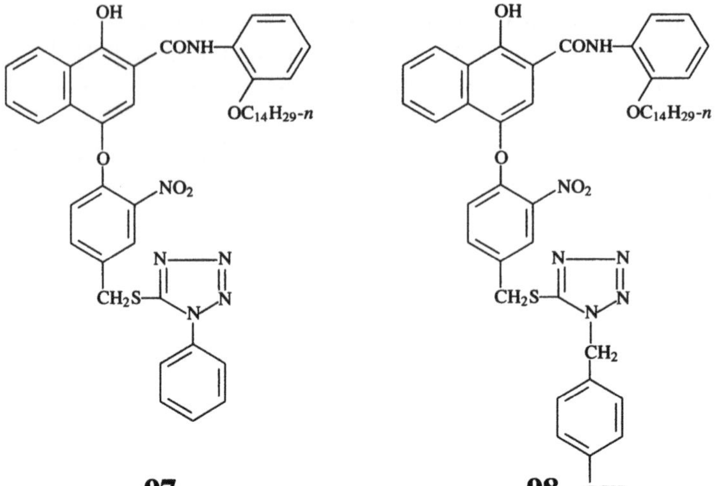

Figure 13.31. Timing DIR couplers releasing a thiolic development inhibitor on the basis of the *p*-quinone-methide mechanism.

Syntheses

A synthetic pathway for preparing **94** is illustrated in Fig. 13.32 [41]. 1,4-Dihydroxy-2-naphthoic acid (**99**) is reacted with 5-chloro-4-formyl-3-methyl-1-phenylpyrazole to give an adduct **100**. The formyl group of **100** is reduced into a hydroxymethyl group, which is in turn converted into a chloromethyl group so as to yield **101**. This intermediate is condensed with the anion of 1-phenyl-5-mercapto-1H-tetrazole to give **102**. Then, this is condensed with a ballast amine by using dicyclohexylcarbodiimide as a dehydrating agent, giving the final product **94**.[14]

13.4.4 Multi-Timing DIR Couplers

The timing DIR mechanisms described in the preceding subsections (i.e., the hydrolysis, the cyclization, and the quinone-methide mechanism) can be combined to construct a multi-timing function (Fig. 13.33). For example, the coupler **103** is based on a combination of the cyclization mechanism and the quinone-methide one [46]. Thus, the coupler reacts with an oxidized color developer to yield the first intermediate, which cyclizes in terms of the cyclization mechanism (cf. Fig. 13.24) to release the second intermediate.[15] This intermediate undergoes elec-

[14] Since 1,4-dihyroxy-2-naphthamide derivatives are used widely in color photography, various methods of preparing them have been reported in patents. In addition to the method described here, they can be synthesized by the reaction of a naphthoyl chloride and an amine and by the reaction of a naphthoic acid phenyl ester and an amine. See Chapters 11 and 12.

[15] In contrast to the scheme of Fig. 13.24, the cyclization leaves a five-membered ring.

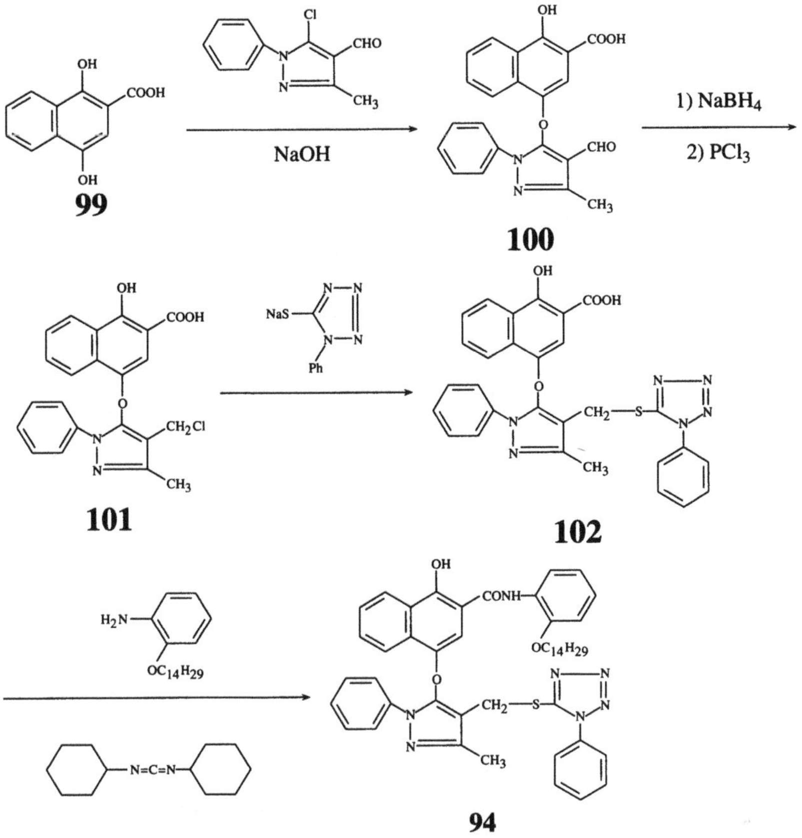

Figure 13.32. Synthetic pathway of the timing DIR cyan coupler **94**.

tron shifts due to the quinone-methide mechanism (cf. Fig. 13.28) so as to release 1-phenyl-5-mercapto-1H-tetrazole as a development inhibitor.

On the other hand, the coupler **104** employs a combination of the quinone-methide mechanism and the hydrolysis one [47]. Thus, the coupler **104** reacts with an oxidized color developer to yield the first intermediate, which releases the second intermediate in accord with the quinone-methide mechanism (cf. Fig. 13.28). The second intermediate undergoes electron shifts based on the hydrolysis mechanism (cf. Fig. 13.21) so as to release 1-(4-methoxyphenyl)-5-mercapto-1H-tetrazole as a DI. It should be noted that the DI derived from **104** is the same as **98**. The tetrazole moiety in **104** is fixed as a keto (thione) form, while the tetrazole moiety in the coupler **98** is incorporated as an enol (thiol) form. They are the same species, when released as an anion.

Another elaborate timing mechanism has been incorporated in **105**, which releases a carbamate anion upon a coupling-off reaction [48]. After decar-

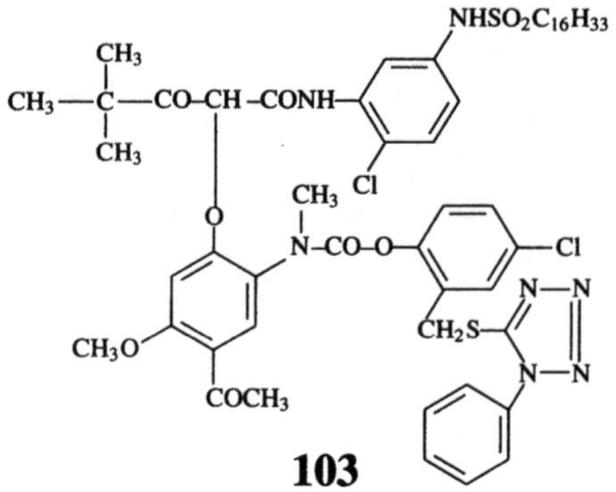

103

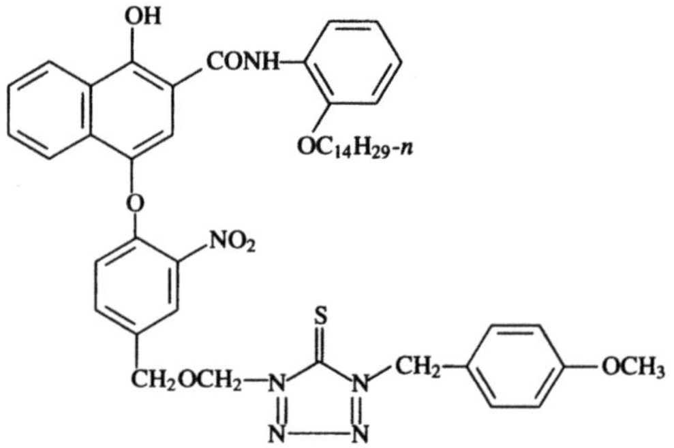

104

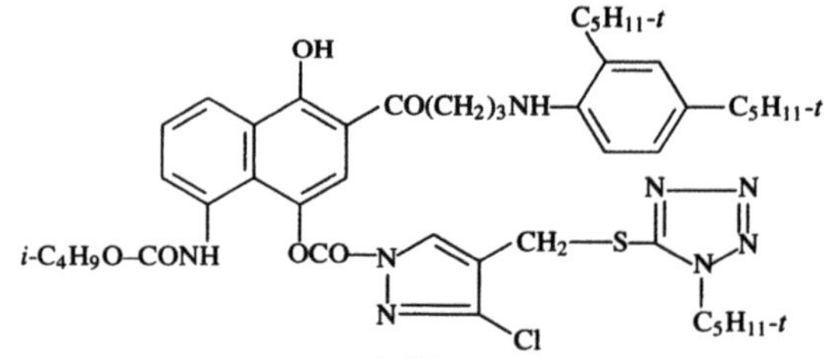

105

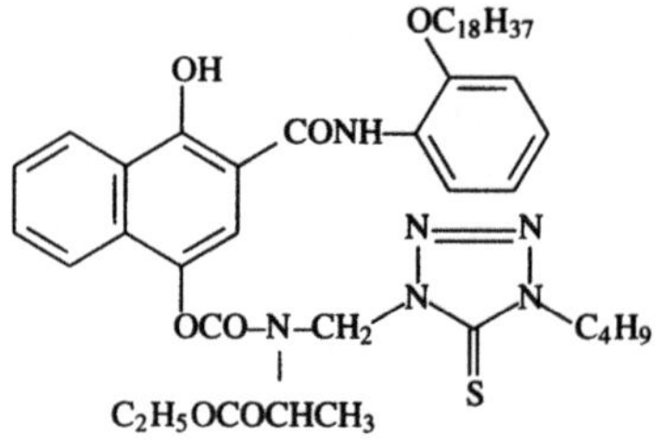

106

Figure 13.33. Multi-timing DIR couplers releasing a thiolic development inhibitor on the basis of the cyclization mechanism, the quinone-methide mechanism and/or the hydrolysis mechanism.

boxylation, the resulting heterocyclic anion undergoes a quinone-methide-like elimination to release a DI.

A further multi-timing mechanism has been incorporated in **106**, which is reacted with an oxidized color developer to release an intermediate carbamate anion ($^{-}OCONR–CH_2–DI$). This is hydrolyzed into another intermediate ($^{-}NR–CH_2–DI$), which finally releases a development inhibitor (DI) [49]. A homolog of **106** has been used in a more recent embodiment described in Chapter 1 (Fig. 1.36).

13.5 DAR Couplers

In contrast to DIR (development *inhibitor* releasing) couplers that involve a development inhibitor moiety, DAR (development *accelerator* releasing) couplers contain a development accelerator moiety. Because there are several mechanisms for accelerating the development of silver halide, DAR couplers have been invented in agreement with the mechanisms.

One of the mechanisms is an electron-donating reaction into silver halide crystals.[16] This can be carried out by using hydrazine derivatives, as exemplified by the DAR coupler (**107**) [50]. As a result of the formation of the corresponding indoaniline cyan dye upon color development, the coupler releases the species represented by $^{-}OCH_2CH_2SCH_2CONH–C_6H_4–NHNHCHO$. This species is adsorbed on the surface of a silver halide crystal, where its NHNH function works as a fogging agent so as to provide a fogging center that accelerates the further development of the silver halide crystal.[17] When the sensitivity for the case in which *N*-[γ-(2,4-di-*t*-amylphenoxy)propyl]-1-hydroxyl-2-naphthamide (the four-equivalent coupler corresponding to the coupler **73**) only is used as a main cyan coupler has been selected as a standard (1.00), the relative sensitivity due to the addition of the coupler **107** (10 mol %) was determined to be 4.20.

Another mechanism is a different type of fogging reaction, where a thiourea moiety incorporated in **108** generates silver sulfide on the surface of a silver halide crystal when released from the coupler by the coupling-off reaction [51]. The silver sulfide works as a fogging center accelerating the further development of the crystal.[18] The relative sensitivity due to the addition of the coupler **108** (10 mol %) has been determined to be 1.80. From this point of view, two-equivalent couplers releasing a thiocyanate anion can be regarded as DAR couplers, though they have been originally disclosed as couplers for releasing a silver halide solvent (^{-}SCN) [52].

To ensure the action of a released development accelerator, its adsorption on the surface of a silver halide crystal is most important to be taken into account.

[16]This mechanism is very important to understand the action of direct positive emulsions used in instant color photography. This point will be discussed later in detail (Chapter 16).

[17]Remember the reduction sensitization.

[18]Remember the sulfur sensitization.

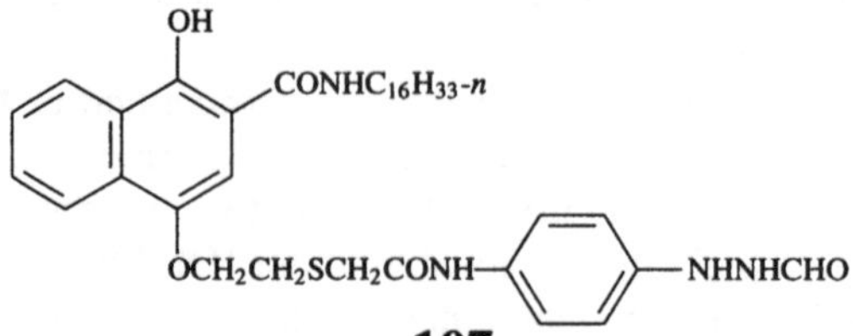

107

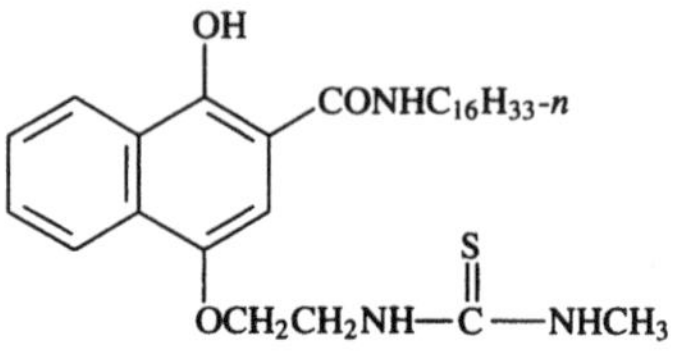

108

Figure 13.34. DAR couplers based on two mechanisms: the generation of fogging centers by reduction and the deposit of silver sulfide.

This is realized by designing a development accelerator having an adsorption site, as exemplified by the DAR coupler (**109**) [53].[19]

The DAR coupler (**110**) has been disclosed to exhibit a development accelerating effect in the presence of hydroxylamine in the color developer solution [55]. Note that the development accelerating effect stems from the pyrazolidone nucleus, which works as an electron transfer agent. Compare the hydrazide group of **109** with the pyrazolidone nucleus of **110**. The pyrazolidone nucleus has been used as a mother skeleton for auxiliary developments. For example, 1-(4-benzoyloxyphenyl)-4,4-dimethyl-3-pyrazolidone has been incorporated in photographic materials, where residual amounts can be deactivated by alkaline hydrolysis, even if they season into the developer solution [56].

The coupler (**109**) is reacted with an oxidized developer (**32**) to give the corresponding yellow dye and a development accelerator (**111**). The development accelerator (**111**) involves a benzotriazole nucleus as an adsorption site and a hydrazine function as a development accelerator site. It should be noted that the benzotriazole nucleus also works as a development inhibitor, which is in disagreement with the development accelerator site. Hence, the balance of the two functions may be important to design such DAR couplers.

[19]Color negative films employing a DAR coupler have been placed on the market under the name Fujicolor HR series (1984), where the DAR coupler has been called "A-coupler", though its precise structure has not been disclosed [54]. The letter A of the term "A-coupler" may stem from the word "accelerator".

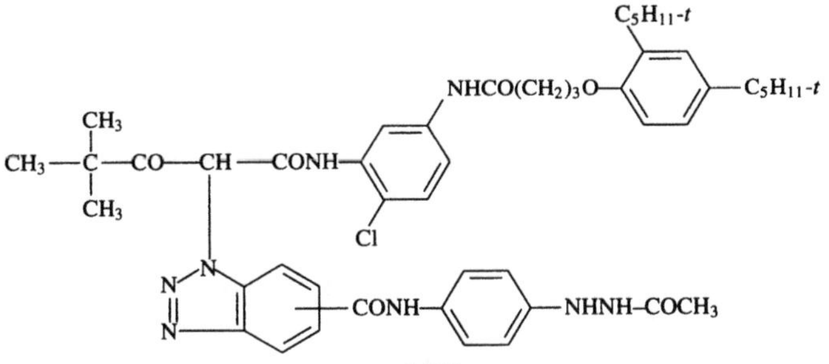

109

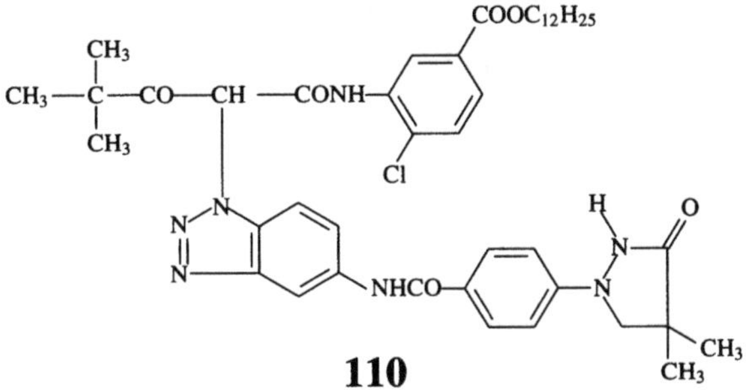

110

Figure 13.35. DAR couplers releasing a development accelerator with an adsorption site. The structural formulas are faithfully cited from the patents. The substitution mode for the benzotriazole nucleus of the DAR coupler **109** may be the same as that for **110** because of their synthetic methods.

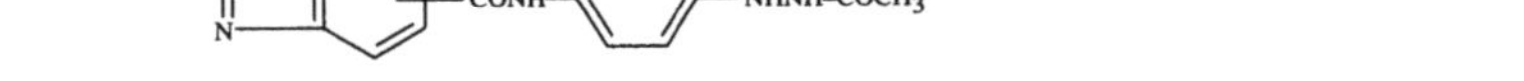

109

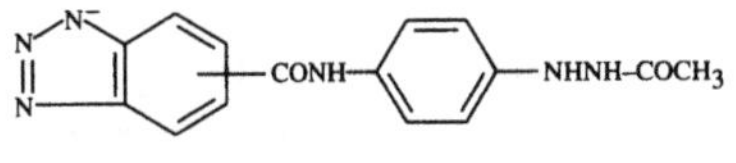

32

– Dye

111

Figure 13.36. DAR coupler (**109**) and the release of a development accelerator (**111**). The benzotriazole nucleus of **111** works as an adsoption site, while the hydrazine function works as a development accelerator as a result of generating a fogging center.

13.6 BAR Couplers

To speed up the processing (color development, bleach, fixation, and washing) of color negative films, the acceleration of the bleach has been extensively investigated. BAR (bleach accelerator releasing) couplers such as **112** have been disclosed to have satisfactory effects. Upon the color development step, the coupler (**112**) incorporated in the red-sensitive layer of a film releases a thiolic species ($^{-}SCH_2CH_2COOH$) as a bleach accelerator [57]. The species is adsorbed in the film and carried over to the next bleaching step, where the acceleration occurs. The effect of another BAR coupler (**113**) has been compared with that of **112**, where the former is more efficient to bleach acceleration [58]. BAR couplers (e.g., **112**) have been used in combination with timing DIR couplers (e.g., **95**) of quinone-methide type, where both the improved granularity and the bleach acceleration have been realized [59].

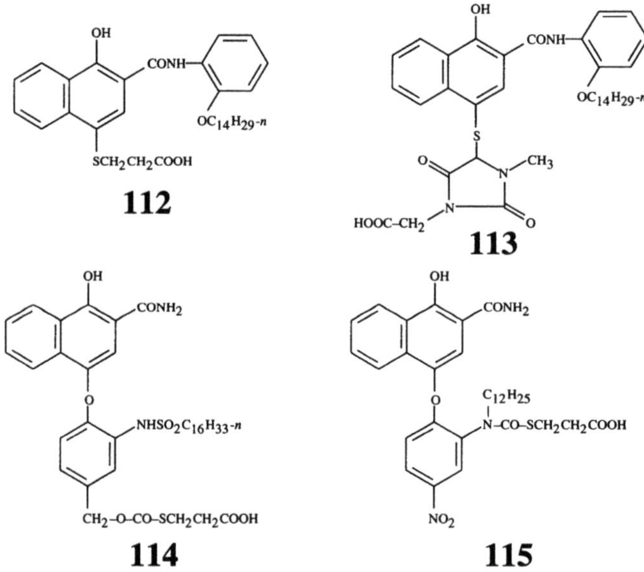

Figure 13.37. BAR (bleach accelerator releasing) couplers.

BAR couplers such as **114** are capable of forming washout cyan dyes and of releasing bleach accelerators [60]. Moreover, the BAR coupler **114** incorporates a timing function based on the quinone-methide and the subsequent hydrolysis mechanism, which finally releases a bleach accelerator ($^{-}SCH_2CH_2COOH$). The bleach accelerator moiety of the formula (**114**) is further varied: $-SCH_2CH_2COOH$, $-SCH_2COOH$, $-OCO–SCH_2CH_2N(CH_2H_5)_2$, $-OCO–SCH_2CH_2N(CH_2H_2)_2O$, and so on. BAR couplers of this type (e.g., **114**) have been used in combination with DIR couplers [61].

BAR couplers of timing releasing type (e.g., **115**) have been used in color reversal films [62]. For example, the BAR coupler **115** releases a bleach accelerator ($^-SCH_2CH_2COOH$) in terms of the cyclization mechanism. The bleaching process of color reversal films, unlike that of color negative films, must oxidize silver generated throughout the image in both regions of maximum image dye density and regions of minimum dye density. In order that a BAR coupler uniformly assists in silver bleaching throughout a reversal film, an additional layer containing non-imaging silver halide and the BAR coupler (**115**) has been coated below the light-sensitive silver halide layers.

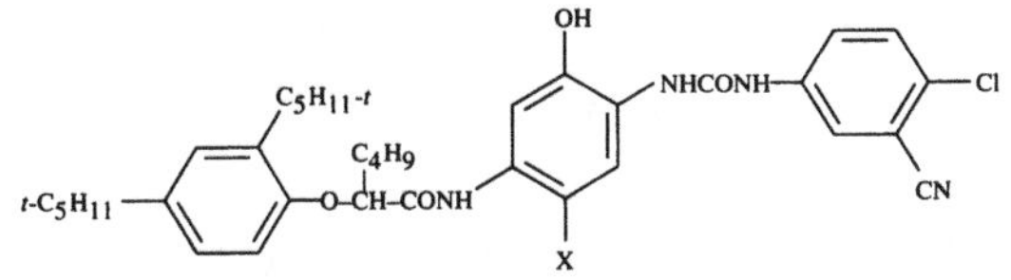

116 ($X = SCH_2CH_2N(CH_2)_2$)
117 ($X = SCH_2COOH$)

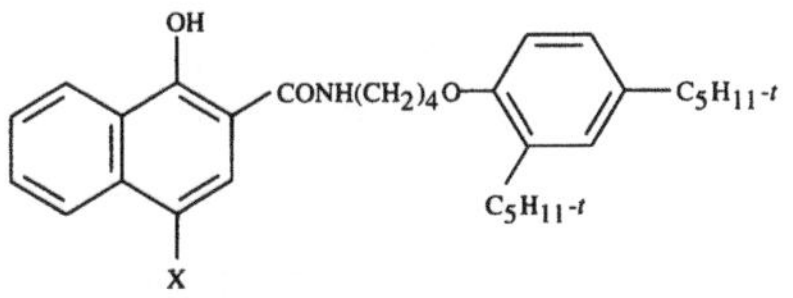

118 ($X = SCH_2CH_2COOH$)
119 ($X = SCH_2CH_2CH_2COOH$)

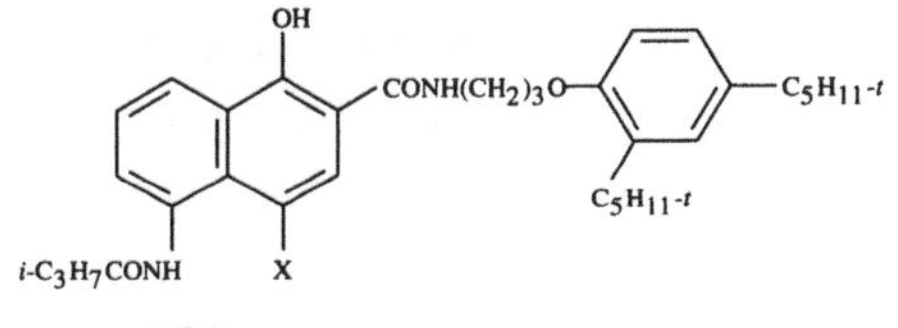

120 ($X = SCH_2CH_2COOH$)
121 ($X = SCH_2CH_2CH_2COOH$)

Figure 13.38. BAR (bleach accelerator releasing) couplers of various types. Each leaving group (X) is removed during chromogenic development to give the sulfide anion (X^-) adsorbed on the surface of a silver crystal.

A combined use (e.g., [63]) of 5-amido-1-naphthol cyan couplers with yellow-colored cyan couplers has been already described in Chapter 12. This use has been further combined with BAR cyan couplers of phenol type (**116** and **117**) and of naphthol type (**118** and **119**) [64].

In more recent embodiments [65,66], BAR cyan couplers of 5-amido-1-naphthol type (e.g., **120** and **121**) have been used in place of those of 1-naphthol type.

References

[1] Mueller FWH (1970) In: The Photographic Image Formation and Structure (Proc Intl Congr Photogr Sci Tokyo, 1967), Focal Press, London, pp. 91–106
[2] Evva F (1967) Z Wiss Photogr. 60:145
[3] Futaki K, Oohashi M (1986, 2002) In: Functionalized Organic Chemicals for Silver Halide Color Photographic Materials, CMC, Tokyo. Chapter 2, Section 3.4
[4] Newmiller RJ, Pontius RB (1961) Photogr Sci Eng. 5:283
[5] Bell P, Ly C (1993) US Patent 5 270 157
[6] Fujita S (1984) Sci Pub Fuji Photo Film. 29:18
[7] Hammond HA, Humphlett WJ, Salminen IF (1977) US Patent 4 009 029
[8] Fujita S (1981) Kagaku no Ryoiki. 39:617
[9] Vargas JR, Dickinson DA (2000) US Patent 6 150 077
[10] Welter TR, Dickinson DA, Chen KT (1996) US Patent 5 567 577
[11] Barr CR, Thirtle JR, Vittum PW (1969) Photogr Sci Eng. 13:74
[12] Itoh I (1986, 2002) In: Functionalized Organic Chemicals for Silver Halide Color Photographic Materials, CMC, Tokyo. Chapter 2, Section 5
[13] Ichijima S (1989) J Soc Photogr Sci Technol Jpn. 52:145
[14] Kida S (1989) J Soc Photogr Sci Technol Jpn. 52:150
[15] Whitmore KE, Staud CJ, Barr CR, Williams J (1964) US Patent 3 148 062
[16] Barr CR, Williams J, Whitmore KE (1966) US Patent 3 227 554
[17] Barr CR, Williams J, Whitmore KE (1972) US Patent 3 701 783
[18] Abbott JR, Coffey WF (1971) US Patent 3 615 506
[19] Fujiwara M, Endo T, Sato R, Masukawa T, Uozumi T (1976) Jpn Patent S51-16142
[20] Sawdey GW (1971) US Patent 3 617 291
[21] Sueyoshi T, Furutachi N, Okumura A, Shishido T (1978) US Patnt 4 095 984
[22] Tanaka M, Yagihara M, Aono T, Hirose T (1979) US Patent 4 149 886
[23] Öhlshläger H, Griesel U, Odenwälder H (1986) US Patent 4 579 816
[24] Odenwälder H, Vetter H, Berghaller P, Hübner (1993) US Patent 5 200 306
[25] Begley WJ, Coms FD, Chen T-H (1999) US Patent 6 007 974
[26] Webb TC (1982) US Patent 4 362 878
[27] Ichijima S, Sakanoue K, Kobayashi H, Adachi K (1984) US Patent 4 477 563
[28] Motoki M, Ichijima S, Saito N, Kamio T, Mihayashi K (1993) US Patent 5 213 958
[29] DeSelms RC, Kapecki JA (1988) US Patent 4 782 012

[30] Bergthaller P (1991) Angew Chem Int Ed Engl. 30:1704
[31] Yokota Y, Aono T, Hirose T (1979) US Patent 4 146 396
[32] Sakanoue K, Ichijima S, Kishimoto S (1989) US Patent 4 812 389
[33] Ichijima S, Sato S, Ono M, Sasaki N (1987) US Patent 4 652 516
[34] Berghaller P, Odenwälder H, Matejec R (1985) US Patent 4 546 073
[35] Lau PTS (1981) US Patent 4 248 962
[36] Szajewski RP, Poslusny JN, Slusarek W (1990) US Patent 4 962 018
[37] Szajewski RP, House GL, Brust TB, Harsell DL, Black DL, Bohan AE, Merrill JP (1994) US Patent 5 356 764
[38] Sato R, Hotta Y, Matsuura K (1983) US Patent 4 409 323
[39] Poslusny JN, Anderson LG, Mooberry JB, Wu ZZ (1996) US Patent 5 576 167
[40] Poslusny JN, Anderson LG, Mooberry JB, Slusarek W K, Wu ZZ (1997) US Patent 5 670 306
[41] Uemura M, Kishi K, Nakagawa S, Kida S, Sugita H (1983) US Patent 4 421 845
[42] Nakagawa S, Sugita H, Kida S, Uemura M, Kishi K (1984) US Patent 4 482 629
[43] Begley WJ, Chen T-H, Coms FD, Singleton D (1994) US Patent 5 306 607
[44] Szajewski RP, Taber TR (1991) US Patent 5 021 555
[45] Merkel PB, Kestner MM, Zengerle PL (1995) US Patent 5 451 496
[46] Burns PA, Taber TR (1989) US Patent 4 861 701
[47] Ohkawa A, Motoki M, Mihayashi K (1994) US Patent 5 360 709
[48] Ohkawa A, Motoki M, Mihayashi K (1994) US Patent 5 286 620
[49] Ohkawa A, Obayashi T, Mihayashi K (1994) US Patent 5 326 680
[50] Kobayashi H, Takahashi T, Hirano S, Hirose T, Adachi K (1983) US Patent 4 390 618
[51] Kobayashi H, Takahashi T, Hirano S, Hirose T, Adachi K (1985) US Patent 4 518 682
[52] Loria A, Salminen IF (1966) US Patent 3 253 924
[53] Mihayashi K, Kobayashi H, Takada S (1988) US Patent 4 746 601
[54] Itoh I, Takada S, Ikenoue S (1984) Nikkakyo Geppo. 1984(12):18
[55] Tsoi S-C, Twist PJ, Southby DT (1999) US Patent 5 958 664
[56] Tsoi SC, Twist PJ (2000) US Patent 6 087 084
[57] Hall JL, Kilminster KN, Romanet RF (1986) Jpn Kokai S61-201247
[58] Sakanoue K, Ichijima S (1989) US Patent 4 842 994
[59] Sakanoue K, Kobayashi H, Ichijima S, Ueda S (1990) US Patent 4 959 299

[60] Begley WJ, Singer SP, Southby DT, Singleton Jr D (1994) US Patent 5 358 828

[61] Begley WJ, Singer SP, Southby DT, Singleton Jr D (1994) US Patent 5 300 406

[62] Bowne AT (1996) US Patent 5 561 031

[63] Ohkawa A, Kamio T, Motoki M, Mihayashi K (1992) US Patent 5 112 730

[64] Ishii Y, Fujita Y, Mihayashi K (1995) US Patent 5 459 022

[65] Matsumoto K, Ito T (2002) US Patent 6 337 176 B1

[66] Omae N, Aida S (2003) US Patent 6 537 740 B1

Chapter 14

DIR Hydroquinones and Related Compounds

14.1 DIR Hydroquinones

14.1.1 DIR Hydroquinones for Reversal Color Process

The color reversal process described in Chapter 8 (Fig. 8.6) consists of exposure, first development (B&W development), color development after exposure (or reversal action with a foggant), and bleach & fixation. Because the color development uses all the silver halide remained after the first B&W development, such DIR couplers as described in the preceding chapter cannot be employed to improve sharpness, granularity, and color reproduction. In other words, interlayer effects (for improved color reproduction) and intralayer effects (for improved granularity and sharpness) should take place during the first B&W development of color reversal films. DIR hydroquinones, which were proposed for instant color films (e.g., [1]) and for color negative films (e.g., [2]), have been employed to bring about the inter- and intralayer effects in the color reversal process [3,4].[1]

The reactions occurring in the first B&W development of the color reversal process in the presence of a DIR hydroquinone are illustrated in Fig. 14.1 [3]. The developer solution for this step contains the potassium salt of hydroquinone-monosulfonic acid (**3**) as a main developer, 4-hydroxymethyl-4-methyl-1-phenyl-3-pyrazolidone (**1**) as an auxiliary developer, and several additives.

---

[1]Color reversal films using DIR hydroquinones have been placed on the market from Fiji Photo Film under the names of Velvia (1990), Provia (1994), and Astia (1997).

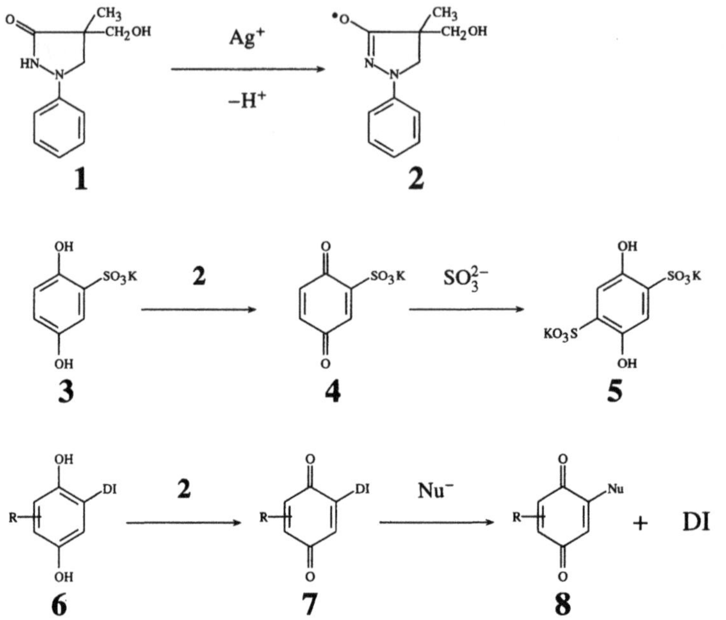

(14.1)

(14.2)

(14.3)

Figure 14.1. Releasing reactions of a DIR hydroquinone in the first B&W development of the color reversal process. The symbol DI represents a developer-inhibitor moiety or molecule. The symbol Nu represents a nucleophilic species.

The following consecutive reactions occur in the emulsion layer containing the DIR hydroquinone (**6**) and liberate development inhibitor molecules.

1. The auxiliary developer (**1**) reduces silver halides in the emulsion layer containing the DIR hydroquinone so as to cause the imagewise generation of the corresponding oxidized form (**2**), as shown in eq. 14.1.

2. The oxidized auxiliary developer (**2**) reacts with the main developer (**3**) to produce **4**, which is further reacted with a sulfite ion, as shown in eq. 14.2. The revived **1** is turned over into further redox reactions.

3. In competition with eq. 14.2, the oxidized auxiliary developer (**2**) reacts with the DIR developer (**6**) to produce the corresponding quinone form (**7**). Then, the quinone intermediate (**7**) is attacked by a nucleophile (Nu^-) such as a hydroxide ion and a sulfite ion so that a developer inhibitor molecule (DI) is released, as shown in eq. 14.3. The former redox reaction is a rate-determining step, while the latter step is very fast.

4. The released DI diffuses and inhibits the development of silver halide in its neighborhood or in other layers.

Although several general formulas of DIR hydroquinones and their properties have been reported in the review by Ishimaru [3], their concrete structures have not been fully specified. One of the general formulas is a hydroquinone with DI at the 2-position and RS at the 5-position. This may be assigned to **9** or its analog which is covered by the patent [2], since this patent has been cited in the review [3].

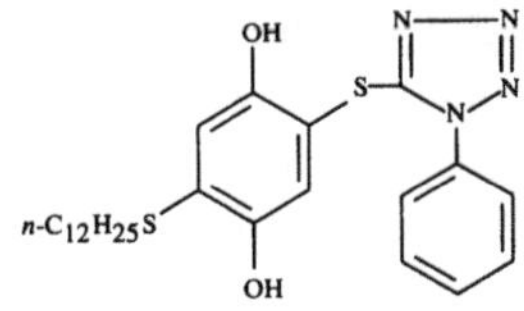

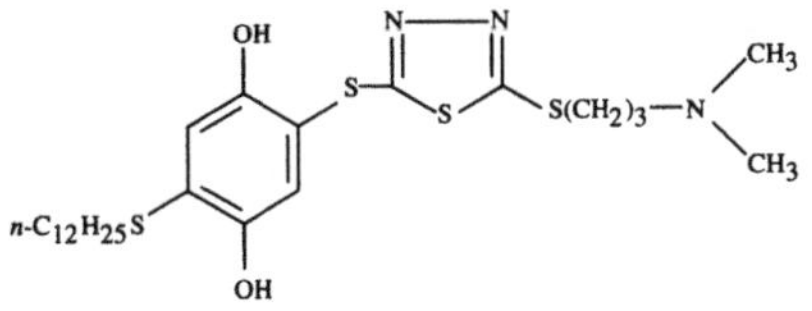

Figure 14.2. DIR hydroquinones based on monocyclic nuclei. For the releasing mechanism, see Fig. 14.1.

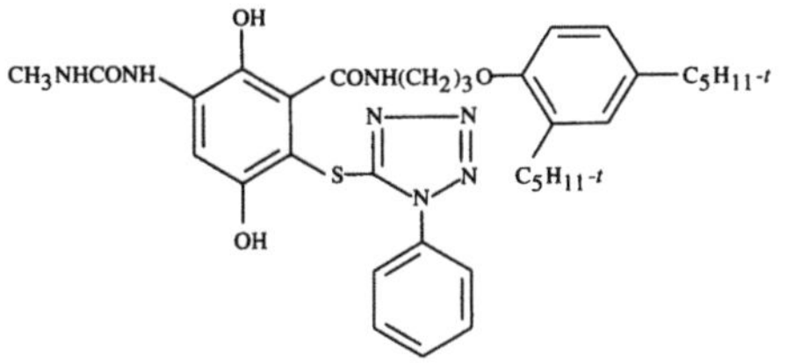

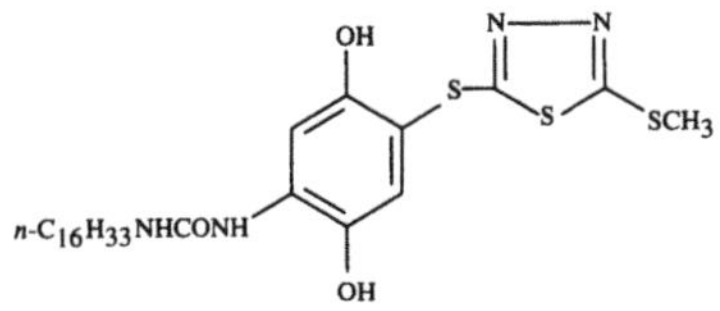

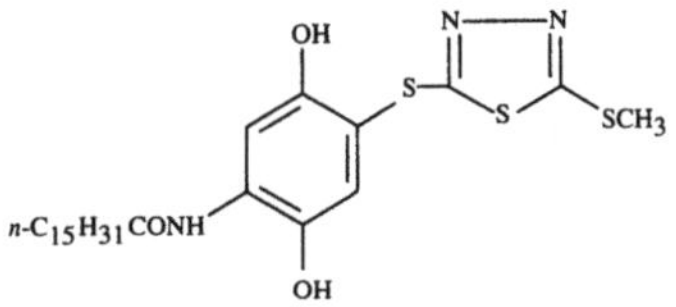

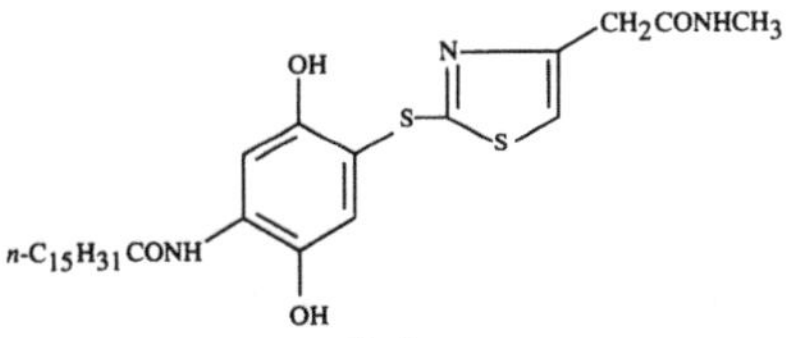

Another DIR hydroquinone **10**, which is also covered within the general formula, has been compared with **9**, where the former is superior to the latter in granularity and sharpness [5].

A further mode of substitution has been disclosed in a patent [6], e.g., **11**, after extensive studies on the influence of substituents to the redox potential. The DIR hydroquinone **11** has been compared with another DIR hydroquinone **12** having a different mode of substitution [7], where the latter is superior to the former in sharpness (edge effect) and interlayer effect. The DIR hydroquinone (**13**) has been disclosed to be effective in a photographic material containing a relatively small amount of silver halide, which has been proposed to reduce the amount of the desilvering solution [8]. The DIR hydroquinone (**13**) contains a thiazole inhibitor in place of the thiadiazole inhibitor of **14**, where this replacement has provided improved sharpness and enhanced interlayer effect [9].

Another type of general formulas are fused hydroquinone nuclei [3], which may be assigned to **15**, **16**, or their analogs, all of which are covered by the patent [6]. The DIR hydroquinones (**15** and **16**) and the n-$C_{16}H_{33}$S-homolog of **9** have been compared with a DIR hydroquinone (**17**) based on a hetero-bicyclic nucleus [7]. The latter DIR hydroquinone (**17**) has been reported to be superior to the former ones in sharpness (edge effect) and interlayer effect.

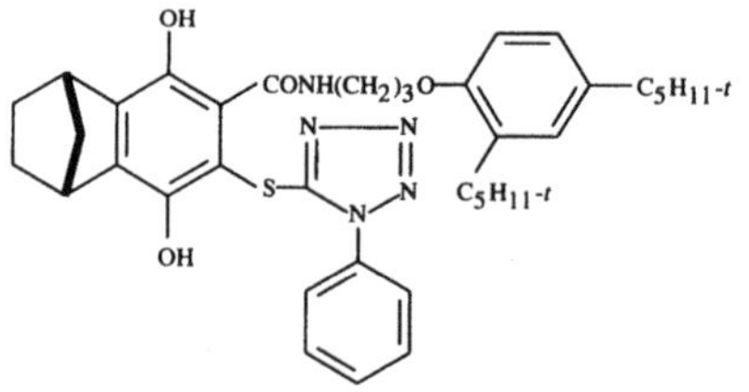

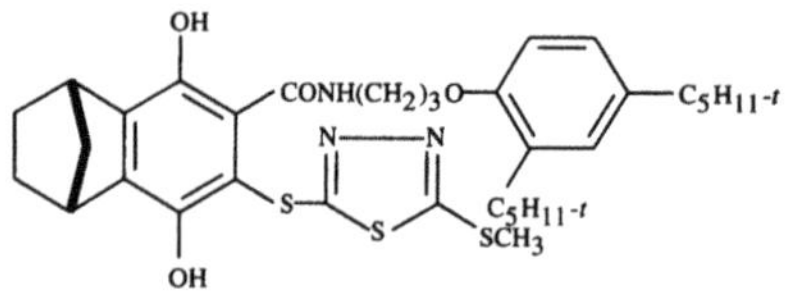

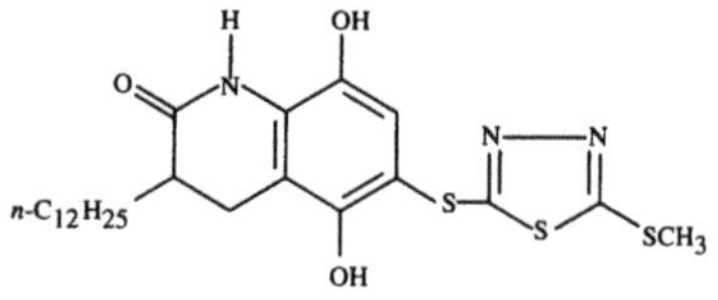

Figure 14.3. DIR hydroquinones based on fused nuclei. For the releasing mechanism, see Fig. 14.1.

In more recent embodiments of color reversal films disclosed in a patent [10], the DIR hydroquinones **12** and **14** have been used in red-, green-, and blue-sensitive emulsion layers.

Syntheses

A synthetic pathway for preparing **12** is illustrated in Fig. 14.4 [7]. 2,5-Dimethoxyaniline (**18**) is reacted with phenyl chloroformate in the presence of pyridine to give 1,4-dimethoxy-2-phenoxycarbonamidobenzene (**19**), which is condensed with 1-hexadecylamine to give the ureido intermediate (**20**). After removing the protective methoxy groups, the resulting hydroquinone derivative (**21**) is oxidized with 2,3-dichloro-5,6-dicyano-1,4-benzoquinone to give the quinone intermediate (**22**). Then, 2-mercapto-5-methylthio-1,3,4-thiadiazole is added to the quinone nucleus of **22** by a 1,4-addition mechanism to give the final product (**12**). This type of 1,4-additions to quinone derivatives has been extensively investigated by using 5-mercapto-1-phenyltetrazole as a nucleophile [11].

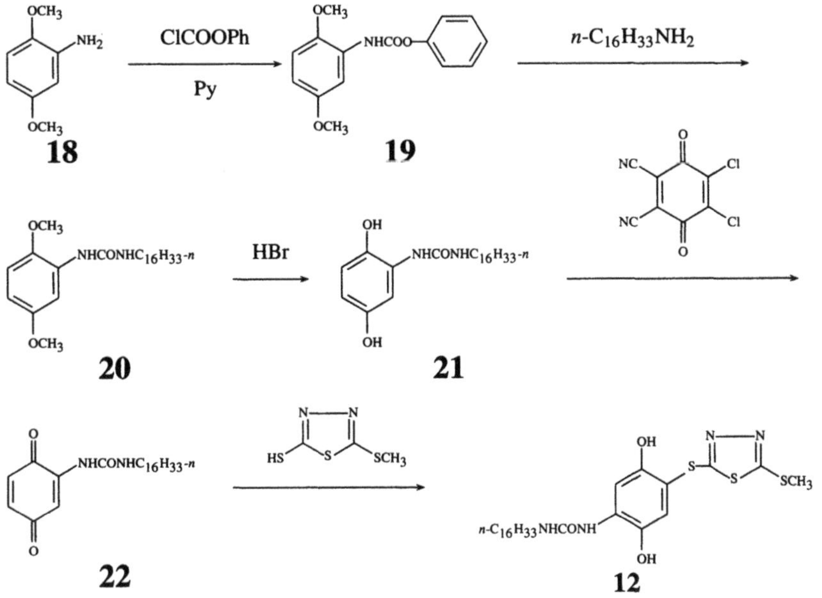

Figure 14.4. Synthetic pathway of the DIR hydroquinone **12**.

14.2 DIR Hydroquinones with Additional Functions

14.2.1 Timing DIR Hydroquinones

DIR hydroquinones can be combined with additional timing functions. Here these DIR hydroquinones are called "timing DIR hydroquinones".[2] For example, the DIR hydroquinone **23** has the function of timing releasing as an additional function [12]. Thus, the DIR hydroquinone releases a precursor that further releases a development inhibitor on the basis of the cyclization mechanism. Comparison data between this type of DIR hydroquinones and such DIR hydroquinones as **9** have been disclosed in this patent.

On the other hand, the DIR hydroquinone **24** releases another DIR hydroquinone, which is further oxidized to release a development inhibitor [13]. The DIR hydroquinone **24** has been used as a comparison compound to show the superiority of such DIR couplers as **17** [7].

14.2.2 DIRR Couplers

The function of DIR hydroquinones can be combined with the function of DIR couplers described in Chapter 13. Such couplers as having combined functions

[2]Remember the discussions on timing DIR couplers and multi-timing DIR couplers, which have appeared in Chapter 13.

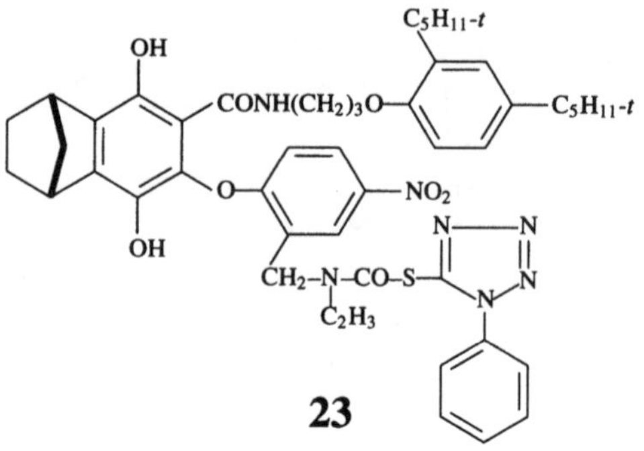

23

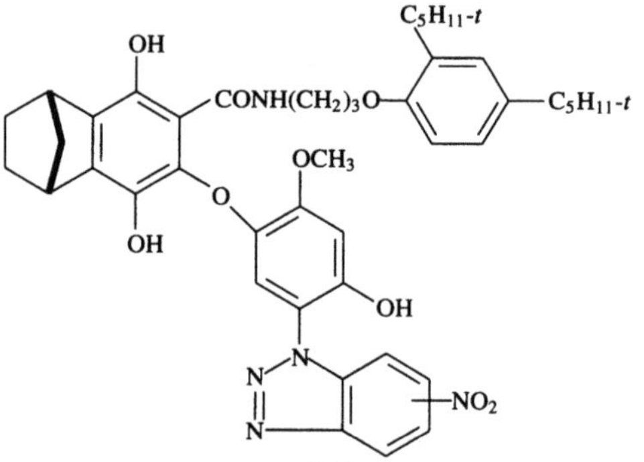

24

Figure 14.5. DIR hydroquinones with additional functions. The DIR hydroquinone **23** has the function of timing releasing based on the cyclization mechanism. The DIR hydroquinone **24** is based on a double oxidization.

are called "DIRR couplers", where DIRR is the acronym of "development inhibitor releaser releasing" [14,15].[3] For example, the DIRR cyan coupler (**25**) reacts with an oxidized color developer in a red-sensitive layer to release a DIR hydroquinone, which diffuses to another emulsion layer and reacts with an oxidized color developer to release a development inhibitor [16].

Since the released DIR hydroquinone requires the action of an oxidizing agent (e.g., the oxidized color developer), it does not interact with unexposed silver halide. Hence the DIR hydroquinone runs through the green-sensitive layer and the cyan-sensitive layer[4] without any photographic interactions and arrives at the blue-sensitive layer, at which the releasing of a development inhibitor occurs from the DIR hydroquinone.

The amount of such a DIRR coupler has been specified to be at the most 1 mol % based on the amount of silver halide in a unit layer of the highest sensi-

[3]A color negative film using a DIRR coupler has been placed on the market under the name of "Fujicolor REALA" (1988).

[4]The cyan-sensitive layer is a fourth (additional) layer sensitive to light of wavelength of ca. 520 nm [14,15].

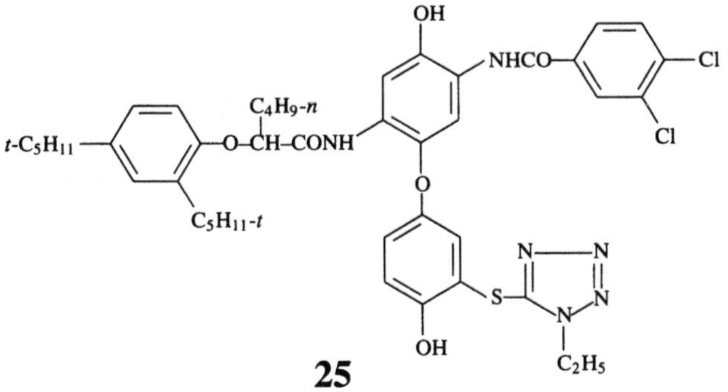

25

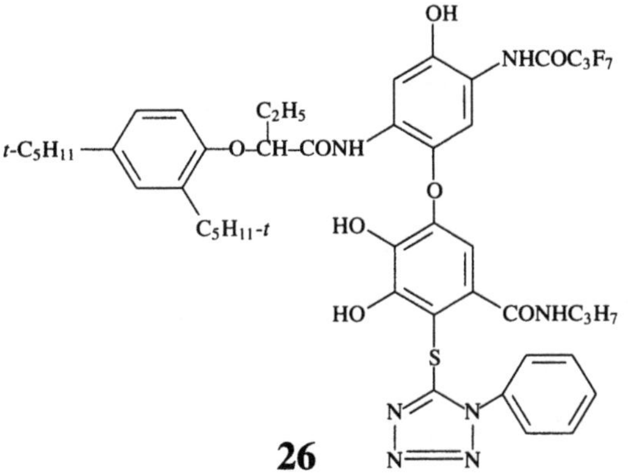

26

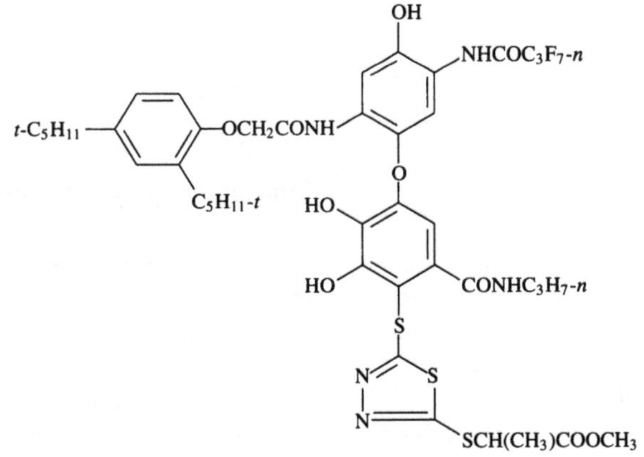

27

Figure 14.6. DIRR (development-inhibitor-releaser-releasing) couplers.

tivity, because the development inhibiting effect is too high to lower the sensitivity excessively, or to make the adjustment of the gradation remarkably difficult [17].

Similar but more complicated DIRR couplers (e.g., **26**) have been disclosed to release a gallic acid amide substituted by a DI moiety [18].

The coupler **27** is a more functionalized DIRR coupler which releases an imagewise precursor of a deactivatable development inhibitor [19]. The same compound has been used in a recent embodiment, as described in Chapter 1 (Fig. 1.36). The behavior of the DIRR coupler (**27**) is illustrated in Fig. 14.7 by using a multilayer photographic film that contains a so-called fourth emulsion layer of cyan sensitivity (Layer 7) [14,20].[5]

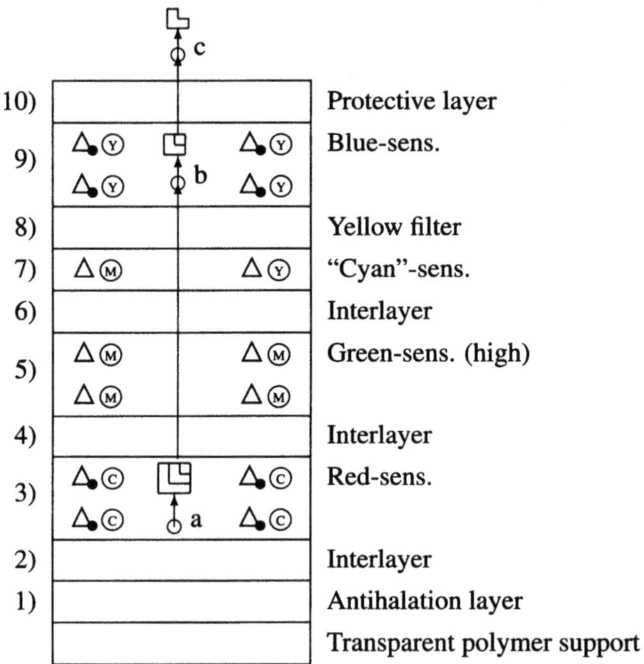

Figure 14.7. Schematic cross-section of multilayer structure with an additional layer of cyan sensitivity, where a DIRR coupler is incorporated in the red-sensitive layer. Δ: Silver halide grain; Δ•: Exposed silver halide grain with a latent image; Ⓨ: Yellow coupler; Ⓜ: Magenta coupler; ©: Cyan coupler; [symbol] : DIR hydroquinone released from the DIRR coupler; [symbol] : DI released from the DIR hydroquinone; and [symbol] : deactivated DI. Step a: Coupling-off reaction releasing the DIR; Step b: Redox reaction releasing the DI; and Step c: Hydrolysis deactivating the DI.

[5]For the fourth emulsion layer, see Subsection 7.6.2 of Chapter 7. However, the effect of the present DIRR coupler does not depend on whether the cyan-sensitive layer is present or not.

In Fig. 14.7, a DIRR coupler (e.g., **27**) is incorporated in the red-sensitive layer (Layer 3). Suppose that the red-sensitive layer (Layer 3) and the blue-sensitive layer (Layer 9) are irradiated in a region of an exposed film so as to generate latent images (△•), whereas the other photo-sensitive layers are not irradiated. The DIRR coupler (e.g., **27**) reacts with an oxidized color developer in the red-sensitive layer (Step a), where the DIR hydroquinone (⊡) is released.[6] The DIR hydroquinone diffuses through the green-sensitive layer, the cyan-sensitive layer, and the interlayers without releasing a DI, so that it reaches the blue-sensitive layer. There in Step b, it is oxidized by an oxidized developer to release the DI (⊡), which inhibits the development of the blue-sensitive emulsion. The DI is dissolved into the development solution and deactivated by hydrolysis (Step c), releasing the deactivated species (⊡).

The effect of the DIRR coupler (e.g., **27**) shown in Fig. 14.7 results in color correction for a cyan dye having a side absorption in the blue region of light. This should be compared with another color correction method using a yellow-colored cyan coupler, which has been discussed in Chapter 12 (Subsection 12.2.2).

Syntheses

A synthetic pathway for preparing the DIRR coupler **26** is illustrated in Fig. 14.8 [18]. The *o*-aminophenol moiety of one starting material is protected by the formation of a benzoxazole ring in 5-chloro-6-nitro-2-*t*-butylbenzoxazole (**28**). The two phenolic hydroxyl groups of a gallic acid ester is protected by the formation of a ketal ring in propyl 3,4-(diphenylmethylenedioxy)-5-hydroxybenzoate (**29**).[7] After these protections, the benzoxazole (**28**) and the protected gallic acid ester (**29**) are condensed in the presence of cuprous chloride to give **30**. Note that the chlorine atom of **28** is activated by the substitution of the nitro group at the *o*-position. The ester group and the oxazole ring of the intermediate (**30**) are hydrolyzed to give **31**. The amino group of **31** is reacted with heptafluorobutanoic acid anhydride and then the nitro group is reduced into an amino group to give another intermediate **32**. The amino group of **32** is reacted with the acyl chloride function of a ballast group, i.e., α-(2,4-di-*t*-amylphenoxy)butyryl chloride, so as to produce **33**. When the carboxylic acid group of the intermediate (**33**) is chlorinated with thionyl chloride and is reacted with propylamine, the amide intermediate is obtained. The ketal protective ring of the amide intermediate is then removed under an acidic condition. The resulting intermediate is reacted with 1-phenyl-1H-tetrazole-5-sulfenyl chloride which is derived from 1-phenyl-5-mercapto-1H-tetrazole. Thereby, the target DIRR coupler (**26**) is obtained.

---

[6]In this example, the DIR hydroquinone represented by the symbol is a gallic acid amide derivative having a deactivatable DI moiety. The DIR hydroquinone corresponds to the DIR (development inhibitor releaser) of the acronym DIRR so that it does not release the DI but it is capable of releasing the DI when oxidized in agreement of the definition.

[7]For an alternative name, we can use propyl 2,2-diphenyl-1,3-benzo[*d*]dioxole-5-benzoate.

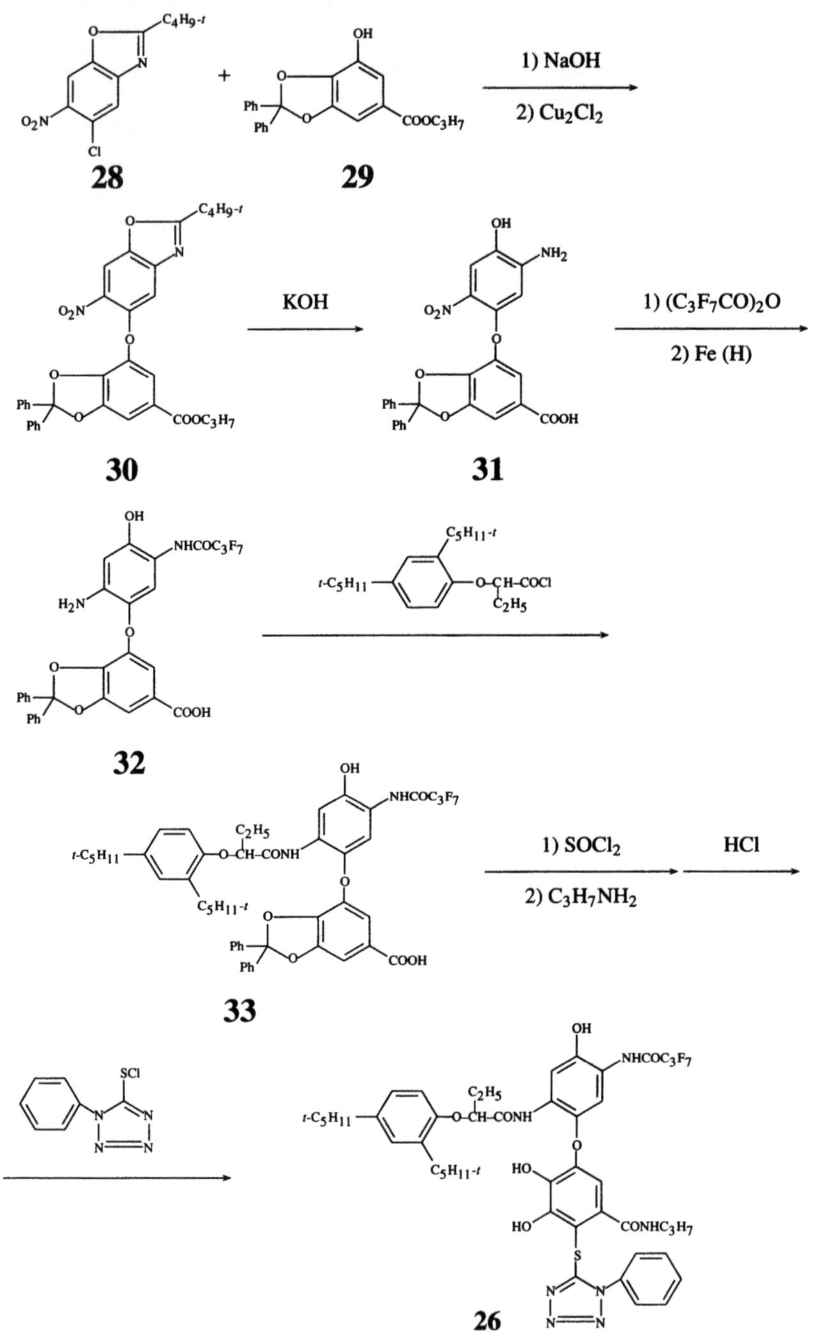

28 + **29**

1) NaOH
2) Cu_2Cl_2

30

KOH

31

1) $(C_3F_7CO)_2O$
2) Fe (H)

32

33

1) $SOCl_2$
2) $C_3H_7NH_2$

HCl

26

Figure 14.8. Synthetic pathway of the DIRR coupler **26**.

14.3 DIR Hydrazines

14.3.1 Applications to Chromogenic Photography

In addition to hydroquinone derivatives, hydrazines have been well-known as another type of B&W developing agents. Their derivatives such as hydrazides have been also known to have reductive power so as to work as nucleating or fogging agents. Hence it has been natural that they have been examined to embody DIR functions. For example, the DIR hydrazines (**34** and **35**) have been disclosed to release a development inhibitor (an indazole or a benzotriazole derivative) upon oxidation [21]. A tetrazole-type inhibitor (1-phenyl-5-mercapto-1H-tetrazole) has been reported to be released from the DIR hydrazine (**36**) and to improve sharpness just as a DIR coupler does [22].

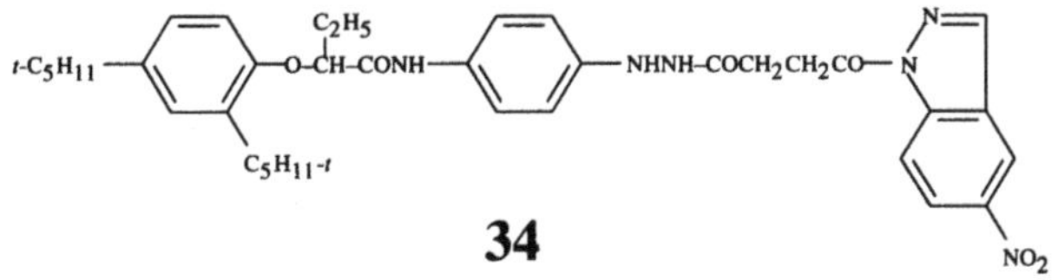

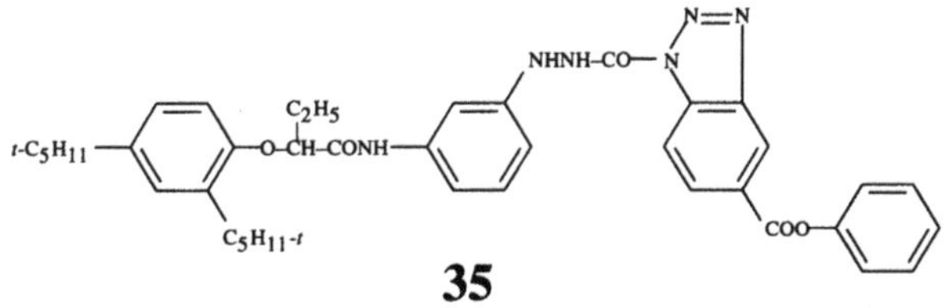

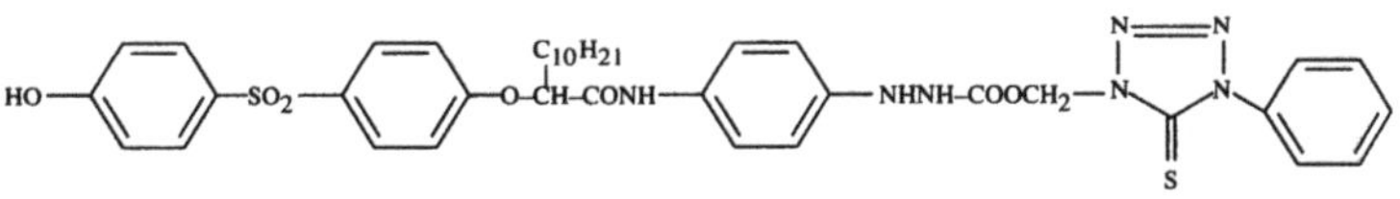

Figure 14.9. DIR hydrazines.

14.3.2 Application to Graphic Arts Films

DIR hydrazines have been successfully used in the hydrazide-nucleated infectious development of films for graphic arts [23–26]. To understand the effect of the DIR hydrazines, we shall start with essential items on the hydrazide-nucleated infectious development.

Films for graphic arts must discriminate both large and small dots from the colorless background as sharply as possible.[8] For this purpose, nucleating agents have been incorporated in films for graphic arts [27,28,29]. Thereby, the films can be developed by using a superadditive developer solution containing methol-hydroquinone along with sulfite stabilizers.[9] This technique is called "hydrazide-nucleated infectious development".

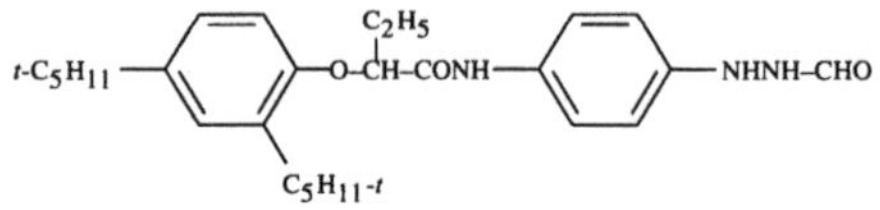

37

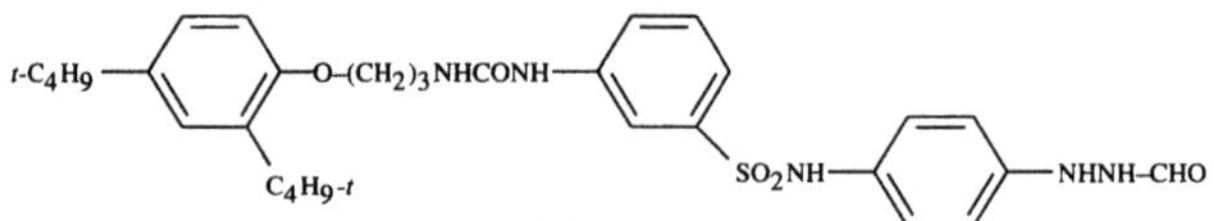

38

Figure 14.10. Nucleating agents.

Representative nucleating agents are hydrazide compounds (e.g., **37** [30] and **38** [25]), where they react with silver halide according to the following equations:

$$\mathrm{HQ + Ag^{+} \rightarrow Q + Ag} \quad (14.4)$$

$$\mathrm{Q + Ar{-}NHNH{-}CHO \rightarrow HQ + Ar{-}N{=}N{-}CHO} \quad (14.5)$$

$$\mathrm{Ar{-}N{=}N{-}CHO + OH^{-} \rightarrow Ar{-}N{=}N{-}H + HCOO^{-}} \quad (14.6)$$

$$\mathrm{(AgX)_n + Ar{-}N{=}N{-}H \rightarrow (AgX)_{n-2}(Ag_2) + Ar{-}H + N_2} \quad (14.7)$$

An exposed silver halide oxidizes hydroquinone (HQ) into benzoquinone (Q), which in turn oxidizes the nucleating agent (Ar–NHNH–CHO) into a formyl-diazene (Ar–N=N–CHO). This is attacked by a hydroxide ion to be hydrolyzed into a diazene Ar–N=NH, which acts as a nucleating agent and produces silver metal on the surface of an unexposed silver halide grain represented as $(AgX)_n$. The silver metal (i.e., $(AgX)_{n-2}(Ag_2)$ on the right-hand side of eq. 14.7) works

[8]Contrasty negative films are essential to printing, since halftone images in graphic arts are converted into dot images. Moreover, photocomposed letters contained in the originals with the halftone images also require contrasty negative films. Although lith development had long been used until the hydrazide-nucleated infectious development was embodied, the lith developer solutions were unstable and difficult in long-term processing because sulfite stabilizers could not be used to maintain contrasty development.

[9]Without nucleating agents, sulfite stabilizers cannot be added, since sulfite species attack benzoquinones and stop lith-type infectious development. As a result, the developer solution is so unstable as to give unreliable results. In the presence of the nucleating agents, a stable developer solution containing the sulfite stabilizers can be used.

as a development center so that the silver halide can be developed although it has been unexposed.

Although the hydrazide-nucleated infectious development has given a more satisfactory dot image quality than the lith development, there has been a demand for a system having a broader latitude to maintain reproducibility of halftone gradation during enlargements or reduction of a dot print. In particular, a counter of typeface and a small-sized colorless area surrounded with black images (e.g., large dots) should remain colorless during the reduction of print size.

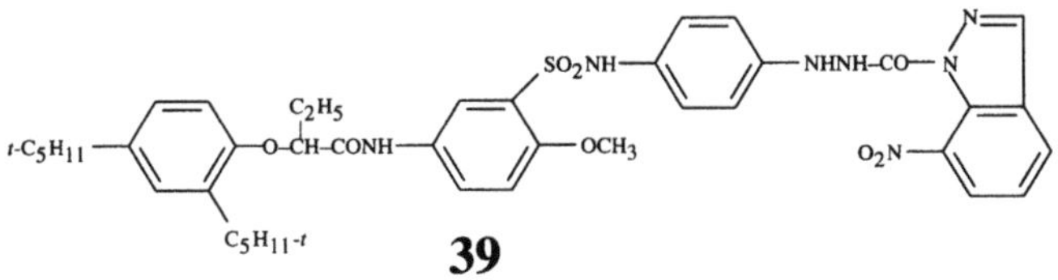

Figure 14.11. DIR hydrazines.

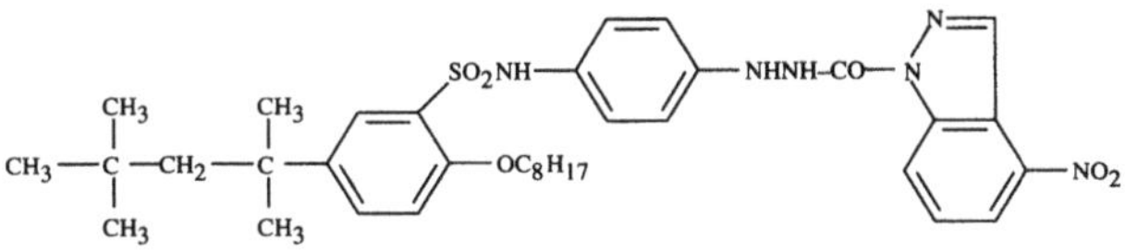

DIR hydrazines such as **39** have been used for this purpose, as disclosed in a patent [31]. Since 4- and 7-nitroindazole have been found to quench the diazene (eq. 14.6) by oxidation, either of them has been incorporated into a DIR hydrazine to inhibit the nucleation represented by eq. 14.7. Thereby, the subsequent nucleating development has not occurred.

A more specific structural formula **40**, which has been covered by the patent [31], has been reported in a review by Katoh [25]. The DIR hydrazine (**40**) releases 4-nitroindazole as a development inhibitor (more precisely a nucleation inhibitor). The 4-nitroindazole is found to be superior to other nitroindazoles, because it can be smoothly deactivated by reduction and sulfonation [25].[10] The resulting sodium 4-aminoindazole-7-sulfonate has no nucleation effect so as to prevent the seasoning of the developer solution.

Figure 14.12 shows a schematic cross-section of a film and its top view during hydrazide-nucleated infectious development. The film comprises an image layer containing nucleating agents (Ⓝ) and silver halide emulsions of high sensitivity, a timing layer, a DIR layer containing DIR hydrazine molecules (Ⓗ) and silver halide emulsion of low sensitivity, and a protective layer. The central area corresponds to the white area surrounded by the dotted area in the top view. This white

[10]The same type of deactivation occurs in the case of 7-nitroindazole. However, the incorporation of the 7-nitroindazole into a DIR hydrazine has been reported to be difficult synthetically [25].

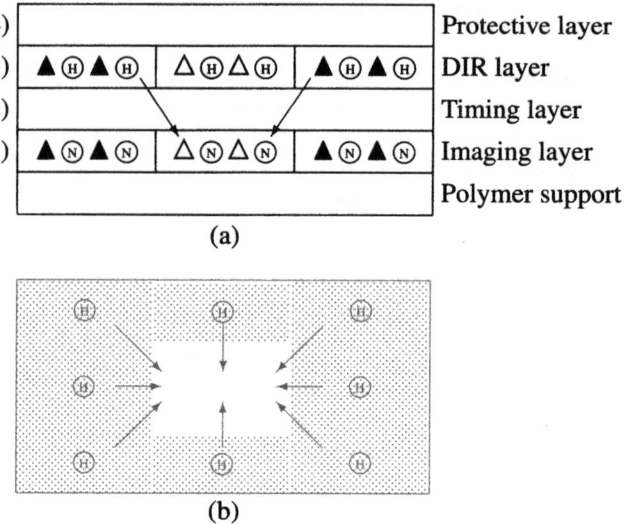

Figure 14.12. Schematic cross-section of a film of graphic arts (a) and its top view (b), where a model counter of type affected by development inhibition (arrows) is represented by a central white box. The symbol Ⓗ denotes a DIR hydrazine, from which an arrow is depicted to represent the release of a development inhibitor (or a nucleation inhibitor). The symbol Ⓝ denotes a nucleating agent.

area represents a model counter of typeface, which is affected by a nucleating development inhibitor released from DIR hydrazine molecules (Ⓗ). The released inhibitor diffuses into the counter area, as represented by arrows.

Recently, the pH of the developer solution has been required to be lowered from 11.9 to 10.6, where a more active DIR hydrazine even under pH 10.6 has been developed, as reported in a review [26].

References

[1] Porter RF, Schwan JA, Gates Jr JW (1968) US Patent 3 379 529

[2] Shiba K, Hirose T, Aono T, Ohi R, Shishido T (1976) US Patent 3 930 863

[3] Ishimaru S, Ikeda H, Sakagami M, Miyazaki K, Hirano S, Tamano J (1992) J Soc Photogr Sci Technol Jpn. 55:174

[4] Shuto S, Kuramitsu M, Kuwashima S, Bando S (1998) Fujifilm Res & Dev. 43:1

[5] Kojima T, Hirano S, Usui H (1988) US Patent 4 791 049

[6] Nakamura K, Hirano S, Ikeda T, Mihayashi K, Ono M, Takahashi T, Kuwabara K, Yagihara M, Itoh I (1988) US Patent 4 740 453

[7] Ono M, Hanaki K, Sakanoue K, Hirano S, Yamamoto M (1993) US Patent 5 210 012

[8] Shuto S, Ozawa T (1994) US Patent 5 310 638

[9] Takeuchi H, Ichikawa S, Matsumoto K (2001) US Patent 6 184 386 B1

[10] Tateishi K, Mikoshiba H, Matsuda N (2002) US Patent 6 399 291 B1

[11] Porter RF, Rees WW, Frauenglass E, Vilgus III HS, Nawn GH, Chiesa PP, Gates Jr JW (1964) J Org Chem. 29:588; Vilgus III HS, Frauenglass E, Jones ET, Porter RF, Gates Jr JW (1964) J Org Chem. 29:594

[12] Nakamura K, Hirano S, Takahashi O (1988) US Patent 4 770 990

[13] Ichijima S, Hirano S, Nakamura K, Mihayashi K (1988) US Patent 4 770 982

[14] Ichijima S (1989) J Soc Photogr Sci Technol Jpn. 52:145

[15] Sasaki N, Takahasi K, Ikoma H (1989) J Soc Photogr Sci Technol Jpn. 52:41

[16] Ichijima S, Usui H, Deguchi N (1986) US Patent 4 618 571

[17] Mihayashi K, Ichijima S (1990) US Patent 4 948 716

[18] Ichijima S, Mihayashi K (1991) US Patent 4 985 336

[19] Matsumoto K, Ito T (2002) US Patent 6 337 176 B1

[20] Ikegawa A, Ohashi Y, Okazaki M (1993) US Patent 5 198 332

[21] Itoh I, Ichijima S, Hirano S (1986) Jpn Kokai S61-213847

[22] Harder W (1987) US Patent 4 684 604

[23] Okamura H (1992) J Soc Photogr Sci Technol Jpn. 55:211

[24] Katoh K (1994) J Soc Photogr Sci Technol Jpn. 57:308

[25] Katoh K (1997) Nippon Kagaku Kaishi. 467

[26] Yasuda S, Ezoe T, Taniguchi M, Ito T, Yamada K (2003) J Soc Photogr Sci Technol Jpn. 66:179

[27] Mifune H, Takada S, Akimura Y, Shishido T (1979) US Patent 4 166 742

[28] Takada S, Akimura Y, Mifune H, Tsujino N (1979) US Patent 4 168 977

[29] Takada S, Akimura Y, Mifune H, Tsujino N (1980) US Patent 4 224 401

[30] Katoh K, Yagihara M (1990) J Soc Photogr Sci Technol Jpn. 53:495

[31] Okamura H, Kawamoto H, Matsumoto K, Katoh H (1993) US Patent 5 204 214

Part IV

Diffusion Transfer Photography

Chapter 15

Silver-Salt Diffusion Transfer Photography

15.1 B&W Diffusion Transfer Process

The present chapter is devoted to the B&W diffusion transfer reversal (DTR) process and its applications before the next chapter on color DTR will be started.

15.1.1 DTR Copy Print

The diffusion transfer reversal (DTR) process was disclosed by Rott (1939) and by Weyde (1941), where a physical development was applied to silver image formation [1]. The DTR process was first applied to copy printing so that it was placed on the market under the names Copyrapid (Agfa, 1949), Gevacopy (Gevaert, 1950), Hishirapid (Mitsubishi Paper Mill, 1956), etc. However, their distributions have now been stopped because of the advance of electronic copying systems.

The DTR copy print gives a positive image according to the following steps:

1. **Exposure**: A negative film is exposed by reflection light from an original picture.

2. **Processing**: The film is dipped into a development fluid.

3. **Transfer**: The wet film is superimposed on a transfer sheet in a face-to-face fashion. Since the transfer sheet is coated with a silver halide solvent such as sodium thiosulfate, a solution physical development takes place to give a positive image on the transfer sheet.

4. **Peeling apart**: The positive image on the transfer sheet is obtained by peeling apart the transfer sheet from the negative film.

Since the mechanism of the solution physical development is primarily the same as that of instant B&W photography, it will be described in the next subsection.

15.1.2 Instant B&W Photography

Procedures and Mechanisms

The DTR process was next applied to instant B&W photography by Land so that the system based on a so-called Land's camera was placed on the market in 1948 under the name TYPE 41 (Polaroid Black and White speed 100) [2,3]. The procedures for obtaining photographs with the system are shown in Fig. 15.1.

1. **Exposure**: Expose the film with a Land's camera (Fig. 15.1b). For the sake of simplicity, the left region of the film is exposed ($h\nu$), while the right region is unexposed, although the intensity of light has gradation according to a real object.

2. **Processing**: Pull the tag attached to the exposed film so that the film is superimposed onto a receiving paper with a pod including an activator fluid (Fig. 15.1c). Pull further the combined unit through a pair of rollers so as to spread the activator fluid between the film and the paper of the combined unit. Then processing is allowed to proceed during about 1 min and a silver image appears on the receiving paper (Fig. 15.1d).

3. **Peeling apart**: Peel the combined unit apart into the film and the receiving paper, the latter of which includes the positive image of silver (Fig. 15.1e).

The processing shown in Fig. 15.1c and 15.1d takes the following steps:

1. **Chemical development**: Negative development occurs in the film. Exposed silver halide (△•) is reduced into silver metal (▲), while unexposed silver halide remains undeveloped (△), as shown in Fig. 15.1c.

2. **Silver salt dissolution**: The unexposed silver halide is dissolved by a silver halide solvent (usually, sodium thiosulfate):

$$2AgCl + 2Na_2S_2O_3 \rightarrow 4Na^+ + Ag(S_2O_3)_2^{3-} + Cl^- \qquad (15.1)$$

3. **Diffusion transfer**: The dissolved salt $Ag(S_2O_3)_2^{3-}$ diffuses through the spread fluid (as shown by the symbol ↑ in Fig. 15.1c) and reaches the receiving paper.

4. **Physical development**: The salt on the receiving paper is developed around physical development nuclei (···) to deposit positive silver images (•), as

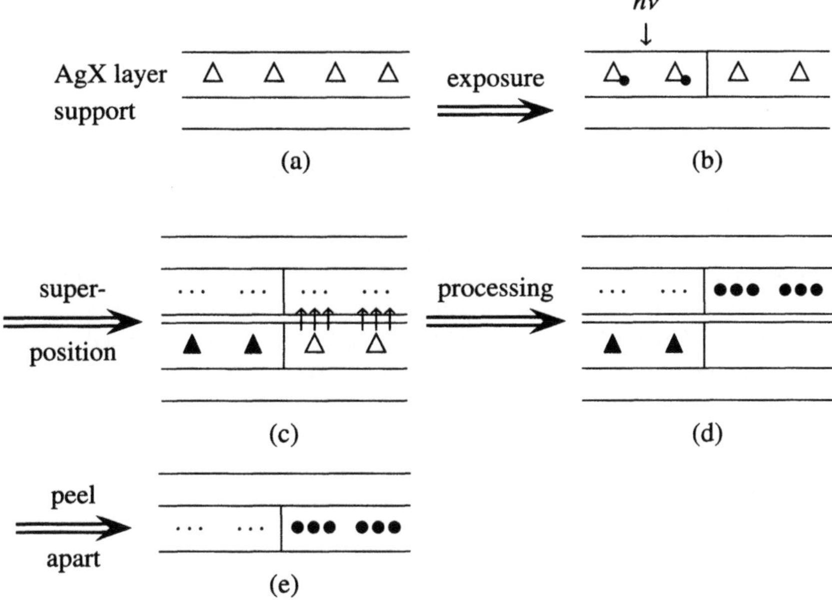

Figure 15.1. Film cross-sections during B&W DTR process. (a) B&W DTR film set for exposure, (b) exposure of the film, (c) superposition of a receiving paper on the exposed film and the beginning of the processing, (d) the end of the processing, and (e) the receiving paper is peeled apart, where the layers contain silver halide and related substances represented by the following symbols, △: unexposed AgX, △•: exposed AgX with a latent image, ▲: developed silver, ···: physical development nuclei, and •: silver image physically developed.

shown in Fig. 15.1d:

$$\mathrm{Ag(S_2O_3)_2^{3-}} + \mathrm{HO{-}C_6H_4{-}OH} + \mathrm{Na_2CO_3}$$

$$\longrightarrow \mathrm{Ag} + \mathrm{O{=}C_6H_4{=}O} + 2\mathrm{Na^+} + 2\mathrm{S_2O_3^{2-}} + \mathrm{H_2O} + \mathrm{CO_2} \qquad (15.2)$$

As found in Fig. 15.1, the silver image deposited is positive, since the silver image is generated in unexposed regions of the receiving paper. Moreover, the image is proper with respect to the left-and-right relationship; i.e, no reflection occurs as a result of the peel-apart operation.

Silver Halide Solvents

One of the most commonly employed silver halide solvents is sodium thiosulfate as shown in eq. 15.1. Thiocyanates (e.g., potassium and sodium thiocyanate), mercaptanes (e.g., mercaptoacetic acid and cysteine), alkali metal cyanides (e.g., potassium cyanide), and cyclic imides (e.g., barbituric acid and uracil) are well-known as other types of silver halide solvents. Compounds containing sulfonyl groups have been disclosed as further types of silver halide solvents. For example, β-disulfones (e.g., 1,1-bis(methanesulfonyl)ethane) [4], sulfinyl-sulfonylalkanes (e.g., $CH_3SO–CH_2–SO_2CH_3$) [5], piperazines having a β-disulfone moiety (e.g, **1**) [6], and sulfonyl ylides (e.g, **2**) [7] have been proposed to be used in instant B&W photography. Cyclic β-disulfones (e.g., **3**) have been also reported to be useful [8].

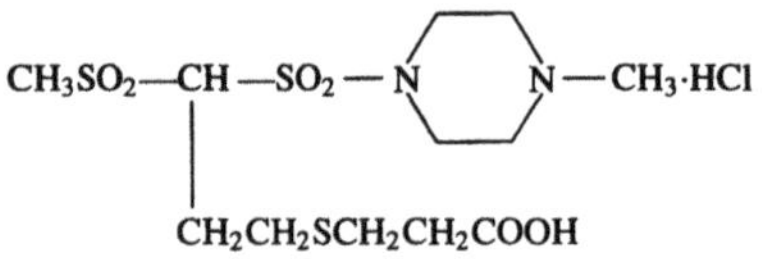

CH_3SO_2—CH—SO_2—N N—CH_3·HCl
$CH_2CH_2SCH_2CH_2COOH$
1

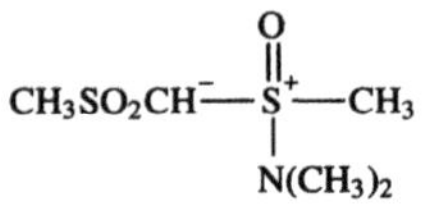

$CH_3SO_2CH^-$—S^+(=O)—CH_3
$N(CH_3)_2$
2

H SCH_3
O_2S SO_2
3 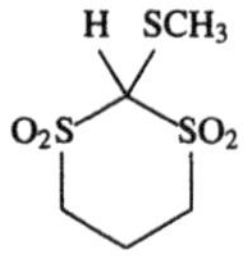

Figure 15.2. Silver halide solvents.

15.1.3 DTR Plate

The DTR process has been applied to a planographic printing plate [9,10]. Planographic printing plates based on the DTR process (especially direct positive sheets) have been placed on the market under the names Supermaster (Agfa), Silvermaster (Mitsubishi Paper Mills), and others.[1]

The procedures for a planographic printing system due to the DTR process are illustrated in Fig. 15.3.[2]

1. **Exposure**: Expose the film (Fig. 15.3b). For the sake of simplicity, the left region of the film is exposed (*hv*), while the right region is unexposed.[3]

2. **Development**: Negative development (chemical development) occurs in the exposed area to give silver metal (▲), which is so hydrophilic as to repel

[1]The planographic printing plates are frequently called "offset printing plates". As such printing plates, there are other predominant ones, which this book does not deal with. For example, a presensitized plate (PS plate) is a representative, which is based on photosensitized polymers coated on a aluminum support.

[2]The film is of one-sheet type. A film of two-sheet type is also disclosed.

[3]In the printing of letters, the intensity of light for such a planographic printing plate has no gradation.

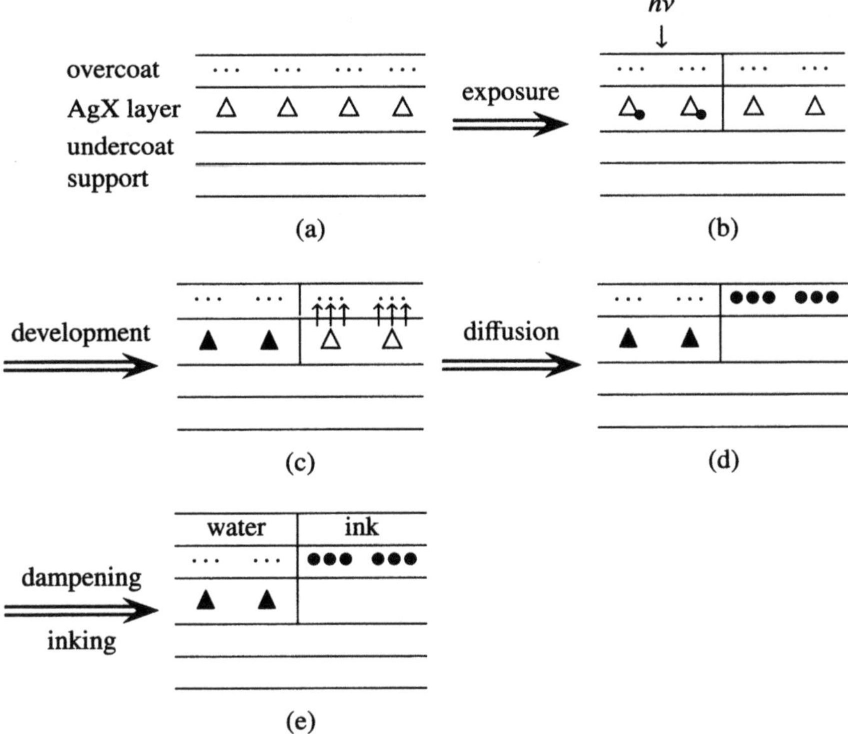

Figure 15.3. Cross-sections for a planographic printing plate based on DTR process. (a) DTR plate set for exposure, (b) exposure of the plate, (c) chemical development and diffusion (d) physical development, and (e) dampening and inking, where the layers contain silver halide and related substances represented by the following symbols, △: unexposed AgX, △•: exposed AgX with latent image, ▲: developed silver, ···: physical development nuclei, and •: silver image physically developed.

oily ink after dampening. The unexposed silver halide (△) is dissolved by sodium thiosulfate and diffuses to the overcoat layer of the film (Fig. 15.3c).

3. **Physical development**: The silver salt reaching the overcoat layer is physically developed by virtue of physical development nuclei (···) so as to give a silver image (• • •), as shown in Fig. 15.3d. The silver image which is naked on the surface has an oleophilic property.

4. **Printing**: After dampening the chemically developed silver, the oleophilic area with the physically developed silver is covered by oily ink, which is transferred onto a sheet of paper (Fig. 15.3e).

15.2 Additive Color Photography

15.2.1 Instant Color Movie

The instant color movie system named Polavision (Polaroid, 1977) has been based on B&W DTR photography and additive color reproduction [11], where a compact multipurpose cassette [12] and a unique projector [13] are characteristic of the motion picture system. Although its commercial lifetime has already been terminated because of the advance of electromagnetic products [14], its technical achievements should be mentioned for further applications to other fields.

Image Formation

The processing of instant color movie film is illustrated in Fig. 15.4 [15,16].

1. Figure 15.4a shows the cross-section of instant color movie film. An additive screen having fine striped structure (B, R, and G) is coated on the bottom support [15]. To guard the screen from an alkali processing solution, an alkaline guard layer is coated on the screen. The positive image receiver layer contains heavy metal salts as nuclei for physical development. The emulsion layer contains silver halide grains for the B&W DTR process. The top overcoat layer contains a bleachable antihalation dye and a precursor of a silver image stabilizer.

2. Figure 15.4b shows the exposure of red light. The upward arrows illustrate that the red light reaches the silver halide emulsion layer in the region of the red filter (△•) but does not reach the silver halide layer in the remaining regions (△).

3. Figure 15.4c shows the application of development agents (long downward arrows), which diffuse into the inner layers but do not attack the additive screen by virtue of the alkali guard layer.

4. First, a chemical development causes the rapid formation of silver metals of low covering power (▲) in the region exposed to the red light (Fig. 15.4d). Then, undeveloped silver halide in the unexposed regions is dissolved with a silver halide solvent so as to diffuse as shown by short arrows.

5. The dissolved silver halide reaches the positive image receiver layer, where it is physically developed to give silver metal images of high covering power.[4]

[4]The structures of negative and positive image silver are compared by using micrographs [11]. The image silver of a grain developed to form a positive deposit has greatly increased its area and hence its covering power. On the other hand, a silver halide grain developed to form negative silver covers roughly the same area as an undeveloped grain.

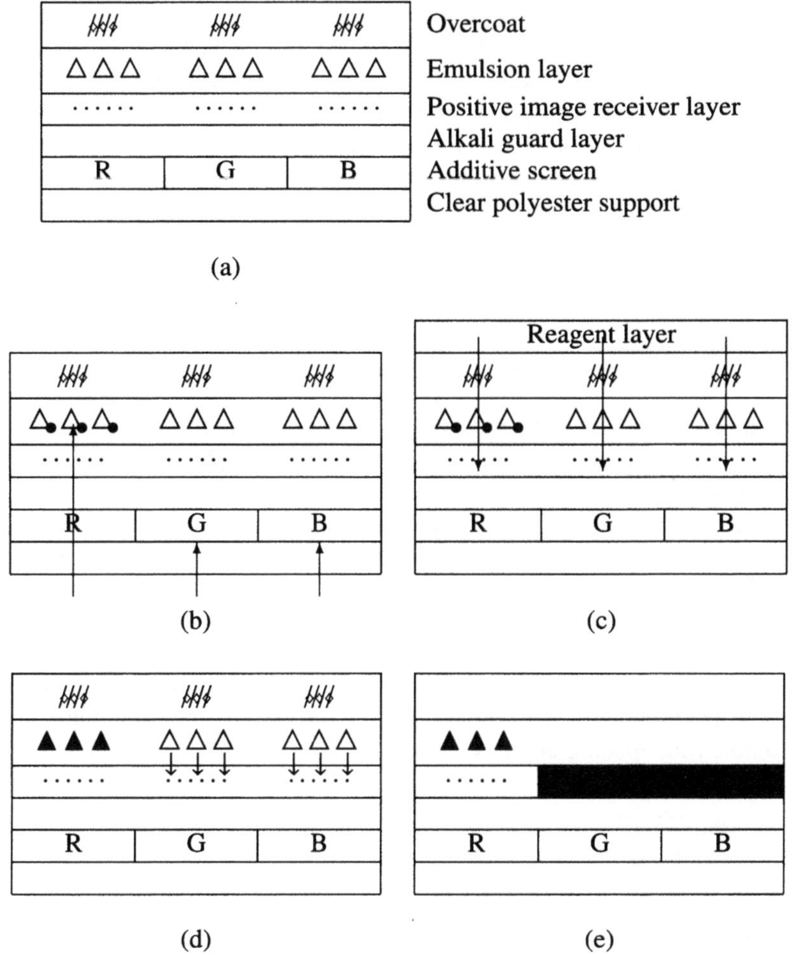

Figure 15.4. Schematic cross-sections during the processing of an instant color movie: (a) film before exposure, (b) exposure to red light, (c) application of development reagents, (d) chemical development and diffusion, and (e) physical development, where the layers contain silver halide and related substances represented by the following symbols, △: unexposed AgX, △•: exposed AgX with latent image, ▲: developed silver, ···: physical development nuclei, and ▬: silver image physically developed. The symbol /// represents an antihalation dye, and ◇ represents a precursor of a silver image stabilizer. The additive screen consists of red (R), green (G), and blue (B) filters.

6. During and after the chemical and the physical development, the antihalation dye is bleached to become colorless and the precursor of a stabilizer (e.g., gold mercaptobenzimidazole) releases a stabilizer to stabilize the resulting silver images [16].

The developed film is loaded in a projector. As found easily in Fig. 15.4e, the projection light is applied from the top (antihalation) layer, where the antihalation dye has been bleached. Because the projected light runs through the low covering silver halide (▲) and through the red filter, red light is observed. On the other hand, the projected light that runs through the regions without ▲ in the emulsion layer is interrupted by the high-covering silver halide (▬) in the positive image receiver layer.[5] Thereby, green and blue lights cannot be observed. Note that the triplet of filters (R, G, and B) shown in Fig. 15.4 corresponds to one color unit for additive color reproduction.

Developing Agents

The invention of nonstaining developing agents has been a crucial point in the instant color movie as well as in instant color photography described in the following chapters. Moreover, the instant color movie has required more elaborate developing agents, because it is based on the difference in covering powers between negative silver images and positive (physically developed) silver images [17]. Hence the development agents should give silver images of low covering power in the first chemical development, whereas they should generate silver images of high covering power in the second physical development. To maximize the difference between the two developments, tetramethylreductic acid (**4**) has been selected as a developing agent [18,19]. This developing agent undergoes an irreversible ring opening upon oxidation:

4 → **5** —[O]→ **6** —alkali→ **7** (15.3)

The positive image silver should maintain neutral tone over its entire density range. In general, however, neutral gray silver image has low covering power, whereas high covering silver image (e.g., colloidal silver) tends to have a red or brown tinge. In order that the positive silver image is neutral gray and has high covering power, the image silver is made aggregated to form a very thin, mattress-

[5] Negative silver images of low covering power have a density of about 0.3 but physically developed silver images have a density of about 3.

like continuum. As a result, it acts like Maxwellian reflectors rather than black absorbers, because the distances between silver aggregates are very short [11].

Striped Additive Screen

The additive screen used in the Polavision film has a striped structure with 1500 triplets (4500 lines) per inch. The procedure for forming the additive screen on an instant movie film is illustrated in Fig. 15.5 [11].

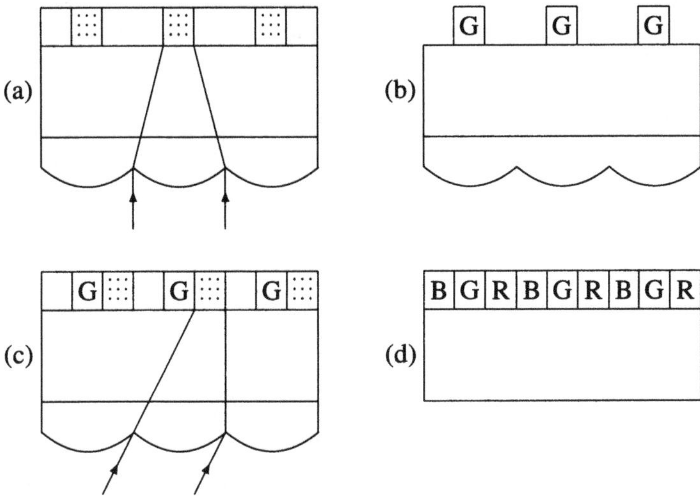

Figure 15.5. Schematic representation of stages in the formation of the striped additive screen.

1. As illustrated in Fig. 15.5a, the back of a transparent film with lenticules is coated with dichromated gelatin. The dichromated gelatin layer of the film is exposed through the lenticules, which are formed to give 1500 stripes per inch (60 stripes per mm). Thereby, 1500 stripes are hardened in the dichromated gelatin layer.
2. The unexposed gelatin is washed away, and the lines that remain are dyed by a coloring substance with a primary color (e.g., green), as illustrated in Fig. 15.5b.
3. The resulting film with the stripes is coated with dichromated gelatin, and the undyed regions of the film are hardened by lenticular exposure, washed, and dyed with another dye (e.g., red), as illustrated in Fig. 15.5c. This procedure is repeated to form stripes dyed with the remaining primary color (e.g., blue).
4. The lenticules are removed to complete an ultrafine array of alternating color stripes in the pattern BGRBGR and so on, as illustrated in Fig. 15.5d.

The resulting film (Fig. 15.5d) is further coated with an alkali guard layer, a positive image receiver layer, an emulsion layer, and an overcoat to complete an instant color movie film shown in Fig. 15.4a.

15.2.2 Instant Color Slide

The instant color slide system placed on the market by Polaroid in 1982 is an application of the instant color movie [20]. A schematic cross-section of the slide film is shown in Fig. 15.6a, which has two additional layers (a release layer and a protector layer) in comparison with the movie film shown in Fig. 15.4. The exposure and processing processes using this film are substantially the same as those of the instant color movie (Fig. 15.4) except that the slide film is separated at the interface between the additional two layers. This means that the chemical developed silver image can be removed at the time of the separation so as to give the completed slide film (Fig. 15.5b), in which the region exposed to red light is depicted.

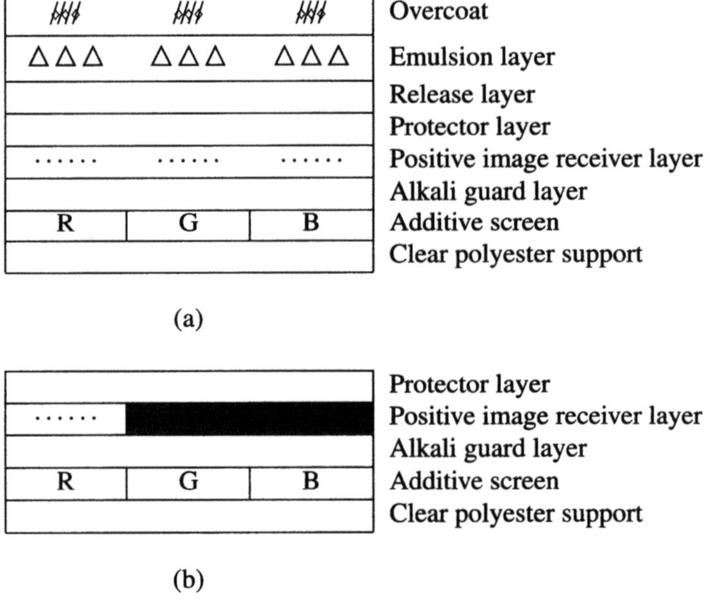

Figure 15.6. Schematic cross-sections of an instant color slide: (a) a slide film before exposure and (b) a slide film after processing. The layers contain silver halide and related substances represented by the following symbols, Δ: unexposed AgX, ···: physical development nuclei, and ▬ : silver image physically developed. The symbol /// represents an antihalation dye, and ◇ represents a silver halide solvent or its precursor. The additive screen consists of red (R), green (G), and blue (B) filters.

References

[1] Rott A, Weyde E (1972) Photographic Silver Halide Diffusion Process, Focal Press, London New York

[2] Land EH (1953) US Patent 2 647 056

[3] Land EH (1954) US Patent 2 698 236

[4] Stewart PH, Heseltine DW (1973) US Patent 3 769 014

[5] Greenwald RB (1977) US Patent 4 047 954

[6] Greenwald RB (1977) US Patent 4 009 167

[7] Greenwald RB (1977) US Patent 4 047 955

[8] Greenwald RB (1976) US Patent 3 958 992

[9] De Haes LM, Gevers HK, Vanheertum JJ (1973) US Patent 3 721 559

[10] Futaki K, Suzuki S, Yoshihiro Y (1973) US Patent 3 728 114

[11] Land EH (1977) Phot Sci Eng. 21:225

[12] Land EH (1971) US Patent 3 615 127

[13] Land EH (1973) US Patent 3 709 588

[14] Kimata H (2002) Shashin Kogyo. 2002(2):85; 2002(3):69

[15] Land EH (1970) US Patent 3 536 488

[16] Land EH, Bloom SM, Farney LC (1972) US Patent 3 704 126

[17] Land EH (1958) US Patent 2 861 885

[18] Bloom SM, Cramer RD (1971) US Patent 3 615 440

[19] Land EH, Bloom SM, Farney LC (1974) US Patent 3 821 000

[20] Chemistry and Engineering News (1982) Nov 8:27

References

[1] [illegible] (1972) Photographic Silver Halide Diffusion Process[illegible] Press, London/New York

[illegible]

Chapter 16

Chemistry in Instant Color Photography

16.1 Instant Color Photography vs. Conventional Color Photography

Instant color photography is called *diffusion transfer reversal color photography* (DTR color photography) or simply *diffusion transfer photography*, since it is based on imagewise dye diffusion (nondiffusion) as a function of exposed silver halide. The beginning of instant color photography was the peel-apart system (Polaroid 108) placed on the market by Polaroid (1963). Afterwards, integrated (mono-sheet) systems called SX-70 (1972, Polaroid), PR-10 (1976, Kodak), and FI-10 (1981, Fuji Photo Film) appeared as more elaborate systems.[1]

Instant color photography is characterized by the immediateness of image appearance as the name contains the word "instant". Then, where does the immediateness come from? In order to answer this question, it is suggestive to compare instant color photography with conventional color photography (chromogenic photography).

Because both instant color photography and conventional color photography are based on silver halide emulsions, they have several common features along with different items, as shown in Table 16.1. The common features (Items 1 to 3) are concerned with physical processes (from light exposure to image storage as a latent image). On the other hand, the remaining items (Items 4 to 9) different

[1] In 2003, the distribution of Kodak's system had already been discontinued.

Table 16.1. Comparison of Instant Color Photography with Conventional Color Photography

		Instant color photography	Conventional color photography
1.	Color reproduction	Sensitizing dyes and multiple layers	Sensitizing dyes and multiple layers
2.	Sensitivity	AgX emulsions	AgX emulsions
3.	Image storage	Latent image	Latent image
4.	Developer	B&W developer	Color developer
5.	Image formation	Dye transfer process	Chromogenic process
6.	Dye image isolation	Dye transfer	Desilverization
7.	Reversal process	Dye developers or direct positive emulsions	—
8.	Dye deposit	Mordant layer	Ballast group
9.	Image dye	Azo dyes, etc.	Azomethine dyes

between the two systems of photography are mainly concerned with chemical processes.

Among these nine items, Items 6 and 7 are mainly concerned with the immediateness of instant color photography.

Dye image isolation Conventional color photography (chromogenic photography) is based on the chromogenic process in which couplers are reacted with an oxidized color developer. In this process, the exposure/nonexposure discrimination due to an original object is converted into the on/off of AgX reduction, which is in turn converted into dye/coupler (color/colorless) discrimination. As a result, a dye image and a silver image are generated at the same site. To isolate the dye image from the silver image, the latter should be removed by an appropriate treatment (desilverization or bleach-fixing). In instant color photography, in contrast, the exposure/nonexposure discrimination is converted into the diffusible/non-diffusible discrimination via the on/off of AgX reduction. Since the diffusible dye is automatically isolated from the silver image, no additional treatments are necessary to obtain the dye image. This means that the mechanism of dye transfer supports the immediateness of instant color photography.

Reversal process To understand this point clearly, let us first examine the negative-positive process and the reversal process in conventional color photography (Fig. 16.1).

a) Color negative-positive process

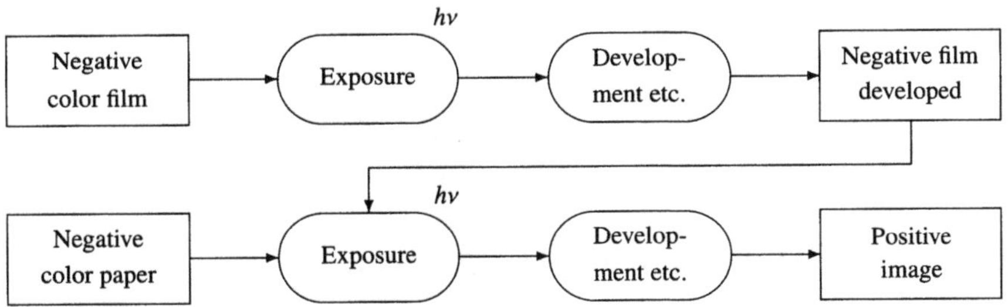

b) Color reversal process

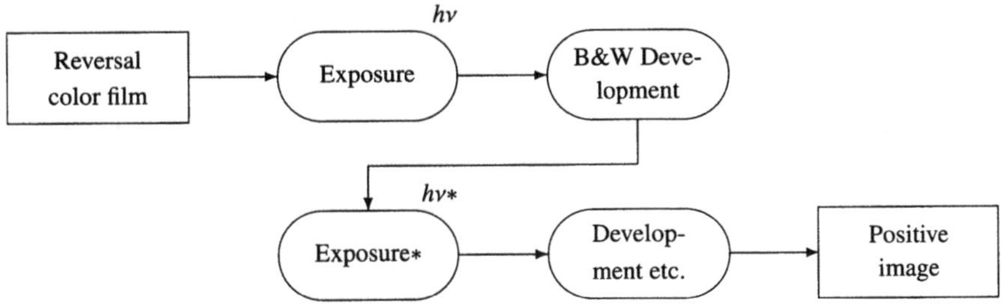

c) Instant color process

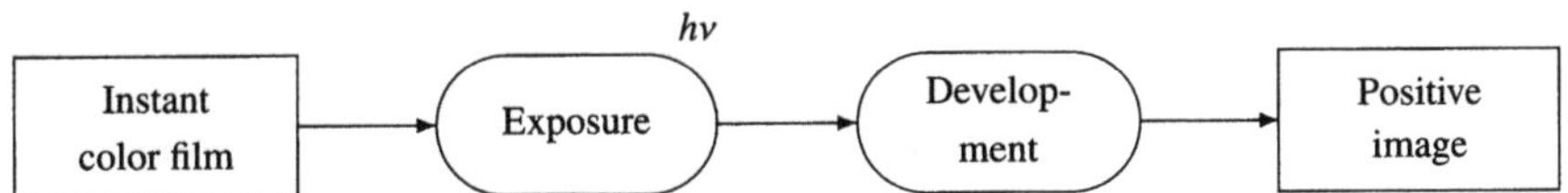

Figure 16.1. Reversal processes in color photography. The Exposure* ($h\nu*$) step in the color reversal process may be a reversal treatment in place of light exposure.

The negative-positive process in conventional color photography uses a negative photographic film as well as a negative color paper, where the former gives a negative image of an original object and the latter converts the negative image into a positive image. The positive image is the negative image of the negative image, as a result of the two development steps illustrated in Fig. 16.1a. See also Chapter 10 (Figs. 10.1 and 10.2).

On the other hand, color reversal film is used as a single photographic material to obtain a positive image of an original object. As found in Fig. 16.1b, the reversal process requires two development steps (i.e., a B&W development and a color development). See also Chapter 8 (Fig. 8.6).

In contrast to the two processes, the instant color process necessitates a "direct" reversal mechanism because of the immediateness (Fig. 16.1c). In other words, a single exposure and a single development are essential to the immediateness.

16.2 Classification of Imaging Compounds for Instant Color Photography

16.2.1 Dye-Stopping and Dye-Releasing

In contrast to the fact that all of the systems for conventional color photography are based on chromogenic development, the systems for instant color photography are based on distinct chemistries, as summarized in reviews by the author (Fujita) [1–8] and others [9–12]. In spite of such distinct chemistries, there are several common features among these systems. This chapter is devoted to discussing such common features.

Imaging compounds for instant color photography are represented by a general formula shown in Fig. 16.2. The FUN moiety denotes a function for changing diffusibility. When we focus our attention on the diffusibility, there are two cases:

1. One case in which a diffusible compound is changed into a non-diffusible one is a *dye-stopping process*. Since this case consists of *dye developers* only, as shown in this chapter, there has been no general name for specifying this case. If necessary, we use the name *imaging compounds of dye-stopping type*.

2. The other case in which a non-diffusible compound is changed into a diffusible one is a *dye-releasing process*. The imaging compounds in this case are called *diffusible dye-releasing compounds* (DDR compounds) or simply *dye releasers*.

The DYE moiety denotes a preformed dye or a dye precursor, which is linked with the FUN through a covalent bond.

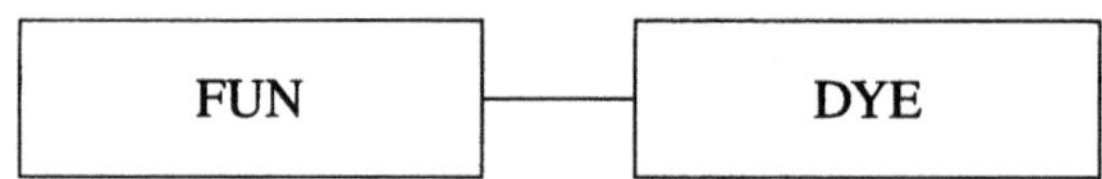

Figure 16.2. Schematic representation of imaging compounds for instant color photography. FUN denotes a function for changing diffusibility and DYE denotes a preformed dye moiety or a dye precursor moiety, where the two moieties are liked with a bond. When diffusible imaging compounds are converted into non-diffusible ones, they are called *imaging compounds of dye-stopping type*. When non-diffusible imaging compounds are converted into diffusible ones, they are called *diffusible dye-releasing compounds* (DDR compounds) or simply *dye releasers*.

16.2.2 Positive- and Negative-Working

Another viewpoint for classification of imaging compounds is whether the FUN works positively or negatively when negative silver halide emulsions are used. This point is important to understand the chemistry of instant color photography, because the immediateness of instant color photography requires a reversal process (Fig. 16.1).

There are two ways to realize such a direct reversal mechanism, i.e., (1) the usage of a positive-working compound (with a negative silver halide emulsion) and (2) the usage of a direct positive silver halide emulsion (with a negative-working compound). Hence it is important whether the FUN works positively or negatively when negative silver halide emulsions are used.

When the FUN gives an image in an unexposed area, the FUN–DYE compound (Fig. 16.2) is regarded as being positive-working. This means that the compound inherently possesses a reversal function. On the other hand, when the FUN gives an image in an exposed area, the FUN–DYE compound is negative-working.

The criterion of positive-working or negative-working mechanism is combined with the former criterion of dye-stopping or dye-releasing mechanism so as to give four categories of functional compounds. Among them, three categories (a, b, and c) have been embodied as FUN–DYE compounds for instant color photography (Fig. 16.3). Dye-stopping and negative-working compounds have been unknown, whereas the other three categories have been disclosed in patents. In this book, the name "dye developer" is used to designate category a (dye-stopping and positive-working), because only dye developers have been known. Category b (dye-releasing and negative-working) are referred to as negative-working dye releasers and category c (dye-releasing and positive-working) are called positive-working dye releasers.

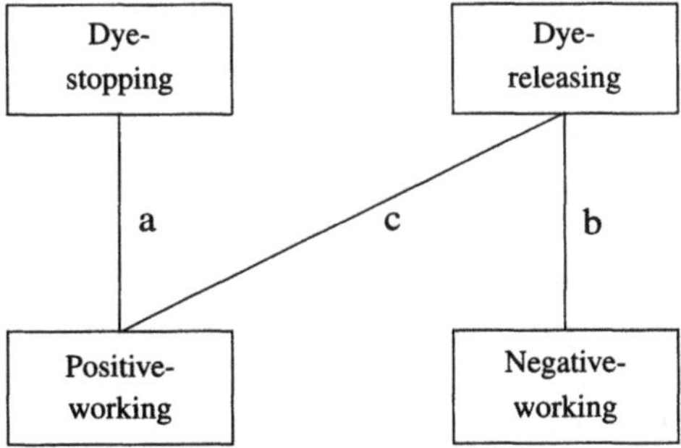

Figure 16.3. Combination of the dye-stopping or dye-releasing criterion with the positive- or negative-working criterion. Three categories (a–c) have been embodied as FUN–DYE compounds.

16.3 Imaging Compounds of Three Categories

16.3.1 Dye Developers as Dye-Stopping and Positive-Working Compounds (Category a)

A representative class of imaging compounds of dye-stopping type are dye developers (**1** in Fig. 16.4), in which the hydroquinone is a FUN shown in Fig. 16.2. They are non-diffusible but change into mobile dianion species (**2**) [13]. The mobile dianion species diffuses to give a dye image in an unexposed area. In an exposed area, they are oxidized into a quinone derivative (**3**) that is non-diffusible so as to give no dye image. This means that the dye developers are positive-working.[2] Since dye developers have been used in both the peel-apart system and the integrated (mono-sheet) system sold from Polaroid, more detailed discussion will appear in Subsection 16.4.1 and Chapter 17.

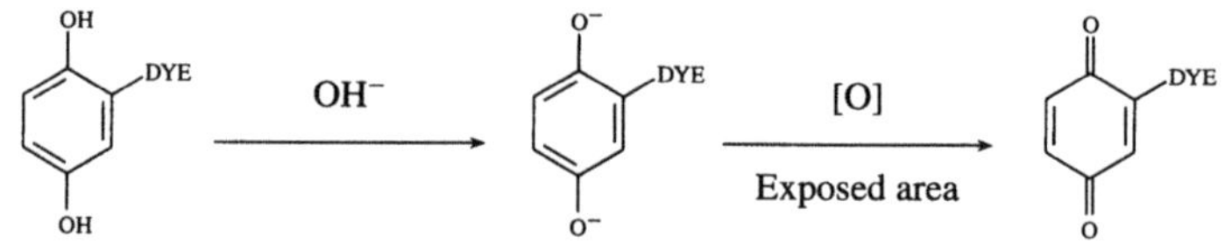

OH, DYE, OH — OH^- → O^-, DYE, O^- — [O], Exposed area → O, DYE, O

1 (immobile) **2** (mobile) **3** (immobile)

Figure 16.4. Imaging compound of dye-stopping type, which works positively. Dye developers shown in this figure are a sole example of the dye-stopping type.

16.3.2 Negative-Working Dye Releasers (Category b)

Dye-Releasing Redox (DRR) Compounds

Imaging compounds of dye-releasing type are called dye releasers, or more formally, diffusible dye releasing redox compounds (DRR compounds).[3] This subsection deals with negative-working dye releasers, as listed in Fig. 16.5.

The first example of this class consists of *p*-sufonamidonaphthol dye releasers (**4**), where the moiety represented by HO–(C=C)$_2$–$NHSO_2R$ can be related to HO–(C=C)$_2$–OH by the isosterism between an OH group and an $NHSO_2R$ group [14,15].[4] These compounds are non-diffusible because of their ballast

[2]Logically speaking, imaging compounds that are both dye-stopping and negative-working are possible. However, no such compounds have been reported. See Fig. 16.3.

[3]They are sometimes abbreviated as "DDR compounds".

[4]For the concept of isosterism, see Ref. [16]. In terms of the isosterism concept, *p*-sulfonamidonaphthols can be related to *p*-naphthalenediols or 1,4-naphthalenediols, which are one of the representative classes of photographic developers.

1) *p*-Sufonamidonaphthol dye releasers

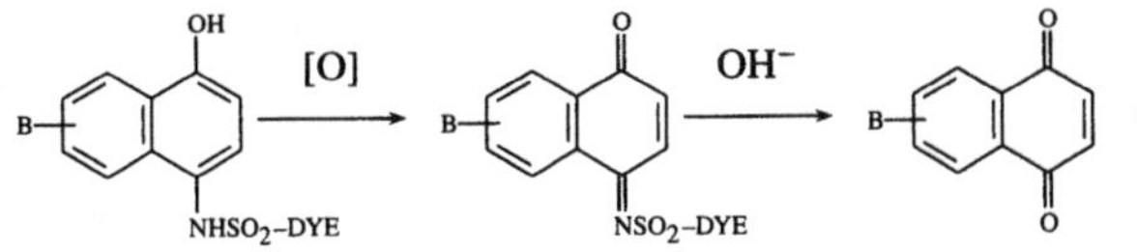

OH; B; $NHSO_2$–DYE → [O] → O; B; NSO_2–DYE → OH^- → O; B; O + $^-NHSO_2$–DYE

4 (immobile) **5** (immobile) **6** (immobile) **7** (mobile)

2) *o*-Sufonamidophenol dye releasers

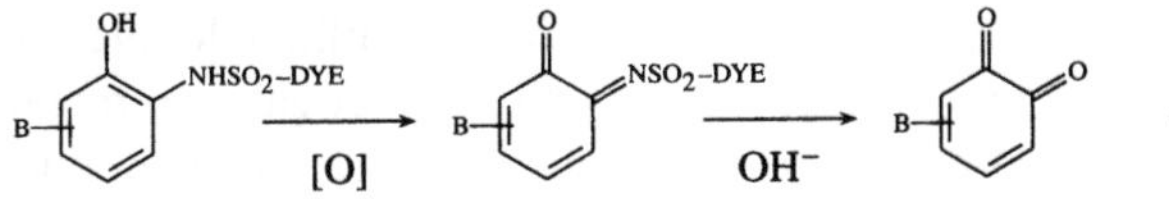

OH; $NHSO_2$–DYE; B → [O] → O; NSO_2–DYE; B → OH^- → O; O; B + $^-NHSO_2$–DYE

8 (immobile) **9** (immobile) **10** (immobile) **7** (mobile)

3) Indole dye releasers

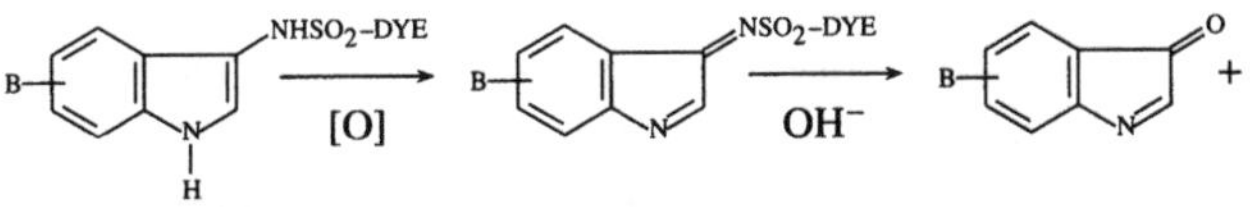

$NHSO_2$–DYE; B; N; H → [O] → NSO_2–DYE; B; N → OH^- → O; B; N + $^-NHSO_2$–DYE

11 (immobile) **12** (immobile) **13** (immobile) **7** (mobile)

4) Dye-releasing hydroquinones

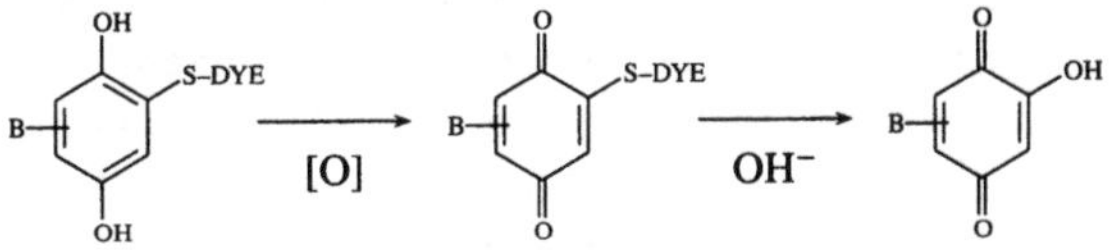

OH; S–DYE; B; OH → [O] → O; S–DYE; B; O → OH^- → O; OH; B; O + ^-S–DYE

14 (immobile) **15** (immobile) **16** (immobile) **17** (mobile)

5) Dye releasers via cyclization

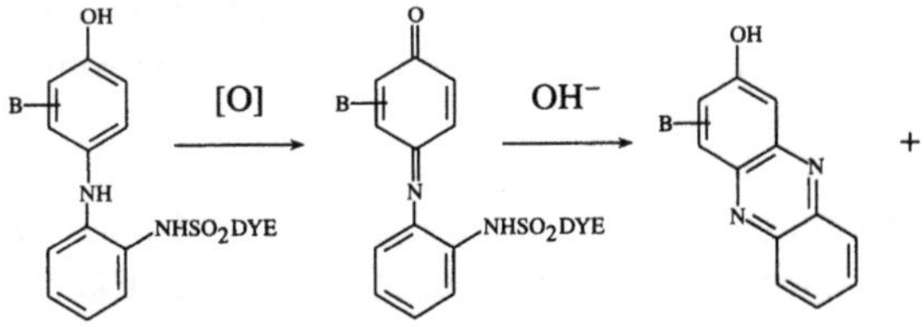

OH; B; NH; $NHSO_2DYE$ → [O] → O; B; N; $NHSO_2DYE$ → OH^- → OH; B; N; N + ^-O_2S–DYE

18 (immobile) **19** (immobile) **20** (immobile) **21** (mobile)

Figure 16.5. Negative-working dye releasers. The symbol B denotes a ballast group. The symbol [O] represents an imagewise oxidation.

groups. Each of them is oxidized into a *p*-naphthoquinone monoimide or 1,4-naphthoquinone monoimide (**5**), which is hydrolyzed into a *p*-naphthoquinone or 1,4-naphthoquinone (**6**) and a diffusible dye having a sulfonamido group (**7**). Since an instant color film using this class has been placed on the market by Kodak, more detailed discussions will appear in Chapter 18. On the other hand, *o*-sulfonamidonaphthol dye releasers have not been utilized in commercialized films, although they have been disclosed as alternative dye-releasers based on a naphthol nucleus [17].

The second example contains *o*-sulfonamidophenol dye releasers (**8**), where the moiety represented by HO–C=C–$NHSO_2R$ can be related to HO–C=C–OH in terms of the isosterism concept [18]. Each of the non-diffusible dye releasers is oxidized into an *o*-benzoquinone monoimide or 1,2-benzoquinone monoimide (**9**), which is then hydrolyzed into an *o*-benzoquinone or 1,2-benzoquinone (**10**) and a diffusible dye (**7**). Since an instant color film using this class has been placed on the market by Fuji Photo Film, more detailed discussions will appear in Chapter 19.

Although the remaining examples listed in Fig. 16.5 have not been marketed until now, it is worthwhile to describe their mechanisms in order to show the scope and limitations of dye releasers of Category b.

The third example consists of indole dye releasers (**11**), where the moiety represented by NH–C=C–$NHSO_2R$ can be related to NH–C=C–OH in terms of the isosterism concept [19]. Among them, indole dye releasers having a 2-hexadecylcarbamoyl group as a ballast group and a methoxy group at the 5-position have been disclosed to give an excellent D_{min}-value [20]. By substituting a multi-oxaalkyloxy group (e.g., $CH_3O(CH_2CH_2O)_2$) for the methoxy group at the 5-position, a higher D_{max}-value has been obtained [21] (cf. Fig. 16.19). A variety of substituents have been examined to obtain more efficient dye releasers of this type [22,23]. Dye moieties in the indole dye releasers (**11**) have been investigated to improve color reproduction and light fastness [24,25].

The fourth example comprises dye-releasing hydroquinones (**14**), where the resulting quinone (**15**) releases a diffusible dye (**17**) [26]. Obviously, the dye releasers are related to the DIR hydroquinones described in Chapter 14. Note that the released species are dyes for the present dye releasers, while the DIR hydroquinones release development inhibitors. Dye-releasing hydroquinones having a dye-sulfonyl moiety have also been disclosed, where a dye-sulfinic acid is released from the oxidized form [27].

The final example comprises dye releasers via cyclization (**18**), where the oxidized aminophenol (**19**) undergoes an intramolecular attack by the sulfonamide anion to form a tricyclic intermediate. The hydrogen at the attacked position of the intermediate is removed to form an azine ring so that a diffusible sulfinate anion linked with a dye moiety (**21**) is released [28].

Dye-Releasing Couplers and Related Compounds

Figure 16.6 lists dye releasing compounds based on chromogenic development. Since stains due to remaining color developers have not been overcome, all of these compounds have not been used in commercialized films.[5] However, they are interesting from a historical point of view and should be mentioned to some extent, because their modes of dye releasing have given versatile hints to developing the dye releasers listed in Fig. 16.5.

Coupler **22** is a diffusible dye-forming coupler, which is immobile because of the ballast group at the 4-position (O–$C_{18}H_{37}$) [29]. Note that the ballast group works as a leaving group during chromogenic development. This coupler is reacted with an oxidized color developer (**23**) to give a cyan dye (**24**), which is mobile because of the two carboxylic acid functions. Magenta and yellow dye forming couplers have been also disclosed in the patent [29].

Dye-releasing couplers represented by **25** have the same structures as colored couplers [30,31]. The coupler, which is non-diffusible because of the ballast group, is reacted with the oxidized color developer (**23**) to give an immobile cyan dye (**26**) and to release a mobile dye (**27**) as a leaving group. The latter mobile dye is used as an image dye when **25** is used in instant color photography. In contrast, the same dye is removed into a development solution when **25** is used as a colored coupler (cf. Chapter 12 on colored couplers).

Dye-releasing couplers via cyclization (e.g., **28**) are related to the dye-releasers (**18**) described in Fig. 16.5 [32,33]. The intermediate (**29**) undergoes cyclization in a similar way to **18** and releases a diffusible dye having a sulfinate anion (**21**).

The last example of Fig. 16.6 shows the dye-releasing mechanism of amidorazone dye releasers (**31**) [34]. More detailed disclosures have appeared to realize direct reversal processes [35,36]. Hydrazide derivatives such as Ballast–CO–NHNH–SO_2DYE have been disclosed as analogous dye-releasing compounds so that they react with *o*-benzoquinone generated by the imagewise oxidation of catechol to release a diffusible dye ($DYESO_2^-$) [37].

As found easily, one of the common problems to use dye-releasing compounds listed in Fig. 16.6 is the increase of stains during the storage of developed films, because unreacted color developers have also diffused to the image-receiving layer and are oxidized by atmospheric oxygen to give serious stains. To reduce the stains, ballasted color developers have been proposed to be employed in place of usual color developers (Fig. 16.7). For example, a ballasted color developer **33** has been used in combination with dye-releasing couplers (**25**) [30]. Another ballasted color developer **34** has been disclosed to be used together with amidorazone dye releasers (**31**) [38]. Since these ballasted color developers are immobile, a diffusible B&W developer (auxiliary developer) is necessary to mediate the oxidative activity of silver halide to the immobile color developers.

---

[5] Although they have not been employed in instant color photography, dye-releasing couplers among them are used as colored couplers in conventional color photography. See Chapter 12.

1) Diffusible dye-forming couplers

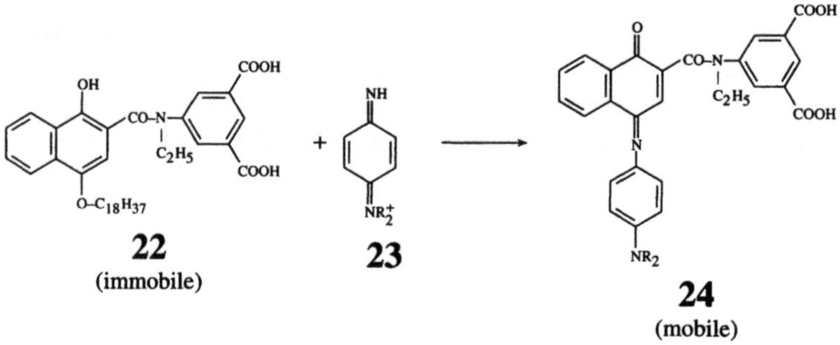

22 (immobile) + **23** → **24** (mobile)

2) Dye-releasing couplers

25 (immobile) + **23** → **26** (immobile) + $^{-}$O–DYE **27** (mobile)

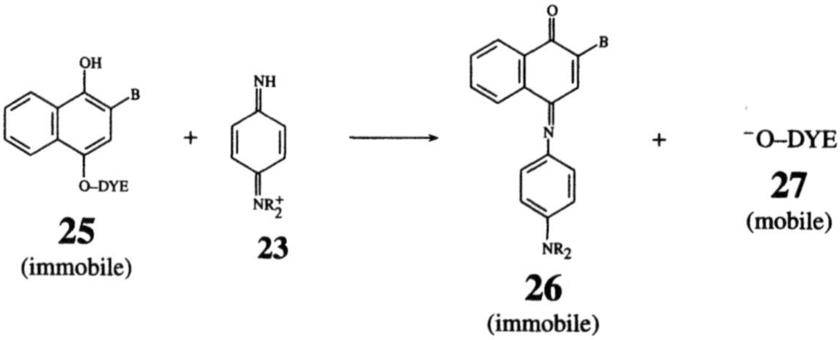

3) Dye-releasing couplers via cyclization

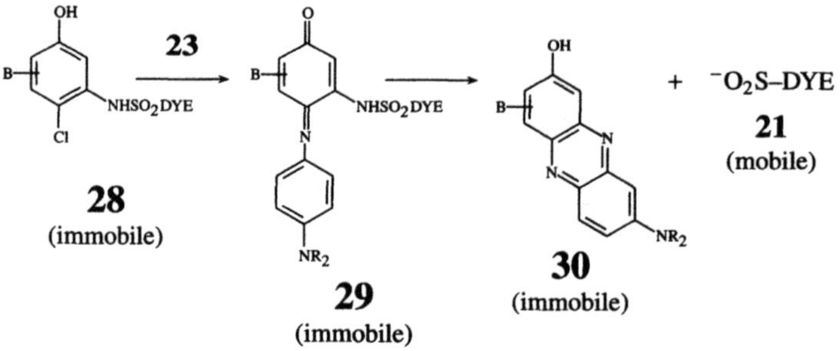

28 (immobile) —**23**→ **29** (immobile) → **30** (immobile) + $^{-}O_2S$–DYE **21** (mobile)

4) Amidorazone dye releasers

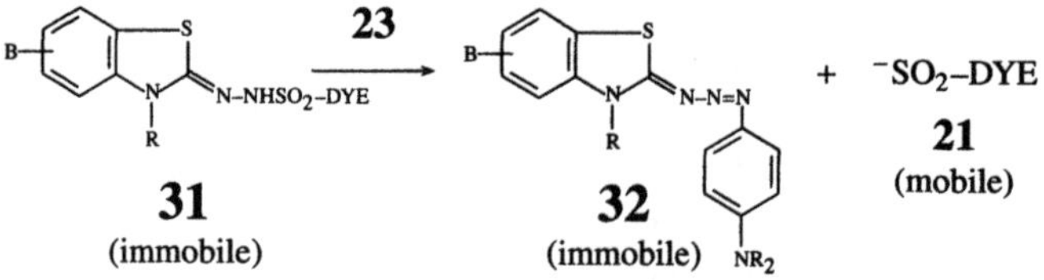

31 (immobile) —**23**→ **32** (immobile) + $^{-}SO_2$–DYE **21** (mobile)

Figure 16.6. Negative-working dye releasers based on chromogenic development. The symbol B denotes a ballast group. The species **23** is generated by the imagewise oxidation of a color developer.

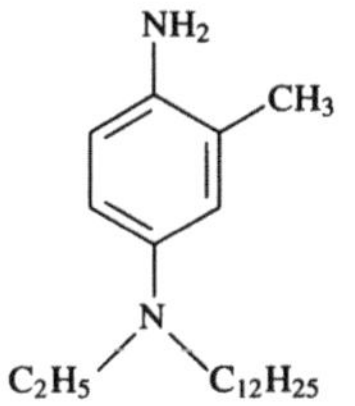

NH$_2$ CH$_3$ N C_2H_5 $C_{12}H_{25}$

33
(immobile)

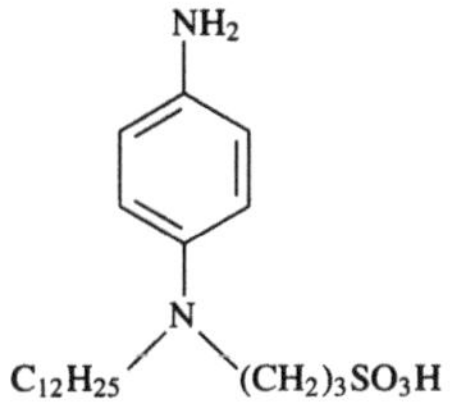

NH$_2$ N $C_{12}H_{25}$ $(CH_2)_3SO_3H$

34
(immobile)

Figure 16.7. Ballasted color developers proposed to reduce stains in instant color photography.

16.3.3 Positive-Working Dye Releasers (Category c)

Category c (positive-working dye releasers) can be further divided into three subclasses, i.e., oxidative-stopping dye releasers, reductive-releasing dye releasers, and silver-ion-induced dye releasers.

Oxidative-Stopping Dye Releasers

Figure 16.8 lists positive-working dye releasers of the oxidative-stopping type. An isoxazole dye releaser such as **35** is immobile and undergoes a ring-opening reaction on the attack of a hydroxide ion [39]. The resulting intermediate **36** is imagewise oxidized into the corresponding nitrone **37**, which is immobile and incapable of releasing a mobile dye. In the area where no oxidation occurs, on the other hand, the intermediate **36** cyclizes into an alternative isoxazole so as to release a diffusible dye **39**.

A ballasted hydroquinone dye releaser such as **40** belongs to this subclass [40]. The immobile hydroquinone releases a diffusible dye **42** upon the intramolecular attack of a phenoxide anion. It is imagewise oxidized into the corresponding benzoquinone **43**, which is immobile and incapable of releasing such a mobile dye.

The image discrimination of dye releasers of this type is not an easy task to be realized, since it depends upon the competition between the oxidation (**36** → **37** or **40** → **43**) and the hydrolysis (**36** → **38** + **39** or **40** → **41** + **42**). Compare this feature of the oxidative-stopping dye releasers with the below-mentioned feature of the reductive-releasing dye releasers.

Reductive-Releasing Dye Releasers

Figure 16.9 lists positive-working dye releasers of the reductive-releasing type. A nitro-type dye releaser **44** has been disclosed to release a mobile dye **47**, where the nitro group of **44** is reduced into a hydroxylamine in the intermediate **45** [41,42]. For a concrete structure, see Fig. 16.23. The reduction occurs as a function of remaining developers which are unreacted during the preceding

1) Isoxazole dye releasers

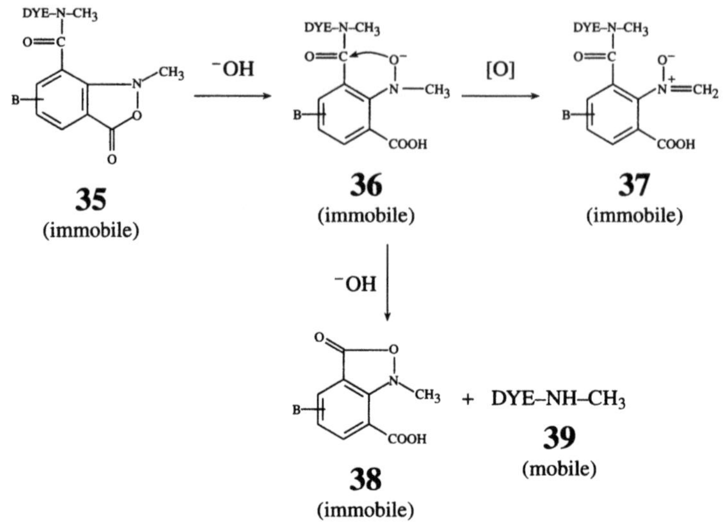

2) Ballasted hydroquinone dye releasers

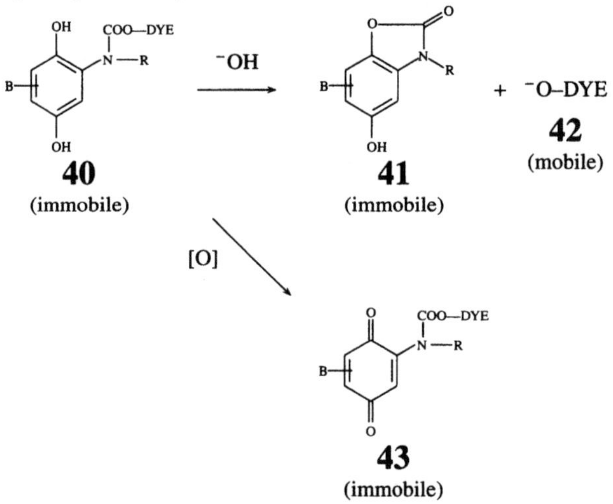

Figure 16.8. Positive-working dye releasers. The symbol B represents a ballast group. Each dye-stopping process is driven by the oxidation reaction.

imagewise development. It should be mentioned that the intermediate **45** resembles **36**. However, a further oxidation of the hydroxylamine of **45** has not been discussed.

A ballasted quinone dye releaser **48** also works as a positive-working reductive-releasing dye releaser [43]. The dye releaser is reduced into the corresponding hydroquinone intermediate **49**, which intramolecularly cyclizes into **50** so as to

1) Nitro-type dye releasers

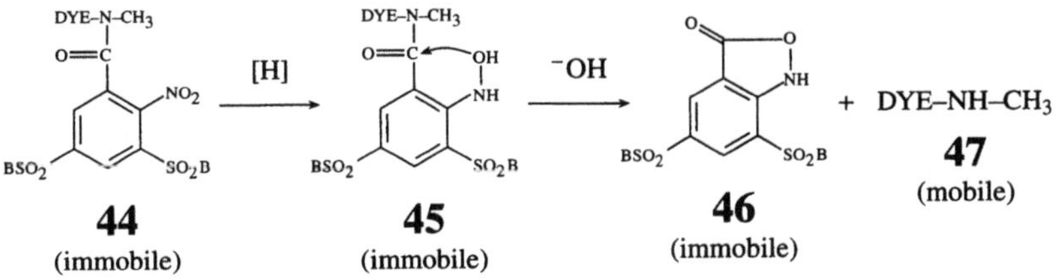

DYE–N–CH$_3$ [H] DYE–N–CH$_3$ $^{-}$OH + DYE–NH–CH$_3$

44 (immobile) 45 (immobile) 46 (immobile) 47 (mobile)

2) Ballasted quinone dye releasers

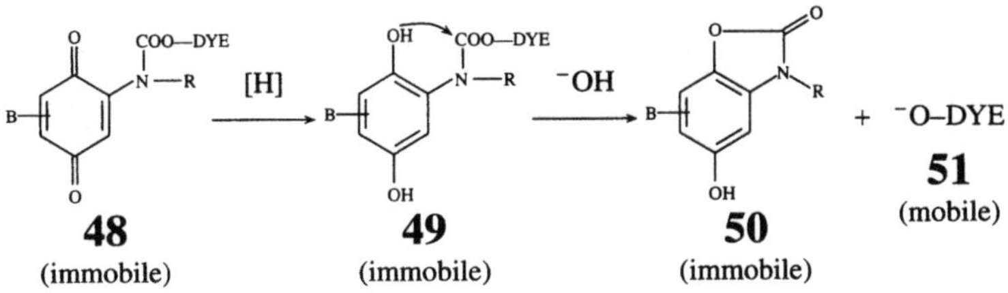

COO—DYE [H] COO—DYE $^{-}$OH + $^{-}$O–DYE

48 (immobile) 49 (immobile) 50 (immobile) 51 (mobile)

3) Carquin dye releasers

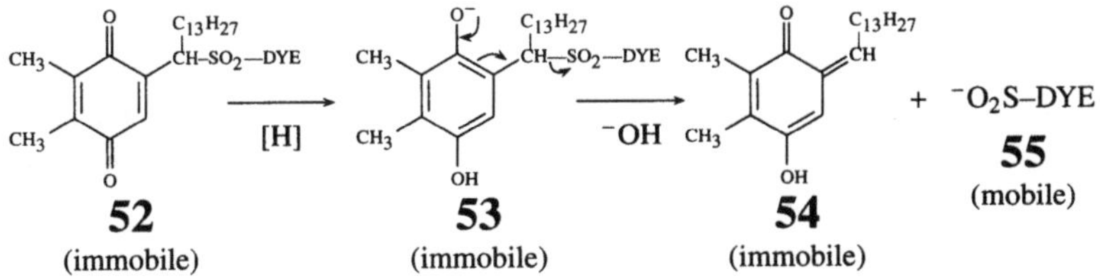

C$_{13}$H$_{27}$ [H] C$_{13}$H$_{27}$ $^{-}$OH C$_{13}$H$_{27}$ + $^{-}$O$_2$S–DYE

52 (immobile) 53 (immobile) 54 (immobile) 55 (mobile)

4) ROSET dye releasers

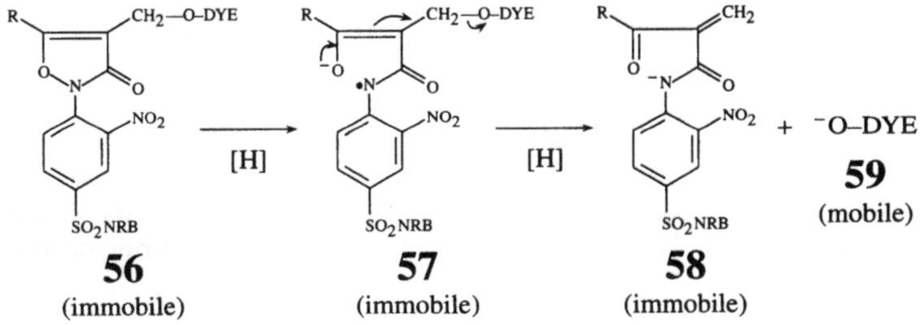

CH$_2$—O–DYE [H] CH$_2$—O–DYE [H] CH$_2$ + $^{-}$O–DYE

56 (immobile) 57 (immobile) 58 (immobile) 59 (mobile)

Figure 16.9. Positive-working dye releasers. The symbol B represents a ballast group. Each dye-releasing process is driven by the reduction reaction represented by [H].

release a diffusible dye **51** in an imagewise fashion. Note that the intermediate (**49**) substantially has the same structure as the ballasted hydroquinone dye releaser (**40**) described above. The employment of the reduction in **48** increases the image discrimination as compared with **40**.

A ballasted quinone dye releaser that has the group CH$_2$–N(R)COO–DYE in place of the group N(R)COO–DYE at the 2-position of **48** has been also disclosed

[43]. In this case, a six-membered ring is formed in place of the five-membered one of **50**.

The original patent [43] has covered both the nitro-type dye releasers (**44**) and the ballasted quinone dye releasers (**48**), where they are represented by a more general formula and referred to as *BEND compounds* (ballasted electron-accepting nucleophilic displacement compounds).

Ballasted quinone dye releasers of another type (e.g., **52**) are referred to as *carquin* [44].[6] The quinone (**52**) is reduced into the corresponding hydroquinone intermediate **53**, which releases a diffusible dye **55** in terms of the quinone-methide mechanism.

4-Isoxazolin-3-one derivatives such as **56** have been reported as a novel type of positive-working reductive-releasing dye releasers, where they require one-electron transfer to undergo a ring opening into the intermediate **57** [45]. The intermediate releases a diffusible dye **59** by virtue of electron transfer. These dye releasers are called *ROSET dye releasers*, where ROSET is the acronym of "Ring Opening by Single Electron Transfer".[7]

Silver-Ion-Induced Dye Releasers

Thiazolidine dye releasers such as **60** are also positive-working, where the dye releasing is based on the opening of a thiazolidine ring induced by the attack of a silver ion [46,47,48]. For a concrete structure, see Fig. 16.25.

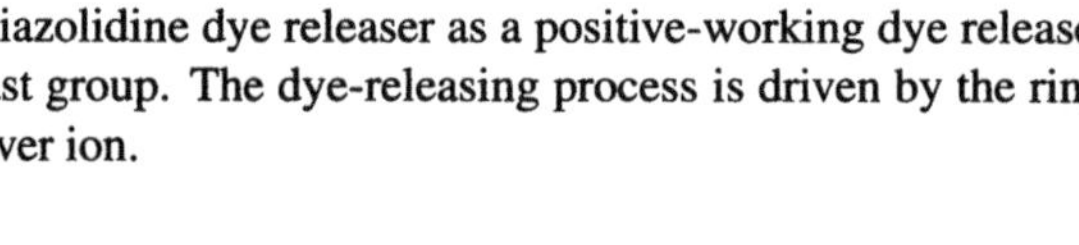

Figure 16.10. Thiazolidine dye releaser as a positive-working dye releaser. The symbol B represents a ballast group. The dye-releasing process is driven by the ring opening due to the attack of a silver ion.

It should be noted that the "silver ion" shown in Fig. 16.10 may be a cationic or anionic species (e.g., $Ag(S_2O_3)^{3-}$) dissolved by an appropriate silver-halide solvent and that the silver ion comes from the residual (unreacted) part of the original silver halide emulsion. The latter fact is a reason for the positive-working mechanism of the thiazolidine dye releaser. Compare this type of dye releasers

[6] The word "carquin" comes from "carrier" and "quinone". This type of dye releasers has been used in COPYCOLOR CCN commercialized by Agfa-Gevaert. See Chapter 20.

[7] A full color copying system based on the ROSET dye releasers has been placed on the market by Fuji Photo Film (1991). See Chapter 20.

with the ones which directly use redox reactions, as listed in Fig. 16.8 an Fig. 16.9.

16.4 Reversal Mechanisms in Instant Color Photography

One of the most remarkable characteristics of instant color photography is its *immediateness*, as the name "instant" suggests. The immediateness mainly stems from the mechanism of dye-image isolation and the direct reversal process, as discussed in Section 16.1. Various mechanisms to realize the direct reversal process have been proposed, as summarized in a review [5]. Among them, several representative mechanisms will be dicussed in this section.

16.4.1 Direct Reversal Mechanism Based on Dye Developers

Dye developers [13], which are positive-working compounds of dye-stopping type (Fig. 16.4), have an intrinsic direct reversal mechanism in themselves, as shown in Fig. 16.11.

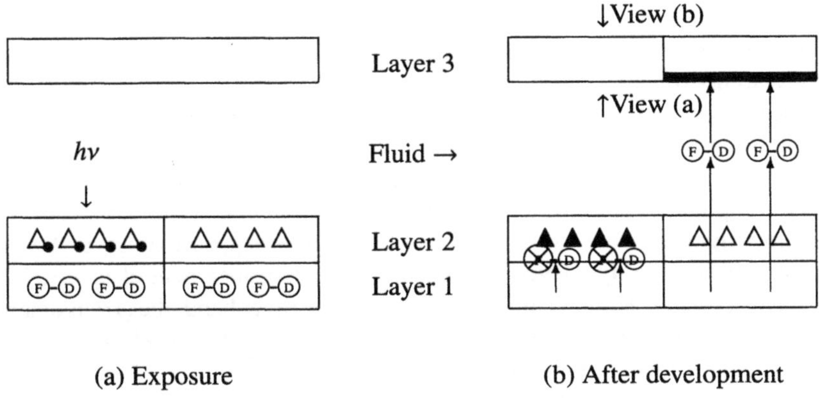

Figure 16.11. Direct reversal process using dye developers during exposure (left) and after development (right). These schematic cross-sections omit polymer supports on which Layers 1 and 2 are coated: (Layer 1) a dye developer layer, (Layer 2) a photo-sensitive layer containing a negative silver halide emulsion, and (Layer 3) an image-receiving layer, where the symbols represent Ⓕ-Ⓓ: dye developer (mobile in an alkaline media), ⊗-Ⓓ: oxidized dye developer (immobile), △: unexposed AgX, △•: exposed AgX with latent image, ▲: developed silver, and ▬▬ : dye image deposited on the image-receiving layer (Layer 3). View (a) represents the direction of view for a peel-apart film, while View (b) indicates that for an integrated (mono-sheet) film.

The left diagram shown in Fig. 16.11 represents a minimum set of layers during the exposure step (a), where Layer 1 is a dye developer layer containing a dye developer (Ⓕ-Ⓓ), Layer 2 is a photo-sensitive layer containing a negative silver halide emulsion, and Layer 3 is an image-receiving layer. The image-receiving layer should be taken off during the exposure of a peel-apart film, but is superimposed onto the film during the development process. If an integrated (mono-sheet) film is employed on the other hand, the exposure must be done through a transparent image-receiving layer.

Suppose that the left half is exposed ($h\nu$) to generate latent images (△•) in Layer 2. Then a development fluid is spread between the exposed film and the image-receiving layer. The fluid develops the exposed silver halide grains in the left half of the right diagram of Fig. 16.11b, where the dye developer molecules are oxidized in accord with the scheme shown in Fig. 16.4 to be prevented from further diffusion (⊗-Ⓓ). Such oxidation does not occur in the right half of the right diagram (b) so that unreacted dye developer molecules (anionic species) diffuse through the development fluid and other possible layers to arrive at the image-receiving layer (Layer 3), where the symbol ▬▬ represents dye images deposited on Layer 3.

Since the dye images appear in the unexposed area (the left part), they are positive in a photographic sense. Although Fig. 16.11 does not specify the light-sensitivity of the silver halide emulsion, respective emulsions undergo this type of direct reversal mechanism. For example, an exposed red-sensitive emulsion layer prevents the diffusion of cyan dye developers so that yellow and magenta dyes reach the image-receiving layer (Layer 3). Since the yellow and magenta dyes absorb blue (400–500 nm) and green lights (500–600 nm), we can observe the remaining red light (600–700 nm).

Since each emulsion (e.g., the red-sensitive emulsion) contains a sensitizing dye of the corresponding complementary color (e.g, cyan), the exposure should be carried out from the side of Layer 2 in Fig. 16.11a. Otherwise, the dye developer of the complementary color in Layer 3 (e.g., the cyan dye developer) absorbs the necessary light (e.g., the red light) so that no effective exposure of Layer 2 occurs. Hence, a film using dye developers should take an arrangement of layers, Layer 1—Layer 2—($h\nu$), as shown in Fig. 16.11a. On the other hand, the arrangement of Layer 1—Layer 2—Layer 3 is essential in order that the dye developer is stopped effectively, as shown in Fig. 16.11b.

These situations of the layer arrangement decide whether the resulting images are proper or reflective [4]. By observing the dye images on Layer 3 from the side of View (a) shown in Fig. 16.11b, we can see a proper image in which the left and right relationship is the same as that of the original object. This case is realized in peel-apart films. On the other hand, the observation from View (b) gives a mirror image in which the left and right relationship is reverse to that of the original object. This case occurs by using integrated (mono-sheet) films, as discussed in detail in Chapter 17.

16.4.2 Direct Reversal Mechanisms for Negative-Working Dye Releasers

Solution Physical Development

Dye-Forming Couplers and Solution Physical Development

Solution physical development, which has been used in instant B&W photography (Chapter 15), can be used to realize a direct reversal mechanism in combination with dye-forming couplers (**22**) described in Fig. 16.6 [31].

Schematic diagrams for the direct reversal process using dye-forming couplers and solution physical development are shown in Fig. 16.12. Layer 1 is a photo-sensitive layer containing a negative silver halide emulsion and a scavenger coupler (ⓢ). Layer 2 (a dye-forming coupler layer) contains development nuclei (· · ·) and a dye-forming coupler (ⓒ). Layer 3 is an image-receiving layer.

After exposure (a), a development fluid is spread between the Layer 2 and Layer 3. Thereby, the subsequent process is initiated, as shown in Fig. 16.12b. First, the exposed silver halide grain with a latent image (△•) is developed by a color developer. The resulting oxidized color developer is scavenged by a scavenger coupler (ⓢ, e.g., **63**) to generate an indoaniline dye (⊗).

On the other hand, the unexposed silver halide in Layer 1 (Fig. 16.12b) is dissolved by an appropriate silver halide solvent, and the resulting silver species (e.g., $Ag(S_2O_3)_2^{3+}$) diffuses to Layer 2, where the physical development takes place by virtue of physical development nuclei (· · ·). This development results in the formation of a diffusible dye (ⓓ) according to the scheme shown at the top of Fig. 16.6. The diffusible dye arrives at Layer 3, where it produces a positive image (▬).

For an integrated (mono-sheet) film, the exposure from *hv* (b) and View (b) can be adopted to keep the proper right-left relationship. For a peel-apart film, the exposure from *hv* (a) and View (a) enable us to observe a proper image. The latter mode of exposure is effective, since the dye-forming coupler in Layer 2 is colorless.

Dye-Releasing Couplers and Solution Physical Development

Solution physical development can also be combined with dye-releasing couplers described in Fig. 16.6 [49]. Thus, the dye-forming coupler (ⓒ) in Layer 2 of Fig. 16.12 is replaced by a dye-releasing coupler such as **25**. A set of such dye-releasing couplers necessary to color reproduction are listed in Fig. 16.13, where the coupler moiety is common to each other [30,31]. Thus, these 1-naphthol couplers couple with an oxidized color developer at the active 4-position at which a respective preformed dye is attached as a leaving group. Thereby, the coupler **64** releases a cyan dye (a Nickel chelate dye) upon coupling with an oxidized

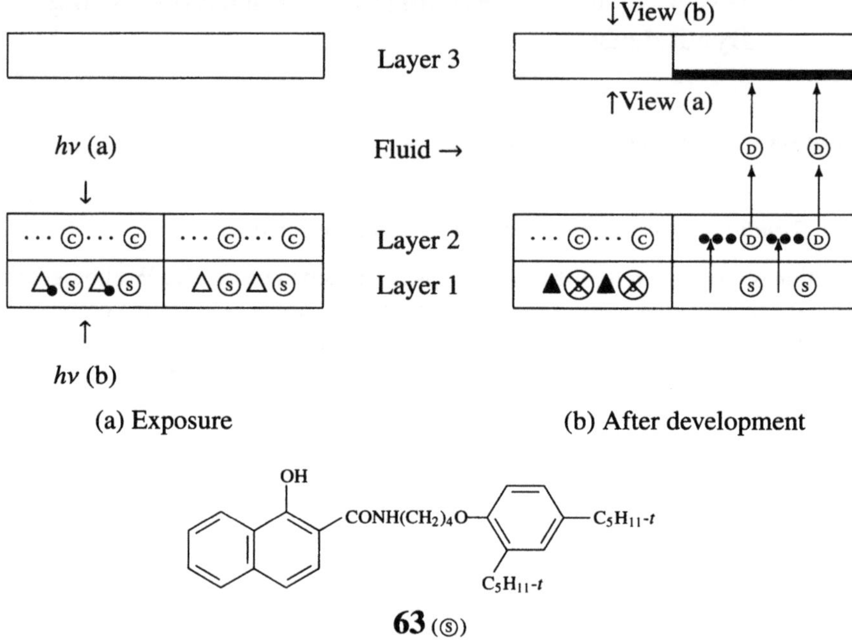

Figure 16.12. Direct reversal process using dye-forming couplers and solution physical development during exposure (left) and after development (right). These schematic cross-sections omit polymer supports on which Layers 1 and 2 are coated: (Layer 1) a photo-sensitive layer containing a negative silver halide emulsion and a scavenger coupler (Ⓢ), (Layer 2) a dye-forming coupler layer containing development nuclei (···) and a dye-forming coupler (Ⓒ), and (Layer 3) an image-receiving layer, where the symbols represent Ⓒ: dye-forming coupler (immobile in an alkaline media), Ⓓ: diffusible dye generated from the dye-forming coupler, Ⓢ: scavenger coupler (e.g., **63**), ⊗: immobile dye generated from the scavenger coupler, △: unexposed AgX, △•: exposed AgX with latent image, ···: physical development nuclei, •••: developed silver and ▬▬: dye image deposited on the image-receiving layer (Layer 3). The exposure from *hν* (a) aims at a peel-apart film, while the exposure from *hν* (b) indicates an integrated (mono-sheet) film. View (a) represents the direction of view for a peel-apart film, while View (b) indicates that for an integrated (mono-sheet) film.

color developer, the coupler **65** releases a magenta azo dye, and the coupler **66** also releases a yellow azo dye.[8]

In contrast to the dye-forming couplers, the dye-releasing couplers can be used in combination with the ballasted color developers listed in Fig. 16.7 [30]. Each dye-releasing coupler and the ballasted color developer (e.g., **33**) are contained in

[8]Although such a set of dye-releasing couplers has not been commercialized, the coupler **65** has been used as a magenta-colored cyan coupler for improving color reproduction in several negative color films commercialized by Kodak and other companies. See Chapter 12 and Fig. 1.25 of Chapter 1.

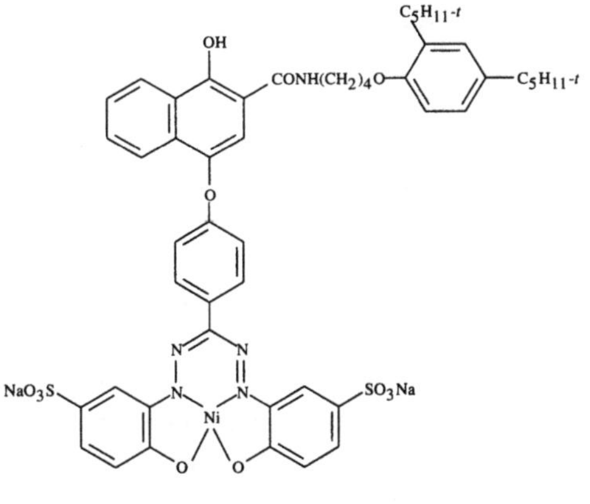

64 (cyan)

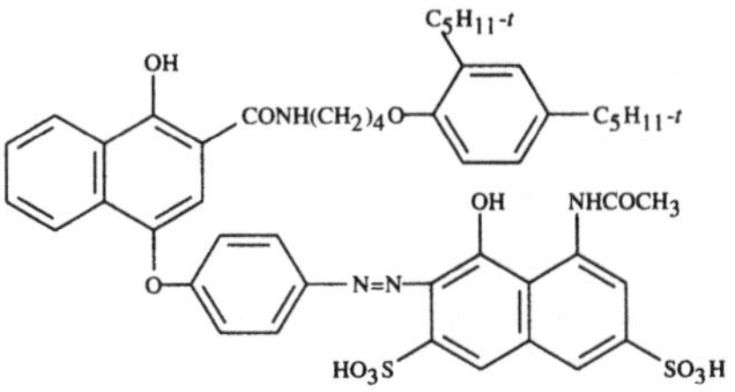

65 (magenta)

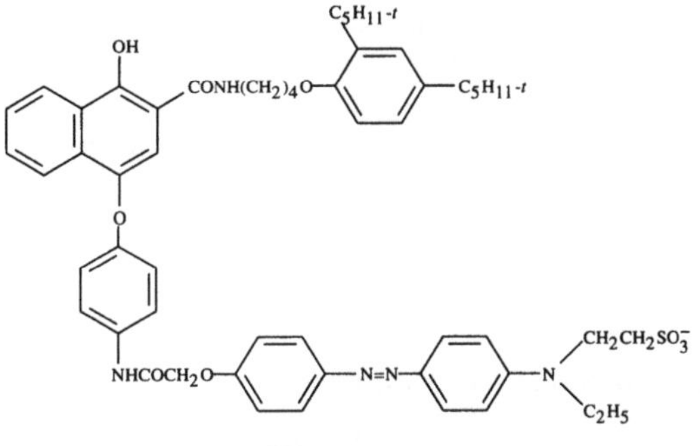

66 (yellow)

Figure 16.13. Dye-releasing couplers for instant color photography. These couplers are called "colored couplers" in conventional color photography. See Fig. 16.6(2).

Layer 2 of Fig. 16.12. The ballasted color developer is also contained in Layer 1 along with a scavenger coupler (e.g., **63**). A diffusible B&W developer (e.g., phenidone) is used in order to mediate the cross-oxidation between silver halide and the ballasted color developer.

In the exposed area of Fig. 16.12b, the B&W developer is reacted with silver halide to generate the corresponding oxidized B&W developer, which oxidizes the ballasted color developer contained in Layer 1. The latter is reacted with the scavenger coupler.

On the other hand, the solution physical development occurs in Layer 2 of the unexposed area, where a dissolved silver species (e.g., $Ag(S_2O_3)_2^{3-}$) is supplied from the unexposed Layer 1. The resulting oxidized B&W developer is reacted with the ballasted color developer to generate the oxidized color developer, which couples with the dye-releasing coupler to release a diffusible dye (see **27** shown in Fig. 16.6). The diffusible dye arrives at Layer 3 of Fig. 16.12b, where it produces a positive image (▬).

It should be emphasized again that, in place of a single diffusible color developer for the dye-forming couplers, the combination of the diffusible B&W developer and the ballasted (non-diffusible) color developer can be employed for the dye-releasing couplers. Remember that the diffusible B&W developer causes less stains than the diffusible color developers.

Other Applications of Solution Physical Development

In principle, the negative-working dye releasers listed in Fig. 16.6 can be combined with solution physical development, if an appropriate modification is applied to the diagram illustrated in Fig. 16.12. For example, the amidorazone dye releasers (**31**) have been used in combination with solution physical development [36] and with the ballasted color developer (e.g., **34**) [50].

Moreover, the negative-working dye releasers listed in Fig. 16.5 can also be combined with solution physical development. In this case, a B&W developer can be used in place of a color developer. For example, an embodiment using an azine-forming dye releaser (e.g., **18**) and solution physical development has been disclosed, where a scavenger coupler shown in Fig. 16.12 is replaced by a scavenger for an oxidized form of the B&W developer [28,33].

Reversal Mechanism by DIR Couplers

Dye-Forming Couplers and DIR Couplers

The interlayer effect of DIR couplers discussed in Chapter 13 can be used to realize a reversal process for instant color photography. For example, an embodiment for using dye-forming couplers is shown in Fig. 16.14 [51]. Layer 1 of Fig. 16.14a is a photo-sensitive layer containing a negative silver halide emulsion and a DIR coupler (Ⓘ, e.g., **67**), Layer 2 is an AgSCN layer containing a dye-forming coupler (©) and development nuclei (e.g., ZnS), and Layer 3 is an image-receiving layer.

The color development of the exposed silver halide (△•) generates an oxidized developer, which couples with the DIR coupler to give a dye (⊗) and a development inhibitor (Fig. 16.14b). The resulting developer inhibitor (e.g., 1-phenyl-5-mercapto-1H-tetrazole released from **67**) diffuses to Layer 2, where it inhibits the AgSCN from further development, as shown by the symbol (△∘).

On the other hand, the AgSCN in the unexposed area of Layer 2 is developed without the influence of the development inhibitor so that the resulting oxidized color developer couples with the dye-forming coupler (Ⓒ) to release the corresponding diffusible dye (Ⓓ). The dye diffuses to Layer 3, where it deposits as an image dye. The exposure from *hv* and the view (View) from the top represent the observation mode for an integrated (mono-sheet) film.

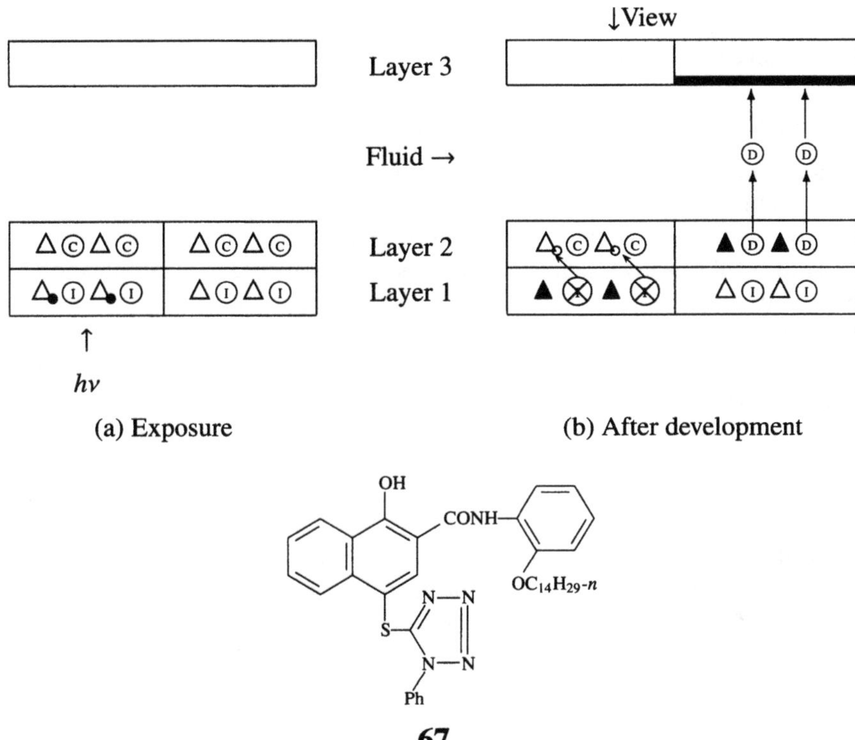

Figure 16.14. Direct reversal process using dye-forming couplers and a DIR coupler during exposure (left) and after development (right). These schematic cross-sections omit polymer supports on which Layers 1 and 2 are coated: (Layer 1) a photo-sensitive layer containing a negative silver halide emulsion and a DIR coupler (Ⓘ), (Layer 2) AgSCN layer containing a dye-forming coupler (Ⓒ) and development nuclei (e.g., ZnS), and (Layer 3) an image-receiving layer, where the symbols represent Ⓒ: dye-forming coupler (immobile in an alkaline media), Ⓓ: diffusible dye generated from the dye-forming coupler, Ⓘ: a DIR coupler (e.g., **67**), ⊗: dye generated from the DIR coupler, △: unexposed AgX, △•: exposed AgX with latent image, △∘: AgSCN inhibited by a development inhibitor, and ▬▬: dye image deposited on the image-receiving layer (Layer 3). The exposure from *hv* and the view (View) from the top represent the observation mode for an integrated (mono-sheet) film.

Other Applications of DIR Couplers

In principle, the negative-working dye releasers listed in Fig. 16.6 can be combined with the reversal process based on DIR couplers, if an appropriate modification is applied to the diagram illustrated in Fig. 16.14. For example, the amidorazone dye releasers (**31**) have been used in combination with DIR releasers [35].

Direct Reversal Emulsions

Core-Shell Emulsions for Direct Reversal Processes

Although direct positive emulsions had been known for a long time, their speeds were too low as to be used in practical color photography [52]. At the beginning of instant color photography, as a result, it seemed to be difficult to realize such direct reversal (positive) emulsions that had practical speeds. This situation had induced the examination of the various direct reversal mechanisms described in the preceding paragraphs, since most high-speed silver halide emulsions were negative ones. From this context, Kodak's PR-10 (1976, Kodak) appeared as an epoch-making product based on a direct reversal emulsion of practical speed, although the distribution of Kodak's PR-10 has now been discontinued (Chapter 18). Later, Fuji's FI-10 (1981, Fuji Photo Film) was placed on the market as another product based on a direct reversal emulsion and its distribution continues at present (Chapter 19).

The direct reversal emulsions of practical speeds are core-shell emulsions that comprise an iodine-rich core and a surrounding shell. The direct reversal mechanism based on such core-shell emulsions is shown in Fig. 16.15, which is cited from the reviews [1,5,11].

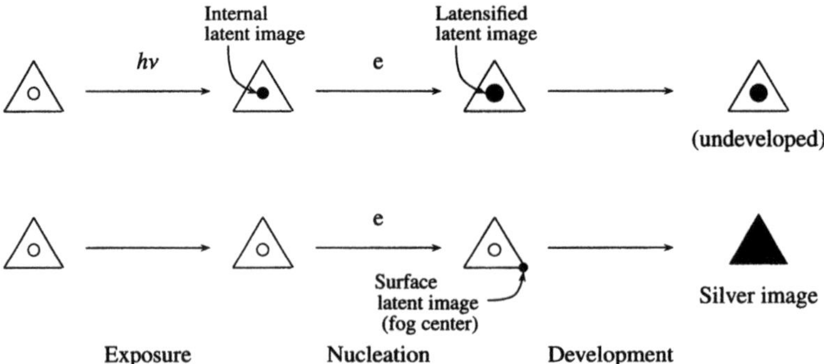

Figure 16.15. Direct reversal mechanism based on core-shell emulsions. The top row represents the fate of a silver halide grain in the exposed area, while the bottom row represents the fate of a silver halide grain in the unexposed area.

The core-shell grain exposed to light (the top row of Fig. 16.15) generates an internal latent image, because each photo-electron is trapped by the core.[9] During the nucleation step, the electron injection due to a nucleating agent reinforces (latensifies) the internal latent image so as to generate no latent images on the surface of the core-shell grain. Because of the absence of surface latent images, the core-shell grain in the exposed area is not reduced during the next development step.

On the other hand, the core-shell grain in the unexposed area (the bottom row of Fig. 16.15) does not generate an internal latent image. During the nucleation step, the electron injection due to the nucleating agent generates latent images on the surface of the core-shell grain, because the absence of an internal latent image does not stimulate the reinforcement (latensification) mechanism. By virtue of the surface latent images, the unexposed silver halides are finally reduced into silver images during the next development step.

It is worthy here to compare the scheme shown in Fig. 16.15 with the color reversal process shown in Fig. 16.1b. Obviously, the nucleation step of Fig. 16.15 has the same functions achieved by the B&W development and the uniform light exposure in the reversal process of Fig. 16.1b. More precisely speaking, the B&W development corresponds to the latensification shown in the top row of Fig. 16.15, while the uniform light exposure corresponds to the surface fogging shown in the bottom row of Fig. 16.15. The two functions are discriminated from each other in the form of separate steps in the color reversal process (Fig. 16.1b), whereas they are discriminated in the form of internal structural separation (core and shell) in the direct reversal emulsions (Fig. 16.15).

Nucleating Agents

One of the most important technologies for realizing an efficient reversal process is the selection of nucleating agents. As found in Fig. 16.15, nucleating agents are fogging agents capable of generating a fog center on the surface of a silver halide grain. Several examples of such nucleating agents are shown in Figs. 16.16 and 16.17.

Hydrazine derivatives (e.g., **68a**) have been well-known as nucleating agents, though they have no aim at the use in instant color photography [53]. These hydrazines and their acyl derivatives (e.g., **68b**) were first examined in instant color photography based on dye-forming couplers (**22**) [54]. A direct reversal emulsion due to the nucleating agent (**68b**) has been successfully used in combination with *p*-sulfonamidonaphthol dye releasers (**4**) [55]. Hydrazides having an adsorption site (e.g., **69**) have been disclosed as nucleating agents [56]. They have been widely used together with hydrazides of another type (e.g., **70**) [57]. Hydrazides having a heterocyclic moiety (e.g., **71**) have been also proposed as nucleating

[9] In contrast, a negative silver halide grain does not suffer from this mechanism and generates a latent image on its surface. Note that such a surface latent image is effective in silver halide development.

agents [58]. Cyclic hydrazides such as **72** have been disclosed as nucleating agents for the use of indole dye releasers such as **11** [59].

CH_3
NHNH–R
68a R = H
68b R = CHO

NH–C–NH–NHNH–CHO
S
69

C_5H_{11}-*t*
NH_2 O C_5H_{11}-*t*
CO
NH– NHNH–$COCH_3$
70

HO S
Ph N O
NHNH–$COCH_3$
71

S
H N N H
H N N H
H C_3H_7-*i*
72

Figure 16.16. Hydrazine and hydrazide nucleating agents for direct reversal core-shell emulsions. These compounds work in the nucleation step shown in Fig. 16.15.

Quaternary ammonium salts such as **73a** and **73b** have been disclosed as nucleating agents for the conventional reversal process [60,61]. Quaternary ammonium salts containing a propargyl or related group (e.g., **74a**) have also been reported [62]. Quaternary salts of the same type (e.g., **74b**) have been used together with such hydrazide nucleating agents as listed in Fig. 16.16 [63]. Pyridinium salts such as **75** work as nucleating agents [64].

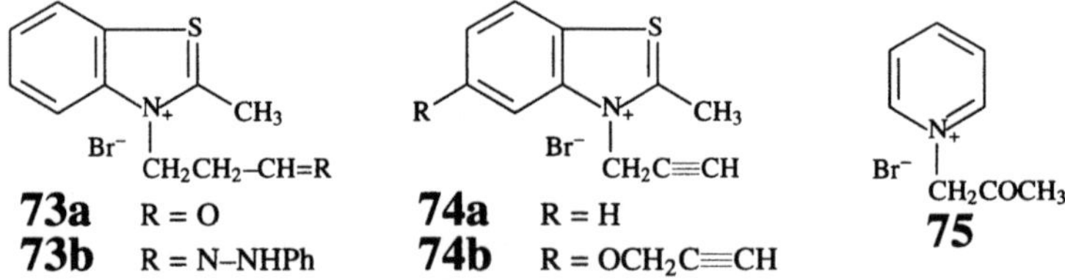

Figure 16.17. Heterocyclic nucleating agents for direct reversal core-shell emulsions. These compounds work in the nucleation step shown in Fig. 16.15.

Dye Releasers and Direct Reversal Emulsions

The direct reversal emulsions (Fig. 16.15) can be combined with any of the dye releasers shown in Fig. 16.5, among which *p*-sulfonamidonaphthol dye releasers

(**4**, Kodak's PR-10, 1976, Chapter 18) and *o*-sulfonamidophenol dye releasers (**8**, Fuji's FI-10, 1981, Chapter 19) have been used in commercialized systems.

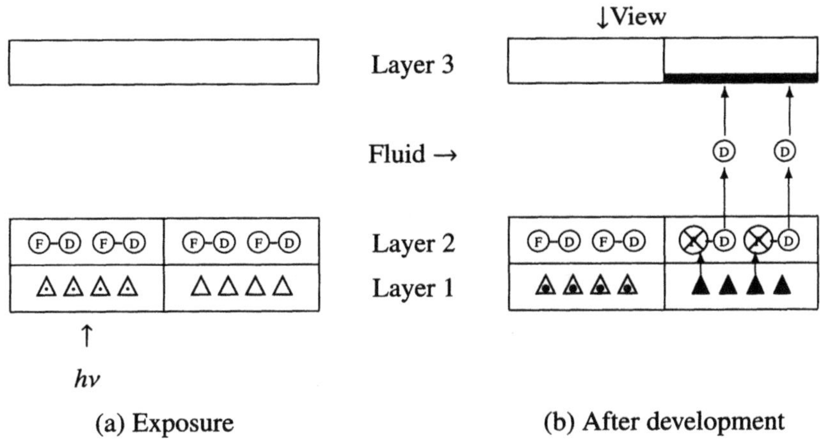

Figure 16.18. Direct reversal process using dye releasers with direct reversal emulsions during exposure (left) and after development (right). These schematic cross-sections omit polymer supports on which Layers 1 and 2 are coated: (Layer 1) a photo-sensitive layer containing a direct reversal emulsion, (Layer 2) a dye-releaser layer, and (Layer 3) an image-receiving layer, where the symbols represent Ⓕ-Ⓓ: a dye releaser (immobile), ⊗-Ⓓ: an oxidized dye releaser (immobile but capable of releasing a mobile dye), Ⓓ: a released dye, △: unexposed AgX, △: exposed AgX with internal latent image, ▲: exposed AgX with latensified latent image, ▲: developed silver, and ▬▬ : dye image deposited on the image-receiving layer (Layer 3).

A minimum set of functions for this type of reversal mechanisms is shown in Fig. 16.18, where Layer 1 of the left diagram is a photo-sensitive layer containing a direct reversal silver halide emulsion, Layer 2 is a dye releaser layer containing a dye releaser (Ⓕ-Ⓓ), and Layer 3 is an image-receiving layer.

The nucleation/development intensifies an internal latent image (△→ ▲) in the exposed area so that no oxidation of an electron transfer agent (ETA fed from an activator fluid) occurs in accord with the scheme in the top row of Fig. 16.15. On the other hand, the nucleation/development causes the cross-oxidation of silver halide and the electron transfer agent (ETA), generating the oxidized species (ETA_{ox}). The oxidized ETA diffuses from Layer 1 to Layer 2, where it oxidizes the dye releaser (Ⓕ-Ⓓ→ ⊗-Ⓓ) to release a dye (Ⓓ). The resulting dye (Ⓓ) diffuses and reaches the image-receiving layer (Layer 3).

The *p*-sulfonamidonaphthol dye releasers (**4**, Kodak) and *o*-sulfonamidophenol dye releasers (**8**, Fuji), which have been used in commercialized systems, will be discussed in detail in following chapters (Chapters 18 and 19). Although the indole dye releasers (**11**, Agfa) have not been placed on the market, a concrete embodiment of them should be described briefly in order to obtain a balanced view on the R&D of dye releasers.

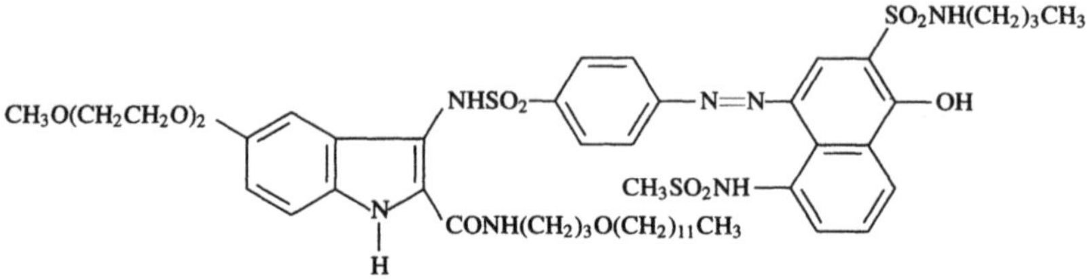

$SO_2NH(CH_2)_3CH_3$

$CH_3O(CH_2CH_2O)_2$ NHSO$_2$ —N=N— OH

CH_3SO_2NH

N CONH$(CH_2)_3O(CH_2)_{11}CH_3$

H

76

Figure 16.19. An indole dye releaser. See Fig. 16.5(3).

The indole dye releaser **76**, which is capable of releasing a diffusible magenta dye according to the scheme shown in Fig. 16.5(3), can be combined with direct reversal emulsions. The dye releaser has a multi-oxaalkyloxy group (e.g., $CH_3O(CH_2CH_2O)_2$) at the 5-position, along with a ballast group at the 2-position and a dye moiety at the 3-position [21]. The replacement of a methoxy group at the 5-position [20] by such a multi-oxaalkyloxy group causes a more increased D_{max}-value. A set of indole dye releasers for full color reproduction based on direct reversal emulsions has been disclosed in a patent [65], although the main claim of the patent has been concerned with acid polymers.

16.4.3 Direct Reversal Mechanisms Based Upon Positive-Working Dye Releasers

Oxidative-Stopping Dye Releasers

This type of dye releasers are frequently called "BEND compounds" (ballasted electron-accepting nucleophilic displacement compounds), as collected in Fig. 16.8. They are positive-working, as found in the left diagram of Fig. 16.20, which employs a negative emulsion.

Layer 1 of the left diagram of Fig. 16.20 is a photo-sensitive layer containing a negative emulsion (Δ), Layer 2 is a dye-releaser layer containing a BEND compound (Ⓕ-Ⓓ), and Layer 3 is an image-receiving layer. After the exposure of the left half of the film (Δ → Δ•), a development fluid is spread between Layer 2 and Layer 3 to initiate the development and diffusion processes (Fig. 16.20b). Thereby, an electron transfer agent (ETA) fed from the development fluid is oxidized by the silver halide emulsions in the exposed area (Δ• → ▲). The resulting oxidized ETA (ETA_{ox}) diffuses to Layer 2 (represented by the upward arrows from Layer 1 and Layer 2 in Fig. 16.20b), where a redox reaction takes place between the ETA_{ox} and the dye releaser (Ⓕ-Ⓓ, e.g., **35** or **40**) so as to

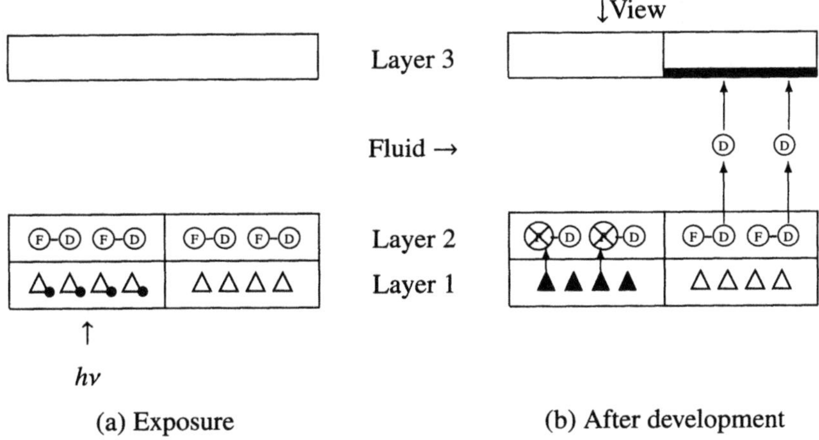

Figure 16.20. Direct reversal process using positive-working dye releasers of oxidative-stopping type during exposure (left) and after development (right). These schematic cross-sections omit polymer supports on which Layers 1 and 2 are coated: (Layer 1) a photo-sensitive layer containing a negative emulsion, (Layer 2) a dye-releaser layer, and (Layer 3) an image-receiving layer, where the symbols represent Ⓕ-Ⓓ: a positive-working dye releaser (immobile but capable of releasing a mobile dye), ⊗-Ⓓ: an oxidized dye releaser (immobile but incapable of releasing a mobile dye), Ⓓ: a released dye, △: unexposed AgX, △•: exposed AgX with latent image, ▲: developed silver, and ▬: dye image deposited on the image-receiving layer (Layer 3).

yield the corresponding oxidized intermediate (⊗-Ⓓ, e.g., **37** or **43**). The latter intermediate is incapable of releasing a diffusible dye.

On the other hand, the dye releaser in the unexposed area is capable of releasing a diffusible dye (Ⓓ, e.g., **39** or **42**) in accord with either of the mechanisms shown in Fig. 16.8. The dye diffuses to Layer 3, as represented by the upward arrows from Layer 2 to Layer 3 (Fig. 16.20b), so as to give a positive dye image (▬).

As a concrete embodiment, let us consider a ballasted hydroquinone having a temporarily shifted dye moiety (**77**). This compound releases a diffusible dye according to the scheme shown in Fig. 16.8(2) [40]. The absorption of the dye incorporated in **77** is shifted to a shorter wavelength region by protecting the hydroxyl group as a 4-chlorobenzoyloxy group. During the development under alkaline conditions, the protective group is removed to regenerate the hydroxyl group so as to exhibit the original color (cyan).

One of the merits of the use of such temporarily shifted dye releasers is its capability of being incorporated into the same layer as the corresponding silver halide emulsion is incorporated. In this case, the shifted dye releaser (**77**) exhibits magenta hue and can be incorporated *in* a red-sensitive emulsion layer. However, the stability of the protected hydroxyl group during storage is frequently difficult to be compatible with the rate of the hydrolyzation during development to generate the original absorption.

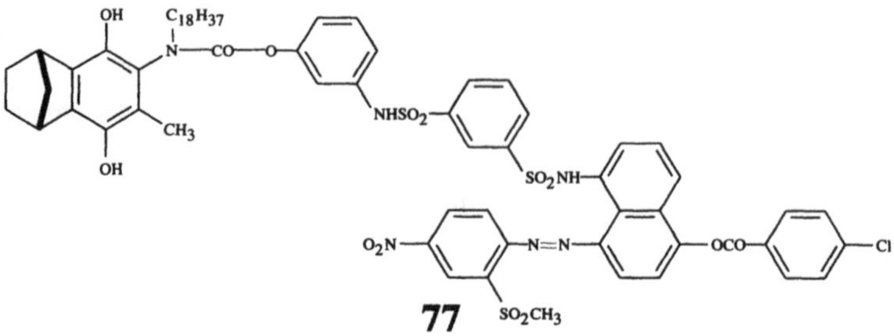

77

Figure 16.21. A ballasted hydroquinone having a temporarily shifted dye moiety. See Fig. 16.8(2).

Reductive-Releasing Dye Releasers

Since dye releasers of reductive-releasing type listed in Fig. 16.9 are positive-working, they can be combined with negative silver halide emulsions to obtain direct positive images. Among them, carquin dye releasers (**52**, Agfa) and ROSET dye releasers (**56**, Fuji) have been used in commercialized photographic materials. Hence, these dye releasers shall be discussed in detail in Chapter 20. In this section, instead, we use another dye releaser (a nitro-type dye releaser) to illustrate essential features of the reversal mechanism at issue since no further discussions on the dye releaser will appear in the remaining part of this book.

A minimum set of functions for obtaining direct positive images is shown in Fig. 16.22, where a nitro-type dye releaser (**44**) is used together with an electron donor [43]. Layer 1 is a photo-sensitive layer containing a negative emulsion, Layer 2 is a dye-releaser layer containing a reductive-releasing dye releaser (Ⓕ-Ⓓ) and a ballasted electron donor (Ⓔ), and Layer 3 is an image-receiving layer.

The left half of the film of Fig. 16.22a is exposed to light (△ → △•). Then a development fluid is spread between Layer 2 and Layer 3 to initiate the development and diffusion processes (Fig. 16.22b). The exposed silver halide with a latent image (△•) in Layer 1 is reduced by an electron transfer agent (ETA) fed from the development fluid (△• + ETA → ▲ + ETA_{ox}). The oxidized ETA (ETA_{ox}) diffuses from Layer 1 to Layer 2 (the upward arrow), where it consumes the electron donor (Ⓔ + ETA_{ox} → ⊗ + ETA) so as to inhibit the redox reaction between the dye releaser (Ⓕ-Ⓓ) and the electron donor (Ⓔ).

On the other hand, an unreacted electron donor in the right half of Layer 2 (Fig. 16.22b) reduces a dye releaser (Ⓔ + Ⓕ-Ⓓ → ⊗ + ⊘-Ⓓ). The reduced species (⊘-Ⓓ) is capable of releasing a diffusible dye (ⓓ), which diffuses to Layer 3, as represented by upward arrows to give dye images (▬▬).

Figure 16.23 shows a nitro-type dye releaser (**78**) and a ballasted electron donor (**79**), which can be used in Layer 2 illustrated in Fig. 16.22 [43]. The dye releaser is capable of releasing a diffusible dye in accord with the scheme shown in Fig. 16.9(1).

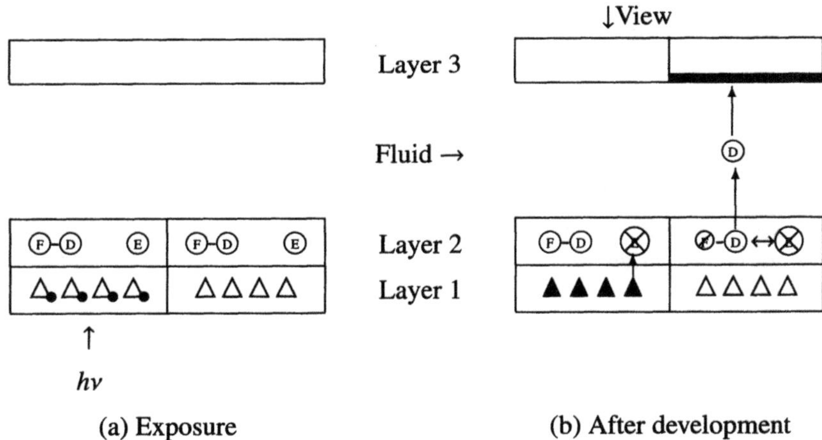

Figure 16.22. Direct reversal process using dye releasers of reductive-releasing type during exposure (left) and after development (right). These schematic cross-sections omit polymer supports on which Layers 1 and 2 are coated: (Layer 1) a photo-sensitive layer containing a negative emulsion, (Layer 2) a dye-releaser layer containing a reductive-releasing dye releaser (Ⓕ-Ⓓ) and a ballasted electron donor (Ⓔ), and (Layer 3) an image-receiving layer, where the symbols represent Ⓕ-Ⓓ: a dye-releasing dye releaser (immobile and incapable of releasing a diffusible dye), Ⓕ̸-Ⓓ: a reduced dye releaser (immobile but capable of releasing a diffusible dye), Ⓓ: a released dye, Ⓔ: an electron donor, ⊗: an oxidized electron donor, △: unexposed AgX, △•: exposed AgX with latent image, ▲: developed silver, and ▬: dye image deposited on the image-receiving layer (Layer 3).

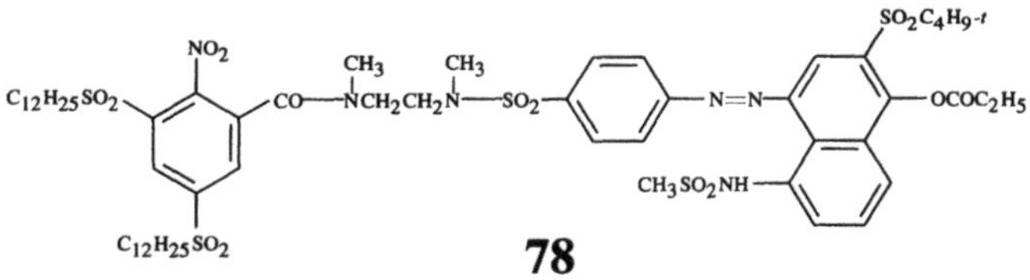

$C_{12}H_{25}SO_2$ NO$_2$ CO–N(CH$_3$)CH$_2$CH$_2$N(CH$_3$)–SO$_2$ N=N SO$_2$C$_4$H$_9$-t OCOC$_2$H$_5$ CH$_3$SO$_2$NH $C_{12}H_{25}SO_2$

78

CO–N(CH$_3$)CH$_2$CH$_2$SO$_2$NHC$_{18}$H$_{37}$ N–CH$_3$ O O $C_{18}H_{37}NHSO_2CH_2CH_2$N(CH$_3$)–CO

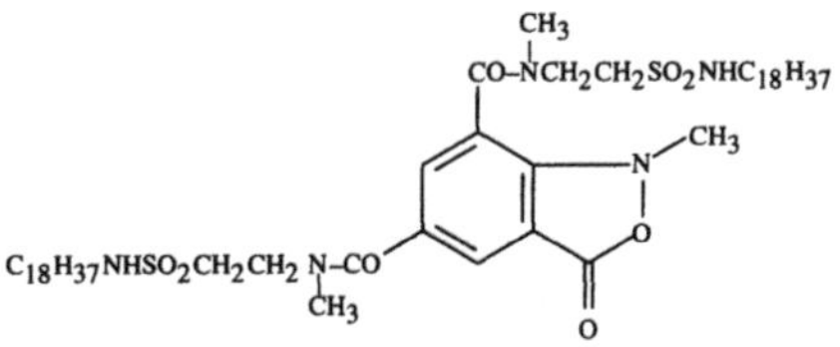

79

Figure 16.23. A nitro-type dye releaser (**78**) having a temporarily shifted dye moiety together with a ballasted electron donor (**79**). See Fig. 16.9(1).

Since the dye moiety is a temporarily shifted dye (yellow), the dye releaser (**78**) can be incorporated in a green-sensitive emulsion layer without lowering film speed. After development, the protective group (COC_2H_5) is removed by hydrolysis to exhibit the original magenta hue. However, this embodiment has not been realized to be placed on the market. It should be noted that the electron donor (**79**) is an application of the dye releasers illustrated in Fig. 16.18(1).

The other dye releasers listed in Fig. 16.9 can be employed in the diagram illustrated in Fig. 16.22. Among them, carquins and ROSET compounds will be discussed in detail (Chapter 20), since they have been used in commercialized photographic materials.

Direct Reversal Process Using Thiazolidine Dye Releaser

Thiazolidine dye releasers shown in Fig. 16.10 work as a kind of positive-working dye releasers. Figure 16.24 illustrates a minimum set of layers using such thiazolidine dye releasers (Ⓕ-Ⓓ): Layer 1 (a photo-sensitive layer) containing a negative emulsion (△), Layer 2 (a dye-releaser layer) containing a thiazolidine dye releaser (Ⓕ-Ⓓ), and Layer 3 (an image-receiving layer).

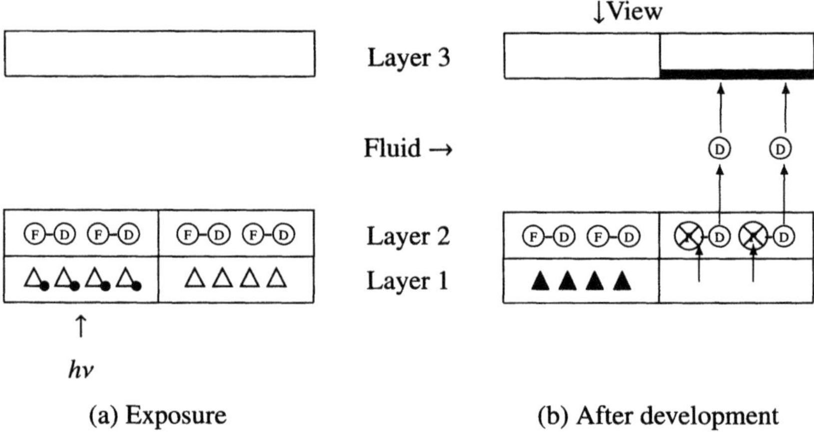

Figure 16.24. Direct reversal process using thiazolidine dye releasers during exposure (left) and after development (right). These schematic cross-sections omit polymer supports on which Layers 1 and 2 are coated: (Layer 1) a photo-sensitive layer containing a negative emulsion, (Layer 2) a dye-releaser layer, and (Layer 3) an image-receiving layer, where the symbols represent Ⓕ-Ⓓ: a dye releaser (immobile), ⊗-Ⓓ: a chelated dye releaser capable of releasing a mobile dye, Ⓓ: a released dye, △: unexposed AgX, △•: exposed AgX with latent image, ▲: developed silver, and ▬▬: dye image deposited on the image-receiving layer (Layer 3).

Upon development (Fig. 16.24b), the unexposed silver halide dissolved by an appropriate silver halide solvent, which diffuses to Layer 2, as represented by the upward arrows from Layer 1 and Layer 2. The dissolved silver species react with the dye releaser (Ⓕ-Ⓓ), the thiazolidine ring of which is cleaved to release a dye having an aldehyde group (Ⓓ). The released dye diffuses to Layer 3 so as to give a positive image (▬).

Figure 16.25 shows the concrete structure of a thiazolidine dye releaser (**80**), which can be incorporated into Layer 2 of Fig. 16.24 [47]. The thiazolidine ring is cleaved so that a yellow azo dye is released according to the scheme shown in Fig. 16.10. Cyan dye releasers which are characterized by two thiazolidine rings (e.g., **81**) have been disclosed recently [66].

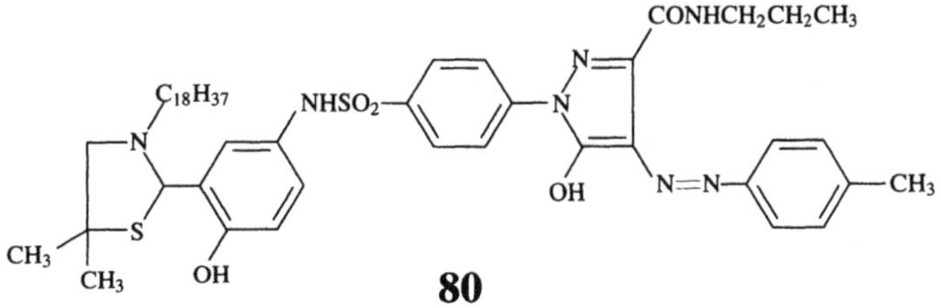

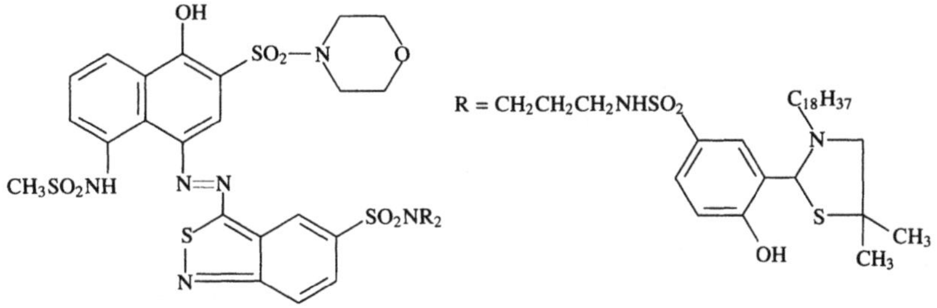

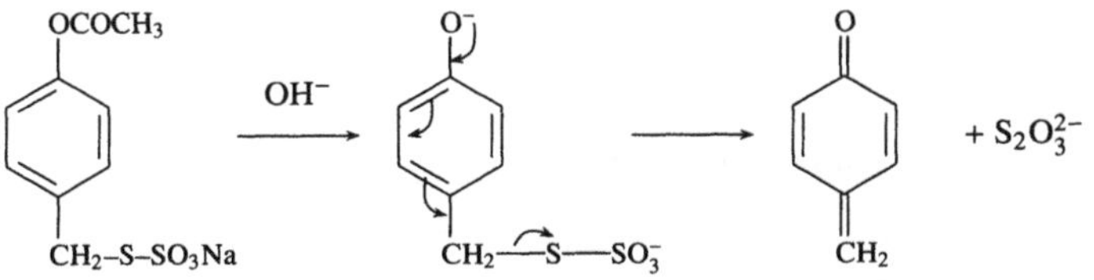

Figure 16.25. A thiazolidine dye releaser and a silver halide solvent precursor. See Fig. 16.10.

The silver halide solvent (e.g., $Na_2S_2O_3$), which dissolves the unexposed silver halide and transports it from Layer 1 to Layer 2, is contained in a development fluid. The solvent can be supplied from Layer 3 in the form of a silver halide solvent precursor (such as **82**), which releases a thiosulfate anion via a quinone-methide mechanism shown in Fig. 16.25 [47].

A more elaborate mechanism in which a dye aldehyde released from a thiazolidine dye releaser further undergoes a β-elimination has been disclosed later

[67]. Several investigations on the dye moieties of thiazolidine dye releasers have appeared: e.g., azomethine dyes [68,69] and azo dyes [70]. Dye releasers in which two thiazolidine rings are incorporated have been proposed [70,71]. More recently, the combined (hybridized) use of dye developers and thiazolidine dye releasers has been disclosed in several patents (e.g., [72]), which will be discussed in Chapter 17.

References

[1] Fujita S (1981) Yuki Gosei Kagaku Kyokaishi. 39:331
[2] Fujita S (1982) Yuki Gosei Kagaku Kyokaishi. 40:176
[3] Fujita S (1982) Kagaku no Ryoiki. 36:617
[4] Fujita S (1983) Kino Zairyo. 3:66
[5] Fujita S (1983) J Soc Photogr Sci Technol Jpn. 46:525
[6] Fujita S (1984) Senshoku Kogyo. 32:167
[7] Itoh I, Fujita S (1985) Denshi Shashin. 24:339
[8] Fujita S (1986, 2002) In: Functionalized Organic Chemicals for Silver Halide Color Photographic Materials, CMC, Tokyo. Chapter 3
[9] Ohyama Y (1970) J Soc Photogr Sci Technol Jpn. 32:243
[10] Land EH (1972) Photogr Sci Eng. 16:247
[11] Hanson WT (1976) Photogr Sci Eng. 20:155
[12] Van de Sande CC (1983) Angew Chem Int Ed Engl. 22:191
[13] Rogers HG (1961) US Patent 2 983 606
[14] Fleckenstein LJ (1975) US Patent 3 928 312
[15] Fleckenstein LJ, Figueras J (1978) US Patent 4 076 529
[16] Larsen AA, Lish PM (1964) Nature. 203:1283
[17] Fleckenstein LJ (1977) US Patent 4 076 529
[18] Koyama K, Maekawa Y, Miyakawa M (1977) US Patent 4 055 428
[19] Vetter H (1976) Jpn Kokai S51-104343
[20] Vetter H, Marx P (1979) US Patent 4 179 291
[21] Jaeken J, Verhecken A, Vetter H, Marx P (1981) US Patent 4 273 855
[22] Vetter H, Püschel W, Melzer A, Peters M (1980) US Patent 4 198 235
[23] Odenwälder H, Vetter H, Janssens W, Jaeken J (1982) US Patent 4 360 581
[24] Vetter H, Püschel W, Marx P (1979) US Patent 4 171 220
[25] Püschel W, Vetter H, Marx P (1980) US Patent 4 242 435
[26] Anderson AE, Lum KK (1973) US Patent 3 725 062
[27] Gompf TE, Lum KK (1972) US Patent 3 698 897

[28] Bloom SM, Stephens RK (1969) US Patent 3 443 940

[29] Whitmore KE, Mader PM (1966) US Patent 3 227 550

[30] Barr CR (1972) US Patent T900 029

[31] Barr CR, Bush WM, Thomas LJ, Cole HE (1973) GB Patent 1 330 524

[32] Bloom SM, Rogers HG (1969) US Patent 3 443 940

[33] Bloom SM (1973) US Patent 3 751 406

[34] Püschel W, Danhäuser J, Kabitzke K, Marx P, Melzer A, Schranz K-W, Vetter H, Perz W (1973) US Patent 3 628 952

[35] Püschel W, Marx P (1973) US Patent 3 736 136

[36] Danhäuser J (1973) US Patent 3 730 718

[37] Püschel W, Vetter H, Odenwälder H (1973) Jpn Kokai S49-64436

[38] Vetter H, Danhäuser J, Kabitzke K, Marx P, Melzer A, Pelz W, Püschel W (1973) GB Patent 1 309 133

[39] Fields DL, Henzel RP, Lau PTS, Chasman RA (1978) US Patent 4 108 850

[40] Fields DL, Henzel RP, Lau PTS, Chasman RA (1976) US Patent 3 980 479

[41] Hinshaw JC, Henzel RP (1979) US Patent 4 139 389

[42] Hinshaw JC, Henzel RP (1980) US Patent 4 218 368

[43] Chasman RA, Dunlap RP, Hinshaw JC (1979) US Patent 4 139 379

[44] Janssens W (1980) US Patent 4 232 107

[45] Nakamura K, Nakamura S (1988) US Patent 4 783 396

[46] Cieciuch RFW, Luhowy RR, Meneghini FA, Rogers HG (1973) US Patent 3 719 489

[47] Cieciuch RFW, Luhowy RR, Meneghini FA, Rogers HG (1977) US Patent 4 060 417

[48] Cieciuch RFW, Luhowy RR, Meneghini FA, Rogers HG (1978) US Patent 4 098 783

[49] Cole HE (1974) Jpn Patent S49-21660

[50] Wingender K, Vetter H, Janssens W (1976) Jpn Kokai S51-77335

[51] Barr CR, Williams J, Whitmore K (1966) US Patent 3 227 551

[52] Bacon RE (1977) Latent Image Effects Leading to Reversal or Desensitization. In: James TH (ed) The Theory of the Photographic Process. Macmillan, New York London, Chapter 7. See especially Section A7

[53] Ives CE (1951) US Patent 2 563 785

[54] Whitmore KE (1966) US Patent 3 227 552

[55] Evans FJ (1975) US Patent 3 923 513

[56] Leone RE, Weber II WW, Wrathall DP (1977) US Patent 4 030 925

[57] Leone RE, Weber II WW, Wrathall DP (1977) US Patent 4 031 127

[58] Leone RE, Elwood E (1978) US Patent 4 080 207

[59] von König A, Odenwälder H, Peters M, Püschel W (1979) US Patent 4 139 387

[60] Lincoln LL, Heseltine DW (1971) US Patent 3 615 615

[61] Lincoln LL, Heseltine DW (1973) US Patent 3 759 901

[62] Adachi K, Ikeda T, Tsujino N (1978) US Patent 4 115 122

[63] Heki T, Inoue N, Hirano S (1989) US Patent 4 880 729

[64] Oishi Y, Hirano S (1982) US Patent 4 324 855

[65] Helling G, Krafft W, Findeisen K, Himmelmann W (1981) US Patent 4 284 708

[66] Arnost MJ, Viski P, Waller DP, Whritenour DC (1998) US Patent 5 716 754

[67] Simon M, Waller DP (1985) US Patent 4 535 051

[68] Arnost MJ, Chinoporos E, McGowan DA, Waller DP (1995) US Patent 5 415 970

[69] Chinoporos E, Pauze RH, Waller DP, Whritenour DC (1995) US Patent 5 414 091

[70] Viski P, Waller DP (1997) US Patent 5 658 705

[71] Arnost MJ (1994) US Patent 5 320 929

[72] Foley JA, Filosa MP, Telfer SJ, Marshall JL, Waller DP (2003) US Patent 6 541 177 B1

Chapter 17

Dye Developers

17.1 Chemical Processes and Multilayer Structure

The first instant color film based on the chemistry of dye developers was placed on the market under the name Polacolor in 1963 from Polaroid Corporation. The Polacolor film has adopted a peel-apart format, which has required a so-called Land's camera. Later, the SX-70 photographic system was introduced by Polaroid as a total system based on an integrated (mono-sheet) film and a specific camera containing a mirror (1972) [1,2]. The chemistry of dye developers has been discussed in an excellent review by Bloom et al. [3]. The beginnings of instant color photography have been looked back upon briefly by Bloom [4] and by Rogers [5].

17.1.1 Chemical Processes of Dye Developers

Dye developers are positive-working compounds of dye-stopping type. The principal reactions of the dye developers have been already discussed in Chapter 16, where the developer (hydroquinone) moiety of each dye developer is tentatively presumed to be oxidized directly by silver halide (Figs. 16.4 and 16.11). This, however, has made a thing too simple, because the dye developer of cyan, magenta, or yellow color is incorporated in a distinct layer of a multilayered film, located separately from the corresponding silver halide layer of red-, green-, or blue-sensitivity.

In order to mediate the redox reaction between the silver halide and the dye developer (**4**) migrating to the silver halide layer, an auxiliary developer such as tolylhydroquinone (**1**) is used as shown in Fig. 17.1. First, the dissociated tolyl-

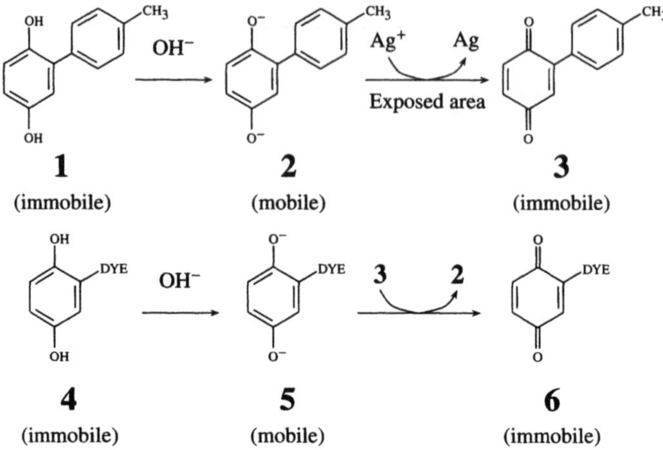

OH⁻ → ; Ag^+ → Ag, Exposed area (17.1)

1 (immobile) **2** (mobile) **3** (immobile)

OH⁻ → ; 3 → 2 (17.2)

4 (immobile) **5** (mobile) **6** (immobile)

Figure 17.1. Consecutive reactions of silver halide, an auxiliary developer (e.g., tolylhydroquinone), and a dye developer.

hydroquinone (**2**) is fed from an activator fluid and diffuses to a respective silver halide emulsion layer, where it is oxidized into tolylbenzoquinone (**3**) by silver halide in the exposed area, as shown in eq. 17.1. Meanwhile, the dye developer (**4**) is dissociated to give the corresponding dianion (**5**), which reaches the silver halide emulsion layer. Then, the tolylbenzoquinone (**3**) oxidizes the dissociated dye developer (**5**) to give the quinone form of the dye developer (**6**), as shown in eq. 17.2 (Fig. 17.1). Thereby, the oxidized dye developer is stopped within the silver halide layer so as to generate no dye images.

17.1.2 Multilayer Structure for Color Reproduction

Three dye developers for color reproduction are incorporated in a multilayered film, which comprises two outer supports, each having several coated layers as illustrated in Fig. 17.2 [1,2].

One of the two outer supports is a black opaque polyester sheet, on which a set of photo-sensitive units are coated. Each of the dye developers is coated on the sheet together with the silver halide emulsion of complementary sensitivity so that the resulting layers construct one of such photo-sensitive units. Thus, a cyan dye developer layer is adjacent to a red-sensitive emulsion layer, a magenta dye developer layer is adjacent to a green-sensitive emulsion layer, and a yellow dye developer layer is adjacent to a blue-sensitive emulsion layer. The three sets of photo-sensitive units are separated from each other by means of interlayers.

The other outer support is a transparent polyester sheet, on which a mordant layer for receiving a dye image is coated along with other important layers such as a timing layer and an acid layer.

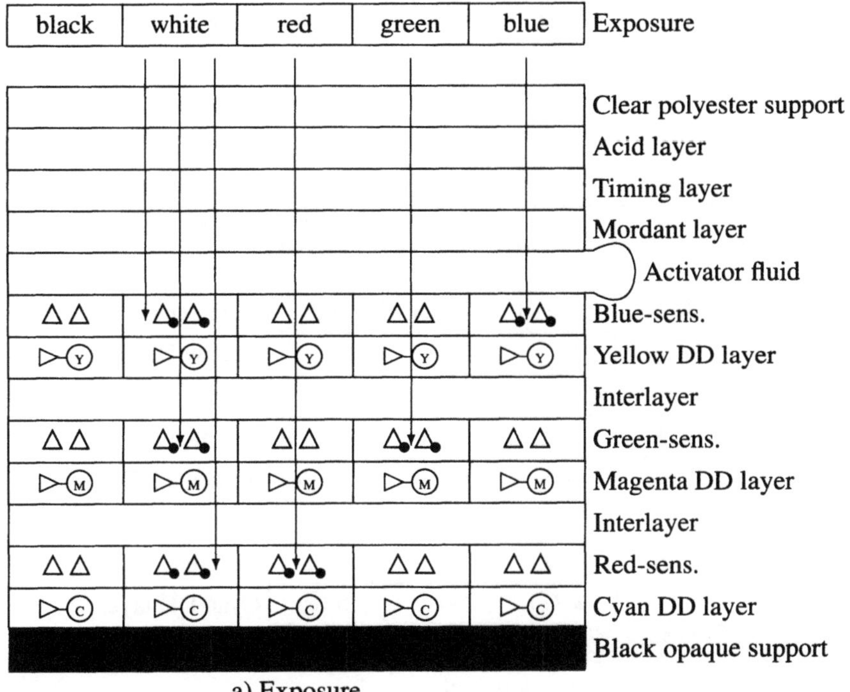

Figure 17.2. Instant color film using dye developers. A schematic cross-section during exposure. △: Silver halide grain, △•: Exposed silver halide grain with a latent image; ▷-(Y): Yellow dye developer, ▷-(M): Magenta dye developer, and ▷-(C): Cyan dye developer.

The two coated sheets are superimposed in a face-to-face fashion so as to construct an integrated (so-called "mono-sheet") film shown in Fig. 17.2, where a pod containing an activator fluid is attached between the two sheets.[1]

An exposure to blue, green, red, or white light (or to no light) is effected through the transparent support as shown in the top of Fig. 17.2 by using a specifically designed camera (e.g., an SX-70 camera), where a downward arrow represents such light as arriving at a silver halide grain and the symbol △• represents the resulting silver halide grain with a latent image.[2]

The integrated film exposed to light (Fig. 17.2) passes through the gap between a pair of rollers equipped in the camera. The pressure due to the rollers ruptures

[1] In instant color photography, the word "pod" is used to refer to a pressure rupturable processing container.

[2] In the case of a peel-apart film, the transparent support is replaced by a paper having a mordant layer (i.e., an opaque support). After the exposure of the photo-sensitive sheet, this sheet is superimposed onto the paper and the activator fluid in a pod is spread between them. Then, they are peeled apart to observe the image-receiving layer.

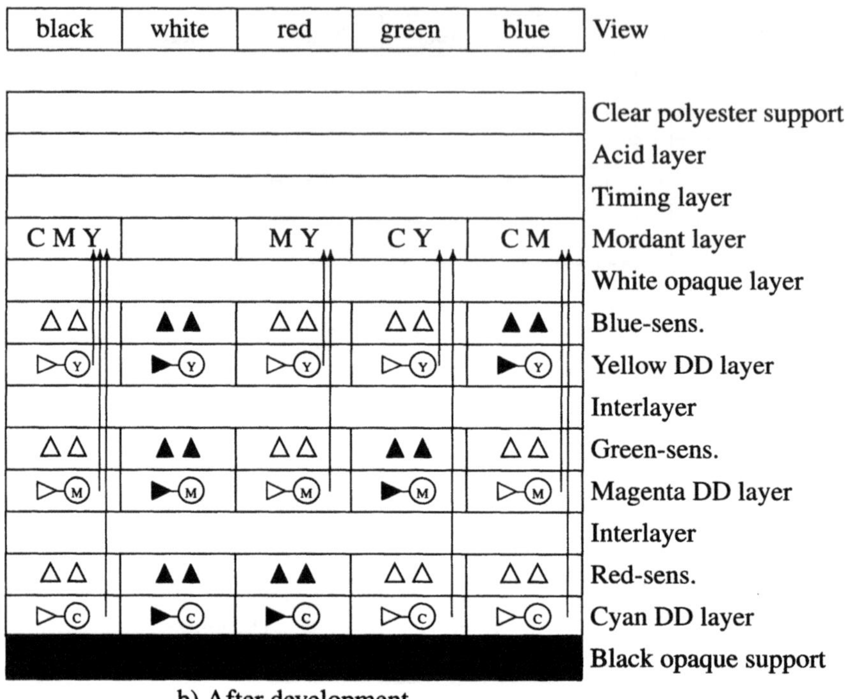

Figure 17.3. Instant color film using dye developers. A schematic cross-section after development. △: Silver halide grain, ▲: Developed silver halide grain; ▷-(Y): Yellow dye developer, ▷-(M): Magenta dye developer, ▷-(C): Cyan dye developer; ▶-(Y): Oxidized yellow dye developer, ▶-(M): Oxidized magenta dye developer, and ▶-(C): Oxidized cyan dye developer. Strictly speaking, the oxidized dye developers are stopped in the corresponding silver halide layers.

the pod to spread an activator liquid between the two outer sheets of the integrated film (Fig. 17.3). Thereby, the development of silver halide emulsions (△ → ▲) is initiated to cause the reactions shown in Fig. 17.1. For example, the silver halide grains in the area exposed to blue light oxidize molecules of a yellow dye developer so that these molecules are stopped within the blue-sensitive layer. In contrast, the molecules of a magenta dye developer and of a cyan dye developer diffuse to the mordant (image-receiving) layer and are captured there, as represented by upward arrows. Hence, the subtractive mixing of cyan and magenta produces blue color (C + M = B) in terms of the dye-stopping mechanism. Similarly, the dye-stopping mechanism reproduces green (C + Y = G), red (M + Y = R), and black (C + M + Y = Black). No captured dyes correspond to white light. The available colors are shown at the top of Fig. 17.3. These colors are observed against the background of the white titanium oxide pigment, which has been contained in the activator fluid and has been spread so as to produce the white background (the white opaque layer).

As discussed in Chapter 16 (Fig. 16.11), the absorption of a dye developer (e.g., 400–500 nm for a yellow dye developer) is located in the same region as that of the corresponding sensitizing dye (e.g., 400–500 nm for a blue-sensitizing dye). The discussion on one photo-sensitive unit in Chapter 16 (Subsection 16.4.1) can be extended to the present case of full color reproduction. Hence, this relationship decides the arrangement of each silver halide layer and the layer of the corresponding dye developer, when we adopt the integrated (mono-sheet) format. As shown in Fig. 17.2, we have an arrangement: a dye developer layer—an emulsion layer—(exposure site). On the other hand, the dye-stopping mechanism requires an arrangement: a dye developer layer—an emulsion layer—(a mordant layer for viewing). The combination of the two arrangements means that we have the mirror image of an original object if no modification is adopted for the mono-sheet format. Hence, the cameras for the SX-70 system have a mirror to obtain a picture having a correct left-right relationship.[3]

17.2 Dye Moieties of Dye Developers

Since the dye moiety of a dye developer molecule is insulated from a developer (hydroquinone) moiety, the former can be discussed independently of the latter. Such dye moieties have been selected from various types of dyes, as summarized in reviews [3,6,7].

17.2.1 Azo and Anthraquinone Dye Developers

A dye developer comprises a dye moiety (chromophore) and a developer moiety in agreement with the general formula (FUN—DYE) discussed in the preceding chapter. Dye developers used in the first Polacolor film (1963, Polaroid) had an azo dye moiety (magenta and yellow) or an anthraquinone dye moiety (cyan), as shown in Fig. 17.4 [8].

The dye moiety of each dye developer has been selected in order that the hue is independent of pH and of the state of oxidation. Moreover, the dye moiety is insulated from the developer moiety by using an appropriate alkylene group. The insulating alkylene links prevent unwanted color shifting by interrupting the conjugation between the two moieties. For example, the cyan dye developer (**7**) has an anthraquinone dye moiety, which is insulated from a developer moiety by a branched propylene group [9]. The magenta dye developer (**8**) and the yellow dye developer (**9**) use an ethylene group as an insulating linkage [10].

[3] The peel-apart format requires no mirror correction. See the discussion in Chapter 16 (Fig. 16.11).

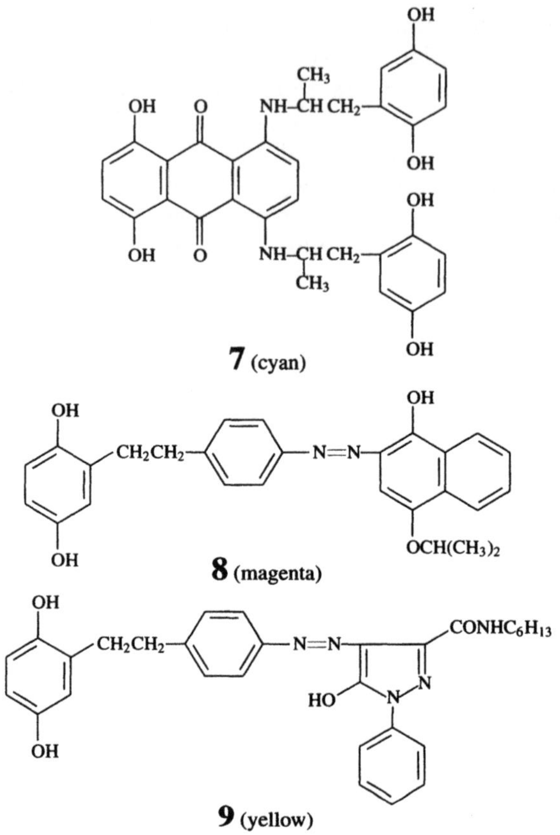

7 (cyan)

8 (magenta)

9 (yellow)

Figure 17.4. A set of cyan, magenta, and yellow dye developers for color reproduction of instant color photography.

Syntheses

Various synthetic pathways for preparing dye developers have been summarized in reviews [3,11]. The cyan dye developer (**7**) has been synthesized according to the scheme shown in Fig. 17.5, where leuco-1,4,5,8-tetrahdyroxyanthraquinone (**10**) is condensed with two moles of 2-(β-aminopropyl)hydroquinone (**11**) [9]. The use of pyridine-*N*-oxide in place of air oxygen has been disclosed to oxidize the leuco intermediate [12].

The syntheses of the magenta and yellow dye developers (**8** and **9**) have used the same amine intermediate, i.e., 2-(4-aminophenyl)ethyl-1,4-diacetoxybenzene hydrochloride (**16**), as an azo component, as illustrated in Fig. 17.6 [10]. 2,5-Dimethoxybenzaldehyde (**12**) is condensed with 4-nitrophenylacetic acid (**13**) in the presence of piperidine to yield a condensed intermediate (**14**), where a decarboxylation together with a dehydration leaves an olefinic double bond. After the hydrogenation of the double bond and the reduction of the nitro group have been

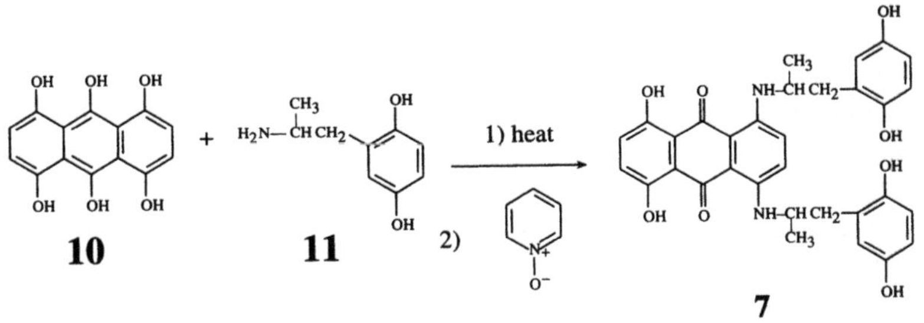

Figure 17.5. Synthetic pathway of the cyan dye developer 7.

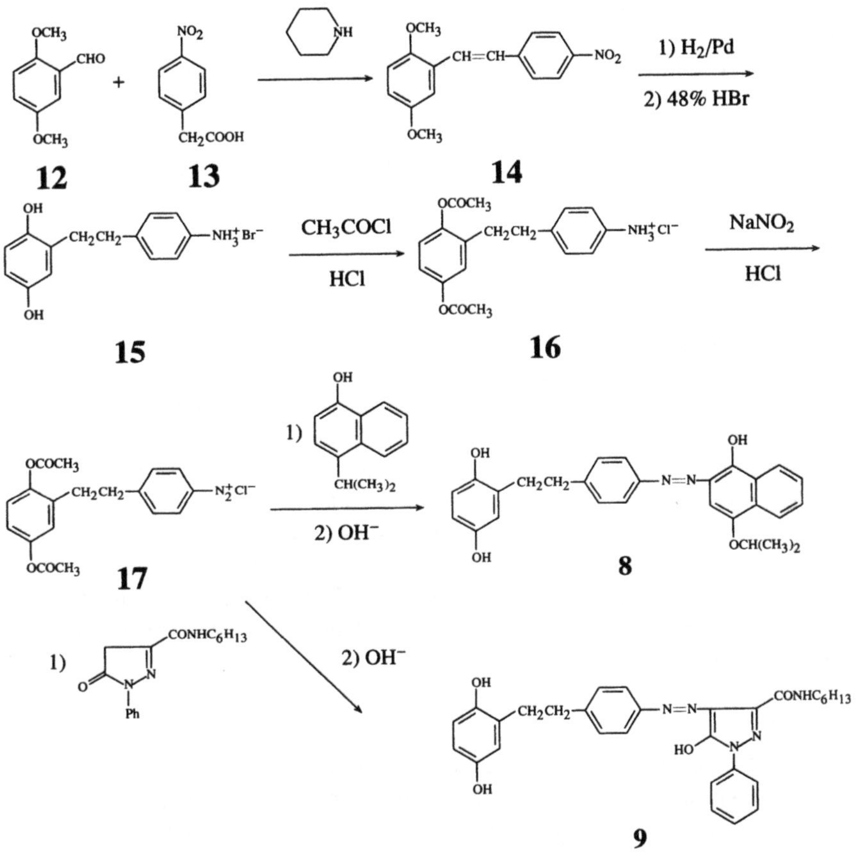

Figure 17.6. Synthetic pathways of the magenta dye developer (8) and of the yellow dye developer (9).

carried out at the same time, the protective methoxy groups are removed by means of concentrated hydrobromic acid so as to produce a hydroquinone intermediate (**15**). The two hydroxyl groups are again protected as acetoxy groups to yield the azo component (**16**).

The azo component (**16**) is diazotized to give the corresponding diazonium salt (**17**), which is coupled with 4-isopropoxy-1-naphthol as a coupling component. The protective (acetyl) groups of the resulting intermediate are removed to yield the magenta dye developer (**8**). Similarly, the use of 1-phenyl-3-(*N*-hexylcarbamolyl)-5-pyrazolone as a coupling component results in the production of the yellow dye developer (**9**).

17.2.2 Temporarily Shifted Dye Developers

There have been several attempts to shift the original absorption of a dye moiety into a shorter wavelength region during exposure and to regenerate the original absorption after development. Such dye developers as having a hypsochromic absorption are called "temporarily shifted dye developers". Thereby, a dye developer layer and a relevant silver halide emulsion layer can be exchanged against the direction of exposure. If sufficiently stable, the dye developer can be incorporated in the emulsion layer.

Acylated Azo Dye Moieties

As examples of temporarily shifted dye developers, Fig. 17.7 shows a cyan (**18**) [13], a magenta (**19**) [14,15], and a yellow dye developer (**20**) [16].

Usually, an azo dye forms an equilibrium (tautomeric) mixture in which a hydrazone form (=N–NH–) is present as a main species together with a less amount of an azo form (–N=N–). In general, the former has a longer absorption than the latter. The conversion of the hydroxyl group into an acetyloxy group fixes an azo form, as found in each of the dye developers shown in Fig. 17.7. As a result, their absorptions are hypsochromically shifted by about 100 nm.

As an example of the shift of a temporarily shifted dye, Fig. 17.8 shows the shifted absorption of the dye developer (**21**) and the regenerated absorption of the hydrolyzed dye developer (**22**) [17].

Syntheses

The shifted magenta dye developer (**19**) has been synthesized according to the scheme shown in Fig. 17.9 [14]. By starting from the magenta dye developer (**8**), the hydroquinone moiety is first oxidized into a benzoquinone. Thereby, the original hydroxyl groups of the hydroquinone moiety are protected from the following acetylation, as found in the resulting key intermediate (**23**). The hydroxyl group conjugate to the azo group in **23** is acetylated by using isopropenyl acetate under acidic conditions. The quinone moiety of the resulting intermediate (**24**) is reduced into the original hydroquinone moiety to yield the shifted magenta dye

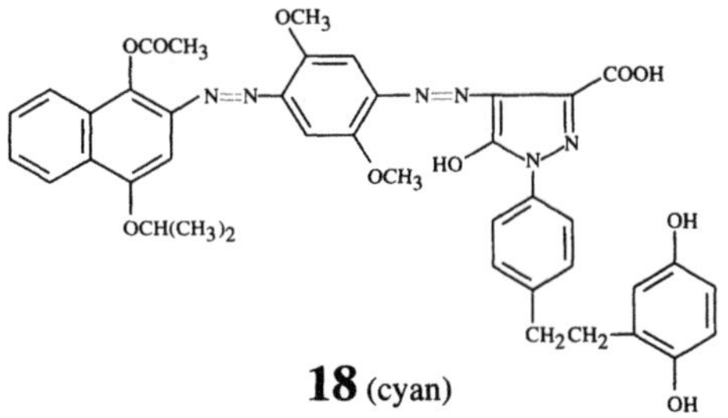

18 (cyan)

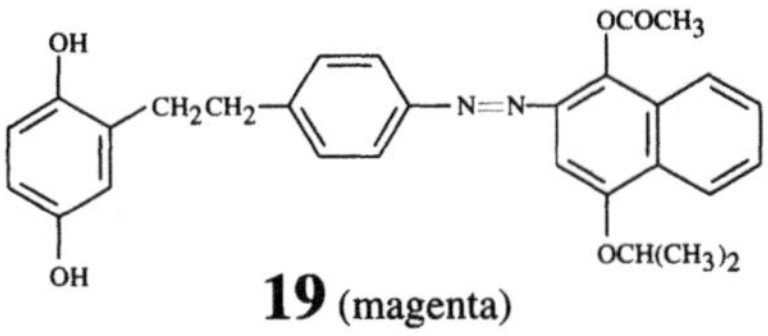

19 (magenta)

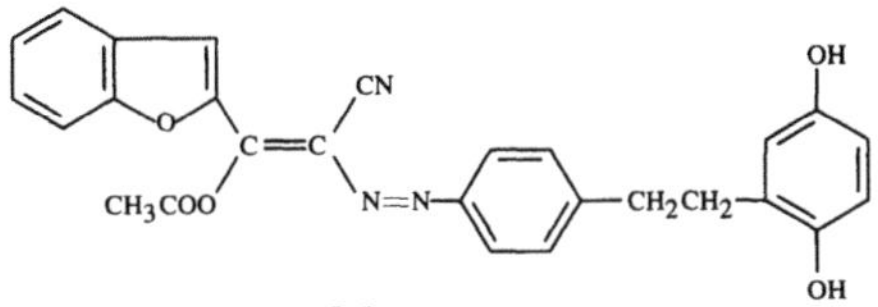

20 (yellow)

Figure 17.7. Temporarily shifted dye developers for color reproduction of instant color photography.

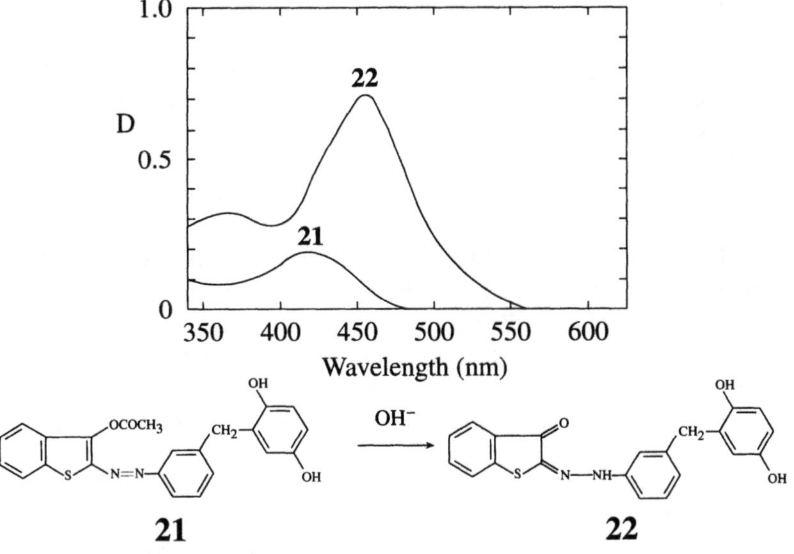

Figure 17.8. Absorption spectra of a temporarily shifted dye developer (**21**) and the regenerated dye developer (**22**) [17].

developer (**19**). This pathway totally requires seven steps if we take account of the azo component (**15**) as a starting material (cf. Fig. 17.6).

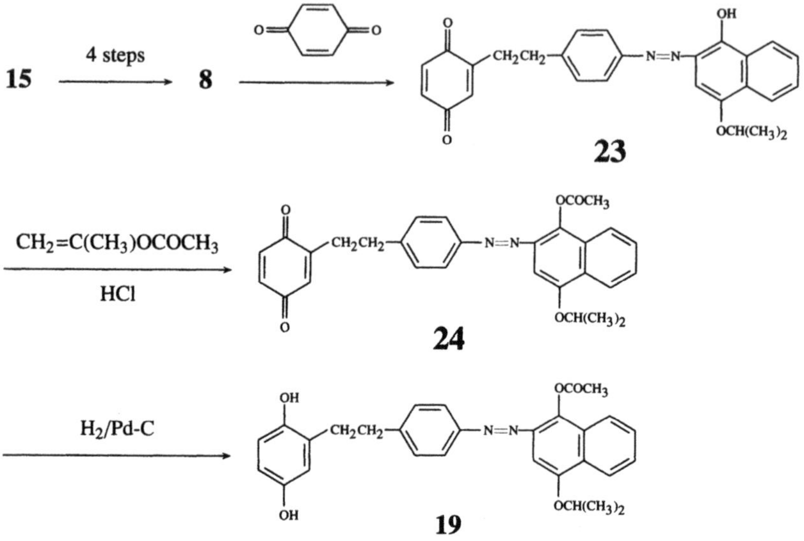

Figure 17.9. Synthetic pathway for preparing the temporarily shifted magenta dye developer (**19**) [14].

An improved pathway for preparing shifted magenta dye developers requires four steps from the azo component (**15**), as shown in Fig. 17.10 [15]. This pathway has employed so-called "diazoxidation," where the hydroquinone moiety of the azo component (**15**) is oxidized into a benzoquinone during the diazotization of the amino group [15]. The resulting intermediate (**25**) is stable under acidic conditions and capable of coupling with a coupling component.

Thereby, the key intermediate (**26**, the same intermediate as **23** if R is equal to isopropyl) is directly obtained. The acetylation and the subsequent reduction produce the target molecule (**29**). Note that **29a** (R = isopropyl) is the same compound as **19**. Compare the two pathways shown in Figs. 17.9 and 17.10.

The reduction of **24** into **19** (Fig. 17.9) has been carried out by using catalytic hydrogenation, which may reduce the azo group if a careful termination is not taken into consideration. On the other hand, *N*,*N*-diethylhydroxylamine is used as a selective reducing agent in the improved pathway shown in Fig. 17.10 [15, 18,19].

The improved pathway shown in Fig. 17.10 can also be regarded as a route to usual (non-shifted) dye developers (**28**). Thus, it requires three steps from the azo component to such non-shifted dye developers (**15** → **25** → **26** → **28**), while the pathway shown in Fig. 17.6 requires four steps (**15** → **16** → **17** → → **8**).

The shifted yellow dye developer (**21**) has also been synthesized by diazoxidation [17] except that 2,5-di-*t*-butylhydroquinone [20] is used as a reducing agent

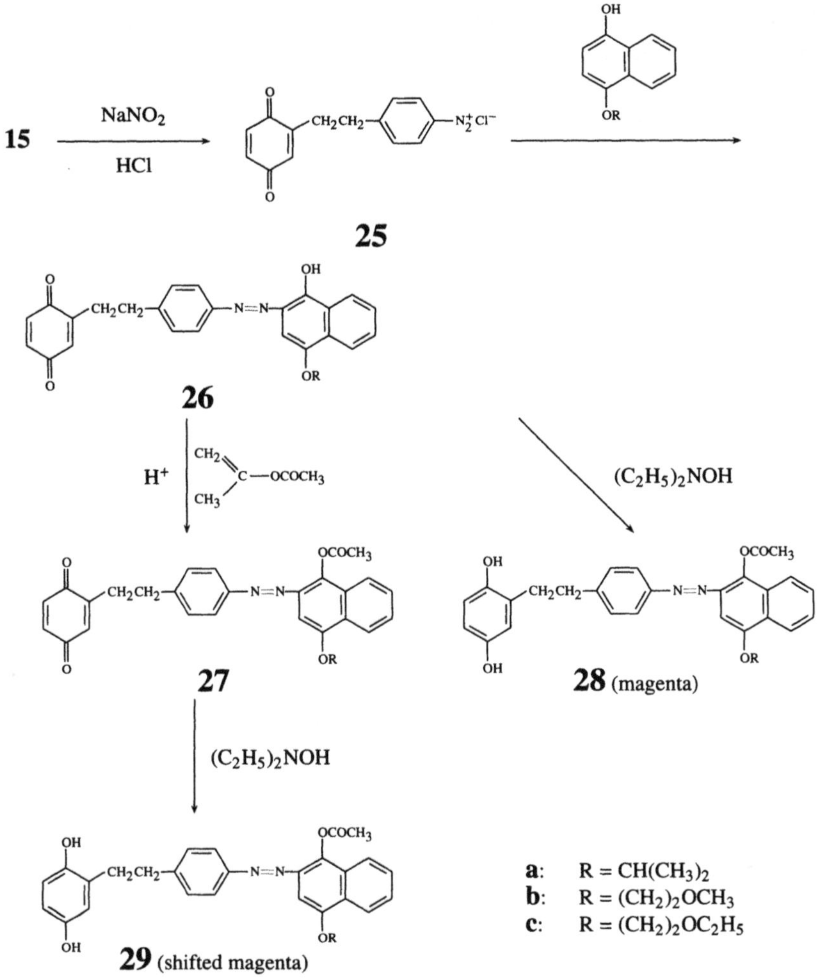

a: R = $CH(CH_3)_2$
b: R = $(CH_2)_2OCH_3$
c: R = $(CH_2)_2OC_2H_5$

Figure 17.10. Improved synthetic pathways for preparing magenta dye developers and the corresponding temporarily shifted dye developers [15].

in place of *N*,*N*-diethylhydroxylamine. The diazoxidation has been applied to the syntheses of other shifted dye developers [16,21]

Leuco Dye Moieties

The leuco forms of azomethine or indoaniline dyes have been incorporated into dye developer molecules [22,23,24]. They have a temporarily shifted absorption, which regenerates the original absorption upon oxidation.[4] Hence, they are called "oxichromic" compounds. The dye developer (**30**) shown in Fig 17.11 is a shifted dye developer of oxichromic type. Appropriate oxidizing agents are necessary to generate original colors; hence, iodoso compounds such as $PhI(OCOCH_3)_2$ [25], monomeric and polymeric *N*-oxides [26,27,28], etc. have been disclosed in patents.

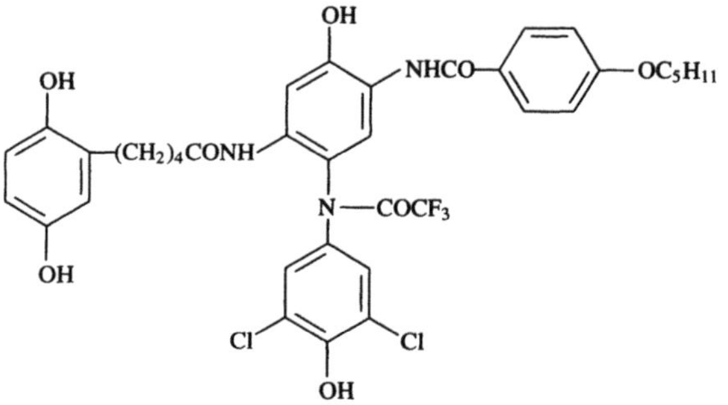

30 (shifted cyan)

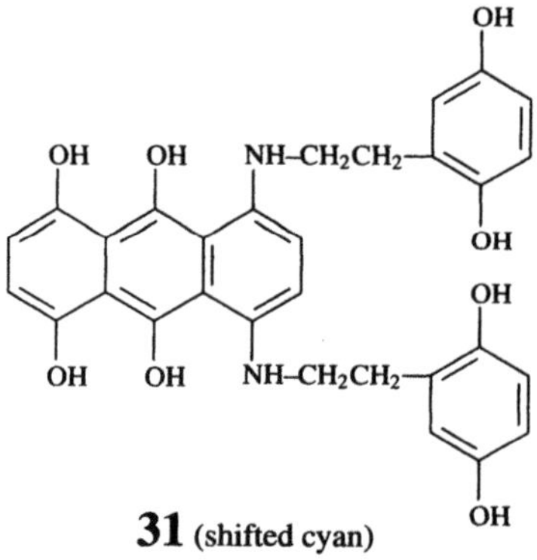

31 (shifted cyan)

Figure 17.11. Leuco dye developers for generating cyan dyes.

The dye developer (**31**) shown in Fig 17.11 has a leuco anthraquinone moiety, which is another shifted dye developer of oxichromic type [29]. The leuco dye moiety is oxidized to an anthraquinone dye similar to **7** and exhibits cyan color.

[4]The regenerated azomethine or indoaniline dyes belong to the same categories as those used in conventional color photography (image dyes generated by chromogenic development). See the previous chapters for further information on the chromogenic development.

17.2.3 Metallized Dye Developers

A Triplet of Metallized Dye Developers for Color Reproduction

The light and dark stabilities of image dyes have been crucial to embody the integrated (mono-sheet) format of the SX-70 system, because the image dyes are tied inside a half-dried film together with various compounds due to an activator fluid. Among possible approaches,[5] the stabilization by metallation has been adopted as a straightforward and powerful strategy. Thereby, the metallized dye developers shown in Fig. 17.12 have been found to exhibit excellent stabilities [3,6].

The cyan dye is a copper phthalocyanine derivative (**32**) [30]; the magenta is a chromium complex of an *o,o'*-dihydroxy azo dye (**33**) [31,32]; and the yellow is a chromed *o,o'*-dihydroxyazomethine (**34**) [33]. These metallized dye developers have long been used as a triplet of dye developers for full color reproduction, as found in several patents of other additives [34,35].[6]

The spectra of the metallized dye developers are shown in Fig. 17.13. The remarkable light stability of these dyes has been reported [36].[7]

Cyan Metallized Dye Developers

Anthraquinone dyes such as **7** had been used as a cyan dye developer in the old-type Polacolor system until Polaroid SX-70 system adopted the copper phthalocyanine derivative (**32**) in 1972 [3]. The copper phthalocyanine nucleus of **32** tolerated more severe conditions of the integrated film of the SX-70 system [30]. Because of the brilliant hue and the excellent light and dark stability, the dye developer (**32**) has seemed to be continuously used until now, as far as we have known from patents (e.g., [37]).

There have been several attempts to develop cyan dye developers of other types. For example, xanthene dyes having a cyan hue have been incorporated in dye developers [38,39].[8] Cyan dye developers having a benzoisothiazole azo moiety have been disclosed in patents [40,41,42], although their main targets have been thiazolidine dye releasers.

---

[5] As discussed in conventional color photography, approaches using stabilizers may be other selections.

[6] The triplet of metallized dye developers shown in Fig. 17.12 has been used from 1972 to the beginning of the 1980s. According to the author's watching of patents, the magenta metallized dye developer was replaced by a xanthene magenta dye developer during the 1980s. Further, the yellow dye developer has now been replaced by a thiazolidine dye releaser. The phthalocyanine dye developer seems to be used still now.

[7] It is to be noted that the magenta dye exhibits a rather large side absorption in the region of 400–500 nm. This is probably one of the reasons to change the metallized magenta into a xanthene dye developer.

[8] Xanthene dye developers having a magenta hue will be discussed later.

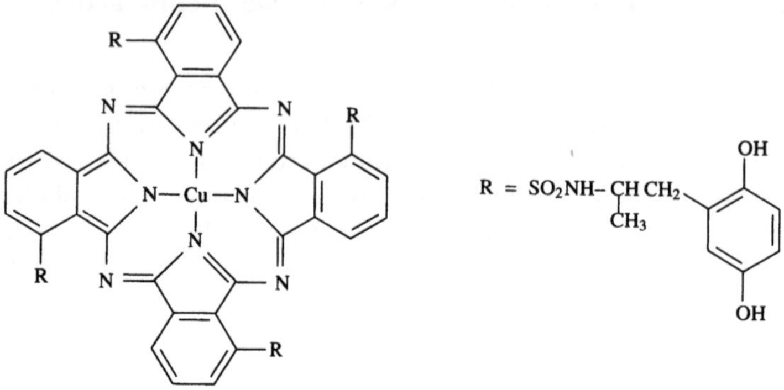

32 (cyan)

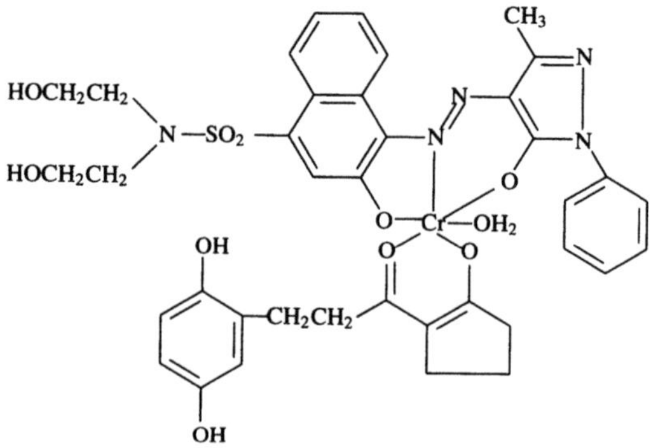

33 (magenta)

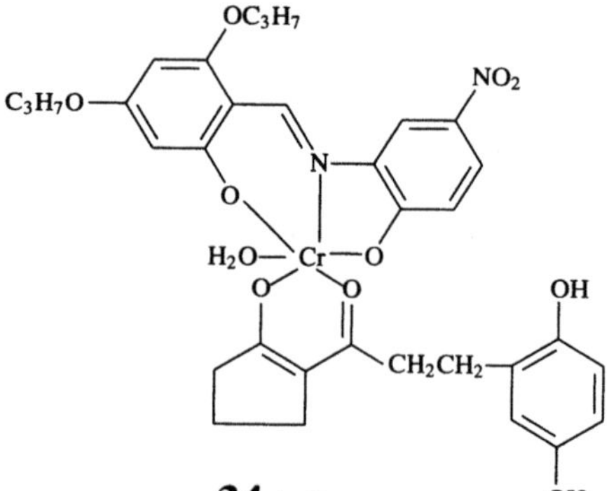

34 (yellow)

Figure 17.12. A set of chelate dye developers for color reproduction of instant color photography. Coordination bonds are not discriminated from other kinds of bonds. The chromium in the magenta metallized dye (**33**) is coordinated to the nitrogen atom attached to the naphthalene nucleus.

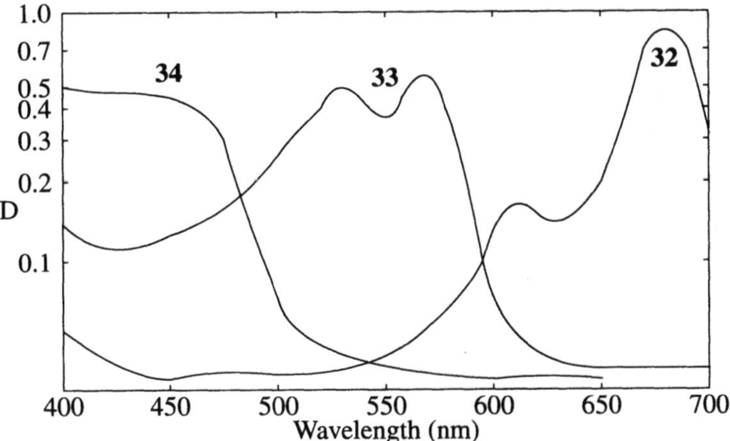

Figure 17.13. Absorption spectra of metallized dye developers: cyan (**32**), magenta (**33**), and yellow (**34**) [3].

Syntheses

The cyan dye developer (**32**) has been synthesized by starting from copper phthalocyanine (**35**), as shown in Fig. 17.14 [30,43]. First, four chlorosulfonyl groups are introduced to the copper phthalocyanine nucleus by using a mixture of $ClSO_3H$ and $SOCl_2$. The resulting intermediate (**36**) is reacted with the corresponding amine (i.e., 1-(2-aminopropyl)-2,5-dimethoxybenzene) to give the condensate (**37**). The methoxy groups of **37** are removed by boron tribromide (BBr_3) to yield the final product (**32**).

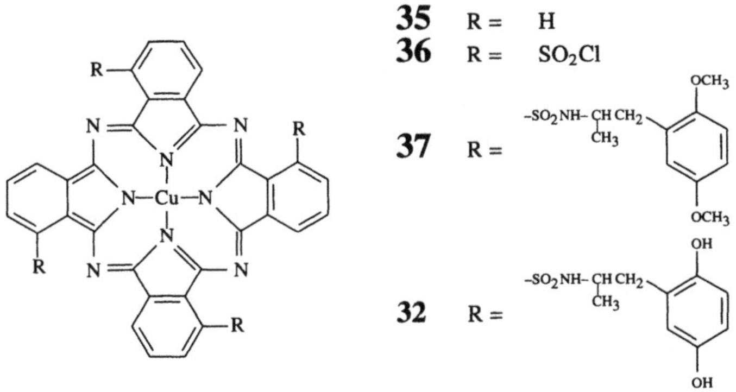

35 R = H

36 R = SO_2Cl

37 R = $-SO_2NH-CH(CH_3)CH_2-$ (2,5-dimethoxyphenyl: OCH_3, OCH_3)

32 R = $-SO_2NH-CH(CH_3)CH_2-$ (2,5-dihydroxyphenyl: OH, OH)

Figure 17.14. Synthetic pathway for preparing the cyan dye developer (**32**).

Magenta Metallized Dye Developers

An *o,o′*-dihydroxy azo dye forms complexes having a dye:chromium ratio of 2:1 or 1:1. In general, the 2:1 complex exhibits a broader absorption spectrum than the 1:1 complex. Hence, the magenta dye developer adopted in Polaroid SX-70 (i.e., **33**) is a 1:1 chromium complex so that the remaining valencies are filled with a colorless ligand linking to a developer (hydroquinone) moiety [31,32].

Table 17.1. Absorption Maxima of Magenta Metallized Dyes (**38**) [3]

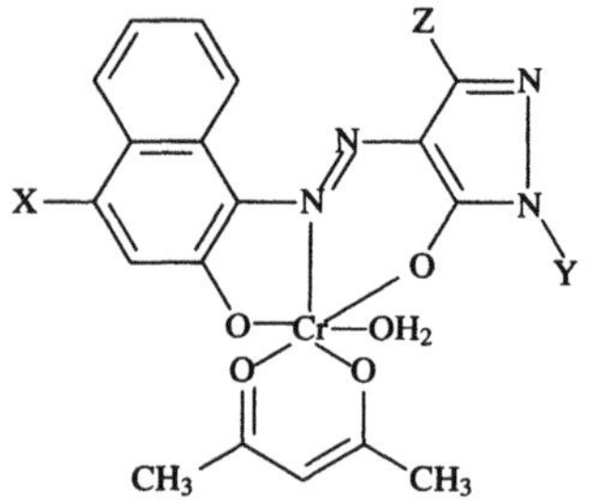

38

Substituents			First band		Second band	
X	Y	Z	λ_{max} (nm)	ϵ_{max} $\times 10^{-4}$	λ_{max} (nm)	ϵ_{max} $\times 10^{-4}$
CN	H	CH_3	528	1.48	568	1.64
CN	CH_3	CH_3	531	1.88	569	2.06
CN	Ph	CH_3	537	2.20	575	2.46
CN	Ph	$COOC_2H_5$	537	1.92	576	2.40
CN	Ph	$CONHCH_3$	535	2.06	573	2.54
CN	Ph	$NHCOCH_3$	546	2.14	585	2.26
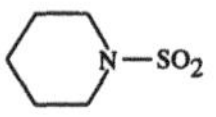 N–SO_2	Ph	CH_3	524	2.14	556	2.20
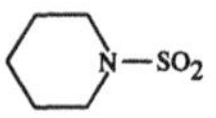 N–SO_2	Ph	$COOC_2H_5$	532	1.72	570	2.00
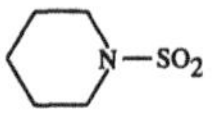 N–SO_2	Ph	$CONHCH_3$	533	1.94	570	2.34

For the final selection of the magenta dye developer (**33**), a vast number of 1:1 complexes have been synthesized, as reported in an article [44] and in a review [3]. The absorption data selected from the table in the review [3] are collected in Table 17.1, where substituents (X, Y, and Z) on the azo ligand of **38** are varied and the other ligand is fixed to be acetylacetone (2,4-pentanedione referred to as acac). For a rough absorption profile, see Fig. 17.13. The X-ray analysis of the crystal of **38** (X = SO_2–N⬡, Y = Ph, Z = CH_3) has clarified that the dye coordinates in the meridional positions of the chromium ion, where the nitrogen atom attached to the naphthalene nucleus (not to the pyrazolone ring) coordinates to the chromium ion [3,45].

By the inspection of the formula (**38**), we can find X, Y, Z, or the ligand (acac) as a linking position of a developer (hydroquinone) moiety. Dye developers with a hydroquinone moiety incorporated in the positions X and Y were first examined, as disclosed in patents [46,47]. The introduction of a hydroquinone moiety to the position Z has also been disclosed [48]. Finally, the linking at the ligand position has been adopted, as found in the formulas of the magenta (**33**) and the yellow dye dveloper (**34**). Thus, the 1,3-diketone ligand linked with a hydroquinone moiety has been selected as a key intermediate after the examination of various types of ligands (e.g, triethylenetetramine, 8-hydroxyquinoline, and salicylic acid) [49,50].

Syntheses

The magenta dye developer (**33**) and the yellow dye developer (**34**) have a common moiety, which contains a hydroquinone moiety and a 1,3-diketone ligand. The common moiety is synthesized according to the scheme shown in Fig. 17.15 [51].

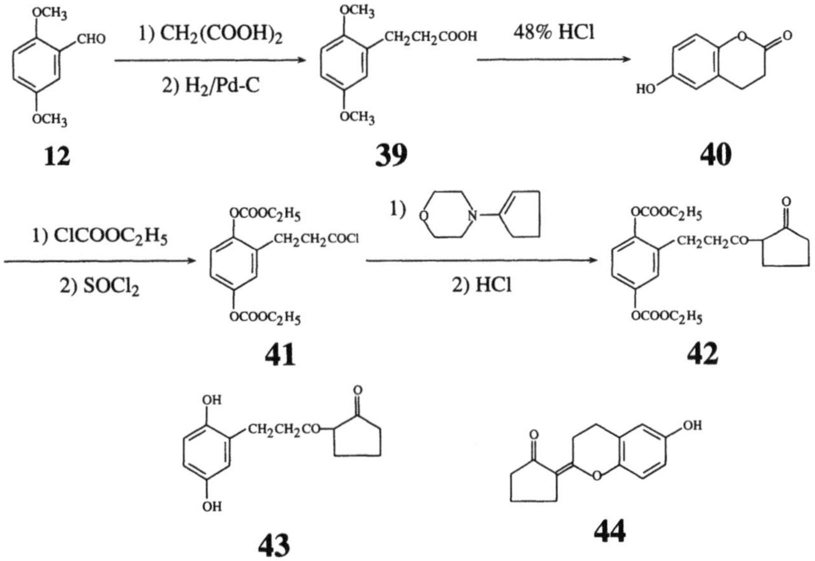

Figure 17.15. Synthetic pathway for the common ligand intermediate (**42**).

The starting material **12** has been already described in Fig. 17.6. It is condensed with malonic acid to give an olefinic compound as a result of decarboxylation, as shown in Fig. 17.15. The olefinic intermediate is hydrogenated catalytically to give **39**. After the protection by a methoxy group in **39** is changed to the protection by a carbonic acid ester, the carboxylic group is converted into an acid chloride (**41**). The latter compound is reacted with an enamine to give the key intermediate (**42**) [51].

As shown in Fig. 17.16, the azo dye moiety of the magenta dye developer (**33**) has been synthesized via such a 1:1-chrome-complexed intermediate as **46** or the corresponding diethylenetriamine (dien) complex,which is reacted with the intermediate (**42**) [31]. The protective groups of the resulting compound (**47**) are removed under alkaline conditions so as to afford the target molecule (**33**).

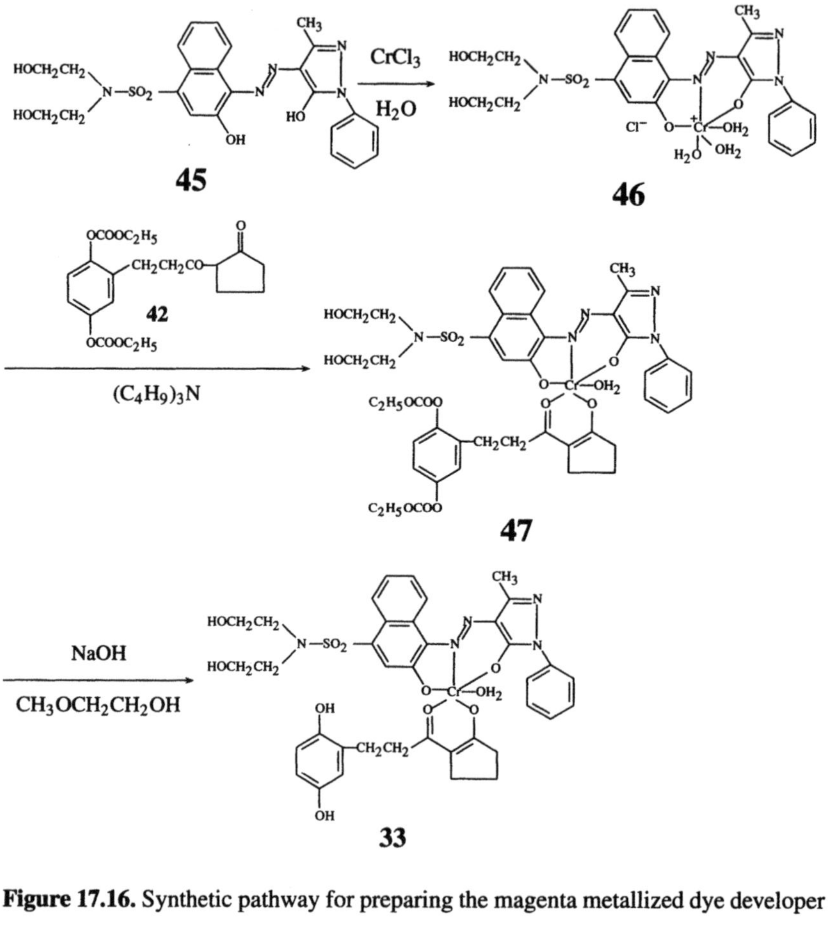

$CrCl_3$ / H_2O

45 → **46**

42

$(C_4H_9)_3N$

47

NaOH / $CH_3OCH_2CH_2OH$

33

Figure 17.16. Synthetic pathway for preparing the magenta metallized dye developer (**33**).

The method shown in Fig. 17.16 has a disadvantage in that the alkali employed in the deblocking step (**47** → **33**) causes some removal of the ligand from the complex as well as the formation of an oxidation product. To eliminate the deblocking step, the unprotected intermediate (**43**) is used in place of the protected intermediate (**42**) [52]. Another intermediate (**44**) has also given a satisfactory result [53].

Yellow Metallized Dye Developers

1:1-Chromed *o,o'*-dihydroxyazomethine dyes have been incorporated in dye developers, as disclosed in patents [33,54]. Within the scope of these patents, dyes in which alkoxy groups are attached to the one benzene ring and a nitro group is attached to the other benzene ring (e.g., **34**) possess good yellow color and satisfactory tinctorial strength [55]. The use of a ligand moiety having two hydroquinone moieties has been disclosed [56]. 2:1-Chromed dye developers have also been disclosed, where each azo dye contains a hydroquinone moiety

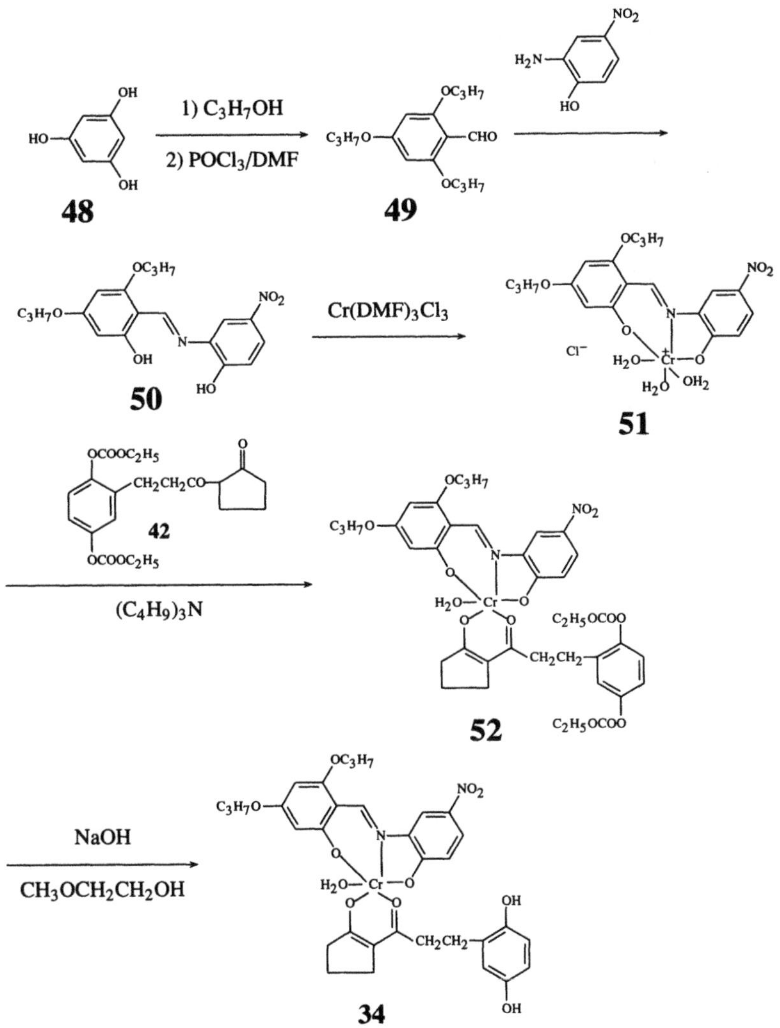

1) C_3H_7OH
2) $POCl_3/DMF$
48 **49**
$Cr(DMF)_3Cl_3$
50 **51**
42
$(C_4H_9)_3N$
52
NaOH
$CH_3OCH_2CH_2OH$
34

Figure 17.17. Synthetic pathway for preparing the yellow metallized dye developer (**34**).

[57]. Azo-azomethine 2:1-chromed complexes have been disclosed as yellow dye developers [58].

Syntheses

The yellow metallized dye developer (**34**) has been synthesized by using the common intermediate (**42**), as illustrated in Fig. 17.17 [33]. The corresponding azomethine dye (**50**) has been prepared by starting from phloroglucinol (**48**) [59]. It has been metallized to give a 1:1-chrome complex (**51**), which has been reacted with the intermediate (**42**) [60]. The final step has been the removal of the protective groups by hydrolysis, giving the target molecule (**34**) [55]. Another method using **44** in place of **42** has also been disclosed in the patent [53], which has been cited above for the alternative method for preparing the magenta dye developer (**33**).

17.2.4 Xanthene Dye Developers

Although the metallized dye developers have excellent light and dark stabilities, their hues are not always satisfactory to realize faithful color reproduction. In particular, magenta metallized dye developers (e.g., **33**) have undesirable absorptions

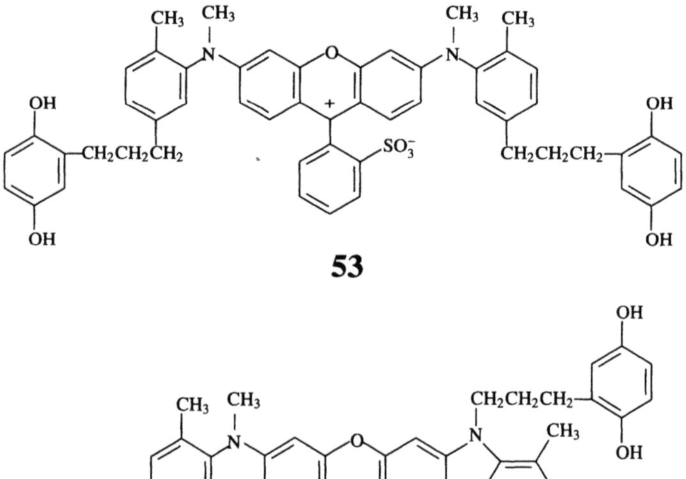

53

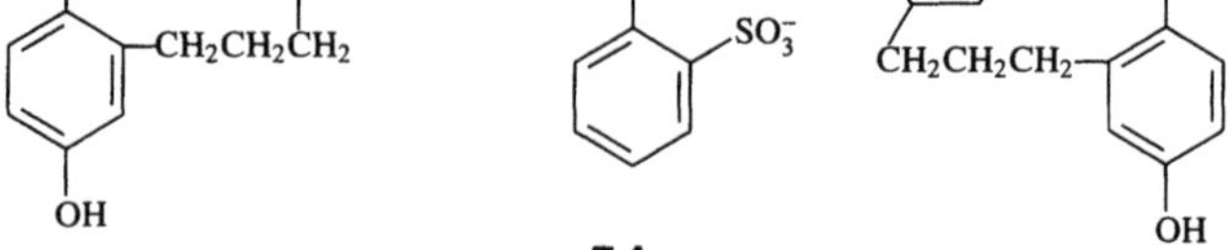

54

Figure 17.18. Xanthene dye developers having magenta hue.

in the blue region of light so that the purity and quality of the "blue image" are impaired.

To improve the hues of magenta dye developers, xanthene dye moieties have been investigated, as disclosed in patents [61,62]. For example, the dye developer (**53**) shown in Fig. 17.18 has been shown to exhibit a sharp-cut short-wavelength profile of its light absorption curve (i.e., high green absorption and high blue transmission). Xanthene dye developers which include three or four hydroquinone moieties (e.g., **54**) have been disclosed in patents [63,64]. The comparison between **53** and **54** has revealed that the former transfers at a much slower rate than the latter [61].

In a typical embodiment, the phthalocyanine cyan dye developer (**32**), the xanthene magenta dye developer **54**, and the chrome-complexed yellow dye developer (**34**) have been employed as a triplet for full color reproduction [61].

Syntheses

For the synthesis of the xanthene dye developer (**54**), Fig. 17.19 shows the preparations of intermediates having a hydroquinone moiety [63].

1) Dimethoxy intermediate:

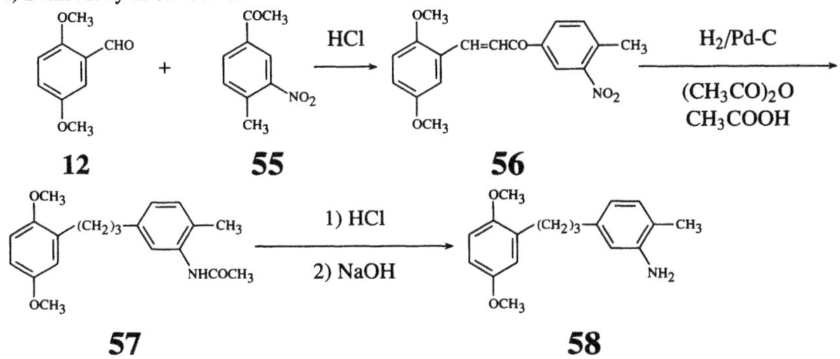

2) Tosylate intermediate:

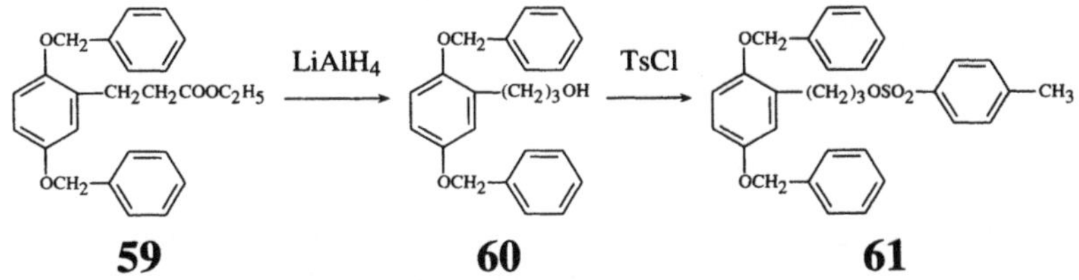

Figure 17.19. Synthetic pathways for preparing intermediates with a hydroquinone moiety. These intermediates are used to synthesize xanthene dye developers. TsCl represents tosyl chloride (*p*-toluenesulfonyl chloride).

The first intermediate (**58**) is an amine derivative having a hydroquinone moiety protected by dimethoxy groups, which come from the common starting material (**12**). This is condensed with an acetophenone derivative (**55**) to yield a chalcone derivative (**56**). The CH=CHCO chain of **56** is exhaustively reduced into a trimethylene chain, where the nitro group is meanwhile reduced into an amine, which is acetylated in the presence of acetic anhydride. The acetyl group is removed to prepare the intermediate (**58**).

The second intermediate (**61**) is a tosylate which includes a hydroquinone moiety protected by dibenzyloxy groups. The ester group in the starting material (**59**) is reduced into an alcohol (**60**), which is converted into the tosylate (**61**).

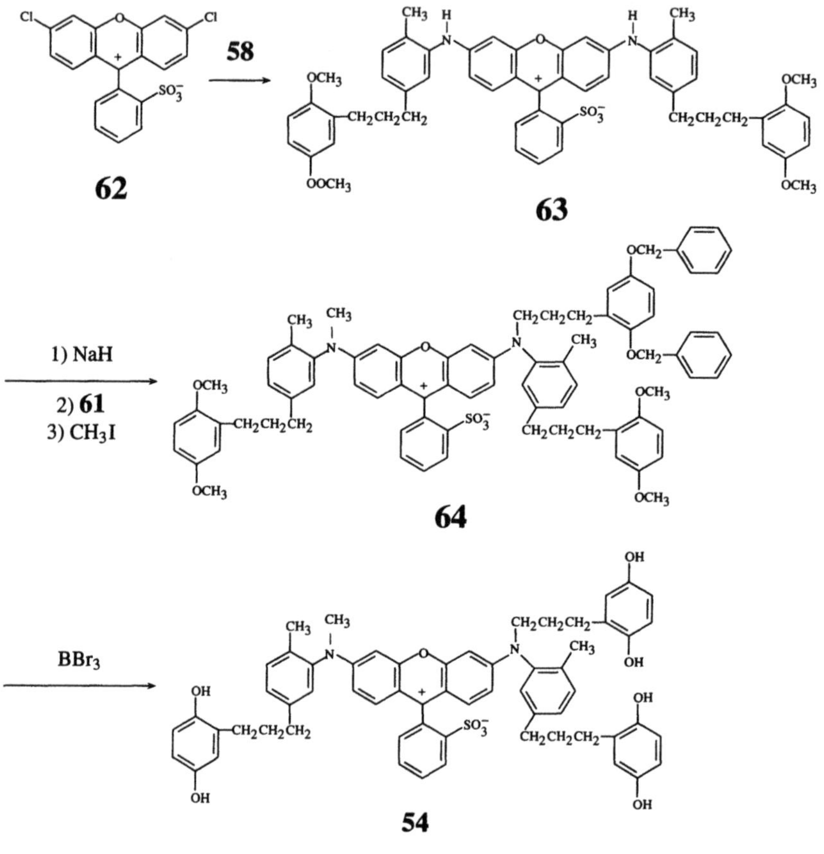

Figure 17.20. Synthetic pathway for preparing the magenta dye developer (**54**). The intermediates **58** and **61** have been prepared according to the schemes shown in Fig. 17.19 [63].

Two moles of the dimethoxy intermediates (**58**) are reacted with 3,6-dichlorosulfonefluoran (**62**), giving a dye intermediate (**63**). After the dye intermediate is treated with sodium hydride dispersion in oil, the resulting amide anion functions are reacted successively with the tosylate intermediate (**61**) and methyl iodide so as to give **64**. The ether groups in **64** are removed by using boron tribromide to yield the final product (**54**). The absorption maximum of **54** appears at 553 nm (ε = 117,500).

17.3 Chemicals for Controlling Dye Transfer Processes

17.3.1 Chemicals Contained in Photo-Sensitive Layers

Auxiliary Developers

The oxidation of a dye developer is mediated by the action of an auxiliary developer, as illustrated in Fig. 17.1. *p*-Tolylhydroquinone (**1**) has been used as an auxiliary developer [65]. The latter has been incorporated into the cyan dye developer layer [34]. It is similar in structure to a dye developer, but capable of moving rapidly through all of the layers of the film, since it does not carry a large dye moiety.

Emulsion Stabilizers

4-Hydroxy-6-methyl-1,3,3a,7-tetrazainden (TAI) has been incorporated in all of the three emulsion layers [34]. This compound is widely used in instant color photography as well as in conventional color photography.

Development Inhibitor Releasers

If a dye developer is stopped (oxidized) at a "wrong" silver halide emulsion layer (i.e., a silver halide emulsion layer that does not correspond to it), this reaction is referred to as "cross-talk." One of the methods that prevents such cross-talk is to make latent images non-effective before "wrong" dye developers reach the emulsion layer. For this purpose, development inhibitors such as 1-phenyl-5-mercapto-1H-tetrazole have once been used in a non-blocked fashion, where the timing of development inhibition is adjusted by changing layers to be incorporated [66,67].

The timing of development inhibition can be adjusted more effectively by blocked development inhibitors, which are listed in Fig. 17.21. They are called *development inhibitor releasers* because of their non-imagewise (i.e., timing) release of development inhibitors. For exmaple, 1-phenyl-5-(*p*-hydroxybenzylthio)tetrazole (**65**) releases 1-phenyl-5-mercapto-1H-tetrazole according to a quinone-methide mechanism [68,69]. This compound has been used in an integrated film based on the triplet of the metallized dye developers shown in Fig. 17.12, where the temperature dependence of D_{max} values has been compensated. The use of such *o*- and *p*-hydroxybenzyl functions has been investigated as blocking groups which are removable with bases [70]. A similar mechanism involved in a six-membered heterocyclic ring works in a blocking group of **66** [71,72]. Five-membered blocking groups (e.g., **67**) have also been incorporated in similar development inhibitor releasers [73,74].

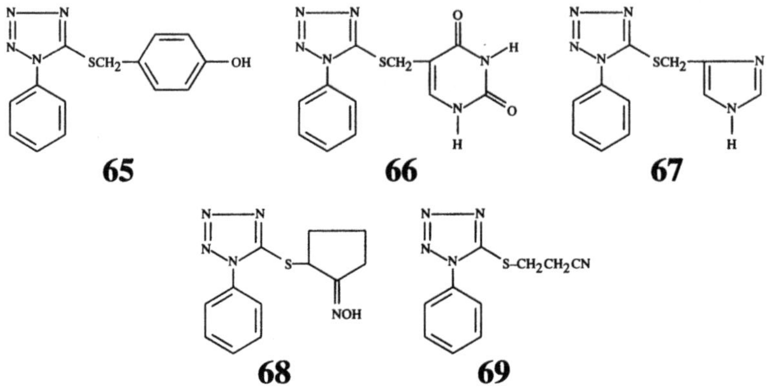

65 **66** **67**

68 **69**

Figure 17.21. Blocked development inhibitors as development inhibitor releasers.

70a R = OH
70b R = $SONH_2$

71

72a R = OH
72b R = $C(=NOH)CH_3$

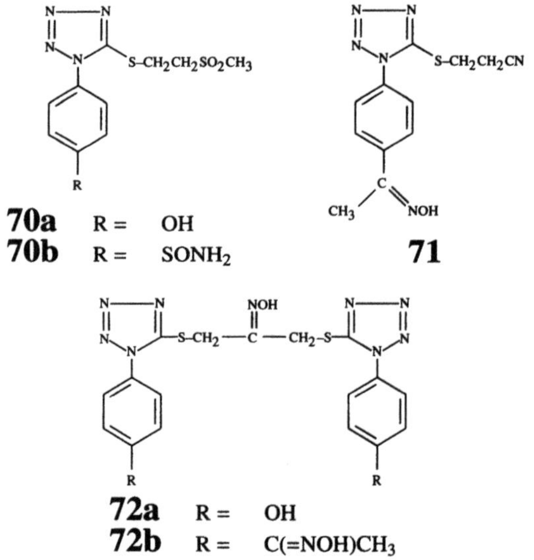

Figure 17.22. Dissociable substituents on the phenyl ring of 1-phenyl-5-mercaptotetrazole.

The compound (**68**) works as a development inhibitor releaser according to another mechanism, although it aims at the application to conventional photography [75]. The compound (**69**) releases a development inhibitor in terms of a retro-

Michael reaction [26]. This patent has disclosed an embodiment in which **69** is employed in a film using oxichromic dye developers (e.g., Fig. 17.11).

A dissociable substituent on the 1-phenyl group of 1-phenyl-5-mercapto-1H-tetrazole influences the capability of development inhibition. For example, such a compound as **70a** or **70b** releases a dianion at a high pH, which acts as a weak silver solvent, thus promoting development [35,76]. On the other hand, the monoanion formed at a low pH gives the silver salt of low solubility, resulting in development inhibition.

The oxime group of **71** can change the dissociation depending upon pH levels. Thereby the compound (**71**) acts as a development inhibitor releaser that works according to the same mechanism as described for **70a** and **70b** [77,78]. The compounds (**72a** and **72b**) employ a combination of this mechanism with another mode of blocking [79,80].

17.3.2 Chemicals Contained in an Activator Fluid

Let us now consider various chemicals contained in the activator fluid shown in Fig. 17.2, which have been employed together with the triplet of metallized dye developers shown in Fig. 17.12 [34]. The chemicals belong to various categories, as listed in Figs. 17.23 and 17.25.

Titanium dioxide and Viscosity-Imparting Reagents

As found by comparing Figs. 17.2 and 17.3, the activator fluid fed from the pod of Fig. 17.2 is spread to give the white opaque layer of Fig. 17.3. This layer comprises a white pigment (titanium dioxide) contained in the fluid.

A thin and square spread of the activator fluid between layers already in place (Fig. 17.2) has been one of the exigencies to embody the integrated film of Polaroid's SX-70 system [2]. To ensure such a thin and square spread, viscosity has been imparted to the fluid by adding sodium carboxymethylhydroxyethylcellulose (**74**) and polyethylene glycol (**81**).

Development Accelerators for Preventing Color Contamination

Since a dye developer (**4**) always has a hydroquinone moiety, it can be improperly stopped in silver halide emulsion layers other than the corresponding proper light-sensitive one ("cross-talk"). This improper oxidation of the dye developer causes color contamination. This type of color contamination is prevented by an interlayer containing appropriate polymers such as a 60:30:4:6 tetrapolymer of butyl acrylate, diacetone acrylamide, styrene, and methacrylic acid (e.g., [34]). The interlayer gives a sufficient timing between the proper oxidation of the corresponding dye developer and the arrival of the other dye developers. Thereby, the undesirable improper oxidation is prevented.

1) Inorganic components:

KOH (5.25), TiO_2 (37.4), SiO_2 (0.55) (Aqueous dispersion of colloidal silca), H_2O (44.26)

2) Organic components:

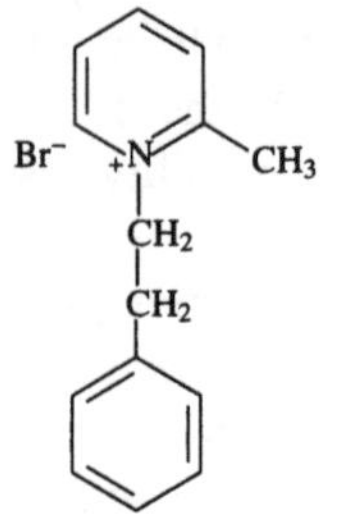

73 (1.27)

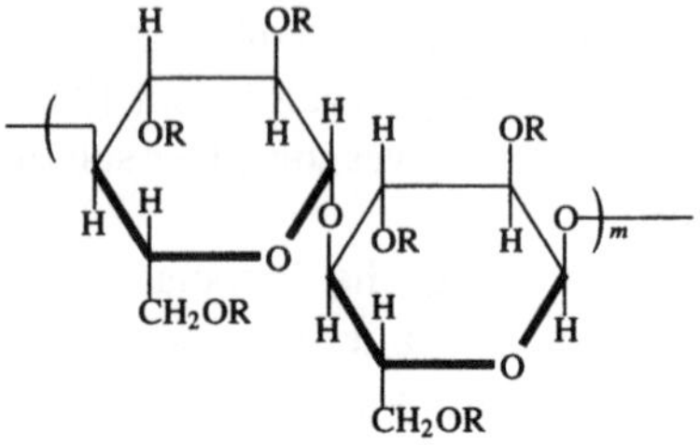

74 (2.0)
R = H, $-CH_2COONa$, or $-CH_2CH_2OH$

$(H_2NCH_2CH_2)_2S$
75 (0.02)

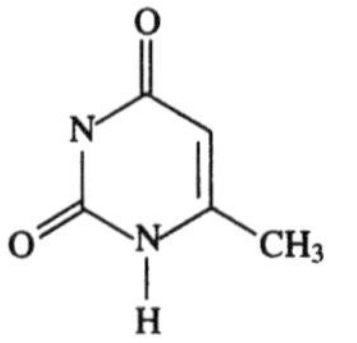

76 (0.7)

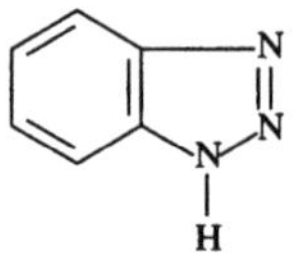

77 (5.48)

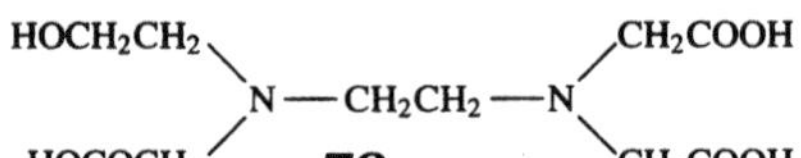

78 (0.75)

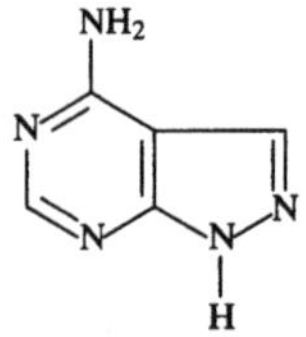

79 (0.25)

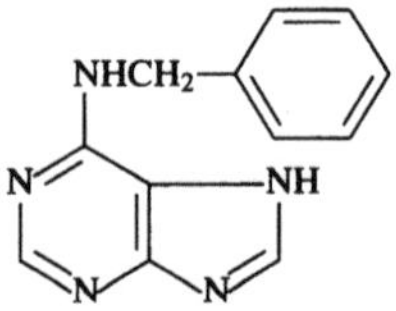

80 (0.45)

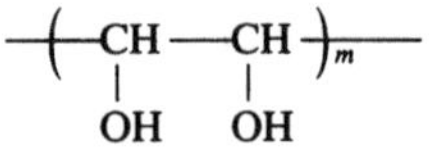

81 (0.45)

Figure 17.23. Chemicals contained in an activator fluid (A) [34]. The value of each pair of parentheses represents a weight percent of a component at issue. The activator fluid also contains the opacifying dyes shown in Fig. 17.25.

There is another type of color contamination "leak," where an undeveloped dye developer in a high-light region may leak so as to give an undesirable dye image. This color contamination can be prevented by using a development accelerator such as *N*-phenethyl-α-picolinium bromide (**73**) [81] together with an auxiliary developer (**1**) [65]. A hydrogen atom of the α-methyl group is removed under alkaline conditions to give a methide derivative (i.e., *N*-phenethyl-2-pyridone methide), which quenches the benzoquinone derivative (**6**) produced by eq. 17.2 of Fig. 17.1. Thereby, the accumulation of the benzoquinone derivative is prevented so as to accelerate a further reaction of eq. 17.2.

Various onium salts, especially 2-alkylpyridinium salts which are capable of giving an exo double bond under alkaline conditions, have been disclosed as development accelerators in patents. For example, *N*-benzyl-α-picolinium bromide (**82**) and related compounds have been reported [67,82]. The gegen ions of such onium salts have been replaced by polymeric anions, e.g., *N*-phenethyl-α-picolinium salt of polyvinyl hydrogen phthalate, which is fed from an overcoat layer [83].

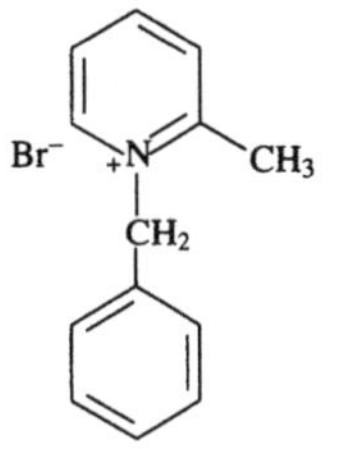

82 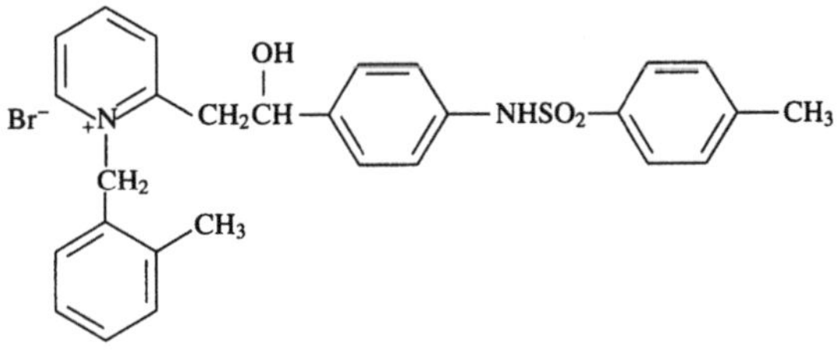 **83**

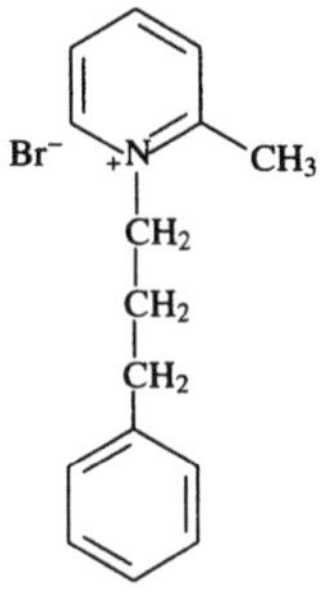

84 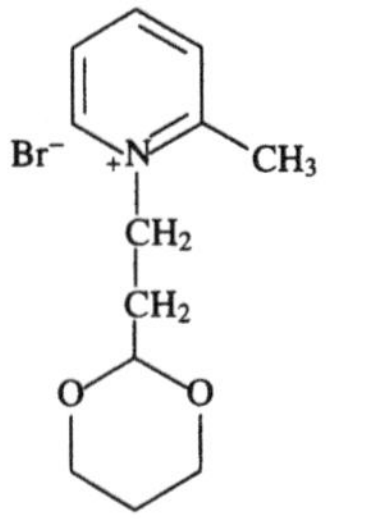 **85**

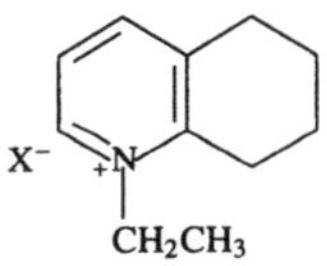

86a $X^- = Br^-$

86b $X^- = {}^-O-C_6H_4-CH_3$

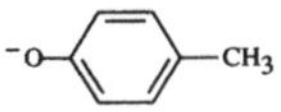

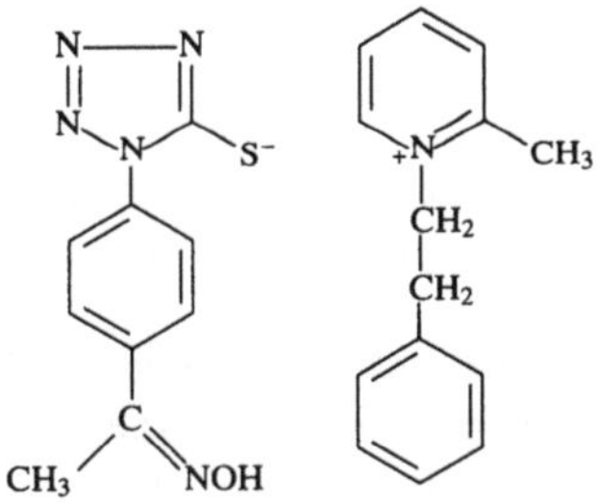

87

Figure 17.24. Pyridinium salts as development accelerators.

A more complicated pyridinium salt (**83**) has been disclosed as a development accelerator for an embodiment that employs the phthalocyanine cyan dye developer (**32**), the xanthene magenta dye developer (**54**), and the chrome-complexed yellow dye developer (**34**) [84]. This compound (**83**) gives the virtually same blue $D_{\min}$ values as **73**. Moreover, the green and blue $D_{\max}$ values due to **83** are higher than the corresponding values due to **73**.

More recently [85,86], the combined use of two pycolinium salts having substituted *N*-alkyl groups (**84** and **85**) has been disclosed for a hybrid film employing dye developers and a thiazolidine dye releaser (see Section 17.4).

Quaternary pyridinium compounds which have a fused 5- to 12-membered saturated carbocyclic ring attached to the 2- and 3-positions of the pyridine ring (e.g., **86a** and **86b**) have also been proposed as development accelerators, where the gegen ions have been varied [87]. As such a gegen ion, the accelerator (**87**) includes an anion of a 1-phenyl-5-mercapto-1H-tetrazole derivative, which is ineffective at a high pH but is capable of acting as a development inhibitor at a lower pH [88].

Temporary Opacifiers

A film ejected from a camera is represented by the schematic cross-section shown in Fig. 17.3, where the silver halide inside the film is still sensitive to light. Although the opaque white layer due to titanium dioxide exhibits some effect of opacification, its capacity is not sufficient, especially on photographing at the seashore on a sunny day and in sunny snow scenes.

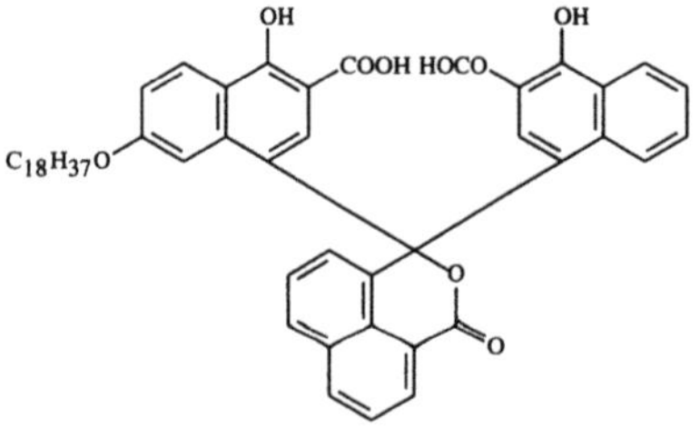

88 (1.35) 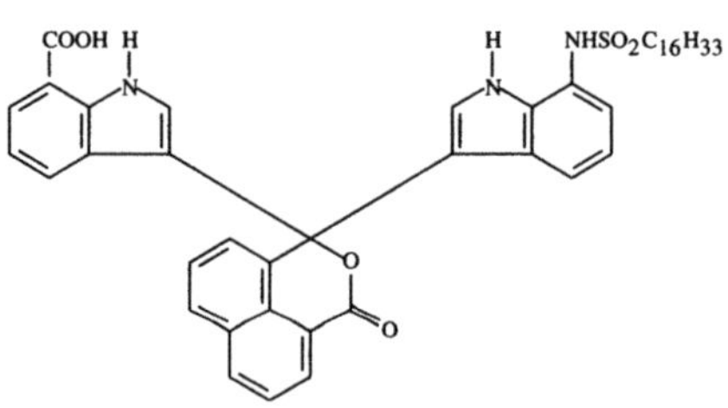 **89** (0.3)

Figure 17.25. Chemicals contained in an activator fluid (B) [34]. Temporary opacifying dyes are illustrated. For other components containing the activator fluid, see Fig. 17.23.

Temporary opacifiers such as **88** and **89** shown in Fig. 17.25 are used to protect the film from exposure during the critical seconds just after ejection from the camera [89]. They have colors at a high pH during the development process, but are converted into colorless derivatives under a low pH after development.

In general, such temporary opacifiers should satisfy the following requirements [6]: (1) they must be stable in caustic solution for at least a year, (2) they must change from colored forms to colorless forms by lowering the pH, (3) their opacifying effect must cover the whole visible spectral region, and (4) they must be photographically inert.

Usual phthaleins such as phenolphthalein are reacted with hydroxide ion to give a colored species at the moderate pH 9. When the pH is further raised, the colored species is faded to a new, colorless form by hydroxylation. The hydroxylation is prevented by introducing such steric hindrance as found in the naphthalein residue

(Table 17.2) [2], where the six-membered lactone of **90** (X = H, Y = H) raises the pK_a about one unit as compared with the corresponding phthalein dye [6].

Table 17.2. pK_a of phthaleins [2]

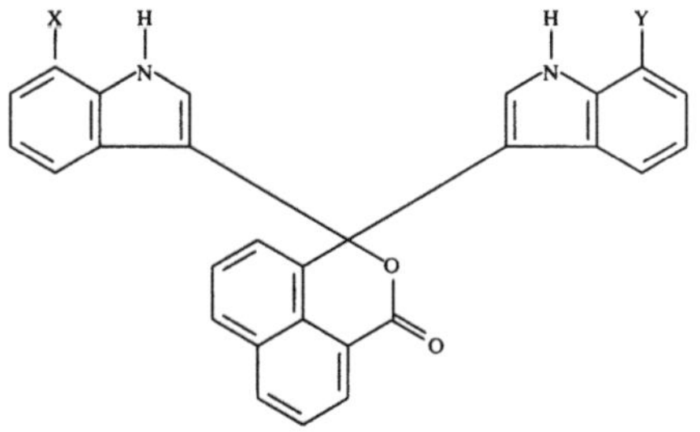

90

Substituents		pK_a	
X	Y		
H	H	11.1	13.8
COOH	COOH	13.2	>15
COOH	$NHSO_2O_{16}H_{33}$	12.9	>15
COOH	$SO_2NH_{18}H_{37}$ (= **89**)	12.9	>15

Hydrogen bonding of the phenolic hydroxyl group with an adjacent carboxylic group can be used to increase pK_a. For example, the pK_a of phenol is 9.99, while salicylic acid exhibits pK_a at 3.00 (for the carboxylic acid group) and at 13.44 (for the hydroxyl group). This function has been incorporated to the naphthaleins, as shown in Table 17.2 [90,91,92]. The combination of these ideas has resulted in the structures of two opacifying dyes, i.e., a naphtholnaphthalein (**88**) and an indolenaphthalein (**89**). They are used in combination to cover the whole visible spectral region.

17.4 Hybrid of Dye Developers and Thiazolidine Dye Releasers

17.4.1 Single Use to Hybrid Use

The first version of dye developers employed azo and anthraquinone dyes, as shown in Fig. 17.4. They were used in peel-apart photographic materials. The second version of dye developers was based on metallized dyes, as shown in Fig. 17.12. They mainly aimed at an integrated (mono-sheet) format. In the third version, the magenta metallized dye developer (**33**) of the second version was replaced by the xanthene dye developer (**54**). Although the dye moieties

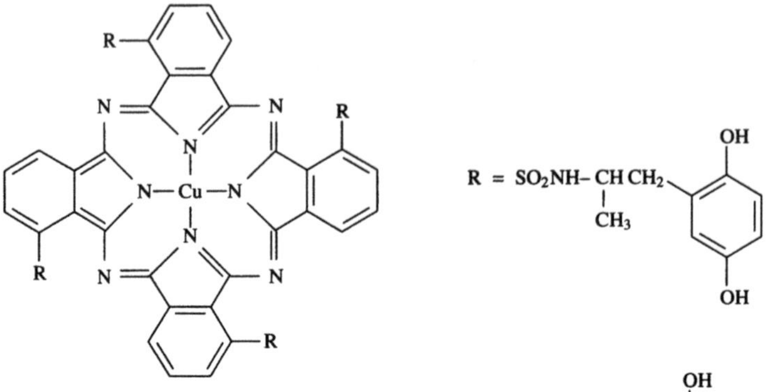

32 (cyan)

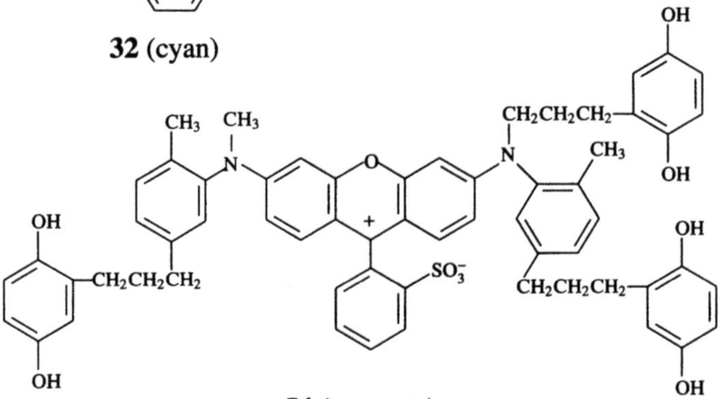

54 (magenta)

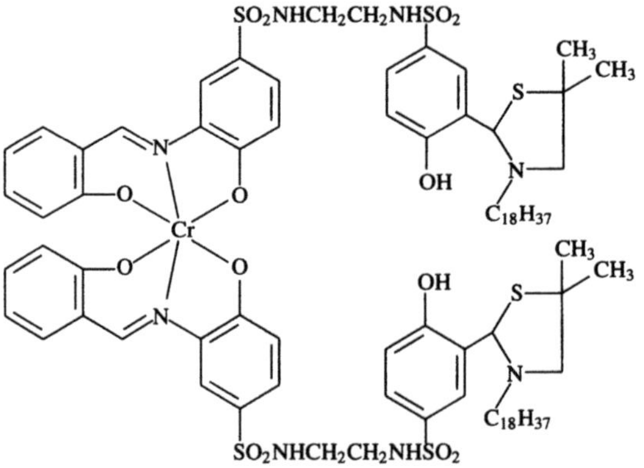

91 (yellow)

Figure 17.26. A hybrid set of dye developers and a thiazolidine dye releaser for color reproduction of instant color photography. Coordination bonds are not discriminated from other kinds of bonds.

were changed, their structures remained within dye developers (i.e., the developer moieties remained unchanged) from the first to the third version.

In the fourth version, the yellow dye developer (**34**) has been replaced by the thiazolidine dye releaser (**91**), as shown in Fig. 17.26 [93]. Thereby, the resulting film employs the thiazolidine dye releaser (**91**) and the dye developers (**32** and **54**) in a hybrid fashion. Although layer structure and additives for such a hybrid film have varied, the triplet listed in Fig. 17.26 have long been used, as still found in a recent patent [86].

The use of such a thiazolidine dye releaser permits the use of much lower coverages of blue-sensitive silver halide than the case used with a yellow dye developer. This means that the magenta dye developer is much less likely to be oxidized by exposed blue-sensitive silver halide, thereby reducing the likelihood of "magenta dropoff" due to "cross-talk."

17.4.2 Multilayered Hybrid Films

A schematic cross-section of a hybrid instant color film using dye developers and a thiazolidine dye releaser is shown in Fig. 17.27. Compare this diagram with that of Fig. 17.2.

Colorless Developers

A new interlayer containing a "colorless" developer has been set up between the blue-sensitive emulsion layer and the yellow dye-releaser layer [93]. The colorless developer (e.g., 2-phenyl-5-*t*-butylhydroquinone) has been selected in order that the oxidized form (e.g., 2-phenyl-5-*t*-butylbenzoquinone) is incapable of oxidizing "wrong" magenta and cyan dye developers [93]. Since the cyan dye developer (**32**) has an $E_{1/2}$ of −275 mV and the magenta dye developer **54** has an $E_{1/2}$ of −300 mV, they can be oxidized by the oxidized form of 2-*p*-tolylhydroquinone ($E_{1/2}$ = −220 mV). However, the oxidized form of 2-phenyl-5-*t*-butylhydroquinone (**92**, $E_{1/2}$ = −362 mV) is incapable of oxidizing the cyan and the magenta dye developer, which can diffuse through the yellow dye releaser layer to give dye images on the mordant layer. In more recent patents, *t*-octylhydroquinone (**93**) [86] and 2-*t*-butyl-5-norbornylhydroquinone (**94**) [37,94] have been used as colorless developers in multicolor films.

Silver Ion Scavengers

On the other hand, silver ion (water-soluble complexed species) which has been generated from an unexposed region of the blue-sensitive emulsion layer should work completely within the yellow thiazolidine dye releaser layer and should not work in other layers. To realize this requirement, the interlayer containing such a scavenger as **95** is set up between the yellow thiazolidine dye releaser layer and the green-sensitive emulsion layer [93].

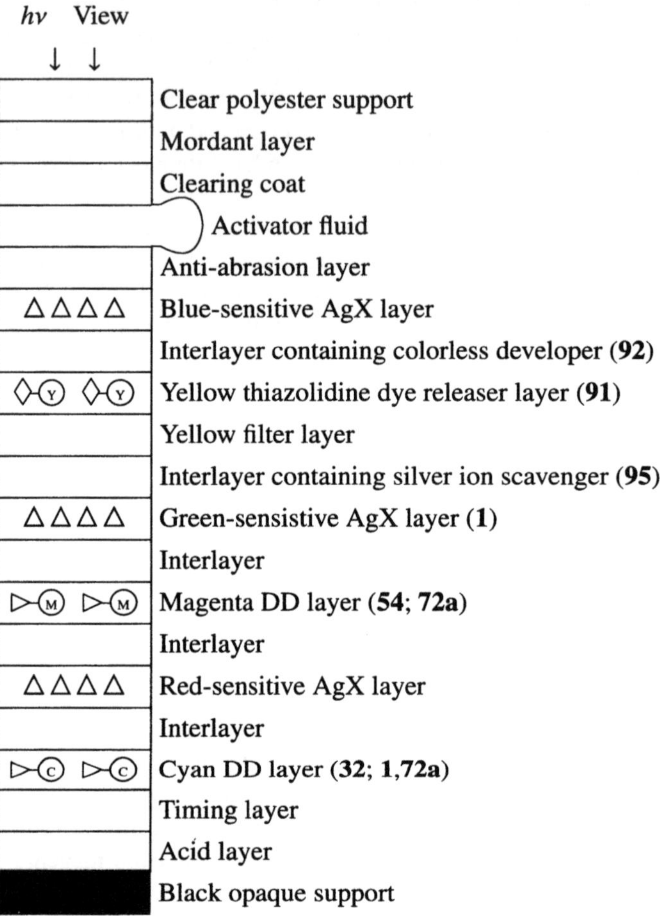

Figure 17.27. Schematic cross-section of a hybrid instant color film using dye developers and a thiazolidine dye releaser. △: Silver halide grain, ◇-Ⓨ: Yellow thiazolidine dye releaser, ▷-Ⓜ: Magenta dye developer, and ▷-Ⓒ: Cyan dye developer.

A polymeric scavenger (**96**) has been disclosed for the single use of thiazolidine dye releasers, where a mercaptotetrazole is incorporated in the polymer chain to capture a silver ion [95].

Silver Halide Solvents and Related Heterocycles

Silver halide solvents have been already discussed in Chapter 15 for instant B&W photography based on a solution physical development. They can be also employed as silver halide solvents for instant color photography based on thiazolidine dye releasers. Mono-*N*-tosylsulfimide derivatives such as **97** have been

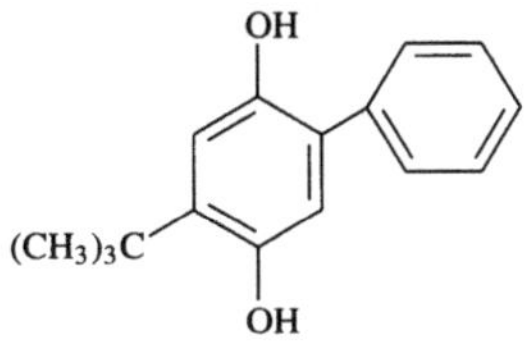

92 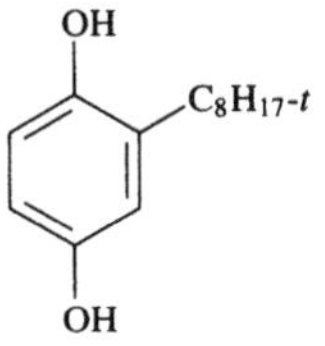 93 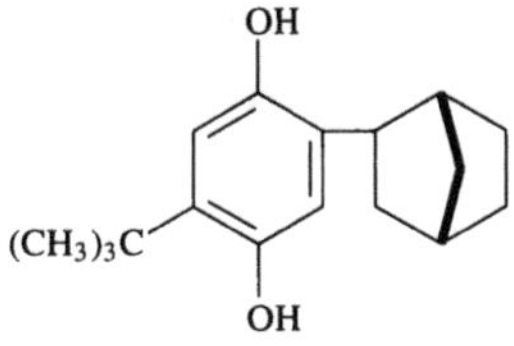 94

Figure 17.28. Colorless developers incapable of oxidizing "wrong" dye developers.

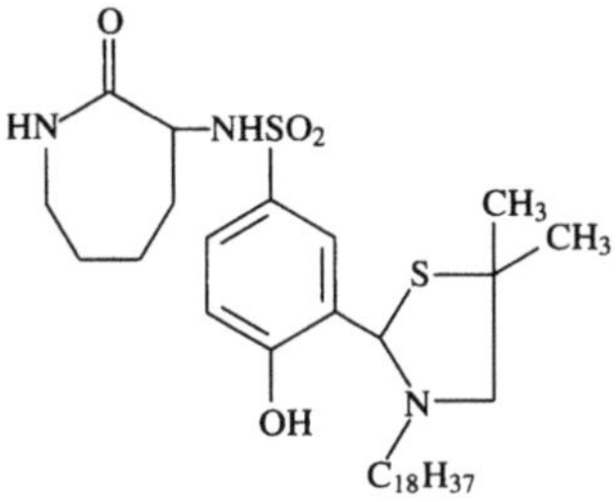

95 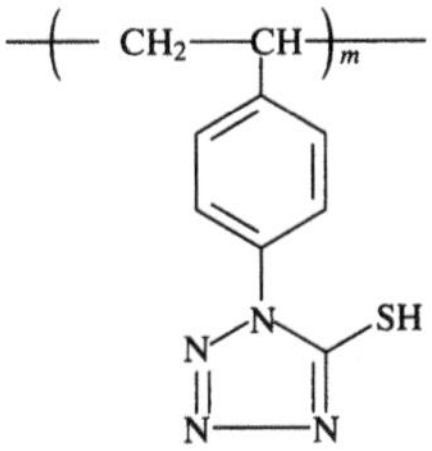 96

Figure 17.29. Silver ion scavengers.

proposed as further examples of silver halide solvents [96]. 2-Methylimidazole (**98**) [93] and 2-ethylimidazole [86,97] disclosed recently in several patents seem to be used as silver halide solvents.

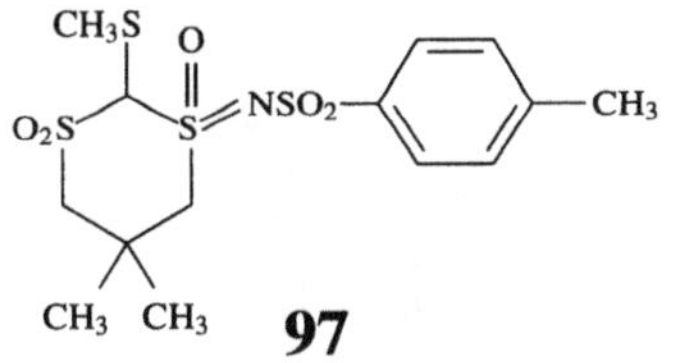

97 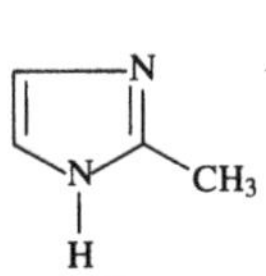 98

Figure 17.30. Silver halide solvents.

As discussed in Chapter 16, a silver halide solvent precursor such as **99** releases a thiosulfate anion ($S_2O_3^{2-}$) via a quinone-methide mechanism [95,98].

Another type of silver halide solvent precursors (e.g., **100**) has been disclosed to be used in instant color photography based on thiazolidine dye releasers [99]. The precursor (**100**) is attacked by a hydroxide ion at the terminal position of the exocyclic double bond so as to release benzaldehyde and the anion of 1,3-dithiane-1,1,3,3-tetraoxide. The latter has been known as a silver halide solvent.

The precursor (**101**) is based on dual blocking [100]. The alkaline attack on **101** produces the subsidiary precursor (**100**) with releasing the anion

($^-$N(Ph)COOCH$_2$Ph). Then, the resulting precursor (**100**) is further hydrolyzed as described above.

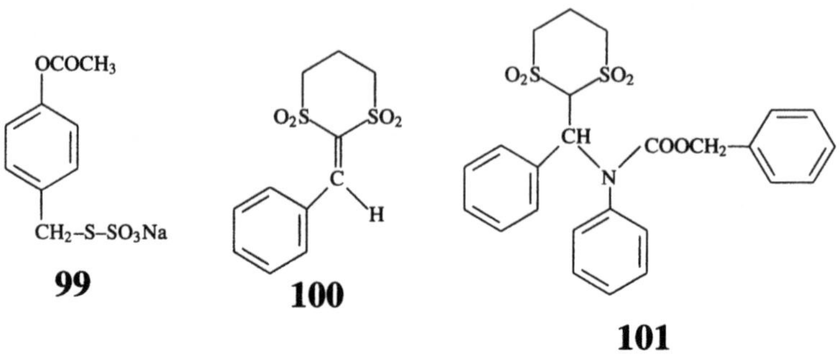

99

100

101

Figure 17.31. Precursors of silver halide solvents.

Although uracil has been known as a silver halide solvent, 6-methyluracil (**76**) involved in an activator fluid for processing the hybrid film is presumed to be added as a stabilizer, not as a silver halide solvent. Note that the 6-methyluracil (**76**) is also used for the single use of dye developers, as shown in Fig. 17.23.

Hypoxanthine (**102**) has been involved in an activator fluid as a stabilizer [86]. The crystallization of hypoxanthine results in visible crystal formation in the final image of a finished photograph. The crystallization can be prevented by adding a compound having a similar structure. Thus, inosine (**103**) has been disclosed as a crystallization inhibitor [97].

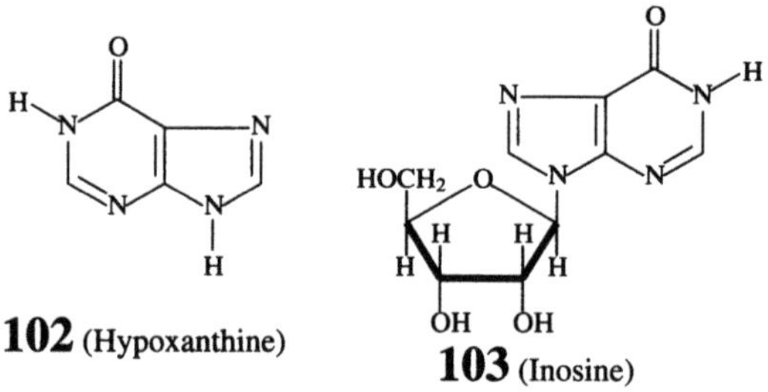

102 (Hypoxanthine)

103 (Inosine)

Figure 17.32. Inosine as a crystallization inhibitor for hypoxanthine.

17.5 Polymers

17.5.1 Polymers for the Mordant Layer

The mordant layer shown in Fig. 17.2 (or Fig. 17.27) works as an image-receiving element, which contains polymeric substances for receiving unreacted dye developers. A mordant polymer for the first version of dye developers shown in Fig. 17.4 has been poly-4-vinylpyridine (**104**) [101]. A mixture of poly-4-vinylpyridine and polyvinyl alcohol has been also used as an image-receiving material [102].

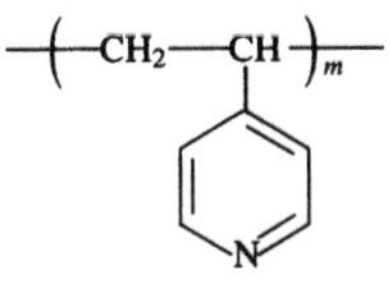

104

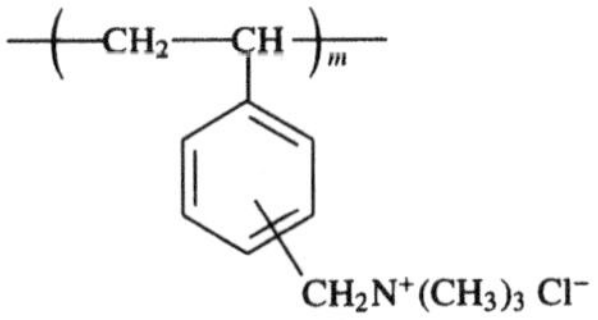

105

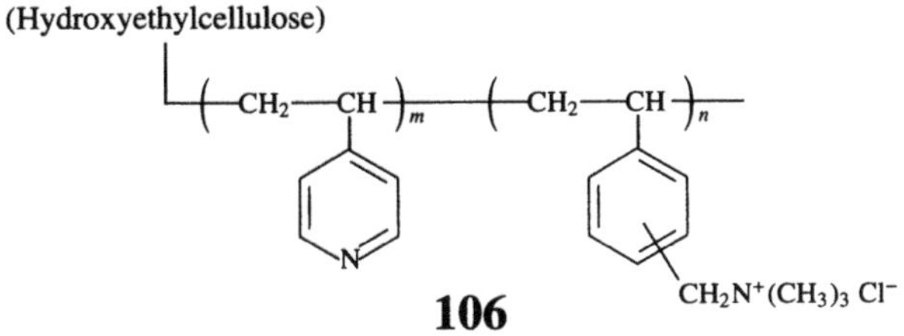

(Hydroxyethylcellulose)

106

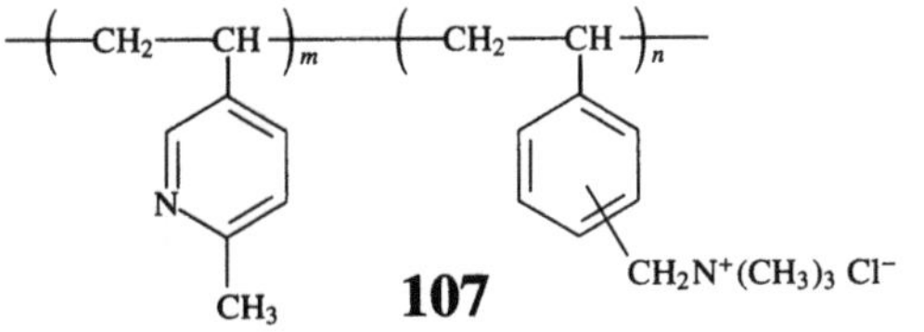

107

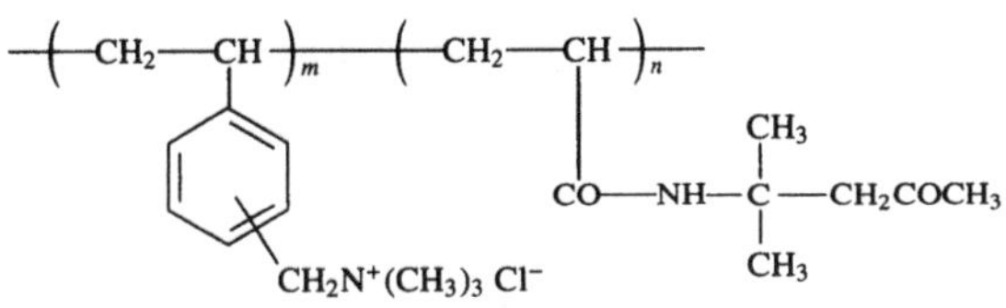

108

Figure 17.33. Mordant polymers for receiving dye developers.

A mixture of poly-4-vinylpyridine (**104**), poly(vinylbenzyltrimethylammonium chloride) (**105**), and polyacrylamide has been employed as a mordant polymer for the second version of dye developers shown in Fig. 17.12 [103]. A graft copolymer (**106**) of 4-vinylpyridine and vinylbenzyltrimethylammonium chloride on hydroxyethylcellulose has been proposed as a mordant polymer for the metallized dye developers shown in Fig. 17.12 [104]. Image-receiving layers comprising such a graft copolymer provide images exhibiting excellent dye densities over a wide temperature range.

Copolymeric mordants such as **107**, in which 5-vinyl-3-methylpyridine is used as a recurring unit in place of 4-vinylpyridine, have been proposed to improve the yellowing of a white background [105]. They have been employed for the third version of a dye developer set (cyan **32**, magenta **54**, and yellow **34**). A copolymeric mordant of another type (**108**) has been used in the form of a mixture with poly(vinylpyrrolidone), where the obtained maximum density is favorable [106].

For the fourth hybrid set shown in Fig. 17.26, copolymeric mordants containing two types of vinylbenzyl quaternary ammonium salts (e.g., **109**) have been disclosed, where good results have been obtained from mixtures of vinylbenzyltrimethylammonium chloride and vinylbenzyldodecyldimethylammonium chloride at ratios from 10:1 to 20:1 [107].

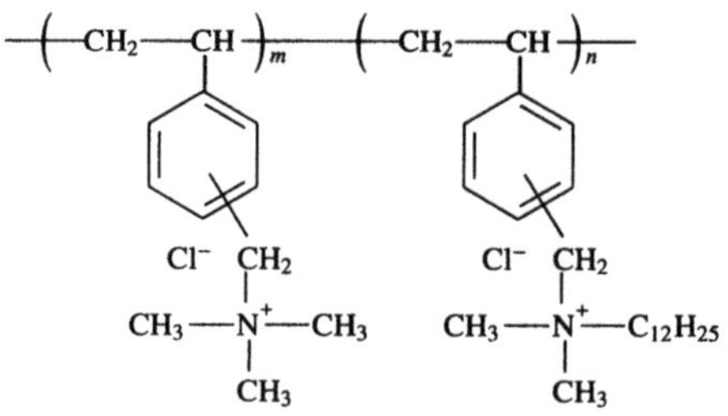

109

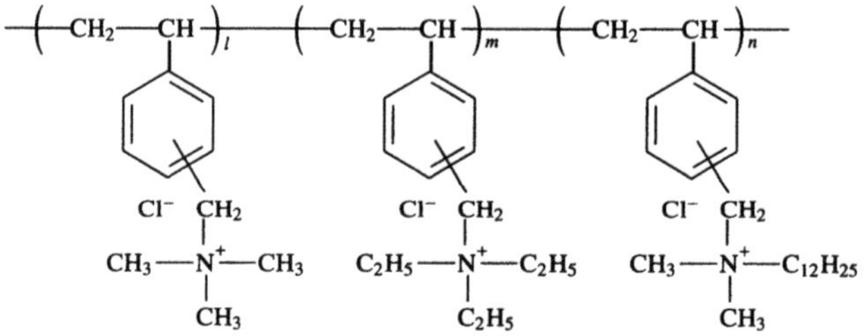

110

Figure 17.34. Mordant polymers for receiving dye developers and dyes derived from thiazolidine dye releasers.

A terpolymer (**110**) comprising vinylbenzyltrimethylammonium chloride, vinylbenzyltriethylammonium chloride, and vinylbenzyldodecyldimethylammonium chloride has been used in more recent patents [37,94,108]. The graft copolymer (**106**) has been also used in an embodiment based on the hybrid set shown in Fig. 17.26 [109].

17.5.2 Polymers for the Acid Layer

The development and the dye transfer take place under alkaline conditions, e.g., pH 12–14, as described above. The alkaline media should be neutralized to assure the stabilities of dye images during storage. Acid polymers such as **111** and

112 have been used for this purpose in the acid layer shown in Fig. 17.2 (or Fig. 17.27) [110]. They can be synthesized by the half esterification of poly(methyl vinyl ether/maleic anhydride) and poly(ethylene/maleic anhydride), which are commercially available. A copolymer of ethyl acrylate and acrylic acid grafted onto polyvinyl alcohol (**113**) has been disclosed to be useful as an acid polymer in an embodiment based on metallized dye developers [111].

A mixture of poly(ethylene/maleic anhydride) and partially (88–90%) hydrolyzed poly(vinyl acetate) has been employed for metallized dye developers [103]. The acid polymer (**111**) has been still employed in examples of recent patents [37,109].

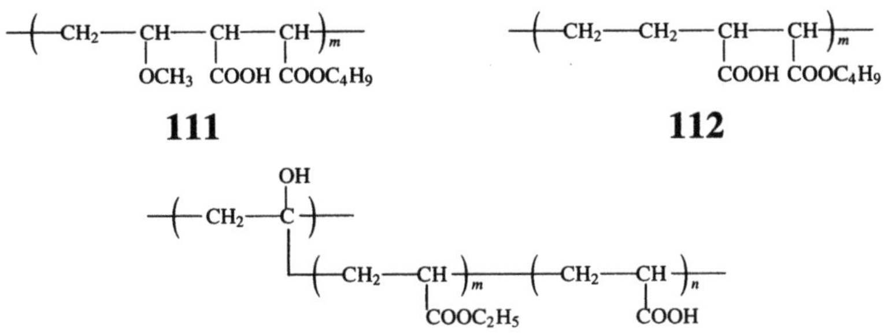

$$-\!\left(CH_2-\underset{OCH_3}{CH}-\underset{COOH}{CH}-\underset{COOC_4H_9}{CH}\right)_m\!-$$

111

$$-\!\left(CH_2-CH_2-\underset{COOH}{CH}-\underset{COOC_4H_9}{CH}\right)_m\!-$$

112

$$-\!\left(CH_2-\overset{OH}{C}\right)\!-\;\;\left(CH_2-\underset{COOC_2H_5}{CH}\right)_m\left(CH_2-\underset{COOH}{CH}\right)_n\!-$$

113

Figure 17.35. Acid polymers.

17.5.3 Polymers for the Timing Layer

To control the action of the acid polymer, a timing layer is positioned between the acid layer and the other layers, as shown in Fig. 17.2 (or Fig. 17.27). The timing layer does not prematurely interfere with the development process so as to ensure that the pH reduction occurs after a sufficient, predetermined period. Such timing layers have been designed to operate in the following ways [112]:

1. **Sieve type**: The timing layer acts as sieves which slowly meter the flow of alkali therethrough, and
2. **Hold-release type**:The timing layer acts as alkali-impermeable barriers for predetermined time intervals before converting to alkali-permeable barriers.

As for timing layers of sieve-type, early instant photographic films have used cellulose acetate, poly(vinyl alcohol), etc. [113]. Acrylic polymers such as a copolymer of diacetone acrylamide and acrylamide (**114**) have been used in the timing layer [114]. A copolymer of diacetone acrylamide and acrylamide grafted onto polyvinyl alcohol (**115**) has been disclosed to give approximately the same temperature-alkali permeability characteristics as polyvinyl alcohol and a copolymer of diacetone acrylamide-acrylamide, even if the former layer is thinner than

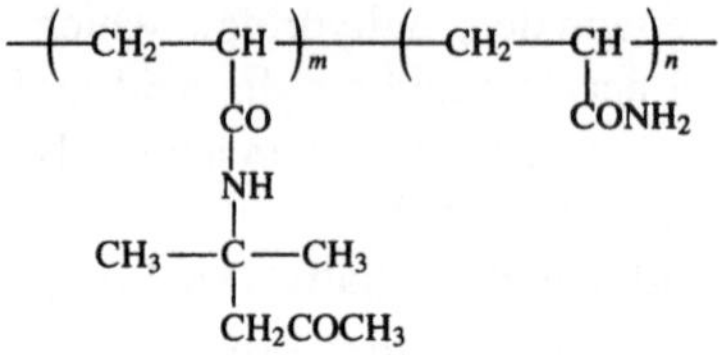

114

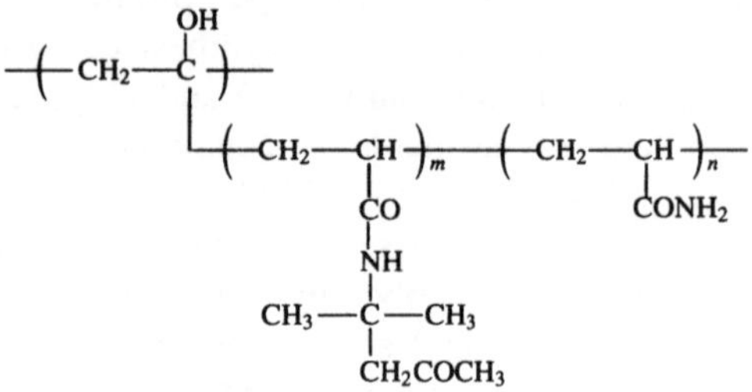

115

Figure 17.36. Timing polymers of sieve-type.

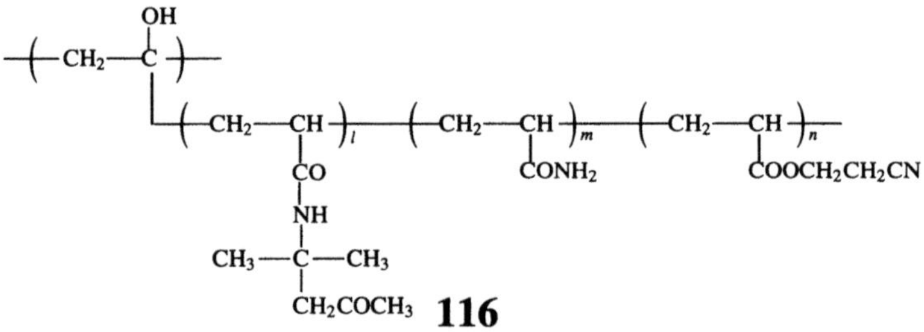

116

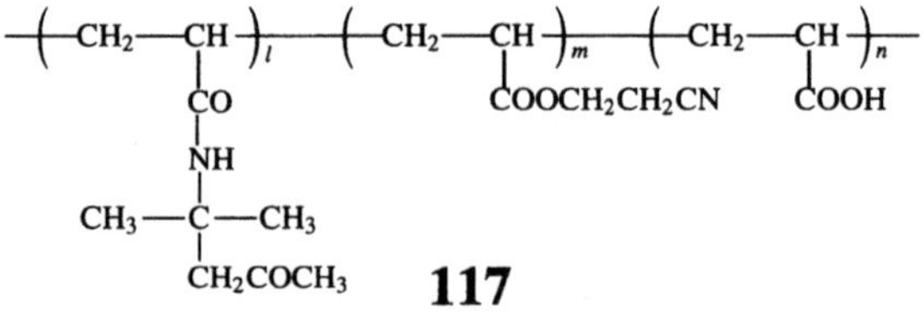

117

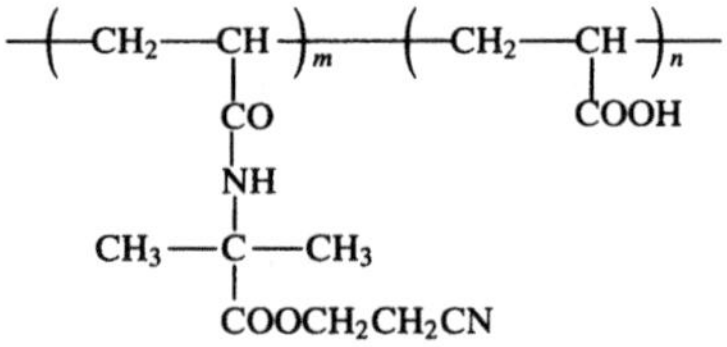

118

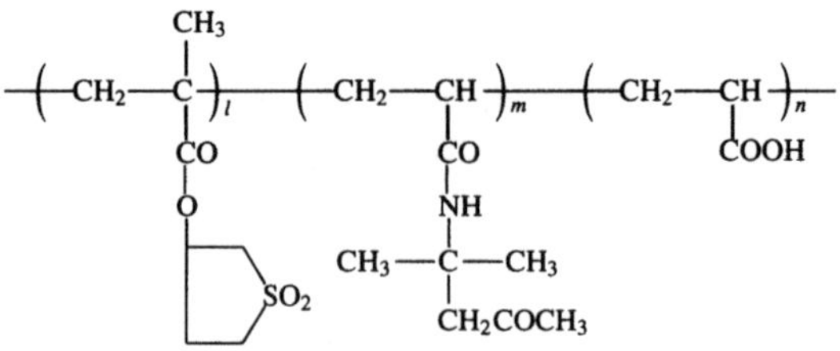

119

Figure 17.37. Timing polymers of hold-release type.

the latter one [115]. A mixture of such a graft copolymer as **115** and an aliphatic polyester urethane resin[9] has been reported to compensate the temperature dependence of D_{max} values for the hybrid set shown in Fig. 17.26 [112]. In particular, the red maximum (D_{max}) after 180 sec processing is 2.08 (at 27℃) vs. 2.00 (at 40℃), while the corresponding values without the aliphatic polyester urethane resin are 2.08 (at 27℃) vs. 1.63 (at 40℃).

Timing polymers of hold-release type have a hydrolyzable group under alkaline conditions so that a timing layer containing them is converted from a condition of impermeability to alkali or related materials (hold) into a condition of substantial permeability thereto (release). For example, the cyanoethyl group in a graft copolymer of diacetone acrylamide/acrylamide/β-cyanoethyl acrylate on poly(vinyl alcohol) **116** undergoes β-elimination to release a carboxyl group, which is capable of converting the permeability of the timing layer at issue [116].

A copolymer of diacetone acrylamide/β-cycanoethyl acrylate/acrylic acid (**117**) also contains a β-cyanoethyl group, which releases a carboxyl group on the polymer chain [117,118]. On the other hand, a copolymer of *O*-β-cyanoethyl-*N*-acrylyl-2-methylalanine/acrylic acid (**118**) undergoes β-elimination to release a carboxyl group on the side chain [119].

A hydrogen atom on the methylene adjacent to a sulfone group is active to cause β-elimination, as found in **119** [120]. Similar polymers have been disclosed as timing polymers of hold-release type [121]. The hold-release mechanism by β-elimination has been incorporated in a condensation polymer chain [122].

More recently, a pentapolymer of butyl acrylate/methyl methacrylate/diacetone acrylamide/ carboxymethoxymethyl acrylate/acrylic acid has been employed as a hold-release timing polymer of another type, which releases a carboxyl group of the direct hydrolysis of the butyl or ethyl ester groups [37]. A sieve timing layer containing the copolymer of diacetone acrylamide and acrylamide grafted onto polyvinyl alcohol and a hold-release timing layer comprising the copolymer of diacetone acrylamide/butyl acrylate/carboxymethoxymethyl acrylate/methacrylic acid have been dually set up in a recent embodiment [109].

References

[1] Land EH (1972) Photogr Sci Eng. 16:247

[2] Land EH, Rogers HG, Walworth VK (1977) One-Step Photography. In: Sturge JM (ed) Neblette's Handbook of Photography and Reprography, 7th edn. Van Nostrand Reinhold, New York, Chapter 12

[3] Bloom SM, Green M, Idelson M, Simon MS (1978) The Dye Developer in the Polaroid Color Photographic Process. In: Venkataraman K (ed) The

[9] An aliphatic polyester urethane resin named "Bayhydrol PU 402A" (commercialized from Bayer) has been used in this patent.

Chemistry of Synthetic Dyes, Vol VIII, Academic, New York London, Chapter VIII
[4] Bloom SM (1977) Polaroid Report, An Annual Journal for Stockholders. 19
[5] Rogers HG (1988) Research Technology Management. September–October. 42
[6] Idelson M (1982) Dyes and Pigments. 3:191
[7] Fujita S (1981) Yuki Gosei Kagaku Kyokaishi. 39:331
[8] Rogers HG (1961) US Patent 2 983 606
[9] Blout ER, Cohler MR, Green M, Simon MS, Woodward RB (1965) US Patent 3 209 016
[10] Blout ER, Green M, Rogers HG (1964) US Patent 3 134 764
[11] Fujita S (1982) Yuki Gosei Kagaku Kyokaishi. 40:176
[12] Kasman S (1965) US Patent 3 173 929
[13] Maekawa Y, Sakanoue S (1976) US Patent 3 982 946
[14] Idelson EM, Rogers HG (1967) US Patent 3 307 947
[15] Fujita S, Sano K (1979) J Org Chem. 44:2647
[16] Fujita S, Maekawa Y, Sano K, Sakanoue S (1977) US Patent 4 014 700
[17] Fujita S, Sakanoue S (1976) US Patent 3 999 991
[18] Fujita S (1979) Yuki Gosei Kagaku Kyokaishi. 37:960
[19] Fujita S, Sano K (1975) Tetrahed Lett. 1695
[20] Fujita S (1975) Jpn Kokai S50-134040
[21] Fujita S, Harada T, Sakanoue S (1979) Jpn Patent S54-21254
[22] Lestina GJ, Bush WM (1975) US Patent 3 880 658
[23] Lestina GJ, Bush WM (1976) US Patent 3 935 262
[24] Lestina GJ, Bush WM (1976) US Patent 3 935 263
[25] Ciurca Jr SJ (1975) US Patent 3 928 043
[26] Hammond HQ, Humphlett WJ, Salminen IF (1977) US Patent 4 009 029
[27] Campbell GA, Cohen HL, Ling HG, Ponticello IS (1975) US Patent 3 868 252
[28] Ciurca SJ, Brault AT (1976) US Patent 3 998 640
[29] Yoshida Y, Imai S, Miyakawa M (1975) Jpn Kokai S50-91324
[30] Idelson EM (1969) US Patent 3 482 972
[31] Idenson EM (1970) US Patent 3 551 406
[32] Idelson EM (1971) US Patent 3 563 739
[33] Idelson EM (1971) US Patent 3 597 200
[34] Baker III TN, Zuckerman B (1981) US Patent 4 302 525

[35] Mehta AC, Nawn GH, Tayor LD (1983) US Patent 4 390 613
[36] Rogers HG, Idelson EM, Cieciuch RFW, Bloom SM (1974) J Photogr Sci. 22:138
[37] Fehervari AF, Manning JJ (2002) US Patent 6 403 278 B1
[38] Locatell Jr L, Zepp CM, Cieciuch RF (1981) US Patent 4 267 251
[39] Locatell Jr L, Zepp CM, Cieciuch RF (1986) US Patent 4 619 784
[40] Arnost MJ, Viski P, Waller DP, Whritenour DC (1997) US Patent 5 691 458
[41] Arnost MJ, Viski P, Waller DP, Whritenour DC (1998) US Patent 5 716 754
[42] Arnost MJ, Viski P, Waller DP, Whritenour DC (1998) US Patent 5 811 530
[43] Bader H, Rickter DO (1975) US Patent 3 888 875
[44] Idelson EM, Karady I, Mark B, Rickter DO, Hooper VH (1967) Inorg Chem. 6:450
[45] Idelson EM, Karady I (1966) J Am Chem Soc. 88:186
[46] Idelson EM (1969) US Patent 3 453 107
[47] Idelson EM (1970) US Patent 3 544 545
[48] Chinoporos E, Idelson EM, King PF (1987) US Patent 4 656 117
[49] Idelson EM (1971) US Patent 3 629 336
[50] Idelson EM (1973) US Patent 3 780 105
[51] Bader H, Feingold MH (1974) US Patent 3 812 191
[52] Bader H, Feingold MH (1976) US Patent 3 970 616
[53] Bader H, Feingold MH (1979) US Patent 4 150 018
[54] Idelson EM (1973) US Patent 3 752 836
[55] Goulston AB, Huyffer PS (1972) US Patent 3 705 184
[56] King PF, Stroud SG (1984) US Patent 4 481 278
[57] Idelson EM (1980) US Patent 4 231 950
[58] Idelson EM (1981) US Patent 4 247 455
[59] Bader H, Jahngen Jr EG (1973) US Patent 3 770 833
[60] Bader H, Jahngen Jr EG (1975) US Patent 3 929 848
[61] Locatell Jr L, Rogers HG, Bilofsky RC, Cieciuch RF, Zepp CM (1981) US Patent 4 264 701
[62] Locatell Jr L, Rogers HG, Bilofsky RC, Cieciuch RF, Zepp CM (1983) US Patent 4 386 216
[63] Borror AL, Cincotta L, Mahoney EM, Feingold MH (1981) US Patent 4 264 507
[64] Borror AL, Cincotta L, Mahoney EM, Feingold MH (1981) US Patent 4 264 704
[65] Weyerts, WJ, Kennard KC, Van Campen JH (1969) Ger Patent 1 422 848

[66] Rogers HG, Lutes HW (1966) US Patent 3 265 498
[67] Weyerts WJ, Salminen WM (1966) US Patent 3 260 597
[68] Grasshoff JM, Taylor LD (1972) US Patent 3 698 898
[69] Grasshoff JM, Taylor LD (1972) US Patent 3 993 661
[70] Taylor LD, Grasshoff JM, Pluhar M (1978) J Org Chem. 43:1197
[71] Bartels-Keith JR, Puttick AJ (1982) US Patent 4 350 754
[72] Bartels-Keith JR, Puttick AJ (1982) US Patent 4 442 290
[73] Bartels-Keith R, Boggs RA, Puttick AJ, Sofen NM (1985) US Patent 4 503 139
[74] Bartels-Keith R, Boggs RA, Puttick AJ, Sofen NM (1989) US Patent 4 847 383
[75] Fujiwara M, Satoh R, Masukawa T, Uozumi T (1976) US Patent 3 938 996
[76] Mehta AC, Nawn GH, Taylor LD (1982) US Patent 4 355 101
[77] Mehta AC, Nawn GH, Taylor LD (1982) US Patent 4 355 092
[78] Mehta AC, Nawn GH, Taylor LD (1986) US Patent 4 593 108
[79] Boggs RA, Mahoney JB, Mehta AC, Schwarzel WC, Taylor LD (1988) US Patent 4 743 533
[80] Boggs RA, Mahoney JB, Mehta AC, Schwarzel WC, Taylor LD (1990) US Patent 4 946 964
[81] Green M, Rogers H (1965) US Patent 3 173 786
[82] Weyerts WJ, Salminen WM (1966) US Patent 3 253 915
[83] December JR, Hass HC, Reid JL (1974) US Patent 3 816 125
[84] Dolphin JM (1987) US Patent 4 677 206
[85] Eckert RD, Huang DD, Pierce FM (1995) US Patent 5 422 233
[86] Foley JA, Filosa MP, Telfer SJ, Marshall L, Waller DP (2003) US Patent 6 541 177 B1
[87] Guarrera DJ, Mattucci NC, Mehta AC, Taylor LD, Warner JC (1998) US Patent 5 705 312
[88] Mehta A, Taylor LD (1985) US Patent 4 543 317
[89] Land EH (1972) US Patent 3 647 437
[90] Bloom SM, Borror AL, Huyffer PS, MacGregor PT (1972) US Patent 3 702 244
[91] Simon MS (1974) US Patent 3 833 614
[92] Simon MS, Waller DP (1974) US Patent 3 833 615
[93] Kleim PO (1988) US Patent 4 740 448
[94] Hasan FB, Huang DD (1999) US Patent 6 001 531

[95] Cieciuch RFW, Luhowy RRL, Meneghini FA, Rogers HG (1977) US Patent 4 060 417

[96] Evans DH, Greenwald RB (1978) US Patent 4 107 176

[97] Eckert RD, Filosa MP, Gomes G, Layne AR, Schwarz MC, Taylor LD (1998) US Patent 5 756 253

[98] Grasshoff JM, Taylor LD (1976) US Patent 3 932 480

[99] Borror AL, Ellis EW (1983) US Patent 4 382 119

[100] Borror AL, Ellis EW (1983) US Patent 4 388 398

[101] Haas HC (1964) US Patent 3 148 061

[102] Rogers HG (1967) US Patent 3 295 970

[103] Taylor LD (1973) US Patent 3 770 439

[104] Bedel SF (1978) US Patent 4 080 346

[105] Land EH, Bronstein-Bonte IY, Taylor LD (1982) US Patent 4 322 489

[106] Bronstein-Bonte IY (1986) US Patent 4 563 411

[107] Grasshoff JM, Simon MS (1988) US Patent 4 794 067

[108] Febonio RL, Foley JA (1998) US Patent 5 747 219

[109] Waterman KC (1997) US Patent 5 633 114

[110] Land EH (1968) US Patent 3 362 819

[111] Bedell SF (1973) US Patent 3 765 885

[112] Lindholm EP, Manning JJ (1997) US Patent 5 593 810

[113] Land EH (1968) US Patent 3 362 821

[114] Taylor LD (1969) US Patent 3 421 893

[115] Taylor LD (1971) US Patent 3 575 701

[116] Bedell SF, Sullivan CL, Taylor LD (1980) US Patent 4 201 587

[117] Sullivan CI (1981) US Patent 4 297 431

[118] Sullivan CI (1984) US Patent 4 426 481

[119] Taylor LD (1981) US Patent 4 288 523

[120] Mehta AC (1984) US Patent 4 461 824

[121] Taylor LD (1984) US Patent 4 458 001

[122] Schwarzel WC, Taylor LD (1983) US Patent 4 391 895

p-Sulfonamidonaphthol Dye Releasers

18.1 Chemical Processes and Multilayer Structure

The instant color film based on *p*-sulfonamidonaphthol dye releasers was marketed by Kodak under the name PR-10 in 1976 [1].[1] This film is now unavailable because Kodak has discontinued the manufacture of the film.[2]

18.1.1 Chemical Processes of *p*-Sulfonamidonaphthol Dye Releasers

p-Sulfonamidonaphthol dye releasers are negative-working compounds of dye-releasing type. The principal reactions of the dye releasers have been already discussed in Chapter 16. Since each dye releaser layer is separated from the corresponding silver halide emulsion layer, the reaction between the dye releaser and the silver halide should be mediated by means of a diffusible electron transfer agent (ETA). The ETA is usually supplied from an activator fluid.

As shown in Fig. 18.1, a phenidone derivative (e.g., 4-hydroxymethyl-4-methyl-1-phenyl-3-pyrazolidone, **1**) is used as an electron transfer agent (ETA), which is dissociated and oxidized into **3** (ETA_{ox}) according to eq. 18.1.[3] The structure of the ETA_{ox} (**3**) exhibits a radical character, after one electron of the

---

[1] Strictly speaking, *p*-sulfonamidonaphthols should be referred to as 4-sulfonamido-1-naphthols.

[2] This is because Kodak lost a patent infringement suit brought on by Polaroid in 1986.

[3] The formal minus charge of the structure **2** is depicted for the sake of convenience, although the symbol $^{-}$O– or :Ö– is sufficient to represent the species.

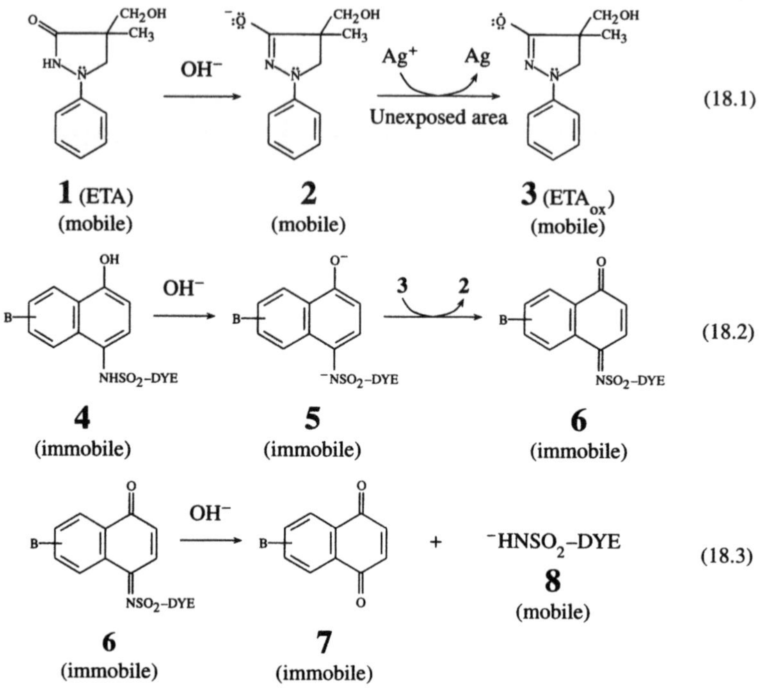

1 (ETA) (mobile) —OH^-→ **2** (mobile) —Ag^+ → Ag, Unexposed area→ **3** (ETA_{ox}) (mobile) (18.1)

4 (immobile) —OH^-→ **5** (immobile) —3 → 2→ **6** (immobile) (18.2)

6 (immobile) —OH^-→ **7** (immobile) + $^-HNSO_2$–DYE **8** (mobile) (18.3)

Figure 18.1. Consecutive reactions of silver halide, an electron transfer agent, and a *p*-sulfonamidonaphthol dye releaser. The letter B represents a ballast group. The symbol DYE denotes a cyan, magenta or yellow dye moiety.

anionic species (**2**) is transferred to a silver cation.[4] Note that the redox reaction (eq. 18.1) occurs in the exposed region of a direct positive emulsion layer.

The ETA_{ox} diffuses into the adjacent dye releaser layer, in which it reacts with the dye releaser dianion (**5**) according to eq. 18.2, giving the oxidized species (**6**).[5] As found in eq. 18.3, the oxidized species is hydrolyzed under alkaline conditions to give a mobile dye (**8**), which diffuses to a mordant layer, giving dye images.

18.1.2 Multilayer Structure for Color Reproduction

Three dye releasers for color reproduction are incorporated in a multilayered film, which comprises two outer supports, each having several coated layers as illustrated in Fig. 18.2 [2].

---

[4] According to the convention of organic chemistry, the symbol •O– is used to represent the oxygen atom of the ETA_{ox}. For the sake of convenience, the oxygen atom of the structure **3** is represented by the symbol :Ȯ–, where lone pairs and an odd electron are fully depicted.

[5] Note that two moles of the ETA_{ox} are necessary to oxidize one mole of the dye releaser dianion (**5**) into the corresponding quinonemonoimide (**6**).

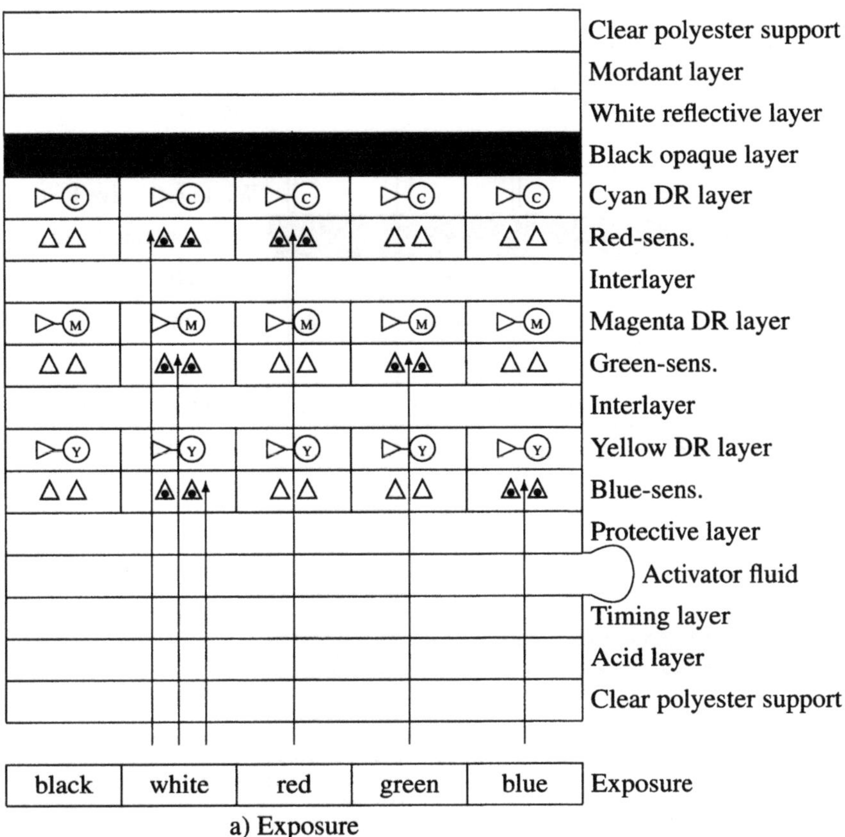

Figure 18.2. Instant color film using negative-working dye releasers. A schematic cross-section during exposure. △: Direct positive silver halide grain, ▲: Exposed silver halide grain with an internal latent image, ▲: Developed silver halide grain; ▷-Ⓨ: Yellow dye developer, ▷-Ⓜ: Magenta dye developer, and ▷-Ⓒ: Cyan dye developer.

One of the two outer supports is a transparent polyester sheet, on which a mordant layer, a white opaque reflective layer, a black opaque layer and a set of photo-sensitive units are coated. This coated sheet is usually called a *light-sensitive sheet*, although it involves even an image-receiving layer along with the light-sensitive layers.[6]

The other outer support is also a transparent polyester sheet, on which an acid layer and a timing layer are coated. This coated sheet is called a *cover sheet*. In order to construct an integrated (so-called "mono-sheet") film shown in Fig. 18.2,

[6]Remember that a receiving layer and a set of light-sensitive layers are coated separately from each other in the multilayer film based on the chemistry of dye developers (Fig. 17.2 in Chapter 17).

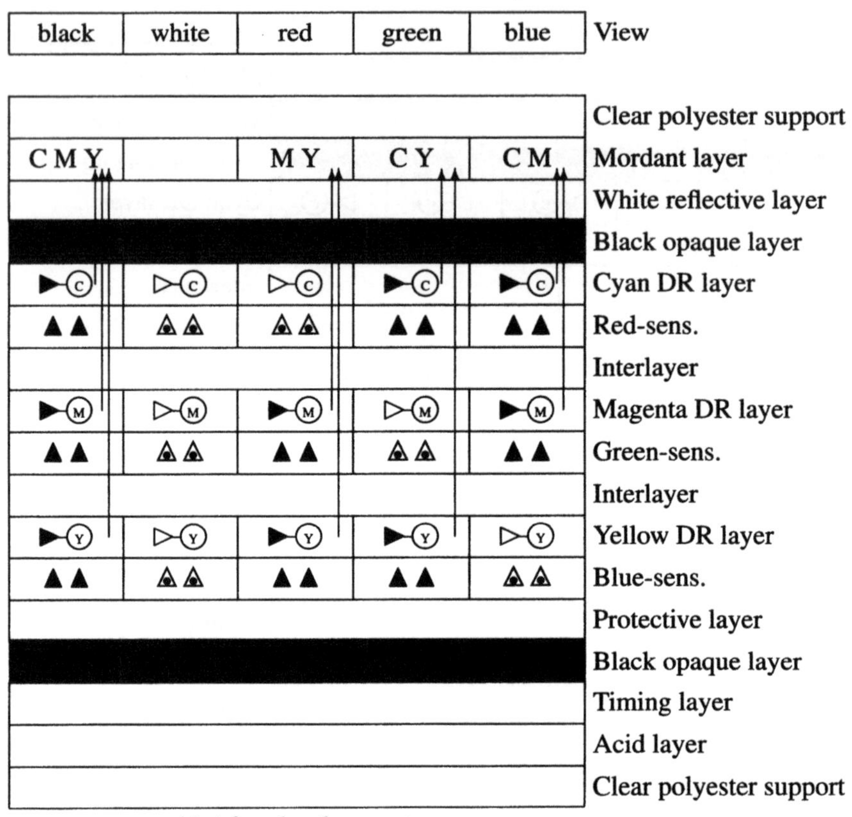

b) After development

Figure 18.3. Instant color film using negative-working dye releasers. A schematic cross-section during and after development. ▲̇: Exposed silver halide grain with an internal latent image, ▲: Developed silver halide grain; ▷-Ⓨ: Yellow dye releaser, ▷-Ⓜ: Magenta dye releaser, and ▷-Ⓒ: Cyan dye releaser; ▶-Ⓨ: Oxidized yellow dye releaser, ▶-Ⓜ: Oxidized magenta dye releaser, and ▶-Ⓒ: Oxidized cyan dye releaser.

the two coated sheets are superimposed in a face-to-face fashion, where a pod containing an activator fluid is attached between the two sheets.

It should be noted that each emulsion layer contains direct positive silver halide emulsions described in Chapter 16 (Subsection 16.4.2), since *p*-sulfonamidonaphthol dye releasers are negative-working.

For an effective capture of light, a unit of photo-sensitive layers should be aligned in the order of (exposure)—a silver halide emulsion layer—a dye releaser layer.[7] The dye releaser system allows the direction of view represented by

[7]This order is the same as that of the film based on dye developers. On the other hand, the dye developer system requires the order represented by (exposure and view)—a silver halide emulsion layer—a dye releaser layer, as discussed in Chapter 17.

(exposure)—a silver halide emulsion layer—a dye releaser layer—(view). Hence, the order of layers shown in Fig. 18.2 is inevitable so that the exposure of light is effected from the bottom of the cross-section.

Figure 18.2 illustrates an exposure to blue, green, red, or white light (or to no light) from the bottom through the transparent support. Each upward arrow represents such light as arriving at a silver halide grain so that the symbol ⚠ beside the arrowhead represents the resulting silver halide grain with an internal latent image.

When passed through the gap between a pair of rollers, the pod (Fig. 18.2) is ruptured to spread an activator fluid between the two outer sheets of the integrated film. Since the activator fluid contains carbon black as an opacifier, a black opaque layer is formed to construct a mini-dark room inside the mono-sheet film. In the mini-dark room, the development of unexposed silver halide emulsions is initiated on the action of a nucleating agent and an ETA (fed from the activator fluid) so as to cause the reactions shown in Fig. 18.1. For example, the silver halide grains (⚠) in the area exposed to blue light do not oxidize an ETA so that the yellow dye releaser remains unreacted. In contrast, the silver halide grains in the corresponding area of the other layers are reacted with an ETA (Δ + ETA → ▲ + ETA_{ox}). The oxidized ETA reacts with a magenta dye releaser or a cyan dye releaser. Each released dye diffuses to the mordant (image-receiving) layer and is captured there, as represented by an upward arrow. Hence, the subtractive color mixing of cyan and magenta produces blue color (C + M = B) in terms of the direct positive emulsion mechanism. Similarly, green (C + Y = G), red (M + Y = R), black (C + M + Y = Black), and white (no dyes) are reproduced. These colors are observed from the top of Fig. 18.3 against the background of the white titanium oxide pigment, which has been originally coated as a white reflective layer.

18.2 *p*-Sulfonamidonaphthol Moieties

18.2.1 Ballasted *p*-Sulfonamidonaphthols

Non-diffusible *p*-sulfonamidonaphthols (4-sulfonamido-1-naphthols) capable of releasing diffusible dyes have been disclosed by Fleckenstein et al. [3,4] as dye releasers for instant color photography.

Although these patents have also disclosed *p*-sulfonamido*phenols*, the *p*-sulfonamido*naphthols* have been finally selected, probably because of their superiority in dye-releasing capability. In particular, the *p*-sulfonamidonaphthol moiety involved commonly in a set of dye releasers shown in Fig. 18.4 has been employed in the PI-10 film marketed by Kodak (1976). As found easily, the ballast group is common to the one used in cyan couplers for conventional color photography. The *p*-sulfonamidonaphthol moiety is combined with a naphthol azo dye having a special diazo component for the cyan dye releaser (**9**) [5,6], with a naphthol azo dye for the magenta dye releaser (**10**) [7], and with a pyrazolone

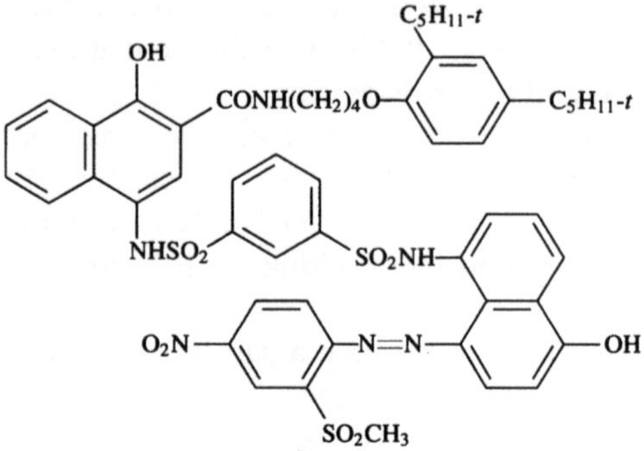

9 (cyan)

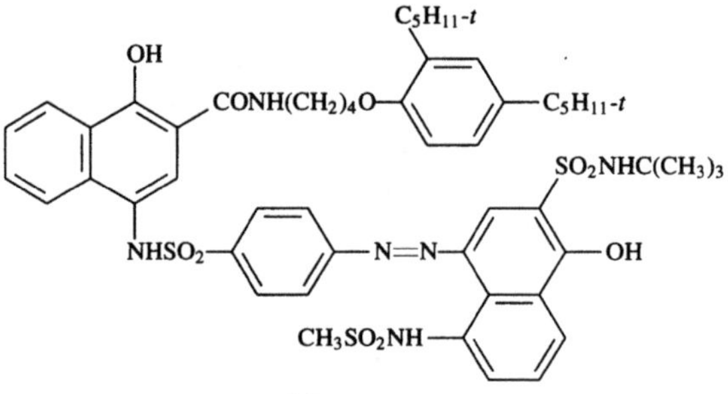

10 (magenta)

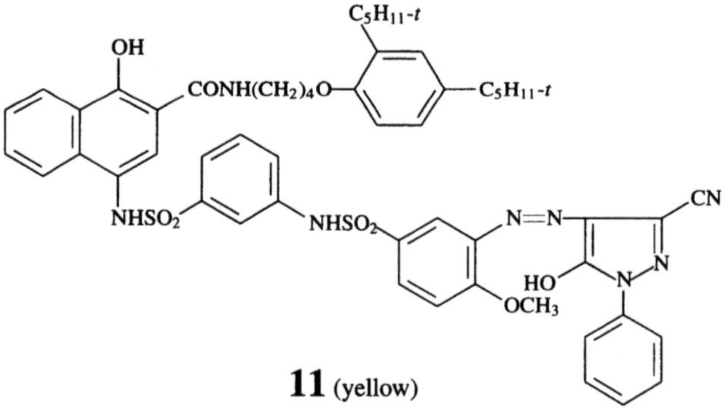

11 (yellow)

Figure 18.4. A set of cyan, magenta, and yellow *p*-sulfonamidonaphthol dye releasers for color reproduction of instant color photography.

azo dye for the yellow dye releaser (**11**) [8]. These dye moieties will be discussed later.

Syntheses

Typical examples of preparing *p*-sulfonamidonaphthol dye releasers have been summarized in a review [9]. To synthesize *p*-sulfonamidonaphthol dye releasers shown in Fig. 18.4, ballasted 4-amino-1-naphthols such as **14** are key intermediates [3]. Such intermediates have been synthesized by starting cyan couplers (ballasted 1-naphthols) for conventional color photography, as illustrated in Fig. 18.5. The amino group at the 4-position is introduced by an azo coupling and the subsequent reduction of the azo group.

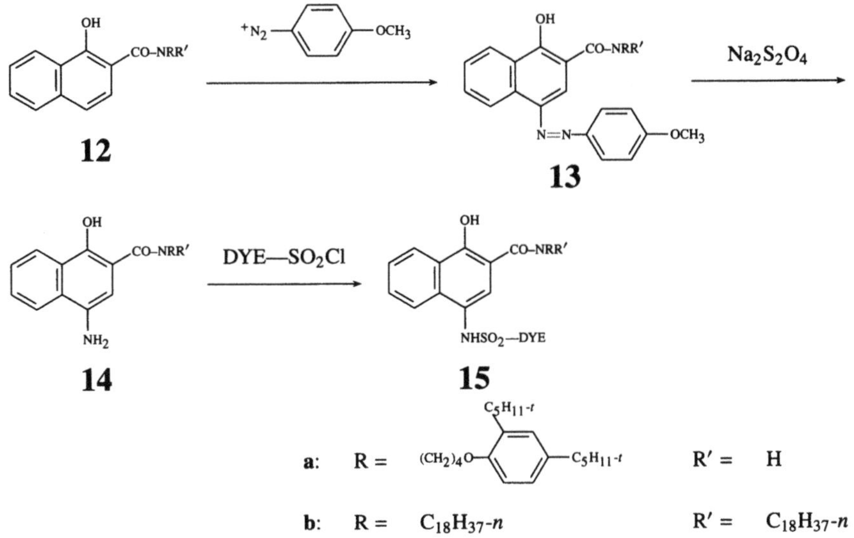

OH CO–NRR' **12** $^{+}N_2$–C$_6$H$_4$–OCH$_3$ → OH CO–NRR' N=N–C$_6$H$_4$–OCH$_3$ **13** $Na_2S_2O_4$ →

OH CO–NRR' NH_2 **14** DYE—SO_2Cl → OH CO–NRR' $NHSO_2$—DYE **15**

a: R = $(CH_2)_4O$–C$_6$H$_3$(C_5H_{11}-*t*)$_2$ R' = H

b: R = $C_{18}H_{37}$-*n* R' = $C_{18}H_{37}$-*n*

Figure 18.5. Synthetic pathways for preparing *p*-sulfonamidonaphthol dye releasers [3, 12].

18.2.2 Improved Dye-Releasing Efficiency

As shown in eq. 18.3 (Fig. 18.1), the quinone (**7**) is formed from the oxidized dye releaser (**6**). The quinone (**7**) is subjected to a hydroxylation reaction (a 1,4-addition) represented by eq. 18.4 (Fig. 18.6), where the position to which the ballast group attaches is altered to match the present explanation. The resulting hydroxylation product (**16**) is then oxidized into a hydroxyl-quinone (**17**), where two moles of ETA_{ox} (**3**) are consumed in vain. This means that such side reactions as the hydroxylation at the 3-position decrease the dye-releasing efficiency.

To prevent such undesired side reactions and to enhance the dye-releasing efficiency, the substitution of an alkyl or phenyl group at the 3-position of the naphthol nucleus (e.g., **18**) has been examined, as shown in Fig. 18.7 [10,11]. These patents have claimed the use of a retained image as well as a transferred image after the bleach-fix process of the photo-sensitive sheet which is peeled apart from the image-receiving sheet. The side reaction represented by eq. 18.4 causes unwanted stain of the retained image. Because the D_{max} with the 3-methyl group for the retained magenta image is lower than that without the 3-methyl group, the dye-releasing efficiency of the former may be concluded to be higher

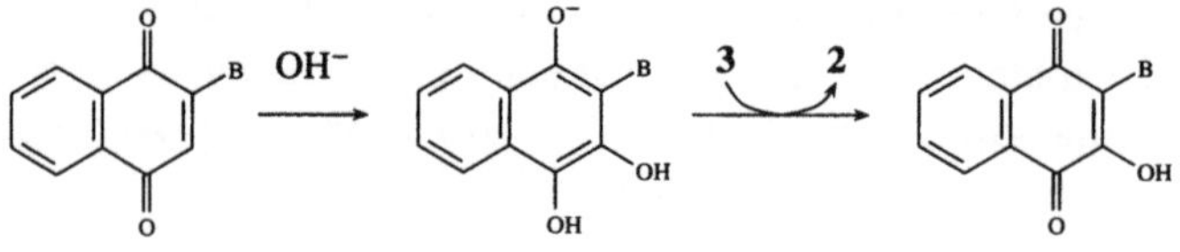

(18.4)

7 (immobile) 16 (immobile) 17 (immobile)

Figure 18.6. Side reactions of *p*-sulfonamidonaphthol dye releasers. The letter B represents a ballast group.

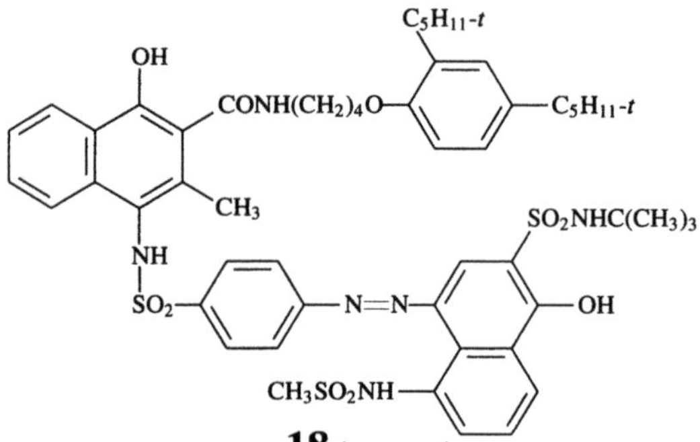

18 (magenta)

Figure 18.7. *p*-Sulfonamidonaphthol dye releaser having a 3-methyl substituent.

than that of the latter. This effect can be explained by the difference between the stoichiometry due to eq. 18.2 and the combined one of eqs. 18.2 and 18.4.

Syntheses

To synthesize *p*-sulfonamidonaphthol dye releasers having a 3-methyl group (e.g., **18**), one must start from the construction of the naphthol nucleus, as shown in Fig. 18.8 [11]. 1-Phenyl-2-propyl *p*-toluenesulfonate (**20**) is condensed with the sodium salt of ethyl malonate to give diethyl 2-(1-phenyl-2-propyl)malonate, which is then hydrolyzed to the corresponding mono ester. This is converted into the acid chloride (**21**), which is cyclized by employing an intramolecular Friedel-Crafts acylation to give 2-ethoxycarbonyl-3-methyl-1-tetralone (**22**). The tetralone ring is catalytically dehydrogenated into a naphthol ring to produce ethyl 1-hydroxy-3-methyl-2-naphthoate (**23**). The ester group ($COOC_2H_5$) of **23** is converted into a ballasted amido group (CO–NRR′). The introduction of a 4-amino group is conducted via an azo derivative in a similar way to the one described in Fig. 18.5 except that the reduction of the azo group is conducted by catalytic hydrogenation in place of the action of sodium dithionite (sodium hydrosulfite).

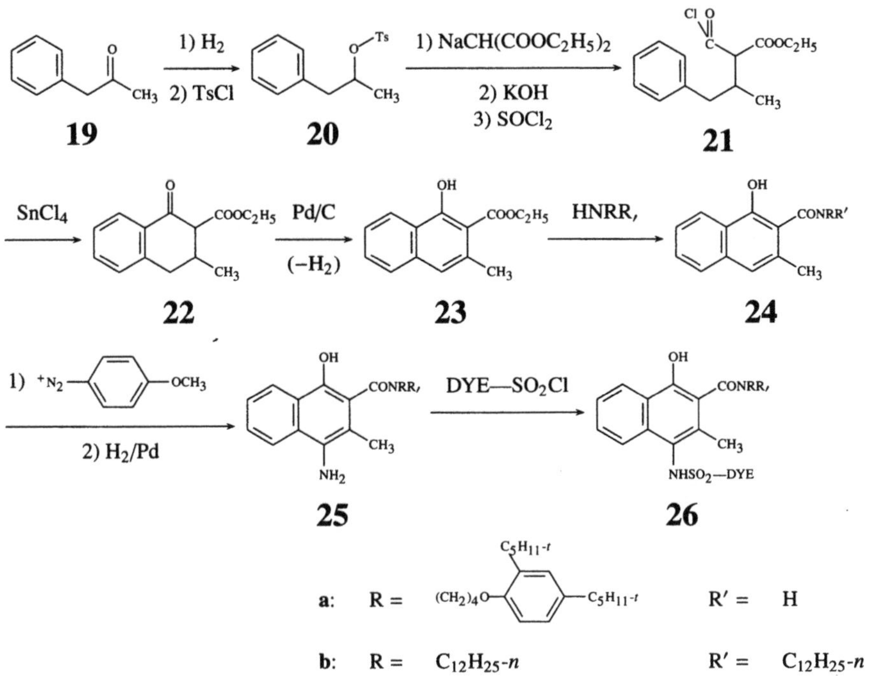

a: R = $(CH_2)_4O$–(2,4-di-C_5H_{11}-*t*-phenyl) R′ = H

b: R = $C_{12}H_{25}$-*n* R′ = $C_{12}H_{25}$-*n*

Figure 18.8. Synthetic pathways for preparing *p*-sulfonamidonaphthol dye releasers having a 3-methyl group [11].

18.2.3 Change of Ballast Groups

Another way of enhancing the dye-releasing efficiency is to use ballast groups based on a secondary amide, e.g. CO–NR_2, where R represents a long-chain alkyl group [12]. Although the change of a ballast group from a primary amide to a secondary amide seems to be trivial, it has practically provided a remarkable improvement of the dye-releasing efficiency. Among them, an *N*,*N*-dioctadecylcarbamoyl group (Fig. 18.9) has been selected to exhibit improved properties. A triplet of such *p*-sulfonamidonaphthol dye releasers for subtractive color reproduction (i.e., **27** for cyan, **28** for magenta, and **29** for yellow color) have been used as control dye releasers in comparison tests for attempted improvements of dye releasers [13].

Ortho- and para-sulfonamidophenol dye releasers that have another sulfonamido group in a conjuaged position with respect to the sulfonamido group have been reported, although they have not been used in commercialized films [13].

Syntheses

The ballasted 4-amino-1-naphthol that is necessary for the syntheses of *p*-sulfonamidonaphthol dye releasers shown in Fig. 18.9 can be prepared according to the scheme illustrated already in Fig. 18.5 [12]. Thus, the amino group at

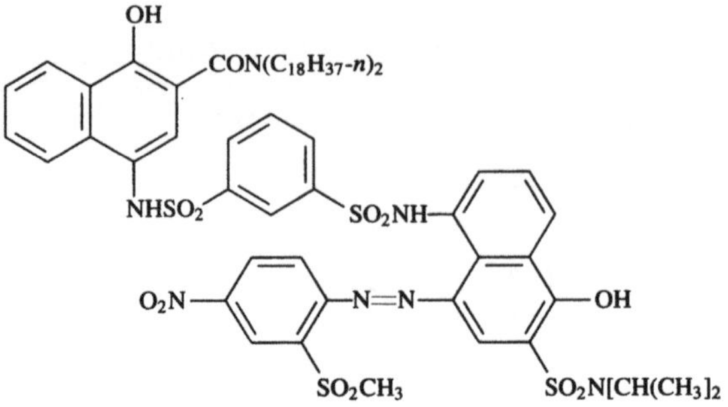

27 (cyan)

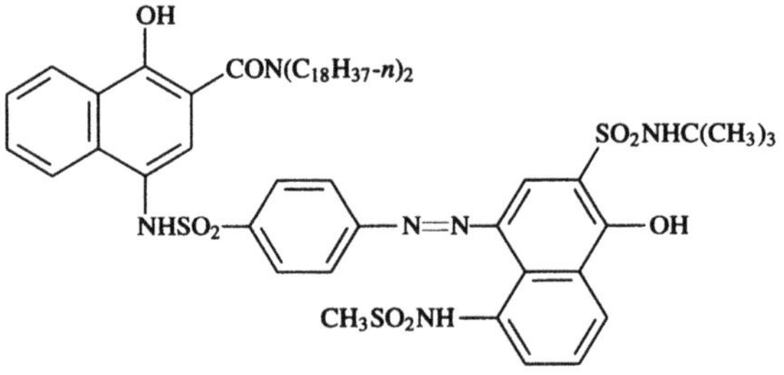

28 (magenta)

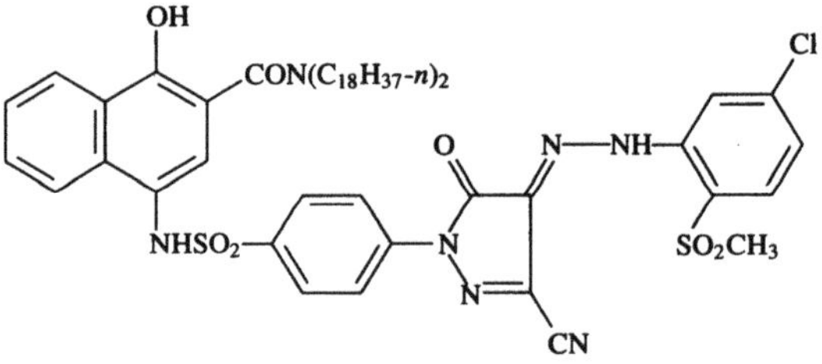

29 (yellow)

Figure 18.9. Another triplet of cyan, magenta, and yellow *p*-sulfonamidonaphthol dye releasers for the color reproduction of instant color photography. The *N,N*-dioctadecylcarbamoyl group of each dye releaser enhances dye-releasing efficiency.

the 4-position is introduced by an azo coupling and the subsequent reduction of the azo group.

18.3 Azo Dye Moieties

18.3.1 Azo Dye as Ionic Species

Dyes released from *p*-sulfonamidonaphthol dye releasers (e.g., those listed in Figs. 18.4 and 18.9) have a terminal sulfonamide group, which enables the dyes to diffuse through layers in a film. In general, an azo dye for a textile dyestuff is present in an equilibrium between an azo form (**30**) and a hydrazone form (**31**), as shown in the top row of Fig. 18.10. In contrast, the dyes from the *p*-sulfonamidonaphthol dye releasers are captured as anionic species by a mordant layer, as shown in the bottom row of Fig. 18.10 (**32** and **33**). This fact requires that the dyes have sufficiently low pK_a's.

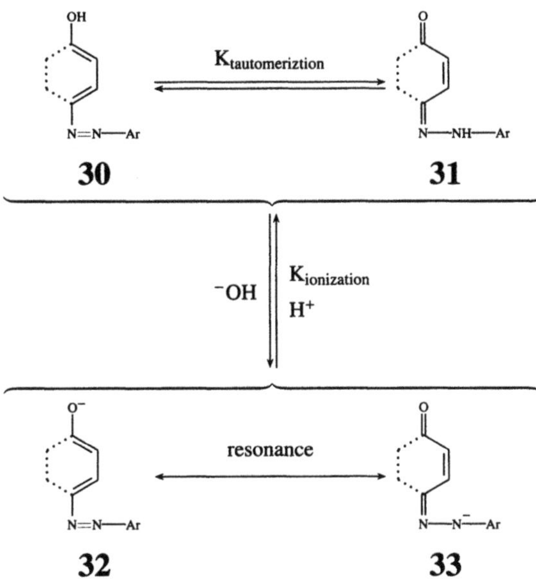

Figure 18.10. Equilibria of monoazo dyes.

18.3.2 Cyan Dye Moieties

The cyan dye released from the dye releaser (**9**) is present as an anionic species (**34**), because its pK_a is about 4 [2]. It has a further unique feature that the hue is shifted on the order of 100 nm, yielding a cyan dye. Note that usual monoazo naphthol dyes have a higher pK_a and a magenta hue. These features come from the structure (**34**), in which a nitro group is present at the 4-position and a methanesulfonyl group at the 2-position (in the 2-methanesulfonyl-4-nitrophenylazo moiety). As shown in Fig. 18.11, the dye (**34**) can be represented by a resonance hybrid (**34a** ↔ **34b**), which allows the minus charge to spread over all the atoms of the chromophoric π-electron system.

Figure 18.12 shows the improvement of dye moieties in which the 1-naphthol nuclei have various substituents. For example, the *N*-*t*-butylsulfamoly group is attached at the 2-position of the 1-naphthol nucleus (**35**) [6]. Compare this compound with a recent example (**27**), which has an *N*,*N*-di(*sec*-propyl)sulfamoly group at the 2-position and a different ballast group. One of the merits due to the

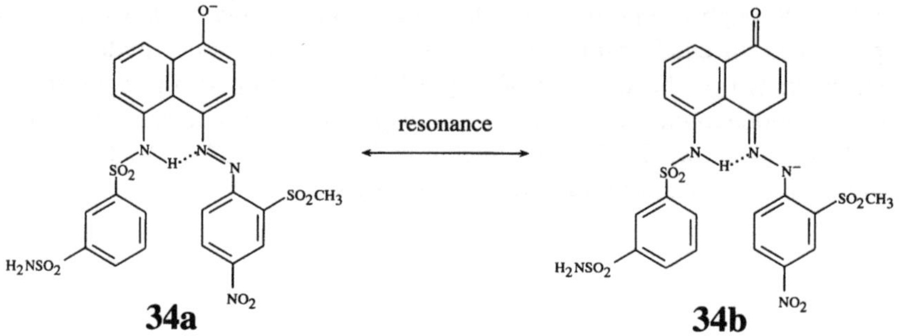

Figure 18.11. Cyan dye released from the dye releaser (**9**), where its anionic species is depicted as a resonance hybrid.

incorporation of such 2-substituents has been reported to be the improvement of light fastness.

An appropriate carbamoyl group at the 2-position, as found in **36**, improves the hue of the corresponding mordanted cyan dye [14,15]. For example, the dye released from **36** has a single band at λ_{max} 656 nm, whereas the dye released from **9** (i.e., **34** of Fig. 18.11) exhibits a shoulder at 615 nm along with λ_{max} 640 nm. As for heat-moisture fastness (60℃, 70% relative humidity), the density loss of the dye released from **36** is measured to be 0.27 (at $D = 1.0$), whereas that of the dye released from **27** is obtained to be 0.48 (at $D = 1.0$).

The incorporation of an alkoxy group on the azo component has been disclosed to improve the light fastness associated with singlet oxygen [16]. For example, the dye released from **37** gives the fading rate constant $k_{ox} = 0.9 \times 10^{-6}\ M^{-1}sec^{-1}$ for bleaching by singlet oxygen in a methanolic solution, whereas the corresponding value of the dye released from **36** is measured to be $k_{ox} = 2.3 \times 10^{-6}\ M^{-1}sec^{-1}$.

The reaction of the cyan dye released from **36** with sulfur dioxide has been investigated to improve the stability of the dye image when employed in peel-apart films [17].[8] The fading has been assigned to the attack of sulfur dioxide to the 3-position of the dye 1-napthol nucleus. Fortunately, however, the original cyan dye can be regenerated simply by washing the film in cold water.

Syntheses

For the synthesis of the cyan dye releaser (**9**, X = H), 5-amino-1-naphthol (**38**, X = H) is used as a starting material, as illustrated in the second part of Fig. 18.13 [5]. The use of 3-fluorosulfonylbenzenesulfonyl chloride (**42**) should be emphasized, where two sulfo functions are differentiated in the form of SO_2Cl and SO_2F. Note that the chlorosulfonyl group is more active than the fluorosulfonyl group. Hence, after the condensation of **38** with **42**, the resulting intermediate having the fluorosulfonyl group (**43**) is coupled with the diazonium ion derived from 2-methanesulfonyl-4-nitroaniline so as to give an azo dye intermediate (**44**).

[8] The peel-apart film was placed on the market under the name "Ektaflex" in 1981 by Kodak.

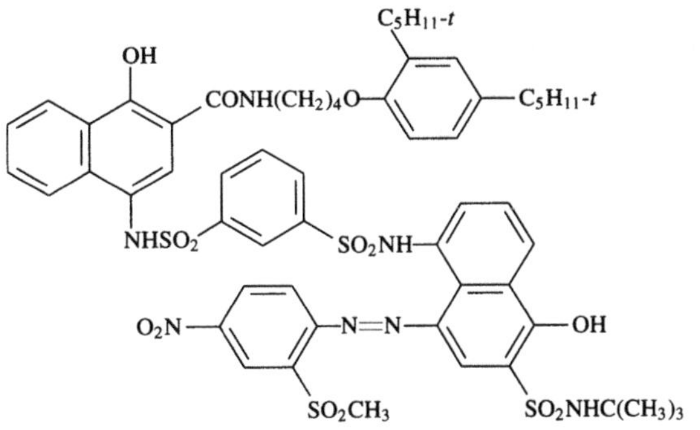

35 (cyan)

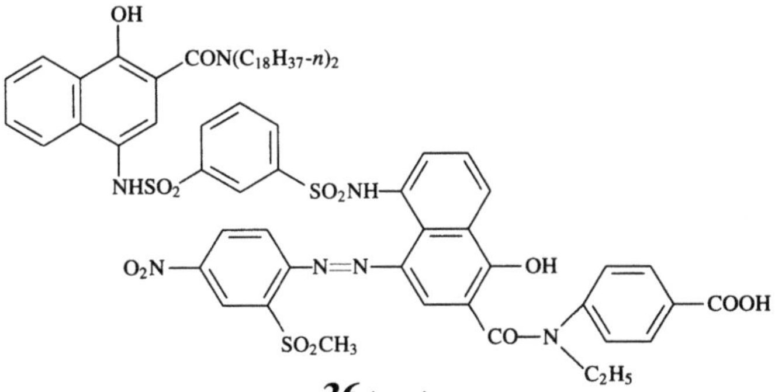

36 (cyan)

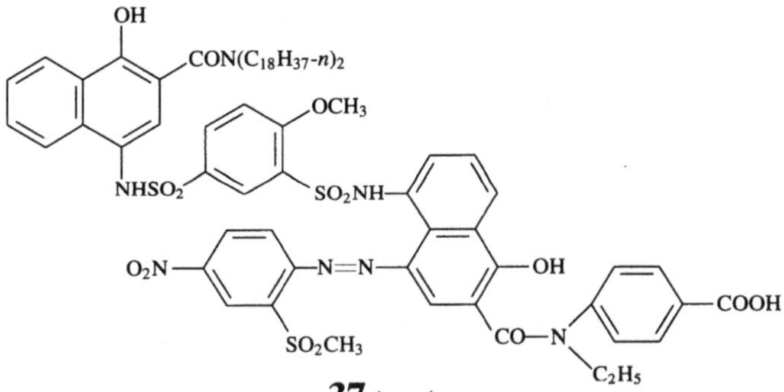

37 (cyan)

Figure 18.12. *p*-Sulfonamidonaphthol dye releasers for releasing a cyan azo dye.

The last step for preparing **9** (X = H) is the condensation of **44** with the ballasted 4-amino-1-naphthol (**15a**) described in Fig. 18.5, where sodium carbonate is used as a base in dimethyl sulfoxide as a solvent.

The cyan dye releaser (**35**, X = $SO_2NHC(CH_3)_3$) requires a substituted 5-amino-1-naphthol (**41**, X = $SO_2NHC(CH_3)_3$) as a starting material [6]. This starting material is synthesized from 5-amino-1-naphthol, as illustrated in the first part of Fig. 18.13. Then, the cyan dye releaser (**35**, X = $SO_2NHC(CH_3)_3$) was synthesized in a similar way according to the second part of Fig. 18.13.

As found easily, the ballasted 4-amino-1-naphthols described in Figs. 18.5 and 18.8 can be used in place of **15a** to synthesize a variety of cyan dye releasers.

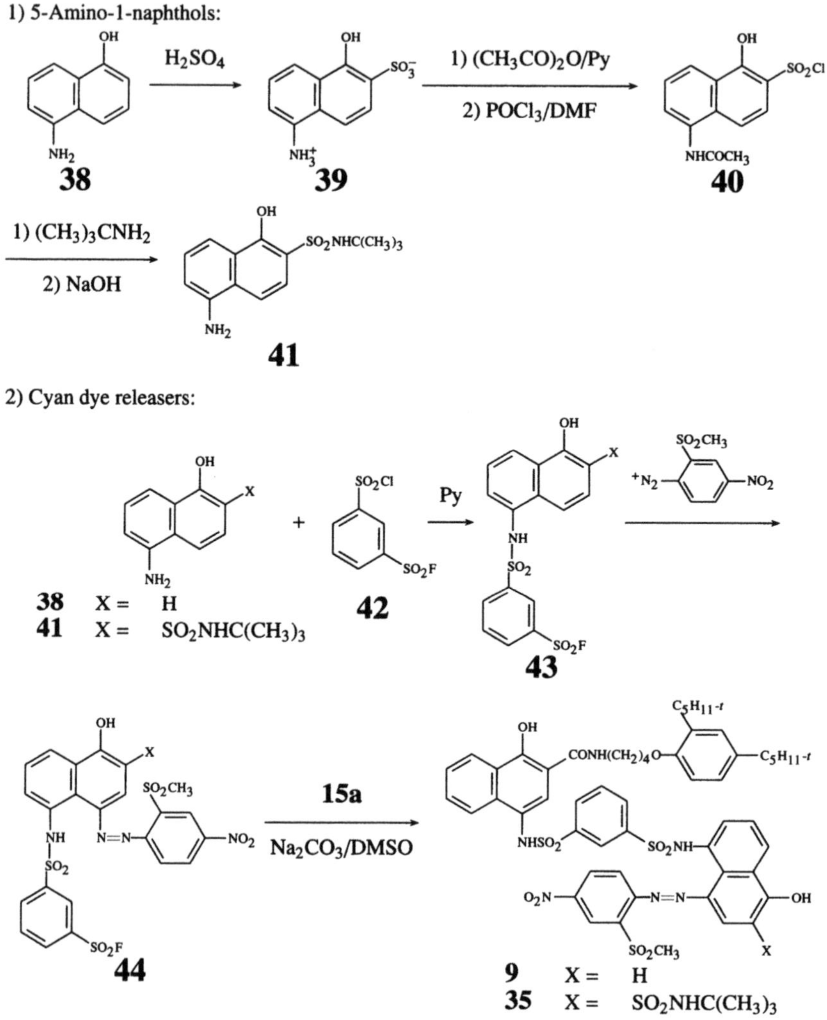

1) 5-Amino-1-naphthols:

38 $\xrightarrow{H_2SO_4}$ **39** $\xrightarrow[\text{2) } POCl_3/DMF]{\text{1) } (CH_3CO)_2O/Py}$ **40** $\xrightarrow[\text{2) NaOH}]{\text{1) } (CH_3)_3CNH_2}$ **41**

2) Cyan dye releasers:

38 X = H
41 X = $SO_2NHC(CH_3)_3$

+ **42** $\xrightarrow{Py}$ **43** $\xrightarrow{^{+}N_2\text{-}C_6H_3(SO_2CH_3)\text{-}NO_2}$ **44** $\xrightarrow[Na_2CO_3/DMSO]{\textbf{15a}}$

9 X = H
35 X = $SO_2NHC(CH_3)_3$

Figure 18.13. Synthetic pathways for preparing cyan dye releasers. The abbreviations have the following meanings: Py—pyridine; DMF—*N*,*N*-dimethylformamide; and DMSO—dimethyl sulfoxide.

18.3.3 Magenta Dye Moieties

The pK_a values of 4-(4-sulfamoylphenylazo)-1-naphthols with various substituents at the 2-position have been correlated to the σ values of the substituents, where positive σ values affect favorably the ease of ionization of the dyes [2]. Thus, the magenta dye released from the dye releaser (**10**) has pK_a of about 4, because it has an electron-withdrawing group ($SO_2NHC(CH_3)_3$) at the 2-position of the 1-naphthol. Hence, it is mordanted as an anionic species (**45**). As shown in Fig. 18.14, the anion (**45**) can be represented by a resonance hybrid (**45a** ↔ **45b**).

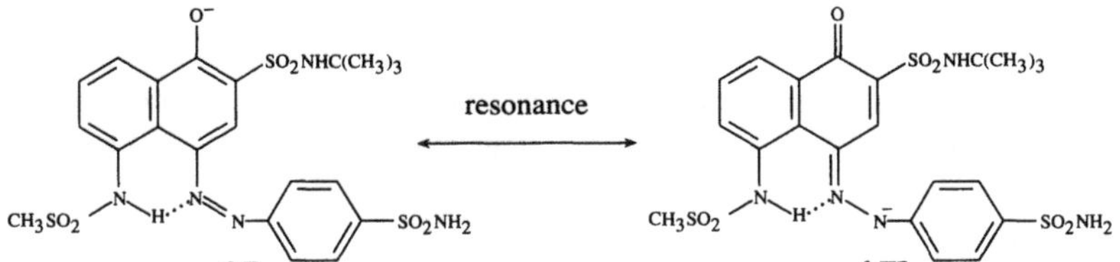

Figure 18.14. Magenta dye anion released from the dye releasers (**10** and **28**).

Table 18.1. Absorption Maxima and Light Stabilities of Magenta Dyes (**46**) [7]

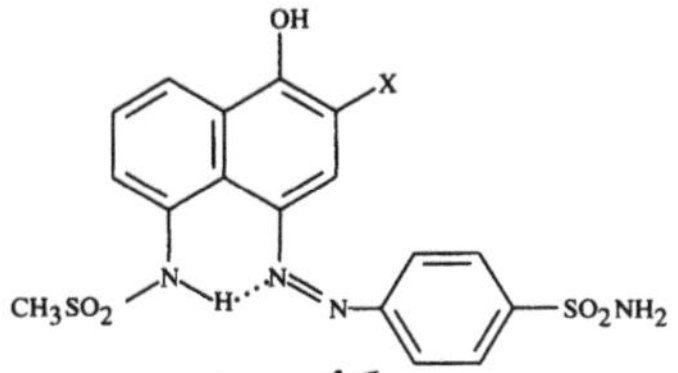

X	λ_{max} (nm)	Half-bandwidth (nm)	Light stability % Loss
$SO_2NHC(CH_3)_2CH_2C(CH_3)$	549	96	16
SO_2NH_2	552	99	10
SO_2NHCH_3	546	98	8
$SO_2NHC(CH_3)_3$	545	103	6
$SO_2NH(CH_2)_2CH(CH_2)_2$	546	97	12
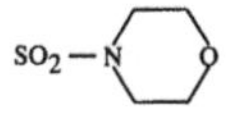 SO_2—N (morpholino ring, O)	551	97	48

The hues of 4-(4-sulfamoylphenylazo)-1-naphthols with various substituents at the 2-position have been also correlated to the σ values of the substituents. The dye hues tend to shift toward yellow as electrons are withdrawn (more positive σ values) [2].

For substituents in the arylazo group, electron-withdrawing groups (high σ values) cause shifts to longer wavelengths. The substituent on the arylazo group can be varied to modify hue without major changes in pKa [2].

By combining the data described above, such dyes as **45** have been selected as magenta dyes (Fig. 18.14). Intramolecular hydrogen bonding between the azo group and the 5-sulfonamido substituent sharpens the absorption of dye [2]. These dyes have been incorporated into such dye releasers as **10** and **28** [7].

The spectral data of several released dyes (**46**), which are cited from the patent [7], are listed in Table 18.1, which also contains data of light stabilities.

Syntheses

The 5-sulfonamido-1-naphthol derivative (**48**) has been synthesized according to the first scheme of Fig. 18.15 [7]. The exhaustive mesylation of **39** and

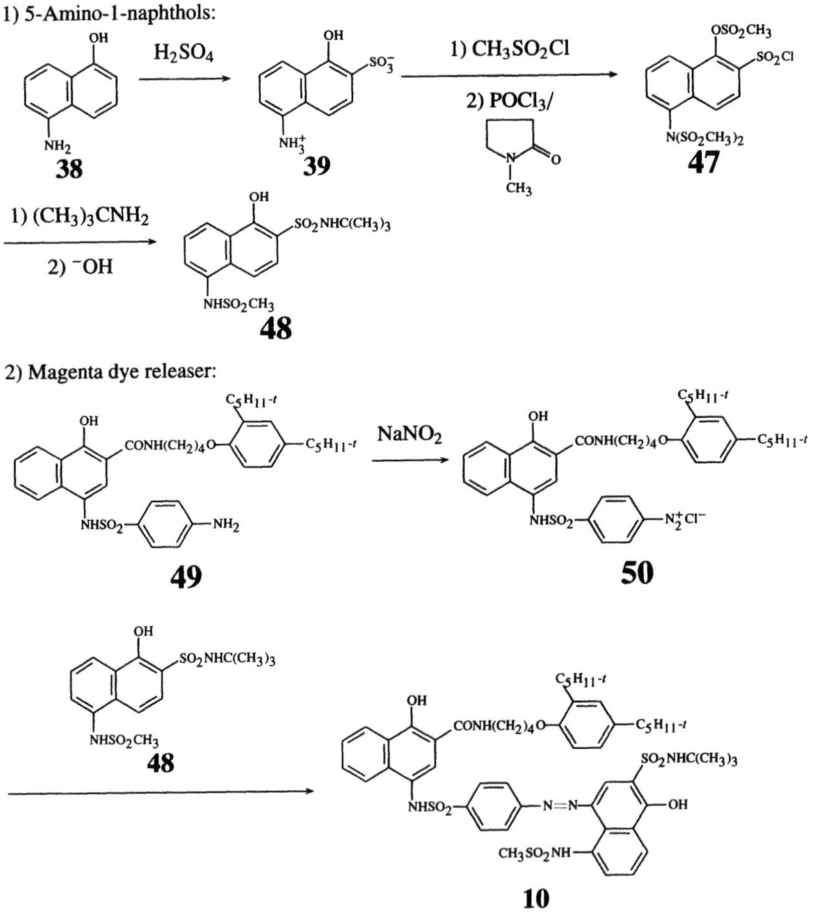

1) 5-Amino-1-naphthols:

H_2SO_4; 1) CH_3SO_2Cl, 2) $POCl_3$/; 1) $(CH_3)_3CNH_2$, 2) ^-OH

38, **39**, **47**, **48**

2) Magenta dye releaser:

$NaNO_2$

49, **50**, **48**, **10**

Figure 18.15. Synthetic pathway for preparing the magenta dye releaser (**10**).

the subsequent chlorination give the sulfonyl chloride (**47**), in which three mesyl (methanesulfonyl) groups are incorporated. The sulfonyl chloride is condensed with *t*-butylamine and hydrolyzed to give 2-(*N*-*t*-butylsulfamoyl)-5-methanesulfonamido-1-naphthol (**48**), which is used as a coupling component in the next scheme.

The second scheme of Fig. 18.15 illustrates a rather unique example in which the formation of an azo group is conducted at the final step [7]. Thus the diazonium salt (**50**) having a *p*-sulfonamidonaphthol moiety is coupled with **48** to give the magenta dye releaser (**10**).

18.3.4 Yellow Dye Moieties

In contrast to the cyan and the magenta dyes described above, arylazopyrozolone yellow dyes do not take anionic forms with sufficiently low pK_a's. Hence, the design of azopyrazolne dyes has been aimed at finding a candidate that shows little hue change in going from the anionic to the neutral (hydrazone) form [2]. As found in the structures of the yellow dye releasers described above (**11** and **29**), 4-arylazo-3-cyano-1-phenyl-5-pyrazolones have been selected as yellow dye moieties [8]. For example, the protonation of the 4-arylazo-3-cyano-1-phenyl-5-pyrazolone released from **11** has little effect on the hue. On the other hand, the protonation of 4-(4-sulfamoylphenylazo)-3-*N*-methylcarbamoyl-1-phenyl-5-pyrazolone, which has been incorporated in an earlier embodiment [4], shifts the hue to shorter wavelength to give the effect of lower density to blue light [8]. Both the yellow dyes have pK_a values of about 5.5 in a methanol-water solution.

Azophenol dyes having electron-withdrawing groups in each of the positions ortho to the phenolic hydroxyl group have been disclosed as yellow dye moieties (e.g., **51**) [18]. Although the ortho substitution may be capable of lowering the pK_a of the dye, it is not clear whether this lowering effect is practically useful or not in this case, because the λ_{max} data disclosed in this patent have been measured at about pH 8.

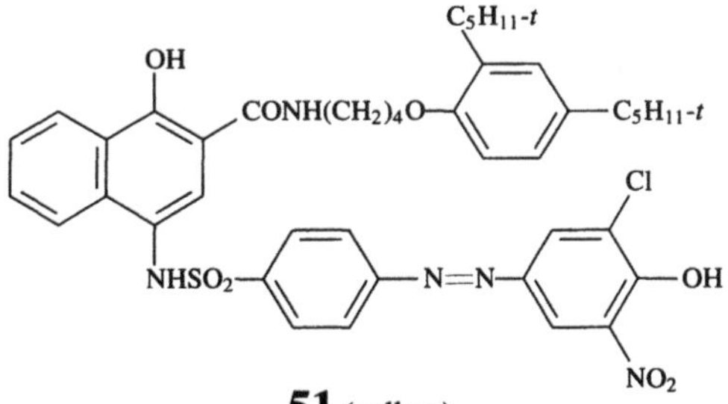

C_5H_{11}-*t*
OH
$CONH(CH_2)_4O$
C_5H_{11}-*t*
Cl
$NHSO_2$
N=N
OH
NO_2

51 (yellow)

Figure 18.16. *p*-Sulfonamidonaphthol dye releaser releasing a yellow azo dye of another type.

Syntheses

To synthesize the yellow dye releaser (**11**), the ballasted *p*-sulfonamidonaphthol intermediate (**53**) has been prepared according to the first scheme of Fig. 18.17 [8]. Thus, the starting material (**14a**) obtained in Fig. 18.5 is condensed with 3-nitrobenzenesulfonyl chloride. The nitro group of the condensate (**52**) is reduced into the amino group of the intermediate (**53**) by catalytic hydrogenation.

The pyrazolone dye moiety has been prepared by a usual azo coupling, as shown in the second scheme of Fig. 18.17. It should be noted that phosphoryl chloride/*N*,*N*-dimethylformamide works as a chlorinating agent converting

1) Ballasted *p*-sulfonamidonaphthol:

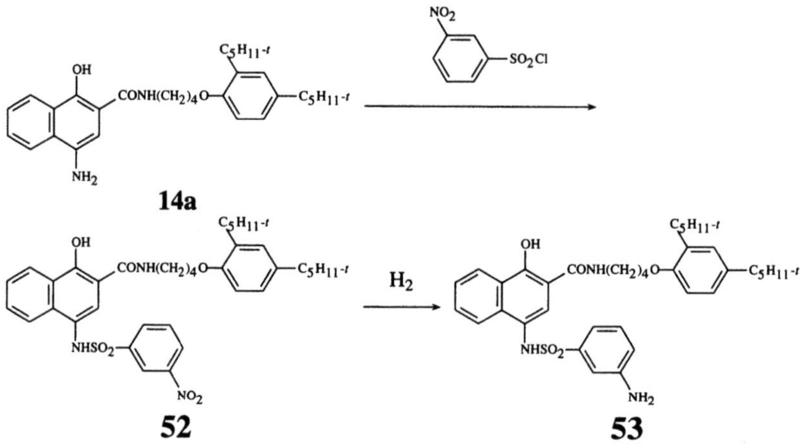

NO_2 SO_2Cl OH $CONH(CH_2)_4O$ C_5H_{11}-*t* C_5H_{11}-*t* NH_2

14a

OH $CONH(CH_2)_4O$ C_5H_{11}-*t* C_5H_{11}-*t* $NHSO_2$ NO_2 H_2 OH $CONH(CH_2)_4O$ C_5H_{11}-*t* C_5H_{11}-*t* $NHSO_2$ NH_2

52 **53**

2) Yellow dye releaser:

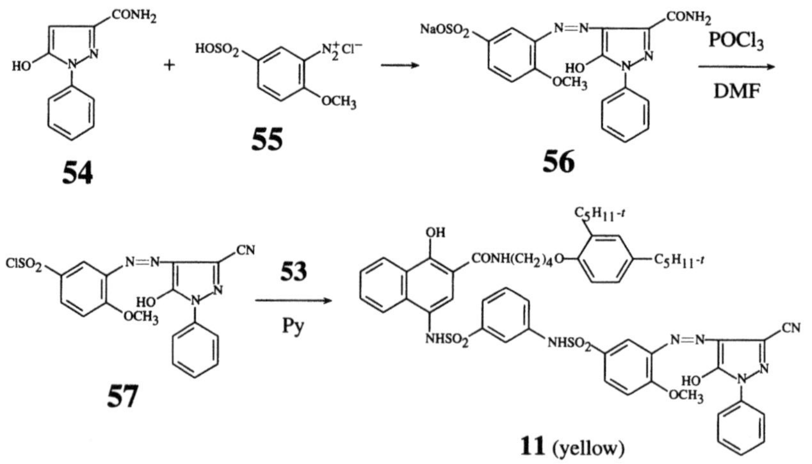

$CONH_2$ HO N N + $HOSO_2$ $N_2^+Cl^-$ OCH_3 $NaOSO_2$ N=N $CONH_2$ HO OCH_3 N N $POCl_3$ DMF

54 **55** **56**

$ClSO_2$ N=N CN HO OCH_3 N N **53** Py OH $CONH(CH_2)_4O$ C_5H_{11}-*t* C_5H_{11}-*t* $NHSO_2$ $NHSO_2$ N=N CN HO OCH_3 N N

57 **11** (yellow)

Figure 18.17. Synthetic pathway for preparing the yellow dye releaser (**11**).

the sulfonic acid into a sulfonyl chloride and at the same time it dehydrates the carbamoyl group to give a cyano group.

18.4 After-Chelation Dye Moieties

During the first half decade of the 1980s, a vast number of patents on after-chelation dye moieties were extensively disclosed from Kodak. Since the manufacturing of Kodak's instant color photographic system was stopped, these patents have never been used in practical films. However, their structures and some properties are summarized briefly, because they would attract the attention of the reader.

18.4.1 Cyan Dye Moieties of After-Chelation Type

In addition to dye releasers releasing pre-metallized dyes, dye releasers having metallizable dye moieties have been disclosed, as collected in Fig.18.18. For example, the dye released from **58** is transferred imagewise to an image-receiving layer, where it is contacted with metal ions to form a metal-complexed azo dye of excellent stability [19,20]. As a result of such after-chelation, the corresponding nickel(II) complex exhibits a λ_{max} of 656 nm and the copper(II) complex exhibits a λ_{max} of 641 nm.

The non-chelated form of the dye released from **59** has a λ_{max} of 629 nm, while the nickel(II) and the copper(II) complex respectively exhibit λ_{max} values of 645 nm and 635 nm [21,22]. The nickel(II) complex of the 6-(2-thienylazo)-3-pyridinol dye released from **60** exhibits a λ_{max} of 637 nm [23,24,25].

The dye releaser (**61**) releases a 1-arylazo-4-isoquinolinol dye, where it exhibits a λ_{max} of 544 nm under a non-metallized condition but the nickel(II) complex shows a λ_{max} of 637 nm [26,27]. *p*-Sulfonamidonaphthol dye releasers having another ballast group (e.g., **62**) have been disclosed to release cyan heterocyclylazo-3-pyridinol dyes [28,29,30]. Pre-metallized dye moieties of cyan hue have been also disclosed to be incorporated in the *p*-sulfonamidonaphthol nucleus [31].

18.4.2 Magenta Dye Moieties of After-Chelation Type

The dye released from **63** has been reported to form a magenta nickel(II) complex, which gives absorption maxima (λ_{max} values) at 539 and 573 nm with the half-bandwidth of 113 nm [32,33,34]. The dye releaser (**64**) releases a metallizable 6-arylazo-3-pyridinol dye, the nickel(II) complex of which exhibits a λ_{max} of 557 nm and a half-bandwidth of 75 nm [35,36,37]. The dye released from **65** shows absorption maxima at 530 (shoulder) and 565 nm after nickel(II) chelation, where the $\lambda_{1/2}$ value at half of the maximum on the short-wavelength side is measured to be 536 nm (half-bandwidth 88 nm) [38,39,40]. The dye

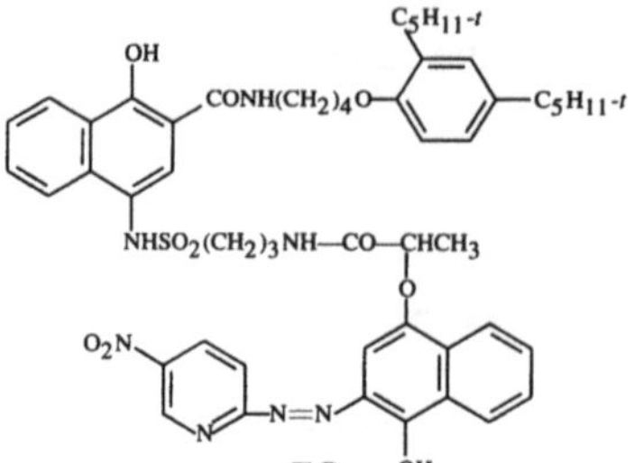

58

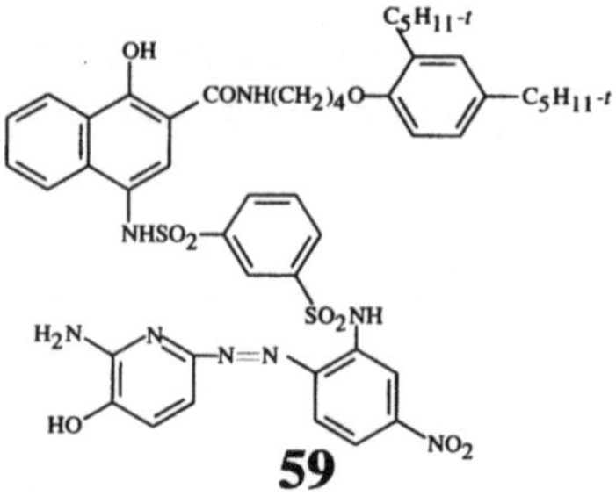

59

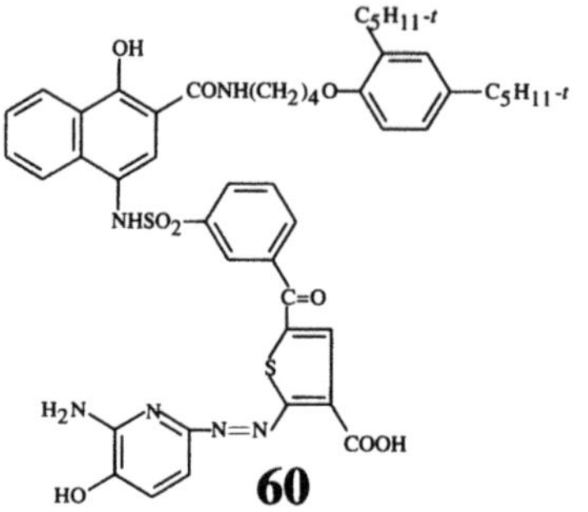

60

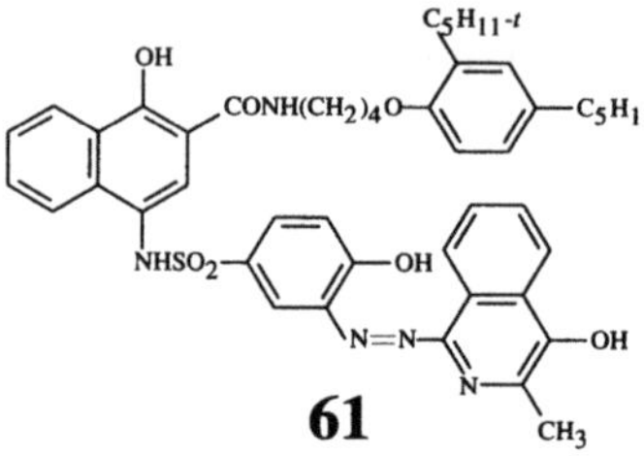

61

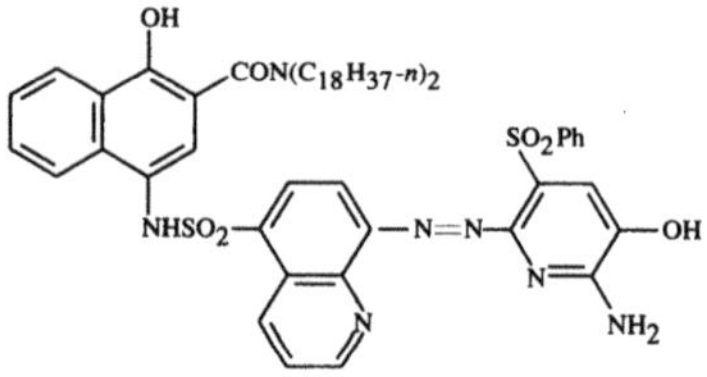

62

Figure 18.18. *p*-Sulfonamidonaphthol dye releasers releasing a cyan azo dye of after-chelation type.

releaser (**66**) releases a similar metallizable magenta dye with a bridged nitrogen, the nickel(II) complex of which exhibits an absorption maximum at 569 nm and the $\lambda_{1/2}$ of 551 nm (half-bandwidth 82 nm) [41,42,43]. The dye releaser (**67**) releases a metallizable 4-(2-heterocyclylazo)phenol dye having a fused heterocyclic ring. The nickel(II) complex of the released dye exhibits the $\lambda_{1/2}$ of 553 nm (half-bandwidth 76 nm), but its absorption maximum has not been reported [44,45,46]. This patent contains comparison data with 2:1 complexes. Dye releasers having a 2:1-nickel(II) metallized dye have been disclosed as magenta dye releasing compounds, where the 2:1-nickel(II) complex corresponding to the dye releaser (**63**) has been compared with the unmetallized **63** [47,48]. Arylazo-pyrazolotriazole dye-releasing compounds have been also proposed as dye releasers of after-chelation type [49,50,51].

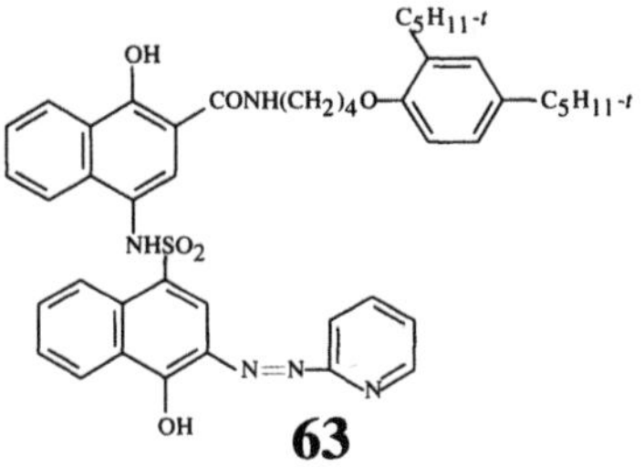

Figure 18.19. *p*-Sulfonamidonaphthol dye releasers releasing a magenta azo dye of after-chelation type.

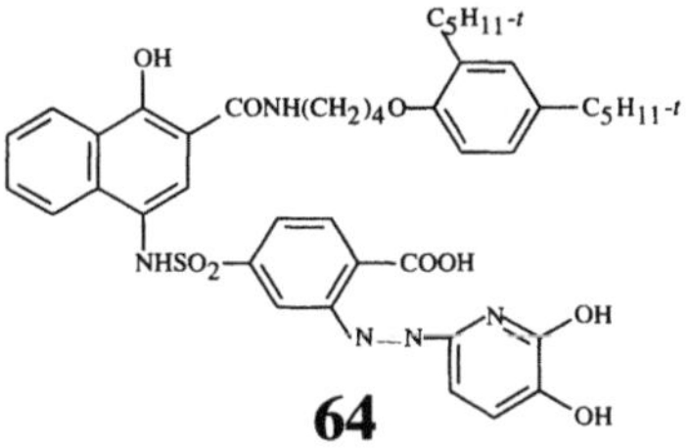

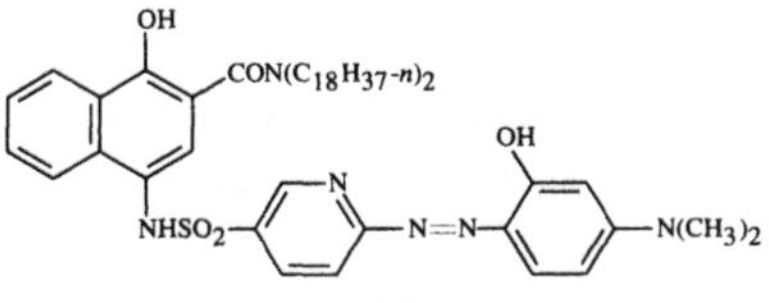

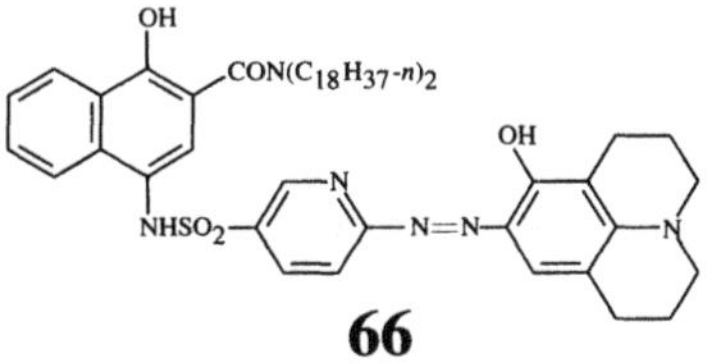

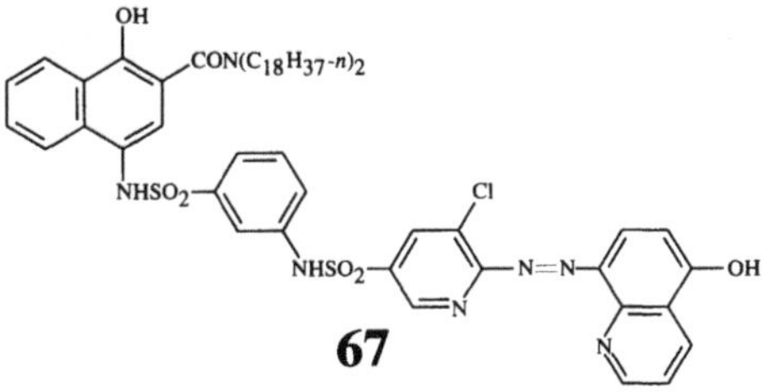

18.4.3 Yellow Dye Moieties of After-Chelation Type

The dye released from **68** shows an absorption maximum at 450 nm after nickel(II) chelation, where the half-bandwidth is determined to be 88 nm [52,53]. The two benzoyloxy groups in the dye moiety of **69** are hydrolyzed to give hydroxyl groups, which form a copper(II) chelate together with the azo group [54,55,56]. The resulting metallized dye exhibits a λ_{max} of 479 nm and a half-bandwidth of 82 nm. Dye releasers having a 2:1-nickel(II) pre-metallized dye have been disclosed as yellow dye releasing compounds [57,58,59].

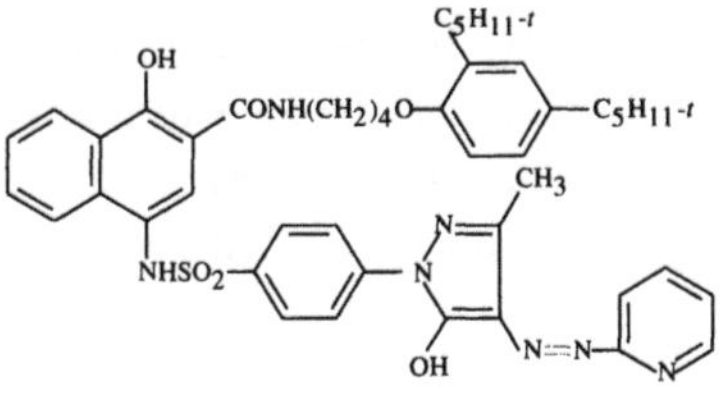

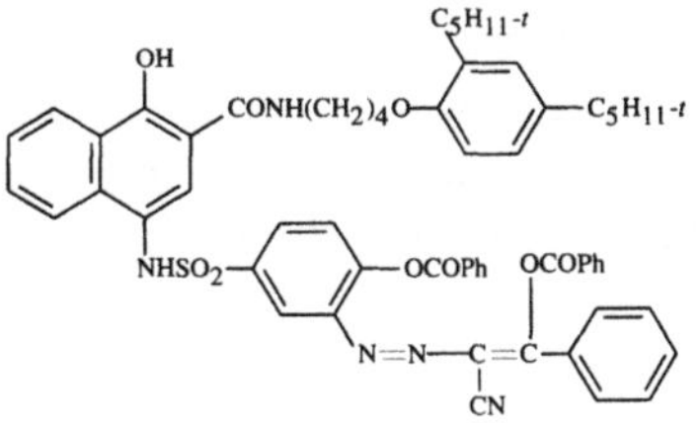

Figure 18.20. *p*-Sulfonamidonaphthol dye releasers releasing a yellow azo dye of after-chelation type.

18.5 Chemicals for Controlling Dye Transfer Processes

18.5.1 Chemicals Contained in Photo-Sensitive Layers

Nucleation Agents and Nucleation Accelerators

Since *p*-sulfonamidonaphthol dye releasers are negative-working, they are combined with direct positive emulsions, as discussed in Chapter 16 (Fig. 16.15). The direct positive emulsions work positively under fogging (nucleation) conditions using nucleating agents, which have been already discussed in Chapter 16 (Figs. 16.16 and 16.17). Such nucleation has been found to be inhibited if very active *p*-sulfonamidonaphthol dye releasers (Fig. 18.9) are used in place of those of low activity (Fig. 18.4). Quinone oxidizing agents such as octadecylbenzoquinone (**70**) incorporated in direct positive silver halide emulsion layers have been disclosed to be useful to accelerate nucleation based on such a nucleating agent as **71** [60].

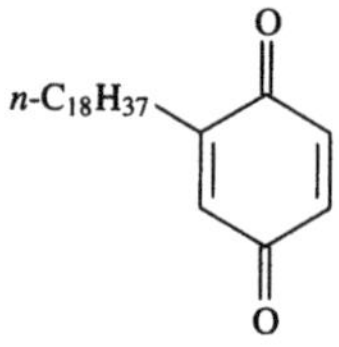

O
n-$C_{18}H_{37}$
O

CH_3NH—C(=S)—NH—C₆H₄—NHNHCHO

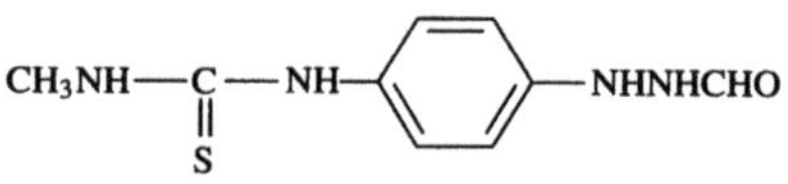

71
(Nucleating agent)

70
(Nucleation accelerator)

Figure 18.21. Nucleation accelerator in combination with a nucleation agent.

Development Accelerators

As shown in eq. 18.1 (Fig. 18.1), such an ETA as **1** reduces silver halide grains in a silver halide emulsion layer. To accelerate this redox reaction, a development accelerator such as **72** has been incorporated in the emulsion layer. An embodiment in which the development accelerator (**72**) is used in combination with the dye releasers shown in Fig. 18.9 has been disclosed [8]. Later, another embodiment coupled with the dye releasers shown in Fig. 18.9 has also disclosed [61].

Compounds for Preventing Color Contamination

The oxidized form of an ETA (eq. 18.2) diffuses within the corresponding photo-sensitive unit and reaches the adjacent emulsion layer so as to give

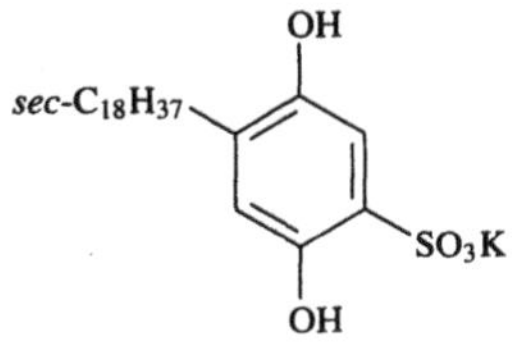

72

Figure 18.22. Development accelerator in combination with electron transfer agents.

correct color reproduction. If the ETA_{ox} diffuses to other units, color contamination would take place. In order to prevent such color contamination, an appropriate reducing agent consuming the ETA_{ox} is involved in each interlayer between two of the units. For example, a ballasted hydroquinone (**73**) [63] or an *o*-sulfonamidophenol (**74**) [64] has been incorporated in each interlayer.

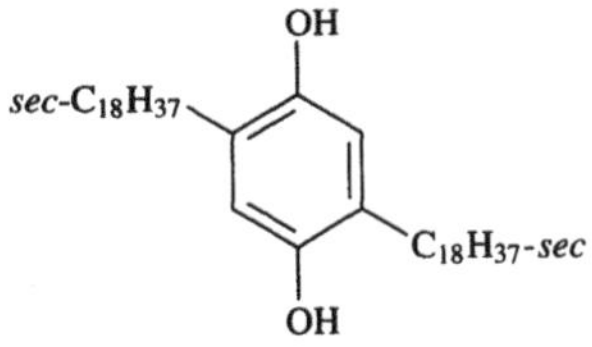

73

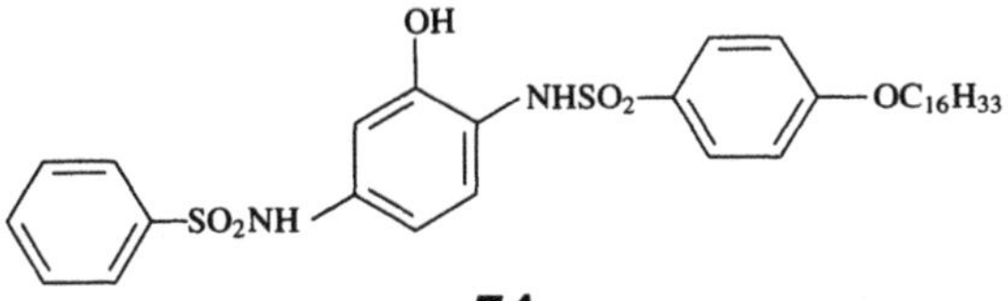

74

Figure 18.23. Compounds for preventing color contamination.

Development Inhibitor Releasers

Development inhibitor releasers have been discussed in detail in Chapter 17 (Subsection 17.3.1), where a variety of blocking groups proposed are listed. All these compounds can be used in instant color films based on *p*-sulfonamidonaphthol dye releaser. As a representative example, 5-(2-cyanoethylthio)-1-phenyl-1H-tetrazole (**75**) is incorporated in the timing layer illustrated in Fig. 18.2 [65]. This compound releases a development inhibitor (**76**) by a retro-Michael reaction shown in Fig. 18.24. On the other hand, a thiocarbamate (**77**), which has been used in a peel-apart format, releases the development inhibitor (**76**) by the hydrolysis of the thiocarbamate function [66].

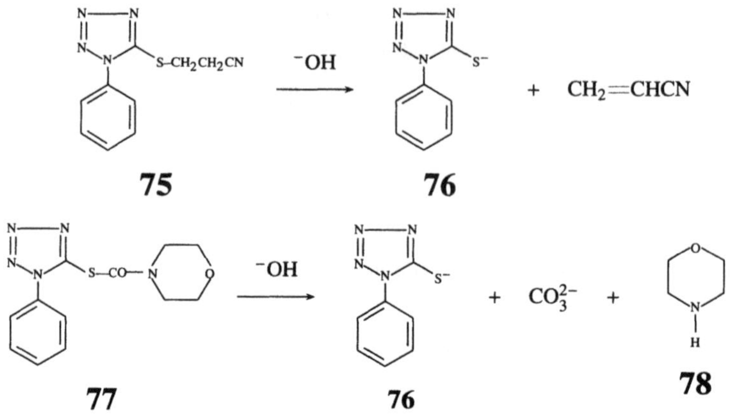

Figure 18.24. 5-(2-Cyanoethylthio)-1-phenyltetrazole and 5-(morpholinocarbonylthio)-1-phenyltetrazole as development inhibitor releasers.

18.5.2 Chemicals Contained in an Activator Fluid

Figure 18.25 lists various chemicals that are contained in the activator fluid shown in Fig. 18.2. The chemicals belonging to various categories have been employed together with the triplet of dye releasers shown in Fig. 18.9 [61,62].

Carbon Black and Viscosity-Imparting Reagents

As found by comparing Fig. 18.3 with Fig. 18.2, the activator fluid contains carbon black, which is spread to give a black opaque layer. Sodium carboxymethylcellulose (**80**) is added to impart viscosity, which ensures a thin and square spread of the activator fluid.

Electron Transfer Agents

In instant color photography based on *p*-sulfonamidonaphthol dye releasers, an electron transfer agent (ETA) supplied from an activator fluid is oxidized as a function of silver halide development, as shown in eq. 18.1 of Fig. 18.1. The oxidized ETA (ETA_{ox}) diffuses to the adjacent dye releaser layer, in which it cross-oxidizes the dye releaser (eq. 18.2). In accordance with these processes, the ETA and its oxidized form (ETA_{ox}) must be soluble and diffusible in alkali media due to the activator fluid. In addition, the cross-oxidations between the silver halide and the ETA and between the ETA_{ox} and the dye releaser should take place smoothly.

Pyrazolidones with various substituents have been disclosed as ETAs in patents. For example, the basic patent on *p*-sulfonamidophenol dye releasers [3] has employed phenidone (1-phenyl-3-pyrazolidone), dimezone (4,4-dimethyl-1-phenyl-3-pyrazolidone), etc., which have long been used in conventional

1) Inorganic components:

KOH (5.25), Na_2SO_3 (103), $KF \cdot H_2O$ (10), H_2O (up to 1 L)

2) Organic components:

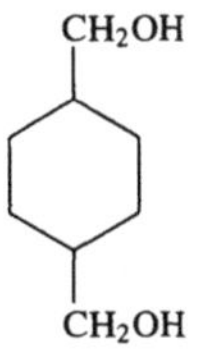

79 (1) 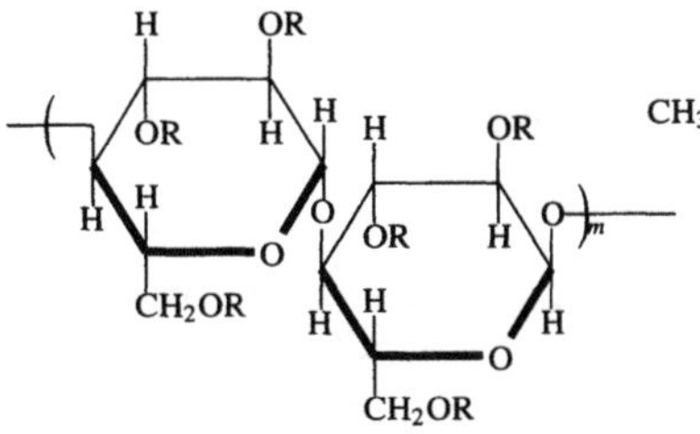 **80** (44.0) R = H, or $-CH_2COONa$ **81** (4)

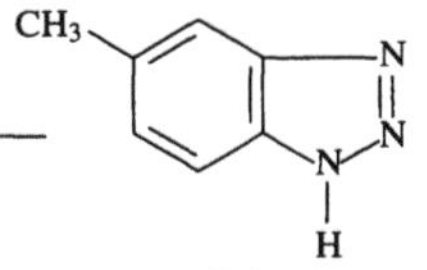

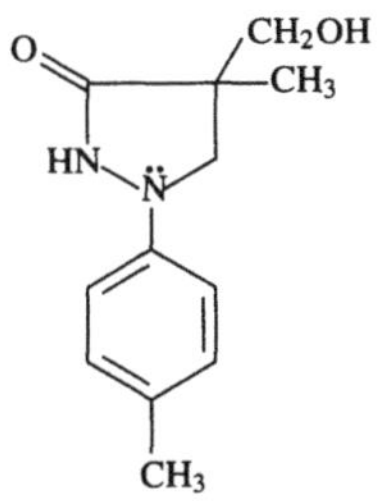

82 (12.0) 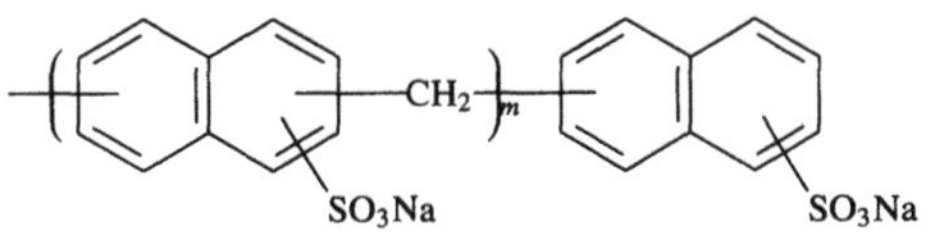 **83** (6.4)

Figure 18.25. Chemicals contained in an activator fluid [61]. The value of each pair of parentheses represents the weight (g) of a component at issue. This fluid has been used in combination with the dye releasers shown in Fig. 18.9 for a peel-apart format. When such an activator fluid is used for a mono-sheet format, carbon black (ca. 190 g) is added as an opacifier [62].

B&W and color photography. Later, 4-hydroxymethyl-4-methyl-1-phenyl-3-pyrazolidone (**84a** = **1**) [67,68] has been widely used as such an ETA [4]. According to the ballast-group change from the dye-developers of Fig. 18.4 to those of Fig. 18.9, a more active ETA has been required. Thus, a methyl group has been introduced as a substituent of the 1-phenyl group, as found in **84b** (= **82**) [61]. Other ETAs having electron-donating groups (e.g., **84c**–**84e**) have been also disclosed in this patent.

ETA-precursors such as **85** have been proposed for the purpose of incorporation in an interlayer of a peel-apart format [69]. This compound releases phenidone under alkaline conditions.

	X	R
84a	X = H	R = H (= **1**)
84b	X = H	R = CH_3 (= **82**)
84c	X = OH	R = CH_3
84d	X = H	R = OCH_3
84e	X = OH	R = OCH_3

85

Figure 18.26. Electron transfer agents.

Other Additives

Improved dye densities are obtained by adding a saturated, aliphatic or alicyclic glycol (e.g., **79**) or a saturated, aliphatic or alicyclic amino alcohol (e.g., 3-amino-1-propanol) to an activator fluid [70].

18.6 Polymers

18.6.1 Polymers for the Mordant Layer

To capture diffusible dyes with a sulfamolyl group which are released from *p*-sulfonamidonaphthol dye releasers, mordant polymers having quaternary ammonium moieties have been proposed in several patents. They are incorporated in the mordant layer shown in Fig. 18.2. For example, copoly[styrene-*N*-benzyl-*N*,*N*-dimethyl-*N*-(3-maleimidopropyl)ammonium chloride] (**86**), which was originally proposed for diffusible dye-releasing couplers [71], has been used in instant color films based on *p*-sulfonamidonaphthol dye releasers [4]. Related copolymers such as **87** have been also disclosed as mordant polymers in patents [72,73].

Polymers and copolymers having quaternary ammonium groups, wherein the total number of carbon atoms in three alkyl groups on the quaternary ammonium group is at least 12, have been disclosed, as exemplified by **88** [74].

More elaborate copolymers which contain divinylbenzene as a repeating unit (e.g., **89**) have been proposed as mordant polymers [75]. A copolymer of this type was formerly used in Kodak's PR-10 film [1,2].

Homopolymers and copolymers having a reactive group which will be covalently bonded to a dye (e.g., a dichlorotriazine in **90**) have been proposed as mordant polymers [76].

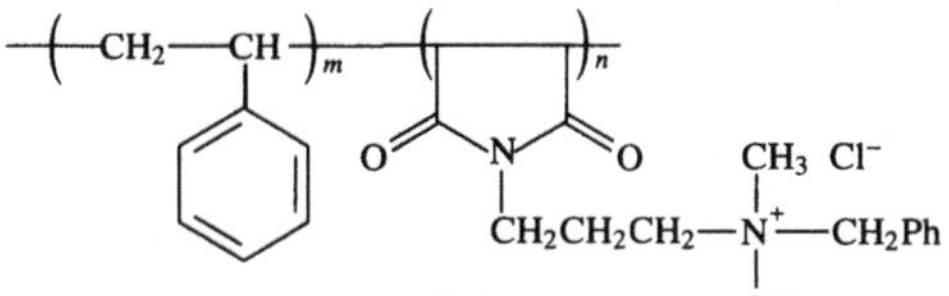

86

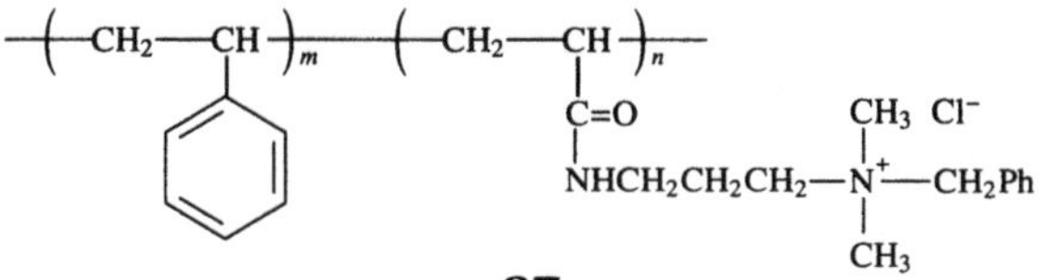

87

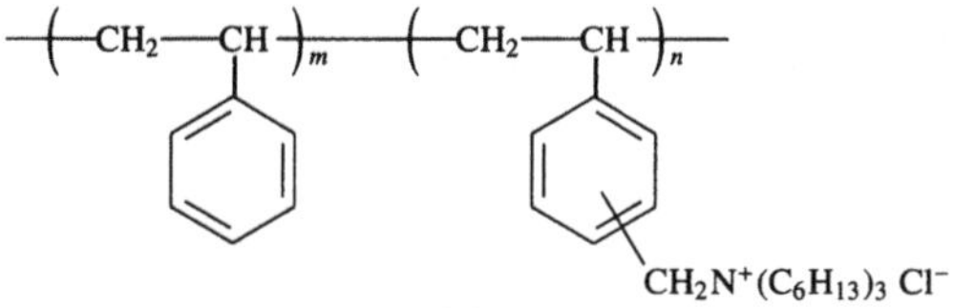

88

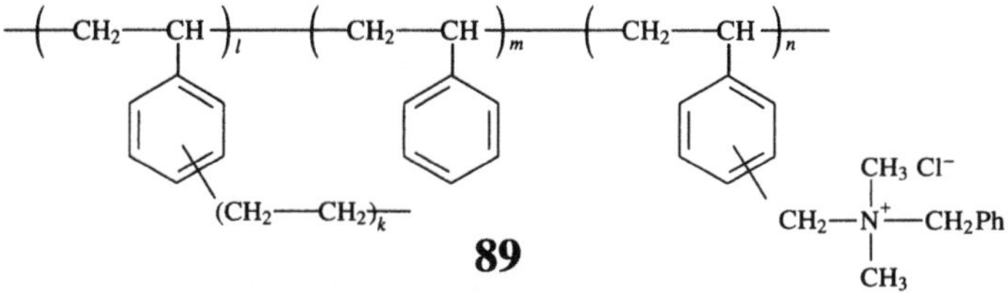

89

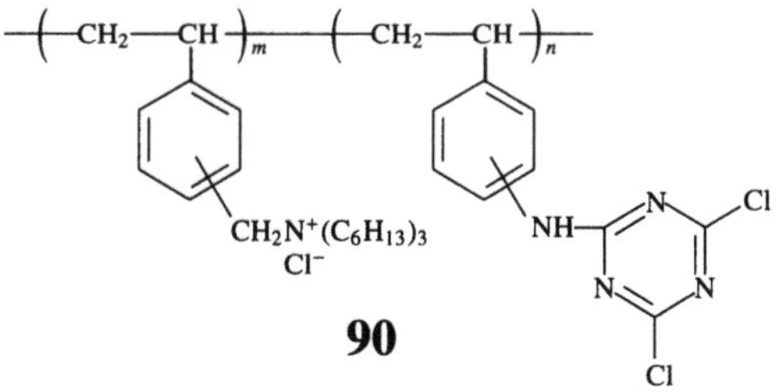

90

Figure 18.27. Mordant polymers for receiving dyes released from dye releasers.

Mordant polymers having imidazole moieties have been disclosed to be useful with chelating dyes [77,78]. Polymers and copolymers derived from *N*-

vinylbenzyliminodiacetic acid disodium salt have been extensively investigated as mordant polymers retaining metal ions, and have been found to be useful for dye releasers of after-chelation type [79,80,81]. Poly(4-vinylpyridine) retaining nickel(II) ions has been proposed as a mordant useful for dye releasers of after-chelation type [82]. Copolymers having terpyridine groups have been reported to retain nickel(II) ions, which are useful as mordant polymers for dye releasers of after-chelation type [83,84].

18.6.2 Polymers for the Acid Layer

After silver halide development and dye transfer under alkaline conditions, the alkaline media are neutralized by an acid polymer incorporated in the acid layer (or the neutralization layer). A copolymer (**91**) of dimeric acrylic acid and acrylic acid has a lower glass transition temperature than a copolymer of acrylic acid butyl ester and acrylic acid (**92**) [85].

Copolymers having acid functions and cross-linkable groups (e.g., **93**) have been used as acid polymers after acid-catalyzed cross-linkage represented by $2\times CONH–CH_2–OCH(CH_3)_2 \longrightarrow CONH–CH_2–NHCO$ [86]. A cross-linked polymer such as **94**, which has been prepared from methyl acrylate and trivinylcyclohexane, has been used for the same purpose [87].

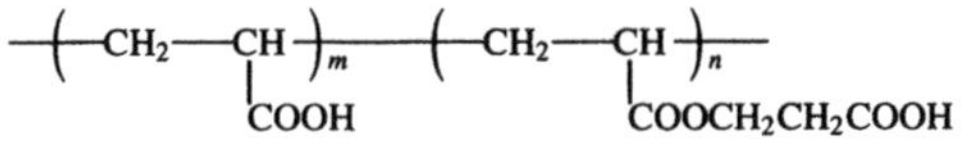

$$-\!\!(CH_2-\underset{COOH}{CH})_m\!\!-\!\!(CH_2-\underset{COOCH_2CH_2COOH}{CH})_n\!\!-$$

91

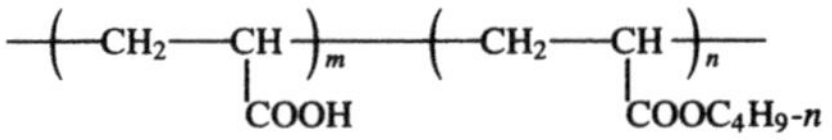

$$-\!\!(CH_2-\underset{COOH}{CH})_m\!\!-\!\!(CH_2-\underset{COOC_4H_9\text{-}n}{CH})_n\!\!-$$

92

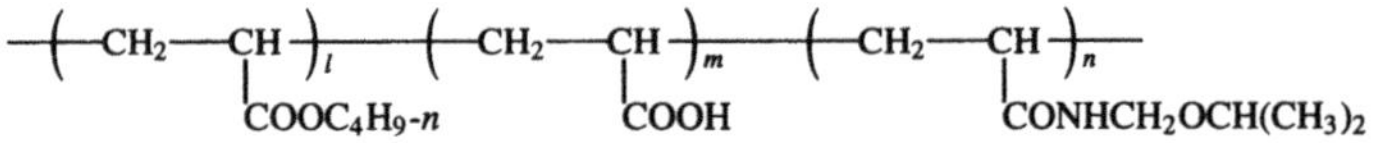

$$-\!\!(CH_2-\underset{COOC_4H_9\text{-}n}{CH})_l\!\!-\!\!(CH_2-\underset{COOH}{CH})_m\!\!-\!\!(CH_2-\underset{CONHCH_2OCH(CH_3)_2}{CH})_n\!\!-$$

93

$$-\!\!(CH_2-\underset{COOH}{CH})_m\!\!-\!\!(CH_2-CH)_n\!\!-$$

$-\!\!(CH-CH_2)\!\!-$

$-\!\!(CH-CH_2)\!\!-$

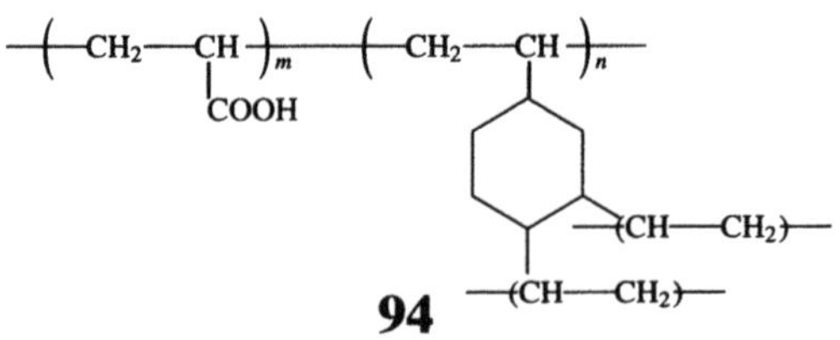

94

Figure 18.28. Acid polymers.

18.6.3 Polymers for the Timing Layer

The timing layer shown in Fig. 18.2 is placed to retard the neutralization by the acid layer. As already discussed in Chapter 17, timing polymers used in the timing layer are classified into sieve type and hold-release type.

A mixture of 95% cellulose acetate (40% acetyl) and 5% poly[styrene-co(maleic anhydride)] (**95**) has been incorporated in a timing layer, the permeability of which does not change before and after incubation [88]. The latter copolymer is regarded as belonging to the hold-release type.

A copolymer containing vinylidene dichloride as a repeating unit (e.g., **96**) has been used in a temporary barrier layer adjacent to a timing layer [89]. This polymer maintains D_{min} values to be constant at high and low temperatures, while D_{max} values are higher at lower temperatures.

A lactone polymer has been prepared by the acid-catalyzed reaction of **97** with butanol, where lactonization and esterification take place. The lactone polymer has been used in the timing layer together with the copolymer **96** [90]. The addition of oxalic acid to the mixture of the lactone polymer and **96** has minimized sensitometric changes that occur with keeping [62].

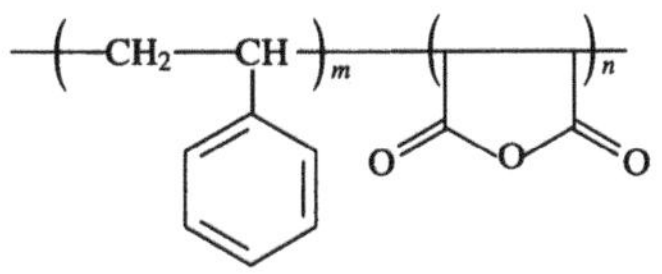

95

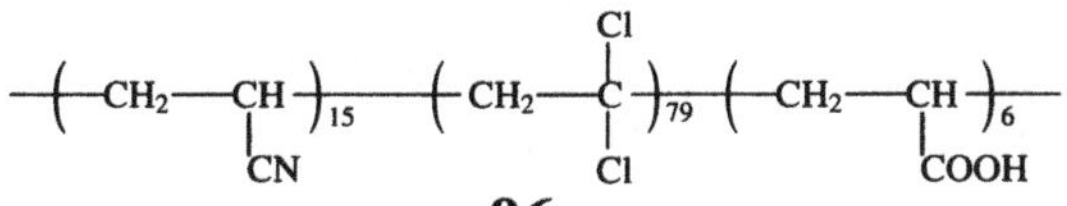

96

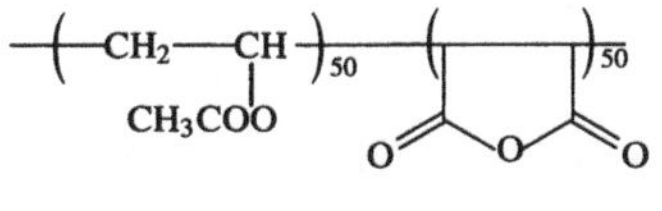

97

Figure 18.29. Timing polymers. The polymer **97** has been used after lactonization and esterifcation.

References

[1] Chem Eng News. (1976) June 21

[2] Hanson Jr WT (1976) Photogr Sci Eng. 20:155; Hanson Jr WT (1977) J Photogr Sci. 25:189; Hanson Jr WT (1979) Interdiscipl Sci Rev. 4:290

[3] Fleckenstein LJ (1975) US Patent 3 928 312

[4] Fleckenstein LJ, Figueras J (1978) US Patent 4 076 529

[5] Landholm RA, Haase JR, Krutak Sr JJ (1975) US Patent 3 929 760

[6] Landholm RA, Haase JR, Krutak Sr JJ (1977) US Patent 4 013 635

[7] Krutak Sr JJ, Hasse JR, Landholm RA (1976) US Patent 3 954 476

[8] Hasse JR, Eldredge CH, Landholm RA (1977) US Patent 4 013 633

[9] Fujita S (1982) Yuki Gosei Kagaku Kyokaishi. 40:176

[10] Gerbal CF, Gompf TE, Collet PD (1981) US Patent 4 258 120

[11] Gerbal CF, Gompf TE, Collet PD (1986) US Patent 4 605 697

[12] Fernandez JM, McCreary MD, Ross RE, Staples JT (1979) US Patent 4 135 929

[13] Ross RE, Fleckenstein LJ, Campbell ME (1983) US Patent 4 407 930

[14] Kilminster KN, Holstead C (1980) US Patent 4 195 993

[15] Kilminster KN, Holstead C (1981) US Patent 4 273 708

[16] Weber LD, Merkel PB, Warren III HC (1985) US Patent 4 524 122

[17] Leyshon LJ, Holstead C (1988) J Photogr Sci. 36:107

[18] Landholm RA, Robbins JM (1979) US Patent 4 156 609

[19] Anderson RB, Hoffmeister EH, Landholm RA (1979) US Patent 4 147 544

[20] Anderson RB, Hoffmeister EH, Landholm RA (1979) US Patent 4 165 238

[21] Chapman DD (1980) US Patent 4 195 994

[22] Chapman DD (1980) US Patent 4 204 993

[23] Krutak JJ, Maleski RJ, Moore WH (1982) US Patent 4 346 161

[24] Krutak JJ, Maleski RJ, Moore WH (1983) US Patent 4 385 104

[25] Krutak JJ, Maleski RJ, Moore WH (1983) US Patent 4 396 546

[26] Chapmann DD, Friday JA, Elwood JK (1979) US Patent 4 148 642

[27] Chapmann DD, Friday JA, Elwood JK (1980) US Patent 4 183 754

[28] Reczek JA, Elwood JK (1983) US Patent 4 419 435

[29] Reczek JA, Elwood JK (1984) US Patent 4 476 207

[30] Reczek JA, Elwood JK (1985) US Patent 4 495 100

[31] Reczek JA (1986) US Patent 4 598 030

[32] Chapman DD, Wu EM (1980) US Patent 4 204 870

[33] Chapman DD, Wu EM (1980) US Patent 4 207 104

[34] Chapman DD, Wu EM (1981) US Patent 4 273 706
[35] Chapman DD, Reczek JA (1981) US Patent 4 287 292
[36] Chapman DD, Reczek JA (1982) US Patent 4 346 155
[37] Chapman DD, Reczek JA (1982) US Patent 4 358 404
[38] Anderson RB, Kalenda NW (1982) US Patent 4 357 410
[39] Anderson RB, Kalenda NW (1983) US Patent 4 367 174
[40] Anderson RB, Kalenda NW (1983) US Patent 4 368 249
[41] Anderson RB, Kalenda NW (1982) US Patent 4 357 412
[42] Anderson RB, Kalenda NW (1982) US Patent 4 366 218
[43] Anderson RB, Kalenda NW (1983) US Patent 4 368 154
[44] Evans S, Elwood K (1983) US Patent 4 420 550
[45] Evans S, Elwood K (1984) US Patent 4 436 799
[46] Evans S, Elwood K (1985) US Patent 4 495 099
[47] Bailey J, Clarke D, Johnston LG (1982) US Patent 4 358 527
[48] Bailey J, Clarke D, Johnston LG (1984) US Patent 4 426 326
[49] Baigrie BD, Bailey J, Johnston LG, Mijovic MV (1979) US Patent 4 142 891
[50] Baigrie BD, Bailey J, Johnston LG, Mijovic MV (1980) US Patent 4 183 753
[51] Baigrie BD, Bailey J, Johnston LG, Mijovic MV (1981) US Patent 4 272 434
[52] Green II A, Kalenda NW (1979) US Patent 4 148 641
[53] Green II A, Kalenda NW (1979) US Patent 4 165 987
[54] Chapman DD, Wu EM (1979) US Patent 4 148 643
[55] Chapman DD, Wu EM (1980) US Patent 4 183 755
[56] Chapman DD, Wu EM (1980) US Patent 4 228 073
[57] Evans S (1983) US Patent 4 407 931
[58] Evans S (1983) US Patent 4 418 131
[59] Evans S (1984) US Patent 4 481 141
[60] Chaffee E, Tuites RC (1982) US Patent 4 341 858
[61] McCreary MD, Farley WC, Erickson WF (1980) US Patent 4 209 580
[62] Wheeler RW (1982) US Patent 4 353 973
[63] Knechel WF (1970) US Patent 3 700 453
[64] Erickson WF, Ross RE (1980) US Patent 4 205 987
[65] Hammond HA, Humphlett WJ, Salminen IF (1977) US Patent 4 009 029
[66] Simmons MJ, Southby DT (1981) GB Patent 2 062 884 B
[67] Mijovic MV, Williams LA (1969) GB Patent 1 093 281
[68] Mijovic MV, Williams LA (1969) GB Patent 1 157 6171
[69] Mooberry JB, Archie Jr WC (1982) US Patent 4 358 525

[70] Kuh AD, Condit PB (1977) US Patent 4 030 920
[71] Abbott TI (1972) US Patent 3 698 896
[72] Cohen HL, King Jr JR, Minsk LM (1973) US Patent 3 709 690
[73] Cohen HL, King Jr JR, Minsk LM (1973) US Patent 3 758 445
[74] Cohen HL, Koeng F, Ponticello I (1975) US Patent 3 898 088
[75] Campbell GA, Cohen H, Hamilton LR (1976) US Patent 3 958 995
[76] Campbell GA, Hamilton LR (1980) US Patent 4 206 279
[77] Petrak KL (1980) US Patent 4 229 515
[78] Petrak KL (1982) US Patent 4 316 972
[79] Campbel GA, Hamilton LR, Brust DP (1980) US Patent 4 193 796
[80] Campbel GA, Hamilton LR, Brust DP (1980) US Patent 4 228 257
[81] Archie Jr WC, Campbel GA (1980) US Patent 4 239 847
[82] Brust DP, Hamilton LR, Wilkes GR (1981) US Patent 4 282 305
[83] Reczek JA, Ponticello IS, Bryan PS (1986) US Patent 4 581 314
[84] Reczek JA, Ponticello IS, Bryan PS (1986) US Patent 4 599 389
[85] Helling G, Krafft W, Findeison K, Himmelmann W (1981) US Patent 4 284 708
[86] Bishop JF, Bowman WA (1983) US Patent 4 389 479
[87] Krafft W, Helling G (1979) US Patent 4 138 260
[88] Abel EP (1977) US Patent 4 009 030
[89] Hannie DE (1977) US Patent 4 056 394
[90] Abel EP (1980) US Patent 4 229 516

o-Sulfonamidophenol Dye Releasers

19.1 Chemical Processes and Multilayer Structure

The instant color film based on *o*-sulfonamidophenol dye releasers was marketed by Fuji Photo Film under the name FI-10 in 1981 [1]. The chemistry of *o*-sulfonamidophenol dye releasers is still effective by continuous improvement so as to be employed in the direct successor "FOTORAMA" system as well as in a recent integral instant photographic system named "Instax" in 1998 [2]. The design, evaluation, and synthesis of *o*-sulfonamidophenol dye releasers have been described in a review by Fujita [3] and in an account written by Fujita, Koyama, and Ono [1].

19.1.1 Chemical Processes of *o*-Sulfonamidophenol Dye Releasers

o-Sulfonamidophenol dye releasers are negative-working compounds of dye-releasing type. The principal reactions of the dye releasers have been already discussed in comparison with *p*-sulfonamidonaphthol dye releasers in Chapter 16. A more detailed profile of the reactions is summarized in Fig. 19.1, in which silver halide, an electron transfer agent (ETA), and a dye releaser participate.

The redox reaction between the silver halide and the ETA (eq. 19.1) occurs in an unexposed area of the corresponding silver halide emulsion layer. Note that each emulsion layer contains a direct positive silver halide emulsion described in Chapter 16, since *o*-sulfonamidophenol dye releasers are negative-working. This reaction for *o*-sulfonamidophenol dye releasers is common to

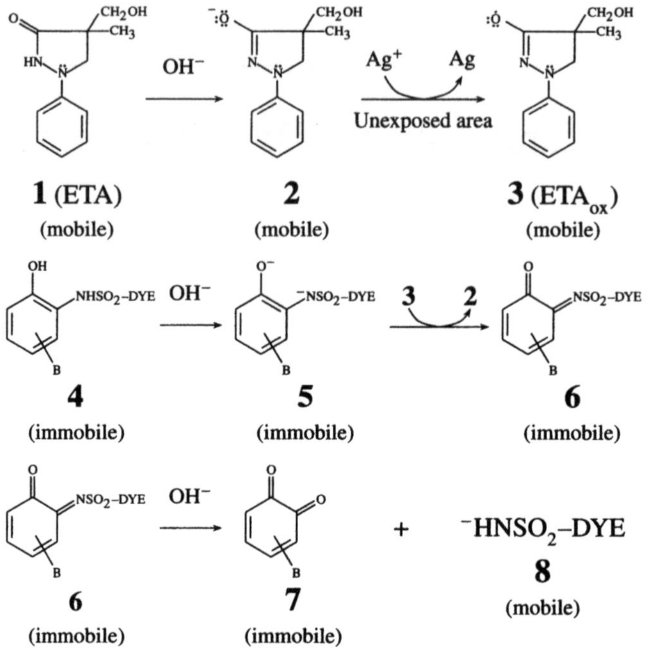

$\xrightarrow{OH^-}$ $\xrightarrow[\text{Unexposed area}]{Ag^+ \;\; Ag}$ (19.1)

1 (ETA) (mobile) **2** (mobile) **3** (ETA_{ox}) (mobile)

$\xrightarrow{OH^-}$ $\xrightarrow{3 \;\; 2}$ (19.2)

4 (immobile) **5** (immobile) **6** (immobile)

$\xrightarrow{OH^-}$ + $^-HNSO_2\text{–DYE}$ (19.3)

6 (immobile) **7** (immobile) **8** (mobile)

Figure 19.1. Consecutive reactions of silver halide, an electron transfer agent, and an *o*-sulfonamidophenol dye releaser. The letter B represents a ballast group. The symbol DYE denotes a cyan, magenta, or yellow dye moiety.

p-sulfonamidonaphthol dye releasers discussed in Chapter 18. The resulting ETA_{ox} (**3**) diffuses to the adjacent dye releaser layer, where the redox reaction represented by eq. 19.2 takes place. The oxidized species (**6**) is then hydrolyzed under alkaline conditions to release a mobile dye (**8**) with leaving an immobile quinone (**7**). The released dye (**8**) diffuses to a mordant layer, giving dye images to be observed (eq. 19.3).

19.1.2 Multilayer Structure for Color Reproduction

The *o*-sulfonamidophenol dye releasers are negative-working in a similar way to the *p*-sulfonamidonaphthol dye releasers described in Chapter 18. Hence, the multilayer structure and the scheme of color reproduction shown in Figs. 18.2 and 18.3 of Chapter 18, which have been illustrated for the *p*-sulfonamidonaphthol dye releasers, can be applied to the *o*-sulfonamidophenol dye releasers without any conceptual modification, as described in several reviews [4–7].

For the sake of convenience, a multilayer structure for using *o*-sulfonamido-phenol dye releasers is depicted in Fig. 19.1, where a light-sensitive sheet coated with at least 12 layers and a cover sheet coated with at least two layers are superimposed in a face-to-face fashion to construct an integrated (so-called mono-sheet) film.

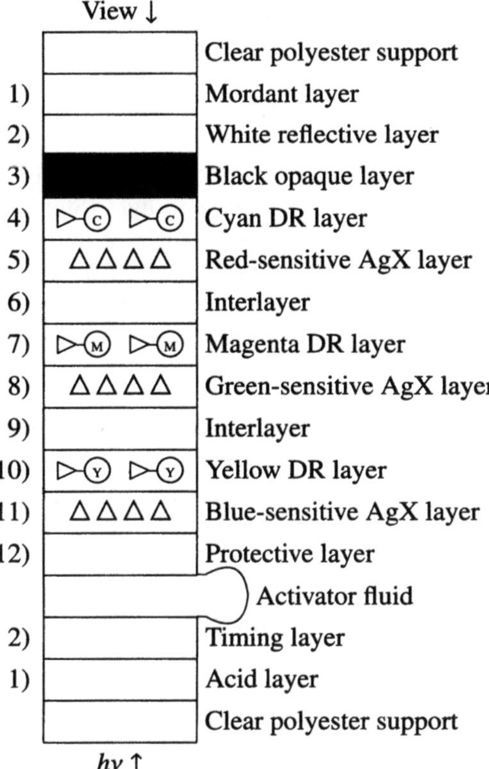

Figure 19.2. Schematic cross-section of an instant color film using *o*-sulfonamidophenol dye releasers. △: Silver halide grain, ▷-(Y): Yellow dye releaser, ▷-(M): Magenta dye releaser, and ▷-(C): Cyan dye releaser.

This film consists of three photo-sensitive units of layers for subtractive color reproduction (Layers 4 and 5; 7 and 8; and 10 and 11; together with interlayers), an image receiving unit (Layers 1 to 3), a neutralizing unit (Layers 1 and 2 on the cover sheet), and an activator fluid. It should be noted that each emulsion contained in Layer 5, 8, or 11 is a direct positive emulsion, which has been already described in Chapter 16. Accordingly, the reactions shown in Fig. 19.1 occur in an unexposed area of Layer 5, 8, or 11 (eq. 19.1) and in the corresponding area of Layer 4, 7, or 10 (eqs. 19.2 and 19.3).

19.1.3 Dynamic Processes of Image Formation

The reactions shown in Fig. 19.1 occur in the specific layers of Fig. 19.2. As a result, diffusion rates through the layers along with reaction rates within respective layers influence dynamic processes of image formation based on the *o*-sulfonamidophenol dye releasers. Schematic illustrations of such dynamic processes are shown in Fig. 19.3, where the respective layer numbers are designated in accordance with those of Fig. 19.2 [4]. These processes (steps) occur consecutively according to the reactivities and the diffusibilities of participant compounds, as shown in the time schedule of Fig. 19.4.

Let us examine the consecutive steps by referring to Fig. 19.3 as well as the time schedule of Fig. 19.4. After exposure to light (Step a), an integrated film (Fig. 19.2) is passed through a gap between a pair of rollers. Thereby, the pod of the film is ruptured to spread an activator fluid between the two outer sheets of the integrated film (Step b). The activator liquid contains carbon black as an opacifier so that a black opaque layer is formed to construct a mini-dark room inside the integrated film.

1. **Step c**: In the mini-dark room, an electron transfer agent (ETA) and an hydroxide ion, which are fed from the activator fluid, penetrate into the layers coated on the light-sensitive sheet (Step c). The first diagram of Fig. 19.3 illustrates Step c.

2. **Step d**: The ETA reaches an emulsion layer (e.g., Layer 8 shown in the second diagram of Fig. 19.3), in which it initiates the development of unexposed silver halide emulsions (eq. 19.1) in the presence of a nucleating agent (Step d). Note again that the silver halide emulsions are direct positive so that the unexposed areas are developed. The resulting ETA_{ox} diffuses to the adjacent dye releaser layer (e.g., Layer 7 shown in the second diagram of Fig. 19.3) or to further separated layers.

3. **Step e**: After development has taken place to a sufficient extent, a development inhibitor (DI) is released from its precursor (DIR) under the alkaline condition.[1] The DIR is usually contained in Layer 2 of the cover sheet. The released DI diffuses to Layer 11 (or others) and stops the development there, as shown in the third diagram of Fig. 19.3.

4. **Steps f and g**: The ETA_{ox} generated in Layer 8 reacts with the magenta dye releaser in Layer 7 (eq. 19.2) to generate a diffusible dye ($^{-}DYE\text{-}SO_2NH^{-}$; eq. 19.3), which diffuses to a mordant layer. An excess ETA_{ox} is returned into an ETA in Layer 6 (and Layer 9), where a ballasted hydroquinone (HQ–B) scavenges the ETA_{ox} to prevent color contamination.[2]

[1] DIRs (development inhibitor releasers) have been already described in Chapter 17 on dye developers. Specific ones for *o*-sulfonamidophenol dye releasers will be discussed in the present chapter.

[2] Ballasted hydroquinones for *p*-sulfonamidonaphthol dye releasers have been already described in Chapter 18. Specific examples for *o*-sulfonamidophenol dye releasers will be discussed in the present chapter (Subsection 19.5.1).

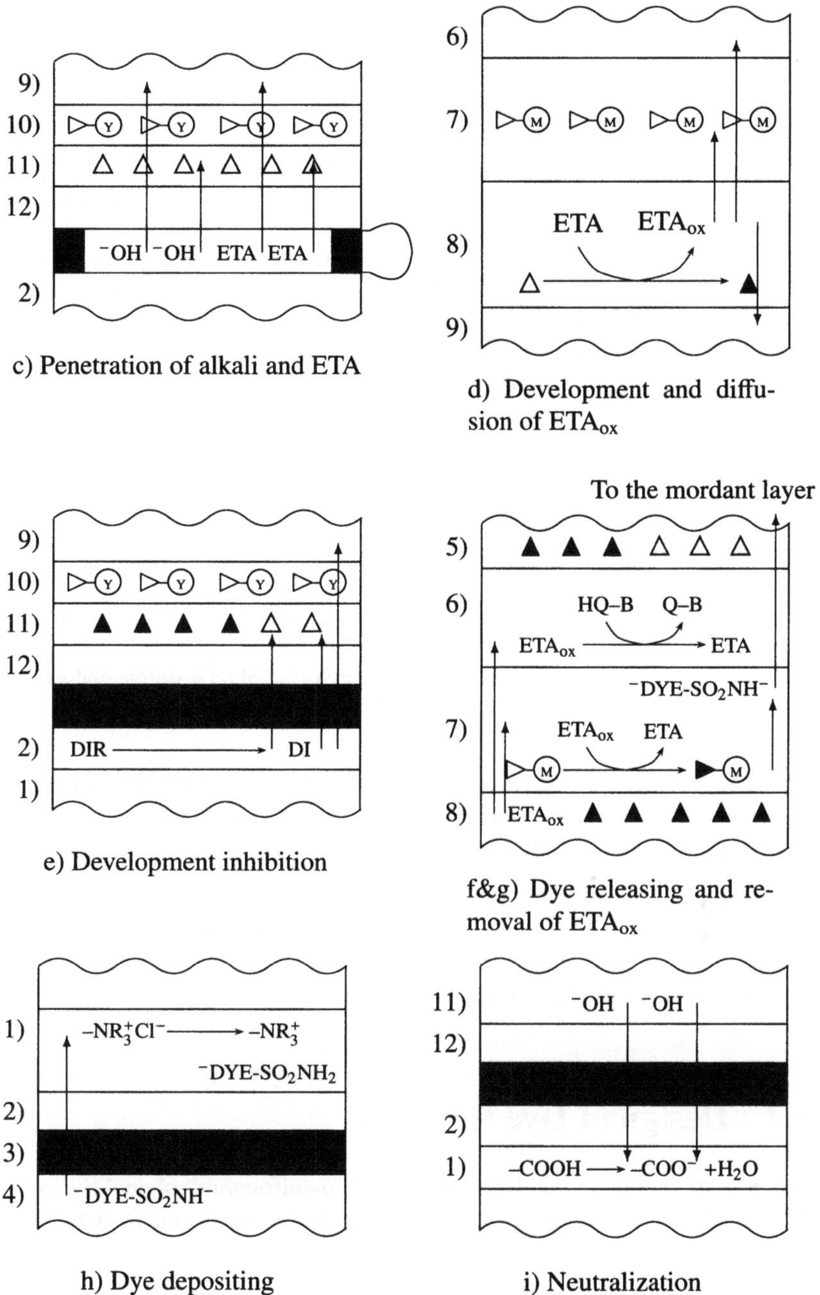

Figure 19.3. Diagrammatic representations of diffusion transfer processes based on *o*-sulfonamidophenol dye releasers. Step a (exposure) and Step b (spreading an activator fluid) are omitted. Each layer number corresponds to that of Fig. 19.2. A schematic time chart is shown in Fig. 19.4. △: Silver halide grain, ▷-(Y): Yellow dye releaser, ▷-(M): Magenta dye releaser, and ▷-(C): Cyan dye releaser.

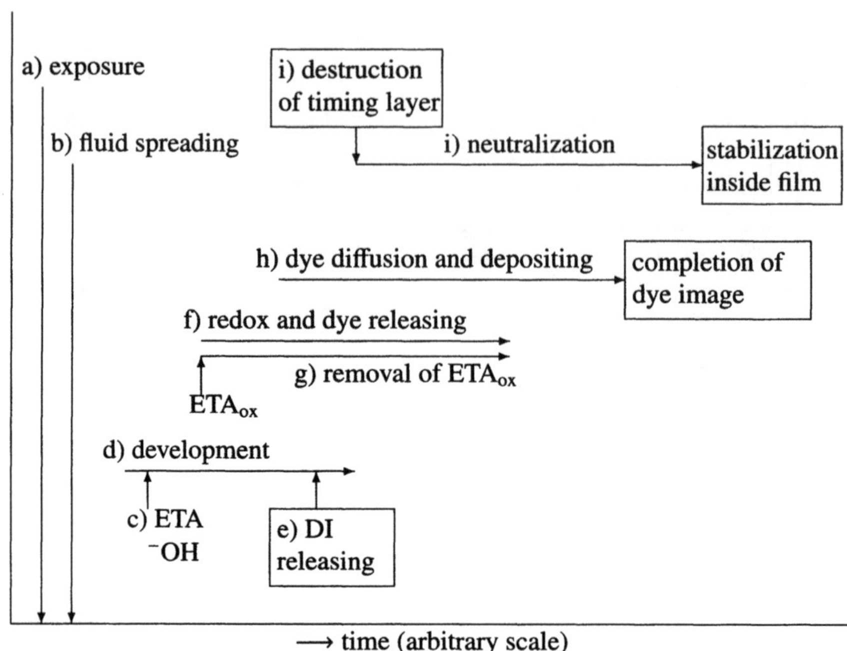

Figure 19.4. Time chart of diffusion transfer processes based on *o*-sulfonamidophenol dye releasers. Steps c to i correspond to the diagrams shown in Fig. 19.3.

5. **Step h**: The dye ($^{-}$DYE-SO_2NH^{-}) diffuses to the mordant layer (Layer 1), where it deposits and forms a dye image on a mordant polymer having a quaternary ammonium ion.[3]

6. **Step i**: Meanwhile, the polymer contained in the timing layer is destructed by alkali. Thereby, the alkaline media due to the activator fluid can contact an acid polymer contained in Layer 1 of the cover sheet. The neutralization assures the stability of the dye image.[4]

19.2 Design of Dye Releasers

According to the preceding discussions, such *o*-sulfonamidophenol dye releasers (FUN—DYE) should be selected so as to satisfy the requirements listed in Table

[3]Such mordant polymers have been already described in Chapter 18 on *p*-sulfonamidonaphthol dye releasers. They can be employed in films based on *o*-sulfonamidophenol dye releasers. Some additional examples will be discussed in the present chapter (Subsection 19.6.1).

[4]Timing polymers and acid polymers have been already described in Chapter 18 on *p*-sulfonamidonaphthol dye releasers. Some additional examples will be discussed for *o*-sulfonamidophenol dye releasers in the present chapter (Subsections 19.6.2 and 19.6.3).

Table 19.1. Properties and Requirements Necessary for Dye Releasers [8,9,10]

Moiety		Properties and requirements
FUN	1)	high S/N ratio of image
	2)	redox potential suitable for cross-oxidation with ETA
	3)	facile cross-oxidation
	4)	facile hydrolysis
	5)	high efficiency of dye releasing
DYE	6)	stability at high pH
	7)	no interaction with chemical species in a film
	8)	high mordantability
	9)	high diffusibility
	10)	appropriate visible absorption (as cyan, magenta, or yellow)
	11)	no changes of visible absorption within wide range of pH
	12)	stability of light
	13)	stability of heat and moisture in dark
FUN-DYE	14)	high solubility in a dispersion oil
	15)	stability during dispersion process
	16)	no crystallization after dispersion
	17)	no diffusion after coating
	18)	appropriate balance between hydrophilic and lipophilic parts
requirements for synthesis	19)	low-priced starting material
	20)	synthetic route with simple and minimum number of steps
	21)	high yield in each step
	22)	applicability to large-scale production
	23)	ease of purification
social requirements	24)	patent-free
	25)	environmentally safe

19.1, where the *o*-sulfonamidophenol moiety (FUN) and the dye moiety (DYE) can be separately taken into consideration [8,9,10].

The first set of requirements (Nos. 1 to 5) in Table 19.1 is concerned with the *o*-sulfonamidophenol moiety (FUN). These requirements come from Steps c, d, f, and g shown in Fig. 19.3.

The second set (Nos. 6 to 13) involves the requirements for the dye moiety (DYE). These requirements are ascribed to Step h as well as to properties of the resulting image dye. Several items of the second set are concerned with the pH change during Step i.

On the other hand, the third set (Nos. 14 to 18) involves the requirements for a dye releaser (FUN–DYE), which are concerned with the production feasibility and the stability of raw photographic films. Note that the dye releaser is dispersed as an oil-in-water dispersion by using a viscous oil and an aqueous gelatin solution. The dispersion is coated to give a layer on a sheet together with other layers.

The fourth set (Nos. 19 to 23) is related to large-scale syntheses of dye releasers (FUN–DYE), where economical and technical requirements are described.

The remaining requirements (Nos. 24 and 25) aim at social accountability, which is becoming increasingly important.

In designing dye releasers, all of these requirements must be taken into consideration. In the following sections, we will deal mainly with the items for the *o*-sulfonamidophenol moieties (FUN) and for the dye moieties (DYE). If necessary, the other items will be involved in the discussions on these items.

19.3 *o*-Sulfonamidophenol Moieties

19.3.1 Designing *o*-Sulfonamidophenol Moieties

The first task is to design an *o*-sulfonamidophenol moiety (FUN) that satisfies the first set of requirements (Nos. 1 to 5) listed in Table 19.1. Figure 19.5 summarizes several targets for the R&D of such *o*-sulfonamidophenol moieties [9,10].

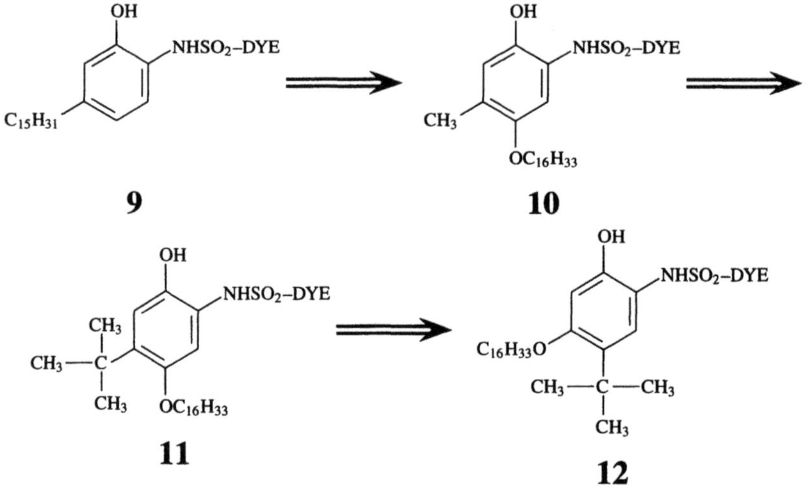

Figure 19.5. Several targets designed for R&D of *o*-sulfonamidophenol dye releasers [9, 10].

Since an *o*-sulfonamidophenol compound **9** had once been reported to be unable to release a diffusible dye under photographic conditions [11], there had appeared no further investigations on compounds of this type until the R&D described here was started in the second half of the 1970s.[5] The inability of **9** to release a dye can be ascribed to the insufficient redox potential of **9** and/or to the insufficient releasing efficiency of its oxidized form. Hence the first working guideline was to adjust the redox potentials of *o*-sulfonamidophenols to be adequate for photographic conditions.

19.3.2 *o*-Sulfonamidophenols Having Alkoxy Groups

To afford the development activity, an additional alkoxy group has been introduced at the 4-position of *o*-sulfonamidophenols (e.g., **10**) [12]. Figure 19.6 shows a set of cyan, magenta, and yellow dye releasers of this type. They have been found to release a dye under photographic conditions so that the working guideline has been verified. However, their dye-releasing efficiencies have been too insufficient to be employed in a practical instant color film.

On the other hand, a detailed investigation on their chemical behaviors has given a foundation to further advances of *o*-sulfonamidophenol dye releasers (e.g., **12**). Moreover, *o*-sulfonamidophenols covered by the patent [12] (homologs of **11**) have recently been employed in thermal photographic materials, as described later in this chapter.

Syntheses

The preparation of *o*-sulfonamidophenol dye releasers has been summarized in a review [8]. The *o*-aminophenol intermediate common to the set of dye releasers shown in Fig. 19.6 (i.e., 2-amino-4-hexadecyloxy-5-methylphenol hydrochloride: **21**) has been originally prepared according to the scheme illustrated in Fig. 19.7 [12].

The pathway shown in Fig. 19.7 contains an elaborate use of protective groups. Thus, during the nitration giving the 2-nitro group of **17**, mesyl (methanesulfonyl) groups are used as protective groups for the hydroquinone function of **16**. The mesyloxy group adjacent to the 2-nitro group is selectively hydrolyzed so as to regenerate a free hydroxyl group, which is in turn protected as a benzyl ether (**18**). Then the other mesyloxy group is hydrolyzed and changed into a hexadecyloxy group (**18** ⟶ **19** ⟶ **20**). Catalytic hydrogenation causes the reduction of the nitro group as well as the reductive cleavage of benzyl ether. The final treatment with hydrochloric acid gives the target molecule **21**.

An improved synthesis of 2-amino-4-hexadecyloxy-5-methylphenol is illustrated in Fig. 19.8 [13]. Catalytic hydrogenation causes the reduction of the nitro group of **23** and the subsequent intramolecular participation of the 1-acetyloxy group. The resulting 2-acetamidophenol (**24**) is treated with *p*-toluenesulfonic

[5] This early history implies a pitfall to be averted in the R&D of functionalized organic compounds, since the very property of a specific compound (e.g., **9**) had been erroneously exploited to hold true for all compounds of that type.

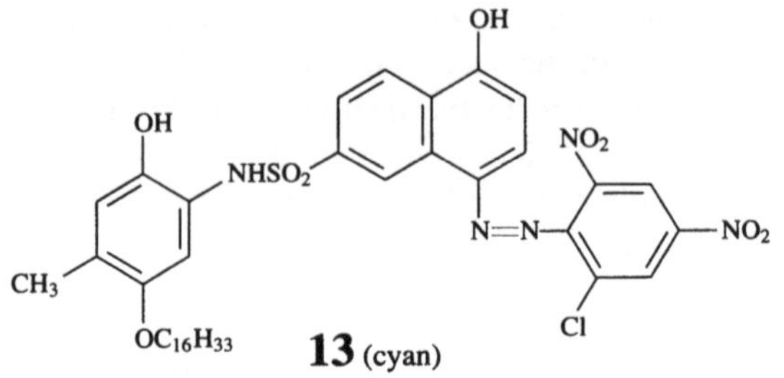

13 (cyan)

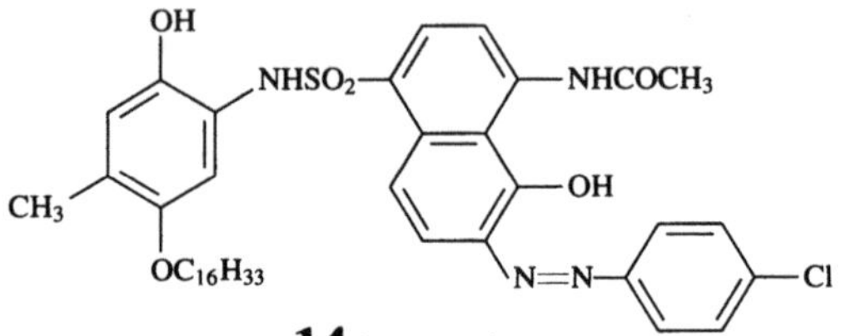

14 (magenta)

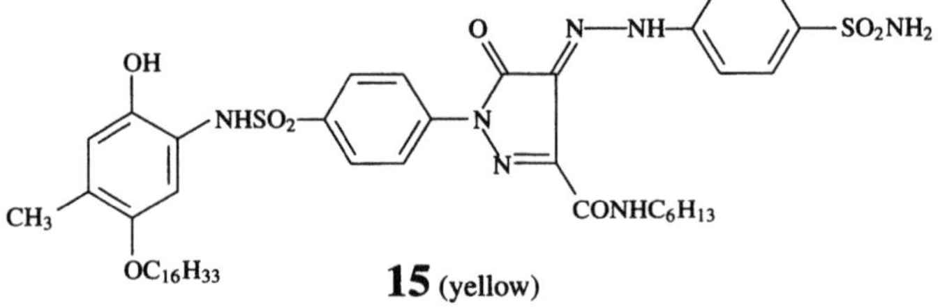

15 (yellow)

Figure 19.6. A set of cyan, magenta, and yellow *o*-sulfonamidophenol dye releasers for color reproduction of instant color photography [12].

acid (TsOH), giving a benzoxazole ring. After the hydrolysis of the remaining acetoxy group, the resulting hydroxyl group of **25** is converted into the hexadecyloxy group of **26**. Finally, a treatment with hydrochloric acid results in an oxazole ring cleavage and an ammonium chloride formation, giving the target compound **21**.

A more general method for preparing benzoxazole intermediates such as **25** has been proposed, where the Beckmann rearrangement is used to form an oxazole ring, as shown in Fig. 19.9 [14]. Thus, an oxime of a 2-acetylphenol (e.g., **29**) undergoes a dehydration and an intramolecular cyclization by introducing hydrogen chloride gas into a solution of the oxime in acetic acid at about 100–120°C. Note that a usual Beckmann rearrangement contains an intermolecular hydration of the cationic intermediate (CH_3C^+=N–Ar + H_2O) to form an acetamide (CH_3CO–NH–Ar), while the present reaction involves an intramolecular attack of the phenolic hydroxyl group to form an oxazole ring.

The conditions of the Beckmann rearrangement employed in Fig. 19.9 (i.e., the use of hydrogen chloride gas as a reagent, the use of acetic acid as a solvent, and a high reaction temperature of about 100–120°C) are unsuitable for a practical large-scale production. A combination of phosphoryl chloride and *N*,*N*-

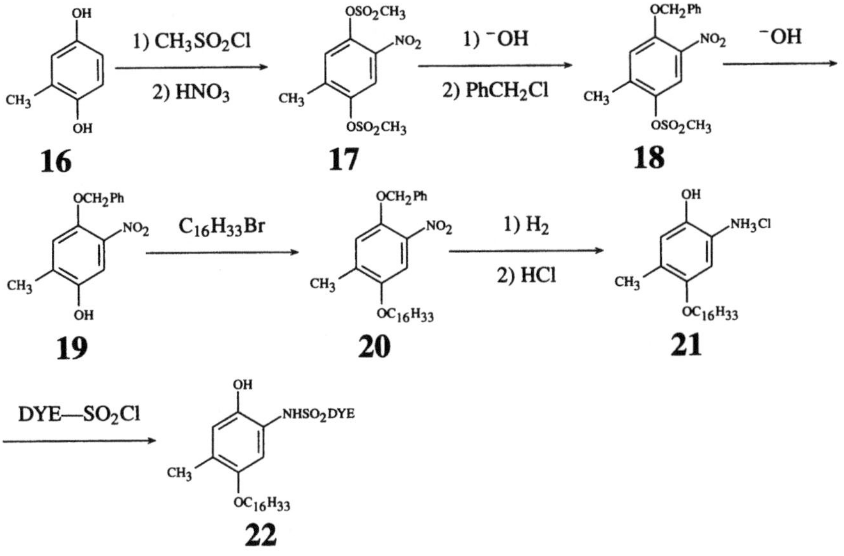

Figure 19.7. Synthesis of 2-amino-4-hexadecyloxy-5-methylphenol by using mesyl and benzyl groups as protecting groups [12].

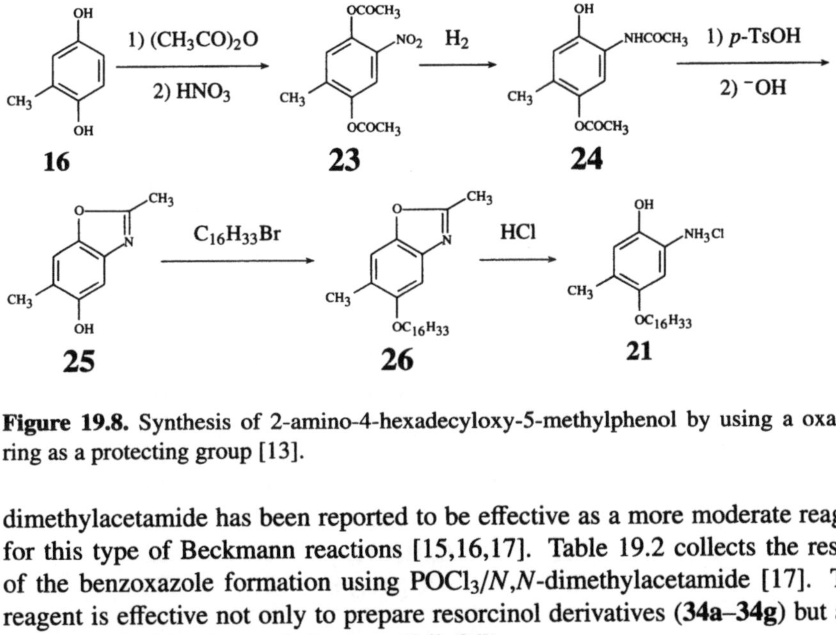

Figure 19.8. Synthesis of 2-amino-4-hexadecyloxy-5-methylphenol by using a oxazole ring as a protecting group [13].

dimethylacetamide has been reported to be effective as a more moderate reagent for this type of Beckmann reactions [15,16,17]. Table 19.2 collects the results of the benzoxazole formation using $POCl_3$/*N*,*N*-dimethylacetamide [17]. This reagent is effective not only to prepare resorcinol derivatives (**34a–34g**) but also to prepare hydroquinone derivatives (**34i–34l**).

The reaction from **33** to **34** may be explained by considering an intermediate akin to a Vilsmeier complex (i.e., $(CH_3)_2N^+{=}C(CH_3)Cl$), which is formed from phosphoryl chloride and *N*,*N*-dimethylacetamide. This intermediate attacks the hydroxyl group of the oxime (HO–N=C<) to form an intermediate

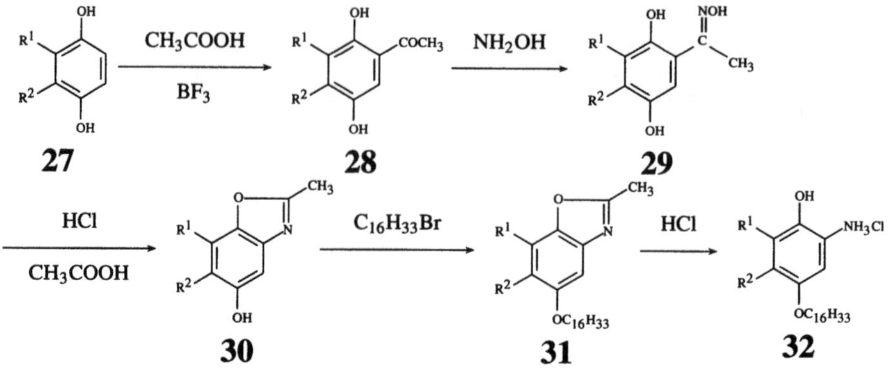

Figure 19.9. Syntheses of various *o*-aminophenols via oxazole intermediates [14]. R^1 and R^2 represents hydrogen, an alkyl group, etc.

Table 19.2. Benzoxazoles Prepared by the Beckmann Rearrangement of Oximes with $POCl_3$/*N*,*N*-Dimethylacetamide [16,17]

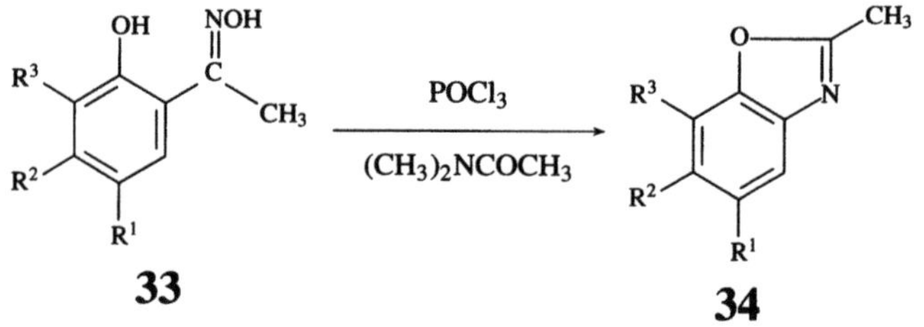

Compd.	R^1	R^2	R^3	Yield (%)
34a	H	OH	H	83
b	CH_3	OH	H	82
c	$C(CH_3)_3$	OH	H	85
d	$C(CH_3)_2C_2H_5$	OH	H	62
e	$C(CH_3)(C_2H_5)_2$	OH	H	87
f	C_6H_{13}-*n*	OH	H	68
g	H	OH	CH_3	72
h	$C(CH_3)_3$	H	CH_3	85
i	OH	H	H	68
j	OH	$C(CH_3)_3$	H	88
k	OH	$C_6H_4CH_3$-*p*	H	92
l	h	$OC_{16}H_{33}$-*n*	H	76

$(CH_3)_2N{-}C(CH_3)Cl{-}O{-}N{=}C<$. Then, this intermediate is converted into $(CH_3)_2N{-}C(CH_3){=}O + Cl^- + {}^+N{=}C<$. The last species undergoes the primary step of the Beckmann reaction.

19.3.3 Improved Dye-Releasing Efficiency

Side Reactions

In order to improve the dye-releasing efficiency of **10**, the hydrolysis of quinone monosulfonimides has been investigated from the viewpoint of side reactions [18,19]. In general, such *o*-quinone monosulfonimides have been reported to be too unstable to be isolated as compared with *p*-quinone monosulfonimides, as summarized in a review [20]. Fortunately, however, a model compound (**36**) can be isolated through the oxidation of **35** (Fig. 19.10) [18,19].

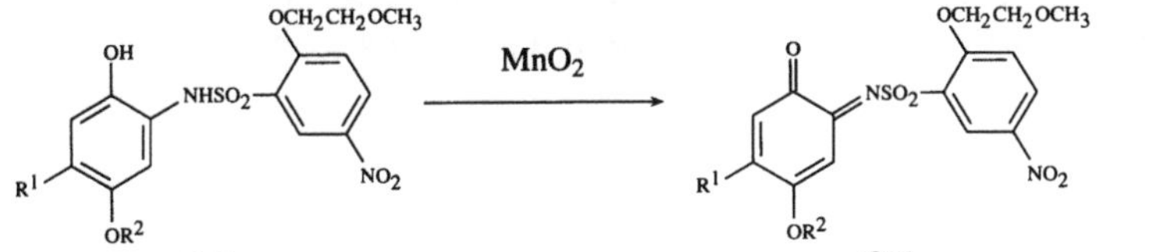

MnO_2

35 **36**

a: R^1 = CH_3 R^2 = CH_3; **b**: R^1 = CH_3 R^2 = $C_{16}H_{33}$
c: R^1 = $C(CH_3)_3$ R^2 = CH_3; **d**: R^1 = $C(CH_3)_3$ R^2 = $C_{16}H_{33}$

Figure 19.10. Syntheses of *o*-quinone monosulfonimide intermediates [18,19].

The alkaline hydrolysis of **36b** in methanol-water has given a *p*-quinone monoacetal (**37**), which can be isolated by quenching in an early stage of the reaction [18,19]. This is a new method for preparing *p*-quinone monoacetals, since electrochemical preparations have been known as standard methods, as summarized in a review [21].

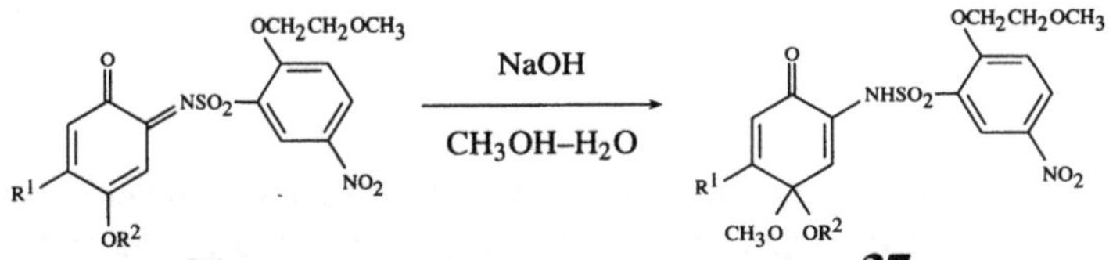

NaOH

$CH_3OH–H_2O$

36 **37**

a: R^1 = CH_3 R^2 = CH_3
b: R^1 = CH_3 R^2 = $C_{16}H_{33}$

Figure 19.11. Isolation of *p*-quinone monoacetals from *o*-quinone monosulfonimide intermediates [18,19].

In a prolonged reaction, the amount of the *p*-quinone monoacetal intermediate (**37b**) reaches a maximum, and then it decreases gradually so that the intermediate is converted into another acetal intermediate (**37a**). As the two intermediates disappear, a hydrolysis product, i.e., 2-(2-methoxyethoxy)-5-nitrobenzenesulfonamide, appears increasingly (final yield 68%).

Although this experiment has not been the same as the reactions occurring in photographic films, it has suggested that nucleophilic attacks at the 4-position cause such side reactions being regarded as 1,4-additons of the RO-C=C-C=O function.

Steric Hindrance

The side reactions described above have been expected to be suppressed by the introduction of steric hindrance, since this is a common methodology in organic chemistry [9,10]. Hence, the second target has been decided to be **11** (Fig. 19.5), where the methyl group of **10** has been replaced by a *t*-butyl group [18,19].

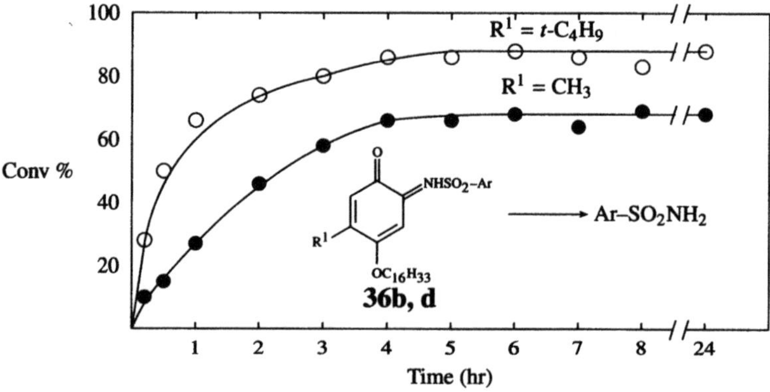

Figure 19.12. Effect of alkyl substituents on hydrolysis [19]. The structure of Ar is shown in Fig. 19.10.

Figure 19.12 shows the effect of alkyl substituents on the efficiency of releasing $Ar\text{-}SO_2NH_2$ under homogeneous conditions [19]. Obviously, the replacement of the methyl group by the *t*-butyl group increases the releasing efficiency to a great extent.

Since an *o*-sulfonamidophenol dye releaser is dispersed as an oil-in-water dispersion by using a viscous oil and aqueous gelatin solution, the dye-releasing reaction occurs under heterogeneous (two-phase) conditions. To simulate such heterogeneous conditions, a combination of ethyl acetate (AcOEt) and water has been selected as a reaction medium. The results are summarized in Fig. 19.13 and Table 19.3, where **38**, **39**, **40**, and **42** have been isolated as main products [19].

Among the main products, **38** and **39** come from the desired 1,2-addition of a hydroxide ion to the imide function, while **40** and **42** are produced by the undesired 1,4-addition. Note that **42** is a rather unusual compound which stems from the 1,4-attack of a sulfonamide anion.

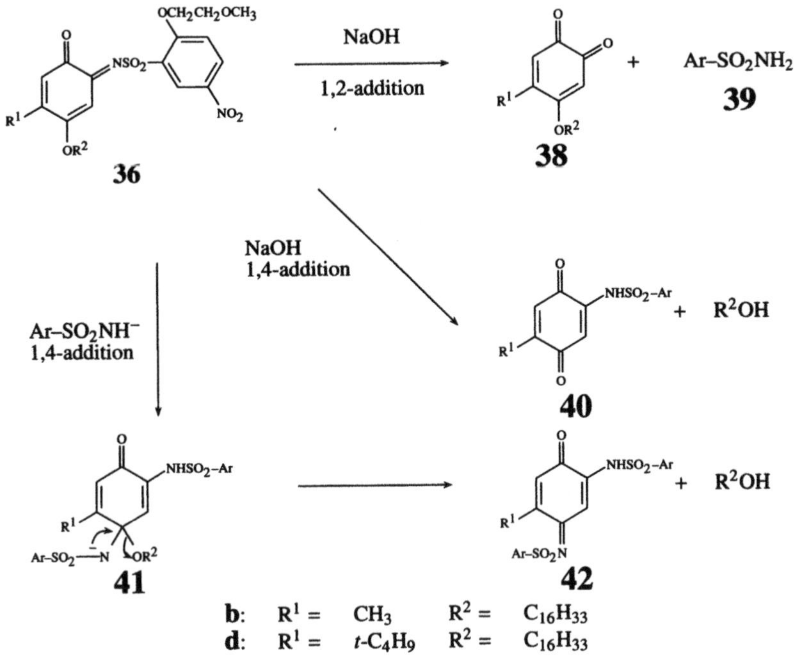

b: R^1 = CH_3 R^2 = $C_{16}H_{33}$
d: R^1 = t-C_4H_9 R^2 = $C_{16}H_{33}$

Figure 19.13. Heterogeneous hydrolysis of **36b** and **36d** [19].

Table 19.3. Heterogeneous Hydrolysis of **36b** and **36d** [19]

Run	Substrate	Condition	Yield (%)			
			38	**39**	**42**	**40**
1	**36b**	2M NaOH/AcOEt	64	81	4	4
2	**36b**	4M NaOH/AcOEt	0	63	12	2
3	**35b**	$K_3Fe(CN)_6$-2M NaOH	53	63	12	3
4	**36d**	2M NaOH/AcOEt	95	87	3	2
5	**36d**	4M NaOH/AcOEt	52	83	5	8
6	**35d**	$K_3Fe(CN)_6$-2M NaOH	93	84	6	0

Table 19.3 shows comparison data, which reveal the superiority of a *t*-butyl group over a methyl group [19]. In particular, compare the yields of the sulfonamide (**39**) between Run 1 and Run 4 and between Run 2 and Run 5.

However, several side reactions due to the 1,4-addition (i.e., giving **40d** and **42d**) still remain even for the *t*-butyl substituent (Run 4 and Run 5). This implies that such side reactions are inevitable as long as dye releasers have an alkoxy group at the para-position of the phenol hydroxyl group.

Electronic Effect

A further working guideline has been obtained by comparing the NMR data of **36** and **38** [19]. These compounds are different only in their substituents (sulfonimido vs. oxo) at the 2-positions. The hydrogen of the group H–C–C=N-SO_2Ar in **36** exhibits an NMR signal in a lower field (by about δ 1 ppm) than the hydrogen of the group H–C–C=O in **38** [19]. This implies that the sulfonimido group (=N-SO_2Ar) is more electron-withdrawing than the oxo group (=O) of the carbonyl group. Thereby, the former is considered to enhance 1,4-additions of **36**, as shown in Fig. 19.13.

$OCH_2CH_2OCH_3$ OH $NHSO_2$ R^2O R^1 NO_2

43

oxidation → No quinone monoimide

$K_3Fe(CN)_6$-
2M NaOH
AcOEt

O O R^2O R^1 + Ar–SO_2NH_2

38 **39**

b: R^1 = CH_3 R^2 = $C_{16}H_{33}$
d: R^1 = t-C_4H_9 R^2 = $C_{16}H_{33}$

Figure 19.14. Oxidation and hydrolysis of **43b** and **43d** [19].

Table 19.4. Oxidation and Hydrolysis of **43b** and **43d** [19]

Substrate	Condition	Yield (%)			
		38	**39**	**42**	**40**
43b	$K_3Fe(CN)_6$-2M NaOH	49	67	–	–
35b	$K_3Fe(CN)_6$-2M NaOH	53	63	12	3
43d	$K_3Fe(CN)_6$-2M NaOH	99	96	–	–
35d	$K_3Fe(CN)_6$-2M NaOH	93	84	6	0

The consideration described above has led to a straight-forward guideline in which the positions of a 4-alkoxy group and a 5-alkyl group are interchanged, as observed in model compounds (**43**) [9,10]. Thereby, a possible side reaction of **43** can be considered to be the 1,4-addition of an α,β-unsaturated carbonyl group, whereas that of **35** is the 1,4-addition of an α,β-unsaturated sulfonimido group. The former 1,4-addition has been expected to be more restrained than the latter by the electronic effect.

As found in Fig. 19.14, a quinone monosulfonimide has not been isolated by the oxidation of **43** even under anhydrous conditions. This implied high dye-releasing efficiency. This expectation has been verified by the data collected in Table 19.4, which contains the results of the simultaneous oxidation and hydrolysis of **43b** and **43d** in comparison with the results of **35b** and **35d**. Thus, the comparisons between **43b** and **35b** as well as between **43d** and **35d** indicate the effects of the exchanges of the substitution positions. On the other hand, the comparison between **43b** and **43d** indicates the effect of a *t*-butyl substituent.

The investigations on side reactions, steric hindrance, and electronic effect have finally led to a target nucleus (**12**) shown in Fig. 19.5 [9,10].

Final Targets

In general, the dye-releasing efficiency of a dye releaser (FUN–DYE) is influenced by other moieties surrounding the functional moiety (FUN). These moieties can be conceptually considered to be a part of the FUN, even if they are released together with the DYE.

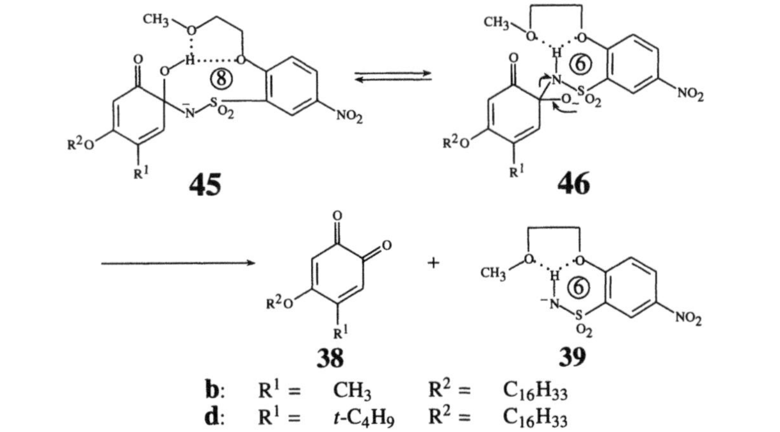

43 → (oxidation) → **44** → (^-OH) → **45** ⇌ **46** → **38** + **39**

b: $R^1 = CH_3$ $R^2 = C_{16}H_{33}$
d: $R^1 = t\text{-}C_4H_9$ $R^2 = C_{16}H_{33}$

Figure 19.15. Speculation for explaining the effect of a 2-methoxyethoxy group [9,10].

In particular, the presence of a 2-methoxyethoxy group has been found to enhance the dye-releasing efficiency. This enhancement has been presumed to stem from an intramolecular chelating effect, as shown in Fig. 19.15 [9,10]. The 1,2-attack of a hydroxide ion to the sulfonimido group of **44** gives an adduct (**45**), where the 2-methoxyethoxy group chelates the hydrogen atom of the hydroxy group in terms of hydrogen bonding. This chelation results in the formation of an eight-membered ring of the intermediate (**45**), as designated by ⑧. The releasing of the sulfonamide (**39**) requires a further intermediate (**46**), which contains a six-membered ring designated by ⑥. The six-membered ring is presumed to be more plausible than the eight-membered ring.

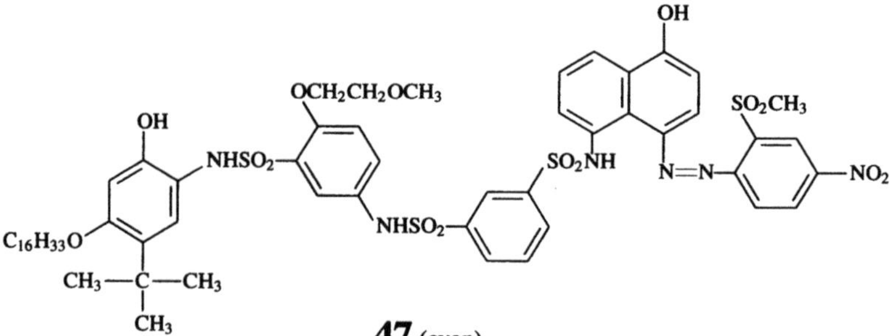

47 (cyan)

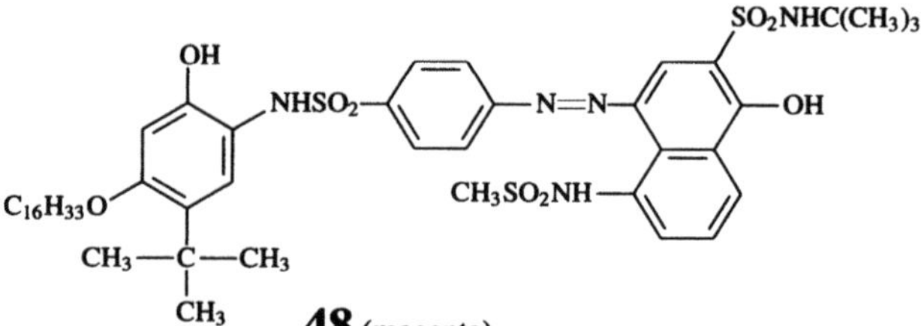

48 (magenta)

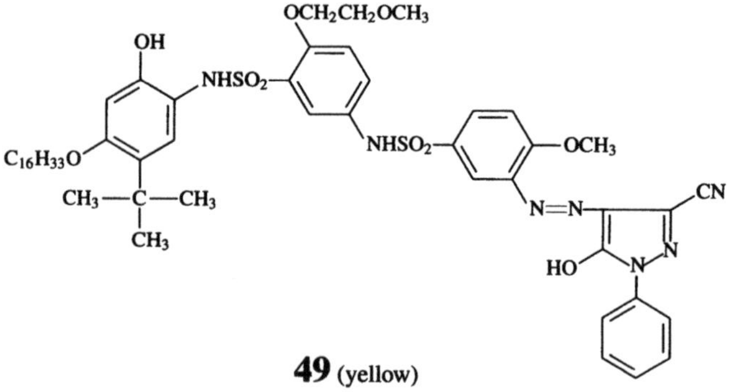

49 (yellow)

Figure 19.16. An improved set of cyan, magenta, and yellow *o*-sulfonamidophenol dye releasers for color reproduction in instant color photography [22].

The design of *o*-sulfonamidophenol moieties by using the model compounds described above has arrived at an improved set of dye releasers, as shown in Fig. 19.16 [22]. The specification of this patent contains experimental comparisons between dye releasers having a *t*-butyl group and the ones having a methyl group by using practical photographic films. Moreover, the effect of 2-alkoxyethoxy groups has been also incorporated in the dye releasers shown in Fig. 19.16 [23]. This type of *o*-sulfonamidophenol dye releasers has been used in the FI-10 film for the FOTORAMA system marketed by Fuji Photo Film in 1981 [1].

Syntheses

As collected in Table 19.2, various benzoxazole intermediates (**34a–34g**) for preparing 2-amino-4-alkyl-5-alkoxyphenols have been easily obtained in high yields. In particular, **34c** serves as a starting material for preparing the dye releasers listed in Fig. 19.16 [16,22].

Another versatile method for preparing 2-amino-4-alkyl-5-alkoxyphenols is the Friedel-Crafts alkylation of 2-acetamido-5-alkoxyphenol, as shown in Fig. 19.17 [24]. Thus, the key step is the conversion of **51** into **52**, where isobutene is used as a carbon source and Amberlite 15 (a synthetic ion-exchange resin from Rohm & Haas Co.) is used as a catalyst.

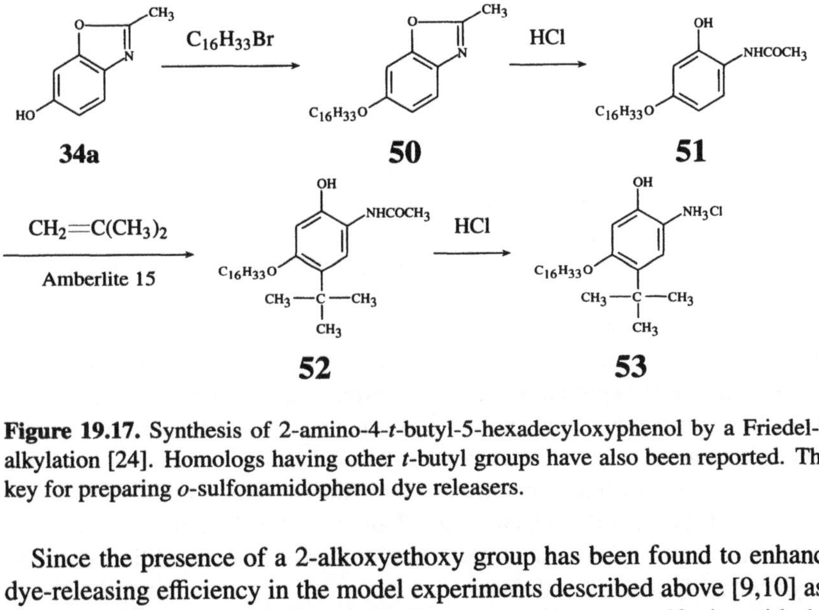

Figure 19.17. Synthesis of 2-amino-4-*t*-butyl-5-hexadecyloxyphenol by a Friedel-Crafts alkylation [24]. Homologs having other *t*-butyl groups have also been reported. This is a key for preparing *o*-sulfonamidophenol dye releasers.

Since the presence of a 2-alkoxyethoxy group has been found to enhance the dye-releasing efficiency in the model experiments described above [9,10] as well as in practical usage [22,23], 2-(2-alkoxyethoxy)benzenesulfonic acids having a nitrogen function have become important as key intermediates for preparing dye releasers of high efficiency. The introduction of such a 2-alkoxyethoxy group at the ortho position of a sulfonyl function is illustrated in Fig. 19.18 [25]. The first method described in one of the patents [23] has used sodium 2-methoxyethylate ($Na^+ \ ^-OCH_2CH_2OCH_3$) prepared by adding sodium hydride to 2-methoxyethanol in the nucleophilic substitution of sodium 2-chloro-5-nitrobenzenesulfonate (**54**). A more convenient method using a combination of sodium hydroxide and manganese dioxide and a further practical method using a combination of sodium hydroxide and sodium silicate ($Na_2O{\cdot}nSiO_2$, where *n* is equal to about 3) have been disclosed in one of the patents [25].

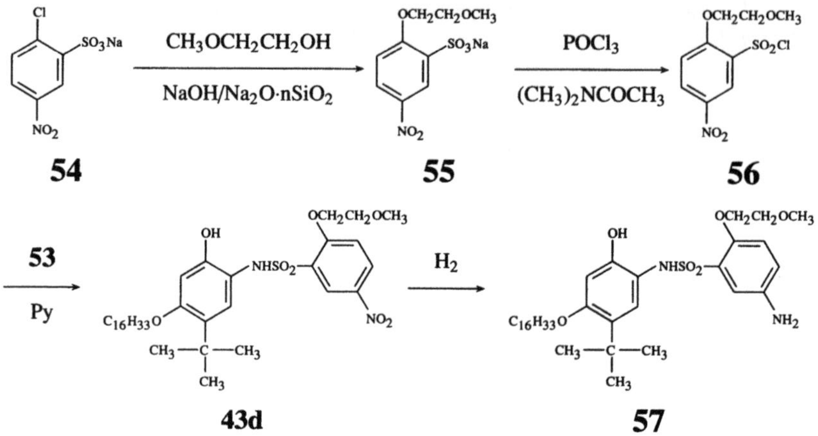

Figure 19.18. Synthesis of an *o*-sulfonamidophenol intermediate having a 2-methoxy-ethoxy group [22,23]. Homologs containing other alkyl groups and other 2-alkoxyethoxy groups have also been prepared in similar ways.

Another key reaction is concerned with the chlorination of sulfonic acids, as found in the reaction of **55**, which produces the corresponding sulfonyl chloride (**56**). To prepare such sulfonyl chlorides, thionyl chloride has been used as a chlorinating agent. However, the reagent evolves sulfur dioxide gas, which is not so easily handled in large-scale production. To avoid the drawback of thionyl chloride, a combination of phosphoryl chloride and *N,N*-dimethylacetamide (DMAC), which has once been described in the preparation of benzoxazoles (Table 19.2), has been reported to be also effective in this chlorination [23].

Further, a combination of phosphoryl chloride and sulfolane has been reported as a more convenient reagent for preparing sulfonyl chlorides, as summarized in Table 19.5 [26]. As for the preparation of the chloride (**56**), Conditions A (with sulfolane) and B (with sulfolane and a small amount of *N,N*-dimethylacetamide) have provided higher yields in shorter reaction times than Condition C (*N,N*-dimethylacetamide without sulfolane).

The condensation of the sulfonyl chloride (**56**) with the *o*-aminophenol (**53**) has been conducted in the presence of a week base (e.g., pyridine), as shown in Fig. 19.18 [22,23]. The use of a strong base (e.g., triethylamine) in the condensation of *p*-toluenesulfonyl chloride (tosyl chloride) with *o*-aminophenol has been reported to result in an *O*-tosylation, while the use of pyridine has preferred an *N*-tosylation [27].

Table 19.5. Sulfonyl Chlorides Prepared with $POCl_3$/Sulfolane [26]

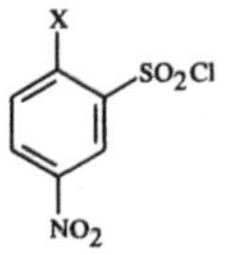

58 X = Cl
59 X = H
56 X = $OCH_2CH_2OCH_3$

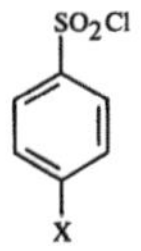

60 X = CH_3
61 X = Cl

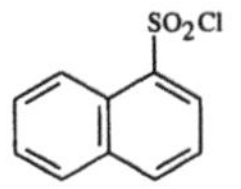

62

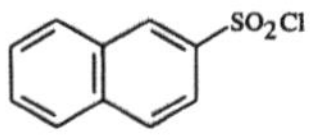

63

Prodct	Condn	Yield (%)
58	A	96
	B	96
	C	–
	D	43
59	A	95
	B	95
	C	89
56	A	96
	B	95
	C	85
60	A	87
	B	84
61	A	96
	B	95
62	A	94
	B	90
63	A	85
	B	91

Reagent	Condition			
	A	B	C	D
$POCl_3$ (mL)	7.3	7.3	7.3	7.3
Sulfolane (mL)	10	10	0	0
Acetonitrile (mL)	10	10	20	20
DMAC (mL)	0	0.5	0.5	7.3
Temperature (℃)	68–72	68–72	70–75	50–55
Time (min)	40–60	15–30	120	90

19.4 Azo Dye Moieties

In addition to the *o*-sulfonamidophenol moieties of high dye-releasing efficiency, dye moieties incorporated in dye releasers are also important to obtain pictures of high quality, as summarized in Table 19.1. Their hue, fastness (light and dark), diffusibility, etc. must be taken into consideration.

19.4.1 Cyan Dye Moieties

4-(4-Nitro-2-methanesulfonylbenzeneazo)-1-naphthols and 4-(2-chloro-4,6-dinitrobenzeneazo)-1-naphthols as cyan azo dyes have already been described in Chapter 18 in combination with *p*-sulfonamidonaphthol moieties. When *o*-sulfonamidophenols are used as functional moieties, the resulting dye releasers (e.g., **13**) have been found to be insufficient. Hence, several innovations have been necessary to enhance the dye releasing efficiency, as discussed in the preceding sections.

One of the guiding principles has been to change the environment surrounding the reaction site of the *o*-sulfonamidophenol function (Fig. 19.15). Although naphthylenes have been tested as linking groups between a FUN and a DYE in place of a phenylene linkage [28], their effects have been still insufficient for practical usage. A morpholino group and a piperidino group incorporated in cyan dye releasers (e.g., **64**) have remarkably improved the dye-releasing efficiency [29].[6] However, the synthetic intermediates (e.g., 2-morpholino-5-nitrobenzenesulfonyl chloride) have been found to be rather labile during storage. 2-Alkoxyethoxy groups have been discovered as more adequate substituents for enhancing the dye-releasing efficiency (e.g., **65**) [23]. This effect has been postulated to come from the intramolecular chelation shown in Fig. 19.15. The cyan dye moiety and the linking group (environment-changing group) of **65** have been combined with a more efficient *o*-sulfonamidophenol moiety to create the cyan dye releaser (**47**) of high dye-releasing efficiency (Fig. 19.16) [22,30].

To expand the structural variety of cyan dyes, various dye moieties have been tested. For example, phthalocyanine dyes such as **66** have been proposed as cyan dye moieties [31,32].

Azo dyes having other azo components have been also disclosed in patents, as shown in Fig. 19.21. The cyan dye moiety of a 4-arylazo-1-naphthol (**67**) comprises a 2,4-dimethanesulfonylphenyl azo component [33]. In order to give longer absorption, an amido group has been introduced at the 2-position of the 4-arylazo-1-naphthol nucleus and one of the methanesulfonyl groups on the azo component is replaced by a benzenesulfonyl group (e.g., **68**). Such cyan dye releasers of longer absorption have been used in heat-developable photographic films [34]. To adjust cyan hue, the azo components derived from 2,4-dialkylsulfonylanilines in such 2-amido-4-arylazo-1-naphthols have been replaced by the counterparts derived from 5-chloro-2-cyano-4-nitroaniline [35]. More recently, heterocyclic azo components have been disclosed to be incorporated in heat-developable *o*-sulfonamidophenol dye releasers, as exemplified by **69** [36].

Syntheses

As a typical example of cyan dye releasers of *o*-sulfonamidophenol type, Fig. 19.22 illustrates a synthetic pathway for preparing the cyan dye releaser (**47**) [22]. It should be noted that this scheme contains 3-[*N*-(1-hydroxy-5-naphthyl)sulfamoyl]benzenesulfonyl chloride (**72**) and the resulting sulfonyl chloride having an azo dye moiety (**75**), whereas the corresponding sulfonyl fluorides have been used in the synthesis of *p*-sulfonamidonaphthol dye releasers, as described in Chapter 18 (Fig. 18.13). A procedure using benzene-1,3-disulfonyl

[6] An intramolecular effect similar to Fig. 19.15 is possible to enhance the dye-releasing efficiency. Although the dye releasers shown in Fig. 19.19 contain the *o*-sulfonamidophenol moiety of an earlier type (**10**) as a functional moiety (FUN), the chemistries of dye moieties discussed in this and the following sections are effective commonly to the improved FUNs (**11**, **12** and related FUNs).

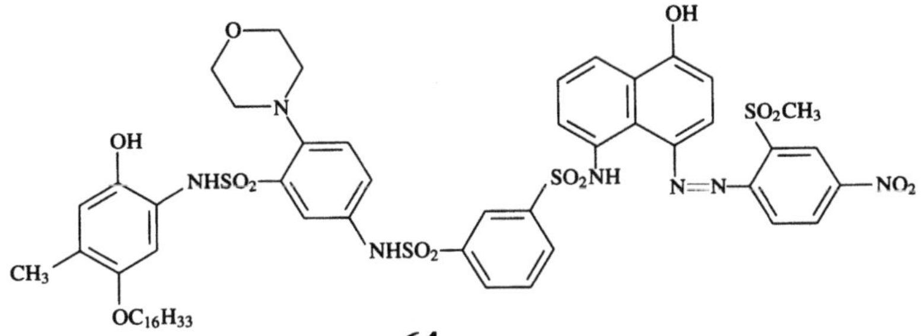

64 (cyan)

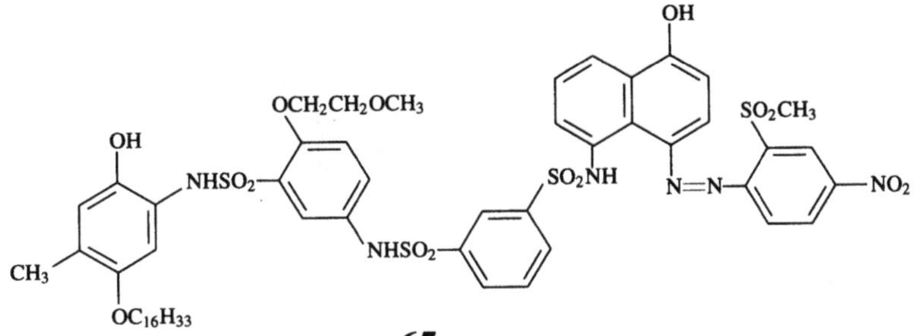

65 (cyan)

Figure 19.19. Cyan dye moieties for *o*-sulfonamidophenols dye releasers. The morpholino group in **64** and the methoxyethoxy group in **65** enhance dye-releasing efficiency according to the scheme shown in Fig. 19.15.

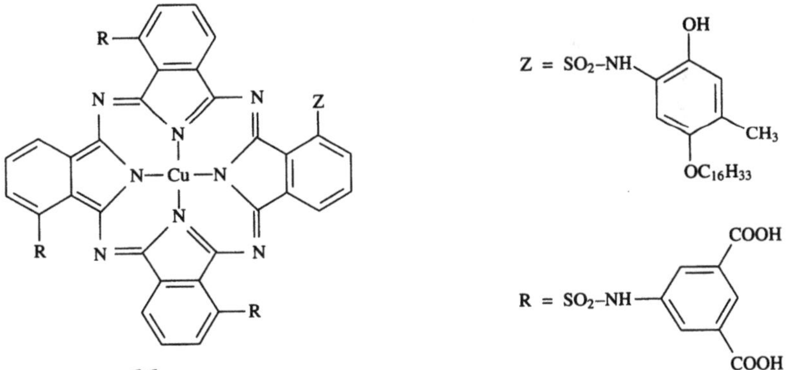

66 (cyan)

Figure 19.20. Phthalocyanine dye moiety for *o*-sulfonamidophenol dye releasers.

chloride (**71**) has been reported to be more practical than the procedure using 3-fluorosulfonylbenzenesulfonyl chloride, where reaction conditions under which one of the two sulfonyl chloride functions in **71** reacts selectively with 5-amino-1-naphthol has been examined [37].

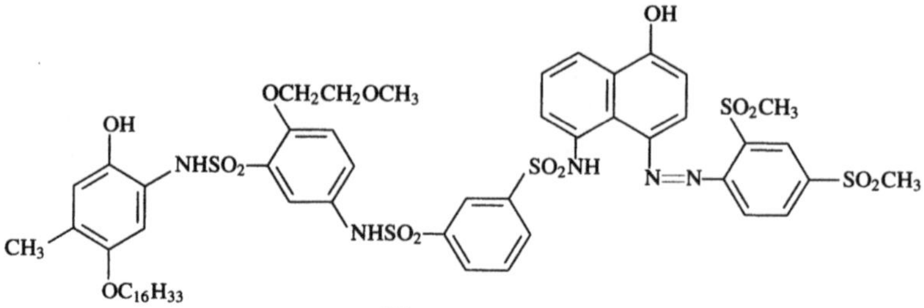

67 (cyan)

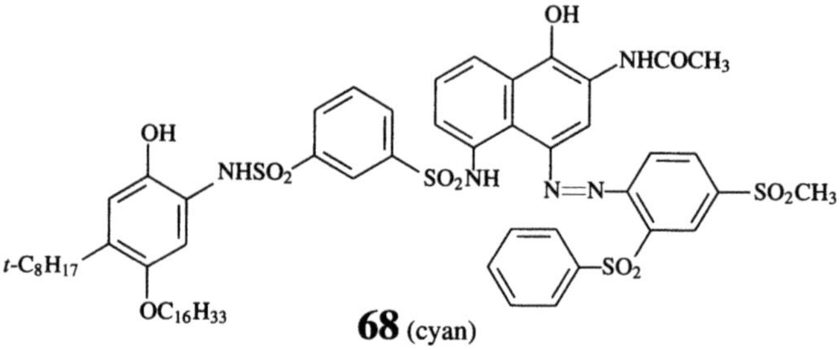

68 (cyan)

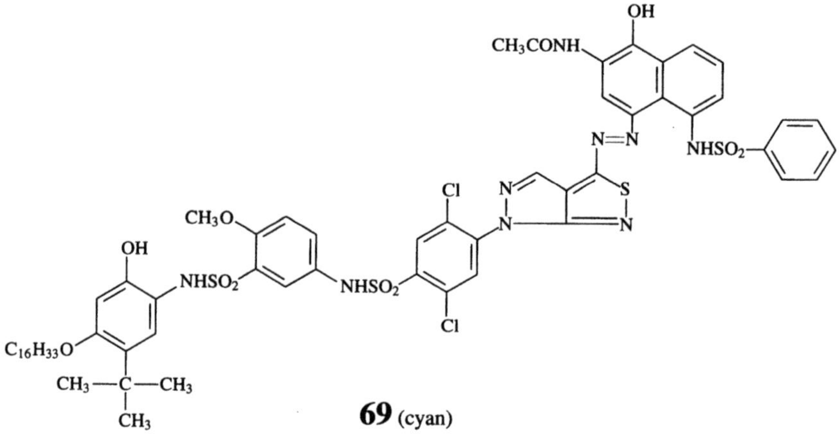

69 (cyan)

Figure 19.21. Cyan dye moieties for *o*-sulfonamidophenol dye releasers.

In an earlier procedure, the sulfonyl chloride (**72**) has been directly coupled with the diazo component (the diazonium salt of 3-methanesulfonyl-4-nitroaniline) to give **75** [23]. Later, the potassium salt (**73**) was selected as a coupling component to give **74** [37], which is in turn chlorinated into **75** by means of phosphoryl chloride/*N*,*N*-dimethylacetamide [22,30]. The final step is the condensation of the dye-sulfonyl chloride (**75**) with the amine intermediate (**57**) prepared in Fig. 19.18 [22,30].

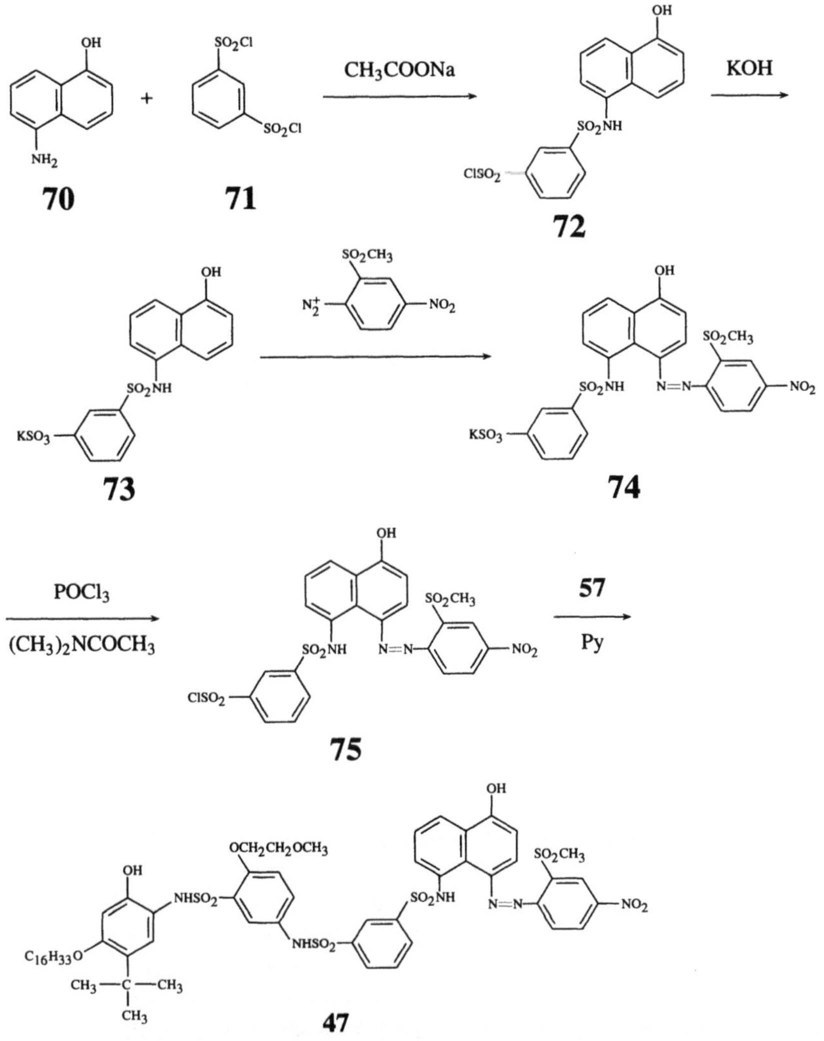

Figure 19.22. Synthesis of the cyan dye releaser (**47**) [22].

19.4.2 Magenta Dye Moieties

Dark Stability

Magenta dyes released from *o*-sulfonamidophenol dye releasers inevitably possess a sulfamoyl group which is a linkage between FUN and DYE. 4-Areneazo-1-naphthol dyes with various substituents on the arene and the naphthol nucleus have been extensively investigated from the viewpoint of dark fastness [38].

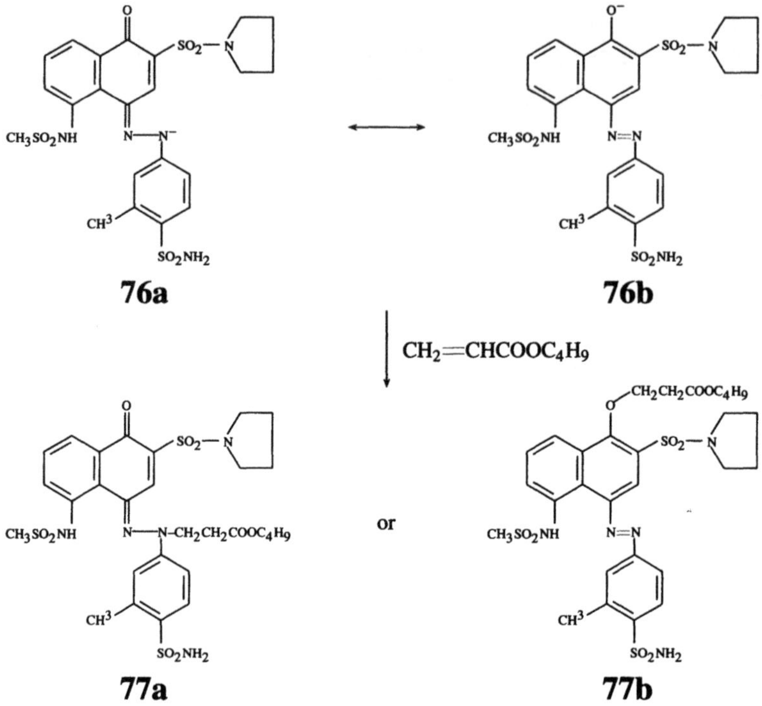

76a ⟷ **76b**

$CH_2{=}CHCOOC_4H_9$

77a or **77b**

Figure 19.23. The mechanism of fading of a magenta dye [38].

Because of the integrated format shown in Fig. 19.2, the inside of a picture ejected from a camera remains wet for several week after the processes of Fig. 19.3. Hence the fastness of dye images under wet conditions (dark stability) is crucially important. The dark stability of the magenta dyes depends on an *N*-substituent at the 2-sulfamoyl group on the naphthol nucleus. The introduction of steric hindrance (e.g., 2-$SO_2NHC(CH_3)_3$) has remarkably restrained the dark fading, as found in the magenta dye released from **48**. However, this level has been still insufficient to be employed in practical photographic films.

The fading of the magenta dyes has been found to be enhanced in the presence of such a vinyl monomer as butyl acrylate.[7] A product due to the Michael reaction shown in Fig. 19.23 has been isolated, although the exact structural formula (**77a** or **77b**) has not been specified on the basis of the resulting data [38].

According to the facts described above, the pseudo-first order rate constants ($\log k'$) of the reactions of the dyes with butyl acrylate can be regarded as indices for dark stability. The $\log k'$ vs. σ plot for various substituents has shown that more electron-donating substituents give lower $\log k'$ values. Table 19.6 shows

[7] Various polymers have been used in instant color films. Unreacted monomers contained in polymer layers can diffuse to other layers during the development processes.

Table 19.6. Rate Constants (k') of the Reaction of Dyes with Butyl Acrylate [38]

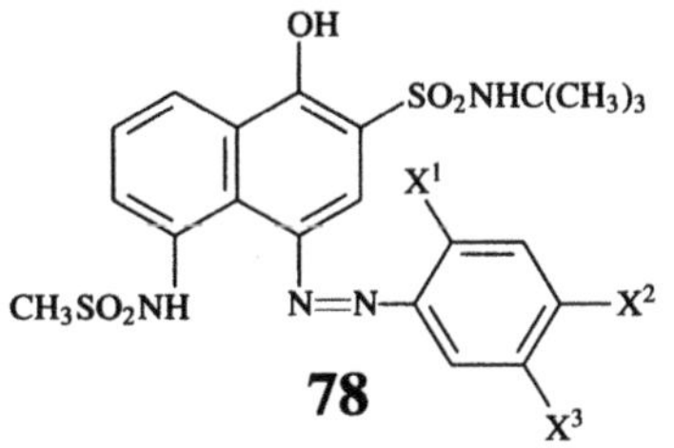

78

dye	X^1	X^2	X^3	k' (day^{-1})
78a	H	SO_2NH_2	H	0.098
78b	H	SO_2NH_2	CH_3	0.099
78c	CH_3	SO_2NH_2	H	0.036
78d	CH_3	SO_2NH_2	CH_3	0.025
78e	OCH_3	SO_2NH_2	OCH_3	0.032
78f	H	H	SO_2NH_2	0.072
78g	H	OCH_3	SO_2NH_2	0.048
78h	H	$OCH_2CH_2OCH_3$	SO_2NH_2	0.027
78i	H	CH_3	SO_2NH_2	0.048

the rate constants of several model released dyes having sulfamoyl groups and other substituents. In particular, **78d** and **78h** have lower $\log k'$ values than the model dye (**78a**) released from **48**.

Dye Releasers Having Improved Dye Moieties

According to the experimental data obtained from the model dyes, dye releasers having such a model dye moiety have been synthesized. In fact, the model dye (**78h**) has been incorporated in an *o*-sulfonamidophenol dye releaser (**79**), which provides a dye image of excellent dark stability [39,40]. Moreover, the model dye (**78d**) of excellent dark stability has also been used as the dye moiety of **80**, as disclosed in patents [41,42].

The modification of dye moieties (DYE) frequently influences the properties of dye releasers (FUN–DYE) such as dye-releasing efficiency. Hence, various attempts to modify dye moieties have been disclosed in patents. For example, 2-*N*-(5-methyl-1,3-dioxan-5-yl)sulfamoyl group is incorporated into the naphthalene nucleus of the 4-areneazo-1-naphthol, as exemplified by **81** [43]. A phenoxy group has been used as a modifier for the arene (benzene) nucleus of the 4-areneazo-1-naphthol, as found in **82** [44]. A morpholino group has been incorporated in the diazo component for preparing **83** [45]. The same morpholine group has been used in the dye releaser (**84**), where 3-sulfamoylbenzenesulfonamido group at the 5-position is a new matter [46]. The 2-sulfonamido-4-cyclohexyl-5-hexadecyloxyphenol moiety of the dye releaser

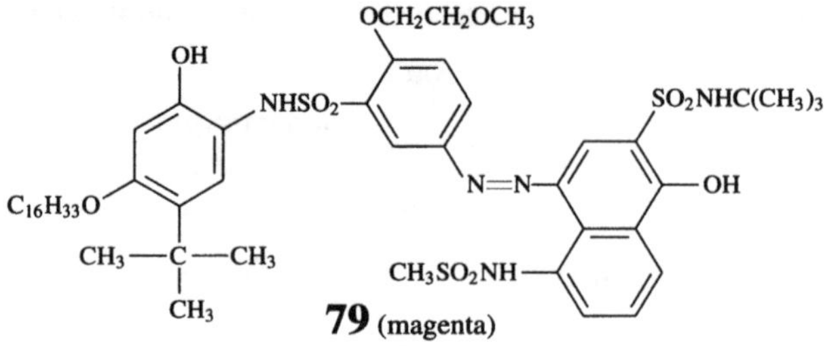

79 (magenta)

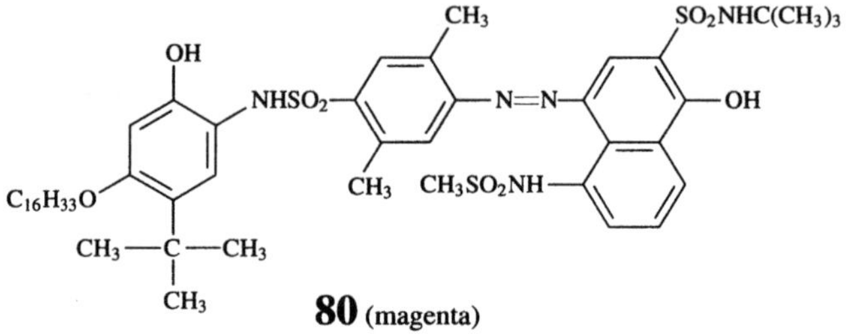

80 (magenta)

Figure 19.24. *o*-Sulfonamidophenol dye releasers releasing magenta dyes of excellent dark stability.

(**84**), which has a cyclohexyl group as a characteristic substituent, is covered by the patent described above [22].

Syntheses

As a typical synthetic pathway for preparing magenta dye releasers of the *o*-sulfonamidophenol type, Fig. 19.26 illustrates the synthesis of the dye releaser (**79**), which is able to release a magenta dye having an excellent dark stability [39,40]. 5-Amino-2-(2-methoxyethoxy)benzenesulfonic acid (**85**) as a diazo component can be prepared by the reduction of sodium 2-(2-methoxyethoxy)-5-nitrobenzenesulfonate (**55**) [47], which has once been used as an environment-changing linking group described in Fig. 19.18. The diazo component (**85**) is diazotized to the corresponding diazonium salt (**86**), which is coupled with 2-(*N*-*t*-butylsulfamoyl)-5-methanesulfonamido-1-naphthol as a coupling component. The resulting azo dye (**87**) is converted into the corresponding sulfonyl chloride (**88**) by means of phosphoryl chloride and *N*,*N*-dimethylacetamide. Finally, the sulfonyl chloride (**88**) is condensed with the *o*-aminophenol intermediate (**53**) to give the dye releaser (**79**).

A synthetic pathway similar to Fig. 19.26 can be applied to the synthesis of the dye releaser (**80**), which is also capable of releasing a magenta azo dye of excellent dark stability [41,42]. For this synthesis, the diazo component (**85**) is replaced by 4-amino-2,5-dimethylbenzenesulfonic acid.

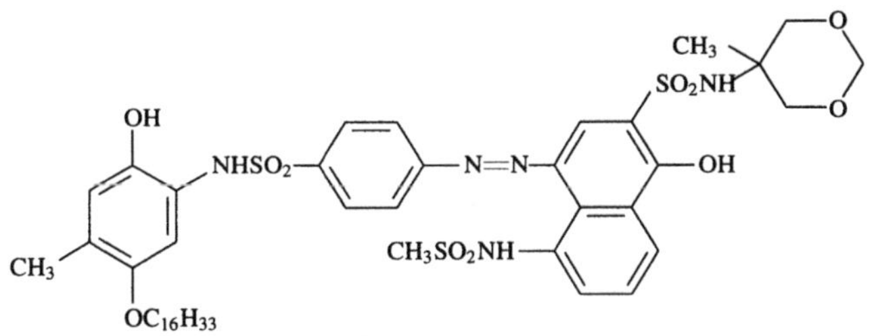

81 (magenta)

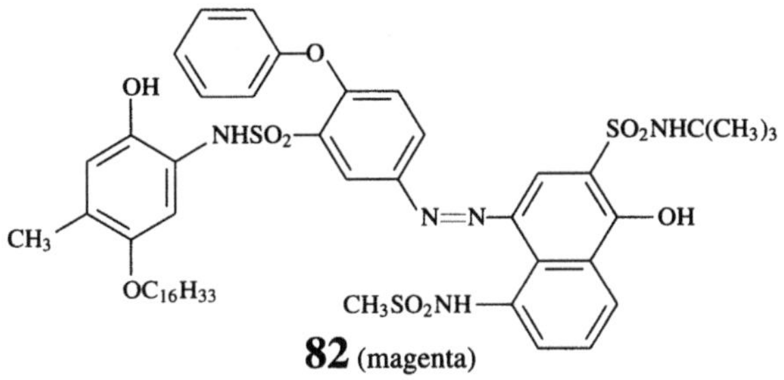

82 (magenta)

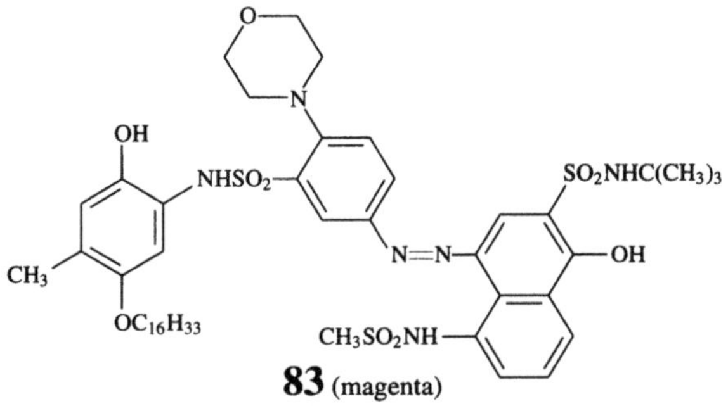

83 (magenta)

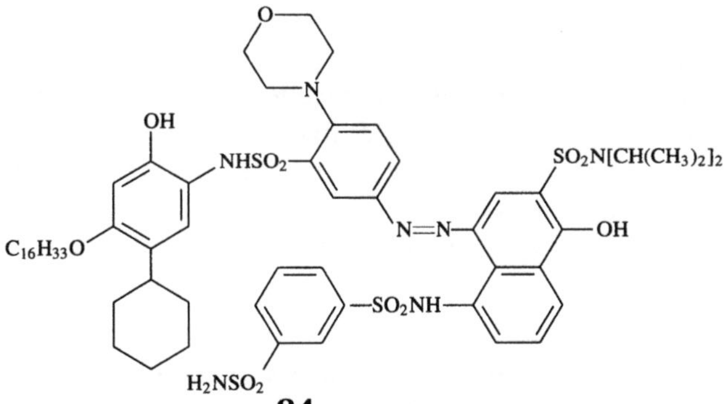

84 (magenta)

Figure 19.25. Variety of substituents for magenta dyes released from *o*-sulfonamidophenol dye releasers.

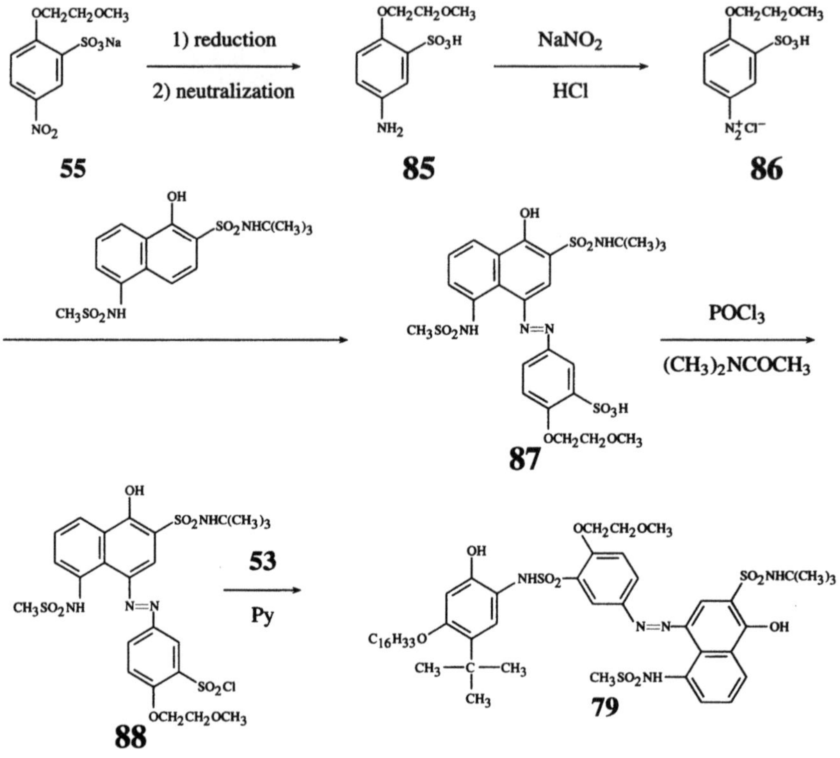

Figure 19.26. Synthesis of the magenta dye releaser (**79**) [39,40].

19.4.3 Yellow Dye Moieties

As exemplified in Fig. 19.16, a 2-methoxyethoxy group is effective to enhance the dye-releasing efficiency of yellow dye releasers (e.g., **49**) [22,23]. A 2-methoxyethoxy group is placed at the para position of the azo group of **89**, where the hue of the released dye is so excellent as to be unchanged over a wide pH range [48]. The same situation holds true in the anilinoazo dye (**90**) [49]. The enhancement of dye-releasing efficiency has been also observed as a result of a 2-methoxyethoxy group in **91** [50]. Further attempts for improving the properties of yellow dye moieties have appeared in several patents, e.g., **92** [51] and **93** [52].

Syntheses

Figure 19.28 illustrates a synthetic pathway for preparing yellow dye releasers of the *o*-sulfonamidophenol type. The target molecule is the dye releaser (**49**) shown in Fig. 19.16 [22,52].

The pyrazolone dye moiety has been prepared by a usual azo coupling of a coupling component (3-cyano-1-phenyl-5-pyrazolone, **94**) and a diazonium salt (**95**) derived from 3-amino-4-methoxybenzenesulfonic acid. The resulting

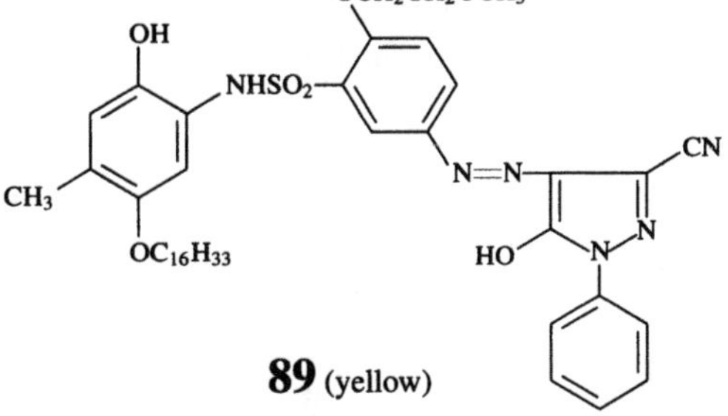

Figure 19.27. *o*-Sulfonamidophenol dye releasers releasing yellow dyes.

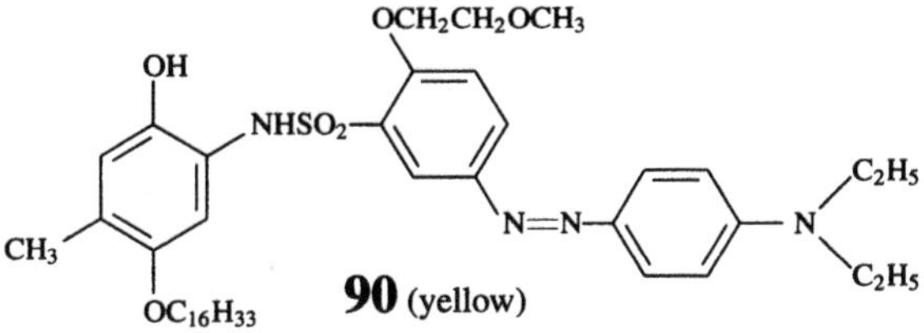

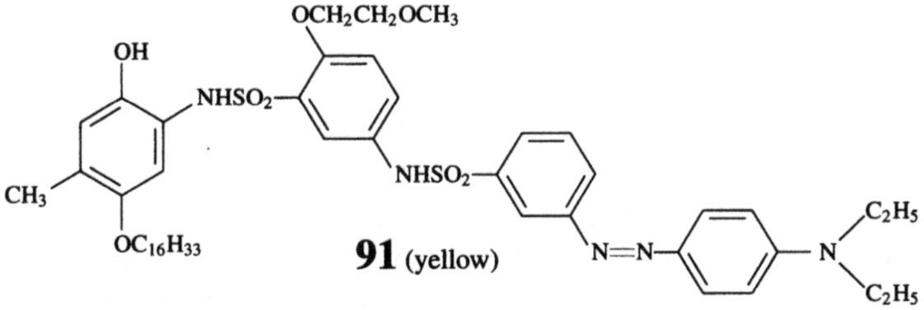

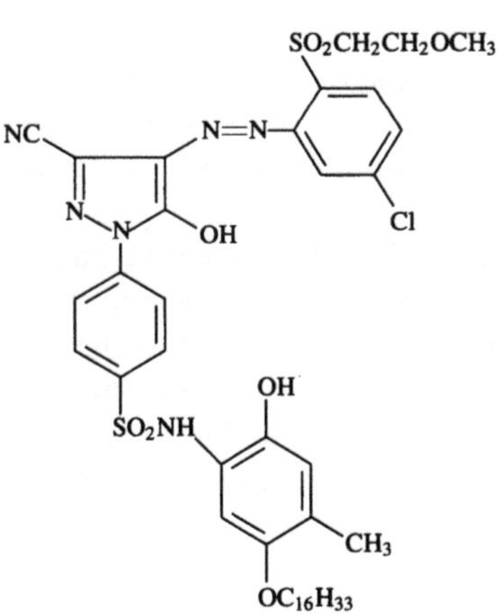

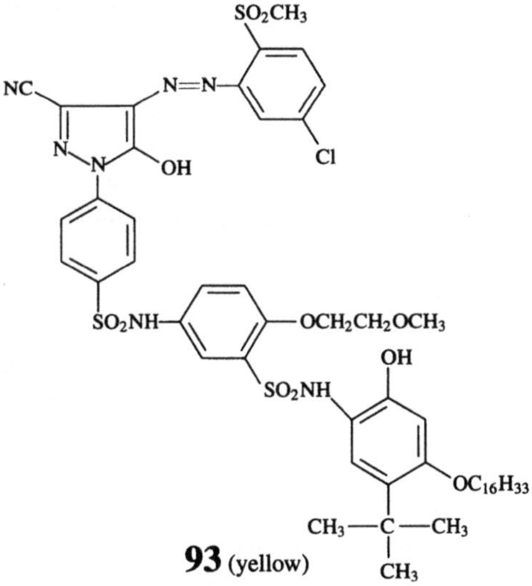

azo-sulfonic acid (**96**) is chlorinated by means of phosphoryl chloride/*N*,*N*-dimethylacetamide so as to give the corresponding dye-sulfonyl chloride (**97**). Then, the sulfonyl chloride (**97**) is condensed with the amine intermediate (**57**), which has been prepared according to the scheme shown in Fig. 19.18.

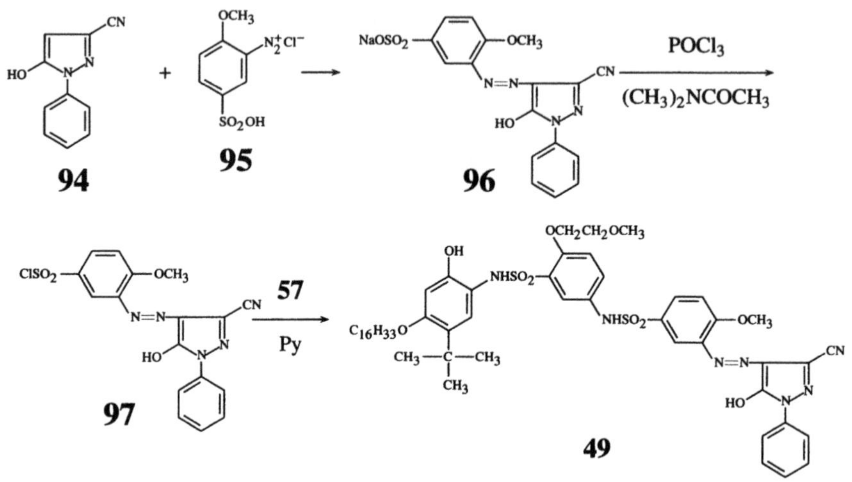

Figure 19.28. Synthetic pathway for preparing the yellow dye releaser (**49**).

Various reagents have been reported to convert dye-sulfonic acids into the corresponding dye-sulfonyl chlorides. However, they have disadvantages for practical use. For example, phosphorus pentachloride is difficult to handle because it is solid. The disadvantage of thionyl chloride/*N*,*N*-dimethylformamide is the evolution of sulfur dioxide gas, as mentioned above. Chlorosulfuric acid is corrosive and fuming in nature. Phosphoryl chloride/*N*,*N*-dimethylformamide has been successfully used to convert dye-sulfonic acids into the corresponding dye-sulfonyl chlorides. However, a more chemoselective reagent has been desirable to be employed in practical preparations.

A combination of phosphoryl chloride and *N*,*N*-dimethylacetamide has been found as a chemoselective reagent for converting dye-sulfonic acids into the corresponding dye-sulfonyl chlorides, as shown in Fig. 19.29 [53].[8] Upon the action of phosphoryl chloride/*N*,*N*-dimethylacetamide, the dye-sulfonic acid (**98**) is converted into the dye-sulfonyl chloride (**99**) with no reaction of the carbamoyl group. On the other hand, phosphoryl chloride/*N*,*N*-dimethylformamide causes the con-

---

[8] Note that this combination has been used in the Beckmann rearrangement for preparing benzoxazole derivatives, as described in Table 19.2. The chlorination described here may be also explained by considering an intermediate akin to a Vilsmeier complex (i.e., $(CH_3)_2N^+{=}C(CH_3)Cl$), which is formed from phosphoryl chloride and *N*,*N*-dimethylacetamide. This intermediate attacks the sulfonic acid group to form an intermediate $(CH_3)_2N{-}C(CH_3)Cl{-}O{-}SO_2Ar$. Then, this intermediate is converted into $(CH_3)_2N{-}C(CH_3){=}O + Cl{-}SO_2Ar$.

version of the carbamoyl group into a cyano group along with the chlorination of the sulfonic acid group, where **97** is obtained.

Table 19.7 shows the preparation of pyrazolone azo dyes having a chlorosulfonyl group by means of the combination of phosphoryl chloride/*N*,*N*-dimethylacetamide [53]. This combination has been widely used as a chlorinating agent for preparing azo-sulfonyl chlorides, as already described in the conversions of **74** into **75** and of **87** into **88**.

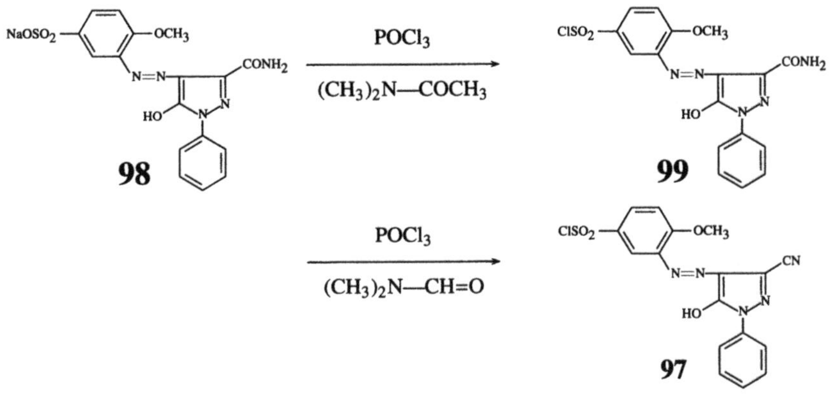

$NaOSO_2$ OCH$_3$ N=N CONH$_2$ HO N N **98** — $POCl_3$ / $(CH_3)_2N—COCH_3$ → $ClSO_2$ OCH$_3$ N=N CONH$_2$ HO N N **99**

— $POCl_3$ / $(CH_3)_2N—CH=O$ → $ClSO_2$ OCH$_3$ N=N CN HO N N **97**

Figure 19.29. Chemoselectivity of phosphoryl chloride/*N*,*N*-dimethylacetamide as compared with phosphoryl chloride/*N*,*N*-dimethylformamide [53].

Table 19.7. Dye-Sulfonyl Chlorides Derived from Dye-Sulfonic Acids by Means of Phosphoryl Chloride/*N*,*N*-Dimethylacetamide [53]

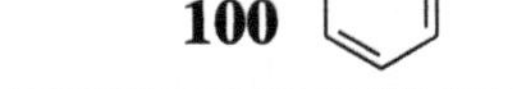

X^3 X^4 X^2 X^1 N=N HO N N **100**

Dye	X^1	X^2	X^3	X^4	yield (%)
100a (= **99**)	$CONH_2$	OCH_3	H	SO_2Cl	93
100b	$CONHCH_3$	OCH_3	SO_2Cl	H	94
100c	$CONHCH_3$	OCH_3	H	SO_2Cl	90
100d	$CONHCH_3$	CH_3	SO_2Cl	H	70
100e	$CONHCH_3$	H	SO_2Cl	CH_3	95
100f	CN	H	$OCH_2CH_2OCH_3$	SO_2Cl	95
100g (= **97**)	CN	OCH_3	H	SO_2Cl	81

19.5 Chemicals for Controlling Dye Transfer Processes

Chemicals for controlling dye transfer processes can be fed from photo-sensitive and related layers, from layers coated on a cover sheet, or from an activator fluid.

19.5.1 Chemicals Contained in Photo-Sensitive Layers

Nucleation Agents and Nucleation Accelerators

Since *o*-sulfonamidophenol dye releasers are negative-working, they are combined with direct positive emulsions, as discussed in Chapter 16. Such direct positive emulsions contain nucleating agents such as **101** [54] and **102** [55]. Another nucleating agent in which an sulfonamido group is involved in place of the amido group of **102** has also been disclosed in one of the series of patents [56].

The nucleating agent (**102**) has been widely used even in recent embodiments, where a hydroxylamine derivative such as **103** or a hydrazine derivative such as **104**[9] has been added as a nucleation accelerator in each silver emulsion layer [57]. The nucleation accelerator is presumed to participate in a redox reaction producing a surface latent image, because of its reducing activity. Thereby, the combination of the nucleating agent with the nucleation accelerator reduces image emergence time.

Development Accelerators

To accelerate the redox reaction between silver halide and an ETA (eq. 19.1 of Fig. 19.1), an additional (auxiliary) developer has been used as a development accelerator. For example, earlier embodiments which have been described as experimental examples based on *o*-sulfonamidophenol dye releasers have employed sodium dodecylhydroquinonesulfonate [49,50] and sodium pentadecylhydroquinonesulfonate (**105**) [46]. More recent embodiments have employed a combination of potassium 2-*sec*-octadecylhydroquinone-5-sulfonate (**106**) and a 2-aryl-4-isoxazolin-3-one derivative (**107**) or a combination of the three development accelerators (**105**, **106**, and **107**) [57].

Dye-Releasing Accelerators

The formation of **42** (Fig. 19.13) strongly suggests that arenesulfonamide anions can be used as nucleophiles in place of hydroxide ions. In fact, the nucleophilic displacement mechanism has been incorporated into dye releasers shown

[9]The mercaptotriazole moiety of **104** may be used as a adsorption site. It should be noted that 5-mercapto-1-phenyl-1H-tetrazole itself works as a development inhibitor.

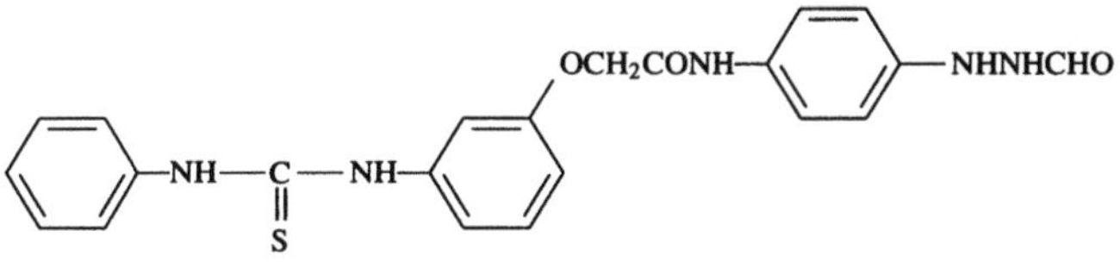

101
(Nucleating agent)

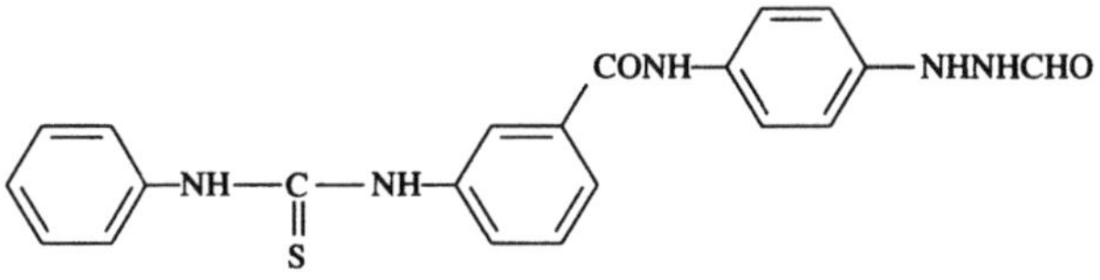

102
(Nucleating agent)

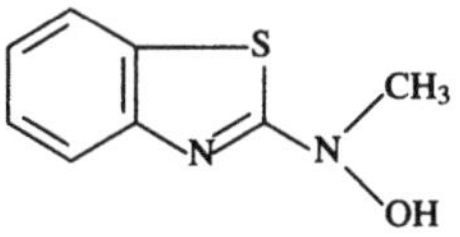

103
(Nucleation accelerator)

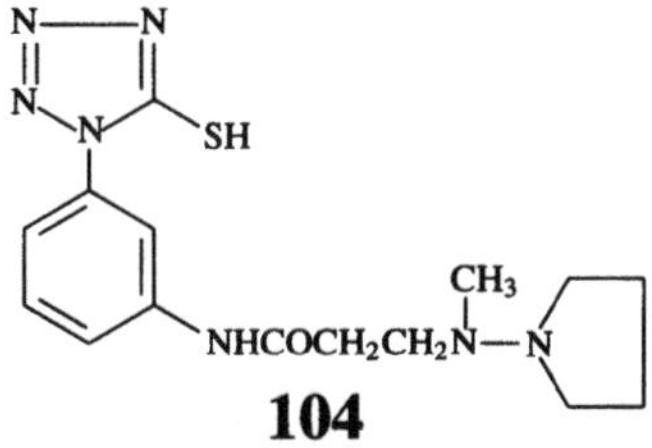

104
(Nucleation accelerator)

Figure 19.30. Nucleating agents and nucleation accelerators. These compounds are incorporated in silver halide emulsion layers.

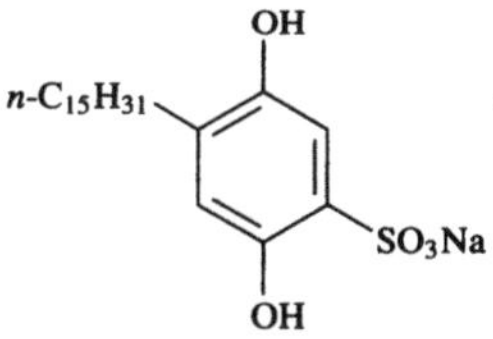

105 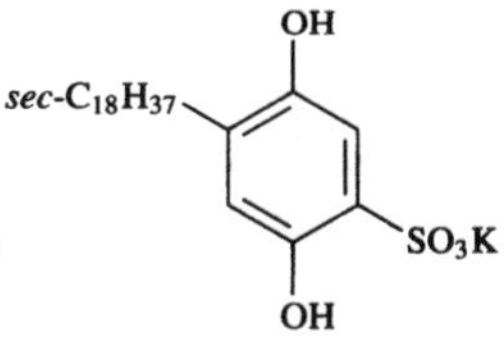 **106** 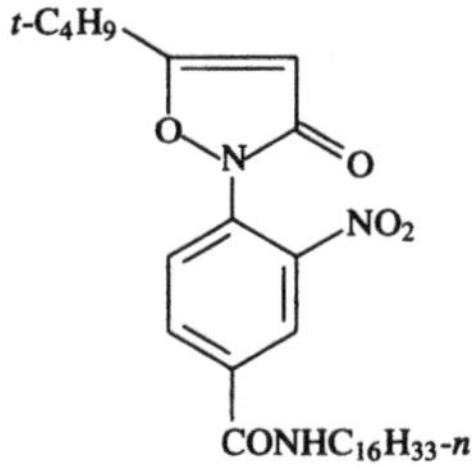 **107**

Figure 19.31. Development accelerator in combination with electron transfer agents. These compounds are incorporated in silver halide emulsion layers.

in Fig. 19.32 [58]. Thus, the *p*-sulfonamidonaphthol dye releaser (**108**) undergoes imagewise oxidation to give the corresponding quinone monoimide (**109**), which in turn releases a dye-sulfonamide (DYE–SO_2NH_2) in terms of intramolecular nucleophilic displacement giving a cyclized product (**110**). The intramolecular nucleophilic displacement works similarly in *o*-sulfonamidophenol dye releasers such as **111** [58].

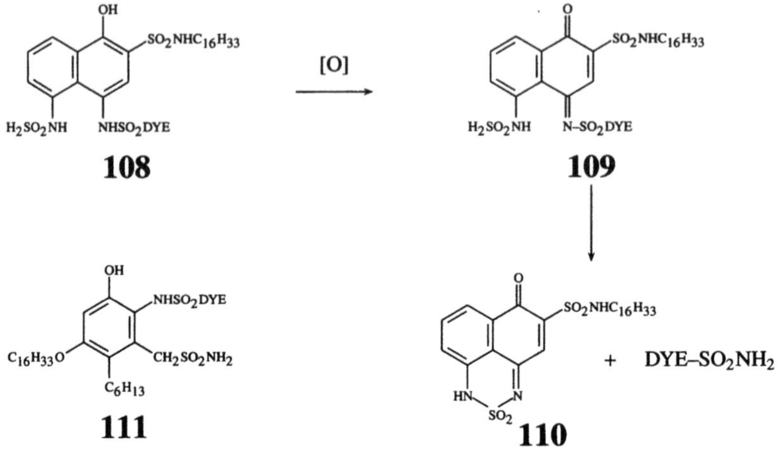

Figure 19.32. Dye releasers based on intramolecular nucleophilic displacement [58].

The nucleophilic displacement of such an arenesulfonamide anion can be applied to accelerate a dye-releasing process shown in eq. 19.3 if the anion fed from an activator fluid attacks the C=N double bond instead of the hydroxide ion. 4-Chlorobenzenesulfonamide has been added to a dye releaser layer to improve the dye-releasing efficiency of a thermal development system [59].

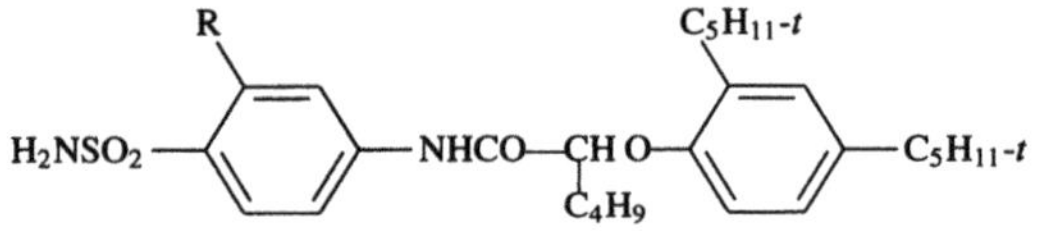

112 R = H
113 R = CH_3

Figure 19.33. Dye-releasing accelerators [60,61]. These compounds are incorporated in dye releaser layers.

A more elaborate embodiment has been employed after arenesulfonamides are ballasted as exemplified by **112** and **113** [60,61]. A merit of using such ballasted dye-releasing accelerators is that they improve the temperature dependence of dye-releasing efficiency so as to maintain D_{max} values at sufficient levels even at low temperatures. For example, the D_{max} values at 5°C vs. 25°C in the presence of **112** have been reported as follows: 1.96 vs. 2.04 for blue (yellow dye), 2.37 vs. 2.45 for green (magenta dye), and 2.44 vs. 2.52 for red (cyan dye) [60]. The dye-releasing accelerator (**112**) has been used in combination with a tertiary amine (e.g., $(C_8H_{17})_2N$-$(CH_2)_{10}$-CONH-C_6H_4-OCH_3-*p*) [62,63].

The merit of such ballasted dye-releasing accelerators is most remarkable when they are incorporated in a yellow dye-releaser layer, since this layer is farthest from the mordant layer, as shown in Fig. 19.2. Thus, a combination of **112** and **113** has been added to a yellow dye-releaser layer in a recent embodiment [57]. The acceleration mechanism of this type has been employed in a recent instant color film marketed under the name "Instax" by Fuji Photo Film [2,64].

Compounds for Preventing Color Contamination

The ETA_{ox} (eq. 19.2) diffuses within and beyond the corresponding photosensitive unit, as shown in Fig. 19.3d. The ETA_{ox} which reaches the adjacent interlayer(s) should be quenched to prevent color contamination, as shown in Layer 7 of Fig. 19.3f&g. To quench the excess ETA_{ox}, the interlayer contains a ballasted hydroquinone such as **114** (HQ–B in Fig. 19.3f&g), which is cited in test examples selected at random from several patents (e.g., [58,62]). More recent embodiments (e.g., those described in [57,65]) have employed a hydroquinone (**115**) having lower *t*-alkyls as ballast groups.

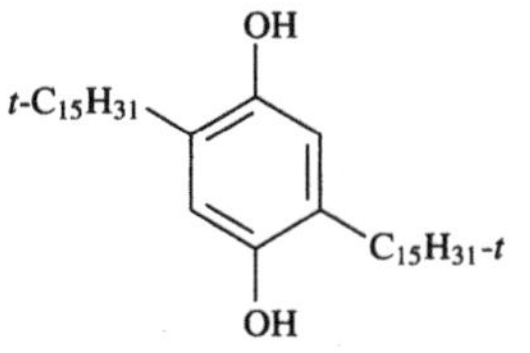

OH, *t*-$C_{15}H_{31}$, $C_{15}H_{31}$-*t*, OH — **114**

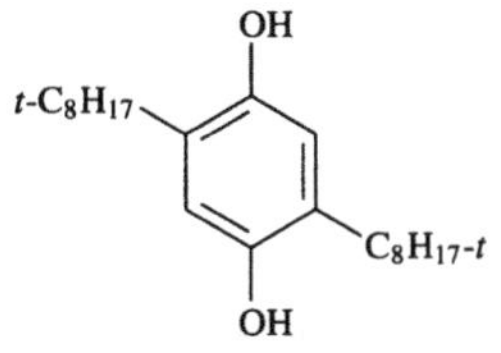

OH, *t*-C_8H_{17}, C_8H_{17}-*t*, OH — **115**

Figure 19.34. Ballasted hydroquinones for preventing color contamination.

Development Inhibitor Releasers

To terminate the development of silver halide, a development inhibitor (DI) is released timely from a development inhibitor releaser (DIR), as shown in Fig. 19.3e (DIR → DI). In usual embodiments, the DIR as a timing precursor (e.g., **116**) is incorporated in a timing layer coated on a cover sheet (e.g., Layer 2 of Fig. 19.2).

Various development inhibitor releasers have been disclosed as timing precursors. For example, the compound (**116**), which is one of such timing precursors used in a recent embodiment [57], releases two moles of 5-mercapto-1-phenyl-1H-tetrazole as the result of a retro-Michael reaction producing a vinylsulfonyl compound.

On the other hand, the development inhibitor releaser (**117**) undergoes the imagewise release of 5-mercapto-1-phenyl-1H-tetrazole, which works as a development inhibitor to form an excellent image with less fog in white background

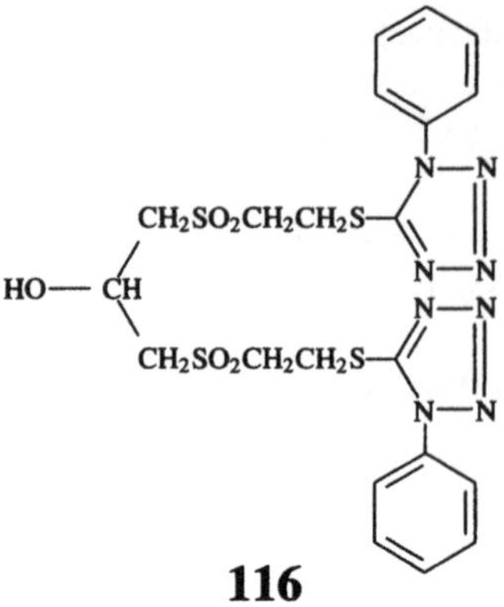

Figure 19.35. Development inhibitor releasers of timing and imagewise types. The timing precursor (**116**) is incorporated in a timing layer coated on a cover sheet (e.g., Layer 2 of Fig. 19.2). The imagewise precursor (**117**) is incorporated in each emulsion layer (e.g., Layer 5, 8, and 11 of Fig. 19.2).

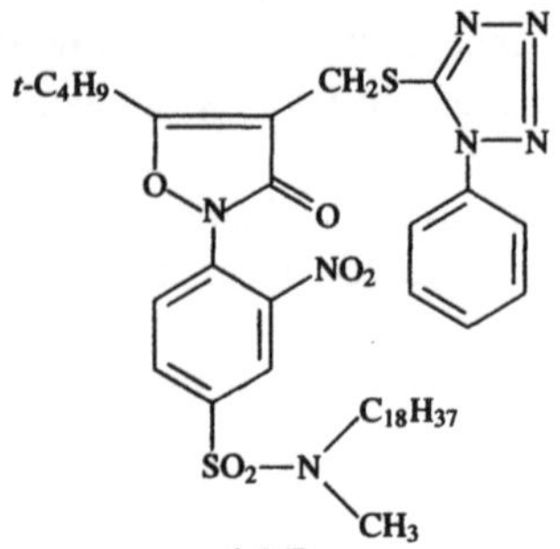

area [66,67]. The mechanism of the imagewise release already has been discussed in Chapter 16 (as dye releasers named ROSET compounds).[10] Another precursor in which a ballast group $CONHC_{16}H_{33}$-*n* is substituted for the sulfamoyl ballast group of **117** has been employed in more recent embodiments [57,65].

19.5.2 Chemicals Contained in an Activator Fluid

Figure 19.36 lists various chemicals contained in an activator fluid, which is used for such an integrated (mono-sheet) film as shown in Fig. 19.2 [57].[11]

Carbon Black and Viscosity-Imparting Reagents

The activator fluid shown in Fig. 19.2 contains carbon black, which gives a black opaque layer as found in the diagrams of Fig. 19.3 (e and i). Sodium carboxymethylcellulose (**119**) is added to impart viscosity, which ensures a thin and square spread of the activator fluid.

Electron Transfer Agents

The electron transfer agents (ETAs) described for processing *p*-sulfonamidonaphthol dye releasers (Chapters 16 and 18) can be used for processing

[10] A photographic material based on ROSET dye releasers will be discussed in Chapter 20.

[11] This patent has disclosed a more elaborate film format than Fig. 19.2, as will be described later (Fig. 19.40). However, the composition of this activator fluid can be regarded as a typical example to be explained.

1) Inorganic composents:

$AgNO_3$ (0.10), Carbon black (160), K_2SO_3 (1.90), $Al(NO_3)_3$ (0.60), $Zn(NO_3)_2$ (0.60), KOH (56.0),

2) Organic components:

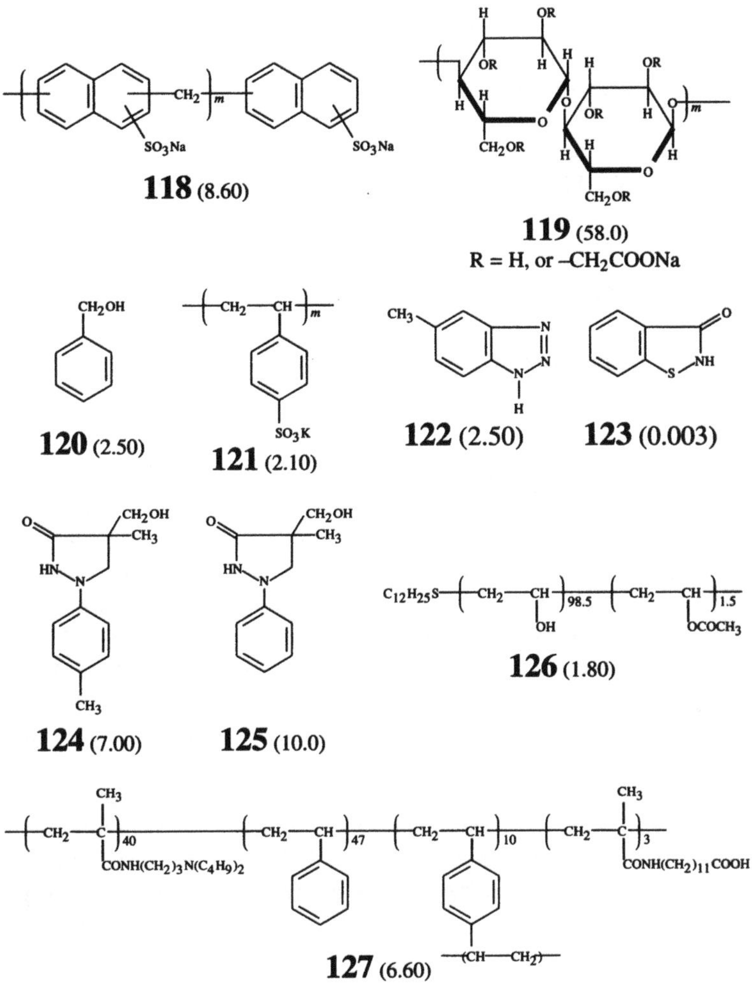

118 (8.60)

119 (58.0)
R = H, or $-CH_2COONa$

120 (2.50) **121** (2.10) **122** (2.50) **123** (0.003)

124 (7.00) **125** (10.0) **126** (1.80)

127 (6.60)

Figure 19.36. Chemicals contained in an activator fluid. The value of each pair of parentheses represents the weight (g) of a component at issue [57].

o-sulfonamidophenol dye releasers. For example, the activator fluid shown in Fig. 19.36 contains 3-pyrazolidone derivatives (**124** and **125**) as ETAs. The redox reaction between silver halide and the ETA(s) (eq. 19.1 of Fig. 19.1) corresponds to the diagram shown in Fig. 19.3d. The redox reaction between the ETA_{ox} and the dye releaser (eq. 19.2 of Fig. 19.1) is illustrated in Fig. 19.3f&g.

Other Additives

A specific tertiary amine polymer (e.g., **127**) has been added in an activator fluid and used together with a dye-releasing accelerator of the arenesulfonamide type (e.g., **112** and/or **113**) which is in turn incorporated in a dye releaser layer. This combined usage reduces the change of D_{max} values upon exposure to heat and moisture after processing [62]. Further, a tertiary amine latex polymer such as **127** has been used in combination with an oligomer or polymer having surface activity [68].

Several heterocyclic compounds are contained in the activator fluid of Fig. 19.36. For example, **123** is an antiseptic and **122** is a stabilizer (antifoggant).

19.6 Polymers

19.6.1 Polymers for the Mordant Layer

In a completed instant color film, a transferred image dye stays in a mordant layer (Layer 1 of the photo-sensitive sheet in Fig. 19.2 and Fig. 19.3h) by means of ionic, dipole-dipole, and/or dispersion-force interaction between the dye anion and the cationic (or heterocyclic) mordant polymer. A polymer latex comprising **128** has been disclosed to be suitable as a mordant for capturing dyes released from dye releasers of the type shown in Fig. 19.6 [69]. Copolymers comprising an imidazole repeating unit (e.g., **129**) have been proposed as a mordant for capturing dyes released from dye releasers of the type shown in Fig. 19.16 [70]. A copolymer (**130**) covered by this patent has been employed in a film of peel-apart format [57].

More complicated copolymers such as **131** have shown excellent mordancy in embodiments employing dye releasers of the type shown in Fig. 19.16 [71]. The mordant polymer (**131**), which contains a quaternary piperidium cation, has been used still in recent embodiments [57,72]. Mordant polymers involving a piperidine moiety in a different manner (e.g., **132**) have also been disclosed [73]. Copolymers containing an imidazolium cation (e.g., **133**) have been proposed as mordant polymers [74].

19.6.2 Polymers for the Acid Layer

In the final step of processing (Fig. 19.3i), an acid polymer involved in the acid layer (Layer 1 of the cover sheet) neutralizes the alkaline media fed from the activator fluid. For this purpose, copolymers comprising acrylic acid and butyl acrylate as repeating monomer units have long been used. Thus, the embodiments of patents published in 1981 [75], in 1982 [22], in 1989 [76], and in 1995 [60] have employed the copolymer represented by **134**, although these patents are selected randomly. A similar copolymer represented by **135** has been used in a more recent embodiment disclosed in 2002 [57].

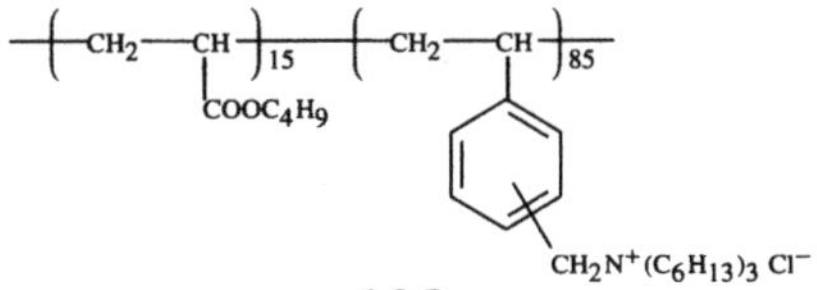

128 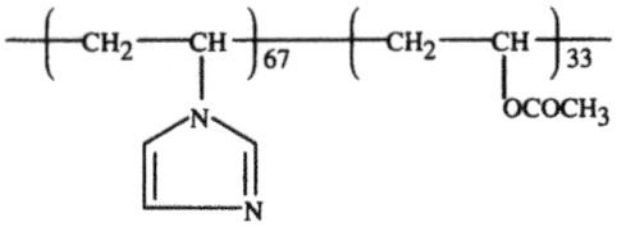 **129**

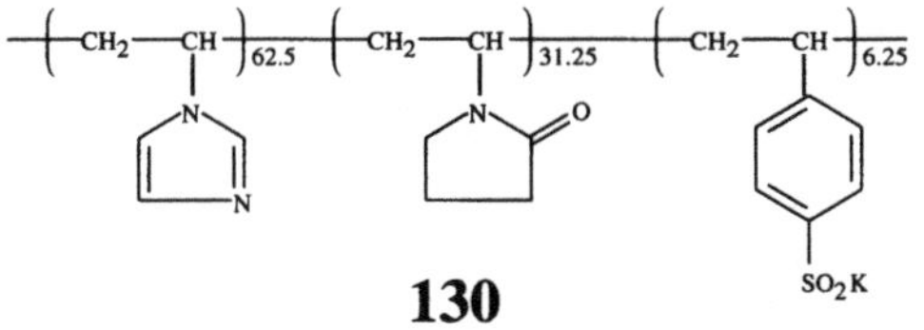

130

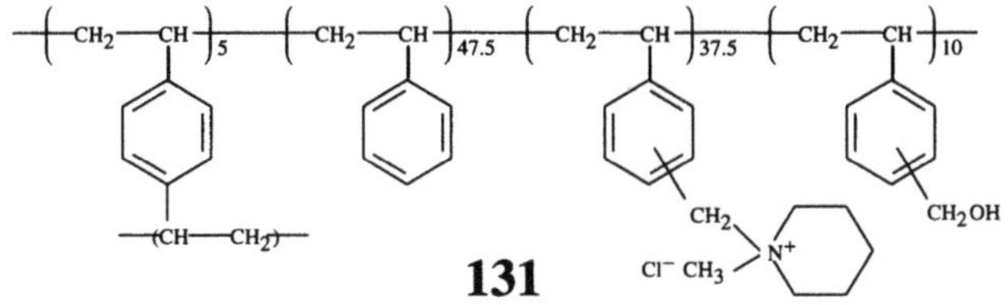

131

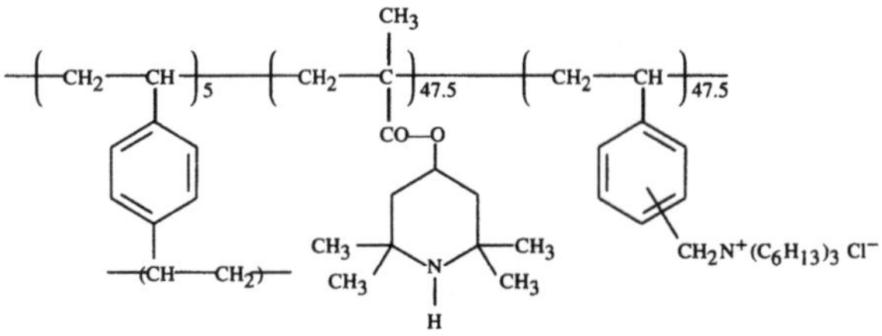

132

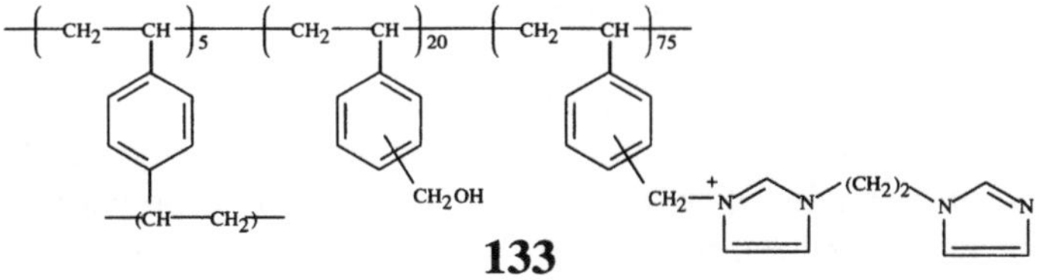

133

Figure 19.37. Mordant polymers used in combination with *o*-sulfonamidophenol dye releasers.

An acid polymer is coated together with a polyhydric alcohol (e.g., glycerol) or cellulose acetate of acetylation degree 45% in an acid layer. Then a hardener such as **136** is added to cause cross-linkage between the carboxylic acid group of the former and the hydroxyl group of the latter [77]. The addition of the hardener (**136**) to the acid layer has been still employed in a recent embodiment [57].

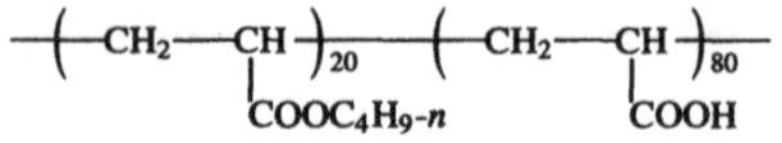

$$-(CH_2-CH(COOC_4H_9\text{-}n))_{20}-(CH_2-CH(COOH))_{80}-$$

134

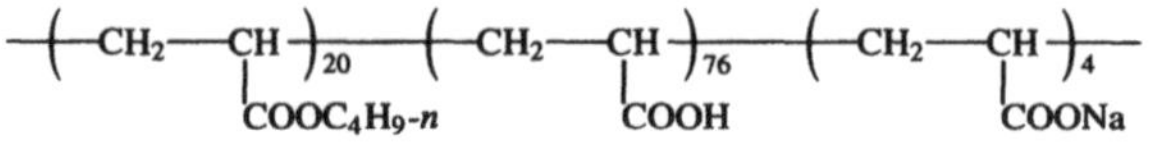

$$-(CH_2-CH(COOC_4H_9\text{-}n))_{20}-(CH_2-CH(COOH))_{76}-(CH_2-CH(COONa))_{4}-$$

135

$$\text{(epoxy)}-CH_2OCH_2CH_2CH_2CH_2OCH_2-\text{(epoxy)}$$

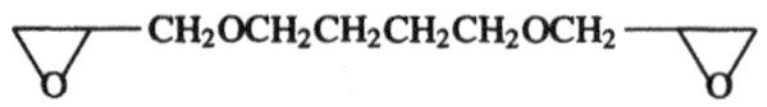

136 (hardener)

Figure 19.38. Acid polymers and a hardener.

19.6.3 Polymers for the Timing Layer

To retard the neutralization by the acid layer (Fig. 19.3i), or in other words, in order to discriminate the process of Fig. 19.3h from that of Fig. 19.3i, one or two timing layers are coated on a cover sheet (Layer 2) of Fig. 19.2 or of Fig. 19.3i. As already discussed in Chapter 17, timing polymers used in the timing layer are classified into sieve type and hold-release type. Most embodiments based on *o*-sulfonamidophenol dye releasers have employed two timing layers, where the first timing layer (adjacent to the acid layer) sometimes called "alkali barrier layer" seems to comprise timing polymers of sieve-type (e.g., cellulose acetate) together with a small amount of timing polymers of hold-release type, while the second one (coated over the first timing layer) sometimes called "temperature compensating layer" seems to comprise timing polymers of hold-release type.

An earlier patent has disclosed that the first timing layer comprises a mixture of cellulose acetate (52.1% acetyl) and poly[styrene-co(maleic anhydride)] and the second timing layer a latex copolymer of styrene/butyl acrylate/acrylic acid/*N*-hydroxymethylacrylamide (50:40:3:7) [75]. The copolymer of the second timing layer has been replaced by a 1:1 mixture of a latex copolymer of styrene/butyl acrylate/acrylic acid (52.8:43.2:4) and a latex copolymer of methyl methacrylate/acrylic acid (96:4) in another earlier patent [78]. A mixture of copolymers **137** and **138** which comprise vinylidene dichloride as a repeating monomer unit has been used in the second timing layer [78]. These earlier embodiments have been concerned with films based on *p*-sulfonamidonaphthol dye releasers. The copolymer **137** has been later used together with a latex polymer of methyl methacrylate in the second timing layer of an embodiment based on *o*-sulfonamidophenol dye releasers [22].

In a cover sheet described in a patent [76], the first timing layer has comprised a mixture of cellulose acetate (51% acetyl) and a copolymer of vinyl ether/maleic monomethyl ester (95:5) and the second timing layer has comprised a mixture of copolymers **139** and **140**, where an additional (auxiliary) acid layer has been incorporated between the the two timing layers. A similar auxiliary acid layer has also been disclosed [79].

As a more recent example, a mixture of latex copolymers represented by **141** and **142** has been used in the second timing layer, where *N*-(2-hydroxyethyl)acrylamide is employed as a monomer unit [80]. This mixture has also been used in a further recent embodiment [57].

$$-(CH_2-CCl_2)_{85}-(CH_2-CH(COOCH_3))_{12}-(CH_2-CH(COOH))_3-$$

137

$$-(CH_2-CCl_2)_{84}-(CH_2-CH(CN))_{10}-(CH_2-C(CH_3)(COOCH_3))_3-(CH_2-CH(COOH))_3-$$

138

$$-(CH_2-CH(C_6H_5))_{49.7}-(CH_2-CH(COOC_4H_9\text{-}n))_{42.3}-(CH_2-CH(CONHCH_2OH))_5-$$

139

$$-(CH_2-C(CH_3)(COOCH_3))_{93}-(CH_2-CH(COOH))_4-(CH_2-CH(CONHCH_2OH))_3-$$

140

$$-(CH_2-C(CH_3)(COOC_4H_9\text{-}n))_{70}-(CH_2-CH(COOH))_{10}-(CH_2-CH(CONHCH_2CH_2OH))_{20}-$$

141

$$-(CH_2-C(CH_3)(COOC_2H_5))_{70}-(CH_2-CH(COOH))_{10}-(CH_2-CH(CONHCH_2CH_2OH))_{20}-$$

142

Figure 19.39. Timing polymers of various types.

19.7 Recent Innovations

19.7.1 Integrated Instant Color Films

Sophisticated Multilayer Structures

Although the standard multilayer format of a mono-sheet instant color film has been shown in Fig. 19.2, several auxiliary layers are added to the multilayer structure from a practical point of view. Moreover, one layer having two or more functions is sometimes separated into layers of a respective function.

For example, Fig. 19.40 shows the schematic cross-section of a recent format disclosed in several patents [57,65,80], where one emulsion layer of each light-sensitive unit (Fig. 19.2) is duplicated into two emulsion layers (high- and low-sensitive layers). In addition, Fig. 19.40 comprises duplicated interlayers (a color mixing prevention layer and an interlayer), duplicated protective layers (a protective layer and a UV absorption layer), and duplicated timing layers (an alkali barrier layer and a temperature compensating layer).

The format shown in Fig. 19.40 seems to be akin to the format of the Instax system marketed by Fuji Photo Film, since the formal report on the system has described that a direct positive emulsion of hexagonal tabular internal latent image type is used in each high-sensitive emulsion layer, while a direct positive emulsion of octahedral internal latent image type is used in each low-sensitive emulsion layer [64]. This means that each emulsion of Fig. 19.2 is divided into a high-sensitive layer and a low-sensitive one, as found in Fig. 19.40. Moreover, such emulsions as comprising hexagonal tabular and octagonal crystals have been disclosed in the patent in combination with the multilayer structure shown in Fig. 19.40 [65].

The report [64] has also described that new timing polymers were developed. This description seems to correspond to the patent [80], which has claimed **141** and **142** together with an embodiment based on the multilayer film shown in Fig. 19.40.

The other technical report on the Instax system [2] has described sulfonamides as dye-releasing accelerators. The sulfonamides **112** and **113**, which have been claimed by the patents [60,61], have been contained in the format shown in Fig. 19.40 [57,65,80]. These dye-releasing accelerators are incorporated in Layer 17 (yellow dye releaser layer) of Fig. 19.40.

Improved Sets of Dye Releasers

An improved set of cyan, magenta, and yellow *o*-sulfonamidophenol dye releasers, is shown in Fig. 19.41, where it is taken from a patent claiming dye-releasing accelerators [61] and from a patent claiming timing polymers [80]. The cyan shown in Fig. 19.41 is the same as **47**, which is covered by the patent cited above [22,30]. The magenta dye releaser **143** comprises a cyclohexyl-substituted

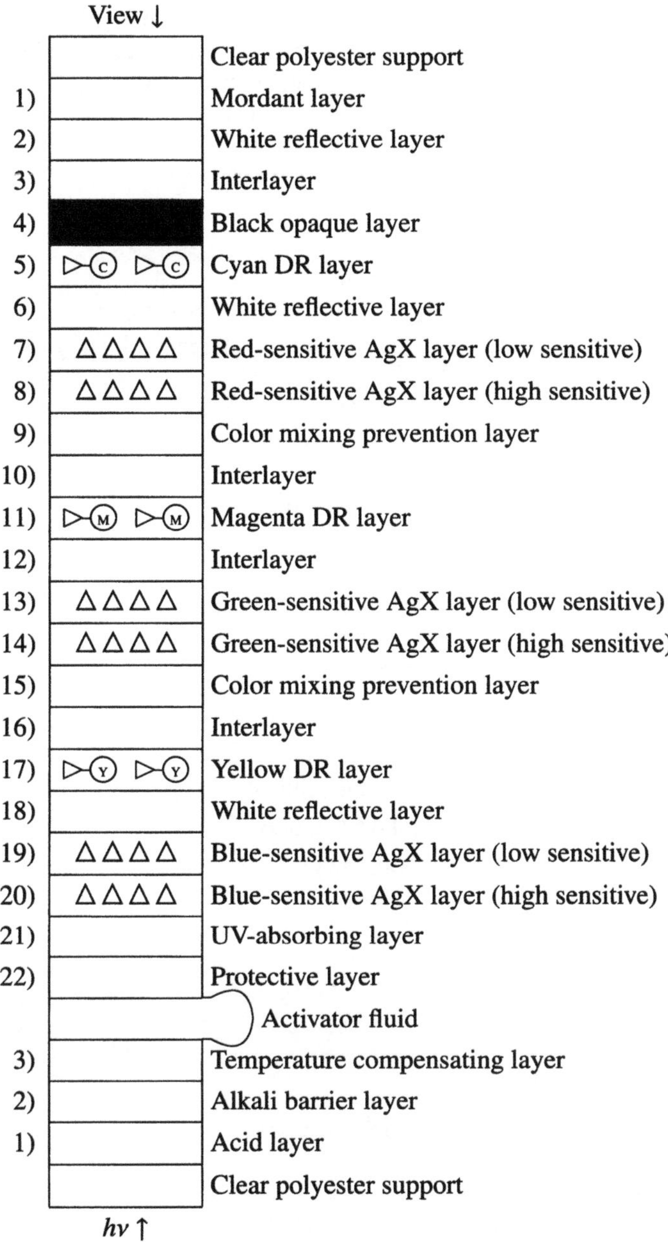

Figure 19.40. Schematic cross-section of a sophisticated multilayer structure of mono-sheet type for instant color photography based on *o*-sulfonamidophenol dye releasers and direct positive emulsions. Δ: Silver halide grain, ▷-Ⓨ: Yellow dye releaser, ▷-Ⓜ: Magenta dye releaser, and ▷-Ⓒ: Cyan dye releaser.

o-sulfonamidophenol (FUN moiety) covered by the patent [22] and an azo dye (DYE) covered by the patent [45]. The FUN moiety substituted by a *t*-butyl group in the yellow dye releaser **144** is covered by the patent [22,52], while the methoxyethoxy linkage is covered by the patent [23].

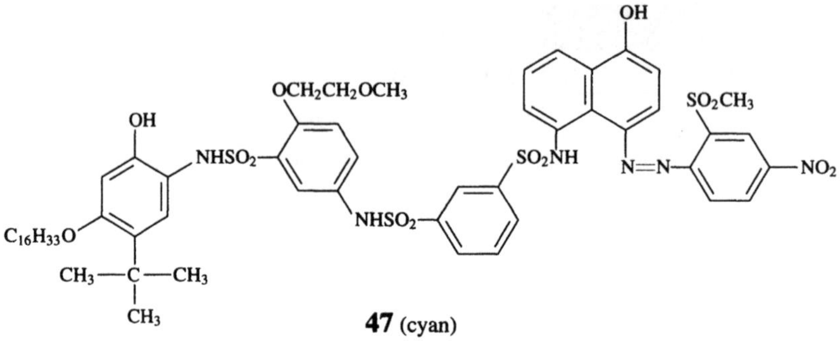

47 (cyan)

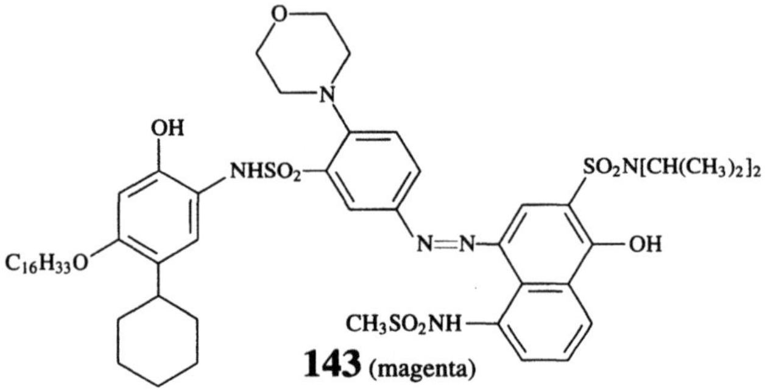

143 (magenta)

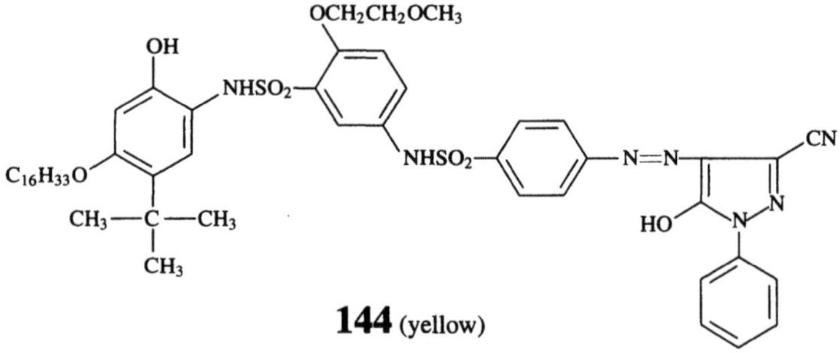

144 (yellow)

Figure 19.41. An improved set of cyan, magenta, and yellow *o*-sulfonamidophenol dye releasers for color reproduction of instant color photography. This set is cited from a patent claiming dye-releasing accelerators [61] and from a patent claiming timing polymers [80].

Another improved set of cyan, magenta, and yellow *o*-sulfonamidophenol dye releasers, is shown in Fig. 19.42, where it is taken from a patent claiming hydrazine additives [57] and from a patent claiming direct-positive emulsions [65]. Two cyan dye releasers have been used as shown in Fig. 19.42, where the one is

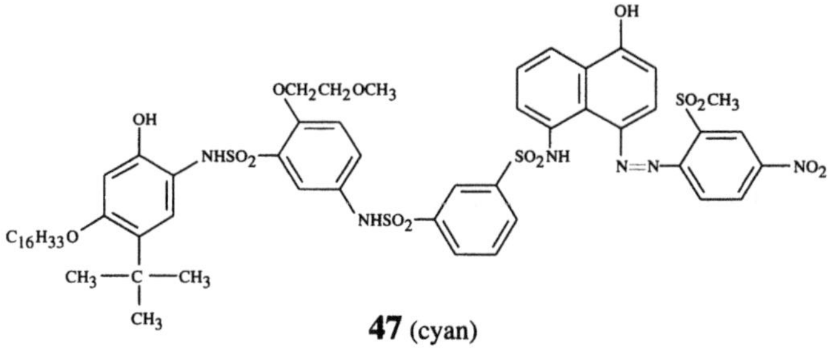

47 (cyan)

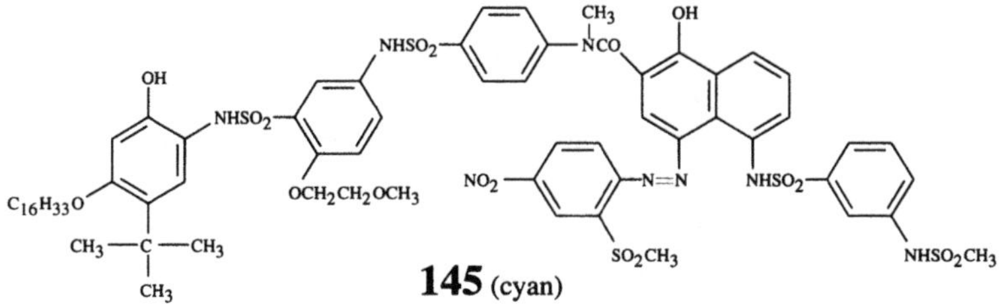

145 (cyan)

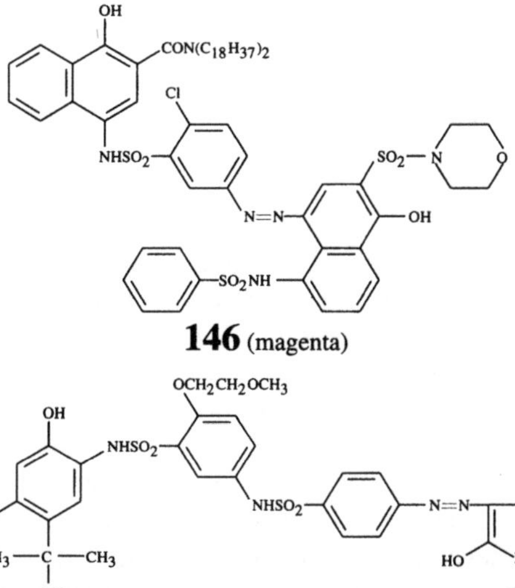

146 (magenta)

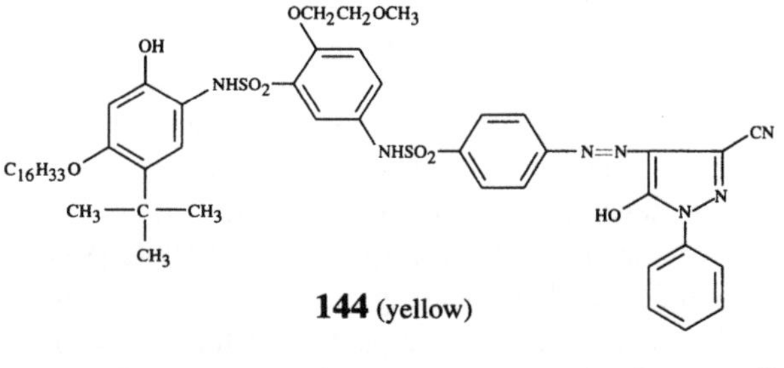

144 (yellow)

Figure 19.42. Another improved set of cyan, magenta, and yellow *o*-sulfonamidophenol dye releasers for color reproduction of instant color photography. This set is cited from a patent claiming hydrazine additives [57] and from a patent claiming direct-positive emulsions [65].

the same as **47**. Their common FUN moiety is covered by the patent cited above [22,30].

The magenta dye releaser **146** comprises a *p*-sulfonamidonaphthol moiety (FUN moiety) covered by an earlier patent [81] and an azo dye (DYE) covered by another earlier patent [82]. These patents [81,82] related to the magenta dye releaser (**146**) have been assigned to Eastman Kodak, although they have already expired. Hence, it is not clear whether or not the set shown in Fig. 19.42 has been really employed in the Fuji Instax system. However, examples appearing in recent patents from Fuji Photo Film involve the set of dye releasers shown in Fig. 19.42.

The yellow dye releaser **144** shown in Fig. 19.42 is the same compound as the one listed in Fig. 19.41.

19.7.2 Peel-Apart Instant Color Films

o-Sulfonamidophenol dye releasers in combination with direct positive emulsions have been also used in peel-apart instant color films.

The schematic cross-section of a recent peel-apart film is depicted in Fig. 19.43 [83]. After the upper sheet (the photo-sensitive sheet having Layers 1 to 11) is exposed to light ($h\nu\uparrow$) from the bottom side, it is superposed onto the lower sheet (the receiving sheet having Layers 1 to 5) in a face-to-face fashion, where a pod containing an activator fluid is placed between the two sheets. Then the superposed sheets pass through the gap between a pair of rollers. Thereby, the pressure due to the rollers ruptures the pod to spread the activator fluid between the two sheets. After the development and the dye transfer are completed, the two sheets are separated from each other so that we can observe a completed picture on the receiving sheet (↓View).[12]

It should be noted that the peel-apart film has an arrangement of layers: a dye releaser layer—a silver halide emulsion layer—($h\nu$, fluid & View)—a mordant layer. This means that the direction of the dye diffusion is inverse to the one of the initial activator flow.[13] On the other hand, The mono-sheet film shown in Fig. 19.2 has an arrangement of layers: (View) a mordant layer—a dye releaser layer—a silver halide emulsion layer—(fluid)—($h\nu$). In this case, the direction of the dye diffusion is the same as the one of the activator flow. At any rate, the chemistry of *o*-sulfonamidophenol dye releasers requires no mirror both in the mono-sheet format and in the peel-apart format. This is a merit as compared with the chemistry of dye developers described in Chapter 17.[14]

A technical report on the new peel-apart film FP100C [83][15] has described the use of sulfonamides as dye-releasing accelerators. The patent claiming the sulfonamide (**113**) [61] has described a peel-apart format like Fig. 19.43, which

[12]The right-and-left relationship for a peel-apart film using dye developers has already been discussed in Chapter 16 (Subsection 16.4.1). The discussion holds true for the peel-apart film using *o*-sulfonamidophenol dye releasers.

[13]This arrangement is essentially the same as the one employed in the peel-apart film using dye developers (Chapter 17). The direction of the dye diffusion is inverse to the one of the initial activator flow.

[14]Remember that the SX-70 camera has a mirror to inverse the right-and-left relationship. See Chapter 17.

[15]A new peel-apart instant color film FP100C has been released from Fuji Photo Film in 2002.

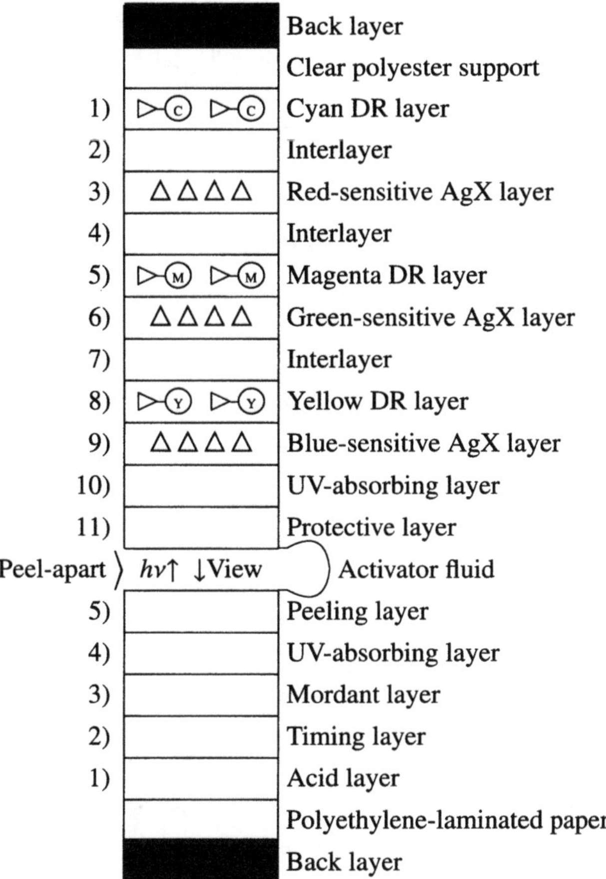

Figure 19.43. Schematic cross-section of a peel-apart photographic material based on *o*-sulfonamidophenol dye releasers and direct positive emulsions. △: Silver halide grain, ▷-Ⓨ: Yellow dye releaser, ▷-Ⓜ: Magenta dye releaser, and ▷-Ⓒ: Cyan dye releaser.

contains a set of dye releasers shown in Fig. 19.44. The same set has appeared in a more recent patent [57], although the employed format takes a more complicated multilayer structure.

The cyan dye releaser **147** has the FUN moiety covered by the patent cited above [22,30]. The DYE moiety has a characteristic 2-carbamoly group.

The magenta dye releaser is the same as **143**, where the cyclohexyl-substituted *o*-sulfonamidophenol (FUN moiety) is covered by the patent [22] and an azo dye (DYE) is covered by the patent [45].

The yellow dye releaser **148** shown in Fig. 19.44 comprises a FUN moiety covered by the patent [22] and a DYE moiety covered by the patent [48].

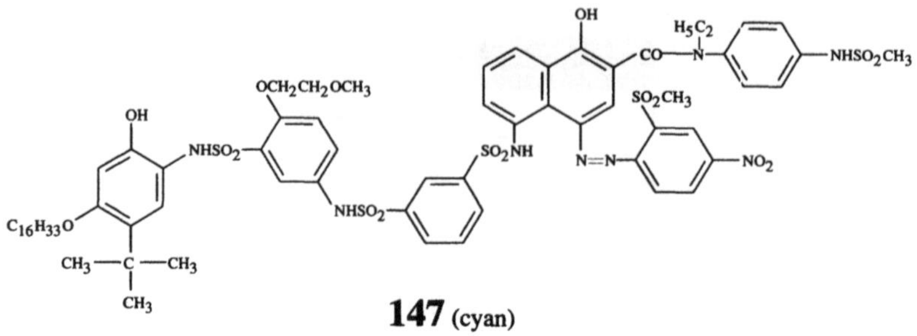

147 (cyan)

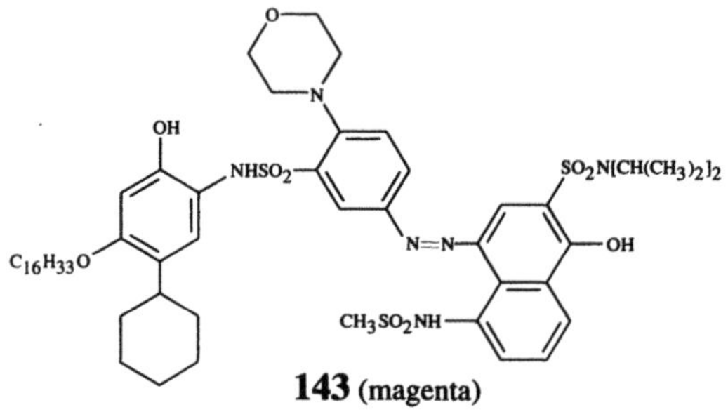

143 (magenta)

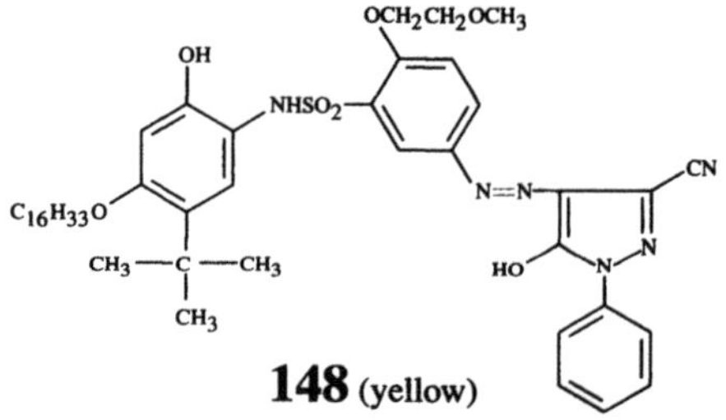

148 (yellow)

Figure 19.44. A set of cyan, magenta, and yellow *o*-sulfonamidophenol dye releasers for a peel-apart instant color film. This set is cited from a patent claiming hydrazine additives [57], from a patent claiming dye-releasing accelerators [61], and from a patent claiming mordant polymers with heterocyclic rings [74].

19.8 Heat Development for Digital Printing

19.8.1 Incorporation of Activators in Layers

A heat-developable color photographic material based on *o*-sulfonamidophenol dye releasers has been disclosed as a new device for digital printing [84]. From the viewpoint of reaction mechanisms, this photographic material can be regarded as a kind of diffusion transfer photographic material of peel-apart type. However, it differs from the usual peel-apart materials described in the preceding section (cf. Subsection 19.7.2) in that an activator fluid is not involved. In other words, it has aimed at a substantially dry process, where only a small quantity of damping water is fed from the outside. This means that activators necessary to initiate silver salt development and dye transfer should be incorporated in a photo-sensitive sheet or a receiver sheet, but not in an activator fluid. Various compounds have been extensively investigated for the purpose of such incorporation, as described in reviews [85,86].[16]

For example, aromatic ethers [87] and esters [88,89,90] have been incorporated as stabilizers; amides have been incorporated to enhance D_{max} values [91]; sulfonamides have been involved as dye-releasing accelerators [59]; base precursors have been used to liberate guanidine bases [92,93]; polyvinyl alcohols have been used to maintain a surface gloss [94]; polymers having an imidazole moiety have been investigated as mordants involved in a receiving sheet [95]; among others.

Among these compounds, guanidinium picolinate (**149**) should be mentioned as an effective base precursor, which liberates guanidine (**151**) according to the equation shown in Fig. 19.45 [93]. Thus guanidinium picolinate (**149**) reacts with zinc hydroxide in the presence of a small amount of water (and heat) to form a zinc chelate species (**150**) and guanidine (**151**).

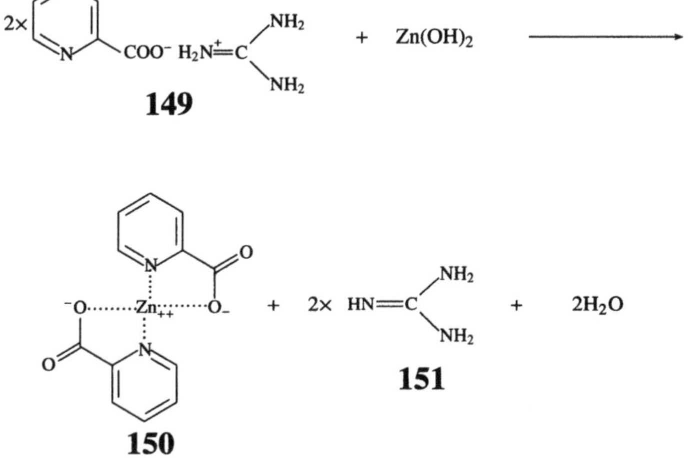

Figure 19.45. Liberation of guanidine from guanidinium picolinate [93].

All of the patents cited in the preceding paragraphs have described dye releasers having a 2-sulfonamido-4-hexadecyloxy-5-*t*-octylphenol as a FUN moiety, which

[16]A heat-developable color photographic material based on *o*-sulfonamidophenol dye releasers has been marketed under the name "Pictrography 1000" by Fuji Photo Film in 1987. Later, it was improved and commercialized under the name "Pictrography 3000" in 1993 and "Pictrography 4000" in 1997.

is covered by a basic patent for *o*-sulfonamidophenol dye releasers [12].[17] As typical examples, a set of dye releasers for a heat-developable photographic material is shown in Fig. 19.46, where they have been cited from a patent claiming saccharide additives [96].

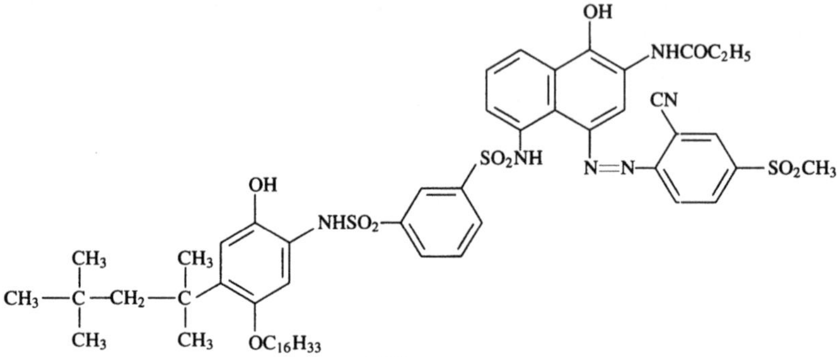

152 (cyan)

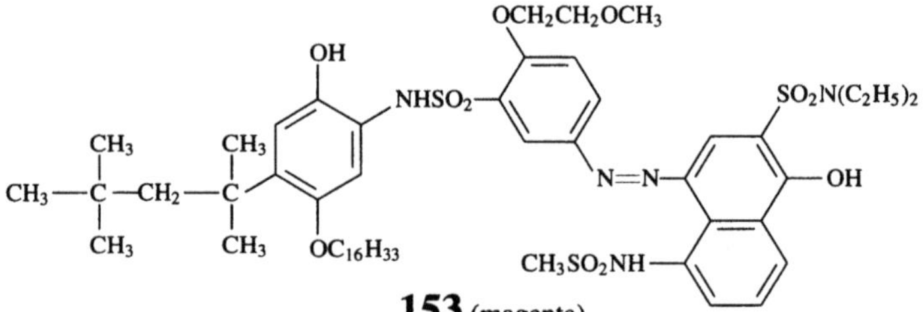

153 (magenta)

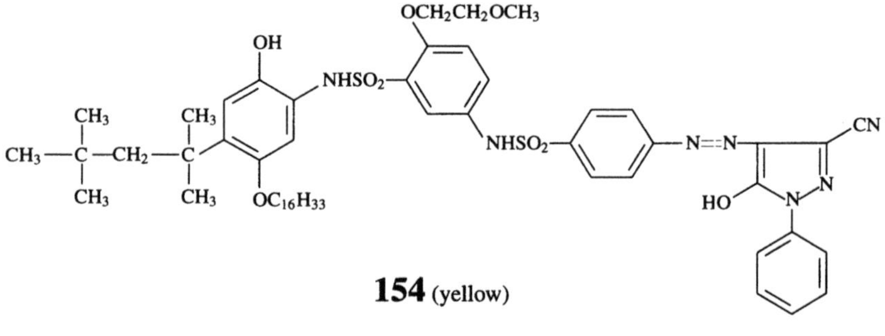

154 (yellow)

Figure 19.46. A set of cyan, magenta, and yellow *o*-sulfonamidophenol dye releasers for heat-developable instant color photography. This set is cited from a patent claiming saccharide additives [96].

The dye moiety of the cyan dye releaser (**152**) shown in Fig. 19.46 is covered by the patent [34], which has been described for discussing the improvement of cyan dye moieties (Fig. 19.21). The FUN moiety of **152** is covered by the patent

[17]The term "*t*-octyl" usually denotes a 1,1,3,3-tetramethyl-1-butyl group in photographic chemistry.

described above [12]. Various cyan dye moieties have been investigated for heat-developable photographic materials in patents, e.g., [97,98,99], although detailed discussions on these dye moieties are omitted in this book.

The magenta dye releaser (**153**) shown in Fig. 19.46 comprises an azo dye (DYE) covered by the patent [45] and the same FUN moiety. Compare this dye releaser with **79**. Another type of magenta dye moieties has been reported [100].

The methoxyethoxy linkage of the yellow dye releaser (**154**) shown in Fig. 19.46 is covered by the patent [23] and the same FUN moiety is used. Compare this dye releaser with **144**. Another yellow dye moiety has been proposed in combination with the *o*-sulfonamidophenol moiety having a *t*-octyl group [101].

19.8.2 False Color Reproduction

Such *o*-sulfonamidophenol dye releasers as listed in Fig. 19.46 have been used in heat-developable photographic materials, where so-called "false color reproduction" has been employed to meet requirements for digital printing.

As already described in this book, the usual subtractive color reproduction for color photography requires a blue-, a green-, and a red-sensitive emulsion layer, as shown, e.g., in Fig. 19.43. In contrast, digital printing is able to employ any set of lights of three colors, because computers can pre-treat image data to meet the set of lights. Usually, laser diodes (LD) with light emissions of 670, 750, and 810 nm are scanned to write digital image data on a photographic material. Hence, silver salt emulsions of the photographic material are required to be sensitive to these lights, so that they bring about false color reproduction,[18] where they are combined with a magenta dye releaser, a cyan dye releaser, and a yellow dye releaser, respectively. Thereby, each emulsion can be coated together with the corresponding dye releaser to give a single coated layer. Moreover, each emulsion layer may be negative, because image data are pre-treated by computers.

Figure 19.47 illustrates the schematic cross-section of a heat-developable photographic material based on *o*-sulfonamidophenol dye releasers, which has a red-sensitive emulsion layer (670 nm), an infrared-sensitive emulsion layer (750 nm), and another infrared-sensitive emulsion layer (810 nm); and each emulsion layer contains a dye releaser of the corresponding false color. A set of dye releasers shown in Fig. 19.46 has also been used in combination with the multilayer photographic material shown in Fig. 19.47 in a patent claiming a dye fixing element [102], except that an additional cyan dye releaser has been admixed.

A plausible set of three dye releasers listed in Fig. 19.48 has seemed to be used in a next generation of heat-developable photographic materials in combination with the film format shown in Fig. 19.47.[19]

[18] For this purpose, infrared-sensitizing dyes have been investigated, as described in Chapter 7.

[19] A dye set similar to the set of Fig. 19.48 has been reported as the previous set in a technical report published by Fuji Photo Film [86].

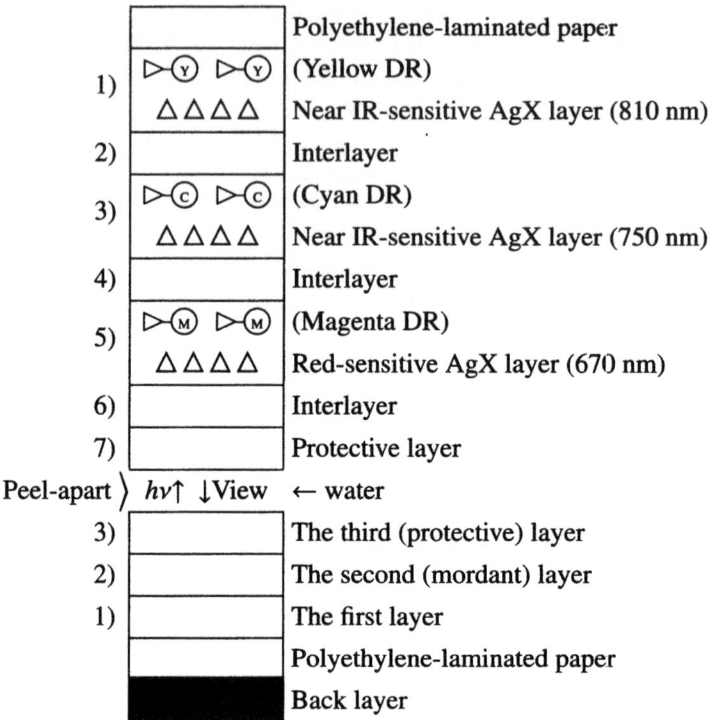

Figure 19.47. Schematic cross-section of a heat-developable photographic material based on *o*-sulfonamidophenol dye releasers and negative silver salt emulsions for false color reproduction. Δ: Silver halide grain, ▷-Ⓨ: Yellow dye releaser, ▷-Ⓜ: Magenta dye releaser, and ▷-Ⓒ: Cyan dye releaser.

The cyan dye releaser (**152**) is the same as above. Another cyan dye releaser (**155**) is used as a co-releaser, where the dye moiety is covered by the patent [35]; the FUN moiety of **155** is covered by the patent described above [12].

The magenta dye releaser (**156**) shown in Fig. 19.48 comprises an azo dye (DYE) covered by the patent [82] and the same FUN moiety. The linkage group containing a morpholino group is covered by the patent [29] (cf. **64**).

The yellow dye releaser is the same as **154** illustrated in Fig. 19.46.

19.8.3 Improvement of Light Stability

Recent progress in heat-developable photographic materials has been concerned with the light stability of images, as summarized in a techincal report [86,103].[20]

[20] A new photographic material with improved light stability has been marketed by Fuji Photo Film under the name "Ever-Rich" in 2000.

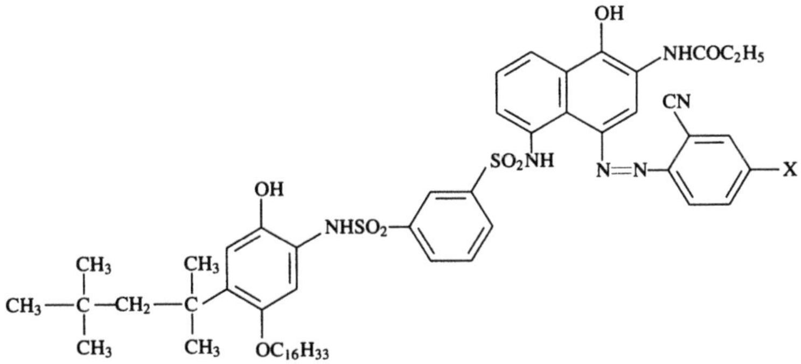

152 (cyan) X = SO_2CH_3
155 (cyan) X = NO_2

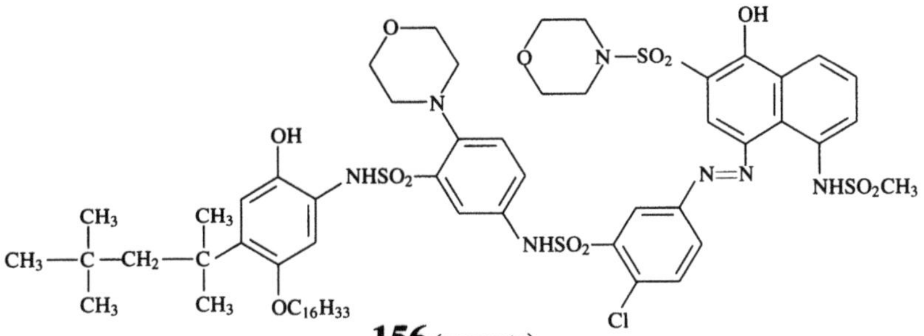

156 (magenta)

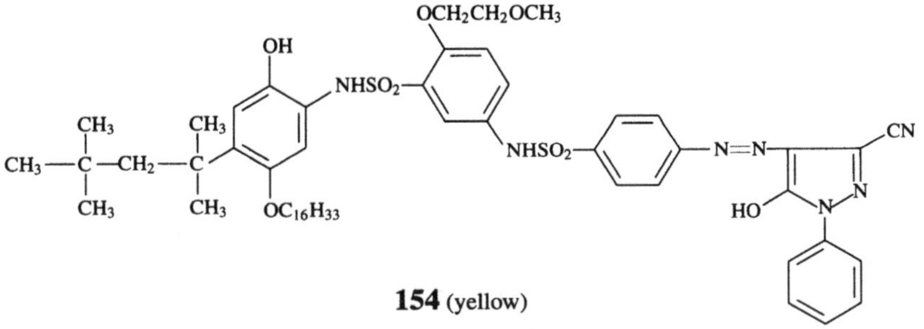

154 (yellow)

Figure 19.48. A more recent set of cyan, magenta, and yellow *o*-sulfonamidophenol dye releasers for heat-developable instant color photography. This set is cited from a patent claiming heterocyclic cyan dye releasers [36].

Two approaches have contributed to the improvement of the light stability. One approach has been the introduction of new heterocyclic azo dye moieties of cyan and magenta colors, while the other has been to develop a new stabilizer [104].

In general, the light fading of azo-dyes has been ascribed to self-sensitized photo-oxidation caused by singlet oxygen [105]. Heterocyclic azo dyes having a pyrazole, an isothiazole, or a thiadiazole as a diazo component have been reported to exhibit excellent stability against singlet oxygen [86]. They have been incorporated into dye releasers for heat-developable photographic materials. For example, a set of such dye releasers for color reproduction is shown in Fig. 19.49. They are cited from a recent patent claiming hydroquinone derivatives [106], since the dye moieties of the dye releasers (**157**, **158**, and **159**) are substantially the same as the heterocyclic azo dyes enumerated in the report [86].

The cyan dye releaser (**157**) comprises an azo dye moiety (DYE) covered by the patent [36]; the FUN moiety covered by the patent described above [12]; and the morpholino-containing linkage group covered by the patent [29] (cf. **64**).

The magenta dye releaser (**158**) shown in Fig. 19.49 comprises an azo dye (DYE) covered by the patent [107]; the FUN moiety covered by the patent described above [12]; and the morpholino-containing linkage group covered by the patent [29] (cf. **64**).

The azo dye moiety of the yellow dye releaser **159** is based on a pyrazolotriazole coupling component and covered by patents [108,109]. The FUN moiety is covered by the patent described above [12]; and the morpholino-containing linkage group is covered by the patent [29] (cf. **64**). The other yellow dye releaser (**154**) shown in Fig. 19.49 is the same as the one illustrated in Fig. 19.46.

In spite of the introduction of the new azo dyes, light stability has remained insufficient under UV light exposure. Dye fading mechanisms other than singlet oxygen have been investigated to bring about further improvement of light stability. Thereby, imidazole polymers such as **130** (Fig. 19.37) have been reported to inhibit the fading reaction of a magenta dye [104]. More sufficient results have been obtained by the addition of stabilized nitroxides as stabilizers [110]. Mordant polymers having a nitroxide radical (e.g., **160**) have been disclosed to give more remarkable results [111]. The mordant (**160**) has been used in a recent embodiment [106], where the nitroxide moiety is presumed to be introduced into an imidazole polymer in the form of $ClCH_2CONH$-(nitroxyl moiety) [86,103,112].

19.8.4 Direct Digital Color Proofs

Heat-developable photographic materials have been applied to produce direct digital color proofs (DDCP) [113], which have become more and more important in accord with the progress of so-called CTP (computer-to-plate publishing).[21] Although dyes for printing based on CMYK (cyan, magenta, yellow, and black) are different in hue from dyes for color photography based on CMY, some attempts to mimimize the difference have appeared, where two cyan dye releasers having absorption maxima of about 630 and 660 nm are used together [114,115]. In addition to such material technologies, color matching technology has been disclosed to cover both software and hardware [116,117].

[21] A DDCP system named "Pictro Proof System" has been marketed by Fuji Photo Film in 1998.

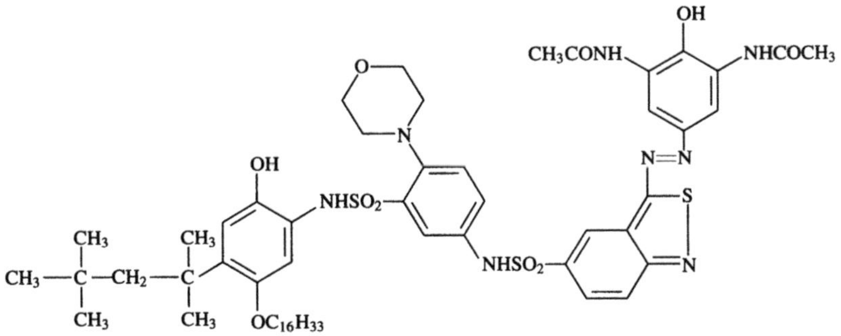

157 (cyan)

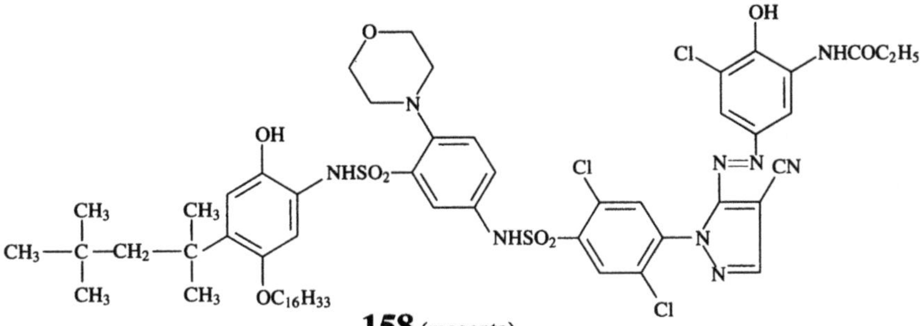

158 (magenta)

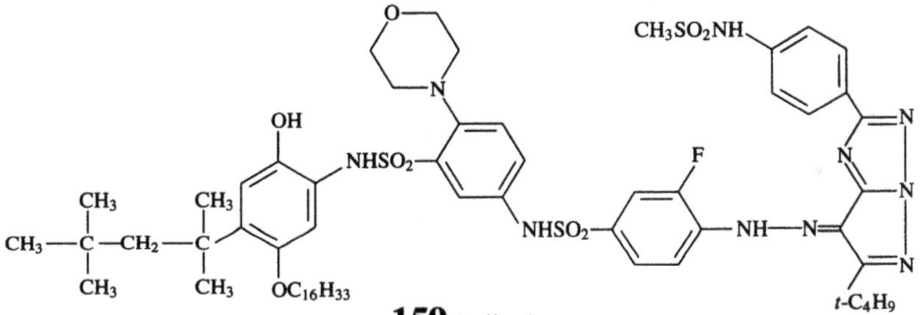

159 (yellow)

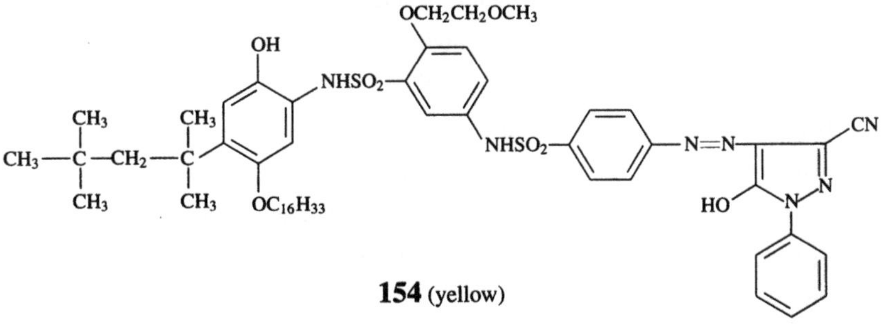

154 (yellow)

Figure 19.49. Another recent set of cyan, magenta, and yellow *o*-sulfonamidophenol dye releasers for heat-developable instant color photography. This set is cited from a patent claiming hydroquinone derivatives [106].

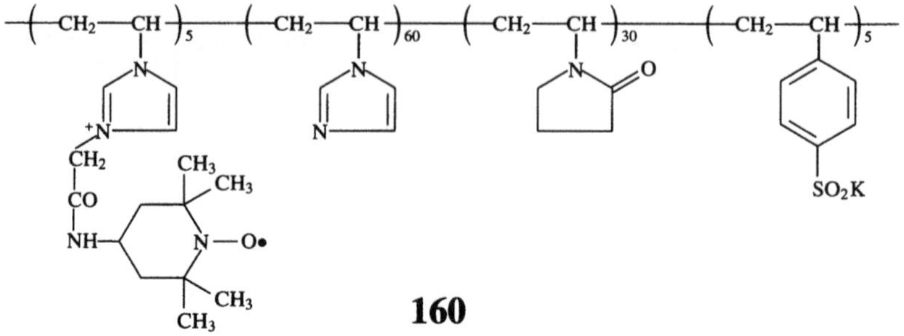

Figure 19.50. Mordant polymer having a nitroxide group.

References

[1] Fujita S, Koyama K, Ono S (1982) Nikkakyo Geppo. 35(11):29

[2] Ishimaru S, Kihayashi K, Takeuchi K, Sawada S, Ookura T (1999) J Soc Photogr Sci Technol Jpn. 62:434

[3] Fujita S (1992) R&D of Dye Releasers for Instant Color Photography: Design, Evaluation, and Synthesis. In: Noyori R (ed) Organic Synthesis in Japan. Past, Present, and Future, Tokyo Kagaku Dozin, Tokyo. pp 89–96

[4] Fujita S (1982) Kagaku no Ryoiki. 36:617

[5] Fujita S (1983) Kino Zairyo. 3:66

[6] Itoh I, Fujita S (1985) Denshi Shashin. 24:339

[7] Fujita S (1986, 2002) In: Functionalized Organic Chemicals for Silver Halide Color Photographic Materials, CMC, Tokyo. Chapter 3

[8] Fujita S (1982) Yuki Gosei Kagaku Kyokaishi. 40:176

[9] Fujita S, Koyama K, Ono S (1991) Nippon Kagaku Kai Shi. 1

[10] Fujita S, Koyama K, Ono S (1992) Rev Heteroatom Chem. 7:229

[11] Fleckenstein LJ (1967) US Patent 4 053 312

[12] Koyama K, Maekawa Y, Miyakawa M (1977) US Patent 4 055 428

[13] Aoki K (1977) Jpn Kokai S52-153923

[14] Koyama K (1980) Jpn Kokai S55-153775

[15] Fujita S, Koyama K, Inagaki Y (1981) Jpn Kokai S56-100771

[16] Fujita S, Koyama K, Inagaki Y (1986) US Patent 4 594 426

[17] Fujita S, Koyama K, Inagaki Y (1982) Synthesis. 68

[18] Fujita S (1981) J Chem Soc Chem Commun. 425

[19] Fujita S (1983) J Org Chem 48:177

[20] Fujita S (1982) Yuki Gosei Kagaku Kyokaishi. 40:307

[21] Fujita S (1985) Yuki Gosei Kagaku Kyokaishi. 43:153

[22] Fujita S, Koyama K, Inagaki Y, Waki K (1982) US Patent 4 336 322

[23] Fujita S, Hayashi H, Ono S, Yoshida Y, Harada T (1981) US Patent 4 268 625

[24] Fujita S, Koyama K, Inagaki Y (1982) US Patent 4 310 693

[25] Fujita S, Ono S, Hayashi H (1981) US Patent 4 246 414

[26] Fujita S (1982) Synthesis. 423

[27] Kurita K (1974) Chem Ind. 345

[28] Inoue K, Miyakawa M (1978) Jpn Kokai S53-47823

[29] Itoh I, Okazaki M, Fujita S, Yoshida Y (1978) Jpn Kokai S53-143323

[30] Fujita S, Koyama K, Inagaki Y, Waki K (1981) Jpn Kokai S56-71061

[31] Fujita S, Sakanoue S, Miyakawa M (1977) Jpn Kokai S52-8827

[32] Fujita S, Sakanoue S, Miyakawa M (1977) Jpn Kokai S52-8827; Fuji Photo Film (1979) GB Patent 1 551 138

[33] Fujita S, Itoh I, Ono S, Harada T, Yoshida Y (1982) US Patent 4 358 526

[34] Sato K, Fujita S, Nakamura K, Naito H (1985) US Patent 4 556 632

[35] Nakamura K, Ito T, Tsukase M, Taguchi T (1991) US Patent 5 066 577

[36] Tateishi K, Kamio T, Kamosaki T, Ishiwata Y, Naruse H (2000) US Patent 6 124 084

[37] Ueda H, Sawaguchi H (1979) US Patent 4 176 134

[38] Katoh K, Harada T, Ono S, Fujita S (1983) Photogr Sci Eng. 27:102

[39] Fujita S, Harada T, Yoshida Y (1981) US Patent 4 268 624

[40] Fujita S, Koyama K, Inagaki Y, Waki K (1981) Jpn Kokai S56-73057

[41] Ono S, Harada T, Fujita S, Yoshida Y (1981) US Patent 4 255 509

[42] Fujita S, Koyama K, Inagaki Y, Waki K (1981) Jpn Kokai S56-71060

[43] Fujita S, Ono S, Yoshida Y (1980) Jpn Kokai S55-36804

[44] Fujita S, Inagaki Y (1980) Jpn Kokai S55-134850

[45] Itoh I, Yoshida Y (1981) US Patent 4 250 246

[46] Ohkawa A, Mikoshiba H, Tsukase M, Ishiwata Y, Naruse H (1997) US Patent 5 665 529

[47] Harada T, Fujita S (1980) US Patent 4 233 237

[48] Fujita S, Harada T, Endo K (1981) US Patent 4 245 028

[49] Ono S, Fujita S (1982) US Patent 4 340 661

[50] Ono S, Fujita S (1983) US Patent 4 368 251

[51] Harada T, Fujita S, Tanaka M, Toriuchi M (1981) Jpn Kokai S56-25737

[52] Fujita S, Koyama K, Inagaki Y, Waki (1981) Jpn Kokai S56-71072

[53] Fujita S (1982) J Chem Soc Perkin I. 1519

[54] Hirano S, Adachi K, Tsujino N (1981) US Patent 4 255 511

[55] Adachi K, Hirano S, Tsujino N (1981) US Patent 4 266 013
[56] Tsujino N, Hirano S, Adachi K (1981) US Patent 4 245 037
[57] Fukagawa N, Ichikawa S, Yoshikawa M (2002) US Patent 6 489 088 B2
[58] Koyama K, Fujita S (1982) US Patent 4 358 532
[59] Hirai H, Yabuki Y, Takeuchi M, Aono T (1986) US Patent 4 590 154
[60] Naruse H, Tsukase M (1995) US Patent 5 418 111
[61] Ohkawa A, Iwanaga H, Naruse H (1996) US Patent 5 496 680
[62] Naruse H, Naito H, Tsukahara J (1995) US Patent 5 447 818
[63] Naruse H, Seto N, Morigaki M, Negoro M (1995) US Patent 5 419 996
[64] Hukuda H, Ōi, Kataoka H, Aosaki K, Takeuchi K, Mihayashi K, Ishimaru S (1999) Fuji Film Res & Dev. 44:1
[65] Matsunaga A, Hara T (2001) US Patent 6 194 134 B1
[66] Nakamura K, Nakamura S (1988) US Patent 4 783 396
[67] Nakamura K, Nakamura S (1990) US Patent 4 950 764
[68] Naito H, Sasaki H (1997) US Patent 5 607 812
[69] Yamamoto T, Katoh K (1981) US Patent 4 308 335
[70] Nakamura T, Nakamura K, Hayashi H (1988) US Patent 4 766 052
[71] Yamanouchi J, Toriuchi M (1989) US Patent 4 861 852
[72] Nakamura K, Takeuch K, Katsumata T, Taguchi T (1999) US Patent 5 976 756
[73] Yamanouchi J, Shiratsuchi K, Karino Y, Shibata T (1991) US Patent 5 023 162
[74] Tsukahara J, Arakatsu H (1996) US Patent 5 498 505
[75] Noguchi Y, Sakaguchi S, Yoshida T (1981) US Patent 4 256 827
[76] Tomiyama H, Horie I, Satake M (1989) US Patent 4 833 063
[77] Sera H, Horie I (1980) US Patent 4 190 448
[78] Yoshida T, Sakaguchi S (1981) US Patent 4 268 604
[79] Tomiyama H (1990) US Patent 4 916 044
[80] Sasaki H, Watanabe H, Asakura T (1999) US Patent 5 972 557
[81] Fernandez JM, McCreary MD, Ross RE, Staples JT (1979) US Patent 4 135 929
[82] Krutak Sr JJ, Haase JR Landholm RA (1976) US Patent 3 954 476
[83] Hara T, Asakura T, Takeuchi K (2003) Fuji Film Res & Dev. 48:1
[84] Naito H, Hara H, Aono T, Sato K, Fujita S (1985) US Patent 4 500 626
[85] Hara H, Sato K (1989) Fuji Film Res & Dev. 34:10
[86] Kosugi T, Uehara K, Irita K, Yokokawa T, Set N, Kamio T (2001) J Soc Photogr Sci Technol Jpn. 64:292

[87] Sakaguchi Y, Aono T, Fujita S (1985) US Patent 4 536 466

[88] Sakaguchi Y, Aono T, Fujita S (1985) US Patent 4 536 467

[89] Sakaguchi Y, Aono T, Fujita S (1985) US Patent 4 555 476

[90] Sakaguchi Y, Aono T, Fujita S (1986) US Patent 4 587 206

[91] Sakaguchi Y, Aono T, Fujita S (1986) US Patent 4 599 296

[92] Sato K, Yabuki Y, Hirai H, Kawata K (1987) US Patent 4 657 848

[93] Hirai H, Yabuki Y, Sato K (1988) US Patent 4 740 445

[94] Aono T, Nakamura K (1989) US Patent 4 818 662

[95] Aono T, Nakamura K, Nakamura T (1986) US Patent 4 619 883

[96] Taguchi T, Hirai H (1991) US Patent 5 051 349

[97] Watanabe K, Taguchi T (1991) Jpn Kokai H03-114042

[98] Tsukase M, Sato K, Hirai H (1993) US Patent 5 223 387

[99] Nakamura K, Tsukase M (1995) Jpn Kokai H07-219180

[100] Sato K, Tsukase M, Shibata T (1988) US Patent 4 741 997

[101] Kitaguchi, Sato K, Fujita S, Naito H (1984) US Patent 4 473 632

[102] Nakamura Y, Arakatsu H (1995) US Patent 5 437 956

[103] Irita K, Arakatsu H, Hiyoshi H, Yokokawa T, Seto N, Kamio T, Watanabe H, Onishi J (2001) Fuji Film Res & Dev. 46:15

[104] Taguchi K, Watanabe H, Seoka Y (1998) Fuji Film Res & Dev. 43:17

[105] Merkel PB, Smith Jr WF (1979) J Phys Chem. 22:2834

[106] Yamazaki T (2001) US Patent 6 329 129

[107] Kamio T, Seto N, Hiyoshi H (1998) US Patent 5 731 140

[108] Yamada M, Matsuda N, Ishiwata Y, Uchida O, Ono M (1996) US Patent 5 585 231

[109] Yamada M, Matsuda N, Ishiwata Y, Uchida O, Ono M (1996) US Patent 6 127 088

[110] Taguchi T, Shibata T, Seto N, Naruse H, Mitsui A (2000) US Patent 6 106 991

[111] Arakatsu H, Seto N, Taguchi T (2001) US Patent 6 268 105

[112] Kosugi T (2001) US Patent 6 261 746 B1

[113] Uehara K, Yokokawa T, Irita K, Kamio T, Tamagawa K, Inoue K, Kondo H, Shimazaki O (1999) Fuji Film Res & Dev. 44:18

[114] Kamosaki T (1998) US Patent 5 817 452

[115] Nakagawa H, Taguchi T (2001) US Patent 6 265 142 B1

[116] Noguchi T, Dounomae Y, Ito W (1998) US Patent 5 734 801

[117] Usami Y, Ohkubo A, Yoda A (1998) Fuji Film Res & Dev. 43:81

Chapter 20

Positive-Working Dye Releasers

20.1 Carquin Dye Releasers

A diffusion transfer photographic system based on ballasted quinone dye releasers (so-called carquin)[1] has been marketed by Agfa-Gevaert under the name "COPY-COLOR CCN" in 1983 [1]. This system has aimed at copying or color proofing but not at amateur use.

Chemical Processes of Carquin Dye Releasers

Carquin (ballasted quinone) dye releasers are positive-working compounds of reductive-releasing type [2]. Their reactions with electron donors (ED), which are associated with imagewise redox reactions between electron transfer agents (ETA) and silver halide, have been briefly discussed in Chapter 16 (Subsection 16.4.3). Since the discussions in Chapter 16 have taken a nitro-type dye releaser as an example of reductive-releasing dye releasers, the same things should be reissued by using carquin dye releasers, as shown in Fig. 20.1.

Equation 20.1 of Fig. 20.1 shows the redox reaction between the anionic species (**2**) of an electron transfer agent (ETA: **1**) and a negative silver halide emulsion, which occurs in a light-exposed area of the corresponding emulsion layer.[2] The

[1] The name "carquin" is the abbreviation of "carrier quinone", since the FUN moiety is sometimes called "carrier".

[2] In the scheme of Fig. 20.1, 4-methyl-1-phenyl-3-pyrazolidone is depicted as a representative of ETA and 2,5-bis(1′,1′,3′,3′-tetramethylbutyl)hydroquinone is depicted as an ED.

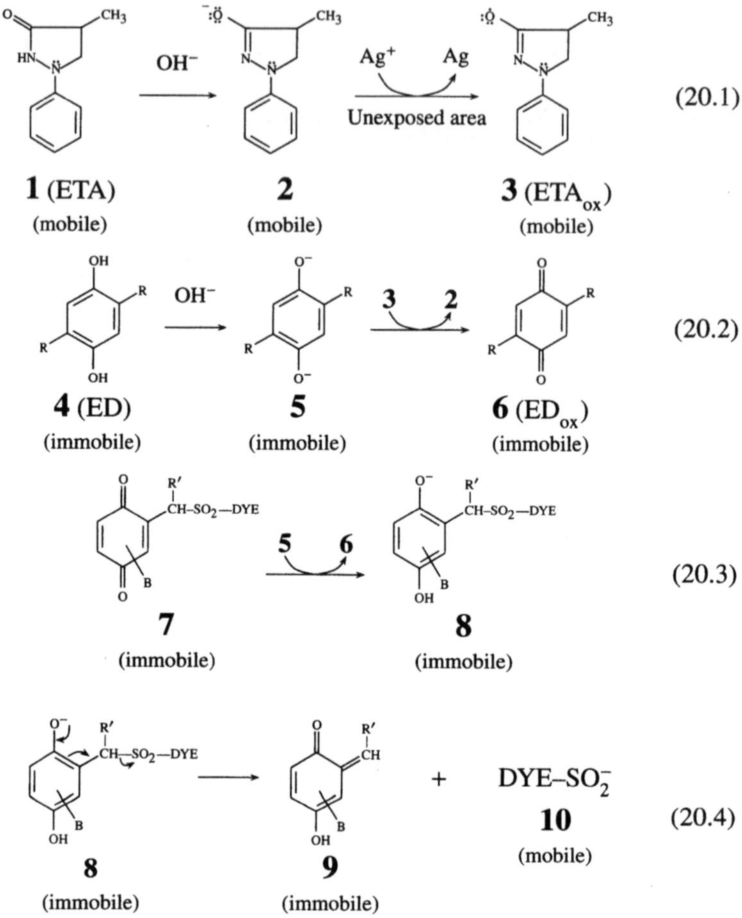

OH⁻ → ; Ag^+ → Ag, Unexposed area (20.1)

1 (ETA) (mobile) **2** (mobile) **3** (ETA_{ox}) (mobile)

OH⁻ → ; **3** → **2** (20.2)

4 (ED) (immobile) **5** (immobile) **6** (ED_{ox}) (immobile)

5 → **6** (20.3)

7 (immobile) **8** (immobile)

→ + $DYE\text{–}SO_2^-$ (20.4)

8 (immobile) **9** (immobile) **10** (mobile)

Figure 20.1. Consecutive reactions of silver halide, an electron transfer agent (ETA), an electron donor (ED), and a carquin dye releaser. The letter R represents a *t*-octyl group, i.e., a 1,1,3,3-tetramethylbutyl group, which is a ballast group. The letter R′ may be an alkyl or a ballast alkyl group. The letter B represents a ballast group. The symbol DYE denotes a cyan, magenta, or yellow dye moiety.

resulting ETA_{ox} (**3**) oxidizes the anionic species (**5**) of an electron donor (ED: **4**), where the ED becomes ineffective, as shown in eq. 20.2. The remaining (effective) ED (or its anionic species) reduces a carquin dye releaser (**7**) to produce the corresponding hydroquinone form (or its anionic species such as **8**), as shown in eq. 20.3. The intermediate (**8**) releases a dye-sulfinate anion (**10**) in terms of a quinone-methide mechanism, as shown in eq. 20.4.

In an exposed area, the reaction shown in eq. 20.2 (after eq. 20.1) should occur, since a negative silver halide emulsion is employed. Instead, the reaction shown in eq. 20.3 must be suppressed so that no diffusible dye is released. In an unexposed area, on the other hand, the reaction shown in eq. 20.3 should take place to release diffusible dye; and the reaction shown in eq. 20.2 must be inhibited. In other words, the ED should reduce the ETA_{ox} (**3** in eq. 20.2) more rapidly than the carquin dye releaser (**7** in eq. 20.3).

20.1.1 Multilayer Structure for Color Reproduction

Multilayer Structure for Carquin Dye Releasers

The schematic cross-section of a photographic material based on positive-working carquin dye releasers is illustrated in Fig. 20.2, where carquins for providing cyan and magenta dyes are used together with a ballasted hydroquinone dye releaser for providing a yellow dye. This multilayer structure is cited from a patent claiming photographic additives [3]. The combined use will be discussed again in Fig. 20.5.

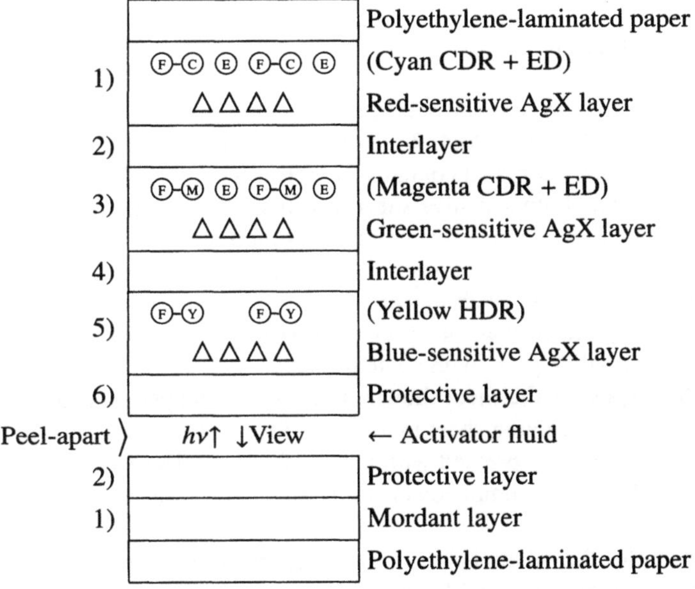

Figure 20.2. Schematic cross-section of a photographic material based on carquin dye releasers (and a hydroquinone dye releaser) and negative silver salt emulsions for color reproduction. Δ: Silver halide grain, Ⓕ-Ⓨ: Yellow hydroquinone dye releaser (yellow HDR) , Ⓕ-Ⓜ: Magenta carquin dye releaser (magenta CDR), Ⓕ-Ⓒ: Cyan carquin dye releaser (cyan CDR), and Ⓔ: Electron donor (ED).

Since this photographic material for copying or color proofing does not require as high speed as materials for amateur use,[3] each dye releaser is coated together with a negative emulsion so as to construct a single layer (Layer 1, 3, or 5). Since each dye releaser absorbs the light of the same region as the corresponding sensitizing dye, the loss of photo-sensitivity is inevitable.

Each of the green- and red-sensitive layers contains such an appropriate electron donor (ED) as 2,5-bis(1′,1′,3′,3′-tetramethylbutyl)hydroquinone in addition to the cyan or magenta carquin dye releaser. Since the yellow dye releaser is used in the hydroquinone form,[4] an electron donor is unnecessary to be involved.

Each interlayer (Layer 2 or 4) contains such a ballasted hydroquinone as potassium 2-*sec*-octadecylhydroquinone-5-sulfonate for preventing color contamination. Such an electron transfer agent (ETA) as 4-methyl-1-phenyl-3-pyrazolidone (**1**), which is involved in Layers 4 and 6, spreads over the layers during development.

Embodiments which comprise a yellow carquin dye releaser (and an ED) in addition to cyan and magenta carquin dye releasers have also been disclosed in several patents (e.g., [4]). Attempts to develop photographic materials of mono-sheet type have been described in patents (e.g., [5]), but they seem not to have been marketed.

Color Reproduction

A brief explanation of color reproduction on the basis of carquin dye releasers is shown in Figs. 20.3 and 20.4. Note that we do not use the multilayer structure shown in Fig. 20.2, because it contains dye releasers of two different mechanisms. Instead, we use a photo-sensitive sheet in which all of the photo-sensitive layers contain carquin dye releasers for the sake of simplicity and of consistency.

Figure 20.3 illustrates an exposure to blue, green, red, or white light (or to no light) from the bottom. Each upward arrow represents such light as arriving at a silver halide grain so that the symbol △• beside the arrowhead represents the resulting silver halide grain with a latent image.

After the exposure, the sheet shown in Fig. 20.3 is dipped in an activator liquid and attached to a receiving sheet in a face-to-face fashion, as shown in Fig. 20.4. The attached sheets are pressed with a pair of rollers. Thereby, the development of exposed silver halide emulsions is initiated so as to cause the reactions shown in Fig. 20.1.

For example, the silver halide grain (△•) in the exposed area of the blue-sensitive layer oxidizes an ETA (eq. 20.1) so that the resulting ETA_{ox} consumes

---

[3] The COPYCOLOR CCN marketed from Agfa-Gavaert was designed to have no sensitivity to the light of about 580 nm, where development work could be carried out under a sodium lamp [1].

[4] This is a positive-working compound of oxidative-stopping type. From those patents that the author is aware of, it is not clear what merits are brought about by the combined use of the reductive-releasing type (cyan and magenta) and the oxidative-stopping type (yellow).

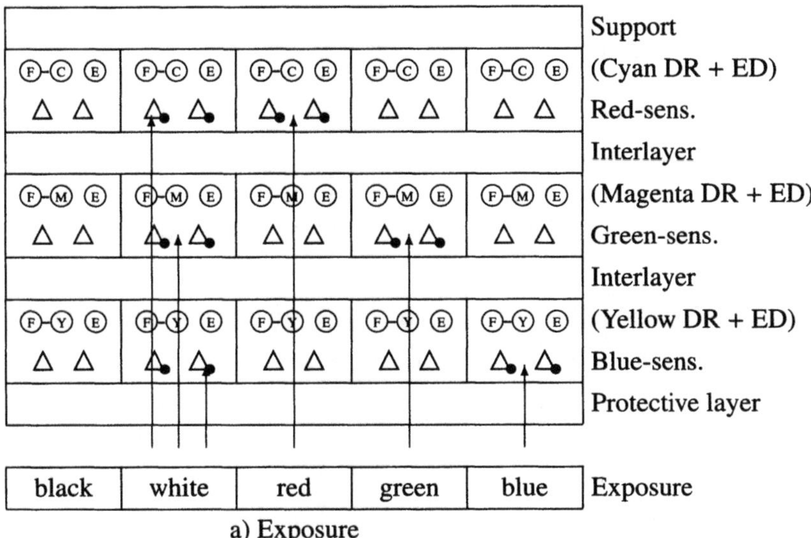

Figure 20.3. Peel-apart color film using negative-working dye releasers of reductive-releasing type. A schematic cross-section during exposure. △: Negative silver halide grain, △•: Exposed silver halide grain with a latent image, ▲: Developed silver halide grain; Ⓕ-Ⓨ: Yellow dye releaser, Ⓕ-Ⓜ: Magenta dye releaser, and Ⓕ-Ⓒ: Cyan dye releaser.

an ED (Ⓔ) to yield an oxidized ED (⊗). Thereby, a carquin (Ⓕ-Ⓨ) in this area remains unchanged so as not to release a yellow dye. Note that the ETA is omitted in the diagram shown in Fig. 20.4. In contrast, the silver halide grain in the corresponding area in each of the other layers does not react with an ETA so that an ED does not suffer from oxidation. Thereby, the reaction between the ED and the carquin (Ⓕ-Ⓜ or Ⓕ-Ⓒ) takes place to give the corresponding reduced carquin (the hydroquinone form: ⊘-Ⓜ or ⊘-Ⓒ), which is capable of releasing a magenta (M) or cyan dye (C). Each released dye diffuses to the mordant (image-receiving) layer and is mordanted there, as represented by an downward arrow. As a result, the subtractive mixing of cyan and magenta produces blue color (C + M = B) in terms of the negative-working mechanism of the reductive-releasing type. This is in agreement with the original blue light exposed. Similarly, green (C + Y = G), red (M + Y = R), black (C + M + Y = K), and white (no dyes) are reproduced.

It should be added that the processes represented by Figs. 20.3 and 20.4 are essentially common to photographic materials based on any positive-working dye releasers of reductive-releasing type. In fact, Figs. 20.3 and 20.4 hold true for ROSET dye releasers described in the next section. Hence, the captions of Figs. 20.3 and 20.4 describe "a peel-apart color film using negative-working dye releasers of negative-releasing type" in a general fashion.

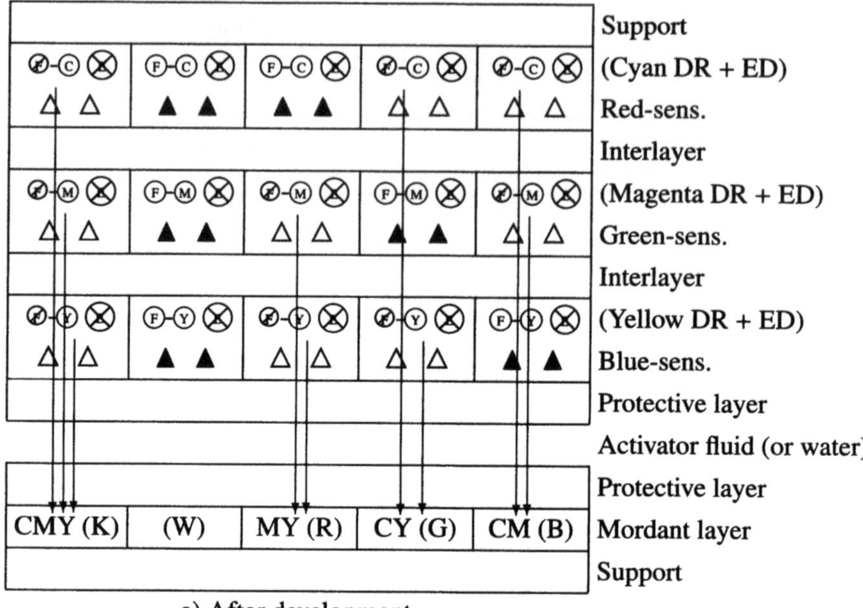

Figure 20.4. Peel-apart color film using negative-working dye releasers of negative-releasing type. A schematic cross-section during and after development. △: Silver halide grain, ▲: Developed silver halide grain; Ⓕ-Ⓨ: Yellow dye releaser, Ⓕ-Ⓜ: Magenta dye releaser, and Ⓕ-Ⓒ: Cyan dye releaser; Ⓕ̸-Ⓨ: Reduced yellow dye releaser, Ⓕ̸-Ⓜ: Reduced magenta dye releaser, and Ⓕ̸-Ⓒ: Reduced cyan dye releaser; Ⓔ: Electron donor, and ⊗: Oxidized electron donor. The letters Y, M, and C represent yellow, magenta, and cyan dyes deposited on the mordant layer. The letters B, G, R, W, and K represent black, green, red, white, and black. The action of an ETA is omitted for the sake of simplicity.

20.1.2 Ballasted Quinone Moieties

The basic patent of carquin dye releasers has disclosed ballasted hydroquinone dye releasers along with ballasted quinone dye releasers (carquins), where the ballasted quinone moiety involved in **11** and its hydroquinone counterpart have been described as examples [2]. The tridecyl group at the quinone-methide position works as a ballast group. Moreover, such a branching has improved dye-releasing efficiency.

Later, ballasted quinones having a 3-aryl substituent have been disclosed, as exemplified by the magenta dye releaser (**12**) [6]. A representative 3-aryl group is the 4-hexadecyloxyphenyl group contained in **12**. The branching at the methine position remains in **12**, in which the tridecyl group is changed into a methyl group. In this patent [6], hydroquinone forms such as involved in the yellow dye releaser (**13**) have also been described as synthetic intermediates for preparing carquins.

In a summary of the patent inspection, a plausible set of three dye releasers is listed in Fig. 20.5, which is an embodiment described in the patent [3]. In

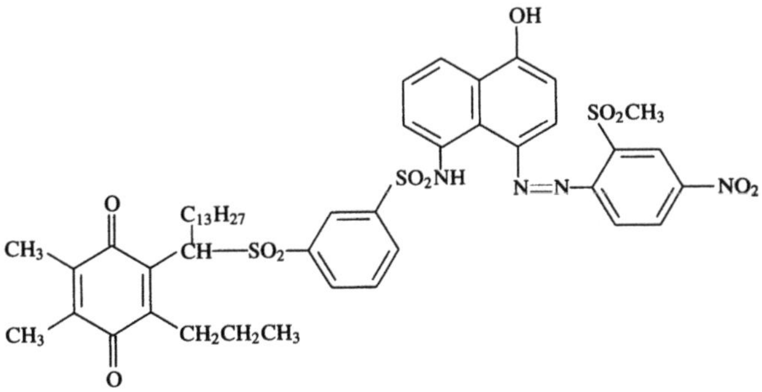

11 (cyan)

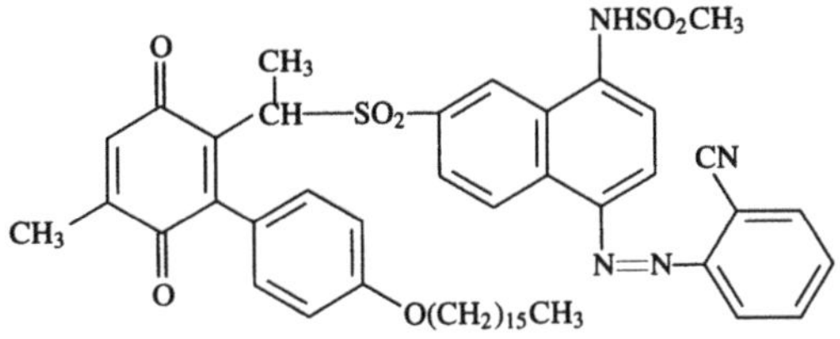

12 (magenta)

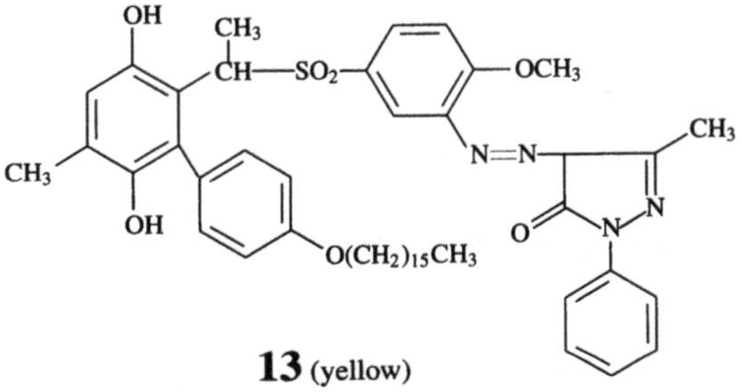

13 (yellow)

Figure 20.5. A set of a cyan carquin dye releaser, a magenta carquin dye releaser, and a yellow hydroquinone dye releaser for full color reproduction. This set is cited from a patent claiming photographic additives [3].

agreement with the multilayer structure shown in Fig. 20.2, the cyan (**11**) and the magenta (**12**) belong to the category of carquin dye releasers, while the yellow (**13**) is a ballasted hydroquinone dye releaser.

Another set of dye releasers having a fused carquin nucleus has appeared in a more recent embodiment, which has been described in a patent claiming mordants, as shown in Fig. 20.6 [7]. The cyan (**14**) and the magenta (**15**) are carquin dye releasers, while the yellow dye releaser (**16**) belongs to the category of ballasted hydroquinone dye releasers. The branching at the methine position remains in these dye releasers.

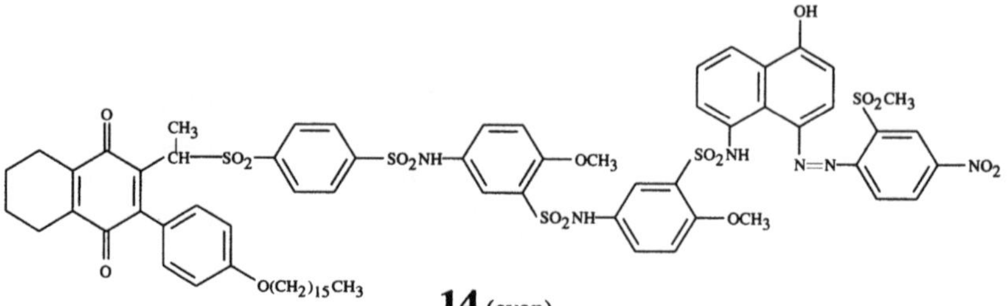

14 (cyan)

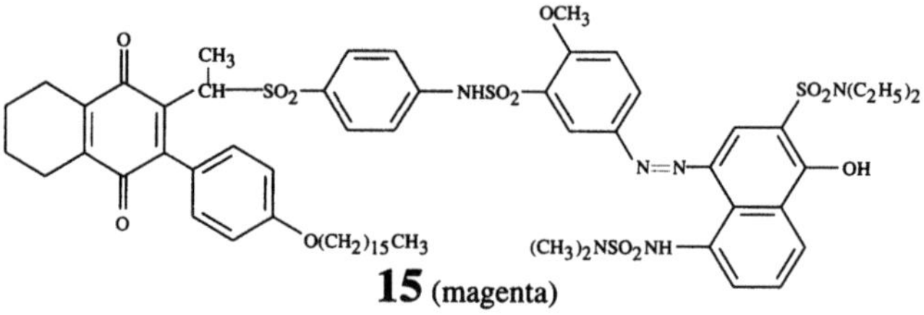

15 (magenta)

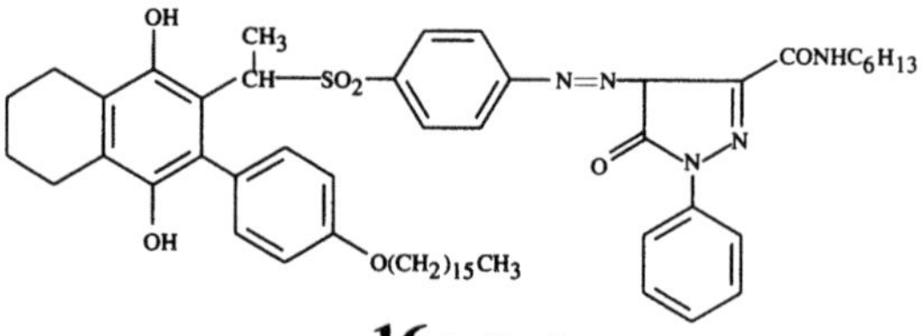

16 (yellow)

Figure 20.6. A set of a cyan carquin dye releaser, a magenta carquin dye releaser, and a yellow hydroquinone dye releaser for full color reproduction. This set is cited from a patent claiming mordants [7].

Linking groups containing three aromatic nuclei have been incorporated in carquin dye releasers such as **14** [8]. Other attempts to incorporate such linking groups have been described in several patents [9,10,11]. Carquin dye releasers with multiple dye units have been disclosed as further examples [12,13].

Syntheses

Figure 20.7 illustrates a synthetic pathway for preparing the magenta dye releaser (**12**) [6]. The first part shows the synthesis of the ballasted hydroquinone moiety (**22**). The introduction of an aryl group at the 3-position of 2-acetyl-5-methylbenzoquinone (**19**) is conducted under acidic conditions. Thus, hexadecyloxybenzene is condensed with 2-acetyl-5-methylbenzoquinone (**19**) in the presence of trifluoroacetic acid so as to give 2-acetyl-3-hexadecyloxy-5-methylbenzoquinone (**20**). The quinone function and the 2-acetyl function are successively reduced, finally giving the ballasted hydroquinone moiety (**22**).

On the other hand, an azo dye moiety having a sulfinyl group (**25**) has been prepared according to the scheme shown in the second part of Fig. 20.7. The

1) Ballasted quinone moiety:

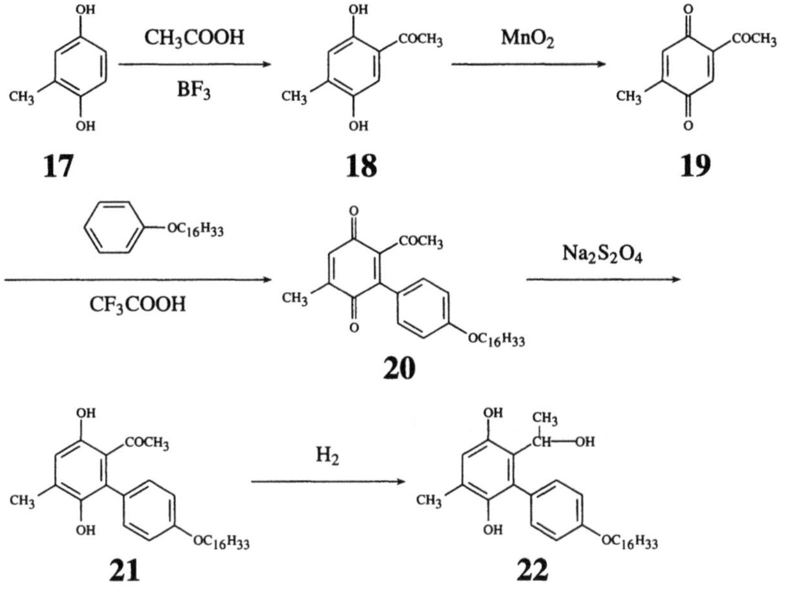

2) Carquin dye releaser:

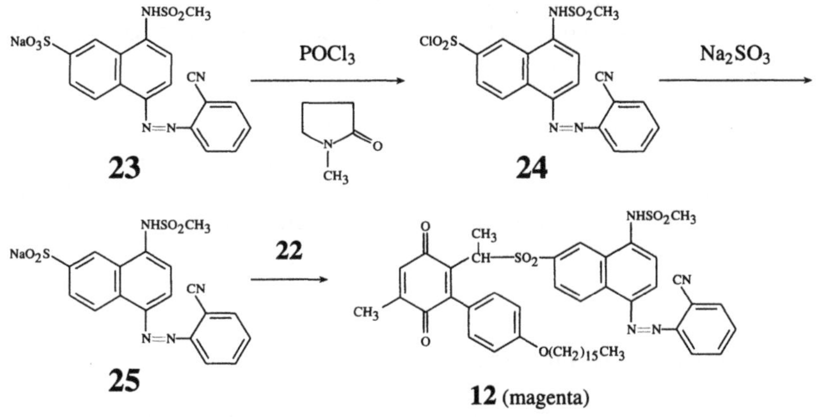

Figure 20.7. Synthetic pathway for preparing the magenta dye releaser (**12**) [6].

conversion of a chlorosulfonyl group (in **24**) into a sulfinic acid function (in **25**) is effected by sodium sulfite. Finally, the condensation between **22** and **25** produces the target molecule (**12**).

20.1.3 Dye Moieties for Carquin Dye Releasers

Various attempts to improve dye moieties in carquin dye releasers have been disclosed in patents. For example, chelatable 4-thienyl imidazole azo dyes have been proposed as cyan dye moieties [14]. Thiazole azo dyes [15], 6-arylazo-2-amino-3-pyridinol dyes [16], thiophene azo dyes [17], and pyridine azo or pyridazine azo dyes [18] have been proposed as cyan dye moieties incorporated in carquin dye releasers.

As for magenta dye releasers, chelatable imidazole azo dyes [19] and pyridine azo dyes [20] have been proposed.

Carquin dye releasers liberating arylazocyanoacetic ester dyes [21], pyridone azo dyes [22], and pyrazolone azo dyes [23] have appeared to exhibit improved properties.

20.1.4 Chemicals Contained in Photo-Sensitive Layers

Electron Donors for Carquin Dye Releasers

An electron donor (ED) reduces a carquin dye releaser according to eq. 20.3 (Fig. 20.1). The earliest example of ED used for carquin dye releasers is ascorbyl palmitate, which has been used in combination with 4-methyl-1-phenyl-3-pyrazolidone (**1**) as an ETA [2]. The sets of carquin dye releasers shown in Figs. 20.5 and 20.6 have been combined with 2,5-bis(1′,1′,3′,3′-tetramethylbutyl)hydroquinone (**4**) as an ED (Fig. 20.8), which has also been used together with the ETA (**1**) [3,7].

ED precursors such as **26** have a hydroquinone group protected as a lactone, which is capable of opening to regenerate a hydroquinone function [24]. Another type of ED precursors involves a CO–CH(OCOR)–CO group, which is hydrolyzed to a CO–CH(OH)–CO group as a reducing agent, as exemplified by **27** [25]. Ballasted indoles such as **28** can be used as an ED [26].

Electron Transfer Agents for Carquin Dye Releasers

An electron transfer agent (ETA) undergoes a redox reaction with silver halide according to eq. 20.1 (Fig. 20.1). Then the resulting ETA_{ox} oxidizes an ED according to eq. 20.2. Such an ETA (or its precursor) has been usually incorporated in an interlayer or a protective layer; but it may be added to an activator fluid.

The ETA (**1**) has been used in combination with the ED (**4**) for treating the sets of carquin dye releasers shown in Figs. 20.5 and 20.6 [3,7]. A protected form such as **29** (Fig. 20.9) has been used together with the ED (**27**) [25]. Another protected ETA such as **30** has been used in combination with the ED (**26**) [27].

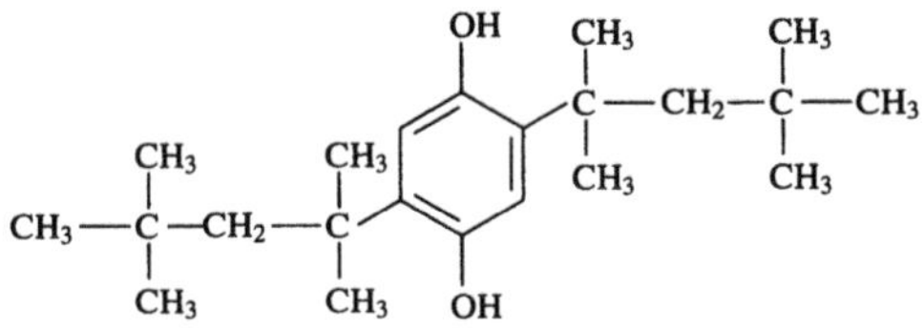

Figure 20.8. Electron donors (ED) for carquin dye releasers.

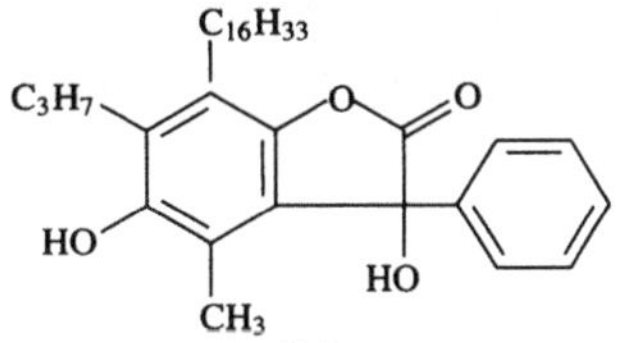

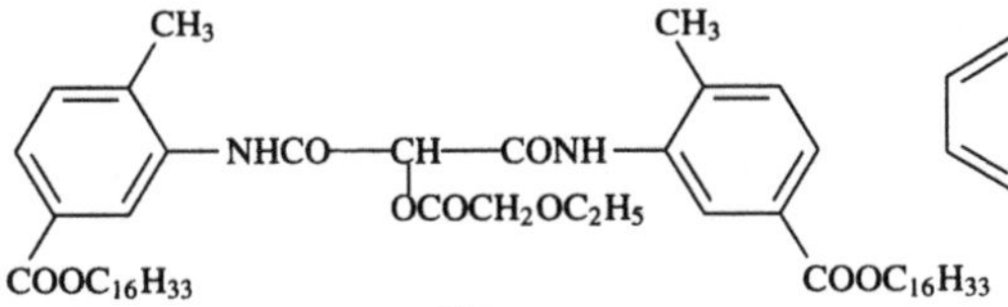

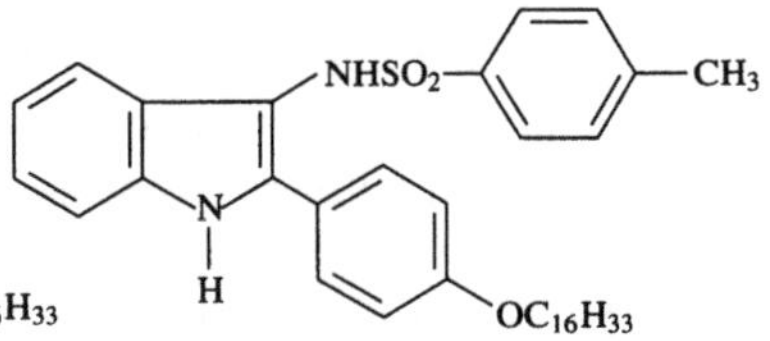

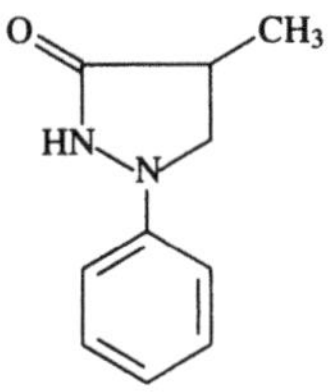

Figure 20.9. Electron transfer agents (ETAs) for carquin dye releasers.

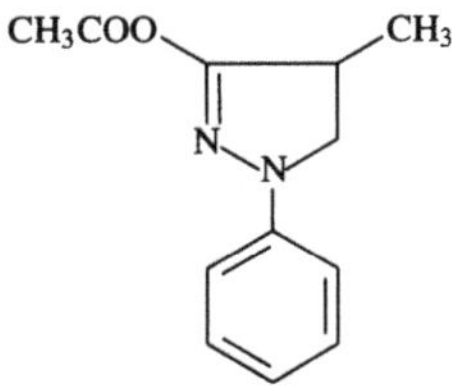

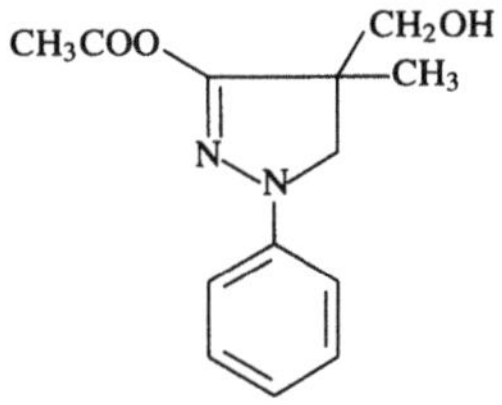

20.1.5 Chemicals Contained in an Activator Fluid

An activator fluid which has been combined with the set of carquin dye releasers shown in Fig. 20.6 comprises sodium hydroxide, sodium orthophosphate, cyclohexane-1,4-dimethanol, sodium thiosulfate, potassium iodide, and 2,2-dimethylpropane-1,3-diol [7]. These components have been covered by the following patents.

The addition of cyclohexane-1,4-dimethanol and 2,2-methylpropane-1,3-diol to an activator fluid has been disclosed to enhance D_{max} values in photographic materials based on carquin dye releasers [27]. Triisopropanolamine and related aminoalcohols have been proposed to increase D_{max} values [4].

Another way to increase D_{max} values is the presence of free iodide ions [3]. Potassium iodide is added for this purpose. This effect has been tested for the set of carquin dye releasers shown in Fig. 20.5.

A silver halide solvent such as sodium thiosulfate has been disclosed to cause an interlayer effect of the resulting silver complex [28]. Thus the physical development of the silver complex diffused to the adjacent green-sensitive emulsion layer decreases the undesired releasing of magenta dye in an area exposed to blue light so that more bright green image is obtained.

20.1.6 Mordant Polymers for Carquin Dye Releasers

Condensation polymers such as **31** have been proposed as mordant polymers for the image-receiving layer for diffusion transfer photographic materials [29,30]. Condensation polymers of another type (e.g., **32**) have been proposed as mordant polymers and used in combination with the set of dye releasers shown in Fig. 20.6 [7]. This patent has reported the comparison with the polyurethane (**31**) and the quaternary ammonium polymer (**32**). More recently, addition polymers having a phosphonium mordant (e.g., **33**) have been disclosed to be used in combination with the set of dye releasers shown in Fig. 20.5 [31,32].

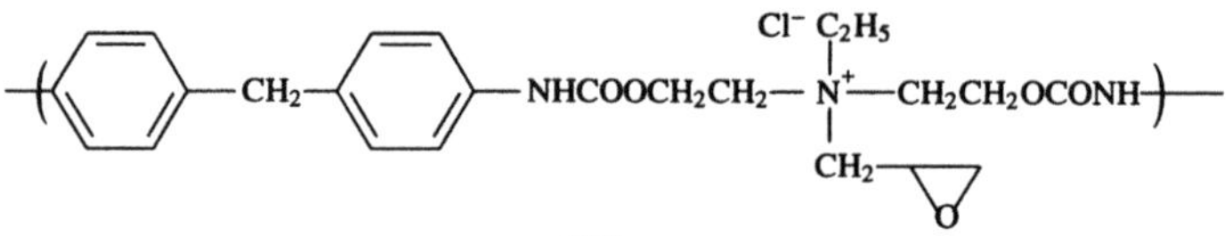

Cl^- C_2H_5 CH_2 NHCOOCH$_2$CH$_2$—N$^+$—CH$_2$CH$_2$OCONH CH_2 O

31

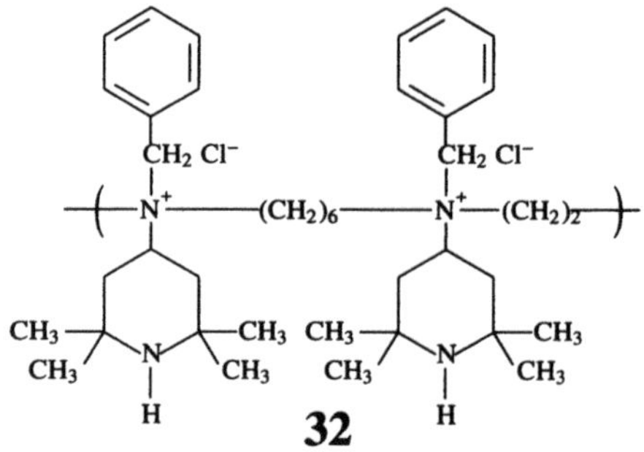

CH_2 Cl^- N$^+$ $(CH_2)_6$ N$^+$ $(CH_2)_7$ CH_3 CH_3 N H

32

CH_2 CH 48.8 0.9 7.5 39.7 CN $CH_2P^+(C_4H_9)_3\ Cl^-$ CH_2Cl CH_2OH

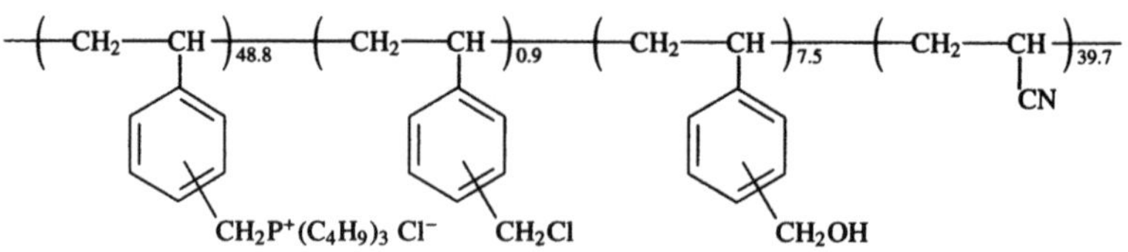

33

Figure 20.10. Mordant polymers for carquin dye releasers.

20.2 ROSET Dye Releasers

A heat-developable diffusion transfer photographic system based on ROSET dye releasers[5] has been placed on the market by Fuji Photo Film under the name "Fujix Pictrostat 100" in 1991 [33,34]. This system has employed heat development where an activator fluid is replaced by a small amount of water.[6] This means that the system can be regarded as a substantial dry system, which is suitable for office-use beyond laboratory-use.

As for heat-developable systems using positive-working dye releasers of reductive-releasing type, a predecessor has been disclosed in a patent [35]. This patent has extensively claimed heat-developable systems using positive-working dye releasers of reductive-releasing type.[7] However, the commercialization of such systems had not been realized until the ROSET dye releasers were developed.

20.2.1 Chemical Processes of ROSET Dye Releasers

ROSET dye releasers are positive-working compounds of reductive-releasing type [36]. The reaction scheme of the ROSET dye releasers is shown in Fig. 20.11. Although this reaction scheme seems to be obtained by reissuing Fig. 20.1 for carquin dye releasers, the employment of heat development provides substantial differences in the selection of a basic agent,[8] an electron transfer agent (ETA, e.g., **34**), and an electron donor (ED, e.g., **37**).

The ETA appeared in eq. 20.5 (Fig. 20.11) is 1,5-diphenyl-3-pyrazolidone, which is more suitable for heat development than 4-methyl-1-phenyl-3-pyrazolidone appeared in eq. 20.1 (Fig. 20.1) [37]. The reaction of eq. 20.5 occurs on the action of an exposed negative silver halide grain. The resulting ETA_{ox} (**36**) oxidizes the anionic species (**38**) of an electron donor (ED: **37**), as shown in eq. 20.6.

The remaining (effective) ED (or its anionic species) reduces a ROSET dye releaser (**40**), where one electron transfer from the ED to the ROSET results in the ring opening at the N—O bond of the five-membered ring. The resulting intermediate (**41**) releases a diffusible dye (**43**), as shown in eq. 20.7. Finally, the radical intermediate (**42**) is reduced into the corresponding anionic species.[9]

---

[5] The name "ROSET" is an acronym of "Ring Opening by Single Electron Transfer."

[6] Another heat-developable system based on *o*-sulfonamidophenol dye releasers has been marketed by Fuji Photo Film under the name "Pictrography", as already discussed in Chapter 19. The two systems are called "Pictrocolor" together.

[7] Although the examples have been limited within BEND compounds, carquins, etc., the first claim of the patent covers embodiments using ROSET dye releasers in a broader sense.

[8] In this scheme, a basic agent is represented by $[OH^-]$, although a precise species is not clarified.

[9] Although one electron shift is sufficient to release a dye, one more electron is necessary to obtain the final stable anionic species.

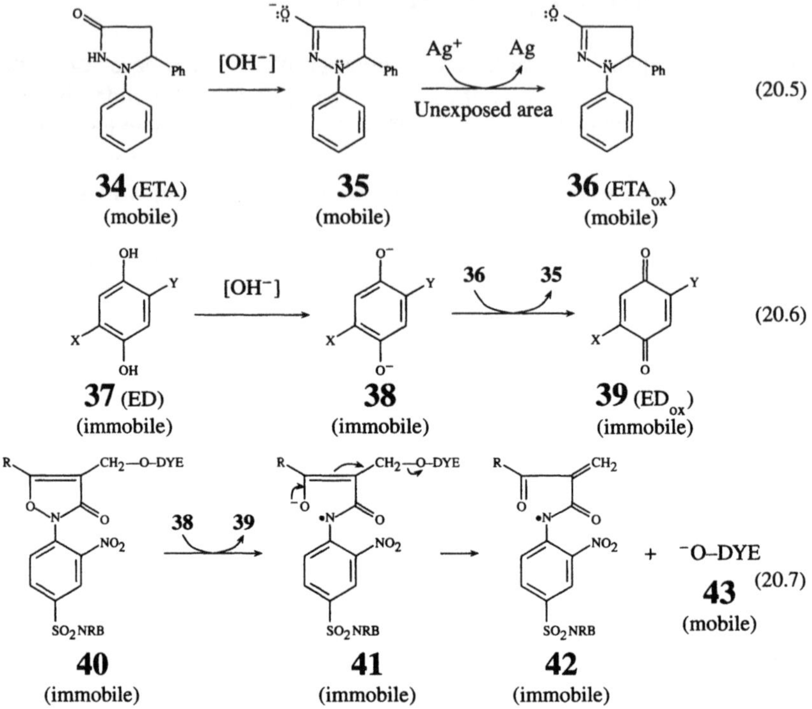

[OH⁻] → Ag⁺ Ag, Unexposed area → (20.5)

34 (ETA) (mobile) **35** (mobile) **36** (ETA_{ox}) (mobile)

[OH⁻] → 36 35 → (20.6)

37 (ED) (immobile) **38** (immobile) **39** (ED_{ox}) (immobile)

38 39 → → + ⁻O–DYE (20.7)

40 (immobile) **41** (immobile) **42** (immobile) **43** (mobile)

Figure 20.11. Consecutive reactions of silver halide, an electron transfer agent (ETA), an electron donor (ED), and a ROSET dye releaser. The letters X and Y represent appropriate substituents, which may be ballast groups. The letter R denotes an alkyl group. The letter B represents a ballast group. The symbol DYE denotes a cyan, magenta, or yellow dye moiety.

20.2.2 Multilayer Structure for Color Reproduction

Multilayer Structure for ROSET Dye Releasers

The schematic cross-section of a photographic material based on positive-working ROSET dye releasers is illustrated in Fig. 20.12. This multilayer structure is cited from a patent claiming photographic additives [39].

Since the photographic material shown in Fig. 20.12 aims at the use of copying, it does not require as high speed as materials for amateur use. Accordingly, each ROSET dye releaser (Ⓕ-Ⓒ, Ⓕ-Ⓜ, or Ⓕ-Ⓨ) is coated together with a negative emulsion (△) so as to construct a single layer (Layer 1, 3, or 5). Moreover, an electron donor (ED: Ⓔ) as well as an electron transfer agent (ETA: [E]) are incorporated in each of the layers (Layere 1, 3, or 5). Each interlayer (Layer 2 or

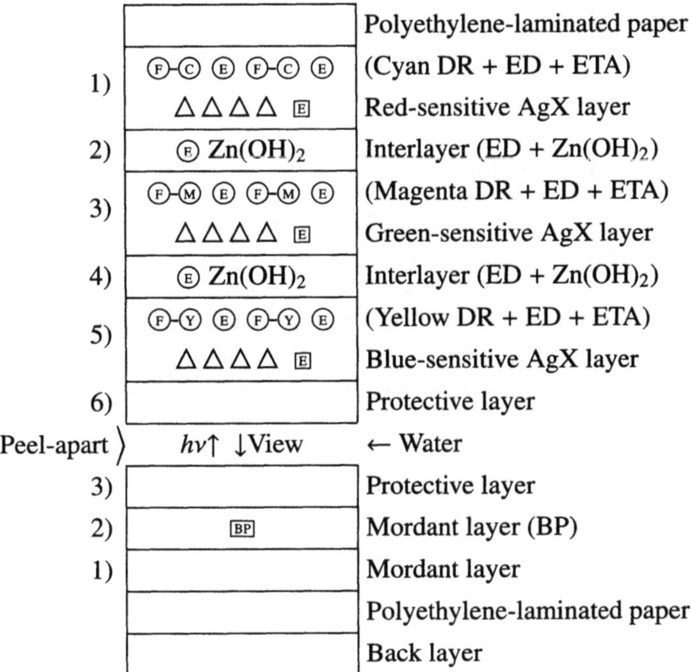

Figure 20.12. Schematic cross-section of a heat-developable photographic material based on ROSET dye releasers and negative silver salt emulsions for color reproduction. △: Silver halide grain, Ⓕ-Ⓨ: Yellow ROSET dye releaser (yellow DR) , Ⓕ-Ⓜ: Magenta ROSET dye releaser (magenta DR), Ⓕ-Ⓒ: Cyan ROSET dye releaser (cyan DR); Ⓔ: Electron donor (ED); [BP]: Base precursor; and [E]: Electron transfer agent.

4) contains the ED (Ⓔ) for preventing color contamination. One of the mordant layers contains a base precursor ([BP]) such as guanidinium picolinate.[10]

Each photo-sensitive layer (Layer 1, 3, or 5) contains all components participating in the reactions shown in Fig. 20.11, i.e., silver halide, an electron transfer agent ([E]: **34**), an electron donor (Ⓔ: **37**), and a ROSET dye releaser (Ⓕ-Ⓒ, Ⓕ-Ⓜ, or Ⓕ-Ⓨ: **40**). This means that the discrimination between eq. 20.6 (after eq. 20.5) and eq. 20.7 is crucial to assure imagewise formation of dyes.

Color Reproduction

Color reproduction by using the film shown in Fig. 20.12 is substantially the same as the scheme shown in Figs. 20.3 and 20.4. As a result, subtractive color

[10] The base-forming mechanism of guanidinium picolinate and zinc hydroxide has been discussed in Chapter 19 (Subsection 19.8.1).

mixings for blue (C + M = B), green (C + Y = G), red (M + Y = R), black (C + M + Y = K), and white (no dyes) are obtained in terms of the negative-working mechanism of the reductive-releasing type.

20.2.3 ROSET Moieties

ROSET moieties are 4-isoxazoline-3-one derivatives which have an electron-accepting aromatic nucleus at the 1-position [36]. The electron-accepting aromatic nucleus is substituted by a 2-nitro group along with another electron-withdrawing group, as found in the structure (**40**). Thereby, it is capable of accepting an electron to give an anion radical species having a closed ring, which undergoes a rapid ring opening to generate a open-chain anion radical species (**41**) according to eq. 20.7 [33,38].

ROSET compounds based on a 4-isoxazoline-3-one nucleus have been clarified to have the following excellent properties [33]:

1. 5-Substituents influence the rates of releasing. Comparison data of a *t*-butyl, a phenyl, and a 4-methoxyphenyl group has shown a 100-times acceleration.

2. The released moiety at the 4-position (i.e., X of 4-X-CH_2) can be selected from a variety of leaving groups such as a phenoxy (ArO^-), a sulfinic acid group ($ArSO_2^-$), and an arylthio (ArS^-).

3. The release occurs almost quantitatively.

4. The releasing reaction occurs even at low pH, since the pKa of the intermediate radical anion (**41**) is low.

5. ROSET dye releasers are thermally stable in themselves.

A set of three ROSET dye releasers for full color reproduction is illustrated in Fig. 20.13, where it is cited from patents claiming electron donors [39] and electron transfer agents [40]. It should be noted that the cyan dye releaser (**44**) and the yellow dye releaser (**46**) release dyes having a phenoxide anion (DYE–O^-), while the magenta dye releaser (**45**) releases a dye having a sulfinate anion (DYE–SO_2^-).

ROSET dye releasers having two dye moieties per single ROSET moiety have been proposed to give high D_{max} values [41]. Compounds having a leaving group –N(SO_2CH_3)—N(Ar)– in place of the group –O—N(Ar)– of the usual ROSET dye releasers have been disclosed as another type of ROSET dye releasers [42].

Another set of ROSET dye releasers for full color reproduction have appeared in more recent embodiments, as shown in Fig. 20.14. This set is cited from a patent claiming cyan dye releasers [43] and a patent claiming yellow dye releasers [44].

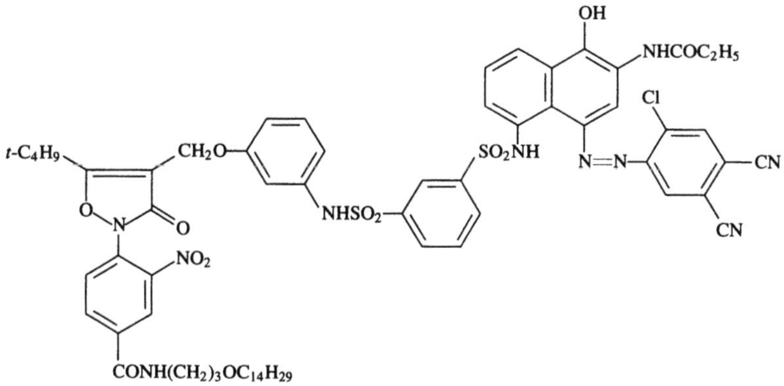

44 (cyan)

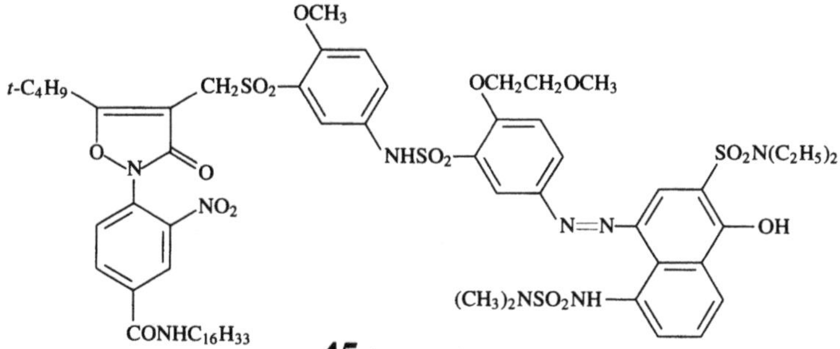

45 (magenta)

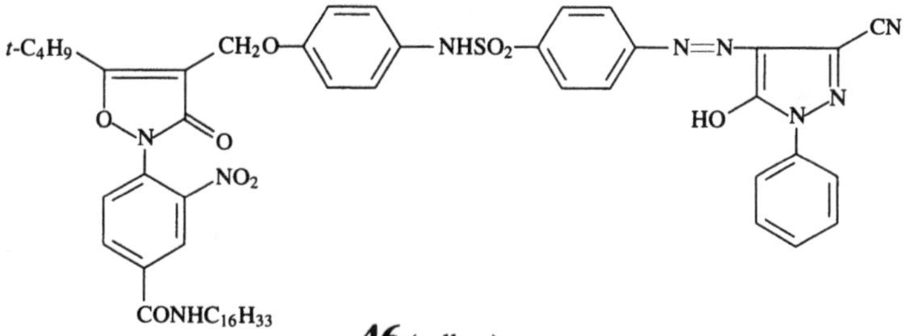

46 (yellow)

Figure 20.13. A set of ROSET dye releasers for full color reproduction. This set is cited from a patent claiming electron donors [39] and a patent claiming electron transfer agents [40].

Syntheses

As an example of syntheses of ROSET dye releasers, Fig. 20.15 illustrates a synthetic pathway for preparing the yellow dye releaser (**59**) [36].

A key intermediate is a ROSET moiety (**54**), which is prepared by starting from 4-chloro-2-nitrobenzoic acid (**50**) [45]. After the introduction of a ballast group, the intermediate (**51**) is condensed with 5-*t*-butyl-3-hydroxyisoxazole (**52**)

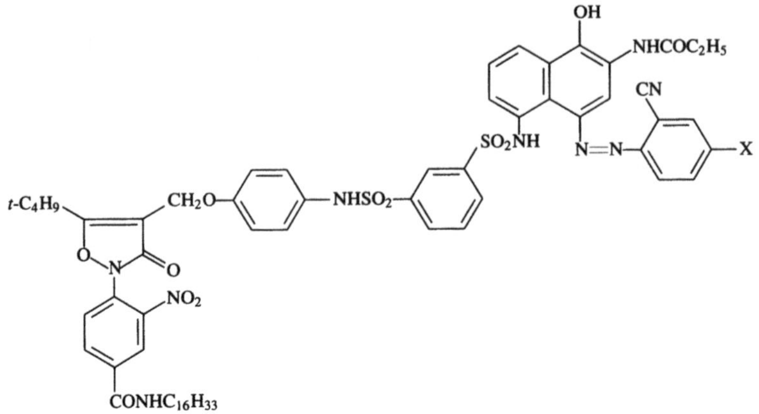

47 X = NO_2 (cyan)
48 X = SO_2CH_3 (cyan)

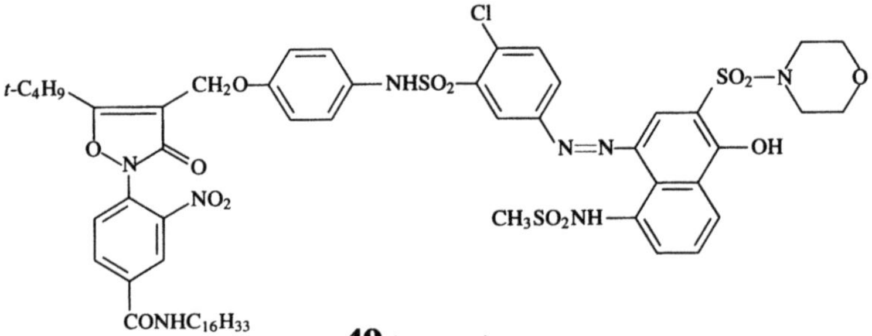

49 (magenta)

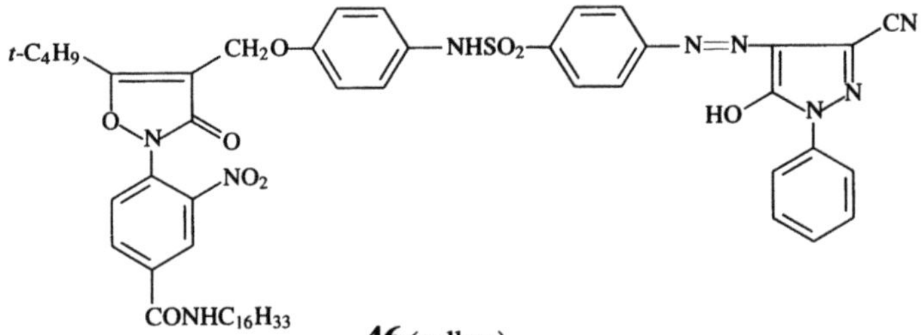

46 (yellow)

Figure 20.14. Another set of ROSET dye releasers for full color reproduction. This set is cited from a patent claiming cyan dye releasers [43] and a patent claiming electron yellow dye releasers [44].

to give a ROSET nucleus (**53**), which undergoes a chloromethylation by means of formaldehyde and hydrogen chloride in the presence of zinc chloride. The resulting key intermediate (**54**) is further condensed with a protected aminophenol (**55**) to give a condensed product (**56**), from which the protecting group is removed to generate the corresponding free amine (**57**).

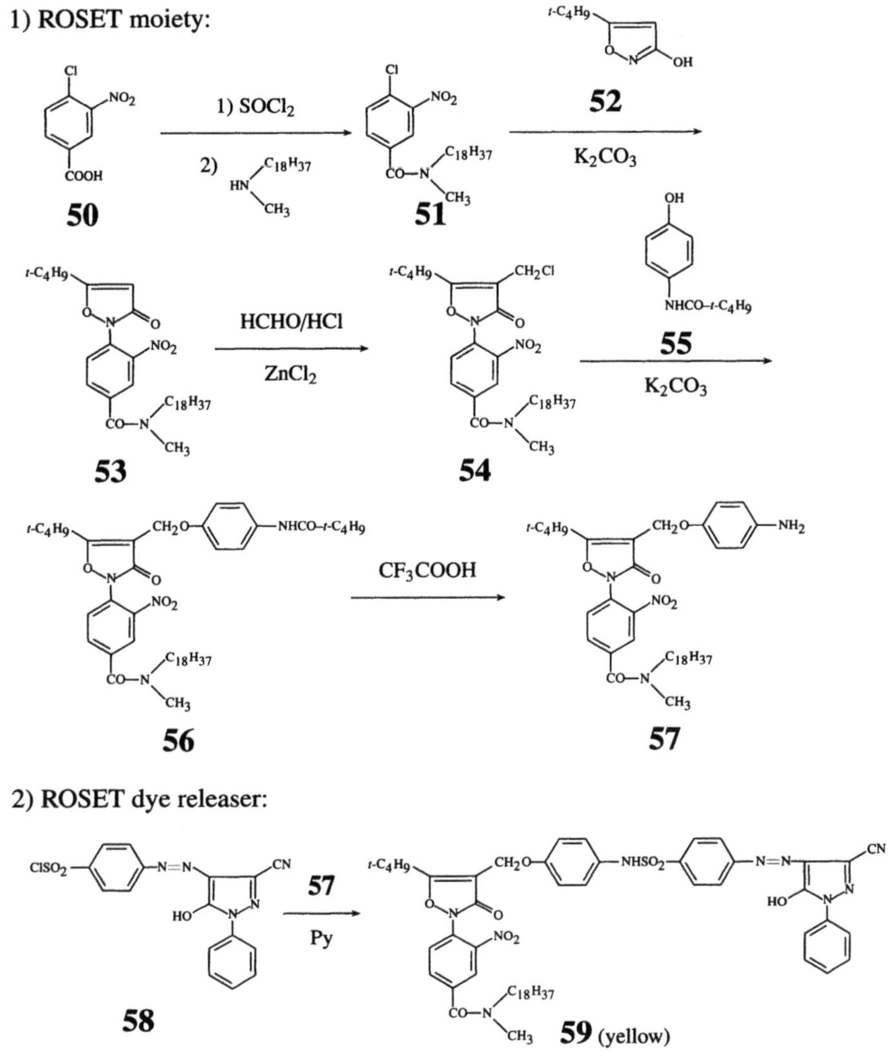

1) ROSET moiety:

2) ROSET dye releaser:

Figure 20.15. Synthetic pathway for preparing the yellow dye releaser (**59**) [36,45].

The final step is the condensation of the dye-sulfonyl chloride (**58**) and the amine intermediate (**57**) in the presence of pyridine so as to produce the yellow ROSET dye releaser (**59**).

20.2.4 Dye Moieties for ROSET Dye Releasers

The dye moieties of the *o*-sulfonamidophenol dye releasers for the heat-developable photographic materials (Chapter 19) can also be employed as dye moieties for ROSET dye releasers. Thus, the dye moieties which are contained in the ROSET dye releasers listed in Figs. 20.13 and 20.14 or related dye moieties have already been described in Chapter 19.

For example, the cyan dye moiety incorporated in the ROSET dye releaser (**44**) is covered by a patent which claims *o*-sulfonamidophenol dye releasers [46]. The cyan dye moieties for the ROSET dye releasers (**47** and **48**) are covered by a patent which also claims *o*-sulfonamidophenol dye releasers [48]. Recently, dye moieties having a heterocyclic diazo component have been proposed as cyan dye moieties for heat-developable dye releasers, as exemplified by **60** [43].

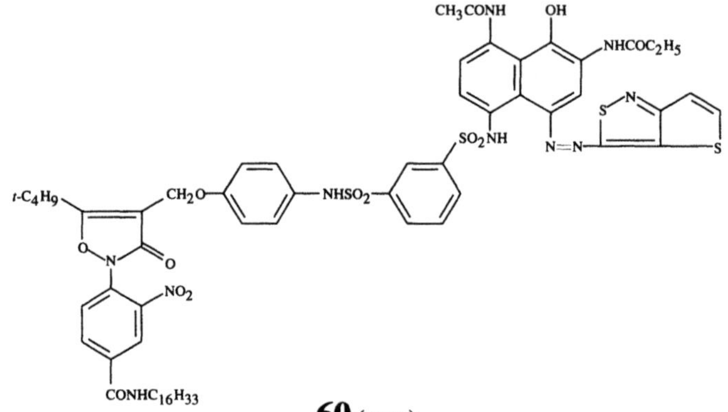

60 (cyan)

Figure 20.16. Cyan ROSET dye releaser for a heat-developable photographic material.

With respect to magenta dye moieties for heat-developable dye releasers, dye moieties having a heterocyclic diazo component have been proposed, as exemplified by **61** [47]. This type of dye moieties has already been described in Chapter 19, where its incorporation in heat-developable *o*-sulfonamidophenol dye releasers has been discussed.

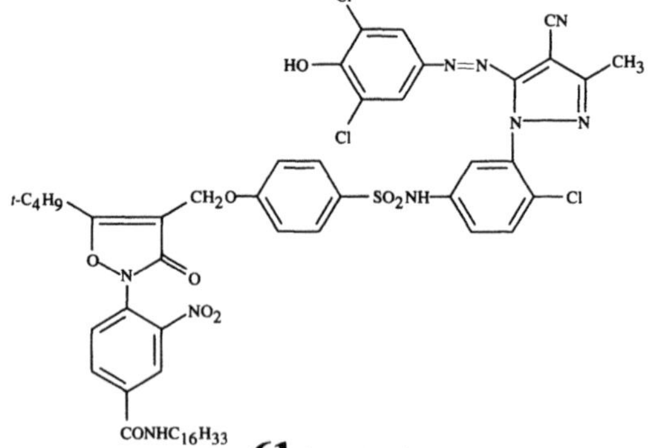

61 (magenta)

Figure 20.17. Magenta ROSET dye releaser for a heat-developable photographic material.

As for yellow ROSET dye releasers, Fig. 20.18 illustrates a recent proposal of yellow dye moieties based on 1H-pyrazolo[1,5-*b*][1,2,4]triazole coupling components [44]. This nucleus has already been described in Chapter 19, where its incorporation in heat-developable *o*-sulfonamidophenol dye releasers has been discussed. In addition, remember that this nucleus has been described in Chapter 10 with respect to its application to magenta couplers for conventional color photography.

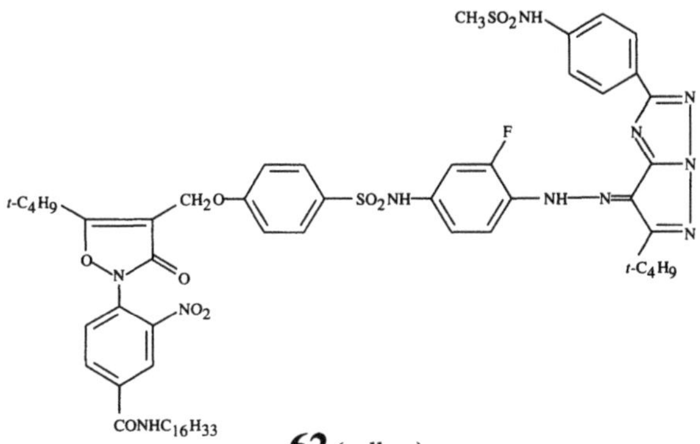

62 (yellow)

Figure 20.18. Yellow ROSET dye releaser for a heat-developable photographic material.

20.2.5 Chemicals Contained in Photo-Sensitive Layers

Electron Donors for ROSET Dye Releasers

An electron donor (ED) reduces a ROSET dye releaser according to eq. 20.7 (Fig. 20.11). Earlier examples of EDs used for ROSET dye releasers are **63** and **64**, which have been used in combination with 4-hydroxymethyl-4-methyl-1-phenyl-3-pyrazolidone as an ETA [36]. 2,5-Dialkylhydroquinones such as **65** have been claimed as EDs [39], where they have been used for reducing the dye releasers shown in Fig. 20.13 in combination with 1,5-diphenyl-3-pyrazolidone as an ETA. 2,5-Dialkylhydroquinones of a similar type (e.g., **66**) have been proposed as EDs [49]. The EDs **65** and **66** have been cited as preferred compounds in the technical report described above [33]. Compounds having two hydroquinone moieties (e.g., **67**) have been also proposed as EDs [49]. In addition, 2-(2′-hydroxyphenyl)hydroquinones have been disclosed to be EDs [50]. Electron donors such as **68** (R = CH_3) have been disclosed to reduce ROSET dye releasers [51]. A homolog **69** (R = $C_{11}H_{23}$-*n*), which is covered by the patent [51], has been reported in recent embodiments, where **69** has been used together with **67** and/or **64** [43,44].

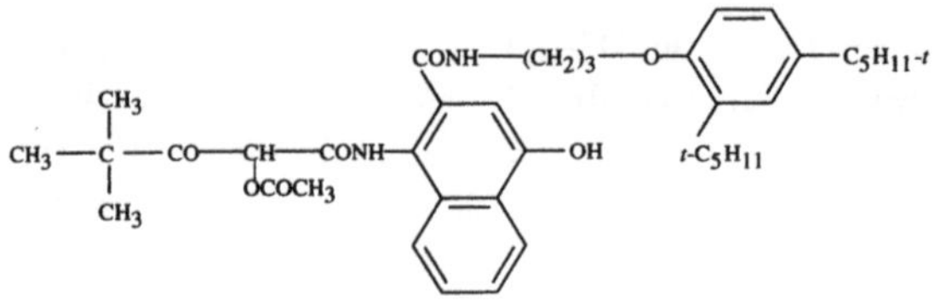

63

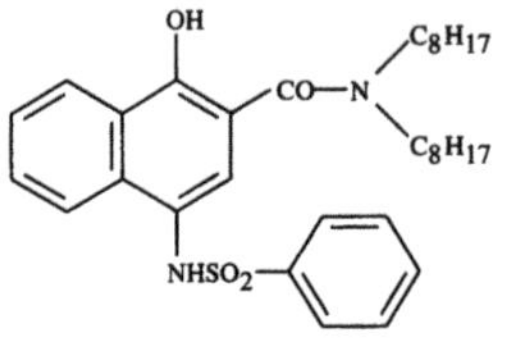

64

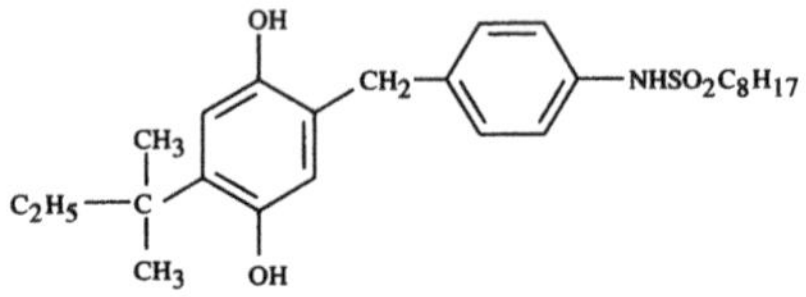

65

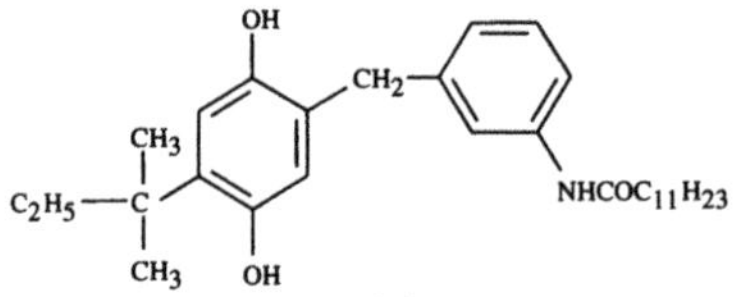

66

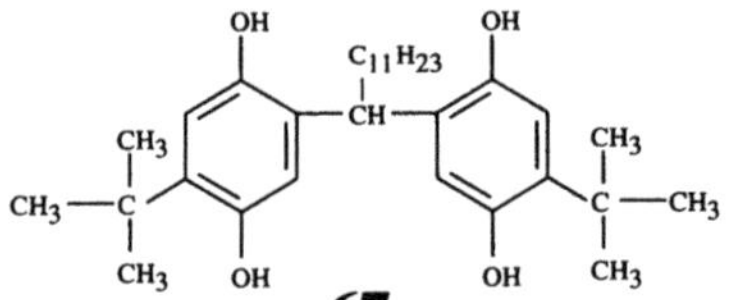

67

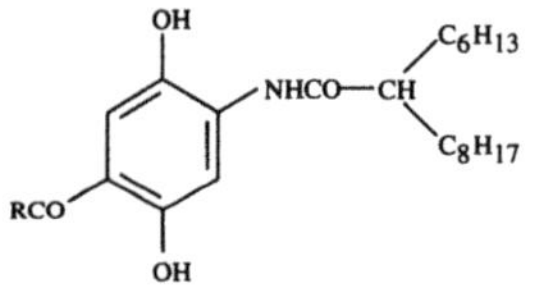

68 R = CH_3
69 R = $C_{11}H_{23}$-*n*

Figure 20.19. Electron donors (EDs) for ROSET dye releasers.

Electron Transfer Agents for ROSET Dye Releasers

An electron transfer agent (ETA, e.g., **34**) undergoes a redox reaction with silver halide according to eq. 20.1 (Fig. 20.11) so as to generate the corresponding oxidized form (ETA_{ox}). To realize faithful color reproduction, the ETA_{ox} should quench an ED effectively according to eq. 20.6. As found in Fig. 20.12 which shows that the ETA, the ED, and each ROSET dye releaser are incorporated in the same layer, the balance between eq. 20.2 and eq. 20.3 is crucial. For example, the ETA (**34**) has been used in combination with the ED (**65**), where the ROSET dye releasers shown in Fig. 20.13 have been used [39]. The ETA (**70**) has been combined with the ED (**66**) [52]. An ETA precursor (**71**) has been used together with the ETA (**34**) in an embodiment which contains the ROSET dye releasers shown in Fig. 20.13 [40]. Such a combined use in addition to a solid particle dispersion technique is effective in order to enhance the stability during storage [37].

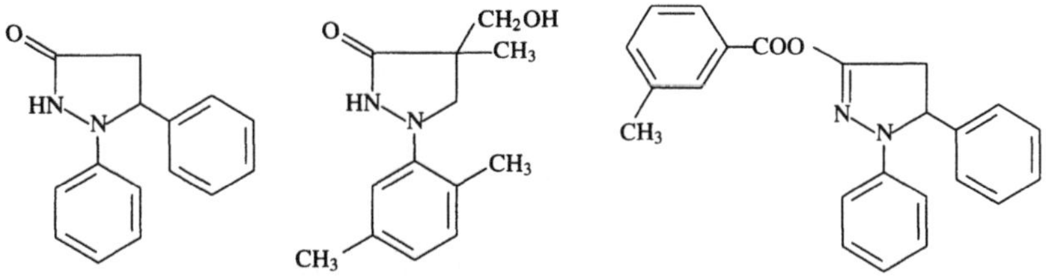

34 **70** **71**

Figure 20.20. Electron transfer agents (ETAs) for ROSET dye releasers.

20.2.6 Mordant Polymers for Dyes from ROSET Dye Releasers

Mordant polymers for ROSET dye releasers have been investigated in a parallel direction with the heat-developable *o*-sulfonamidophenol dye releasers described in Chapter 19. The mordant polymer (**72**) has been used in several embodiments which have employed the ROSET dye releasers shown in Fig. 20.13 [39] and Fig. 20.14 [44].

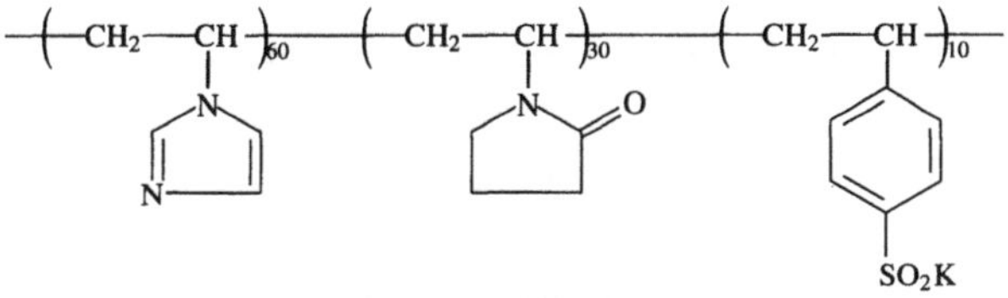

72

Figure 20.21. Mordant polymer.

References

[1] IIzuka H (1984) Insatsu Zasshi. 67(3):45

[2] Janssens W (1980) US Patent 4 232 107

[3] Janssens W, Claeys DA (1985) US Patent 4 496 405

[4] Janssens W, Claeys DA (1982) US Patent 4 353 975

[5] Boie I, Wingender K (1983) US Patent 4 407 929

[6] Van de Sande CC, Janssens W, Lässig W, Meier E (1983) US Patent 4 371 604

[7] Vanmaele LJ, Janssens W (1989) US Patent 4 814 255

[8] Van de Sande CC, Van den Bergh AM, Janssens W, Vetter H (1986) US Patent 4 605 613

[9] Vanmaele LJ, Van de Sande CC, Janssens W, Vetter H (1988) US Patent 4 777 124

[10] Vanmaele LJ, Janssens W, Van de Sande CC (1989) US Patent 4 855 223

[11] Vetter H, Van de Sande CC, Vanmaele LJ, Van den Bergh AM, Janssens W (1991) US Patent 5 037 731

[12] Van de Sande CC, Van den Bergh AM, Verhecken A (1987) US Patent 4 663 273

[13] Vanmaele LJ, Van de Sande CC (1989) US Patent 4 871 654

[14] Schenk G, Bergthaller P, Heidenreich H, Volfrüm G (1985) US Patent 4 524 123

[15] Bergthaller P, Stolzenburg R, Marx P, Hemprecht R (1985) US Patent 4 524 124

[16] Bergthaller P, Volfrüm G, Heidenreich H (1986) US Patent 4 579 817

[17] Bergthaller P, Fürstenwerth H, Stolzenburg R, Marx P, Heidenreich H (1986) US Patent 4 579 805

[18] Bergthaller P, Stolzenburg R, Marx P (1986) US Patent 4 767 698

[19] Bergthaller P, Schenk G, Volfrüm G (1984) US Patent 4 461 827

[20] Stolzenburg R, Bergthaller P, Volfrüm G, Strauss J (1985) US Patent 4 521 506

[21] Schenk G, Bergthaller P, Volfrüm G, Stolzenburg R (1983) US Patent 4 393 132

[22] Bergthaller P, Rosenhahn L, Herd KJ, Fuerstenwerth H, Stolzenburg R (1985) US Patent 4 559 296

[23] Van de Sande CC, Kok P (1988) US Patent 4 748 108

[24] Lässig W, Meier E, Sdhleger S (1982) US Patent 4 366 240

[25] Odenwälder H, Marx P (1983) US Patent 4 420 557

[26] Odenwälder H, Vetter H, Janssen W, Jaeken J (1982) US Patent 4 360 581

[27] Krafft W (1984) US Patent 4 440 849

[28] Janssens W, Claeys DA (1983) US Patent 4 396 699

[29] Vermeulen LL, Vervloet LH, De Smedt WP (1988) US Patent 4 772 536

[30] Bergthaller P, von Bonin W, Helling G, Marx P (1980) US Patent 4 186 014

[31] Janssens W, Timmerman DM, Claeys DA (1989) US Patent 4 855 211

[32] Timmerman DM, Claeys DA, Janssens W (1995) US Patent 5 380 619

[33] Nakamura K (1992) J Soc Photogr Sci Technol Jpn. 55:185

[34] Yokokawa T, Nakamura K, Matsumoto N (1992) Fuji Film Res & Dev. 37:49

[35] Sawada S, Fujita S, Koyama K (1985) US Patent 4 559 290

[36] Nakamura K, Nakamura S (1988) US Patent 4 783 396

[37] Taguchi T, Nakamine T, Kawata K, Hirai H (1994) J Soc Photogr Sci Technol Jpn. 57:340

[38] Nakamura K, Koya K, Sato K (1989) Yuki Gosei Kagaku Kyokai Shi. 47:629

[39] Nakamura K, Hirai H (1990) US Patent 4 968 598

[40] Taguchi T, Nakamine T, Kawata K (1992) US Patent 5 139 919

[41] Matsuda N, Yamada M, Ono M, Ishiwata Y, Uchida O (1996) US Patent 5 543 279

[42] Nakamura K (1990) US Patent 4 891 304

[43] Tateishi K, Kamio T, Komosaki T, Ishiwata Y, Naruse H (2000) US Patent 6 124 084

[44] Yamada M, Matsuda N, Ishiwata Y, Uchida O, Ono M (2000) US Patent 6 127 088

[45] Nakamura K, Nakamura S (1991) US Patent 5 071 994

[46] Tsukase M, Sato K, Hirai H (1993) US Patent 5 223 387

[47] Kamio T, Seto N, Hiyoshi H (1998) US Patent 5 731 140

[48] Nakamura K, Ito T, Tsukase M, Taguchi T (1991) US Patent 5 066 577

[49] Ono M, Hirai H, Ohki N, Hanaki K, Nakamura K (1991) US Patent 5 026 634

[50] Ono M, Motoki M, Hirai H (1991) US Patent 5 032 487

[51] Ono M, Nakamine, T, Hirai H (1993) US Patent 5 236 803

[52] Nakamine T, Kawata K, Taguchi T (1991) US Patent 5 017 454

Part V

Dye Bleach Photography

Chapter 21

Silver Dye Bleach Photography

21.1 Chemical Processes and Multilayer Structure

The beginning of silver dye bleach photography can be ascribed to Cristenson's patent on the reductive cleavage of an azo dye, which was caused by sodium hydrosulfite, stannous chloride etc. in the presence of image silver [1]. Extensive efforts to apply such silver dye bleach reactions to practical photography had been made by Gaspar so that a vast number of patents had appeared, as summarized in several reviews [2,3].[1] For example, he (or his coworker) disclosed thiocarbamide (thiourea), semicarbazide ($NH_2NHCONH_2$), etc. [4], amidol (2,4-diaminophenol dihydrochloride) [5], and 2,3-dimethylquinoxaline [6] for use as mediators of the redox reaction between silver metal and an azo dye; and hydrochloric acid and sodium chloride [7] and sulfuric acid, ethansulfonic acid, etc. [8] for use in the dye bleach compositions. Later, a photographic material based on the silver dye bleach reactions was placed on the market by Ciba-Geigy under the name "Cibachrome" in 1964.[2]

21.1.1 Chemical Processes of Dye Bleach Photography

Although a dye bleach reaction is simply represented by a formula: $Ar{-}N{=}N{-}Ar' + 4H^+ + 4Ag \rightarrow A{-}NH_2 + Ar'{-}NH_2 + 4Ag^+$, many additives are necessary in

[1] Honoring his efforts, silver dye bleach photography is frequently called "Gaspar Color".

[2] The "Cibachrome" is now called "Ilfochrome" because economic relationships between Ciba-Geigy and Ilford have been changed.

order that this reaction is employed in a practical use. One of the most essential additives is a proton carrier, which is usually called a *bleach catalyst*. Figure 21.1 shows a scheme involving an azine derivative as a bleach catalyst.

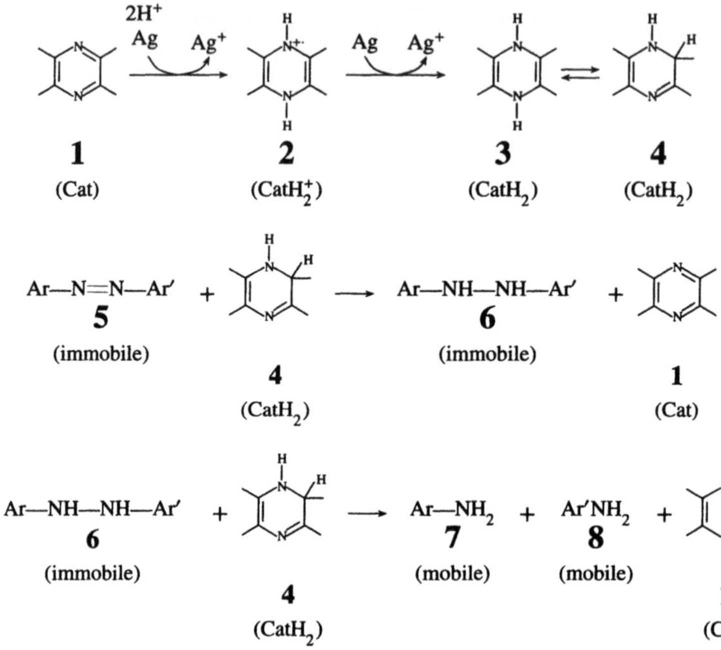

$2H^+$ Ag Ag^+ Ag Ag^+ (21.1)

1 (Cat) **2** ($CatH_2^+$) **3** ($CatH_2$) **4** ($CatH_2$)

Ar—N=N—Ar′ + **4** ($CatH_2$) ⟶ Ar—NH—NH—Ar′ + **1** (Cat) (21.2)

5 (immobile) **6** (immobile)

Ar—NH—NH—Ar′ + **4** ($CatH_2$) ⟶ $Ar{-}NH_2$ + $Ar'NH_2$ + **1** (Cat) (21.3)

6 (immobile) **7** (mobile) **8** (mobile)

Figure 21.1. Consecutive reactions of silver metal, a bleach catalyst, and a dye.

The azine derivative (Cat: **1**) as a bleach catalyst is reduced by silver metal, which is in turn generated by the imagewise development (reduction) of silver halide. This reduction produces **3**, which is changed into a more stable intermediate ($CatH_2$: **4**) according to eq. 21.1. One mole of the intermediate ($CatH_2$) reduces the azo dye (**5**) into the corresponding hydrazine (**6**), as shown in eq. 21.2). The hydrazine is further reduced and decomposed into amines (**7** and **8**) by additional one mole of the intermediate ($CatH_2$), as shown in eq. 21.3. Totally, one mole of the azo dye is fully reduced by two moles of $CatH_2$ or, in other words, by four moles of silver metal.

21.1.2 Multilayer Structure for Color Reproduction

Multilayer Structure for Silver Dye Bleach Photography

The schematic cross-section of a silver dye bleach photographic material is illustrated in Fig. 21.2. This multilayer structure is cited from a patent claiming a masking method [9] and from a patent claiming a method of acid addition

[10]. The inverse arrangement of Layer 1 and Layer 2 has also been reported in a technical report [11].

Layer	Content	Description
9)		Protective layer
8)	△△	Blue-scns. (no dye)
7)	△Y△Y	Blue-sens.
6)		Yellow filter
5)	△△	Green-sens. (no dye)
4)	△M△M	Green-sens.
3)		Interlayer
2)	△△	Red-sens. (no dye)
1)	△C△C	Red-sens.
		White opaque support

Figure 21.2. Schematic cross-section of multilayer structure of a silver dye bleach photographic material. This scheme illustrates silver halide grains and dyes only. △: Silver halide grain; Y: Yellow azo dye; M: Magenta azo dye; C: Cyan azo dye.

To reach optimum speed, double layers per each photo-sensitive unit has been incorporated, where a dyed silver halide emulsion (Layer 1, 4, or 7) is combined with a non-dyed silver halide emulsion (Layer 2, 5, or 8), as found in Fig. 21.2. Note that each of the non-dyed emulsion layers (Layer 2, 5, or 8) causes higher sensitivity and each of the dyed emulsions aims at superior granularity [11].

21.1.3 Color Reproduction

To discuss the features of silver dye bleach processing and color reproduction, let us here use a simple material (Fig. 21.3) for the sake of convenience, because the practical multilayer structure shown in Fig. 21.2 is too complicated to briefly summarize the essential features.

1. The dye bleach material is exposed by light (downarrows), as shown in Fig. 21.3a. The symbols △ and $\Delta_{\bullet}$ represent an unexposed silver halide grain and an exposed one with a latent image. The letters C, M, and Y express a cyan, a magenta, and a yellow azo dye, respectively.

2. The material is processed by silver development as shown in Fig. 21.3b to produce silver metal grains in a negative imagewise fashion.

3. In the dye-bleaching step (Fig. 21.3c), the silver metal grains are utilized to bleach dyes according to the reactions shown in Fig. 21.3d. The resulting amines are removed by dissolving them into the acidic dye-bleaching solution.

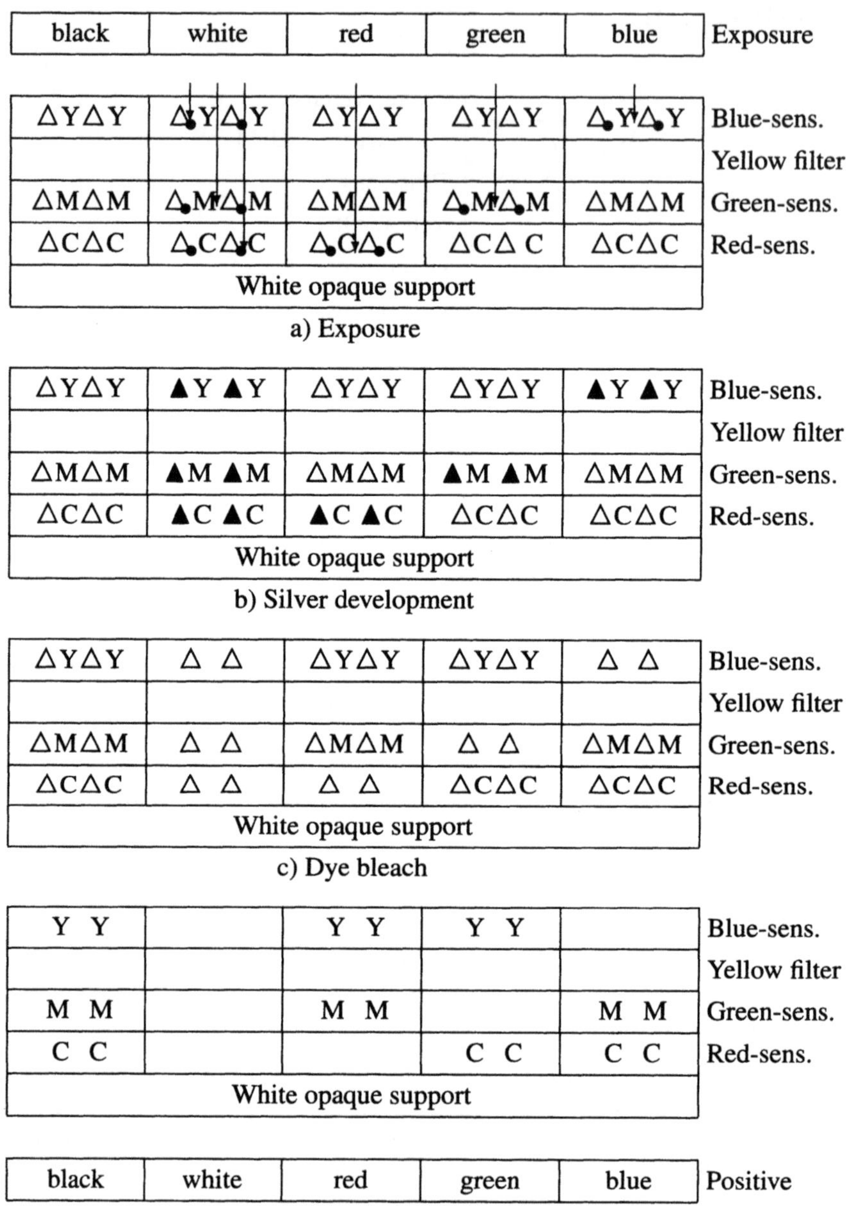

Figure 21.3. Schematic processing of silver dye bleach photographic materials. △: Silver halide grain; △•: Exposed silver halide grain with a latent image; ▲: Developed silver halide grain; Y: Yellow dye; M: Magenta dye; C: Cyan dye.

4. Finally, the developed silvers are converted (bleached) to silver salts if remained. The unreacted silver halides and the regenerated silver salts are removed (fixed) from the layers in a thiosulfate solution (Fig. 21.3d). As a result, the material reproduces blue, green, red, white, and black in a direct positive fashion.

21.2 Azo Dyes for Silver Dye Bleach Photography

21.2.1 Bisazo Dyes

Azo dyes for silver dye bleach photography should be capable of being decomposed (bleached) according to the scheme shown in Fig. 21.1, where the resulting amines should be dissolvable in the bleach solution.

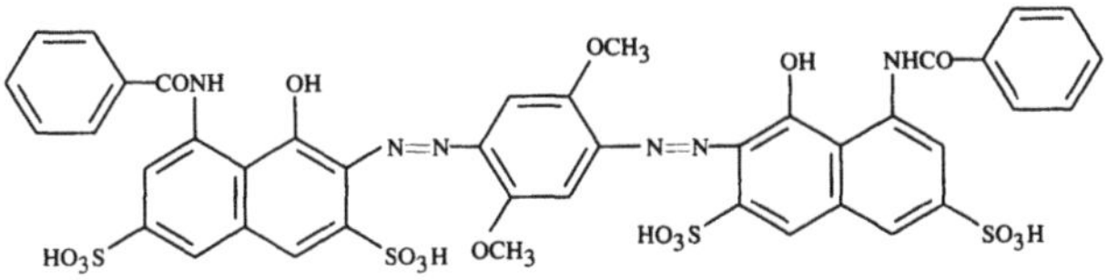

9 (cyan)

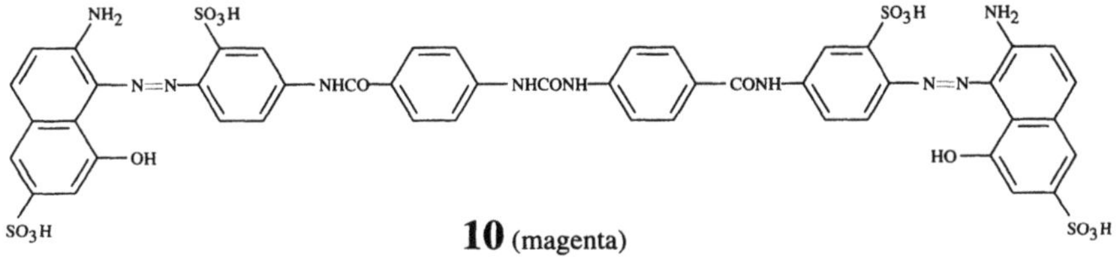

10 (magenta)

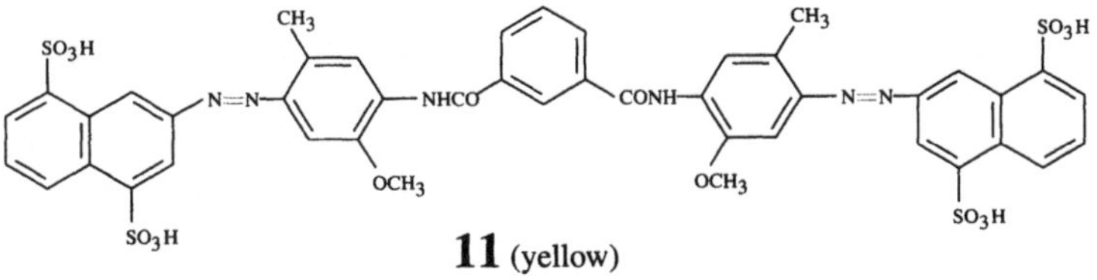

11 (yellow)

Figure 21.4. A set of azo dyes for the full color reproduction of silver dye bleach photography. This set is cited from a patent claiming a masking method [9] and from a patent claiming a method of acid addition [10].

Moreover, the azo dyes themselves are non-diffusible and remain immobile within respective dye layers. This feature is essentially different from that of monoazo dyes used in diffusion transfer photography, since the latter dyes are diffusible or capable of becoming diffusible during processing [12].[3] Hence, bisazo dyes with appropriate molecular weights have been preferred in most silver dye bleach photographic materials. A set of bisazo dyes shown in Fig. 21.4 (**9** to **11**) exemplifies this feature, since this set has appeared commonly in many patents (e.g., [9,10]).

21.2.2 Cyan Bisazo Dyes

Monoazo dyes usually exhibit yellow or magenta color, but not cyan color under undissociated conditions.[4] Hence, in order to obtain cyan dyes, it is inevitable to use bisazo dyes in which the two azo groups are conjugated and incorporated in a single chromophore. Figure 21.5 illustrates examples of bisazo dyes for silver dye bleach photography. A bisazo dye having a benzoylamino group (**12**) has been reported in a patent [13]. A patent belonging to the same series has disclosed bisazo dyes having a characteristic linking unit (e.g., **13** and **14**) [14].

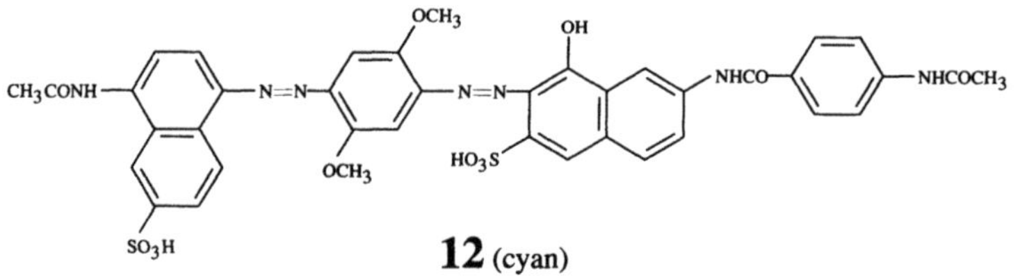

12 (cyan)

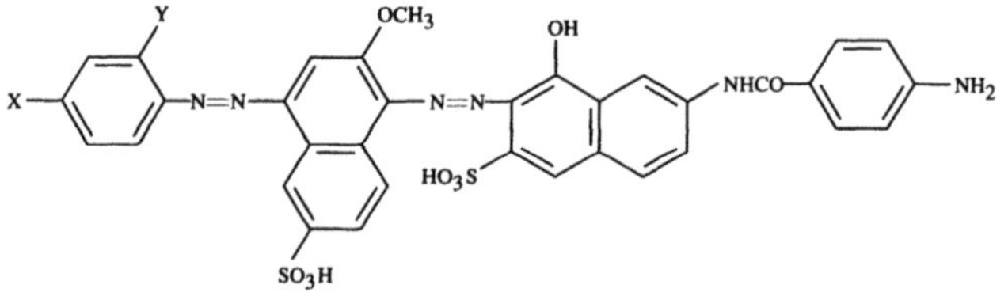

13 X = CH_3O Y = H (cyan)
14 X = H Y = $N(CH_3)_2$ (cyan)

Figure 21.5. Cyan bisazo dyes for silver dye bleach photography.

Trisazo dyes have also been disclosed to be used for silver dye bleach photography. For example, the dye (**15**) has been reported to have blue hue [15].

Polymeric azo dyes have been prepared by azo coupling reactions between a bifunctional coupling component and a bifunctional diazo component [16,17, 18]. An example (**16**) is shown in Fig. 21.7. Since the chromophores in **16** are

[3]This review has compared three types of dyes used in color photography, i.e., azomethine dyes for conventional color photography, monoazo dyes for instant color photography, and bisazo dyes for silver dye bleach photography.

[4]Cyan dyes for diffusion transfer photography are monoazo dyes, as described in Chapters 18 and 19. It should be noted, however, that they are used under dissociated conditions, i.e., in anionic forms.

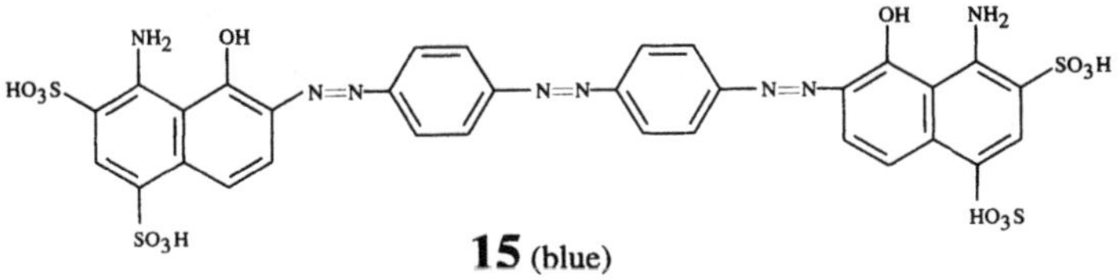

15 (blue)

Figure 21.6. Trisazo dyes for silver dye bleach photography.

separated from each other by a tetramethylene group (in $NHCO(CH_2)_4CONH$), it exhibits magenta hue. Polymeric azo dyes of another type have been prepared by using phenol-formaldehyde condensations and proposed as dyes for silver dye bleach photography [19].

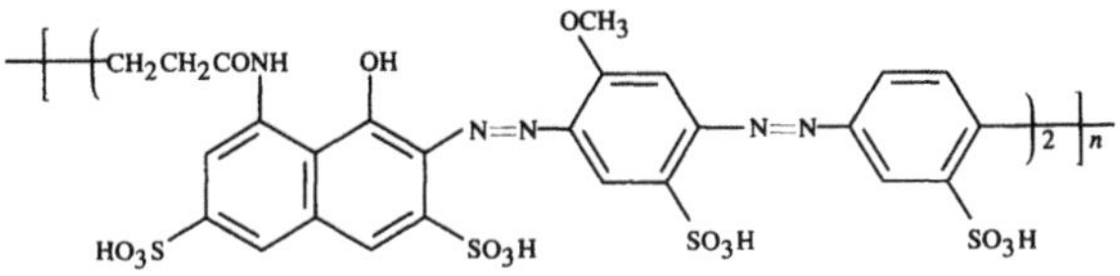

16 (magenta)

Figure 21.7. Polymeric azo dyes for silver dye bleach photography.

21.2.3 Magenta Bisazo Dyes

As for magenta dyes, a monoazo structure is sufficient to exhibit magenta color. A remaining problem is to give non-diffusibility to a dye molecule. This can be accomplished by the incorporation of ballast functions. A most promising methodology is that two dye moieties of the same structure are linked with a non-conjugative group. A predecessor of this methodology has appeared in an earlier stage of the history of silver dye bleach photography [20].[5] However, it had taken a long time until practically usable magenta dyes such as **10** were developed [21]. More complicated linking groups have also been reported. For example, the linking group represented by the formula $NHCO–C_6H_4–NHCONH–C_6H_4–CONH$ of **10** has been replaced by other linking groups such as $NHCONH–C_6H_4–NHCO–C_6H_4–CONH–C_6H_4–NHCONH$ [22] and $NHCO–C_6H_3Cl–O(CH_2)O–C_6H_3Cl–CONH$ [23]. On the other hand, another example (**17**) shown in Fig. 21.8 has the same linking group but different terminal naphthalene nuclei [24].

An amino group on the terminal naphthalene nuclei has been replaced by a substituted phenylamino group to give a magenta dye of another type (**18**) [25]. In

[5] This patent has used a simple methylene group as a linkage in place of $NHCO–C_6H_4–NHCONH–C_6H_4–CONH$ of **10**.

the magenta dye (**19**), the amido linkages have been replaced by ureido linkages [26]. Various linking groups have been disclosed to modify the properties of magenta dyes of this type (e.g., **20**) [27,28]. The magenta dye (**21**) has a linking group containing a thiophene ring [29].

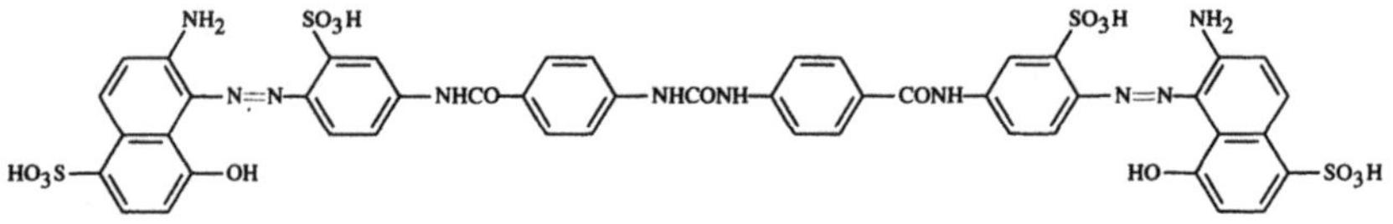

17 (magenta)

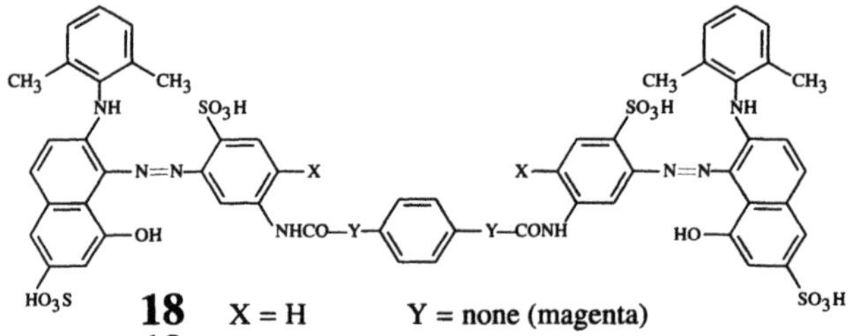

18	X = H	Y = none (magenta)
19	X = H	Y = NH (magenta)
20	X = OCH_3	Y = none (magenta)

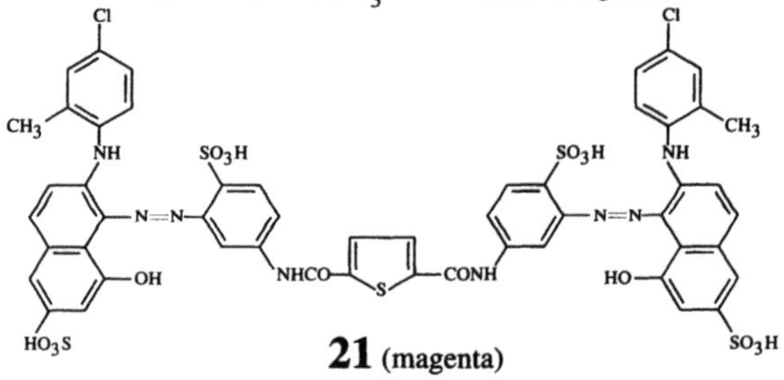

21 (magenta)

Figure 21.8. Magenta azo dyes for silver dye bleach photography.

21.2.4 Yellow Bisazo Dyes

Various linking groups have been investigated in order to improve photographic properties. For example, the linking group of the yellow dye **11** has been modified by fixing the chromophore. The *m*-phenylene of **11** has been replaced by a group containing a 1,3,5-triazine nucleus in a yellow dye (**22**) [30]. The *m*-phenylene has been replaced by a thiophene nucleus in another yellow dye (**23**) [31]. Yet another yellow dye (**24**) has been disclosed, where a central sulfone group is characteristic of the linking group [32]. This patent also contains the application of the linking group to magenta dyes.

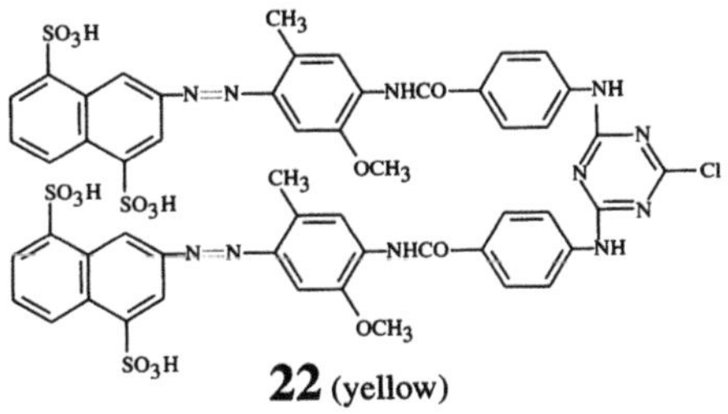

22 (yellow)

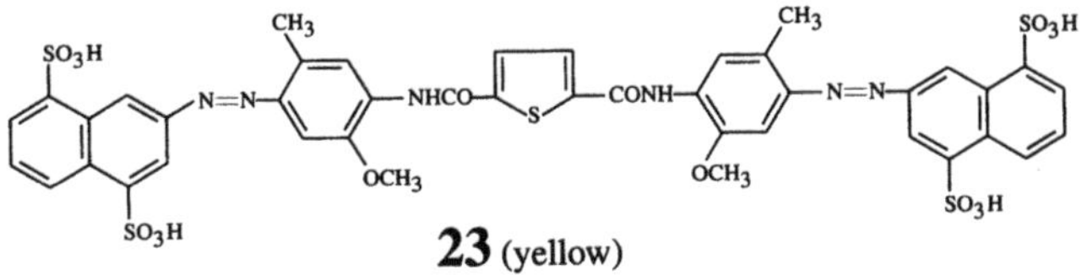

23 (yellow)

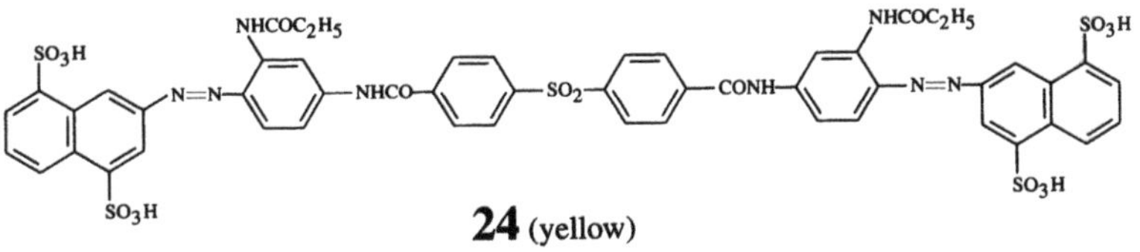

24 (yellow)

Figure 21.9. Yellow azo dyes for silver dye bleach photography.

21.2.5 Colloidal Dispersion of Azo Dyes

In general, bisazo dyes for silver dye bleach photography are used in the form of their alkali metal salts, although the structures depicted above are shown as free acid forms. The use of such bisazo dyes as alkali earth metal salts (e.g., barium salts) [33] or as complexed forms of metal oxides (e.g., cupper oxide) [34] has been proposed in an earlier stage of the history of silver dye bleach photography. More recently, the use of alkali earth metals has been revived to assure effective dispersion of dyes in gelatin [35]. By the use of a colloidal dispersion of water-insoluble salts of water-soluble azo dyes, the ratio of azo dye to gelatin can be controlled between 1:1 to 1:10. Thereby, the thickness of the resulting layer coated with the colloidal dispersion has been reduced so as to result in shorter processing times and in increased image sharpness.

Since this patent [35] seems to disclose the state-of-the-art at that time, the set of dyes described in the example of the patent is cited, as illustrated in Fig. 21.10. These dyes as lanthanum salts have been reported to show absorption maxima under coated conditions as follows: 617 nm and 766 nm (subsidiary max) for **25**, 570 nm for **10**, and 436 nm and 460 nm (shoulder) for **26**. In a practical embodiment, they have been coated as colloidal dispersions of the calcium salts.

The comparison between the set of dyes listed in Fig. 21.4 and that in Fig. 21.10 indicates that only slight changes of substituents have been done. Thus, the cyan dye **25** has an extra methanesulfonyl group at one of the terminal naphthalene nuclei, as compared with the dye **9**; the magenta dye **10** has not been changed;

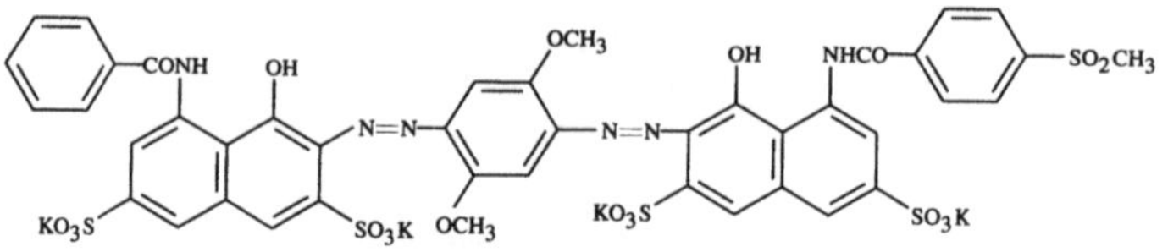

25 (cyan)

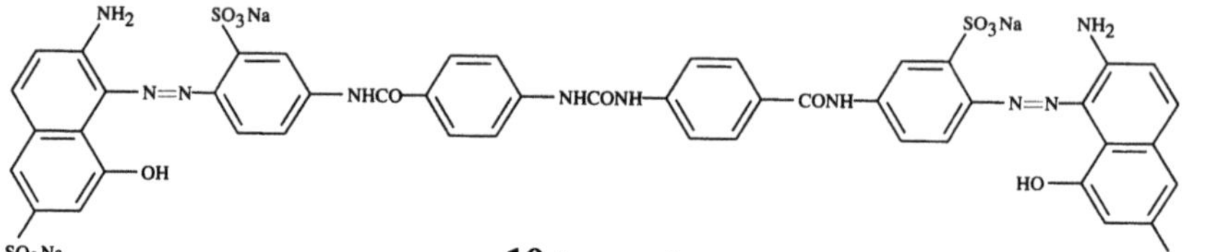

10 (magenta)

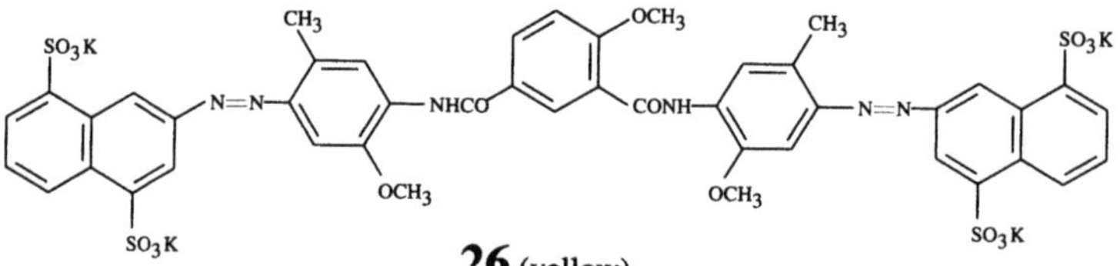

26 (yellow)

Figure 21.10. Another set of azo dyes for the full color reproduction of silver dye bleach photography. This set is cited from a patent claiming a dispersion method [35].

and the yellow dye **26** has an extra methoxy group at the central linking unit as compared with the dye **11**.

21.2.6 Masking by Development Inhibitors

In general, magenta dyes have an undesirable absorption at a blue-light region along with a necessary absorption at a green-light region. This undesirable absorption has been compensated by a masking technique based on an interlayer development inhibition [9]. The masking technique has used the development inhibiting activity of an iodide anion [9,36]. A green-sensitive silver bromide/silver iodide emulsion (which contains or does not contain a magenta bleachable dye) is coated in the green-sensitive layers (Layers 4 and 5) shown in Fig. 21.2. On the other hand, a blue-sensitive *iodide-free* silver bromide emulsion which contains a yellow bleachable dye is coated in the green-sensitive layer (Layer 7) and a blue-sensitive silver bromide/silver iodide emulsion (low iodide content) is coated in the dye-free green-sensitive layer (Layer 8). Moreover, the interlayer (Layer 6) contains nuclei for physical development (e.g., colloidal silver).

Let us consider exposure to green light to understand essential features of this masking method. In an exposed region of the green-sensitive layer (Layer 7 and/or 8), silver development occurs to liberate iodide ions, which migrate into the adjacent interlayer and there inhibit the physical development due to the physical-development nuclei. Thereby, dissolved silver ions form a masked silver image which is the inverse of the negative images formed in Layer 7 and/or 8. In the subsequent dye-bleaching step, the silver metal generated in the exposed region of the green-sensitive layer bleaches the magenta dye, whereas the inverse pattern of the silver metal generated in the interlayer causes a remote action to control the bleaching of the yellow dye according to the inverse pattern. Hence, the more the magenta dye is bleached, the less the yellow dye is bleached. In other words, the more magenta dye remains, the less yellow dye remains in the exposed region to green light. This means that the undesirable blue-absorption of the magenta dye has been compensated by the less remains of the yellow dye.

The experimental details of the masking method have been disclosed in the original patent [36]. A more effective method has been proposed, where a silver halide emulsion is added to the interlayer (Layer 6) in addition to the colloidal silver [9]. Further improvements have also been published in related patents [37, 38].

21.3 Chemicals for Dye Bleach

21.3.1 Bleach Catalysts

As found in Fig. 21.1, bleach catalysts (Cat, e.g., **1** shown in eq. 21.1) play essential roles in a dye bleaching step. Quinoxalines, phenazines, anthraquinones, and naphthoquinones have been proposed as such bleach catalysts. For example, 2,3-dimethyl- (**27**), 2,3-diphenylquinoxaline (**28**), 2,3-diamino- (**29**), and 2-hydroxy-3-aminophenazine (**30**) have been disclosed to accelerate the bleach reactions [39]. Diphenylmethane derivatives such as 2,2′,4,4′-tetramino-5,5′-dimethyldiphenylmethane have been claimed as bleach catalysts [40]. Alloxazine (**31**) and its derivative have been disclosed to be effective as bleach catalysts [41]. *N*-Substituted isoalloxazine derivatives such as 7,8,10-trimethyl- (**32**) and 7,8-dimethyl-10-hydroxyethylisoalloxazine (**33**) have been proposed as bleach catalysts of another type [42]. Indophenazine (**34**) and its methyl and methoxy derivatives have been disclosed as further bleach catalysts [43]. Benzofuro[2,3-b]quinoxaline (**35**), benzothieno[2,3-b]quinoxaline (**36**), and their derivatives have also been disclosed as bleach catalysts [44]. Phenazine-*N*-oxide (**37**) and its methyl derivative (**38**) have been reported as novel bleach catalysts [45]. Cinnoline and its 4-methyl (**39**) and 4-ethoxy derivative (**40**) have been proposed as another series of bleach catalysts [46]. 5,8-Dioxobenzopyrazines (e.g., **41**) [47], imidazo[4,5-b]pyrazine, and its quaternary ammonium salts (e.g., **42**) [48] have been proposed as bleach catalysts. Benzo[c]cinnolines (e.g., **43** and **44**) have been disclosed as bleach catalysts [49,50].

By the inspection of the compounds listed in Fig. 21.11, a reducible group (N=N or N=C–C=N) involved in the heterocyclic ring is essential to exhibit a bleach-catalytic activity. As summarized in a review [3], the redox potential of the reaction of $AgI + e \rightleftharpoons Ag + I^-$, is measured to be −300 to 0 mV against

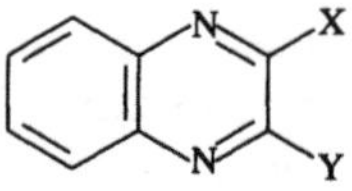

27 X = CH_3 Y = CH_3
28 X = Ph Y = Ph

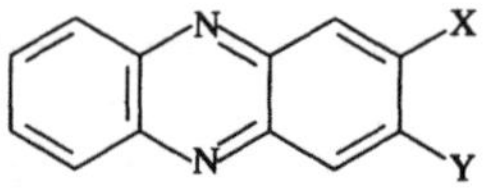

29 X = NH_2 Y = NH_2
30 X = NH_2 Y = OH

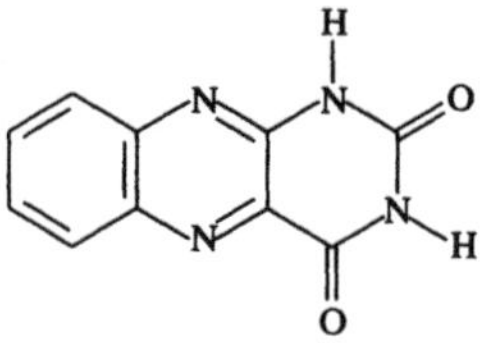

31

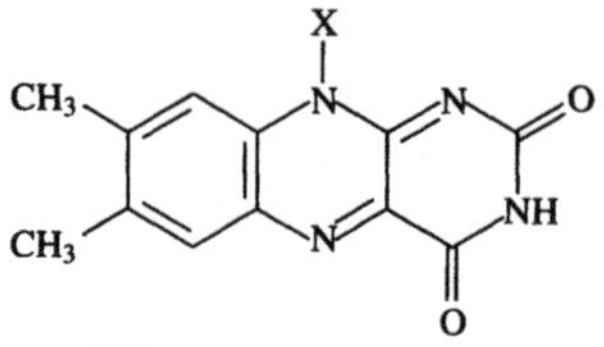

32 X = CH_3
33 X = CH_2CH_2OH

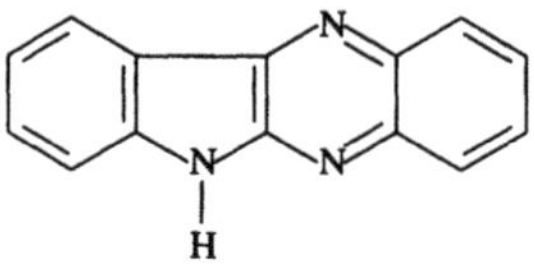

34

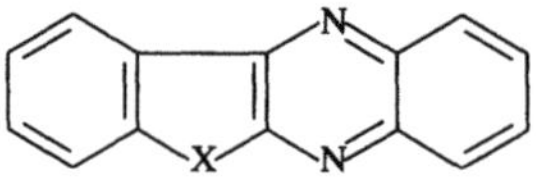

35 X = O
36 X = S

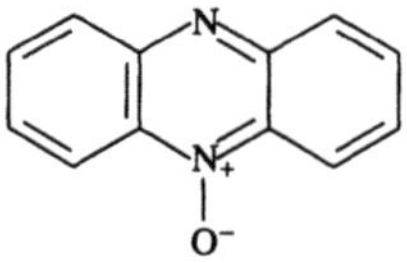

37 X = H
38 X = CH_3

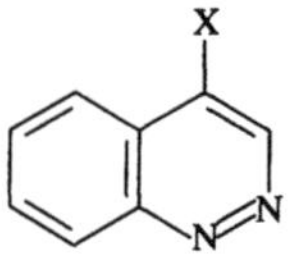

39 X = CH_3
40 X = OC_2H_5

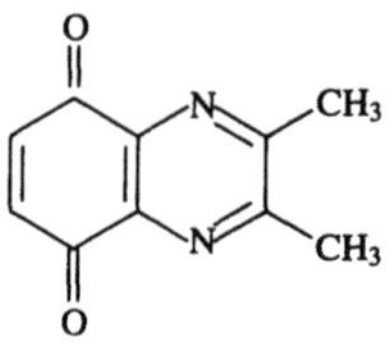

41

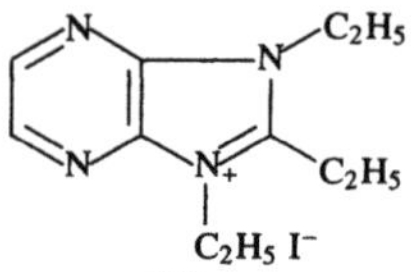

42

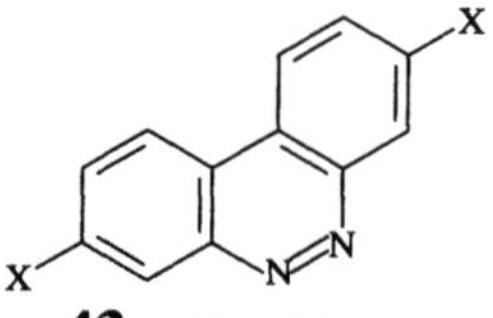

43 X = OH
44 X = OCH_3

Figure 21.11. Various bleach catalysts.

the standard hydrogen electrode. On the other hand, the redox potential of the reaction, Ar–N=N–Ar′ + $4H^+$ + 4e $\rightleftharpoons$ A–NH_2 + Ar′–NH_2 + $4Ag^+$, is measured to be 300 to 400 mV. In order to mediate the two reactions, the redox potential of a bleach catalyst (Cat) should be selected to be −200 to 350 mV according to the reaction represented by Cat + 2H + 2e $\rightleftharpoons$ $CatH_2$. A combined use of a catalyst with a redox potential between −30 and +105 mV (preferably −30 and +60 mV) and another catalyst with a redox potential between −125 and −30 mV (preferably −100 and −30 mV) has been disclosed to improve color balance among three color gradations [51]. For example, a quinoxaline (**45**) of the former category and another quinoxaline (**46**) of the latter category are used together as bleach catalysts for the dye bleaching of the dyes listed in Fig. 21.10.

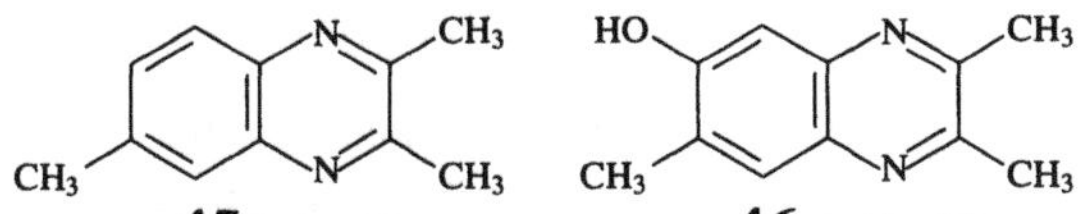

45 (+2 mV) **46** (−46 mV)

Figure 21.12. Combination of bleach catalysts.

21.3.2 Other Additives for Bleach Baths

In addition to the bleach catalysts described in the preceding subsection, a bleach bath contains at least: (a) a strong acid, (b) a water-soluble iodide, (c) a water-soluble oxidizing agent, and (d) an antioxidant.

Strong acids Although sulfuric acid is usually used as the strong acid, a more convenient method using solid acids has been disclosed [10,52]. For example, a 1:1 or 2:1 adduct of urea and sulfuric acid has been used successfully in place of sulfuric acid.

Water-soluble iodides Potassium iodide is a typical example of water-soluble iodides, which work as silver ligands [6,11].

Water-soluble oxidizing agent Nitrobenzenesulfonic acid salts are added as water-soluble oxidizing agents [53,54]. In particular, sodium 3-nitrobenzenesulfonate has been widely used as long as patent literature is inspected intimately (e.g., [10,55]).

Antioxidants Since the potassium iodide described above tends to undergo aerial oxidation, it is stabilized against such aerial oxidation by the addition of an antioxidant selected from reductones and mercapto compounds [56]. For example, 4-mercaptobutyric acid ($HSCH_2CH_2CH_2COOH$) has been widely used in recent embodiments.

Yet other additives Organic developing agents (such as 4-hydroxylmethyl-4-methyl-1-phenyl-3-pyrazolidinone) or benzotriazole derivatives have been proposed as bleaching accelerators [55]. Pyridinium salts such as *N*-methylpyridinium iodide and ethylenedipyridinium dibromide have also been disclosed as bleaching accelerators [55].

References

[1] Christensen JH (1924) US Patent 1 517 049

[2] Friedman JS (1968) History of Color Photography. 2nd ed, Focal, London, Chapter 24

[3] Nakamura T (1986, 2002) In: Functionalized Organic Chemicals for Silver Halide Color Photographic Materials, CMC, Tokyo. Chapter 4

[4] Gaspar B (1935) US Patent 2 020 775

[5] Gaspar B (1935) US Patnet 2 00 4625

[6] Goldfinger P (1940) US Patent 2 217 544

[7] Gaspar B (1941) US Patent 2 255 463

[8] Gaspar B (1955) US Patent 2 699 394

[9] Oetiker A, Chylewski C, Marthaler M (1980) US Patent 4 197 123

[10] Buser H, Morand A (1982) UA Patent 4 366 232

[11] Meyer A (1974) Photogr Sci Eng. 18:530

[12] Fujita S (1984) Senshoku Kogyo. 32:167

[13] Taylor GJ (1947) US Patent 2 420 630

[14] Taylor GJ (1947) US Patent 2 420 631

[15] Chechak JJ (1942) US Patent 2 286 714

[16] Gaspar B, Dreyfuss PD (1958) US Patent 2 844 574

[17] Gaspar B (1949) US Patent 2 470 769

[18] Gaspar B (1944) US Patent 2 356 759

[19] Young RV (1943) US Patent 2 331 755

[20] Carroll B, Chechak JJ (1942) US Patent 2 294 892

[21] Dreyfuss P (1965) US Patent 3 211 554

[22] Freytag K-H, Kabitzke K, Bockly E, Danhauser (1971) US Patent 3 598 594

[23] Baettig K, Jan GC (1991) US Patent 5 043 257

[24] Nickel H, Suckfüll F, Seidel B, Böckly E (1970) US Patent 3 506 450

[25] Anderaú W, Mory R, Piller B (1969) US Patent 3 454 402

[26] Piller B, Lenoir J (1974) US Patent 3 808 194

[27] Piller B, Lenoir J, Foehlich A, Stauner T, Tshopp P (1978) US Patent 4 118 232

[28] Lenoir J, Tschopp P, Loeffel H, de Montmollin R (1976) US Patent 3 931 142

[29] Piller B (1973) US Patent 3 749 576

[30] Mory R, Anderau W (1965) US Patent 3 178 291

[31] Anderau W (1969) US Patent 3 443 952

[32] Piller B (1974) US Patent 3 787 215

[33] Carroll BH (1942) US Patent 2 304 884

[34] Gáspár B (1936) US Patent 2 055 407

[35] Steiger R, Schellenberg M (1989) US Patent 4 803 151

[36] Marthaler M (1977) US Patent 4 046 566

[37] Mollet H, Oetiker A, Boragine C (1982) US Patent 4 310 617

[38] Mollet H, Wyrsch D (1983) US Patent 4 368 256

[39] Gáspár B (1942) US Patent 2 270 118

[40] Gáspár B (1946) US Patent 2 410 025

[41] Mueller FWH (1951) GB Patent 657 374

[42] Mueller FWH (1951) US Patent 2 541 884

[43] Friedman JS (1953) US Patent 2 627 461

[44] Mueller FWH (1954) US Patent 2 669 517

[45] Watanabe S, Ohi R, Sugiyama M, Kondo H (1971) US Patent 3 615 494

[46] Kendall JD, Fry DJ, Robinson RR (1954) GB Patent 711 247

[47] Nakamura T, Watanabe S, Ogawa A, Shishido T (1979) Jpn Patent S54-35931

[48] Nakamura T, Watanabe S, Ogawa A (1979) Jpn Patent S54-35930

[49] Jan G (1981) US Patent 4 266 011

[50] Jan G (1983) US Patent 4 369 318

[51] Marthaler M, Jan G (1981) US Patent 4 304 846

[52] Buser H, Morand A (1989) US Patent 4 879 413

[53] Young RV (1942) GB Patent 539 190; Young RV (1942) US Patent 2 304 987

[54] Seymour MW, Young RV (1943) GB Patent 539 509; Young RV, Seymour MW (1943) US Patent 2 322 084

[55] Libicky A, Schär M (1985) US Patent 4 546 069

[56] Kramp E, Lenoir J, Marthaler M, Schaller R (1976) US Patent 3 961 957

Index

A

B

C

D

E

F

G

H

I

J

K

L

M

N

O

P

Q

R

S

T

U

V

www.ingramcontent.com/pod-product-compliance
Ingram Content Group UK Ltd.
Pitfield, Milton Keynes, MK11 3LW, UK
UKHW021933200726
13853UKWH00010B/1380
* 9 7 8 3 6 6 2 0 9 1 3 1 9 *